Orogenic Andesites and Crustal Growth

Geological Society books refereeing procedures

The Society makes every effort to ensure that the scientific and production quality of its books matches that of its journals. Since 1997, all book proposals have been refereed by specialist reviewers as well as by the Society's Books Editorial Committee. If the referees identify weaknesses in the proposal, these must be addressed before the proposal is accepted.

Once the book is accepted, the Society Book Editors ensure that the volume editors follow strict guidelines on refereeing and quality control. We insist that individual papers can only be accepted after satisfactory review by two independent referees. The questions on the review forms are similar to those for *Journal of the Geological Society*. The referees' forms and comments must be available to the Society's Book Editors on request.

Although many of the books result from meetings, the editors are expected to commission papers that were not presented at the meeting to ensure that the book provides a balanced coverage of the subject. Being accepted for presentation at the meeting does not guarantee inclusion in the book.

More information about submitting a proposal and producing a book for the Society can be found on its website: www.geolsoc.org.uk.

It is recommended that reference to all or part of this book should be made in one of the following ways:

GÓMEZ-TUENA, A., STRAUB, S. M. & ZELLMER, G. F. (eds) 2014. *Orogenic Andesites and Crustal Growth*. Geological Society, London, Special Publications, **385**.

ELBURG, M. A., SMET, I. & De Pelsmaeker, E. 2014. Influence of source materials and fractionating assemblage on magmatism along the Aegean Arc, and implications for crustal growth. *In*: GÓMEZ-TUENA, A., STRAUB, S. M. & ZELLMER, G. F. (eds) *Orogenic Andesites and Crustal Growth*. Geological Society, London, Special Publications, **385**, 137–160. First published online May 30, 2013, http://dx.doi.org/10.1144/SP385.1

GEOLOGICAL SOCIETY SPECIAL PUBLICATION NO. 385

Orogenic Andesites and Crustal Growth

EDITED BY

A. GÓMEZ-TUENA
Universidad Nacional Autónoma de México, México

S. M. STRAUB
Lamont Doherty Earth Observatory of Columbia University, USA

and

G. F. ZELLMER
Massey University, New Zealand

2014
Published by
The Geological Society
London

THE GEOLOGICAL SOCIETY

The Geological Society of London (GSL) was founded in 1807. It is the oldest national geological society in the world and the largest in Europe. It was incorporated under Royal Charter in 1825 and is Registered Charity 210161.

The Society is the UK national learned and professional society for geology with a worldwide Fellowship (FGS) of over 10 000. The Society has the power to confer Chartered status on suitably qualified Fellows, and about 2000 of the Fellowship carry the title (CGeol). Chartered Geologists may also obtain the equivalent European title, European Geologist (EurGeol). One fifth of the Society's fellowship resides outside the UK. To find out more about the Society, log on to www.geolsoc.org.uk.

The Geological Society Publishing House (Bath, UK) produces the Society's international journals and books, and acts as European distributor for selected publications of the American Association of Petroleum Geologists (AAPG), the Indonesian Petroleum Association (IPA), the Geological Society of America (GSA), the Society for Sedimentary Geology (SEPM) and the Geologists' Association (GA). Joint marketing agreements ensure that GSL Fellows may purchase these societies' publications at a discount. The Society's online bookshop (accessible from www.geolsoc.org.uk) offers secure book purchasing with your credit or debit card.

To find out about joining the Society and benefiting from substantial discounts on publications of GSL and other societies worldwide, consult www.geolsoc.org.uk, or contact the Fellowship Department at: The Geological Society, Burlington House, Piccadilly, London W1J 0BG: Tel. +44 (0)20 7434 9944; Fax +44 (0)20 7439 8975; E-mail: enquiries@geolsoc.org.uk.

For information about the Society's meetings, consult *Events* on www.geolsoc.org.uk. To find out more about the Society's Corporate Affiliates Scheme, write to enquiries@geolsoc.org.uk.

Published by The Geological Society from:
The Geological Society Publishing House, Unit 7, Brassmill Enterprise Centre, Brassmill Lane, Bath BA1 3JN, UK

The Lyell Collection: www.lyellcollection.org
Online bookshop: www.geolsoc.org.uk/bookshop
Orders: Tel. +44 (0)1225 445046, Fax +44 (0)1225 442836

British Library Cataloguing in Publication Data

A catalogue record for this book is available from the British Library.
ISBN 978-1-86239-369-1
ISSN 0305-8719

Distributors
For details of international agents and distributors see:
www.geolsoc.org.uk/agentsdistributors

Typeset by Techset Composition India (P) Ltd., Bangalore and Chennai, India.
Printed by Berforts Information Press Ltd, Oxford, UK

Contents

Temporal evolution and crustal growth

Foreword

The editors of this volume invited a brief retrospective about how its 16 chapters reflect progress since my 1981 book *Orogenic Andesites and Plate Tectonics* (*OAPT*) on a similar topic. The epilogue of *OAPT* concluded that the most frequent mechanism, and therefore the most fundamental process, of andesite genesis is low-pressure crystal fractionation of anhydrous phenocryst phases from primary basalt that resulted from partial melting of peridotite mantle that had been metasomatized by fluids or melts from the subducted slab. That is, Bowen, Kuno and Osborne got most of it right before plate tectonics. The last page of *OAPT* concluded that the chief 'andesite problems' for the future were the relative proportion of slab-derived components in the mantle source, the details of the crystal fractionation process, the frequency of other differentiation mechanisms, and the spatial and temporal variations in these parameters. It may be instructive to compare the results of the 16 papers of the present volume with those conclusions.

In the introduction, the editors discuss six topics and especially the relative proportion of slab-derived components v. peridotite in the mantle source of arc magmas (their Fig. 1). They summarize two end-member options as (1) the 'basalt-input model', in which andesites are crustal-level differentiates of partial melts of mantle peridotite (attributed to *OAPT* by the editors as 'an elegant explanation that linked old-school petrology with the conceptual revolution of plate tectonics'), v. (2) 'slab/mantle models', in which andesites are primary melts of sub-crustal (slab-derived and peridotite) mantle sources. To me, these two options lie within a continuum of sources in which the percentage of slab-derived input to the mantle wedge ranges upward from $<1\%$.

The present volume shows that there is indeed no consensus about the mass fraction of slab-derived sources: no one number applies to all subduction zones and all times in Earth history. I note, however, that even the maximum amount invoked in this volume is $<18\%$ and applies to 'hot-subduction zones', such as Mexico and SW Japan. The slab surface temperature beneath the volcanic front of such arcs is estimated to be $>900\ ^{\circ}C$ using the '80 km decoupling' model of Syracuse *et al.* (2007). The only papers in this volume that deal with the more usual case are about Guatemala and the Marianas, and they discuss rocks more felsic than andesite. In both, the felsic rocks are interpreted as direct or indirect differentiates of mafic melts of peridotite (Singer *et al.* and Stern *et al.* respectively). Most current papers about these more typical subduction zones debate whether the mass fraction of slab-derived material is closer to 1 or 5%; whether the slab component is added to the mantle wedge as a solid, dilute low-temperature fluid, high-temperature silicate melt or something in between; how the mass fraction of slab component relates to the percentage of mantle flux melting; and the mass fraction of basaltic crust v. sediment in the slab source.

It may therefore be concluded that few if any orogenic andesites are primary melts of purely mantle peridotite plus water, the possible exceptions bearing names like boninite and high-magnesium andesite. In this volume, adakite seems to be distinguished from the above by lower Mg–Ni–Cr contents, and three papers regard the most mafic of them as primary partial melts of young pyroxenitic mantle formed by recent reaction between slab melts and mantle peridotite (i.e. modified mantle wedge). All three cases are from hot subduction zones. To me, this topic reprises the evolution of the topic from the seminal papers by T. H. Green & Ringwood (1968) to Ringwood (1974), both from the Australian National University (ANU). The first argued that typical calcalkaline andesite is a primary melt of dry subducted eclogite whereas the latter argued that parental arc magmas are mafic melts of modified mantle wedge. My PhD at the ANU lay between the two papers. As part of my thesis, I created a forward model of arc magma genesis to test the ability of the Green and Ringwood hypothesis to explain the suite of rocks from which Green selected the andesite for his experimental study. I concluded that the suite could not be explained as a product of partial melting of the slab (Gill 1974), and presented this to Ringwood in his office. He quickly understood the implications and modified his story to emphasize the reaction products between peridotite and felsic melts at 2–3 GPa. Others soon explored the implications further (e.g. Kelemen 1990), but Ringwood (1974) presented the first heuristic version and cartoon image of the 'slab/mantle model' of this volume.

It appears that this model is a reasonable interpretation of some high-magnesium andesites in hot subduction zones, even when an obvious garnet signature is absent. Indeed, there is a continuum between high-magnesium andesite, more typical calcalkaline basalt and arc tholeiitic basalt as the mass fraction and character of slab-derived component in the mantle source changes, and how this affects the degree of flux melting. The most fundamental modification of the 'basalt-input model' since *OAPT* is that current geodynamic models of subduction zones (e.g. Kelemen *et al.* 2003; Syracuse *et al.* 2007) and current understanding of trace element partitioning between eclogite, fluids and melts (e.g. Johnson & Plank 1999; Kessel *et al.* 2005) agree that at least the sediments, and often some of the mafic rocks, of subducted ocean crust *melt* beneath most and perhaps all arcs. Reaction between that melt and the overlying mantle results in flux melting of the modified mantle wedge in which the percentage melting of the wedge is proportional to, and greater than or equal to, the mass fraction of slab flux added (e.g. Kimura *et al.* 2010). Consequently, arc magmas come from recently modified mantle as Ringwood foresaw. This implies that slab melting usually occurs at the water-saturated solidus of sediment and basalt, which in turn often requires an external source of water and some degree of focused fluid flow through the slab. It also opens the potential for relatively small degree slab melts in which accessory minerals might be stable and buffer trace elements.

On the other hand, I found no evidence in papers of this volume that whole suites of rocks from one volcano or one arc are related to one another by partial melting of the modified mantle wedge. Rather, I found evidence in most papers that the suites are related by differentiation processes within the crust (i.e. within the stability field of plagioclase or hornblende or both). Even parental andesitic magmas with >5% slab components seem restricted to hot subduction zones.

Modern micro-analytical techniques indicate that not all the crystals in some (and perhaps most) orogenic andesite crystallized shortly before eruption from a magma with the composition of the host rock. The 'basalt-input model' is not often that simple, but was never claimed to be. To me, the most fundamental point about this topic is that most arc crust, and most of the diversity in trace element and isotope ratios of arc magmas, come from the modified mantle wedge and not old unrelated crust. That arc magmas contain crystals or melts or both from multiple magma batches over 10^2–10^5 years has no consequence for these two conclusions. Cannibalizing the products of earlier differentiation events, or recharging the system with heterogeneous younger magma just adds scatter to differentiation trends.

The introductory chapter concludes that papers favouring the origin of orogenic andesite by fractional crystallization of parental basalt may be qualitatively convincing but lack quantitative models of purported liquid lines of descent even for major elements. Three reasons are given for this failure of the 'classical model': difficulty finding rocks that are strictly related to one another even in one volcano; difficulty for even current quantitative models to handle the range of likely variables (pressure, temperature, f_{O2}, volatile content) during magma evolution; and the potential for the 'crystal cargo' of minerals to be unrelated to the host magma. *OAPT* discussed the first two of these, and micro-analyses of minerals and their inclusions increasingly demonstrate the third. To me, this is a question of the glass being half-full or half-empty. What mass fraction of crystals in a rock lies outside the range that can be related to the host by varying the magmatic variables? What mass fraction is assimilated from much older and unrelated crustal rocks, or mixed from partial melts thereof? From my own experience with arc basalts, basaltic andesites and even most andesites, the answer to both questions usually is 'small' (minority and perhaps <10%) – especially in oceanic arcs or where continental crust is <30 km thick. If so, then most of the mass of arc crust, and most of the diversity in trace element and isotope ratios in arc magmas, comes from the modified mantle wedge.

To me, the greatest difference from the *OAPT* viewpoint is in the chapter by Kent in which he, following Reubi & Blundy (2009), shows a paucity of andesitic melt inclusions in arc volcanoes despite the abundance of andesitic lavas. Although this may be biased by selection of the host minerals, he argues that eruptible andesite melt is rare, and that andesite lavas are instead physical mixtures of more mafic recharging magma plus older more felsic magma or mush that stalled in the crust. Although this idea still requires differentiation of basalt to andesite melt in arc crust, it adds that erupted andesite magma usually is not this melt. An example might be the 1968–2010 eruption of Volcan Arenal in Costa Rica that is the subject of a journal special issue. Its crystal-rich basaltic andesite lavas contain few andesitic melt inclusions but the temporal change in the lavas can be interpreted quantitatively as being magmas along a liquid line of descent that was a largely closed system for the first half of the eruption by volume and a recharging open system for the second half. The mass fraction of 'unrelated' crystals and melts (i.e. those >500 years older, or that could not precipitate from host magma compositions) may be small: see the Introduction to that issue (Gill *et al.* 2006).

Four other major topics are raised in the chapters of this book: the timescale of slab recycling through the mantle and the residence time of magma in the crust, the temporal evolution of arcs over tens of millions of years, and the need to lose mafic cumulates from the crust if primary arc magmas are basaltic. These were also discussed in *OAPT* and much has been learned about all of them since then. To me, the principal change has been the lack of confirmation about *OAPT*'s model of temporal evolution from tholeiitic to calcalkaline ± shoshonitic suites.

The concluding chapter of *OAPT* posed several other problems and questions for future research, such as the following: are the differences between tholeiitic and calcalkaline suites due to differences in primary magma composition, water content and f_{O2} control, or are the latter just more affected by magma mixing or crustal assimilation? By what processes do melt and crystals separate in arcs and how are the crystals in rocks related to the fractionating phases during chemical evolution of erupted magmas? How can one distinguish between effects of crustal-level 'assimilation' v. mantle-level 'metasomatism'? These and other topics continue to be active research areas in a field that remains vigorous three decades on because of its fundamental importance for understanding what makes our planet so unique and its volcanoes so hazardous.

References

GILL, J. 1974. Role of underthrust oceanic crust in the genesis of a Fijian calc-alkaline suite. *Contributions to Mineralogy and Petrology*, **43**, 29–45.

GILL, J., REAGAN, M., TEPLEY, F. & MALAVASSI, E. 2006. Introduction to Arenal Volcano, Costa rica: magma genesis and volcanological processes. *Journal of Volcanology Geothermal Research*, **157**, 1–8.

GREEN, T. H. & RINGWOOD, A. E. 1968. Genesis of the calc-alkaline igneous rock suite. *Contributions to Mineralogy and Petrology*, **18**, 105–162.

JOHNSON, M. C. & PLANK, T. 1999. Dehydration and melting experiments constrain the late of subducted sediments. *Geochemistry Geophysics Geosystems*, **1**, http://dx.doi.org/10.1029/1999GC000014

KELEMEN, P. B. 1990. Reaction between ultramafic wall rock and fractionating basaltic magma: Part I. Phase relations, the origin of calc-alkaline magma series and the formation of discordant dunite. *Journal of Petrology*, **31**, 51–98.

KELEMEN, P. B., RILLING, J. L., PARMENTIER, E. M., MEHL, L. & HACKER, B. R. 2003. Thermal structure due to solid-state flow in the mantle wedge beneath arcs. *In*: EILER, J. (ed.) *Inside the Subduction Factory*. American Geophysical Union, Washington, DC, 293–311.

KIMURA, J-I., KENT, A. *ET AL*. 2010. Origin of cross chain geochemical variation in Quaternary lavas from the northern Izu arc: using a quantitative mass balance approach to identify mantle sources and mantle wedge processes. *Geochemistry Geophysics Geosystems*, **11**, Q10011, http://dx.doi.org/10.1029/2010GC003050

KESSEL, R., SCHMIDT, M. W., ULMER, P. & PETTKE, T. 2005. Trace element signature of subduction-zone fluids, melts, and supercritical liquids at 120–180 km depth. *Nature*, **437**, 725–727.

REUBI, O. & BLUNDY, J. 2009. A dearth of intermediate melts at subduction zone volcanoes and the petrogenesis of arc andesites. *Nature*, **461**, 1269–1273.

RINGWOOD, A. E. 1974. The petrological evolution of island arc systems. *Journal of the Geological Society, London*, **130**, 183–204.

SYRACUSE, E. M., VAN KEKEN, P. E. & ABERS, G. A. 2007. The global range of subduction zone thermal models. *Physics of the Earth and Planetary Interiors*, **183**, 73–90. http://dx.doi.org/10.1016/j.pepi.2010.02.004

JAMES GILL
Earth and Planetary Sciences
University of California Santa Cruz, USA

An introduction to orogenic andesites and crustal growth

ARTURO GÓMEZ-TUENA[1]*, SUSANNE M. STRAUB[2] & GEORG F. ZELLMER[3]

[1]*Centro de Geociencias, Universidad Nacional Autónoma de México, Querétaro 76230, Mexico*

[2]*Lamont Doherty Earth Observatory of Columbia University, 61 Route 9W, Palisades, NY 10964, USA*

[3]*Department of Soil and Earth Sciences, IAE, Massey University, Palmerston North 4442, New Zealand*

**Corresponding author (e-mail: tuena@geociencias.unam.mx)*

Abstract: This chapter provides an overview of the current state of research on orogenic andesites. While their importance as proxies to the evolution of the continental crust has long been recognized, andesite genesis has remained highly controversial with a broader consensus yet to be reached. The controversy is fuelled by the question of whether orogenic andesites are primary melts of slab and mantle materials, or instead derivative products of basaltic mantle melts that differentiate in the overlying crust. These hypotheses are addressed in three sections of the book devoted to slab–mantle processes, the complexities of melt differentiation at crustal levels, and models pertaining to arc crustal growth. We believe that cross-fertilization and discussion among seemingly opposite and irreconcilable hypotheses will smooth the pathway towards a holistic communal model of andesite petrogenesis.

Among the terrestrial planets, an andesitic continental crust is unique to Earth. Representing only *c.* 1 vol% of the silicate Earth, continents cover *c.* 40% of its surface and, by being permanently emerged above sea-level, constitute the foundation of human life and habitat. While the rate and mechanisms of continental growth have varied through time (Taylor & McLennan 1995; Keller & Schoene 2012), the discovery of Hadean detrital zircons demonstrates that a felsic continental crust must have existed as early as *c.* 4.4 Ga (Wilde *et al.* 2001; Bell *et al.* 2011) and, since then, it has constituted an enduring feature during planetary evolution (Harrison 2009). Yet, and despite its importance for habitability and as an archive of Earth's history, the formation of andesitic continents has long intrigued geoscientists because their very existence presents a paradox. Andesitic rocks have high abundances of silica, aluminium, sodium and potassium, but are relatively depleted in iron, magnesium, calcium and titanium. Hence, unlike basalts, they cannot be direct partial melts of the Earth's peridotite mantle. However, the andesitic crust is a major repository of incompatible elements in the silicate Earth and complements the depleted mantle, both of which were presumably separated from the Earth's primitive mantle (Hofmann 1988; Rudnick 1995; Rudnick & Gao 2003). Yet how did an andesitic continental crust form? By which processes did these elements segregate to form a separate reservoir that is nearly as old as Earth, but fundamentally different from the crust produced by mantle melting? To date, many models of andesite crust formation have been proposed that range from sialic meteoritic infall during the early history of the Earth (Donn *et al.* 1965) to the gradual or episodic extraction of continental crust from the mantle through geological history (Harrison 2009; Cawood *et al.* 2013).

Much of the early continental crust and its genetic information have been destroyed by tectonism and erosion over the course of geological time. On modern Earth, however, andesitic magmas with remarkable compositional similarities to the average continental crust are currently being produced at convergent margins (Gill 1981; Rudnick & Gao 2003). This observation gave rise to the 'andesite model' (Taylor 1967), which postulates continental crust formation by cycles of accretion of andesitic belts that had been created directly above orogenic areas. Over 30 years ago, the significance of orogenic andesites as an offspring of plate tectonics was recognized in the books *Orogenic Andesites and Plate Tectonics* (Gill 1981) and *Andesites: Orogenic Andesites and Related Rocks* (Thorpe 1982). Illustrating developments and state-of-the-art scientific thinking in the field of subduction-related igneous rocks, these publications became landmarks for professional geologists and scholars alike, as they provided an

From: GÓMEZ-TUENA, A., STRAUB, S. M. & ZELLMER, G. F. (eds) 2014. *Orogenic Andesites and Crustal Growth.* Geological Society, London, Special Publications, **385**, 1–13.
First published online October 8, 2013, http://dx.doi.org/10.1144/SP385.16

elegant explanation that linked old-school petrology with the conceptual revolution of plate tectonics. Since then, much progress has been made in refining this approach, stimulated by novel analytical advances as well as by emerging geophysical tools, both assisted by an invigorated computational power and an increasing availability of global data banks.

Most current studies on convergent margins have confirmed the causal link between plate subduction and andesite formation. The re-emergence of the short-lived cosmogenic isotope ^{10}Be in arc volcanoes provided indisputable proof that subducted sediments are recycled at convergent margins (Tera *et al.* 1986; Morris *et al.* 1990). By now, numerous studies have also demonstrated that the budget of fluid-mobile large-ion lithophile elements of arc magmas mostly derive from the various components that constitute the subducted slab, such as oceanic sediments, altered oceanic crust (AOC) and possibly abraded crust from the upper plate (Plank & Langmuir 1993; Miller *et al.* 1994; Elliott *et al.* 1997; Kelemen *et al.* 2003; Plank 2005; Goss & Kay 2006; Tonarini *et al.* 2011). Subducted serpentinite from either within or beneath the subducted oceanic crust contributes water that plays a vital role in mobilizing the elements from the overlying oceanic crust (Schmidt & Poli 1998; Ranero *et al.* 2003; Straub & Layne 2003; Hacker 2008). Importantly, quantifications of slab contributions showed that the elemental flux from the voluminous AOC rivals and exceeds the concomitant flux from the highly enriched, but very thin, sediment layer. Because the ratios of radiogenic isotopes of the AOC are similar to those of subarc mantle, a strong AOC flux thus buffers the flux from the sediment and effectively conceals the magnitude of the total slab flux in Sr–Nd–Pb–Hf isotope space (Miller *et al.* 1994; Straub & Zellmer 2012). The strong link between slab input and arc output is also emphasized by U series studies that suggest recycling from slab to surface within less than a few 100 ka and possibly even within a few thousand years (e.g. Newman *et al.* 1984; Elliott *et al.* 1997; Turner *et al.* 1997; Sigmarsson *et al.* 1998).

While there is little doubt about the efficient recycling of many trace elements, including climatically active volatiles, the origin of the distinct major element composition of arc magmas is less clear (Fig. 1). The major contention is whether andesites are primary melts of the subducted slab or the subarc mantle, or a mixture of both (Kelemen 1995; Straub *et al.* 2011); or alternatively, whether andesites derive from basaltic partial melts of peridotites (the basalt-input model) and are only created by secondary processing in the overlying crust (Hildreth & Moorbath 1988; Plank & Langmuir 1998; Annen *et al.* 2006; Reubi & Blundy 2009). Because major elements make up >99% of the melt mass, the question of slab and mantle v. a crustal origin is of key importance for the mode and rate of arc crustal growth. This is illustrated by means of the arc output equation (equation 1):

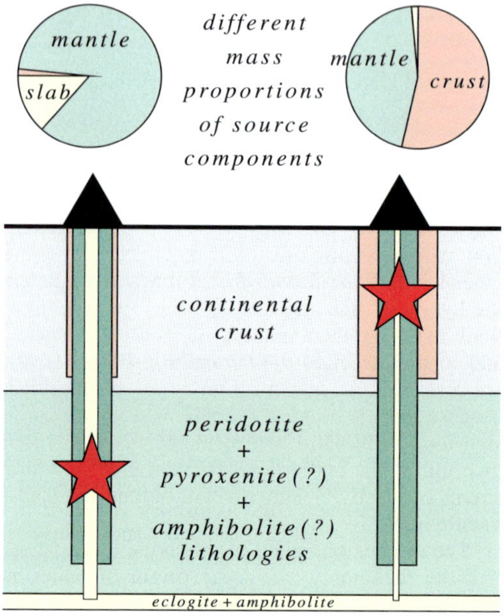

Fig. 1. Schematic diagram of the two contrasting principal locations of andesite formation (red star). Left: andesites form following hybridization of slab and mantle components in the subarc mantle; right: andesites evolve from parental arc basalts in the upper plate crust. Note different mass proportions of source components (estimated) that may result depending on mode of andesite formation.

$$\text{elemental flux} = \text{element abundance} \times \text{arc density} \times \text{arc growth rate} \quad (1)$$

where arc density is the density of the melts, the elemental abundance is the concentration of an element in the primary melt, and the arc growth rate is the mass of melt added per unit time (given in km^3 per km of arc length per myr). Because the range of possible arc densities is limited (*c.* 2.3–2.8 g cm^{-3}), the elemental flux is principally a function of the arc growth rate and the element abundance, and both depend on the composition of the primary arc melts. For example, in a basalt-input model, the calculation of the mode and rate of arc crustal growth requires corrections to lower incompatible element abundances in the basaltic parental magmas (roughly by a factor of 2–3), as well as to the original volumes of the basaltic melts that arrive at the Moho. The latter is difficult to estimate since it needs to account for the mafic residues that must be periodically delaminated from the base of the crust, an inescapable condition in order to stabilize continents with an intermediate composition over geological time (Arndt & Goldstein 1989; Kay & Kay 1993). Clearly, upon this model, formation of an andesitic crust will require time. On the other hand, if the flux to the Moho was an andesitic hybrid of slab and mantle components, corrections to lower incompatible element abundances, and to the effective melt volume added to crust, are less incisive and may not even be required, because there is no need for delamination or extensive melt differentiation in the crust. This would then imply fast and efficient crustal growth with a strong connectivity between arc output and slab–mantle input.

The articles compiled in this book demonstrate that the dichotomy about the origin of andesitic melts from crust v. mantle is very much alive and far from being resolved. Rather than favouring a singular view, it was the goal of the editors to incite discussion to a point that may open a pathway for new research on the petrogenesis of andesite, as well as to identify its role in the context of the broader geochemical cycles of Earth.

The slab–mantle connection

Slab fluids, slab melts, slab diapirs and melt-rock reactions

At convergent margins, the downgoing sedimentary and igneous oceanic crust gradually transforms with increasing depth to greenstone, amphibolite and finally to eclogite at *c.* 80 km depth. Concomitantly, the slab expels aqueous fluids that with increasing depth grade into solute-rich fluids, partial hydrous silicic melts or supercritical fluids, and potentially 'slab diapirs' (Rapp *et al.* 2003; Kessel *et al.* 2005; Savov *et al.* 2005; Gómez-Tuena *et al.* 2008; Behn *et al.* 2011). 'Slab diapirs', or thermochemical slab plumes, are mechanical mixtures of hydrous mantle and partially molten basalts and sediments (Gerya *et al.* 2004) that may detach buoyantly from the slab and rise into the hot core of the mantle wedge (Gerya & Yuen 2003; Behn *et al.* 2011; Marschall & Schumacher 2012), and that may 'relaminate' to the base of the overriding crust (Hacker *et al.* 2011). Recent experimental data suggest that melting of the slab and diapirism may be the rule rather than the exception beneath arc front depths in most subduction zones (Hermann & Spandler 2008; Klimm *et al.* 2008; Plank *et al.* 2009; Behn *et al.* 2011), and either mechanism can produce melts that strongly resemble andesitic arc magmas and that may add to arc crustal growth (Castro *et al.* 2013; Behn *et al.* 2011; and see also **Gómez-Tuena *et al.* 2013**). Thus, the simplest concept is that andesite arc crust grows through addition of such slab components that may or may not be additionally modified in the mantle wedge (Ringwood 1974).

Arc volcanic rocks with the characteristics of slab melts, such as $SiO_2 > 56$ wt%, fractionated heavy rare earth element (HREE) and high Sr/Y ratios (Defant & Drummond 1990), have been found in many arc settings, and are broadly referred to as 'adakites' after their type locality of Adak Island in the Aleutians (Kay 1978). In a landmark study, Defant & Drummond (1990) proposed that adakitic melts were linked to the subduction of young (<25 Ma) and warm oceanic crust, for it was only there that the thermal structure of the modern Earth allowed for melting of subducted metabasalts. Later studies, however, showed that adakitic rocks are found in a variety of settings that depart from warm slab geotherms, or from convergent margins (see Castillo 2012 for a recent overview). Other authors have also shown that adakitic compositions can be produced by crustal differentiation (Müntener *et al.* 2001; MacPherson *et al.* 2006; Alonso-Perez *et al.* 2009; Zellmer *et al.* 2012). For instance, Zellmer *et al.* (2012) recently proposed that the adakitic signatures in western Honshu were caused by fractional crystallization in the presence of garnet that was stabilized in the lower crust by the comparatively higher water abundances of the parental melts. However, a contribution by **Shibata *et al.* (2013)** confronts the fractionation hypothesis of Zellmer *et al.* (2012) for the SW Japanese arc, and relates the emergence of arc-front adakites in northern Kyushu to the abrupt change in the age of the subducted slab from the Philippine Sea in the south to the younger and hotter Miocene plate in the north. While these studies demonstrate that the term 'adakite' has indeed lost much of its genetic connotation, it is also true that their petrogenetic significance is still far from being resolved.

All the same, and despite the widespread occurrence of adakites, there is no doubt that the majority of modern arc volcanics lack tell-tale garnet signatures that may be traced to slab melting at eclogite facies conditions. The absence of garnet signatures is a key argument in models that propose most andesites to ultimately derive from basaltic parental melts that contain only a few mass-per cent of slab contributions (Gill 1981; Tatsumi & Eggins 1995). Intriguingly, some experimental evidence shows that peridotite melting under hydrous conditions can produce andesite melts with SiO_2 abundances as high as 60 wt% (Baker *et al.* 1994; Hirose 1997; Gaetani & Grove 1998; Carmichael 2002). However, primitive magnesian andesites are only produced by hydrous melting if (a) the temperatures and pressures of the last equilibration are extremely low (<1000 °C and 1 GPa), (b) water contents are kept at levels close to saturation (>7 wt%) and (c) the residual mantle has the paragenesis of a refractory harzburgite (Hirose 1997; Parman & Grove 2004; Grove *et al.* 2012). Since subtle variations in these parameters will inevitably render the melts basaltic, and in view of the rarity of water contents >7 wt% H_2O in arc magmas (Ruscitto *et al.* 2012; Plank *et al.* 2013), andesite production by hydrous peridotite melting should be exceptionally rare in most arcs.

However, does the lack of a garnet signature in arc magmas indeed preclude a link between andesite genesis and slab melting or diapirism? Two contributions to this volume argue otherwise. First, **Straub *et al.* (2013)** present a study of young magnesian arc basalts to andesites from the central Mexican Volcanic Belt. While these rocks lack a garnet signature and have low, non-adakitic Sr/Y ratios, it had been inferred earlier on the basis of the high-Ni olivines that they contained >15–18% of a silicic slab component consequent to melt-rock reaction in the mantle wedge (Straub *et al.* 2011). Here, Straub *et al.* (2013) show that this reaction process is capable of blending a strongly fractionated slab component (10–30%) with steep HREE pattern with the subarc mantle in such ways that the HREE (Ho–Lu) and TiO_2 remain controlled by the mantle (which produces partial melts with flat HREE and negative TiO_2 anomalies) with all other elements controlled or at least significantly augmented by slab additions.

Second, Gómez-Tuena *et al.* (2013) suggest that the lack of a garnet signature was due to recycling by means of ‘slab diapirs’. In western Mexico, garnet signatures are strongly recorded in arc-front magmas, located about 140 km above the subducted plate (Gómez-Tuena *et al.* 2011), but conspicuously absent in rear-arc andesitic stratovolcanoes, despite their location at >300 km depth above one of the warmest slabs on Earth. This observation defies models of depth-progressive slab transition from amphibolite to eclogite. Moreover, over an area of 2000 km^2, peripheral volcanism exhibits highly variable trace element patterns within a few kilometres of each other, suggesting a highly heterogeneous mantle source region. These observations can be explained by a model in which the subducted materials are detached from the deeper slab as buoyant diapirs and subsequently melt in the hot core of the mantle wedge at relatively shallow depths. A multitude of plumes that may partially melt at different pressures and temperatures may then create the exceptional diversity of erupted melts in western Mexico.

Timescales of melt transfer from slab and mantle to surface

An important question is the timescale of mass transfer from slab to crust and surface, as this provides unique information on the rate of arc crustal growth. The application of uranium-series isotope analysis of arc magmas in the past two decades provides strong arguments in favour of a rapid mass transfer. Uranium-series isotopes enable processes to be traced on the timescales <350 ka ($^{238}U-^{230}Th$), <150 ka ($^{231}Pa-^{235}U$) and <8 ka ($^{226}Ra-^{230}Th$) (Condomines *et al.* 1988; Turner *et al.* 2003). Therefore, uranium-series disequilibria in young arc magmas should trace events that caused the respective parent/daughter elements in the uranium series isotope chain to fractionate. Dependent on the geochemical behaviour of the elements, however, uranium-series isotopes have been proposed to trace either release of slab fluids (Elliott *et al.* 1997; Turner *et al.* 1997; Sigmarsson *et al.* 1998; Zellmer *et al.* 2000) or the onset of slab and/or mantle melting (Avanzinelli *et al.* 2012). Thus, the disequilibria detectable in arc magmas imply time spans between release of the slab component and eruption in the order of at most 10^4-10^5 years (Elliott *et al.* 1997), melt ascent rates of at least several metres per year (Turner *et al.* 2003) and possibly multiple slab dehydration events within less than 1000 years prior to eruption (Turner *et al.* 2000). However, these timescales must still be considered against the background of an ongoing debate on the U-series elemental behaviour. For example, Pa is normally considered as fluid immobile (Elliott *et al.* 1997; Bourdon *et al.* 1999) and, therefore, ^{231}Pa excesses should be controlled by dynamic melting of the mantle (Turner *et al.* 1996). However, Avanzinelli *et al.* (2012) recently argued that a slab component probably includes significant amounts of Th and Pa and thus traces slab melt release at timescales of <150 ka prior to eruption.

Despite substantial advances in this field over the last few decades, high-precision U-series analyses remain challenging and progress is slow. The contribution by **Dosseto & Turner (2013)** serves well to illustrate the promise and complexity of U-series studies. The authors report new high-precision data on previously analysed samples from Kamchatka, which results in a re-interpretation of an earlier genetic model proposed by Dosseto *et al.* (2003). Rather than mixing of slab fluids into the mantle wedge within <150 ka, it is now argued that ($^{238}U/^{230}Th$) disequilibria and ^{226}Ra and

^{231}Pa excesses are best explained by dynamic melting of a recently (<10 ka) metasomatized mantle source followed by late-stage overprinting in the crust. This model is thought to be superior to models of multiple dehydration events (Bourdon *et al.* 2003) or melting under high oxygen fugacity (Beier *et al.* 2010) in producing the observed U-series systematics.

The role of the overriding crust

The basalt-input model implies that orogenic andesites would not exist without the overriding crust. Clearly, the overriding crust is a physical barrier that forces ascending mantle magmas to stagnate, cool and differentiate by various processes, such as fractional crystallization, assimilation of crustal basement and mixing between stagnating and newly ascending magmas. There is also a well-documented global correlation between crustal thickness and melt silica abundances (Leeman 1983) that holds despite the more recent discovery of common silicic rock series in intra-oceanic arcs (Kodaira *et al.* 2007; Haase *et al.* 2011; **Smith 2013**; **Stern *et al.* 2013**). A plethora of petrographic and geochemical studies on arc magmas, often substantiated by the presence of crustal xenoliths, also provide irrefutable evidence of melt modification in the crust (Eichelberger 1978; Davidson & Tepley 1997; Nishimoto *et al.* 2005; Davidson *et al.* 2007; Ortega-Gutiérrez *et al.* 2008; Reubi & Blundy 2009). Moreover, rare mantle xenoliths (Blatter & Carmichael 1998) and high-Mg# basalt magmas that lack garnet signatures demonstrate that peridotite is present in the subarc mantle and contributes to arc magma genesis.

The evolution of silicic andesitic magmas from basaltic parental melts, however, entails one significant problem: the disposal of large volumes of mafic and ultramafic cumulates. Simple mass balance calculations require mafic cumulates and restites to comprise at least half of the basaltic melt volume that crosses the Moho (Jull & Kelemen 2001; Tatsumi *et al.* 2008). Because the average continental crust is andesitic, such mafic residues should be periodically removed and recycled back into the mantle in a process that is often referred to as 'delamination' or 'lithospheric foundering' (Arndt & Goldstein 1989; Kay & Kay 1993; Tatsumi *et al.* 2008; Stern & Scholl 2010). While active foundering has been deduced from seismic profiles (Zandt *et al.* 2004) and interpreted to play a significant role in the magmatic evolution of the Andes (**Kay *et al.* 2013**), evidence from exposed crustal sections indicates that delamination does not seem sufficient to render an andesitic bulk continental crust (Greene *et al.* 2006). Indeed, other studies in active arcs even suggest that lithospheric foundering may be counterbalanced by triggering voluminous basaltic volcanism, and thus not necessarily displace the bulk crust to more felsic compositions (Elkins-Tanton 2007; Mori *et al.* 2009).

Melt evolution in the crust

The importance of intra-crustal processing and differentiation in andesite genesis is reflected in eight contributions to this book. Two studies consider fractional crystallization a principal mechanism of crustal differentiation (**Elburg *et al.* 2013**; Smith 2013), while the other six papers emphasize the importance of melt mixing. This triage agrees with the new wealth of microbeam data that became available in recent years that invariably demonstrate the ubiquity of melt mixing on the microscopic scale of otherwise macroscopically inconspicuous rocks (Davidson & Tepley 1997; Reubi & Blundy 2009; Kent *et al.* 2010; Straub *et al.* 2011). The overwhelming evidence of melt mixing is noteworthy, as fractional crystallization – with or without contamination – was the key concept behind the classical model of andesites as derivative melts of parental basaltic magmas in the overlying crust (Gill 1981; Hildreth & Moorbath 1988; Plank & Langmuir 1988).

Smith (2013) suggests that the common occurrence of high-Mg# andesite in the late Cenozoic Papuan volcanic arc is linked to extensional tectonics, and emphasizes the importance of crustal thickness in defining fractionation assemblages. The two magma series of low-Mg# and high-Mg# andesites are proposed to derive from a primary basaltic magma through either high-pressure fractionation of amphibole (high-Mg# andesites) or low-pressure fractionation of olivine and clinopyroxene (low-Mg# andesites). Elburg *et al.* (2013) investigate basalt to rhyolite series from three Quaternary Aegean volcanoes, Methana, Santorini and Nisyros, that are evenly distributed within a *c.* 400 km-long segment along the Aegean arc volcanic front. They suggest that dacitic and rhyolitic magmas evolve from parental basalt within the crust with only minor crustal contamination. Apatite, titanomagnetite and amphibole are recognized as important fractionating phases, whereby the variable amphibole stability at different crustal thickness is thought to control various levels of depletion of middle and heavy rare earth elements at different volcanoes.

However, while such models are qualitatively convincing, they still require confirmation by modelling of the purported liquid lines of descent (LLD) for the major element oxides. The lack of such models, evident also in other studies that provide otherwise compelling evidence for melt evolution by fractional crystallization (Wade *et al.* 2005) has several reasons. First, there is the problem of how to recognize the 'true' liquid line of descent from magma series that do not represent continuous stratigraphic sections of individual volcanoes (see also Smith 2013). Volcanic time series that should best record the LLD are even more difficult to establish from individual volcanoes that undergo multiple eruptions from various vents (see Elburg *et al.* 2013). Second, even if time series are recognized, mixing of newly ascending magma with stagnating crustal magma bodies (i.e. 'recharge melt mixing', see **Kent 2013**) may overprint and erase the original LLDs. Moreover, in two contributions to this book, **Zellmer *et al.* (2013*a*, *b*)** describe an extreme variant of recharge melt mixing from basalts (SW Japan) and andesites (Chile), in which newly ascending aphyric magmas take up crystals from previous batches of partially crystallized magmas of similar composition within the same magmatic system, that ultimately erupt as porphyritic lavas. Thus, if the crystal cargo in the rocks consists mostly of antecrysts and xenocryst, the mineral compositions may be decoupled from the melt compositions, and thus have little meaning for the LLD along which melts may evolve. Third, despite the continuous development of sophisticated algorithms, such as MELTS and its later improvements (Ghiorso & Sack 1995; Ghiorso *et al.* 2002), quantitative models still fail to capture the complexities of LLDs that evolve along changing temperatures and pressures, melt fO_2, volatile content and mineral/melt partitioning behaviour.

An important aspect of melt evolution by fractional crystallization with minor or negligible crustal contamination is that many key signatures of the basaltic parental melts (e.g. ratios of radiogenic isotopes and incompatible elements) are preserved until andesitic and dacitic compositions are reached. Thus, after correction for mass and abundance level, silicic arc magmas are still able to provide estimates of slab and mantle fluxes. In the alternative model of mixing of basaltic and silicic melts, such assessments are much more difficult and depend on the origin of the silicic end-member (see also Gómez-Tuena *et al.* 2013). The silicic end-member may have different origins, such as the ambient crust (Eichelberger 1978), a derivate melt of an earlier basalt injection into the same magma system or a combination of both (Annen *et al.* 2006; Reubi & Blundy 2009). If the silicic end-member is derived from pre-existent crust, the mass and elemental abundance of juvenile mantle melts may be small, and probably less than half of the melt volume produced.

Several contributions to this book and many other published studies show that bimodality is a common characteristic of arc magmas world-wide. Respective end-members are mafic (basalt to basaltic andesitic) and silicic (dacitic–rhyolitic), and andesites are quite often produced by melt mixing. The mixing process, however, may not be complete. For example, bimodality of dacitic and basaltic andesite magmas at the Holocene Guatemalan Volcán de Santa María is well preserved. Here, **Singer *et al.* (2013)** propose that these dacites formed by crystal fractionation from previous intrusions of basaltic andesite at >5 km crustal depth over tens of thousands of years. Their eruption was ultimately triggered by newly ascending basaltic andesitic melts, which incompletely mingled during ascent, and did not yet produce intermediate andesites. Stern *et al.* (2013) propose a similar scenario in order to account for the strongly bimodal series of basaltic–andesites and dacite–rhyolites erupted at the Quaternary Diamante cross-chain volcanoes, which are part of the 110 km long Anatahan Felsic Province along the southern Mariana Volcanic arc. However, in this model, the silicic series originate from a pre-existing tonalitic middle crust that was reheated by hot and newly ascending mantle-derived magmas.

Mixing of mafic and silicic magmas can be very efficient in creating homogeneous melts. For instance, a study of phenocryst compositions of apparently homogeneous andesites from Mt Hood in the Cascades arc indicated an origin by homogenization of basaltic and silicic melts (Kent *et al.* 2010). In this volume, Kent (2013) further suggests that the cryptic bimodality is a consequence of 'mafic recharge' where hydrous mafic magma intrudes into a pre-existing crustal silicic magma or mush. Such 'eruption filtering' may be instrumental in the preferential eruption of only highly hybridized andesites, whereas the original silicic and mafic parental magmas are retained in the crust. Lastly, **Steiner & Streck (2013)** propose that mixing of tholeiitic basaltic mantle melts with ambient crust may play a key role in the evolution of calc-alkaline andesitic magmas with their typical low FeO/MgO at high SiO_2. This claim is substantiated by the fact that it was reported in a non-subduction setting, namely the intra-continental Miocene Strawberry Volcanic Suite in proximity to the Columbia River flood basalt province in the western USA.

Residence times of magmas in the crust

An important complement to understanding the processes of crustal differentiation is the question of residence times of melts in the crust. Do erupted melts pass rapidly through the crust or is there an extended stagnation period? Within the past two decades, a wide array of geochronometers have become available, which include U-series disequilibrium studies of bulk rocks and mineral separates (e.g. Condomines *et al.* 2003), crystal size distribution analysis (e.g. Cashman & Marsh 1988; Higgins 1996), chemical diffusion profiles of minerals (e.g. Lasaga 1983; Costa *et al.* 2003; Zellmer *et al.* 2003; Morgan & Blake 2006; Ruprecht & Plank 2013), the study of isotopic variations between growth zones within individual crystals (e.g. Davidson & Tepley 1997) and U–Th dating of accessory minerals (e.g. Schmitt 2011).

Although the determination of melt residence times in the crust is still a developing field where results do not always agree, it is evident that the timescales of magma storage in the crust are quite variable. Thus far, the data provide evidence for rapid passage as well as extended time of melt migration and differentiation, as magma bodies may evolve within *c.* 10^3-10^5 years. There are also distinct variations among different lithologies: porphyritic lava domes in intermediate to felsic systems may record short timescales reflecting last-minute mixing at the onset of eruption, but also complex and protracted crystallization episodes of the order of tens of thousands of years or longer (Zellmer 2008). Mafic melts may traverse the crust more rapidly, despite frequently recording uptake of old cumulate crystals (Zellmer *et al.* 2005). Extremely rapid ascent of some arc magmas from great depths to the surface (within a few days) has been suggested on the basis of seismic studies (Blot 1972; Fedotov *et al.* 1983) and the preservation of ultramafic hornblende–peridotite mantle xenoliths in their host magmas (Blatter & Carmichael 1998). The two Zellmer *et al.* (2013*a*, *b*) contributions in this volume provide further, albeit circumstantial, evidence for extremely rapid melt transfer from depth through the crustal lid.

Temporal evolution and crustal growth

Volcanic arcs are active over tens of millions of years, which means that the timescales of arc crustal growth well exceed the timescales of crustal differentiation. Thus, a simple way to assess a causal relation between arc crust thickening and melt SiO_2 abundances in orogenic andesites is to observe arc evolution through time. Three contributions to this book discuss the complex interplay between evolving tectonic parameters and rock compositions through time. Kay *et al.* (2013) review the Miocene to Holocene arc magmatism of the central Argentine and Chilean Andes, **Yoshida *et al.* (2013)** present a detailed summary of the Cenozoic evolution of the NE Japan arc and **Bryan *et al.* (2013)** report the occurrence of early to middle Miocene syn-extensional andesitic magmatism in the context of the Cenozoic evolution of the Gulf of California. At the Andean margin, Neogene magmas show the influence of crustal thickening and foundering, plateau uplift and landward frontal arc migration, subduction erosion and slab–mantle mixing over a continuously changing Benioff zone (Kay *et al.* 2013). The NE Japan arc is characterized by clear secular changes of mode of magmatic activity and style of magma plumbing system, which affect eruption volumes and magmatic compositions in response to geotectonic evolution and mantle wedge convection (Yoshida *et al.* 2013). In the Mexican margin, a middle Miocene phase of calc-alkaline andesite magmatism occurs in an active rift environment without evidence of concurrent plate subduction, further questioning the belief that the orogenic andesites are confined to active convergent margins (Bryan *et al.* 2013; and also see Steiner & Streck 2013).

All of these studies illustrate the complexity of andesite formation that varies strongly within and among different settings. However, they have one remarkable common factor, namely the lack of unidirectional trends of arc chemistry with time that would be interpreted as an effect of increasing crustal thickening and maturing. These findings agree with results from the 50 myr Izu-Bonin Mariana arcs in the NW Pacific, which has been well investigated for temporal trends (Lee *et al.* 1995; Straub 2003; Straub *et al.* 2010; Ishizuka *et al.* 2011). The results suggest that crustal thickness is not a singular controlling factor on the composition of orogenic andesites, but merely one of many controlling factors.

Summary and outlook

The contributions to this book mostly confirm and expand the notion that a fundamental connection exists among convergent margins, andesites and continents (Gill 1981; Thorpe 1982). At the same time, it is also obvious that the genesis of orogenic andesites remains enigmatic and that no consensus model exists. Because of substantial progress with respect to confirming the validity, magnitude and efficiency of slab recycling, however, it is more important than ever to determine mass fluxes associated with convergent margin volcanism. The rate and mode of arc crustal growth is intimately linked with the flux of elements – including climatically active gases – that are perpetually cycled through the subduction interface and, as such, can strongly influence the evolution of global climate (Kennett & Thunell 1977; Huybers & Langmuir 2009; Kutterolf *et al.* 2013).

A hallmark of orogenic andesites is their consistent calc-alkaline composition despite highly variable geological parameters, such as crustal thickness and architecture, convergence rate, thermal structures, age of subducted crust, and so on. In view of such constancy, it seems unlikely that more than one fundamental process produces orogenic andesites and, in particular, not two processes that are so different, such as the slab–mantle models and the crustal differentiation (basalt-input) model outlined above. Does the diversity observed reflect innumerous variations of one of these models, or possibly a complex combination of both? How can such questions be answered?

Although the contributions to this book do not provide a consensus model of orogenic andesite formation, they show what it is needed at this junction to achieve further progress. The core of the problem lies with the fact that the genetic process cannot be directly observed, but needs to be inferred indirectly from the mixed and processed 'products' of andesite formation, namely the erupted magmas. This book highlights the diverse models and hypotheses by which these data are interpreted, and illustrates the lack of consensus about the mechanisms and loci of melt differentiation. For example, while the dominance of melt mixing is obvious, it is not clear where and how the magmas differentiate (i.e. evolve a broader spectrum of basaltic to dacitic–rhyolitic rocks) in the first place. Does this happen below or above the Moho? Is this done by melting of various slab and primary and secondary mantle lithologies under the subduction influence, or does this happen by fractional crystallization and/or contamination in the overriding crust?

In the past decades, the widely supported basalt-input model triggered many studies with a focus on the most mafic series in arc settings, on the assumption that these best reflect initial signals of slab and mantle processes. However, the possibility that much of the geochemical diversity found in arcs is created by sub-Moho processes opens up a different perspective, and emphasizes the need to study equally all mafic and evolved end-members without a preconceived notion of a probable origin. Regardless of a crustal v. mantle origin, silicic arc magmas must carry vital genetic information that is an essential complement to their basaltic counterparts. Basaltic to silicic magma series emitted from individual vents should be ideally suited to unravelling the processes of melt differentiation, and to obtain insights through quantitative models that are essential to understanding the genesis of orogenic andesites and the global continents.

We would like to acknowledge all colleagues who have contributed to our understanding of orogenic andesites. We are grateful for the opportunity to present some of this progress within a dedicated volume. We thank R. Gertisser for a very positive and encouraging review of the initial book proposal, and all reviewers who provided timely and constructive feedback. This chapter benefited from discussions with C. Annen, T. Plank and F. Ortega-Gutiérrez. T. Anderson and A. Hills provided administrative support throughout the editorial process. SMS acknowledges support from the US National Science Foundation for her research on arcs. GFZ acknowledges funding by the Global Volcanism Program for a research fellowship at Smithsonian's Natural Museum of Natural History. AGT's research on andesites was funded by PAPIIT-UNAM project IN107810.

References

ALONSO-PEREZ, R., MÜNTENER, O. & ULMER, P. 2009. Igneous garnet and amphibole fractionation in the roots of island arcs: experimental constraints on andesitic liquids. *Contributions to Mineralogy and Petrology*, **157**, 541–558, http://dx.doi.org/10.1007/s00410-008-0351-8

ANNEN, C., BLUNDY, J. D. & SPARKS, R. S. J. 2006. The genesis of intermediate and silicic magmas in deep crustal hot zones. *Journal of Petrology*, **47**, 505–539, http://dx.doi.org/10.1093/petrology/egi084

ARNDT, N. & GOLDSTEIN, S. 1989. An open boundary between lower continental crust and mantle: its role in crust formation and crustal recycling. *Tectonophysics*, **161**, 201–212.

AVANZINELLI, R., PRYTULAK, J., SKORA, S., HEUMANN, A., KOETSIER, G. & ELLIOTT, T. 2012. Combined $^{238}U-^{230}Th$ and $^{235}U-^{231}Pa$ constraints on the transport of slab-derived material beneath the Mariana Islands. *Geochimica et Cosmochimica Acta*, **92**, 308–328, http://dx.doi.org/10.1016/j.gca.2012.06.020

BAKER, M., GROVE, T. L. & PRICE, R. 1994. Primitive basalts and andesites from Mt. Shasta region, N. California: products of varying melt fraction and water content. *Contributions to Mineralogy and Petrology*, **118**, 111–129.

BEHN, M. D., KELEMEN, P. B., HIRTH, G., HACKER, B. R. & MASSONNE, H.-J. 2011. Diapirs as the source of the sediment signature in arc lavas. *Nature Geoscience*, **4**, 641–646, http://dx.doi.org/10.1038/ngeo1214

BEIER, C., TURNER, S. P., SINTON, J. M. & GILL, J. B. 2010. Influence of subducted components on back-arc melting dynamics in the Manus Basin. *Geochemistry, Geophysics, Geosystems*, **11**, http://dx.doi.org/10.1029/2010GC003037

BELL, E. A., HARRISON, T. M., MCCULLOCH, M. T. & YOUNG, E. D. 2011. Early Archean crustal evolution of the Jack Hills Zircon source terrane inferred from Lu–Hf, $^{207}Pb/^{206}Pb$, and $\delta^{18}O$ systematics of Jack Hills zircons. *Geochimica et Cosmochimica Acta*, **75**, 4816–4829, http://dx.doi.org/10.1016/j.gca.2011.06.007

BLATTER, D. L. & CARMICHAEL, I. S. E. 1998. Hornblende peridotite xenoliths from central Mexico reveal the highly oxidized nature of subarc upper mantle. *Geology*, **26**, 1035–1038.

BLOT, C. 1972. Volcanisme et séismes du manteau supérieur dans l'Archipel des Nouvelles-Hébrides. *Bulletin Volcanologique*, **36**, 446–461, http://dx.doi.org/10.1007/BF02597120

BOURDON, B., TURNER, S. & ALLÈGRE, C. 1999. Melting dynamics beneath the Tonga-Kermadec Island Arc Inferred from $^{231}Pa-^{235}U$ systematics. *Science*, **286**, 2491–2493, http://dx.doi.org/10.1126/science.286.5449.2491

BOURDON, B., TURNER, S. & DOSSETO, A. 2003. Dehydration and partial melting in subduction zones: constraints from U-series disequilibria. *Journal of Geophysical Research: Solid Earth*, **108**, http://dx.doi.org/10.1029/2002JB001839

BRYAN, S. E., OROZCO-ESQUIVEL, T., FERRARI, L. & LÓPEZ-MARTÍNEZ, M. 2013. Pulling apart the Mid to Late Cenozoic magmatic record of the Gulf of California: is there a Comondú Arc? *In*: GÓMEZ-TUENA, A., STRAUB, S. M. & ZELLMER, G. F. (eds) *Orogenic Andesites and Crustal Growth*. Geological Society, London, Special Publications, **385**. First published online July 31, 2013, http://dx.doi.org/10.1144/SP385.8

CARMICHAEL, I. S. E. 2002. The andesite aqueduct: perspectives on the evolution of intermediate magmatism in west-central (105°–99°W) Mexico. *Contributions to Mineralogy and Petrology*, **143**, 641–663.

CASHMAN, K. V. & MARSH, B. D. 1988. Crystal size distribution (CSD) in rocks and the kinetics and dynamics of crystallization II: Makaopuhi lava lake. *Contributions to Mineralogy and Petrology*, **99**, 292–305, http://dx.doi.org/10.1007/BF00375363

CASTILLO, P. R. 2012. Adakite petrogenesis. *Lithos*, **134–135**, 304–316, http://dx.doi.org/10.1016/j.lithos.2011.09.013

CASTRO, A., VOGT, K. & GERYA, T. 2013. Generation of new continental crust by sublithospheric silicic-magma relamination in arcs: a test of Taylor's andesite model. *Gondwana Research*, **23**, 1554–1566, http://dx.doi.org/10.1016/j.gr.2012.07.004

CAWOOD, P. A., HAWKESWORTH, C. J. & DHUIME, B. 2013. The continental record and the generation of continental crust. *Geological Society of America Bulletin*, **125**, 14–32, http://dx.doi.org/10.1130/B30722.1

CONDOMINES, M., HEMOND, C. & ALLÈGRE, C. J. 1988. U–Th–Ra radioactive disequilibria and magmatic processes. *Earth and Planetary Science Letters*, **90**, 243–262, http://dx.doi.org/10.1016/0012-821X(88)90129-X

CONDOMINES, M., GAUTHIER, P. J. & SIGMARSSON, O. 2003. Timescales of magma chamber processes and dating of young volcanic rocks. *In*: BOURDON, B., HENDERSON, G. M., LUNDSTROM, C. C. & TURNER, S. P. (eds) *Uranium-Series Geochemistry*. Reviews in Mineralogy and Geochemistry. The Mineralogical Society of America, Washington, DC, 125–174.

COSTA, F., CHAKRABORTY, S. & DOHMEN, R. 2003. Diffusion coupling between trace and major elements and a model for calculation of magma residence times using plagioclase. *Geochimica et Cosmochimica Acta*, **67**, 2189–2200, http://dx.doi.org/10.1016/S0016-7037(02)01345-5

DAVIDSON, J. P. & TEPLEY, F. J. 1997. Recharge in volcanic systems: evidence from isotope profiles of phenocrysts. *Science*, **275**, 826–829, http://dx.doi.org/10.1126/science.275.5301.826

DAVIDSON, J., TURNER, S., HANDLEY, H., MACPHERSON, C. & DOSSETO, A. 2007. Amphibole 'sponge' in arc crust? *Geology*, **35**, 787–790, http://dx.doi.org/10.1130/G23637A.1

DEFANT, M. & DRUMMOND, M. 1990. Derivation of some modern arc magmas by melting of young subducted lithosphere. *Nature*, **347**, 662–665.

DONN, W. L., DONN, B. D. & VALENTINE, W. G. 1965. On the Early History of the Earth. *Geological Society of America Bulletin*, **76**, 287–306, http://dx.doi.org/10.1130/0016-7606(1965)76[287:OTEHOT]2.0.CO;2

DOSSETO, A. & TURNER, S. 2013. Reappraisal of uranium-series isotope data in Kamchatka lavas: implications for continental arc magma genesis. *In*: GÓMEZ-TUENA, A., STRAUB, S. M. & ZELLMER, G. F. (eds) *Orogenic Andesites and Crustal Growth*. Geological Society, London, Special Publications, **385**. First published online June 11, 2013, http://dx.doi.org/10.1144/SP385.5

DOSSETO, A., BOURDON, B., JORON, J.-L. & DUPRÉ, B. 2003. U–TH–PA–RA study of the Kamchatka arc: new constraints on the genesis of arc lavas. *Geochimica et Cosmochimica Acta*, **67**, 2857–2877, http://dx.doi.org/10.1016/S0016-7037(03)00086-3

EICHELBERGER, J. C. 1978. Andesitic volcanism and crustal evolution. *Nature*, **275**, 21–27, http://dx.doi.org/10.1038/275021a0

ELBURG, M. A., SMET, I. & PELSMAEKER, E. D. 2013. Influence of source materials and fractionating assemblage on magmatism along the Aegean Arc, and implications for crustal growth. *In*: GÓMEZ-TUENA, A., STRAUB, S. M. & ZELLMER, G. F. (eds) *Orogenic Andesites and Crustal Growth*. Geological Society, London, Special Publications, **385**. First published online May 30, 2013, http://dx.doi.org/10.1144/SP385.1

ELKINS-TANTON, L. T. 2007. Continental magmatism, volatile recycling, and a heterogeneous mantle caused by lithospheric gravitational instabilities. *Journal of Geophysical Research*, **112**, B03405, http://dx.doi.org/10.1029/2005JB004072

ELLIOTT, T., PLANK, T., ZINDLER, A., WHITE, W. & BOURDON, B. 1997. Element transport from slab to volcanic front at the Mariana arc. *Journal of Geophysical Research*, **102**, 14 991–15 019.

FEDOTOV, S. A., GOREL'CHIK, V. I., STEPANOV, V. V. & GARBUZOVA, V. T. 1983. The development of the great Tolbachik fissure eruption in 1975 from seismological data. *In*: FEDOTOV, S. A. & MARKHININ, Y. K. (eds) *The Great Tolbachik Fissure Eruption*. Cambridge University Press, Cambridge, 189–203.

GAETANI, G. A. & GROVE, T. L. 1998. The influence of water on melting of mantle peridotite. *Contributions to Mineralogy and Petrology*, **131**, 323–346.

GERYA, T. V. & YUEN, D. A. 2003. Rayleigh–Taylor instabilities from hydration and melting propel 'cold plumes' at subduction zones. *Earth and Planetary Science Letters*, **212**, 47–62, http://dx.doi.org/10.1016/S0012-821X(03)00265-6

GERYA, T. V., YUEN, D. A. & SEVRE, E. O. D. 2004. Dynamical causes for incipient magma chambers above slabs. *Geology*, **32**, 89–92, http://dx.doi.org/10.1130/G20018.1

GHIORSO, M. S. & SACK, R. O. 1995. Chemical mass transfer in magmatic processes IV. A revised and internally consistent thermodynamic model for the interpolation and extrapolation of liquid–solid equilibria in magmatic systems at elevated temperatures and pressures. *Contributions to Mineralogy and Petrology*, **119**, 197–212, http://dx.doi.org/10.1007/BF00307281

Ghiorso, M. S., Hirschmann, M. M., Reiners, P. W. & Kress, V. C. 2002. The pMELTS: a revision of MELTS for improved calculation of phase relations and major element partitioning related to partial melting of the mantle to 3 GPa. *Geochemistry, Geophysics, Geosystems*, **3**, 1–35, http://dx.doi.org/10.1029/2001GC000217

Gill, J. 1981. *Orogenic Andesites and Plate Tectonics*. Springer, Berlin.

Gómez-Tuena, A., Mori, L., Rincón-Herrera, N. E., Ortega-Guitérrez, F., Solé, J. & Iriondo, A. 2008. The origin of a primitive trondhjemite from the Trans-Mexican Volcanic Belt and its implications for the construction of a modern continental arc. *Geology*, **36**, 471–474.

Gómez-Tuena, A., Mori, L., Goldstein, S. L. & Pérez-Arvizu, O. 2011. Magmatic diversity of western Mexico as a function of metamorphic transformations in the subducted oceanic plate. *Geochimica et Cosmochimica Acta*, **75**, 213–241, http://dx.doi.org/10.1016/j.gca.2010.09.029

Gómez-Tuena, A., Díaz-Bravo, B., Vázquez-Duarte, A., Pérez-Arvizu, O. & Mori, L. 2013. Andesite petrogenesis by slab-derived plume pollution of a continental rift. *In*: Gómez-Tuena, A., Straub, S. M. & Zellmer, G. F. (eds) *Orogenic Andesites and Crustal Growth*. Geological Society, London, Special Publications, **385**. First published online June 11, 2013, http://dx.doi.org/10.1144/SP385.4

Goss, A. R. & Kay, S. M. 2006. Steep REE patterns and enriched Pb isotopes in southern Central American arc magmas: evidence for forearc subduction erosion? *Geochemistry, Geophysics, Geosystems*, **7**, http://dx.doi.org/10.1029/2005GC001163

Greene, A. R., DeBari, S. M., Kelemen, P. B., Blusztajn, J. & Clift, P. D. 2006. A detailed geochemical study of Island Arc Crust: the Talkeetna Arc Section, South–Central Alaska. *Journal of Petrology*, **47**, 1051–1093, http://dx.doi.org/10.1093/petrology/egl002

Grove, T. L., Till, C. B. & Krawczynski, M. J. 2012. The role of H_2O in subduction zone magmatism. *Annual Review of Earth and Planetary Sciences*, **40**, 413–439, http://dx.doi.org/10.1146/annurev-earth-042711-105310

Haase, K. M., Krumm, S., Regelous, M. & Joachimski, M. 2011. Oxygen isotope evidence for the formation of silicic Kermadec island arc and Havre–Lau backarc magmas by fractional crystallisation. *Earth and Planetary Science Letters*, **309**, 348–355, http://dx.doi.org/10.1016/j.epsl.2011.07.014

Hacker, B. R. 2008. H_2O subduction beyond arcs. *Geochemistry, Geophysics, Geosystems*, **9**, http://dx.doi.org/10.1029/2007GC001707

Hacker, B. R., Kelemen, P. B. & Behn, M. D. 2011. Differentiation of the continental crust by relamination. *Earth and Planetary Science Letters*, **307**, 501–516, http://dx.doi.org/10.1016/j.epsl.2011.05.024

Harrison, T. M. 2009. The Hadean Crust: evidence from >4 Ga Zircons. *Annual Review of Earth and Planetary Sciences*, **37**, 479–505, http://dx.doi.org/10.1146/annurev.earth.031208.100151

Hermann, J. & Spandler, C. 2008. Sediment melts at sub-arc depths: an experimental study. *Journal of Petrology*, **49**, 717–740.

Higgins, M. D. 1996. Magma dynamics beneath Kameni volcano, Thera, Greece, as revealed by crystal size and shape measurements. *Journal of Volcanology and Geothermal Research*, **70**, 37–48, http://dx.doi.org/10.1016/0377-0273(95)00045-3

Hildreth, W. & Moorbath, S. 1988. Crustal contributions to arc magmatism in the Andes of central Chile. *Contributions to Mineralogy and Petrology*, **98**, 455–489.

Hirose, K. 1997. Melting experiments on lherzolite KLB-1 under hydrous conditions and generation of high-magnesian andesites. *Geology*, **25**, 42–44.

Hofmann, A. W. 1988. Chemical differentiation of the Earth: the relationship between mantle, continental crust, and oceanic crust. *Earth and Planetary Science Letters*, **90**, 297–314, http://dx.doi.org/10.1016/0012-821X(88)90132-X

Huybers, P. & Langmuir, C. 2009. Feedback between deglaciation, volcanism, and atmospheric CO_2. *Earth and Planetary Science Letters*, **286**, 479–491, http://dx.doi.org/10.1016/j.epsl.2009.07.014

Ishizuka, O., Taylor, R. N., Yuasa, M. & Ohara, Y. 2011. Making and breaking an island arc: a new perspective from the Oligocene Kyushu–Palau arc, Philippine Sea. *Geochemistry, Geophysics, Geosystems*, **12**, http://dx.doi.org/10.1029/2010GC003440

Jull, M. & Kelemen, P. B. 2001. On the conditions for lower crustal convective instability. *Journal of Geophysical Research*, **106**, 6423–6446, http://dx.doi.org/10.1029/2000JB900357

Kay, R. 1978. Aleutian magnesian andesites: melts from subduction Pacific Oceanic crust. *Journal of Volcanology and Geothermal Research*, **4**, 117–132.

Kay, R. & Kay, S. 1993. Delamination and delamination magmatism. *Tectonophysics*, **219**, 177–189.

Kay, S. M., Mpodozis, C. & Gardeweg, M. 2013. Magma sources and tectonic setting of Central Andean andesites (25.5–28°S) related to crustal thickening, forearc subduction erosion and delamination. *In*: Gómez-Tuena, A., Straub, S. M. & Zellmer, G. F. (eds) *Orogenic Andesites and Crustal Growth*. Geological Society, London, Special Publications, **385**. First published online August 7, 2013, http://dx.doi.org/10.1144/SP385.11

Kelemen, P. B. 1995. Genesis of high Mg# andesites and the continental crust. *Contributions to Mineralogy and Petrology*, **120**, 1–19, http://dx.doi.org/10.1007/BF00311004

Kelemen, P. B., Yogodzinksi, G. & Scholl, D. W. 2003. Along-strike Variation in the Aleutian Island Arc: genesis of high-mg# andesite and implication for the continental crust. *In*: Eiler, J. (ed.) *Inside the Subduction Factory*. Geophysical Monograph. American Geophysical Union, Washington, DC, 223–276.

Keller, C. B. & Schoene, B. 2012. Statistical geochemistry reveals disruption in secular lithospheric evolution about 2.5 Gyr ago. *Nature*, **485**, 490–493, http://dx.doi.org/10.1038/nature11024

Kennett, J. P. & Thunell, R. C. 1977. On explosive Cenozoic Volcanism and climatic implications.

Science, **196**, 1231–1234, http://dx.doi.org/10.1126/science.196.4295.1231

Kent, A. J. K. 2013. Preferential eruption of andesitic magmas: Implications for volcanic magma fluxes at convergent margins. *In*: Gómez-Tuena, A., Straub, S. M. & Zellmer, G. F. (eds) *Orogenic Andesites and crustal Growth*. Geological Society, London, Special Publications, **385**. First published online September 12, 2013, http://dx.doi.org/10.1144/SP385.10

Kent, A. J. R., Darr, C., Koleszar, A. M., Salisbury, M. J. & Cooper, K. M. 2010. Preferential eruption of andesitic magmas through recharge filtering. *Nature Geoscience*, **3**, 631–636, http://dx.doi.org/10.1038/ngeo924

Kessel, R., Schmidt, M., Ulmer, P. & Pettke, T. 2005. Trace element signature of subduction-zone fluids, melts and supercritical liquids at 120–180 km depth. *Nature*, **437**, 724–727.

Klimm, K., Blundy, J. & Green, T. 2008. Trace element partitioning and accessory phase saturation during H_2O-saturated melting of basalt with implications for subduction zone chemical fluxes. *Journal of Petrology*, **49**, 523–553.

Kodaira, S., Sato, T., Takahashi, N., Ito, A., Tamura, Y., Tatsumi, Y. & Kaneda, Y. 2007. Seismological evidence for variable growth of crust along the Izu intraoceanic arc. *Journal of Geophysical Research: Solid Earth*, **112**, http://dx.doi.org/10.1029/2006JB004593

Kutterolf, S., Jegen, M., Mitrovica, J. X., Kwasnitschka, T., Freundt, A. & Huybers, P. J. 2013. A detection of Milankovitch frequencies in global volcanic activity. *Geology*, **41**, 227–230, http://dx.doi.org/10.1130/G33419.1

Lasaga, A. C. 1983. Geospeedometry: an extension of geothermometry. *In*: Saxena, S. K. (ed.) *Kinetics and Equilibrium in Mineral Reactions*. Springer, New York, 81–114.

Lee, J., Stern, R. J. & Bloomer, S. H. 1995. Forty million years of magmatic evolution in the Mariana arc: the tephra glass record. *Journal of Geophysical Research: Solid Earth*, **100**, 17 671–17 687, http://dx.doi.org/10.1029/95JB01685

Leeman, W. P. 1983. The influence of crustal structure on compositions of subduction-related magmas. *Journal of Volcanology and Geothermal Research*, **18**, 561–588, http://dx.doi.org/10.1016/0377-0273(83)90026-4

MacPherson, C., Dreher, S. & Thirlwall, M. F. 2006. Adakites without slab melting: high pressure differentiation of island arc magma, Mindanao, the Philippines. *Earth and Planetary Science Letters*, **243**, 581–593.

Marschall, H. R. & Schumacher, J. C. 2012. Arc magmas sourced from melange diapirs in subduction zones. *Nature Geoscience*, **5**, 862–867, http://dx.doi.org/10.1038/ngeo1634

Miller, D. M., Goldstein, S. L. & Langmuir, C. 1994. Cerium/lead and lead isotope ratios in arc magmas and the enrichment of Pb in the continents. *Nature*, **368**, 514–520.

Morgan, D. J. & Blake, S. 2006. Magmatic residence times of zoned phenocrysts: introduction and application of the binary element diffusion modelling (BEDM) technique. *Contributions to Mineralogy and Petrology*, **151**, 58–70, http://dx.doi.org/10.1007/s00410-005-0045-4

Mori, L., Gómez-Tuena, A., Schaaf, P., Goldstein, S. L., Pérez-Arvizu, O. & Solís-Pichardo, G. 2009. Lithospheric removal as a trigger for flood basalt magmatism in the Trans-Mexican Volcanic Belt. *Journal of Petrology*, **50**, 2157–2186, http://dx.doi.org/10.1093/petrology/egp072

Morris, J., Leeman, W. & Tera, F. 1990. The subducted component in island arc lavas: cEonstraints from Be isotopes and B–Be systematics. *Nature*, **344**, 31–36.

Müntener, O., Kelemen, P. & Grove, T. 2001. The role of H_2O during crystallization of primitive arc magmas under uppermost mantle conditions and genesis of igneous pyroxenites: an experimental study. *Contributions to Mineralogy and Petrology*, **141**, 643–658.

Newman, S., Macdougall, J. D. & Finkel, R. C. 1984. ^{230}Th–^{238}U disequilibrium in island arcs: evidence from the Aleutians and the Marianas. *Nature*, **308**, 268–270, http://dx.doi.org/10.1038/308268a0

Nishimoto, S., Ishikawa, M., Arima, M. & Yoshida, T. 2005. Laboratory measurement of P-wave velocity in crustal and upper mantle xenoliths from Ichino-megata, NE Japan: ultrabasic hydrous lower crust beneath the NE Honshu arc. *Tectonophysics*, **396**, 245–259, http://dx.doi.org/10.1016/j.tecto.2004.12.010

Ortega-Gutiérrez, F., Elías-Herrera, M. & Dávalos-Elizondo, M. G. 2008. On the nature and role of the lower crust in the volcanic front of the Trans-Mexican Volcanic Belt and its fore-arc region, southern and central Mexico. *Revista Mexicana de Ciencias Geológicas*, **25**, 346–364.

Parman, S. & Grove, T. 2004. Harzburgite melting with and without H_2O: experimental data and predictive modeling. *Journal of Geophysical Research*, **109**, http://dx.doi.org/10.1029/2003JB002566

Plank, T. 2005. Constraints from thorium/lanthanum on sediment recycling at subduction zones and the evolution of the continents. *Journal of Petrology*, **46**, 921–944, http://dx.doi.org/10.1093/petrology/egi005

Plank, T. & Langmuir, C. H. 1988. An evaluation of the global variations in the major element chemistry of arc basalts. *Earth and Planetary Science Letters*, **90**, 349–370, http://dx.doi.org/10.1016/0012-821X(88)90135-5

Plank, T. & Langmuir, C. H. 1993. Tracing trace elements from sediment input to volcanic output at subduction zones. *Nature*, **362**, 739–743, http://dx.doi.org/10.1038/362739a0

Plank, T. & Langmuir, C. 1998. The chemical composition of subducting sediment and its consequences for the crust and mantle. *Chemical Geology*, **145**, 325–394.

Plank, T., Cooper, L. B. & Manning, C. E. 2009. Emerging geothermometers for estimating slab surface temperatures. *Nature Geoscience*, **2**, 611–615, http://dx.doi.org/10.1038/ngeo614

Plank, T., Kelley, K. A., Zimmer, M. M., Hauri, E. H. & Wallace, P. J. 2013. Why do mafic arc magmas contain ~4 wt% water on average? *Earth and Planetary Science Letters*, **364**, 168–179, http://dx.doi.org/10.1016/j.epsl.2012.11.044

Ranero, C. R., Phipps Morgan, J., McIntosh, K. & Reichert, C. 2003. Bending, faulting and mantle serpentinization at ocean trenches. *Nature*, **425**, 367–373.

RAPP, R., SHIMIZU, N. & NORMAN, M. 2003. Growth of early continental crust by partial melting of eclogite. *Nature*, **425**, 605–609.

REUBI, O. & BLUNDY, J. 2009. A dearth of intermediate melts at subduction zone volcanoes and the petrogenesis of arc andesites. *Nature*, **461**, 1269–1273, http://dx.doi.org/10.1038/nature08510

RINGWOOD, A. E. 1974. The petrological evolution of island arc systems: twenty-seventh William Smith Lecture. *Journal of the Geological Society, London*, **130**, 183–204, http://dx.doi.org/10.1144/gsjgs.130.3.0183

RUDNICK, R. L. 1995. Making continental crust. *Nature*, **378**, 571–578, http://dx.doi.org/10.1038/378571a0

RUDNICK, R. & GAO, S. 2003. Composition of the continental crust. *In*: RUDNICK, R. (ed.) *Treatise on Geochemistry*. Elsevier, Oxford, 1–64.

RUPRECHT, P. & PLANK, T. 2013. Feeding andesitic eruptions with a high-speed connection from the mantle. *Nature*, **500**, 68–72, http://dx.doi.org/10.1038/nature12342

RUSCITTO, D. M., WALLACE, P. J., COOPER, L. B. & PLANK, T. 2012. Global variations in H_2O/Ce: 2. Relationships to arc magma geochemistry and volatile fluxes. *Geochemistry, Geophysics, Geosystems*, **13**, http://dx.doi.org/10.1029/2011GC003887

SAVOV, I. P., RYAN, J. G., D'ANTONIO, M., KELLEY, K. & MATTIE, P. 2005. Geochemistry of serpentinized peridotites from the Mariana Forearc Conical Seamount, ODP Leg 125: implications for the elemental recycling at subduction zones. *Geochemistry, Geophysics, Geosystems*, **6**, http://dx.doi.org/10.1029/2004GC000777

SCHMIDT, M. & POLI, S. 1998. Experimentally based water budgets for dehydrating slabs and consequences for arc magma generation. *Earth and Planetary Science Letters*, **163**, 361–379.

SCHMITT, A. K. 2011. Uranium series accessory crystal dating of magmatic processes. *Annual Review of Earth and Planetary Sciences*, **39**, 321–349, http://dx.doi.org/10.1146/annurev-earth-040610-133330

SHIBATA, T., YOSHIKAWA, M., ITOH, J., UJIKE, O., MIYOSHI, M. & TAKEMURA, K. 2013. Along-arc geochemical variations in Quaternary magmas of northern Kyushu Island, Japan. *In*: GÓMEZ-TUENA, A., STRAUB, S. M. & ZELLMER, G. F. (eds) *Orogenic Andesites and Crustal Growth*. Geological Society, London, Special Publications, **385**. First published online August 15, 2013, http://dx.doi.org/10.1144/SP385.13

SIGMARSSON, O., MARTIN, H. & KNOWLES, J. 1998. Melting of a subducting oceanic crust from U–Th disequilibria in austral Andean lavas. *Nature*, **394**, 566–569.

SINGER, B. S., JICHA, B. R. ET AL. 2013. Lying in wait: deep and shallow evolution of dacite beneath Volcán de Santa María, Guatemala. *In*: GÓMEZ-TUENA, A., STRAUB, S. M. & ZELLMER, G. F. (eds) *Orogenic Andesites and Crustal Growth*. Geological Society, London, Special Publications, **385**. First published online June 11, 2013, http://dx.doi.org/10.1144/SP385.2

SMITH, I. E. M. 2013. High-magnesium andesites: the example of the Papuan Volcanic Arc. *In*: GÓMEZ-TUENA, A., STRAUB, S. M. & ZELLMER, G. F. (eds) *Orogenic Andesites and Crustal Growth*. Geological Society, London, Special Publications, **385**. First published online July 31, 2013, http://dx.doi.org/10.1144/SP385.7

STEINER, A. & STRECK, M. J. 2013. The Strawberry Volcanics: generation of 'orogenic' andesites from tholeiite within an intra-continental volcanic suite centred on the Columbia River flood basalt province, USA. *In*: GÓMEZ-TUENA, A., STRAUB, S. M. & ZELLMER, G. F. (eds) *Orogenic Andesites and Crustal Growth*. Geological Society, London, Special Publications, **385**. First published online August 6, 2013, http://dx.doi.org/10.1144/SP385.12

STERN, R. J. & SCHOLL, D. W. 2010. Yin and yang of continental crust creation and destruction by plate tectonic processes. *International Geology Review*, **52**, 1–31, http://dx.doi.org/10.1080/00206810903332322

STERN, R. J., TAMURA, Y. ET AL. 2013. Volcanoes of the Diamante cross-chain: evidence for a mid-crustal felsic magma body beneath the Southern Izu–Bonin–Mariana arc. *In*: GÓMEZ-TUENA, A., STRAUB, S. M. & ZELLMER, G. F. (eds) *Orogenic Andesites and Crustal Growth*. Geological Society, London, Special Publications, **385**. First published online June 27, 2013, http://dx.doi.org/10.1144/SP385.6

STRAUB, S. M. 2003. The evolution of the Izu Bonin–Mariana volcanic arcs (NW Pacific) in terms of major elements. *Geochemistry, Geophysics, Geosystems*, **4**, 1018, http://dx.doi.org/10.1029/2002GC000357

STRAUB, S. M. & LAYNE, G. D. 2003. The systematics of chlorine, fluorine, and water in Izu arc front volcanic rocks: implications for volatile recycling in subduction zones. *Geochimica et Cosmochimica Acta*, **67**, 4179–4203, http://dx.doi.org/10.1016/S0016-7037(03)00307-7

STRAUB, S. M. & ZELLMER, G. F. 2012. Volcanic arcs as archives of plate tectonic change. *Gondwana Research*, **21**, 495–516, http://dx.doi.org/10.1016/j.gr.2011.10.006

STRAUB, S. M., GOLDSTEIN, S. L., CLASS, C., SCHMIDT, A. & GÓMEZ-TUENA, A. 2010. Slab and mantle controls on the Sr–Nd–Pb–Hf isotope evolution of the post 42 Ma Izu–Bonin Volcanic Arc. *Journal of Petrology*, **51**, 993–1026, http://dx.doi.org/10.1093/petrology/egq009.

STRAUB, S. M., GÓMEZ-TUENA, A., STUART, F. M., ZELLMER, G. F., ESPINASA-PERENA, R., CAI, Y. & IIZUKA, Y. 2011. Formation of hybrid arc andesites beneath thick continental crust. *Earth and Planetary Science Letters*, **303**, 337–347, http://dx.doi.org/10.1016/j.epsl.2011.01.013

STRAUB, S. M., ZELLMER, G. F., GÓMEZ-TUENA, A., ESPINASA-PEREÑA, R., MARTIN-DEL POZZO, A. L., STUART, F. M. & LANGMUIR, C. H. 2013. A genetic link between silicic slab components and calc-alkaline arc volcanism in central Mexico. *In*: GÓMEZ-TUENA, A., STRAUB, S. M. & ZELLMER, G. F. (eds) *Orogenic Andesites and Crustal Growth*. Geological Society, London, Special Publications, **385**. First published online September 12, 2013, http://dx.doi.org/10.1144/SP385.14

TATSUMI, Y. & EGGINS, S. 1995. *Subduction Zone Magmatism*, HOFFMAN, P., JEANLOZ, R. & KNOLL, A. (eds) Blackwell Science, Cambridge.

TATSUMI, Y., SHUKUNO, H., TANI, K., TAKAHASHI, N., KODAIRA, S. & KOGISO, T. 2008. Structure and growth of the Izu–Bonin–Mariana arc crust: 2. Role of crust–mantle transformation and the transparent Moho in arc crust evolution. *Journal of Geophysical Research: Solid Earth*, **113**, http://dx.doi.org/10.1029/2007JB005121

TAYLOR, S. R. 1967. The origin and growth of continents. *Tectonophysics*, **4**, 17–34, http://dx.doi.org/10.1016/0040-1951(67)90056-X

TAYLOR, S. R. & MCLENNAN, S. M. 1995. The geochemical evolution of the continental crust. *Reviews of Geophysics*, **33**, 241–265, http://dx.doi.org/10.1029/95RG00262

TERA, F., BROWN, L., MORRIS, J., SACKS, I. S., KLEIN, J. & MIDDLETON, R. 1986. Sediment incorporation in island arc magmas: inferences from Be-10. *Geochimica et Cosmochimica Acta*, **50**, 535–550.

THORPE, R. S. 1982. *Andesites. Orogenic Andesites and Related Rocks* (1st edn). John Wiley & Sons, Chichester.

TONARINI, S., LEEMAN, W. P. & LEAT, P. T. 2011. Subduction erosion of forearc mantle wedge implicated in the genesis of the South Sandwich Island (SSI) arc: evidence from boron isotope systematics. *Earth and Planetary Science Letters*, **301**, 275–284, http://dx.doi.org/10.1016/j.epsl.2010.11.008

TURNER, S., HAWKESWORTH, C., VAN CALSTEREN, P., HEATH, E., MACDONALD, R. & BLACK, S. 1996. U-series isotopes and destructive plate margin magma genesis in the Lesser Antilles. *Earth and Planetary Science Letters*, **142**, 191–207, http://dx.doi.org/10.1016/0012-821X(96)00078-7

TURNER, S., HAWKESWORTH, C. ET AL. 1997. $^{238}U-^{230}Th$ disequilibria, magma petrogenesis, and flux rates beneath the depleted Tonga–Kermadec island arc. *Geochimica et Cosmochimica Acta*, **61**, 4855–4884, http://dx.doi.org/10.1016/S0016-7037(97)00281-0

TURNER, S. P., GEORGE, R. M. M., EVANS, P. J., HAWKESWORTH, C. J. & ZELLMER, G. F. 2000. Time-scales of magma formation, ascent and storage beneath subduction-zone volcanoes. *Philosophical Transactions of the Royal Society of London. Series A: Mathematical, Physical and Engineering Sciences*, **358**, 1443–1464, http://dx.doi.org/10.1098/rsta.2000.0598

TURNER, S. P., BOURDON, B. & GILL, J. 2003. Insights into magma genesis at convergent margins from U-series isotopes. *In*: BOURDON, B., HENDERSON, G. M., LUNDSTROM, C. C. & TURNER, S. P. (eds) *Uranium Series Geochemistry*. Reviews in Mineralogy and Geochemistry. The Mineralogical Society of America, Washington, DC, 255–315.

WADE, J. A., PLANK, T. ET AL. 2005. The May 2003 eruption of Anatahan volcano, Mariana Islands: geochemical evolution of a silicic island-arc volcano. *Journal of Volcanology and Geothermal Research*, **146**, 139–170, http://dx.doi.org/10.1016/j.jvolgeores.2004.11.035

WILDE, S. A., VALLEY, J. W., PECK, W. H. & GRAHAM, C. M. 2001. Evidence from detrital zircons for the existence of continental crust and oceans on the Earth 4.4 Gyr ago. *Nature*, **409**, 175–178, http://dx.doi.org/10.1038/35051550

YOSHIDA, T., KIMURA, J.-I. ET AL. 2013. Evolution of the Late Cenozoic Magmatism and the Crust-Mantle structure in the NE Japan Arc. *In*: GÓMEZ-TUENA, A., STRAUB, S. M. & ZELLMER, G. F. (eds) *Orogenic Andesites and Crustal Growth*. Geological Society, London, Special Publications, **385**. First published online September 24, 2013, http://dx.doi.org/10.1144/SP385.15.

ZANDT, G., GILBERT, H., OWENS, T. J., DUCEA, M., SALEEBY, J. & JONES, C. H. 2004. Active foundering of a continental arc root beneath the southern Sierra Nevada in California. *Nature*, **431**, 41–46, http://dx.doi.org/10.1038/nature02847

ZELLMER, G. F. 2008. Some first-order observations on magma transfer from mantle wedge to upper crust at volcanic arcs. *In*: ANNEN, C. & ZELLMER, G. F. (eds) *Dynamics of Crustal Magma Transfer, Storage and Differentiation*. Geological Society, London, Special Publications, **304**, 15–31, http://dx.doi.org/10.1144/SP304.2

ZELLMER, G., TURNER, S. & HAWKESWORTH, C. 2000. Timescales of destructive plate margin magmatism: new insights from Santorini, Aegean volcanic arc. *Earth and Planetary Science Letters*, **174**, 265–281, http://dx.doi.org/10.1016/S0012-821X(99)00266-6

ZELLMER, G., SPARKS, R., HAWKESWORTH, C. & WIEDENBECK, M. 2003. Magma emplacement and remobilization time scales beneath Montserrat: insights from Sr and Ba zonation in plagioclase phenocrysts. *Journal of Petrology*, **44**, 1413–1431.

ZELLMER, G. F., ANNEN, C., CHARLIER, B. L. A., GEORGE, R. M. M., TURNER, S. P. & HAWKESWORTH, C. J. 2005. Magma evolution and ascent at volcanic arcs: constraining petrogenetic processes through rates and chronologies. *Journal of Volcanology and Geothermal Research*, **140**, 171–191, http://dx.doi.org/10.1016/j.jvolgeores.2004.07.020

ZELLMER, G. F., IIZUKA, Y., MIYOSHI, M., TAMURA, Y. & TATSUMI, Y. 2012. Lower crustal H_2O controls on the formation of adakitic melts. *Geology*, **40**, 487–490, http://dx.doi.org/10.1130/G32912.1

ZELLMER, G. F., FREYMUTH, H., CEMBRANO, J. M., CLAVERO, J. E., VELOSO, E. A. E. & SIELFELD, G. G. 2013*a*. Altered mineral uptake into fresh arc magmas: insights from U–Th isotopes of samples from Andean volcanoes under differential crustal stress regimes. *In*: GÓMEZ-TUENA, A., STRAUB, S. M. & ZELLMER, G. F. (eds) *Orogenic Andesites and Crustal Growth*. Geological Society, London, Special Publications, **385**. First published online July 18, 2013, http://dx.doi.org/10.1144/SP385.9

ZELLMER, G. F., SAKAMOTO, N., IIZUKA, Y., MIYOSHI, M., TAMURA, Y., HSIEH, H.-H. & YURIMOTO, H. 2013*b*. Crystal uptake into aphyric arc melts: insights from two-pyroxene pseudo-decompression paths, plagioclase hygrometry, and measurement of hydrogen in olivines from mafic volcanics of SW Japan. *In*: GÓMEZ-TUENA, A., STRAUB, S. M. & ZELLMER, G. F. (eds) *Orogenic Andesites and Crustal Growth*. Geological Society, London, Special Publications, **385**. First published online June 11, 2013, http://dx.doi.org/10.1144/SP385.3

Along-arc geochemical variations in Quaternary magmas of northern Kyushu Island, Japan

TOMOYUKI SHIBATA[1]*, MASAKO YOSHIKAWA[1], JUN'ICH ITOH[2], OSAMU UJIKE[3], MASAYA MIYOSHI[1,4] & KEIJI TAKEMURA[1]

[1]*Institute for Geothermal Research, Graduate School of Science, Kyoto University, Noguchibaru, Beppu, Oita, 874-0903, Japan*

[2]*Geological Survey of Japan, The National Institute of Advanced Industrial Science and Technology, Higashi 1-1-1, Tsukuba, Ibaragi 305-8566, Japan*

[3]*Graduate School of Science and Engineering, University of Toyama, Gofuku 3190, Toyama, Toyama 930-8555, Japan*

[4]*Geological Laboratory, Faculty of Education and Regional Studies, University of Fukui, Bunkyo 3-9-1, Fukui, Fukui 910-8507, Japan*

**Corresponding author (e-mail: tomo@bep.vgs.kyoto-u.ac.jp)*

Abstract: We analysed major and trace elements and Sr–Nd–Pb isotope ratios of Himeshima, Futagoyama and Aso volcanoes, northern Kyushu, Japan. The purpose of the study was to investigate the petrogenesis of Quaternary arc magmas of the northern Kyushu arc, Japan, through along-strike variations. The northern Kyushu arc is underlain by the Philippine Sea Plate, which is divided by the Kyushu–Palau Ridge into the Palaeogene West Philippine Basin segment (60–40 Ma) and the Miocene Shikoku Basin segment (26–15 Ma). The Sr/Y ratios of the arc lavas decrease from north to south along the volcanic front. Mixing relations in Sr–Nd–Pb isotope space suggest recycling of the subducted slab materials from the Philippine Sea Plate to the arc. The $^{87}Sr/^{86}Sr$ ratios decrease with increasing Sr/Y and SiO_2, which argues against a genetic link of fractional crystallization between adakites and basaltic magmas. The observations further suggest that partial melting of the hot and young Shikoku Basin slab produces the high Sr/Y component visible in the arc magmas in the north, whereas dehydration of the older West Philippine Basin slab produces the low Sr/Y arc magmas in the south.

Since Defant & Drummond (1990) suggested that adakitic magmas could be generated by the partial melting of hot subducted slab material under mantle depth conditions, it has been argued that both dehydration (e.g. Ishikawa & Nakamura 1991; Miller *et al.* 1994; Shibata & Nakamura 1997) and partial melting (Defant & Drummond 1990) of the subducting slab can feed slab components to the mantle wedge, with these processes being dependent on slab temperature (e.g. Seghedi *et al.* 2004; Castillo 2006). Adakites are characterized by high Al_2O_3 (>15 wt%) and Sr (>300 ppm) abundances but low Y (<15ppm) and Yb (<2 ppm) concentrations and, therefore, have high Sr/Y ratios (>50), which indicates a role of residual garnet in the source of the arc magmas (Defant & Drummond 1990). The Sr/Y ratio and Y concentration can decrease by fractional crystallization and/or assimilation of crustal materials so that the Sr/Y field of adakite is extended down to 20–40 (e.g. Defant & Drummond 1993). Because of the higher temperature required for slab melting, the melting of young and hot subducted slab was first related to the origin of adakite. Later studies have related adakite formation to special slab geometries (e.g. Defant & Kepezhinskas 2001; Seghedi *et al.* 2004), including remnant slab left in the upper mantle (e.g. Desonie 1992; Pe-Piper & Piper 1994), oblique or fast subduction (e.g. Sajona *et al.* 1993), arc–arc collision (e.g. Petford & Atherton 1996), subduction initiation (Peacock *et al.* 1994), slab tearing and the formation of slab windows (e.g. Yogodzinski *et al.* 2001) and flat subduction (e.g. Gutscher *et al.* 2000). However, it has also been proposed that magmas with the chemical properties of adakites can be generated without slab melting, for example during partial melting of garnet-bearing lower crustal material (e.g. Yumul *et al.* 2000) or fractionation of garnet at lower crustal depth due to low water contents in the primary basaltic magmas (Zellmer *et al.* 2012). Accordingly, further research on adakitic formation is required (e.g. Castillo 2006).

Quaternary arc volcanoes are common in SW Japan and the Ryukyu arcs (Fig. 1), and are related

From: GÓMEZ-TUENA, A., STRAUB, S. M. & ZELLMER, G. F. (eds) 2014. *Orogenic Andesites and Crustal Growth*. Geological Society, London, Special Publications, **385**, 15–29.
First published online August 15, 2013, http://dx.doi.org/10.1144/SP385.13

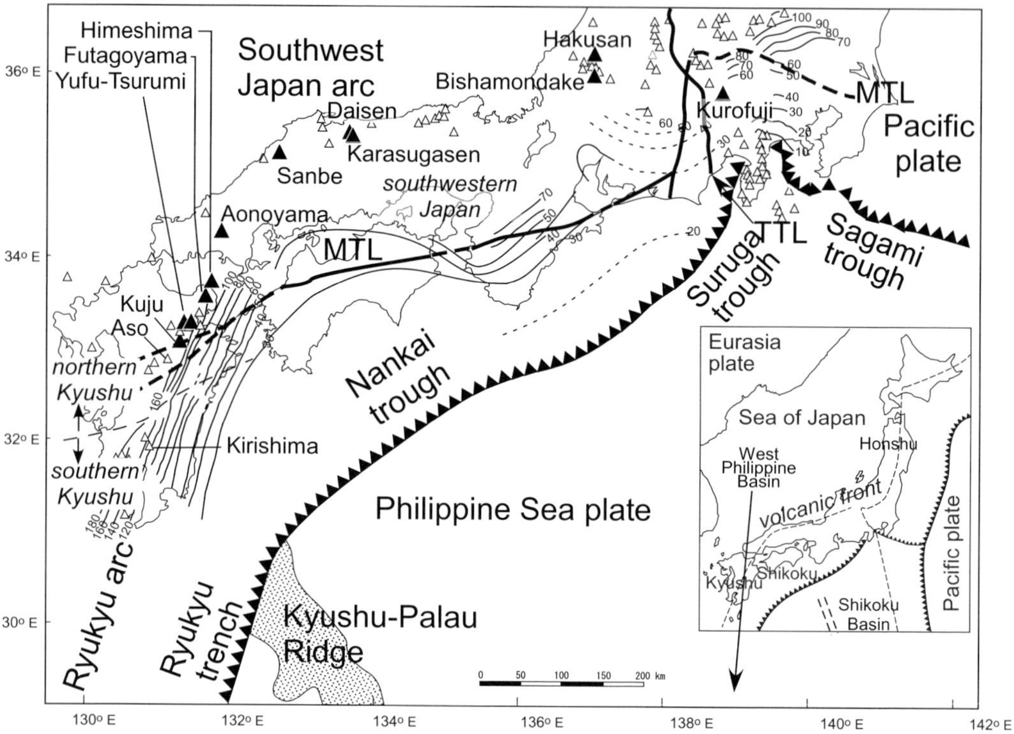

Fig. 1. Quaternary volcanoes and tectonics of the western Japanese Islands. Open and solid triangles indicate the locations of Quaternary volcanoes. Larger and solid triangles indicate sites of adakite eruptions. The line with closed triangles denotes trenches. Thin solid and dashed lines are isodepth contours of the upper boundaries of the seismic plane of PSP, after Noguchi (1998), Uehira *et al.* (2001) and Shibutani (2002). Thick solid and dashed lines are tectonic lines from Taira (2001). MTL and TTL denote the Median and Tanakura tectonic lines, respectively.

to the subduction of the Philippine Sea Plate (PSP) beneath the western and southern parts of Japan along the Nankai Trough and Ryukyu Trench (Fig. 1). Although it is not clearly defined, the arc–arc junction between the Japan and Ryukyu arcs is thought to exist somewhere between the islands of Kyushu and Shikoku.

In the SW Japan arc, adakitic lavas erupted at the Aonoyama, Sanbe, Daisen, Hakusan, Bishamondake and Kurofuji volcanoes of Honshu (e.g. Morris 1995; Kimura *et al.* 2005; Shibata *et al.* 2008; Ujike & Mashimo 2009) (Fig. 1). These adakitic magmas have high Sr/Y ratios (70–300 maximum in individual volcanoes) and low Y (5–10 ppm minimum in individual volcanoes), and are thought to have been generated by partial melting of subducted PSP material. This agrees with the young age of the subducted slab (26–15 Ma; Mahony *et al.* 2011), which is thus presumably hot (Defant & Drummond 1990).

The Quaternary volcanoes on Kyushu Island (Fig. 1) comprise the Himeshima, Futagoyama, Yufu–Tsurumi, Kuju, Aso, Kirishima and Sakurajima volcanoes, from north to south along-strike of the volcanic front (Fukuyama & Ono 1981; Kobayashi 1984; Itoh 1990; Ohta *et al.* 1990; Kamata 1997, 1998; Hoshizumi *et al.* 1998; Imura & Kobayashi 2001; Fujisawa *et al.* 2002; Furukawa *et al.* 2009). Adakitic lavas are restricted to Yufu–Tsurumi and Kuju volcanoes in northern Kyushu (Kita *et al.* 2001; Sugimoto *et al.* 2007), but no Quaternary adakites are observed at Aso (e.g. Hunter 1998) and the volcanoes south of Aso along the entire Ryukyu arc. Sugimoto *et al.* (2007) suggested that the trace element and Sr, Nd and Pb isotopic compositions of adakitic magmas from Yufu–Tsurumi volcano indicate derivation of the magmas by partial melting of the subducting PSP. In contrast, Zellmer *et al.* (2012) suggested that these adakites may have formed by fractional crystallization of mantle-derived mafic magmas within the garnet stability field in the crust.

The Honshu–Kyushu arc transition is a particularly favourable setting in which to assess these controversial models for the origin of the adakitic lavas, because of the potential relationship between the

PSP materials and the along-strike variation of the lava chemistry. The Kyushu–Palau ridge divides the oceanic crust of the PSP into northeastern and southwestern segments with ages of 26–15 Ma (Shikoku Basin) and 60–40 Ma (West Philippine Basin), respectively (Mahony *et al.* 2011). Although there are no clear plate images beneath northern Kyushu, the northern extension of the Kyushu–Palau ridge potentially corresponds to the boundary between the SW Japan and Ryukyu arcs. If adakite genesis was related to the subducted slab rather than the overlying crust, then the spatial distribution of Quaternary adakites should correlate with the age of the subducted PSP.

In order to test this correlation and determine the petrogenesis of the northern Kyushu adakites, we compiled major and trace elements and Sr–Nd–Pb isotope ratios from volcanoes along the arc front that includes the transition from adakitic to non-adakitic arc volcanism. Comprehensive geochemical data, including Sr, Nd and Pb isotope ratios, were restricted to Kuju and Yufu–Tsurumi volcanoes (Kita *et al.* 2001; Sugimoto *et al.* 2007), so we present new geochemical and Sr–Nd–Pb isotope ratios of lavas from Himeshima and Futagoyama volcanoes and Nd–Pb isotope ratios from Aso volcano. This enables us to investigate the along-strike arc variations of the Quaternary lavas in northern Kyushu. On the basis of these variations, and the potential correlation with the down-going PSP, we discuss processes that could generate adakitic magmas in the study area.

Samples

Lava samples were used to determine the geochemical and Sr–Nd–Pb isotopic compositions. The Himeshima volcano lavas are the hornblende–dacites and rhyolites of Itoh (1990), who reported major element oxides and some trace element abundance data. We collected hornblende dacites from Futagoyama lava domes. Post-caldera Aso basalts, andesite, dacite and rhyolites from Miyoshi *et al.* (2011) were used for the analyses of Nd–Pb isotopic ratios, and trace element abundances.

Analytical methods

Major and trace element abundances (Rb, Ba, Sr, Zr, Nb) of whole rocks were determined using a RIGAKU 3070 X-ray fluorescence (XRF) system using the glass beads and pressed pellets methods, respectively. Procedures and instrumental set-up are outlined in Sugimoto *et al.* (2007). Trace elements were analysed by VG Elemental PQ3 and Thermo Scientific X2 Series quadrupole inductively coupled plasma mass spectrometers (Q-ICP-MS), installed at the Institute for Geothermal Sciences (IGS), Kyoto University, Japan. The chemical procedures and analytical protocol are those of Chang *et al.* (2003). The analytical reproducibility for each element was determined by repeat analyses of the JB-3 rock powder (an international reference material from the Geological Survey of Japan) and was less than 3% (1σ, relative standard deviation), with the exception of Y (5%). Whole-rock Pb, Nd and Sr isotope ratios were measured using a Finnigan MAT 262 thermal ionization mass spectrometer (TIMS), installed at IGS, Kyoto University, Japan, following the methods outlined in Yoshikawa & Shibata (2003) and Shibata *et al.* (2007). The normalizing factors used to correct isotope fractionation during mass spectrometry are $^{86}Sr/^{88}Sr = 0.1194$, $^{146}Nd/^{144}Nd = 0.7219$, and 0.001% per atomic mass unit, respectively. The normalizing factor for the Pb isotope ratio was determined using NIST SRM981. The measured isotope ratios of the international standard materials are $^{87}Sr/^{86}Sr = 0.710279 \pm 0.000028$ (2σ) for NIST SRM 987; $^{143}Nd/^{144}Nd = 0.511851 \pm 0.000013$ (2σ) for La Jolla; and $^{206}Pb/^{204}Pb = 16.944 \pm 0.004$ (2σ), $^{207}Pb/^{204}Pb = 15.499 \pm 0.004$ (2σ) and $^{208}Pb/^{204}Pb = 36.721 \pm 0.010$ (2σ) for NIST SRM981.

Results

The major and trace element abundances and Sr–Nd–Pb isotope ratios of lavas from Himeshima, Futagoyama and Aso obtained in this study are listed in Tables 1 and 2, together with previously reported data by Itoh (1990) and Miyoshi *et al.* (2011). Selected geochemical data for the Quaternary magmas of northern Kyushu are also listed in Table 3, including mafic enclaves from Yufu–Tsurumi volcano, based on our data and data from previous studies (Itoh 1990; Ohta *et al.* 1990; Ohta & Aoki 1991; Hunter 1998; Kita *et al.* 2001; Sugimoto *et al.* 2007; Miyoshi *et al.* 2011).

The total alkali and silica (TAS) diagram (Le Maitre *et al.* 1989) is shown in Figure 2. Based on this diagram, andesites, dacites and rhyolites of northern Kyushu can be divided into two groups, with comparatively high and low alkali concentrations, where the alkaline series is only represented by the more evolved magmas from Aso volcano.

Along-arc geochemical variations in the northern Kyushu area

Major elements

Samples from the Himeshima, Futagoyama, Yufu–Tsurumi and Kuju volcanoes (Fig. 1) have SiO_2, Al_2O_3 and MgO abundances (Table 1) within the

Table 1. *Major and trace element concentrations of Himeshima, Futagoyama and Aso volcanoes, northern Kyushu, Japan*

	Himeshima						Futagoyama					
	F0319-1①	950305-2	L0309-2	Kabasu-dake	Lava	0703Dike	9071301	9071302	9071303	9071305	9071307	9071308
SiO_2	65.67*	73.69*	69.44*	14.00*	72.78*	68.43*	64.48*	64.02*	63.57*	64.76*	64.86*	64.89*
TiO_2	1.00*	15.00*	0.30*	1.00*	0.12*	0.38*	0.60*	0.60*	0.61*	0.61*	0.52*	0.55*
Al_2O_3	15.89*	15.26*	15.46*	15.51*	15.14*	16.01*	16.18*	16.64*	16.95*	17.09*	17.09*	16.04*
Fe_2O_3*	4.21*	1.25*	2.97*	2.76*	1.83*	3.15*	5.16*	5.28*	5.20*	5.34*	4.68*	4.46*
MnO	0.09*	0.09*	0.10*	0.08*	0.09*	0.08*	0.09*	0.09*	0.12*	0.12*	0.10*	0.14*
MgO	2.02*	0.25*	1.08*	0.86*	0.48*	1.02*	2.18*	1.95*	1.96*	2.07*	1.34*	1.83*
CaO	4.50*	1.74*	2.81*	2.42*	1.56*	3.22*	4.85*	4.85*	5.54*	4.23*	4.65*	5.49*
Na_2O	4.91*	3.77*	4.76*	4.44*	4.29*	4.69*	3.75*	3.81*	3.57*	3.55*	3.95*	4.26*
K_2O	1.77*	3.55*	2.64*	2.65*	3.35*	2.31*	2.01*	1.98*	1.91*	2.03*	2.06*	1.54*
P_2O_5	0.29*	0.03*	0.19*	0.07*	0.13*	0.18*	0.11*	0.10*	0.12*	0.08*	0.11*	0.12*
Total	99.86*	99.78*	99.75*	29.30*	99.77*	99.47*	99.41*	99.33*	99.54*	99.87*	99.34*	99.32*
Rb	35	35	56	56	74	50	50	48	44	52	49	32
Sr	1090	1100	824	752	286	916	686	681	769	604	719	689
Y	14.3	11.1	12.3	10.1	9.61	12.3	17.1	26.2	14.5	14.3	12.1	14.2
Zr	157	137	116	153	56	134	112	48	81	107	75	127
Nb	10	10	13	13	16	11	13	12	12	15	12	12
Ba	538	559	705	735	850	672	686	654	669	693	671	598
Cs	0.682	0.506	1.38	1.06	1.99	0.945	0.873	0.568	1.66	0.823	0.797	1.05
Pb	10.2	10.1	13.2	14.3	16.2	12.8	7.93	6.70	6.83	8.63	8.27	9.54
Th	3.42	3.38	2.94	2.80	2.60	2.44	5.71	3.64	4.24	6.52	4.56	5.27
U	0.913	0.928	1.32	1.06	1.62	0.753	1.21	0.843	1.01	1.1	0.999	1.04
La	18.2	15.5	14.7	14.3	10.5	17.1	24.1	48.2	21.2	23.9	22.3	21.9
Ce	40.1	31.5	32.6	31.3	23.7	39.0	46.1	45.5	42.3	46.2	43.1	43.7
Pr	5.12	3.81	4.13	3.91	2.87	4.87	5.29	9.46	4.72	4.94	4.76	4.93
Nd	21.4	15.3	16.7	15.4	11.0	19.8	20.2	35.7	18.0	18.2	17.9	19.0
Sm	4.43	3.23	3.64	3.23	2.72	4.14	3.82	6.03	3.41	3.34	3.19	3.61
Eu	1.28	1.18	0.898	0.858	0.459	1.13	1.17	1.53	1.08	1.05	1.03	1.16
Gd	3.85	2.82	3.19	2.76	2.52	3.52	3.9	6.13	3.38	3.4	3.12	3.53
Tb	0.512	0.387	0.445	0.36	0.359	0.461	0.517	0.801	0.442	0.453	0.389	0.454
Dy	2.9	2.26	2.42	2.05	1.88	2.55	2.92	4.33	2.44	2.52	2.07	2.52
Ho	0.55	0.429	0.445	0.36	0.325	0.46	0.584	0.845	0.480	0.501	0.400	0.490
Er	1.57	1.27	1.27	1.06	0.892	1.28	1.76	2.41	1.41	1.49	1.18	1.45
Tm	0.215	0.169	0.174	0.13	0.108	0.161	0.252	0.324	0.196	0.213	0.166	0.209
Yb	1.5	1.26	1.18	0.95	0.776	1.19	1.74	2.12	1.35	1.51	1.14	1.43
Lu	0.22	0.183	0.174	0.123	0.108	0.161	0.261	0.303	0.197	0.224	0.171	0.216

	Futagoyama			Aso							
	9102701	9102702	9102703	AS-OT	AS076	ASK333	AS079	AS041	AS091	AS034	Vb-01
SiO_2	65.55	66.10	65.93	71.10†	69.77†	66.04†	61.15†	58.60†	52.13†	48.18†	53.17†
TiO_2	0.52	0.57	0.57	0.45†	0.44†	0.70†	0.88†	0.84†	0.85†	0.87†	0.94†
Al_2O_3	16.68	15.54	15.69	14.26†	14.83†	16.08†	16.06†	17.64†	19.03†	17.63†	17.65†
Fe_2O_3*	3.95	4.40	4.44	2.25†	2.85†	4.22†	7.56†	6.45†	8.65†	11.52†	9.49†
MnO	0.06	0.07	0.08	0.05†	0.08†	0.10†	0.14†	0.15†	0.12†	0.16†	0.15†
MgO	2.02	2.25	2.35	0.56†	0.57†	1.05†	1.66†	2.53†	3.13†	6.06†	4.55†
CaO	4.61	4.15	4.15	1.66†	1.15†	2.70†	4.58†	5.74†	10.17†	11.78†	8.57†
Na_2O	4.05	4.10	3.98	3.40†	4.79†	4.58†	4.29†	4.10†	2.81†	2.17†	2.97†
K_2O	1.79	2.09	2.10	5.32†	4.75†	3.98†	3.21†	2.51†	1.63†	0.86†	2.05†
P_2O_5	0.09	0.13	0.14	0.08†	0.06†	0.13†	0.58†	0.26†	0.20†	0.18†	0.26†
Total	99.32	99.41	99.42	99.13†	99.27†	99.57†	100.11†	98.82†	98.82†	99.41†	99.80†
Rb	39	43	47	279†	161†	192†	111†	107†	61†	25†	63†
Sr	763	927	921	173†	138†	323†	526†	618†	798†	545†	531†
Y	11.7	14.2	13.9	29†	43†	36†	38†	29†	21†	16†	28†
Zr	107	107	102	344†	369†	350†	287†	216†	111†	69†	197†
Nb	9	8	8	18†	15†	17†	10†	11†	5†	4†	7†
Ba	628	639	643	689†	764†	814†	590†	573†	365†	158†	364†
Cs	0.56	1.18	0.84	16.1†	11.0	9.74†	5.80	5.70†	3.96†	0.656	4.50
Pb	6.93	8.75	8.71	20.4†	7.25	26.6†	5.02	17.8†	10.3†	1.28	3.33
Th	4.72	6.12	6.52	29.4†	20.1	19.8†	14.8	11.5†	6.1†	3.03	8.31
U	1.11	1.53	1.63	7.85†	5.3	5.24†	3.79	3.06†	1.63†	0.812	2.18
La	24.8	25.2	25.0	33.9†	43.73	34.8†	37.7	24.5†	15.8†	9.84	23.7
Ce	36.1	44.9	45.2	74.7†	87.9	78.5†	77.9	55.6†	34.9†	20.7	51.8
Pr	4.59	5.55	5.42	8.36†	9.39	9.17†	9.01	6.73†	4.48†	2.69	6.02
Nd	17.6	22.6	21.9	31.5†	38.2	36.8†	38.3	28.2†	19.5†	12.2	25.9
Sm	3.19	4.4	4.24	6.30†	8.00	7.58†	8.39	6.15†	4.45†	3.07	5.90
Eu	1.1	1.29	1.27	0.967†	1.54	1.69†	2.11	1.72†	1.34†	1.01	1.51
Gd	3.37	4.93	4.7	6.1†	8.08	7.51†	8.60	6.17†	4.60†	3.37	6.10
Tb	0.40	0.51	0.50	0.866†	1.14	1.06†	1.15	0.869†	0.641†	0.485	0.844
Dy	2.17	2.55	2.52	5.1†	6.84	6.22†	6.61	5.08†	3.75†	2.93	4.92
Ho	0.425	0.480	0.472	1.06†	1.41	1.28†	1.33	1.05†	0.760†	0.593	0.990
Er	1.26	1.43	1.43	3.29†	4.42	3.91†	3.93	3.15†	2.25†	1.73	2.95
Tm	0.177	0.196	0.192	0.495†	0.652	0.583†	0.553	0.458†	0.316†	0.2398	0.414
Yb	1.21	1.32	1.3	3.49†	4.51	4.00†	3.67	3.12†	2.10†	1.58	2.77
Lu	0.183	0.193	0.195	0.525†	0.689	0.613†	0.555	0.474†	0.319†	0.235	0.414

*Data from Itoh (1990).
†Miyoshi *et al.* (2011).
Fe_2O_3*: total Fe as Fe_2O_3*.

Table 2. *Sr–Nd–Pb isotopic compositions of the Quaternary magmas from Himeshima, Futagoyama and Aso volcanoes, northeastern Kyushu, Japan*

	$^{87}Sr/^{86}Sr$	$^{143}Nd/^{144}Nd$	$^{206}Pb/^{204}Pb$	$^{207}Pb/^{204}Pb$	$^{208}Pb/^{204}Pb$
Himeshima					
F0319-1①	0.703721 ± 08	0.512843 ± 09	18.198 ± 07	15.542 ± 07	38.254 ± 10
950305-2	0.703724 ± 09	0.512826 ± 09	18.189 ± 10	15.526 ± 09	38.200 ± 20
L0309-2	0.703665 ± 06	0.512862 ± 13	18.247 ± 12	15.547 ± 12	38.312 ± 34
Kabasu-dake	0.703664 ± 10	0.512840 ± 12	18.239 ± 08	15.548 ± 10	38.310 ± 36
Daruma Lava	0.704008 ± 07	0.512732 ± 12			
0703Dike	0.703693 ± 10	0.512813 ± 10	18.221 ± 09	15.535 ± 08	38.257 ± 22
Futagoyama					
F01	0.704012 ± 11	0.512706 ± 06	18.211 ± 01	15.554 ± 01	38.348 ± 03
F02	0.704011 ± 12	0.512679 ± 09	18.229 ± 06	15.573 ± 06	38.421 ± 15
F03	0.703978 ± 09	0.512653 ± 08	18.238 ± 05	15.601 ± 04	38.492 ± 10
F05	0.704096 ± 13	0.512687 ± 08	18.231 ± 03	15.533 ± 02	38.320 ± 05
F07	0.703845 ± 13	0.512711 ± 08	18.171 ± 05	15.517 ± 05	38.224 ± 12
F08	0.703862 ± 12	0.512702 ± 06	18.177 ± 02	15.534 ± 02	38.268 ± 07
F09	0.703890 ± 13	0.512704 ± 12	18.114 ± 04	15.492 ± 03	38.120 ± 09
F10	0.703668 ± 06	0.512878 ± 10	18.103 ± 01	15.499 ± 01	38.083 ± 02
F11	0.703688 ± 11	0.512826 ± 12	18.100 ± 01	15.491 ± 01	38.063 ± 03
Aso					
AS-OT	0.704355 ± 10	0.512739 ± 13	18.404 ± 05	15.618 ± 05	38.593 ± 11
AS076	0.704232 ± 13	0.512710 ± 09	18.343 ± 05	15.585 ± 06	38.457 ± 13
ASK333	0.704242 ± 12	0.512754 ± 13	18.332 ± 05	15.567 ± 05	38.413 ± 10
AS079	0.703996 ± 09	0.512792 ± 11	18.297 ± 02	15.553 ± 02	38.340 ± 04
AS041	0.704249 ± 13	0.512764 ± 11	18.352 ± 05	15.577 ± 05	38.431 ± 11
AS091	0.704000 ± 09	0.512783 ± 12	18.421 ± 04	15.556 ± 04	38.388 ± 10
AS034	0.704036 ± 11	0.512783 ± 11	18.278 ± 06	15.550 ± 05	38.300 ± 13
Vb-01	0.704060 ± 13	0.512599 ± 13	18.302 ± 03	15.558 ± 03	38.360 ± 06

range of adakites ($SiO_2 > 56$ wt%, $Al_2O_3 > 15$ wt% and $MgO < 3$ wt%) generated by the melting of the subducted slab (Table 3; Defant & Drummond 1990; Castillo 2006), although the maximum MgO concentrations from Yufu–Tsurumi (3.6 wt%) and Kuju (3.4 wt%) are slightly above the range of slab-derived adakites (Table 3). The SiO_2, Al_2O_3 and MgO concentrations of most of the samples from Aso volcano fall outside the range of adakites formed by slab melting (Table 3), although those of a post-caldera lava (ASO41; $SiO_2 = 58.6$ wt%, $Al_2O_3 = 17.6$ wt%, $MgO = 2.5$ wt%) are within the range of adakite.

Trace element compositions

Magmas from northern Kyushu are characterized by pronounced negative Nb anomalies and relative enrichments of large ion lithophile elements (LILE) (Fig. 3), features that are characteristic of magmas generated in island-arc settings (e.g. Wood *et al.* 1979; Perfit *et al.* 1980). This suggests contributions of the PSP slab materials to the origin of the Quaternary magmas along the volcanic front of northern Kyushu.

Trace element patterns of magmas from individual volcanoes show along-arc geochemical variations in the northern Kyushu arc (Fig. 3). Heavy rare earth element (HREE) abundances increase from north to south, as indicated by the mantle-normalized Yb values ($[Yb]_N$) of the lava samples from the Himeshima (1.5–3.0), Futagoyama (2.5–3.5), Yufu–Tsurumi (3.0–3.8), Kuju (3.0–4.4) and Aso (4.3–9.9) volcanoes (Table 3). HREE-depleted magmas are indicative of magma generation in the garnet stability field (e.g. Kay 1978), so the influence of garnet decreases towards the south. In addition, positive anomalies of Sr are observed at Himeshima, Futagoyama, Yufu–Tsurumi and Kuju volcanoes, whereas both positive and negative anomalies are found at Aso volcano. This observation may indicate that the petrogenesis or source material of Aso volcano is different from the other volcanoes in northern Kyushu.

Variations in Sr/Y ratios and Y concentrations

The relationship between Sr/Y ratios and Y concentrations within samples from the northern Kyushu area is shown in Figure 4. Sr/Y ratios from individual volcanoes partially overlap with those of the adjacent volcanoes (Table 3 and Fig. 4), but the

Table 3. *Selected geochemical character of Quaternary magmas from northern Kyushu*

	SiO_2 (wt%)	Al_2O_3 (wt%)	MgO (wt%)	Sr (ppm)	Y (ppm)	Yb (ppm)	$[Yb]_N$	Sr/Y	$^{87}Sr/^{86}Sr$
Himeshima	65.9–74.7*	15.0–16.2*	0.1–2.0*	286–1100	9.61–14.3	0.776–1.50	1.5–3.5	30–99	0.70367–0.70400
Futagoyama	63.6–66.1	16.2–17.1	2.0–2.4	600–900	11.7–17.1	1.21–1.76	2.5–3.5	40–66	0.70367–0.70400
Yufu–Tsurumi	58.3–65.8†	16.0–16.7†	1.8–3.6†	436–604†	14.4–26.6†	1.50–1.93†	3.0–3.8	20–42†	0.70389–0.70488†
Kuju	58.3–61.5‡	16.0–16.7‡	2.3–3.4‡	436–511‡	15–24‡	1.5–2.2‡	3.0–4.4	18–34‡	0.70402–0.70427‡
Aso	48.2–71.1§	14.3–19.1§	0.5–6.1§	138–798§	16–35.7§	1.6–4.9§	3.2–9.9	6–38§	0.70420–0.70446§
Slab-derived adakite	>56‖	>15‖	<3 (really up to 6)‖	>300‖	<15‖	<1.9‖	<3.9	>20‖	<0.704‖

*Itoh (1990).
†Sugimoto *et al.* (2007).
‡Kita *et al.* (2001).
§Hunter (1998) and Miyoshi *et al.* (2011).
‖Castillo (2006) and Defant & Drummond (1990).

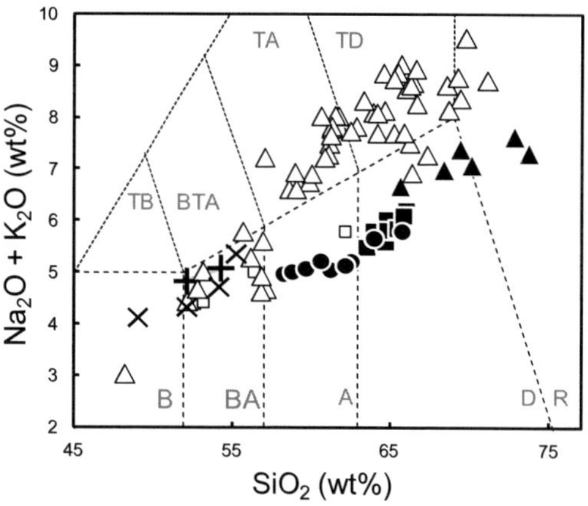

Fig. 2. SiO_2 v. total alkalis ($Na_2O + K_2O$) in wt%. Solid triangles, squares and circles represent data of Himeshima, Futagoyama and Yufu–Tsurumi volcanoes, respectively. Open squares and triangles are Kuju and Aso volcanoes, respectively. Data for mafic enclaves in lavas of Yufu–Tsurumi volcano and basalt of Oninomi volcano are shown as × and +, respectively. Rock types: A, andesite; B, basalt; BA, basaltic andesite; D, dacite; TB, trachy basalt; BTA, basaltic trachy-andesite; TA, trachy-andesite; TD, trachy-dacite; R, rhyolite (Le Maitre *et al.* 1989).

maximum Sr/Y decreases from the high Sr/Y at Himeshima volcano to lower Sr/Y at Futagoyama and Yufu–Tsurumi volcanoes, and to the lowest Sr/Y ratio at Aso. Similar observations have been presented by Mahony *et al.* (2011) and Zellmer *et al.* (2012). Concurrent with the Sr/Y ratios, Y increases from the north (Y = 9.61–14.3 ppm) to the south (Y = 20.8–35.7 ppm) (Fig. 4).

According to the discrimination proposed by Castillo (2006), adakites derived by slab melting have Sr/Y ratios of >20 and Y of <15 ppm. Using these criteria, the northern Kyushu dacite lavas from Himeshima, Futagoyama and Yufu–Tsurumi are classified as adakites. Although some samples from Aso volcano reach a maximum Sr/Y ratio of 38, all samples have Y concentrations of >16 ppm. Therefore, Aso magmas cannot be classified as adakites.

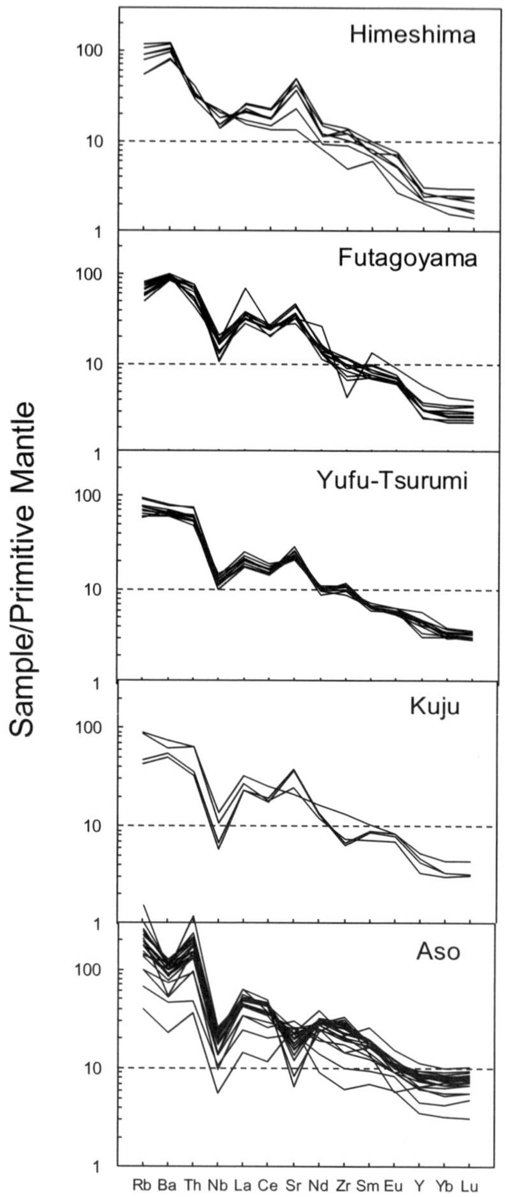

Fig. 3. Primitive mantle-normalized multi-element diagrams for five Quaternary volcanoes. Data for Yufu–Tsurumi, Kuju and part of Aso are from Sugimoto *et al.* (2007), Kita *et al.* (2001) and Hunter (1998), respectively. The values for the primitive mantle are from Sun & McDonough (1989).

Sr, Nd and Pb isotope variations

The isotopic compositions of Quaternary lavas in the northern Kyushu area are shown in $^{87}Sr/^{86}Sr$ v. $^{143}Nd/^{144}Nd$ and $^{208}Pb/^{204}Pb$ v. $^{206}Pb/^{204}Pb$ diagrams (Fig. 5). All these lavas plot close to two-component mixing curves between mid-ocean ridge basalt (MORB)-type mantle and PSP sediments. More specifically, the arc volcanics are bracketed by the Shikoku basin basalts (SBB) and PSP sediments (PSP-S) (Fig. 5). This supports the hypothesis that PSP slab-derived components were source materials for the arc lavas. Some samples have isotopic compositions that plot off the mixing curves, as shown in Nd–Sr isotope space

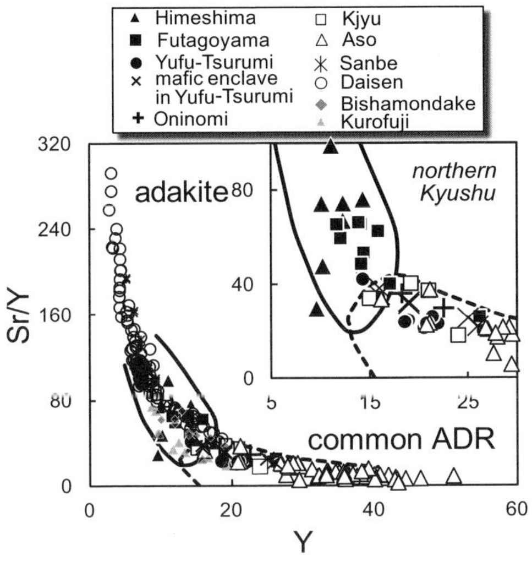

Fig. 4. Y (in ppm) v. Sr/Y of Quaternary magmas from SW Japan and northern Kyushu. ADR, andesite, dacite and rhyolite. Ranges of adakite and ADR are from Defant & Drummond (1990). Data of Yufu–Tsurumi, Kuju and part of Aso are from Sugimoto *et al.* (2007), Kita *et al.* (2001) and Hunter (1998), respectively.

where a significant number of samples are plotting off the mixing curve with lower $^{143}Nd/^{144}Nd$ values (Fig. 5a). Similar deviations are also seen in the Pb–Pb isotope plots, where Pb isotope ratios became more variable at the radiogenic end of the arc data array and deviate from the mixing line of SBB and PSP-S to higher $^{208}Pb/^{204}Pb$ (Futagoyama) and higher $^{206}Pb/^{204}Pb$ (Aso). These observations indicate that the whole-rock Sr–Nd–Pb isotope ratios also document the assimilation of crustal materials. To assess this, the effect of crustal materials is examined by comparing the arc data to the Sr–Nd–Pb isotope data of sedimentary rock from the Shimanto Group (Hosono *et al.* 2003) and the Pb isotope data of a migmatite from the Higo metamorphic belt (Shibata & Kobayashi, unpublished data), which are considered to represent crustal basement in the study area. In Figure 5a, a mixing curve between the adakites of Himeshima volcano and the sedimentary rocks of the Shimanto Group passes close to the samples with the relatively low Sr and Nd isotope ratios. The Pb isotopic compositions of the Shimanto sedimentary rocks (Hosono *et al.* 2003) and the migmatite are plotted in Figure 5b. The Pb isotopic compositions of Aso depart from the mixing line of SBB and PSP-S to those of the migmatite. Near the study area, no crustal materials with higher $^{208}Pb/^{204}Pb$ have been found, which could account for the Pb isotopic trend of the Futagoyama lavas. However, granulites with $^{206}Pb/^{204}Pb$ ratios similar to Quaternary magmas of northern Kyushu also show higher $^{208}Pb/^{204}Pb$ ratios at lower $^{206}Pb/^{204}Pb$ (e.g. Downes *et al.* 1991). Although further investigations are required, lower crustal materials like these granulites may provide a contaminant for the magmas from Futagoyama volcano.

Magma genesis during Quaternary volcanism in the northern Kyushu area

High Sr/Y ratios and low Y concentrations are indicative of the presence of residual garnet during partial melting or fractional crystallization (e.g. Kay 1978; Defant & Drummond 1990). However, as mentioned above, mechanisms other than slab

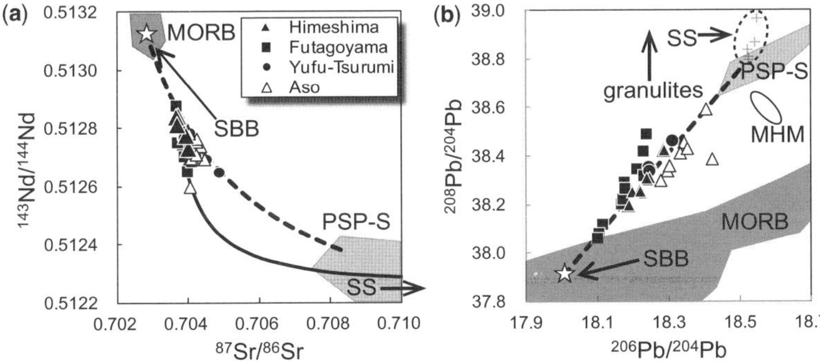

Fig. 5. Sr–Nd–Pb isotope systematics of northern Kyushu magmas. (**a**) $^{87}Sr/^{86}Sr$ v. $^{143}Nd/^{144}Nd$ and (**b**) $^{206}Pb/^{204}Pb$ v. $^{208}Pb/^{204}Pb$. The dashed line represents mixing curves between MORB and PSP sediment, or Shikoku basin basalt (SBB) and PSP sediment (PSP-S). Data for sedimentary rocks of Shimanto group (SS) and migmatites from Higo metamorphic belt (MHM) are from Hosono *et al.* (2003) and Shibata & Kobayashi (unpublished data). Isotope ratios of MORB, SBB and PSP sediment are from Hickey-Vargas (1991), Shibata & Nakamura (1997) and Shimoda *et al.* (1998).

melting could potentially lead to high Sr/Y ratios, including partial melting of garnet-bearing lower crustal material (Atherton & Petford 1993; Arculus *et al.* 1999; Yumul *et al.* 2000) and fractional crystallization of a mantle-derived basalt at lower crustal depth (Zellmer *et al.* 2012).

Zellmer *et al.* (2012) used the plagioclase hygrometer of Lange *et al.* (2009) to infer that the abundance of H_2O in basaltic parental magmas for samples collected from the Aso, Kuju, Yufu and Oninomi (close to Yufu) volcanoes of northern Kyushu increased from north to south. They also suggested that the Sr/Y ratios of lavas decreased from north to south, which is confirmed by this study. A recent experimental study indicates that garnet is stable in basaltic melt under moderate pressure (12 kbar), and that the stability of garnet decreases with increasing H_2O (Müntener *et al.* 2001). On the basis of these studies, Zellmer *et al.* (2012) suggested that the high Sr/Y ratio of magmas from the area may be caused by the evolution of mantle-derived basalt magmas in the garnet-stability field in the lower crust, and that the decrease in garnet stability from north to south resulted in decreasing Sr/Y ratios in northern Kyushu lavas. Despite this hypothesis providing an intriguing explanation for the arc magmas along the Honshu–Kyushu transition, the mechanism that caused the increase in Sr/Y ratios of adakites from the northern Kyushu area remains ambiguous, primarily because of the genetic relationship between basaltic magmas and adakitic magmas.

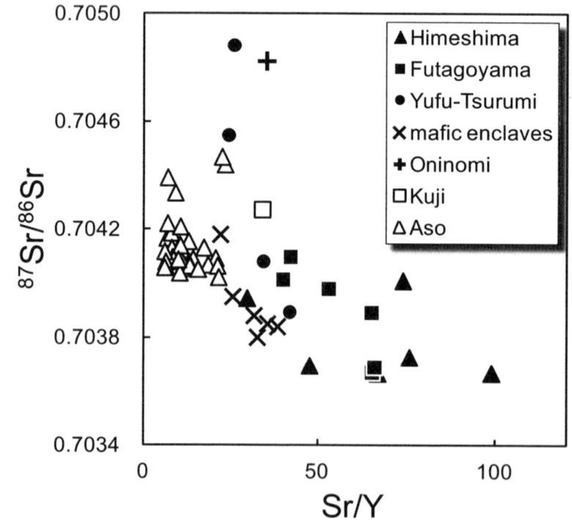

Fig. 6. Relationship between Sr/Y ratios and $^{87}Sr/^{86}Sr$ ratios. Data for Yufu–Tsurumi, mafic enclaves, Kuju and part of Aso are from Sugimoto *et al.* (2007), Ohta & Aoki (1991), Kita *et al.* (2001) and Hunter (1998), respectively. Data for Oninomi are also from Kita *et al.* (2001).

We investigated the relationship between Sr/Y and Sr isotope ratios in order to determine the processes that caused elevated Sr/Y ratios in adakites of the northern Kyushu area (Fig. 6). Figure 6 shows a clear inverse correlation between $^{87}Sr/^{86}Sr$ ratios and Sr/Y values. If the elevated Sr/Y ratios were the result of fractional crystallization of basaltic magmas in the crust, as suggested by Zellmer *et al.* (2012), the more silica-rich magmas would have higher Sr/Y ratios at similar or higher $^{87}Sr/^{86}Sr$, depending on the amount of crustal contamination. Instead, as the Sr/Y and SiO_2 increase in the arc magma, the $^{87}Sr/^{86}Sr$ ratio decreases (Fig. 6). Kita *et al.* (2001), for example, have reported an $^{87}Sr/^{86}Sr$ ratio of 0.70482 for basalts from Oninomi volcano, obtained from within 3 km (north) of Yufu–Tsurumi volcano, which has the significantly higher $^{87}Sr/^{86}Sr$ of the adakitic lavas of Yufu–Tsurumi proper ($^{87}Sr/^{86}Sr = 0.70389$). These data suggest that the high Sr/Y ratios of adakites from northern Kyushu were more likely generated during melting of subducted slab material, rather than during magmatic evolution under crustal conditions.

Figure 7 presents $^{87}Sr/^{86}Sr$ v. SiO_2 (Fig. 7a) and Sr/Y v. SiO_2 (Fig. 7b) for magmas from

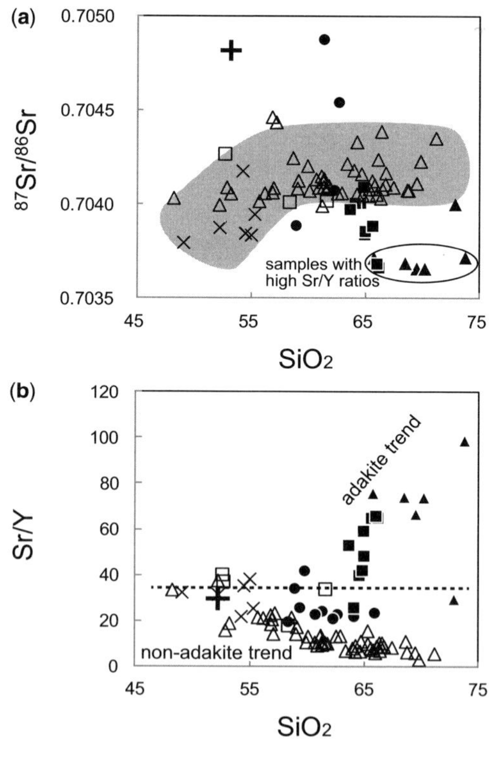

Fig. 7. Variation diagrams of (**a**) SiO_2 v. $^{87}Sr/^{86}Sr$ ratios and (**b**) SiO_2 v. Sr/Y ratios. Symbols as in Figure 4.

northern Kyushu. Data for the Oninomi basalt (Kita *et al.* 2001) and mafic enclaves in lavas of Yufu–Tsurumi (Ohta & Aoki 1991) are also shown in Figure 7. Trends of increasing and decreasing Sr/Y ratios with increasing melt SiO_2 are observed (Fig. 7b). Here, we call the former and latter trends the 'adakite trend' and 'non-adakite trend', respectively. The Sr/Y ratio of adakitic magmas in the study area is >35. The Oninomi basalt and mafic enclaves follow the non-adakite trend, although their Sr/Y ratios are >20. Samples that plot on the mixing curve of SBB and PSP-S towards low Sr–Nd isotopic composition (Fig. 5a) follow the non adakite trend (Tables 1 & 2). Hunter (1998) suggested that assimilation and fractional crystallization (AFC) influenced the geochemical and Sr–Nd isotope compositions of Aso volcano, which belongs to the non-adakite trend. We therefore propose that the non-adakite trend was generated from an 'adakitic' basalt by an AFC process within the crust. The adakite trend crosses to the non-adakite trend at SiO_2 of *c.* 60 wt%. This 'crossing' trend could possibly be produced by mixing of adakitic rhyolite magmas with non-adakitic andesite magmas. From the discussion above, we conclude that there is no direct genetic relationship between the basalts with high Sr/Y ratios and the adakites in the study area. We propose that the high Sr/Y is caused by an increasing abundance of melts sourced from the subducted slab, with along-arc Sr/Y variations indicating differing degrees of slab melt contributions.

Distribution of adakites in the SW Japan arc and its relation to the structure of the subducted plate

The tectonic setting and distribution of adakitic magmatism within the Japanese islands is shown in Figure 1. The subduction of the PSP along the Nankai Trough and Ryukyu Trench is the prime control over the formation of the SW Japan and Ryukyu volcanic arcs, respectively. The Kyushu–Palau ridge now subducting south of Kyushu divides the PSP into two segments. The northeastern section was formed between 15 and 26 Ma by the ocean spreading in the Shikoku Basin. The older southwestern segment (West Philippine Basin) was generated during the period 60–40 Ma (Mahony *et al.* 2011). Although the extension of the subducted Kyushu–Palau ridge is thought to be located somewhere between Kyushu and Shikoku islands, its exact location is still unclear. However, if the geochemistry of Quaternary magmas in this area can be used to identify differences in the subducted PSP, it would be possible to identify the location of the PSP boundary beneath the arc in the present-day or at least in the Quaternary-age arc magmas.

Based on geochemical evidence, Shinjo *et al.* (2001) suggested that the Quaternary magmas of the Ryukyu arc were generated by partial melting of a mantle wedge that was hydrated by fluids derived from dehydration of the older PSP. In comparison, adakites have been reported from several Quaternary volcanoes on Honshu, in the SW Japan arc, including the Sanbe, Aonoyama, Daisen, Karasugasen, Hakusan and Kurofuji volcanoes (Fig. 1), and these are thought to have formed by melting of subducted PSP slab materials (e.g. Morris 1995; Kimura *et al.* 2005; Shibata *et al.* 2008; Ujike & Mashimo 2009). Given that adakites are not found within the Ryukyu arc (e.g. Shinjo *et al.* 2001), the inboard extension of the Kyushu–Palau ridge can be defined using the distribution of adakitic magmas by assuming melting of the young and hot Shikoku Basin PSP plate.

As outlined above, the southwestern tip of slab melting-related magmas within northern Kyushu is represented by Kuju volcano. This would suggest that the boundary between the older West Philippine Basin crust and young Shikoku Basin crust of the PSP runs beneath Kuju volcano. In addition, Zellmer *et al.* (2012) documented that the H_2O contents in the primary basalts beneath Kyushu (4–5 wt%) and SW Japan arcs (*c.* 2 wt%) differ by a factor of 2. They emphasized that the lower H_2O contents in the SW Japan arc might be related to loss of water in the forearc due to the shallow (*c.* 60 km) subduction of the young Philippine Sea slab compared to that of northern Kyushu (*c.* 120 km), dilution of the hydrous wedge mantle with dry mantle derived from the back-arc, or a combination of these processes (Zellmer *et al.* 2012). The Oninomi (Yufu–Tsurumi) area is considered to be part of the Kyushu arc by Zellmer *et al.* (2012); however, the slightly lower H_2O suggests transitional characteristics (Zellmer *et al.* 2012, fig. 3). Mahony *et al.* (2011) suggested that an increase in B/Nb ratio between the Kuju (B/Nb < 1) and Aso (B/Nb ≈ 4) volcanoes indicated the addition of aqueous fluids derived from the subducting Kyushu–Palau ridge to the mantle wedge beneath Aso volcano. However, this scenario is inconsistent with the observation that the lavas with high B/Nb ratios continue to Kirishima volcano (Mahony *et al.* 2011), and the *c.* 100 km volcanic gap between the Kirishima and Aso volcanoes. One alternative idea is that the boundary between the younger PSP and the Kyushu–Palau ridge is located beneath Aso volcano, with the boundary between the Kyushu–Palau ridge and the older PSP running beneath Kirishima volcano (Fig. 8). If this is the case, the volcanic gap between Aso and Kirishima could be explained.

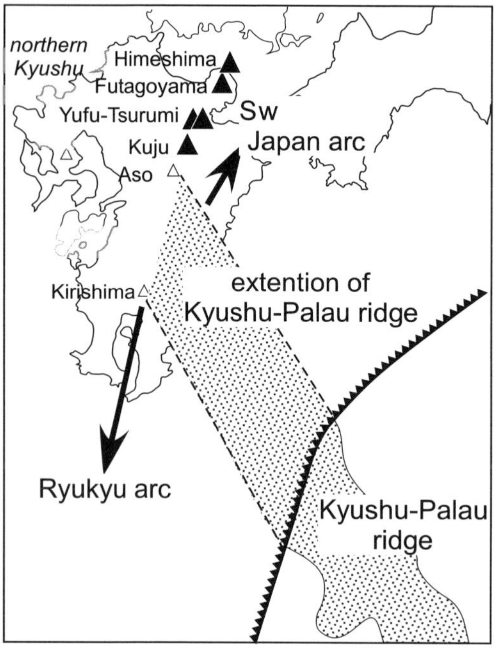

Fig. 8. Schematic map showing the subducted PSP and Kyushu–Palau ridge in relationship to the SW Japan arc.

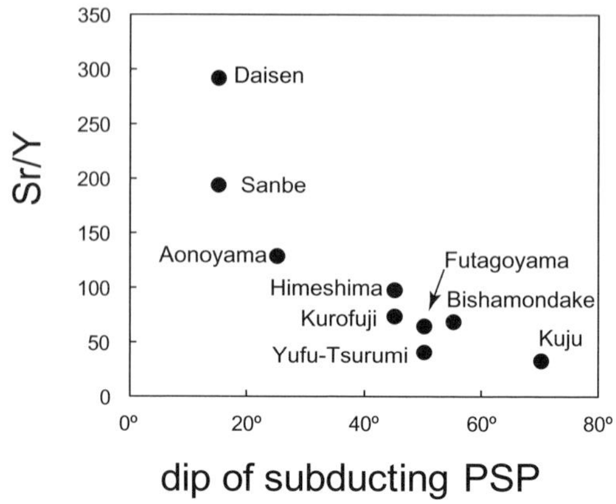

Fig. 9. Relationship between the dip angle of the PSP and the Sr/Y ratio. The maximum Sr/Y values of an individual volcano are plotted. The Sr/Y data are from Daisen and Sanbe, Morris (1995); Aonoyama, Kakubuchi & Nagao (1994); Himeshima and Futagoyama, this study; Kurofuji, Shibata *et al.* (2008); Yufu–Tsurumi, Sugimoto *et al.* (2007); Bishamondake, Ujike *et al.* (1999); and Kuju, Kita *et al.* (2001). The dip angle of the PSP beneath each volcano in NE Kyushu was estimated using the data from the hypocentre provided by the Japan Meteorological Agency (Abe, pers. comm.). The others were estimated from the studies of Nakajima & Hasegawa (2007) and Uchida *et al.* (2010).

Although the exact spatial relation between the subducted Kyushu–Palau ridge and the along-arc geochemical variations of the lavas remains unclear, the western and eastern limits of the adakite distribution along the volcanic front would indicate the down-going young slab formed in the Shikoku Basin. It is thus important to elucidate further the origin of the adakitic rocks in SW Japan and in northern Kyushu.

Along-arc Sr/Y ratio variations of the SW Japan arc

Figure 4 shows the Sr/Y ratios v. Y concentrations for Quaternary lavas from the adakitic volcanoes of the SW Japan arc. As documented by Zellmer *et al.* (2012), Sr/Y increases from south to north in northern Kyushu, with this trend continuing on to Honshu. However, Sr/Y decreases to the east of Daisen volcano (Fig. 1). Zellmer *et al.* (2012) suggested that Sr/Y ratio variations are related to differences in the age and dip angle of the subducting PSP.

We examined the relationship between the dip of the subducting slab and Sr/Y. In Figure 9, we plot the maximum Sr/Y of individual adakitic volcanoes in the SW Japan arc against the dip of PSP. The Sr/Y decreases with increasing dip of PSP. From this observation, we emphasize that the main factor controlling the Sr/Y of adakites from the SW Japan arc may be the dip of the PSP and not intracrustal differentiation.

Conclusions

The major, trace element and Sr, Nd and Pb isotope data of Quaternary magmas from northern Kyushu were investigated. An along-arc variation of decreasing Sr/Y from north (Sr/Y = 30–100) to south (Sr/Y = 6–24) is observed. Major and trace element compositions of magmas from Himeshima, Futagoyama, Yufu–Tsurumi and Kuju are within the ranges that can be linked to partial melting of the subducted slab (e.g. Defant & Drummond 1990; Castillo 2006). The Sr–Nd–Pb isotope ratios of the magmas lie principally on the mixing curves between MORB and PSP-S, or SBB and PSP-S, with minor deviation being attributable to shallow crustal contamination. The decrease of $^{87}Sr/^{86}Sr$ with increasing Sr/Y and SiO_2 argues against a genetic link by fractional crystallization between adakites and basaltic magmas. On the basis of these observations, we conclude that the high Sr/Y ratio of the Quaternary magmas of northern Kyushu reflects partial melting of the subducted

PSP. The boundary of the young PSP and Kyushu–Palau ridge might be located beneath the Aso volcano, indicating that the Kyushu–Palau ridge presumably runs beneath Aso, which also coincides with the southwestern end of the SW Japan arc.

We thank B. Vaglarov, Y. Yonezawa and Y. Kameishi for their valuable help with experiments. We also thank Y. Abe for estimating the dip of the PSP beneath northern Kyushu Island using the hypocentre data provided by the Japan Meteorological Agency. We also greatly appreciate the critical and constructive comments provided by P. Schaaf and J. Kimura. Part of this work was facilitated by KAGI21 and a Grant-in-Aid for Scientific Research (Ministry of Education, Culture, Sports, Science and Technology, Japan) awarded to T. Shibata.

References

ARCULUS, R. J., LAPIERRRE, H. & JAILLARD, E. 1999. Geochemical window into subduction and accretion processes: Raspas metamorphic complex, Ecuador. *Geology*, **27**, 547–550.

ATHERTON, M. P. & PETFORD, N. 1993. Generation of sodium-rich magmas from newly underplated basaltic crust. *Nature*, **362**, 144–146.

CASTILLO, P. R. 2006. An overview of adakite petrogenesis. *Chinese Science Bulletin*, **51**, 257–268.

CHANG, Q., SHIBATA, T., SHINOTSUKA, K., YOSHIKAWA, M. & TATSUMI, Y. 2003. Precise determination for trace elements in geological standard rocks using inductively coupled plasma mass spectrometry (ICP-MS). *IFREE Report for 2001–2002*, 357–360.

DEFANT, M. J. & DRUMMOND, M. S. 1990. Derivation of some modern arc magmas by melting of young subducted lithosphere. *Nature*, **347**, 662–665.

DEFANT, M. J. & DRUMMOND, M. S. 1993. Mount St. Helens: potential example of the partial melting of the subducted lithosphere in a volcanic arc. *Geology*, **21**, 547–550.

DEFANT, M. J. & KEPEZHINSKAS, P. 2001. Evidence suggests slab melting in arc magmas. *EOS*, **82**, 62–70.

DESONIE, D. L. 1992. Geologic and geochemical reconnaissance of Isla San Esteban: post-subduction orogenic volcanism in the Gulf of California. *Journal of Volcanology and Geothermal Research*, **52**, 123–140.

DOWNES, H., KEMPTON, P. D., BRIOT, D., HARMON, R. S. & LEYRELOUP, A. F. 1991. Pb and Os isotope systematics in granulite facies xenoliths, French Massif Central: implication for crustal process. *Earth and Planetary Science Letters*, **102**, 341–357.

FUJISAWA, Y., OKUNO, M., NAKAMURA, T. & KOBAYASHI, T. 2002. Eruptive activities of Tsurumi volcano in Japan during the past 30 000 years. *Journal of Geological Society of Japan*, **108**, 48–58.

FUKUYAMA, H. & ONO, K. 1981. *Geological map of Sakurajima volcano 1:25 000.* Geological Map of Volcanoes, Geological Survey of Japan, **1**, 1–8.

FURUKAWA, K., MIYOSHI, M., SHINMURA, T., SHIBATA, T. & ARAKAWA, Y. 2009. Geology and petrology of the Pre-Aso volcanic rocks distributed in the NW qall of Aso caldera: the eruption and magma plumbing system of the pre-caldera volcanism. *Journal of Geological Society of Japan*, **115**, 658–671.

GUTSCHER, M. A., MAURY, R. C., EISSEN, J. P. & BOURDON, E. 2000. Can slab melting be caused by flat subduction? *Geology*, **28**, 535–538.

HICKEY-VARGAS, R. 1991. Isotope characteristics of submarine lavas from the Philippine Sea: implication for the origin of arc basin magmas of the Philippine tectonic plate. *Earth and Planetary Science Letters*, **107**, 290–304.

HOSHIZUMI, H., ONO, K., MIURA, K. & NODA, T. 1998. *Geology of the Beppu district, with Geological Sheet Map at 1:50 000.* Geological Survey of Japan, **131**.

HOSONO, T., NAKANO, T. & MURAKAMI, H. 2003. Sr–Nd–Pb isotopic compositions of volcanic rocks around the Hishikari gold deposit, southwest Japan: implications for the contribution of a felsic lower crust. *Chemical Geology*, **201**, 19–36.

HUNTER, A. G. 1998. Intracrustal controls on the coexistence of tholeiitic and calc-alkaline magma series at Aso volcano, SW Japan. *Journal of Petrology*, **39**, 1255–1284.

IMURA, R. & KOBAYASHI, T. 2001. *Geological map of Kirishima volcano 1:50 000.* Geological Map of Volcanoes, Geological Survey of Japan, **11**, 1–7.

ISHIKAWA, T. & NAKAMURA, E. 1991. Origin of the slab component in arc lavas from across-arc variation of B and Pb isotopes. *Nature*, **370**, 205–208.

ITOH, J. 1990. Petrology of the Hime–Shima volcanic group. *Journal of Mineralogy, Petrology and Economic Geology*, **85**, 541–558.

KAKUBUCHI, S. & NAGAO, T. 1994. *Are the dacites from Aono volcanic group adakites?.* Abstract, Volcanological Society of Japan, **1994**, 173.

KAMATA, H. 1997. *Geology of the Miyanohara district with Geological Sheet Map at 1:50 000.* Geological Survey of Japan, **127**.

KAMATA, H. 1998. Quaternary volcanic front at the junction of the Southwest Japan Arc and the Ryukyu Arc. *Journal of Asian Earth Sciences*, **16**, 67–75.

KAY, R. W. 1978. Aleutian magnesian andesites: melts from subducted Pacific Ocean crust. *Journal of Volcanology and Geothermal Research*, **4**, 117–132.

KIMURA, J., TATEMO, M. & OSAKA, I. 2005. Geology and geochemistry of Karasugasen lava dome, Daisen–Hiruzen volcano group, southwest Japan. *The Island Arc*, **14**, 115–136.

KITA, I., YAMAMOTO, M., ASAKAWA, Y., NAKAGAWA, M., TAGUCHI, S. & HASEGAWA, H. 2001. Contemporaneous ascent of within-plate type and island-arc type magmas in the Beppu–Shimabara graben system, Kyushu Island, Japan. *Journal of Volcanology and Geothermal Research*, **111**, 99–109.

KOBAYASHI, T. 1984. Geology of Yufu–Tsurumi volcanoes and their latest eruptions. *Memoirs of the Geological Society of Japan*, **24**, 93–108.

LANGE, R. A., FREY, H. M. & HECTOR, J. 2009. A thermodynamic model for the plagioclase-liquid hygrometer/thermometer. *American Mineralogist*, **94**, 494–506.

LE MAITRE, R. W., BATEMAN, P. ET AL. 1989. *A Classification of Igneous Rocks and Glossary of Terms.* Blackwell, Oxford.

MAHONY, S. H., WALLACE, L. M., MIYOSHI, M., VILLAMOR, P., SPARKS, R. S. J. & HASENAKA, T. 2011.

Volcano–tectonic interactions during rapid plate-boundary evolution in the Kyushu region, SW Japan. *Geological Society of America Bulletin*, **123**, 2201–2223.

Miller, M. M., Goldstein, S. L. & Langmuir, C. H. 1994. Cerium/Lead and lead isotope ratios in arc magmas and the enrichment of lead in the continents. *Nature*, **368**, 514–519.

Miyoshi, M., Shibata, T., Yoshikawa, M., Sano, T., Shinmura, T. & Hasenaka, T. 2011. Genetic relationship between post-caldera and caldera-forming magmas from Aso volcano, SW Japan. Constraints from Sr isotope and trace element compositions. *Journal of Mineralogical and Petrological Sciences*, **104**, 114–119.

Morris, P. A. 1995. Slab melting as an explanation of Quaternary volcanism and aseismicity in southwest Japan. *Geology*, **23**, 395–398.

Müntener, O., Kelemen, P. B. & Grove, T. L. 2001. The role of H_2O during crystallization of primitive arc magmas under uppermost mantle conditions and genesis of igneous pyroxenites: an experimental study. *Contributions to Mineralogy and Petrology*, **141**, 643–658.

Nakajima, J. & Hasegawa, A. 2007. Subduction of the Philippine Sea plate beneath southwestern Japan: slab geometry and its relationship to arc magmatism. *Journal of Geophysical Research*, **112**, B08306.

Noguchi, S. 1988. Geometry of the Philippine Sea Slab and the Convergent Tectonics in the Tokai District, Japan. *Zisin*, **49**, 295–325.

Ohta, T. & Aoki, K. 1991. Origin of andesitic magma in Yufu–Tsurumi volcano group – a binary mixing model. *Journal of Mineralogy, Petrology and Economic Geology*, **86**, 1–15.

Ohta, T., Hasenaka, T. & Fujimaki, H. 1990. Geology and petrography of Yufu–Tsurumi volcano group, Oita Prefecture. *Journal of Mineralogy, Petrology and Economic Geology*, **85**, 113–129.

Peacock, S. M., Rushmer, T. & Thompson, A. B. 1994. Partial melting of subducting oceanic crust. *Earth and Planetary Science Letters*, **121**, 227–244.

Pe-Piper, G. & Piper, J. W. 1994. Miocene magnesian andesites and dacites, Evia Greece: adakites associated with subducting slab detachment and extension. *Lithos*, **31**, 125–140.

Perfit, M. R., Gust, A. E., Bence, A. E., Arculus, R. J. & Taylor, S. R. 1980. Chemical characteristics of island-arc basalt: implications for mantle sources. *Chemical Geology*, **30**, 227–256.

Petford, N. & Atherton, M. 1996. Na-rich partial melts from newly underplated basaltic crust: the Cordillera Blanca Batholith, Peru. *Journal of Petrology*, **37**, 1491–1521.

Sajona, F. G., Maury, R. C., Bellon, H., Defant, M. J., Cotton, J., Pubiller, M. & Rangin, C. 1993. Initiation of subduction and the generation of slab melts in western and eastern Mindanao, Philippines. *Geology*, **21**, 1007–1010.

Seghedi, I., Downes, H. *et al.* 2004. Neogene–Quaternary magmatism and geodynamics in the Carpathian–Pannonian region: a synthesis. *Lithos*, **72**, 117–146.

Shibata, T. & Nakamura, E. 1997. Across arc variations of isotope and trace element compositions from Quaternary basaltic volcanic rocks in northeastern Japan: implications for interaction between subducted oceanic slab and mantle wedge. *Journal of Geophysical Research*, **102**, 8051–8064.

Shibata, T., Yoshikawa, M. & Koshimizu, S. 2008. Temporal variations of geochemical characteristics of volcanic rocks from north of Kofu Basin, central Japan, and its relationship with the subduction zone structure. *Magma*, **89**, 1–16.

Shibutani, T. 2002. Crustal and slab structures beneath the southern Kii peninsula inferred from receiver function analysis. *In*: *Proceedings of Meeting 13K-7*, DPRI, Kyoto University, March, 271–277.

Shimoda, G., Tatsumi, Y., Nohda, S., Ishizaka, K. & Jahne, B. M. 1998. Setouchi high-Mg andesites revisited: geochemical evidence for melting of subducting sediments. *Earth and Planetary Science Letters*, **160**, 479–492.

Shinjo, R., Woodhead, J. D. & Hergt, J. M. 2001. Geochemical variation within the northern Ryukyu Arc: magma source compositions and geodynamic implications. *Contributions to Mineralogy and Petrology*, **140**, 263–282.

Sugimoto, T., Shibata, T., Yoshikawa, M. & Takemura, K. 2007. Sr–Nd–Pb isotopic and major and trace element compositions of the Yufu–Tsurumi volcanic rocks: implications for the magma genesis of the Yufu–Tsurumi volcanoes, northeast Kyushu, Japan. *Journal of Mineralogical and Petrological Sciences*, **101**, 270–275.

Sun, S. S. & McDonough, W. F. 1989. Chemical and isotopic systematics of oceanic basalts: implications for mantle composition and processes. *In*: Saunders, A. D. & Norry, M. J. (ed.) *Magmatism in the Ocean Basins*. Geological Society, London, Special Publications, **42**, 313–345.

Taira, A. 2001. Tectonic evolution of the Japanese island arc system. *Annual Review of Earth and Planetary Sciences*, **29**, 109–134.

Uchida, N., Matsuzawa, T., Nakajima, J. & Hasegawa, A. 2010. Subduction of a wedge-shaped Philippine Sea plate beneath Kanto, central Japan, estimated from converted waves and small repeating earthquakes. *Journal of Geophysical Research*, **115**, B07309.

Uehira, K., Shimizu, H., Matsuo, N. & Goto, K. 2001. The configuration and focal mechanism of deep seismic plane beneath western tip of Shikoku-Chugoku and Kyushu island, Japan. *The Monthly Earth*, **23**, 669–673.

Ujike, O. & Mashimo, R. 2009. Preliminary report on the significance of magmatic enclaves in adakite genesis at Hakusan volcano, central Japan. *Journal of Mineralogical and Petrological Sciences*, **104**, 82–85.

Ujike, O., Tanaka, R. & Watanabe, K. 1999. Petrology of adakitic rocks from Quaternary Bishamon-dake volcano, central Japan. *Journal of Mineralogical and Petrological Sciences*, **94**, 315–328.

Wood, D. A., Joron, J. L. & Treuil, M. 1979. A re-appraisal of use of trace elements to classify and discriminate between magma series erupted in different tectonic setting. *Earth and Planetary Sciences Letters*, **45**, 326–336.

Yogodzinski, G. M., Lees, J. M., Churikova, T. G., Dorendorf, F., Woerner, G. & Volynets, O. N. 2001.

Geochemical evidence for the melting of subducting oceanic lithosphere at plate edges. *Nature*, **409**, 500–504.

YOSHIKAWA, M. & SHIBATA, T. 2003. Sr and Nd isotopic ratios and Rb, Sr, Sm and Nd concentrations of JB-2 rock reference material with isotope dilution method. *Annual Report of Institute for Geothermal Sciences*, **FY2002**, 30–31.

YUMUL, G. P., JR, DIMALANTA, C. B., BELLON, H., FAUSTINO, D. V., DE JESUS, J. V., TAMAYO, R. A., JR. & JUMAWAN, F. T. 2000. Adakitic lavas in the Central Luzon back-arc region (Philippines): lower crust partial melting products? *The Island Arc*, **9**, 499–512.

ZELLMER, G. F., IIZUKA, Y., MIYOSHI, M., TAMURA, Y. & TATSUMI, Y. 2012. Lower crustal H_2O controls on the formation of adakitic melts. *Geology*, **40**, 487–490.

A genetic link between silicic slab components and calc-alkaline arc volcanism in central Mexico

SUSANNE M. STRAUB[1]*, GEORG F. ZELLMER[2,3], ARTURO GÓMEZ-TUENA[4], RAMON ESPINASA-PEREÑA[5], ANA LILLIAN MARTIN-DEL POZZO[6], FINLAY M. STUART[7] & CHARLES H. LANGMUIR[8]

[1]*Lamont Doherty Earth Observatory at the Columbia University, 61 Route 9W, Palisades NY 10964, USA*

[2]*Institute of Earth Sciences, Academia Sinica, 128 Academia Road, Sec. 2, Nankang, Taipei 11529, Taiwan, Republic of China*

[3]*Present address: Institute of Agriculture and Environment, Massey University, Palmerston North 4442, New Zealand*

[4]*Centro de Geociencias, Universidad Nacional Autónoma de México, Querétaro 76230, México*

[5]*Centro Nacional de Prevención de Desastres, Secretaría de Gobernación, Av. Delfín Madrigal 665, Col. Pedregal de Santo Domingo C.P. 04360, Delegación Coyoacán, México D.F., Mexico*

[6]*Instituto de Geofísica, Universidad Nacional Autónoma de México, Ciudad Universitaria, Mexico, D.F. 04510, Mexico*

[7]*Isotope Geosciences Unit, Scottish Universities Research and Reactor Centre, East Kilbride G75 0QF, UK*

[8]*Department of Earth and Planetary Sciences, Harvard University, 20 Oxford Street, Cambridge, MA 02137, USA*

**Corresponding author (e-mail: smstraub@ldeo.columbia.edu)*

Abstract: A fundamental question in the formation of orogenic andesites is whether their high melt SiO_2 reflects the recycling of silicic melts from the subducted slab or the processing of basaltic mantle melts in the overlying crust. The latter model is widely favoured, because most arc magmas lack the 'garnet' signature of partial slab melts. Here we present new trace element data from Holocene high-Mg# >64–72 calc-alkaline basalts to andesites (50–62 wt% SiO_2) from the central Mexican Volcanic Belt that crystallize high-Ni olivines with the high $^3He/^4He = 7–8$ of the upper mantle. These magmas have been proposed to be partial melts from 'reaction pyroxenites', which formed by hybridization of mantle peridotite (*c.* 82–85%) and heavy rare earth element-depleted silicic slab melt (>15–18%). Forward and inverse models suggest that the absence of a garnet signature in these melts reflects the efficient buffering of the heavy rare earth elements (Ho to Lu) in the subarc mantle. In contrast, all elements more incompatible than Ho – excepting TiO_2 – are more or less strongly controlled by the silicic slab flux that also directly contributes to the silicic arc magma formation. Our study emphasizes the strong link between slab recycling and the genesis of orogenic andesites.

Supplementary material: Methods, additional data and modelling parameters are available at http://www.geolsoc.org.uk/SUP18686

Deciphering andesite petrogenesis at convergent margins is fundamental to understanding the differentiation of silicate Earth and the growth of the andesitic continental crust (Ringwood 1974; Gill 1981; Rudnick 1995). The plate tectonic model clearly links andesite genesis to plate subduction; however, it remains debated where and how andesitic arc magmas are produced. Low-degree partial melts from the subducted eclogitic slab are silicic (Beard 1995; Rapp & Watson 1995; Skora & Blundy 2010), but display strong garnet signatures (e.g. Pertermann & Hirschmann 2003; Klimm *et al.* 2008). However, clear garnet signatures are comparatively rare in global arc magmas,

From: GÓMEZ-TUENA, A., STRAUB, S. M. & ZELLMER, G. F. (eds) 2014. *Orogenic Andesites and Crustal Growth.* Geological Society, London, Special Publications, **385**, 31–64.
First published online September 12, 2013, http://dx.doi.org/10.1144/SP385.14

and are mostly related to local or transient regional conditions (e.g. Kay 1978; Defant & Drummond 1990; Yogodzinksi *et al.* 2001; Gomez-Tuena *et al.* 2007*a*). The missing garnet signature is therefore a key argument in models that consider formation of arc magmas through partial melting of a peridotite mantle that absorbed only trace amounts (*c.* 1–2 wt%) of a highly enriched slab component. While this slab component may control the budget of the fluid-mobile, highly incompatible elements in arc magmas (Morris *et al.* 1990; Plank & Langmuir 1993; Elliott *et al.* 1997), it imparts no garnet signature. Moreover, the model implies that initial arc melts are typical partial peridotite melts with low SiO_2 (<50 wt%) and high MgO (>10 wt%) abundances (e.g. Langmuir *et al.* 1992). Because the abundances of melt water in arc magmas are generally too low to cause more than a moderate increase of melt SiO_2 in the initial mantle melts (e.g. Baker *et al.* 1994; Ruscitto *et al.* 2012), the evolution to andesitic, dacitic and rhyolitic magmas must then take place in the overlying crust, by fractional crystallization or crustal assimilation, or a mixture of both (e.g. Eichelberger 1978; Hildreth & Moorbath 1988; Plank & Langmuir 1988; Tamura & Tatsumi 2002; Reubi & Blundy 2008, 2009).

While this view of andesite genesis is widely accepted, comprehensive datasets that combine detailed geological information with high-quality geochemical data reveal major problems. An excellent example is provided by the monogenetic volcanoes of the Holocene Sierra Chichinautzin Volcanic Field in the central Mexican Volcanic Belt (MVB) which have been the object of many detailed studies in recent years (e.g. Wallace & Carmichael 1999; Straub & Martin-Del Pozzo 2001; Siebe *et al.* 2004*a*; Schaaf *et al.* 2005; Straub *et al.* 2008, 2011; Agustín-Flores *et al.* 2011). Many of the Sierra Chichinautzin volcanoes erupt olivine-rich, high-Mg# (up to 74) and high MgO (up to 10 wt%) calc-alkaline basalts to andesites that display the arc-typical enrichments of fluid-mobile large-ion lithophile elements (LILE) relative to the high-field-strength elements (HFSE) and rare earth elements (REE). Intriguingly, the overall SiO_2 increase from basalt to andesite cannot be attributed to a fractional crystallization link, as basalt to andesite series have different $^{87}Sr/^{86}Sr$ and $^{143}Nd/^{144}Nd$ isotope ratios (e.g. Fig. 1a, b; Wallace & Carmichael 1999; Schaaf *et al.* 2005). On the other hand, the linear, continuous trends of major and trace elements with melt SiO_2 suggests mixing of basaltic and silicic compositions (Fig. 1a). The magmas erupt through *c.* 45 km-thick continental crust (e.g. Kim *et al.* 2010), and thus the silicic end member may be a silicic component from the ambient crust. If this was the case, huge amounts of silicic crust on the order of tens of per cent must be incorporated in the intermediate andesites (Fig. 1a). Such amounts, however, are improbable, because the Mexican high-Mg# andesites are both unusually primitive in radiogenic isotopes and rich in Ni (up to 250 ppm) and Cr (up to 500 ppm). Moreover, the high-Mg# magmas commonly crystallize fosteritic olivines with the high $^3He/^4He = 7–8\ R_a$ typical of mantle melts (Fig. 1, Straub *et al.* 2011, 2013). In view of the extreme sensitivity of low-He mantle melts towards the assimilation of ^4He-rich crustal material with $^3He/^4He < 1$, these melts can have assimilated only trace amounts of crustal material (<0.0001%, curve b in Fig. 1c; Straub *et al.* 2011, 2013).

The high $^3He/^4He$ olivines also have unusual high-Ni abundances which provide important clues to magma genesis. The Ni of the Mexican olivine exceeds by far the Ni of olivines from partial peridotite melts (Fig. 1d). Straub *et al.* (2011) proposed that such high-Ni olivines may reflect the presence of silica-deficient and silica-excess 'reaction pyroxenites' in the mantle wedge that formed through reaction of silicic slab components with the ambient peridotite mantle (Straub *et al.* 2008, 2011). Upon melting, such pyroxenite segregations may produce a broader range of high-Mg# basaltic and dacitic initial component melts. Mixing of many of these melts during ascent through mantle and crust then produces a series of high-Mg# basaltic to andesitic and dacitic melt that only experienced minor modification (olivine loss and recharge melt mixing) in the overlying crust (Straub *et al.* 2011). In other words, this model suggests that Mexican arc andesites are principally hybrids of slab and mantle components, and do not form through crustal differentiation.

While this 'pyroxenite model' accounts well for the major element characteristics (Straub *et al.* 2008, 2011, 2013), it is not clear how it influences the trace elements. Any 'reaction pyroxenite' that gives rise to silicic mantle melts must contain at least *c.* 15–18% of a silicic slab component. This is the minimum amount required to transform peridotite to 'reaction pyroxenite'. The obvious question is how this large amount (and more) of slab material reconciles with incompatible trace element of the central MVB calc-alkaline magmas that – with rare exceptions – have no garnet signatures (LaGatta 2003; Schaaf *et al.* 2005; Straub *et al.* 2008). The purpose of this paper is to discuss the systematics of incompatible trace elements in the context of the 'pyroxenite model'. While we focus on basaltic to andesite calc-alkaline high-Mg# (>64) magmas that have high-Ni and high $^3He/^4He$ olivines and that are least affected by crustal differentiation, these data are discussed in the context of all central MVB calc-alkaline magmas.

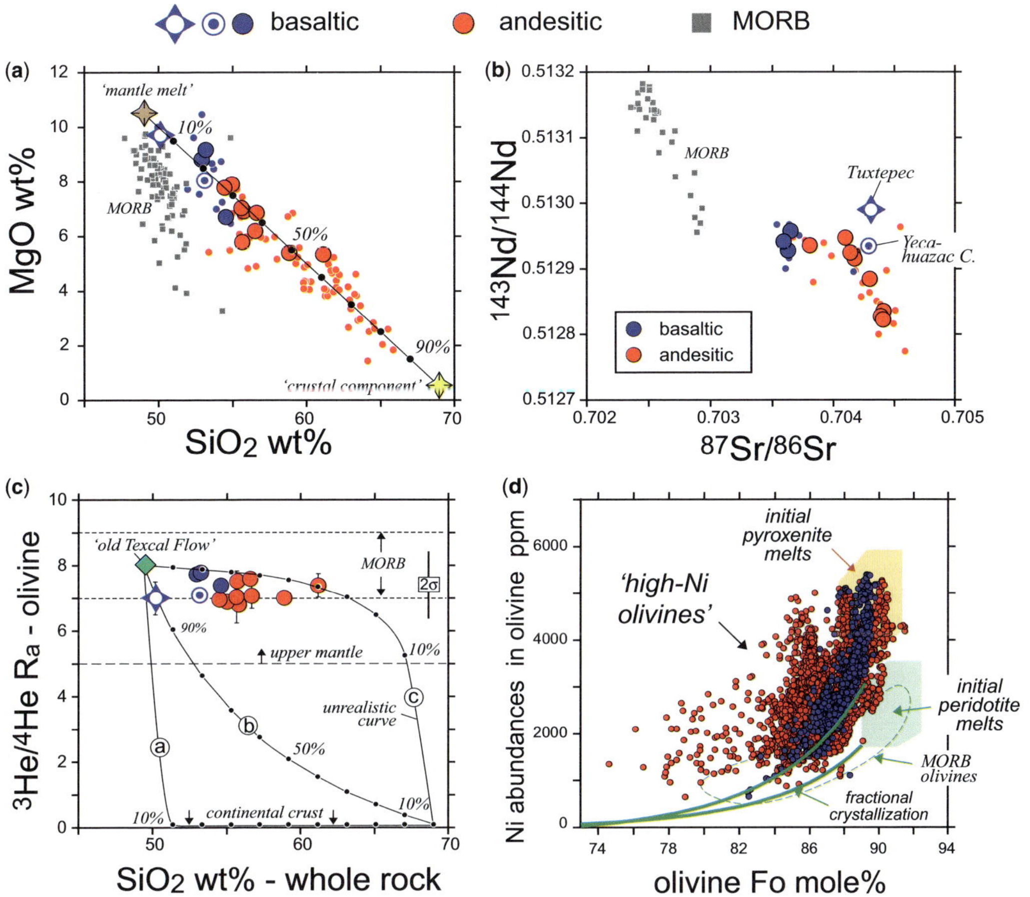

Fig. 1. Summary of previous studies on calc-alkaline magmas in the central Mexican Volcanic Belt (MVB). MORB are from PetDB (2011); for other data sources see text. Basaltic and andesitic magmas are quartz-normative and olivine-normative magmas respectively (as discussed in text). Large symbols in (a–c) denote samples for which $^{3}He/^{4}He$ has been obtained. (**a**) Whole-rock SiO_2 v. MgO (in wt%). Linear mixing curve assumes mixing between a basaltic melt (50 wt% SiO_2, 10 wt% MgO) and a silicic crustal component (69 wt% SiO_2, 0.5 wt% MgO). (**b**) Whole-rock $^{87}Sr/^{86}Sr$ v. $^{143}Nd/^{144}Nd$. (**c**) Whole-rock SiO_2 (wt%) v. olivine $^{3}He/^{4}He$ (R_a). External precision of $^{3}He/^{4}He$ is $\pm 1\ R_a$. 'Old Texcal Flow' from Straub *et al.* (2011) has the highest $^{3}He/^{4}He = 8\ R_a$ in central MVB and is taken as proxy to $^{3}He/^{4}He$ of mantle melt. Because crustal He is by four orders of magnitude higher than He in mantle melts, the mixing curve (curve a) clearly misses the data if only traces of crustal material were assimilated. Curve b assumes loss of 99.99% of the crustal He. Stippled curve c assumed that crustal material has only *c.* 6% of mantle He which is not realistic (Ballentine & Burnard 2002). See Supplementary material for more details on SiO_2–$^{3}He/^{4}He$ mixing curves. (**d**) Olivine Ni (ppm) v. Fo (mole%) of olivines with high $^{3}He/^{4}He = 7–8\ R_a$. Green field indicates olivines in equilibrium with initial mantle melts from peridotite, thick green lines are the liquid lines of descent for olivines crystallizing from peridotite melts. Ochre field indicates olivines in equilibrium with initial mantle melts from 'reaction pyroxenites'. Modified from Straub *et al.* (2011). Data from Straub *et al.* (2011, this study).

Geological background

Central Mexican Volcanic Belt

The geological setting of the Mexican Volcanic Belt has been described in detail elsewhere (e.g. Gómez-Tuena *et al.* 2007*b*; Ferrari *et al.* 2011), and only a short summary is given here. The MVB is a Pliocene–Quaternary volcanic arc that is related to the subduction of the Cocos and Rivera plates along the Middle American Trench (Fig. 2). The Middle American trench runs oblique at an angle of *c.* 17° to the MVB, owing to the near-horizontal subduction of the slab beneath the forearc in the central-eastern arc (Pardo & Suarez 1995). In the central MVB, the arc front is located at *c.* 300–350 km from trench, where the slab steepens abruptly to 65–75° (Perez-Campos *et al.* 2008). The central segment of the MVB between composite volcanoes Popocatepetl and Nevado de Toluca (*c.* 100–98°30′W) is constructed on *c.* 45 km thick sialic crust of Proterozoic granulites overlain by Jurassic metapelites, granites and Mesozoic limestones (Kim *et al.* 2010; Ortega-Gutiérrez *et al.* 2012).

Subduction zone parameters, such as slab age, slab dip and crustal thickness and composition, are constant in central MVB. All the same, Holocene volcanoes Nevado de Toluca, Popocatepetl and the volcanoes of the interjacent Sierra Chichinautzin Monogenetic Field erupt calc-alkaline basaltic to dacitic magmas with highly variable trace element patterns that include a subordinate group of mildly alkaline Nb-rich basalts and basaltic andesites (e.g. Verma 1999; Wallace & Carmichael 1999; LaGatta 2003; Siebe *et al.* 2004*a*; Schaaf *et al.* 2005). The latter are referred to as 'OIB-type' or high-'Nb arc basalts' (HNAB), as their trace element patterns resemble those of intraplate magmas. The far more abundant calc-alkaline series are characterized by strong enrichments of LILE relative to REE and HFSE and strongly resemble average continental crust. Together with the HNABs, the calc-alkaline magmas form a continuum in compositional space with many transitional compositions, but no 'natural' boundary (Verma 1999; Wallace & Carmichael 1999; LaGatta 2003; Schaaf *et al.* 2005). Following LaGatta (2003), all magmas with $<Nb = 16$ ppm are considered as calc-alkaline magmas. The HNAB have >16–35 ppm Nb, and are discussed in detail elsewhere (Straub *et al.* 2013).

Sampling strategy

Samples for this study are from the Sierra Chichinautzin Volcanic Field, in addition to a few samples from Popocatepetl volcano. Sample collection took place during multiple field trips in the years 2004–2009. The field trips were spaced about a year apart, which allowed for intermittent major and trace element analyses that then guided further field work. Originally, we searched for zoned monogenetic volcanoes as prime targets for studying the causes of melt differentiation. While this approach was successful for the HNAB (Straub *et al.* 2013), the calc-alkaline monogenetic volcanoes were either nearly homogeneous (e.g. Pelado, Tlaloc, Siebe *et al.* 2004*a*, 2005; Schaaf *et al.* 2005), or the eruptive sequence could not be, or could only partially be, reconstructed (e.g. Cuatepel, Guespalapa, this study). Several monogenetic volcanoes erupt both calc-alkaline and HNAB magmas, but without significant zoning in either group (e.g. Yecahuazac, Suchiooc, this study).

Calc-alkaline magmas comprise basalts, basaltic andesites, andesites and dacites. Many calc-alkaline basalts to andesites have high values of Mg# >64, fairly high Ni >100 ppm and Cr >155 ppm, and olivine as single, or dominant, phenocryst regardless of the broad SiO_2 range. For simplicity, the calc-alkaline series are divided into two groups based on their normative composition: a 'basaltic' (olivine-normative) and 'andesitic' (quartz-normative) group (Table 1, Straub *et al.* 2011). Olivine phenocrysts in both basaltic and andesitic magmas have been analysed for $^3He/^4He$ (Straub *et al.* 2011). These 'high-$^3He/^4He$ samples' are denoted by large, filled blue and red circles in the figures, and originate from monogenetic volcanoes Cuatepel (two samples), Guespalapa (seven), Suchiooc Cone (two), Tuxtepec (one) and Yecahuazac (one). One sample is from Popocatepetl. All other calc-alkaline basalts and andesites are shown as small blue and red circles.

The samples displayed are either from our own field work, or have been published by Straub *et al.* (2008, 2011, 2013) and LaGatta (2003). Sample locations are shown in Figure 2 and geographical coordinates are given in Table 1. This sample set fully represents the spectrum of the central MVB magmas and is free from interlaboratory bias. For these reasons and for clarity, other published compositional data on central MVB volcanics were omitted from the plots (e.g. Wallace & Carmichael 1999; Siebe *et al.* 2004*a*, 2005; Schaaf *et al.* 2005; Meriggi *et al.* 2008; Agustín-Flores *et al.* 2011).

All new data are listed in Table 1. Many samples were taken from the same volcanoes that have been selected for the $^3He/^4He$ work, which includes Popocatepetl (six additional samples), Cuatepel (two), Guespalapa (12), Suchiooc (10), Yecahuazac (three) and Tuxtepec (one). In addition, we report data for samples from Sierra Chichinautzin monogenetic volcanoes Cuatzin (one sample), Ocusacayo (two), Pelado (three), Teuthli (two), Tlaloc (two), Muñeco (one) and a lava flow in proximity to (but not from) Tuxtepec Cone (one). Subvolcanic rocks were analysed from Zempoala Ridge (two) and from Cerro Magdalena (one).

Sample locations

In the following, the volcanoes with the high $^3He/^4He$ samples are briefly described. *Volcan Cuatepel* is located just east of the village of Juchitepec and halfway between the central Sierra Chichinautzin and Popocatepetl volcano. Cuatepetl magmas are quartz-normative, and produced a scoria cone of olivine-phyric, dark-grey andesite bombs

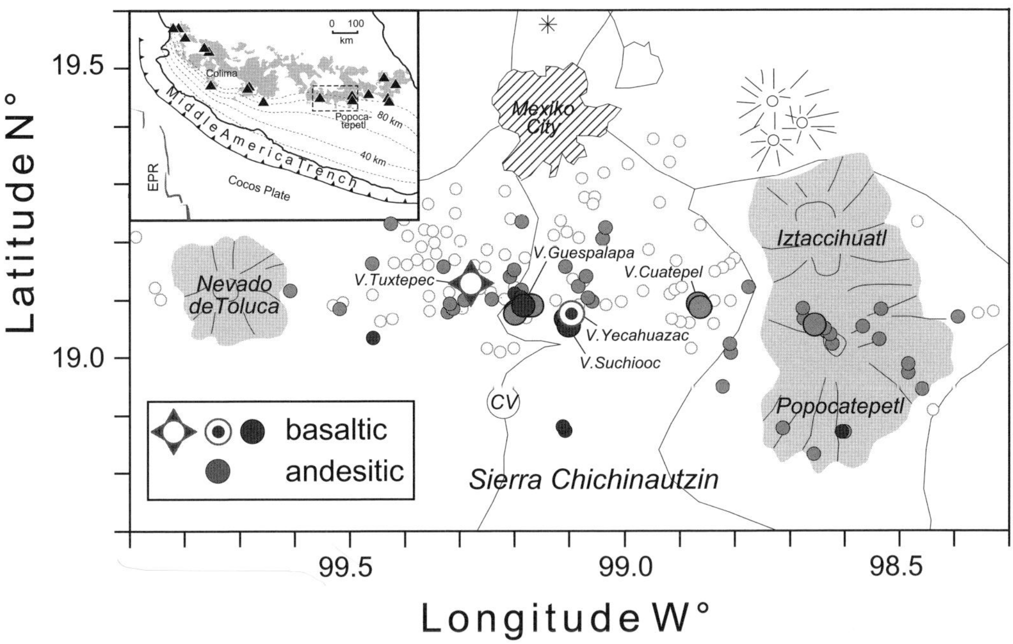

Fig. 2. Geological setting of the central Mexican Volcanic Belt. Inset shows position of study area within the Mexican subduction zone. Slab contours after Pardo & Suarez (1995). The enlargement shows the monogenetic volcanoes (small open grey circles) of the Sierra Chichinautzin that is flanked by Quaternary composite volcanoes Nevado de Toluca and Popocatepetl. Only calc-alkaline compositions (Nb < 16 ppm) are shown. Large symbols are samples with high-$^{3}He/^{4}He = 7-8\ R_a$ olivine; small symbols are other samples (this study and published data). See text for further details.

(58 wt% SiO_2), and an olivine-rich basaltic andesite flow (56 wt% SiO_2) at the base of the western flank. The youthful appearance suggests a Holocene age. There is no information about the relative succession of cone and flow.

The *Guespalapa Volcanic Complex* formed between 2.8 and 4.7 ka BP (Siebe *et al.* 2004*a*, *b*). All magmas are olivine-phyric. It is a broad shield volcano that is topped by three scoria cones: El Caballito, with multiple vents, and cones Manteca and El Hoyo. The Guespalapa shield consists of quartz-normative basaltic andesites to andesites with *c.* 55.4–61.2 wt% SiO_2 and low Nb (= 4.1–8.8 ppm). The three scoria cones erupt quartz-normative basaltic andesites, with the exception of El Caballito, which produces both quartz-normative (55.4 wt% SiO_2) and olivine-normative basaltic andesites (54.4–54.8 wt% SiO_2). El Caballito has also relatively high Nb = 10.8–13.6 ppm that is more typical for transitional rather than calc-alkaline compositions. Manteca (56.7 wt% SiO_2) and El Hoyo (54.9 wt% SiO_2) are the most silicic Guespalapa magmas. El Hoyo (Mg# = 69.8) and Manteca (Mg# = 66.2) have higher values of Mg# than El Caballito (Mg# = 65.0), but much lower Nb = 3.5–4.2 ppm. There is no information about the temporal succession except that Manteca and El Caballito formed after the Guespalapa lava shield.

Volcan Suchiooc is a scoria cone that is associated with lavas flows of up to 10 km length (Espinasa-Perena 2006). All magmas are olivine-phyric. The lava flows were emplaced during the Brunhes Chron and hence are younger than <0.73–0.79 Ma (Urrutia-Fucugauchi & Martin del Pozzo 1993). Calc-alkaline basaltic andesites (53.0–54.2 wt% SiO_2, Nb = 10.2–12.9 ppm) can be found at Suchiooc Cone and proximal lava flows within a *c.* 1 km perimeter are olivine–normative. Olivine–normative basaltic HNABs (>16 ppm Nb) seem to have been emplaced after the calc-alkaline basalts. They extrude within a few hundred metres from Suchiooc Cone and make up all distal lava flows.

Like Suchiocc, *Volcan Yecahuazac* consists of an olivine-normative calc-alkaline cone (53.1 wt% SiO_2; Nb = 8.4 ppm) and olivine-normative HNABs basaltic lava flows that breach through the southwestern flank of the cones and seem to postdate cone formation. All magmas are olivine-phyric. Our field observations and data show that a lava flow emplaced in front of the

Table 1. *Major and trace element abundances of calc-alkaline and high-Nb magmas of the central Mexican Volcanic Belt*

Volcano	Stratigraphic Unit	Sample_ID	Longitude W	Latitude N	Laboratory[a]	Group[b]	Group[c]	$^3He/^4He$[d]	Data source[e]	SiO_2	TiO_2	Al_2O_3	$Fe_2O_3^t$	MnO	MgO	CaO	Na_2O	K_2O	P_2O_5	LOI	Sum	Mg#[f]	Qtz[g]
Popocatepetl	Ventorillo aphyric dacite	**CH-09-18**	98°38′04″	19°02′39″	WSU/CGEO	Andesitic	Calc-alk.		This study	62.85	0.78	15.78	5.39	0.09	3.69	4.90	4.23	2.08	0.23		100.02	60.9	13.6
Popocatepetl	Dyke in upper B.Nexpayantla	**CH-09-19**	98°38′04″	19°02′39″	WSU/CGEO	Andesitic	Calc-alk.		This study	64.07	0.47	15.98	3.87	0.06	1.44	3.89	4.24	1.99	0.15		96.16	45.9	20.8
Popocatepetl	Volcaniclast in upper B.Nexpayantla	**CH-09-20**	98°38′04″	19°02′39″	WSU/CGEO	Andesitic	Calc-alk.		This study	63.41	0.74	16.42	4.84	0.08	2.93	4.79	4.38	1.75	0.19		99.52	58.0	15.7
Popocatepetl	Volcaniclast in upper B.Nexpayantla	**CH-09-21**	98°38′04″	19°02′39″	WSU/CGEO	Andesitic	Calc-alk.		This study	59.76	0.82	15.85	5.51	0.09	4.10	5.30	4.02	1.70	0.20		97.35	62.8	11.9
Popocatepetl	Dyke in lower B. Nexpayantla	**CH-09-22**	98°38′04″	19°02′39″	WSU/CGEO	Andesitic	Calc-alk.		This study	56.54	0.89	16.72	6.30	0.10	4.92	7.07	4.07	1.22	0.21		97.41	64.0	5.7
Popocatepetl	Nealtican flows	**Popo5**	98°32′14″	19°01′51″	Harvard	Andesitic	Calc-alk.		This study	61.36	0.90	16.29	5.68	0.09	3.84	5.23	4.33	1.75	0.24		99.72	60.6	11.8
Cuatepel	Cuatepel lava flow	**CH-08-26**	98°51′33″	19°05′25″	WSU/CGEO	Andesitic	Calc-alk.		This study	55.59	0.94	16.26	7.21	0.12	7.46	7.01	3.55	1.25	0.19		98.84	70.2	3.1
Cuatepel	Cuatepetel Cone	**CH-08-25**	98°51′28″	19°05′08″	WSU/CGEO	Andesitic	Calc-alk.		This study	58.77	0.91	16.21	6.24	0.11	5.40	5.89	3.97	1.49	0.21		98.57	66.3	8.4
Guespalapa	Guespalapa shield lava	**CH-09-1**	99°11′55″	19°04′45″	WSU/CGEO	Andesitic	Calc-alk.		This study	55.43	1.20	17.09	7.42	0.11	5.86	6.79	4.31	1.10	0.26		99.56	64.2	2.0
Guespalapa	Guespalapa shield lava	**CH-07-8**	99°10′26″	19°05′26″	WSU/CGEO	Andesitic	Calc-alk.		This study	55.49	0.98	17.19	7.32	0.12	6.72	7.53	3.90	0.98	0.18		99.69	67.6	2.1
Guespalapa	Guespalapa shield lava	**CH-09-2**	99°11′52″	19°04′49″	WSU/CGEO	Andesitic	Calc-alk.	*	Straub *et al.* (2011); this study	55.71	1.20	17.19	7.28	0.11	5.79	6.80	4.33	1.12	0.25		99.78	64.4	2.3
Guespalapa	Guespalapa shield lava	**S12A**	99°12′21″	19°04′18″	Harvard	Andesitic	Calc-alk.		This study	55.80	1.18	17.12	7.09	0.11	5.88	6.66	4.26	1.13	0.28		99.52	65.4	2.9
Guespalapa	Guespalapa shield lava	**CH-07-9**	99°10′20″	19°05′26″	WSU/CGEO	Andesitic	Calc-alk.	*	Straub *et al.* (2011); this study	55.81	0.98	17.24	7.34	0.12	6.92	7.58	3.87	1.01	0.17		100.31	68.2	2.1
Guespalapa	Guespalapa shield lava	**CH-05-11**	99°11′18″	19°05′22″	WSU/CGEO	Andesitic	Calc-alk.		This study	55.86	1.27	17.26	7.31	0.12	6.00	6.78	4.41	1.12	0.27	0.17	100.57	65.1	1.8
Guespalapa	Guespalapa shield lava	**CH-05-14**	99°12′07″	19°04′33″	WSU/CGEO	Andesitic	Calc-alk.		This study	56.09	1.19	17.26	7.16	0.11	5.96	6.83	4.33	1.13	0.25	0.42	100.73	65.4	2.3
Guespalapa	Guespalapa shield lava	**CH-08-3**	99°11′56″	19°04′34″	WSU/CGEO	Andesitic	Calc-alk.	*	Straub *et al.* (2011); this study	61.18	0.80	15.71	5.84	0.10	5.35	5.45	4.20	1.60	0.22		100.45	67.6	10.3
Guespalapa	El Caballito bomb	**MCH-06-8**	99°11′02″	19°05′21″	WSU/CGEO	Basaltic	Calc-alk.		This study	54.38	1.47	16.97	8.09	0.13	6.72	7.24	4.20	1.17	0.32	0.39	99.87	65.4	
Guespalapa	El Caballito bomb	**MCH-06-9**	99°11′02″	19°05′25″	WSU/CGEO	Basaltic	Calc-alk.	*	Straub *et al.* (2011); this study	54.58	1.48	17.04	8.10	0.13	6.71	7.28	4.26	1.14	0.32	0.25	100.22	65.3	
Guespalapa	El Caballito bomb	**CH-05-10**	99°11′11″	19°05′27″	WSU/CGEO	Basaltic	Calc-alk.		This study	54.82	1.47	17.18	8.02	0.12	6.48	7.00	4.40	1.18	0.33	0.08	100.19	64.8	
Guespalapa	El Caballito bomb	**CH-07-5**	99°10′45″	19°05′26″	WSU/CGEO	Andesitic	Calc-alk.		This study	55.40	1.37	17.20	7.55	0.12	6.09	6.81	4.43	1.15	0.30		99.66	64.7	1.0
Guespalapa	Manteca bomb	**CH-07-6**	99°10′38″	19°05′26″	WSU/CGEO	Andesitic	Calc-alk.	*	Straub *et al.* (2011); this study	56.58	1.01	17.48	7.15	0.11	6.20	7.12	4.09	1.02	0.18		100.24	66.4	3.3
Guespalapa	Manteca bomb	**CH-07-7**	99°10′38″	19°05′26″	WSU/CGEO	Andesitic	Calc-alk.		This study	56.87	1.00	17.48	7.01	0.11	5.98	7.02	4.22	1.03	0.19		100.22	66.0	3.5
Guespalapa	El Hoyo crater rim lava	**CH-07-11**	99°10′11″	19°05′28″	WSU/CGEO	Andesitic	Calc-alk.		This study	54.86	0.94	16.77	7.72	0.13	7.80	7.84	3.74	0.87	0.16		100.06	69.7	0.6
Guespalapa	El Hoyo crater rim lava	**CH-07-12**	99°10′10″	19°05′29″	WSU/CGEO	Andesitic	Calc-alk.	*	Straub *et al.* (2011); this study	54.99	0.95	16.76	7.77	0.13	7.92	7.86	3.73	0.88	0.16		100.37	69.9	0.6
Guespalapa	El Hoyo crater rim lava	**CH-07-13**	99°10′08″	19°05′31″	WSU/CGEO	Andesitic	Calc-alk.		This study	55.00	0.95	16.82	7.77	0.13	7.93	7.91	3.71	0.90	0.15		100.50	69.9	0.5
Guespalapa	El Hoyo crater rim lava	**CH-07-10**	99°10′11″	19°05′26″	WSU/CGEO	Andesitic	Calc-alk.		This study	55.22	0.95	16.90	7.78	0.13	7.74	7.88	3.78	0.89	0.16		100.66	69.4	0.7

Guespalapa	El Hoyo crater vesicular lava	**CH-07-14**	99°10′02″	19°05′25″	WSU/CGEO	Andesitic	Calc-alk.	*	Straub *et al.* (2011); this study	54.50	0.95	16.76	7.63	0.12	7.79	7.79	3.62	0.86	0.16		99.42	69.9	1.0
Suchiooc	Suchiooc Cone lava	**CH-05-2**	99°06′16″	19°04′06″	WSU/CGEO	Basaltic	Calc-alk.		This study	53.67	1.28	16.10	8.21	0.13	8.48	7.92	3.76	1.11	0.29	0.21	100.13	70.1	
Suchiooc	Suchiooc Cone bomb	**CH-08-8**	99°06′18″	19°04′03″	WSU/CGEO	Basaltic	Calc-alk.	*	Straub *et al.* (2011); this study	52.97	1.25	15.94	8.20	0.13	8.33	7.83	3.68	1.03	0.28		99.31	71.0	
Suchiooc	Suchiooc Cone bomb	**CH-08-7**	99°06′18″	19°04′03″	WSU/CGEO	Basaltic	High-Nb		This study	53.68	1.46	16.35	8.32	0.13	7.93	7.11	4.09	1.22	0.34		99.80	68.4	
Suchiooc	Suchiooc distal lavas	**CH-08-11**	99°03′58″	18°57′52″	WSU/CGEO	Basaltic	High-Nb	*	Straub *et al.* (2011); this study	51.60	1.78	16.80	9.21	0.15	7.50	7.93	4.15	1.25	0.43		99.99	65.3	
Suchiooc	Suchiooc distal lavas	**CH-08-10**	99°05′20″	19°01′02″	WSU/CGEO	Basaltic	High-Nb		This study	51.70	1.70	16.89	8.99	0.14	7.74	7.79	3.99	1.14	0.42		99.59	66.2	
Suchiooc	Suchiooc distal lavas	**CH-08-9**	99°05′20″	19°01′02″	WSU/CGEO	Basaltic	High-Nb		This study	52.03	1.69	16.58	9.13	0.14	7.33	7.69	4.06	1.21	0.42		99.87	66.1	
Suchiooc	Suchiooc distal lavas	**CH-08-12**	99°06′24″	18°58′55″	WSU/CGEO	Basaltic	High-Nb		This study	52.17	1.78	16.52	9.40	0.15	7.18	7.51	3.87	1.35	0.52		99.52	63.5	
Suchiooc	Suchiooc proximal lava	**CH-08-6**[h]	99°06′01″	19°03′20″	WSU/CGEO	Basaltic	Calc-alk.	*	Straub *et al.* (2011); this study	53.23	1.26	16.11	8.26	0.13	9.18	7.70	3.73	1.04	0.28		100.10	71.7	
Suchiooc	Suchiooc proximal lava	**CH-08-13**	99°06′18″	19°04′06″	WSU/CGEO	Basaltic	Calc-alk.		This study	53.33	1.26	15.89	8.26	0.13	9.02	7.77	3.69	1.05	0.28		99.87	71.3	
Suchiooc	Suchiooc proximal lava	**CH-08-14**	99°05′58″	19°04′48″	WSU/CGEO	Basaltic	Calc-alk.		This study	53.12	1.25	15.74	8.22	0.13	9.27	7.67	3.57	1.21	0.29		99.64	72.0	
Suchiooc	Suchiooc proximal lava	**CH-07-1**	99°06′02″	19°03′33″	WSU/CGEO	Basaltic	Calc-alk.		This study	54.17	1.33	16.16	8.17	0.13	8.75	7.59	3.82	1.07	0.30		100.68	70.9	
Suchiooc	Suchiooc proximal lava	**CH-05-1**	99°08′59″	19°03′58″	WSU/CGEO	Basaltic	High-Nb		This study	53.07	1.59	16.61	8.62	0.14	7.12	7.59	4.10	1.24	0.41	0.17	99.63	65.3	
Suchiooc	Suchiooc proximal lava	**CH-07-2**	99°06′00″	19°03′26″	WSU/CGEO	Basaltic	High-Nb		This study	52.17	1.67	16.51	8.84	0.14	7.70	7.62	4.07	1.21	0.41		99.46	66.4	
Tuxtepec	Tuxtepec bomb	**S15**[i]	99°16′51″	19°07′41″	Harvard	Basaltic	Calc-alk.	*	Straub *et al.* (2011); this study	50.16	1.00	14.87	8.76	0.14	9.73	9.80	3.20	1.52	0.51	0.00	99.71	71.6	
Tuxtepec	Tuxtepec bomb	**S14**	99°16′51″	19°07′41″	Harvard	Basaltic	Calc-alk.		This study	49.91	0.98	14.66	8.75	0.14	10.02	9.62	3.15	1.55	0.51	0.01	99.30	72.3	
Yecahuazac	Yecahuazac Cone bomb	**CH-08-15**	99°05′41″	19°04′26″	WSU/CGEO	Basaltic	Calc-alk.	*	Straub *et al.* (2011); this study	53.13	1.14	16.03	7.92	0.13	8.05	8.17	3.67	1.45	0.39		99.29	69.8	
Yecahuazac	Yecahuazac lava flow	**CH-08-16**	99°05′46″	19°05′46″	WSU/CGEO	Basaltic	High-Nb		This study	53.24	1.50	16.53	8.26	0.13	7.54	7.21	4.11	1.20	0.37		99.27	67.5	
Yecahuazac	Yecahuazac lava flow?	**MCH-06-6**[j]	99°04′44″	19°05′20″	WSU/CGEO	(Basaltic)	High-Nb		This study		1.41								0.37				
Yecahuazac	Yecahuazac lava flow?	**CH-05-3**[j]	99°05′43″	19°05′05″	WSU/CGEO	(Basaltic)	High-Nb		This study		1.50								0.40				
C. Magdalena	Shallow intrusive	**CH-05-19**	99°11′09″	19°14′07″	WSU/CGEO	(Andesitic)	Calc-alk.		This study		0.55								0.17				n.a.
Cuatzin	Cuatzin lava flow	**CH-09-16**	99°06′22″	19°09′28″	WSU/CGEO	Andesitic	Calc-alk.		This study	62.76	0.78	16.09	5.12	0.09	2.67	4.65	4.18	1.87	0.24		97.95	54.3	16.5
Ocusacayo	Ocusacayo distal lava	**CH-05-6**	99°04′59″	19°07′25″	WSU/CGEO	Andesitic	Calc-alk.		This study	64.45	0.71	16.11	4.71	0.08	2.85	4.63	4.29	2.01	0.24		99.61	57.9	16.8
Ocusacayo	Ocusacayo lava	**CH-05-7**	99°04′08″	19°08′27″	WSU/CGEO	(Andesitic)	Calc-alk.		This study		0.66								0.24				n.a.
Pelado	Pelado Cone bomb	**CH-07-3**	99°12′29″	19°08′24″	WSU/CGEO	Andesitic	Calc-alk.		This study	61.95	0.94	15.83	5.70	0.10	4.40	5.35	4.14	1.79	0.28		99.91	63.7	12.5
Pelado	Pelado shield lava	**ASC42**	99°11′12″	19°07′02″	WSU/CGEO	(Andesitic)	Calc-alk.		This study		1.14								0.34				n.a.
Pelado	Pelado shield lava	**CH-07-4**	99°12′00″	19°09′08″	WSU/CGEO	Andesitic	Calc-alk.		This study	60.24	1.01	16.05	6.08	0.10	4.36	5.67	4.07	1.73	0.28		98.98	62.0	10.8
Teuthli	V.Teuthli shield lava	**CH-05-9**	99°02′19″	19°12′19″	WSU/CGEO	(Andesitic)	Calc-alk.		This study		0.95								0.21				n.a.
Teuthli	V.Teuthli shield lava	**CH-05-8**	99°02′02″	19°13′29″	WSU/CGEO	(Andesitic)	Calc-alk.		This study		1.29								0.21				n.a.
Tlaloc	V.Tlaloc shield lava	**CH-05-4**	99°03′30″	19°05′50″	WSU/CGEO	(Andesitic)	Calc-alk.		This study		0.87								0.23				n.a.
Tlaloc	V.Tlaloc shield lava	**CH-05-5**	99°03′57″	19°06′13″	WSU/CGEO	(Andesitic)	Calc-alk.		This study		0.88								0.22				n.a.
Zempoala Ridge	Shallow intrusive	**CH-05-18**	99°19′19″	19°04′44″	WSU/CGEO	(Andesitic)	Calc-alk.		This study		0.74								0.19				n.a.
Zempoala Ridge	Shallow intrusive	**CH-05-17**	99°18′42″	19°05′07″	WSU/CGEO	(Andesitic)	Calc-alk.		This study		0.62								0.18				n.a.
flow near Tuxtepec	Lava flow	**S13**	99°16′30″	19°07′16″	Harvard	Andesitic	Calc-alk.		This study	53.32	1.36	16.60	9.94	0.15	5.44	7.35	3.59	1.05	0.38	0.01	99.20	55.5	2.9
Muneco	Lava flow	**S16**	99°19′45″	19°09′26″	Harvard	Andesitic	Calc-alk.		This study	60.14	0.81	17.09	5.83	0.10	4.07	5.48	4.21	1.65	0.23	0.01	99.61	61.4	10.2

(Continued)

Table 1. *Major and trace element data of calc-alkaline and high-Nb magmas of the central Mexican Volcanic Belt (Continued)*

Volcano	Stratigraphic Unit	Sample_ID	Oliv[g]	Neph[g]	Ni	Sc	V	Cr	Co	B	Li	Be	Cs	Ba	U	Tl	Rb	W	Th	Nb	Ta	La	Ce	Pb
Popocatepetl	Ventorillo aphyric dacite	**CH-09-18**			63	12.31	94.73	119.90	15.65		24.68	1.80	3.32	500	2.13	0.37	59.87	0.39	5.94	6.84	0.50	20.30	43.95	9.87
Popocatepetl	Dyke in upper B.Nexpayantla	**CH-09-19**			17	5.76	52.56	37.65	7.54	10	21.11	1.62	2.06	520	1.81	0.33	51.34	0.20	4.57	5.72	0.43	16.05	33.06	9.43
Popocatepetl	Volcaniclast in upper B.Nexpayantla	**CH-09-20**			39	11.83	85.10	72.88	12.94		24.63	1.56	2.60	456	1.58	0.41	49.00	0.26	4.25	5.26	0.38	15.92	33.49	8.90
Popocatepetl	Volcaniclast in upper B.Nexpayantla	**CH-09-21**			77	14.25	107.96	162.87	18.35		17.78	1.70	2.82	456	1.60	0.40	47.41	0.30	4.47	6.82	0.46	17.93	38.50	8.76
Popocatepetl	Dyke in lower B. Nexpayantla	**CH-09-22**			76	16.98	140.93	164.88	23.35		12.71	1.37	0.53	352	0.91	0.15	28.37	0.18	2.58	5.45	0.34	14.43	32.35	4.86
Popocatepetl	Nealtican flows	**Popo5**			53	16.00	113.32	125.30	17.22	16	23.97	1.69	2.64	428	1.60	0.36	51.78	0.33	4.58	9.08	0.60	17.46	37.95	8.07
Cuatepel	Cuatepel lava flow	**CH-08-26**																						
Cuatepel	Cuatepetel Cone	**CH-08-25**																						
Guespalapa	Guespalapa shield lava	**CH-09-1**			103	17.40	117.17	147.88	26.43		12.79	1.48	0.83	263	0.66	0.16	20.74	0.13	1.93	7.54	0.47	13.19	30.35	4.63
Guespalapa	Guespalapa shield lava	**CH-07-8**			126	19.45	129.89	279.94	29.03		12.61	1.09	0.83	234	0.57	0.15	19.38	0.13	1.68	4.16	0.25	9.60	22.08	4.20
Guespalapa	Guespalapa shield lava	**CH-09-2**			100	17.61	120.62	155.40	26.39		13.79	1.56	0.85	271	0.67	0.16	21.65	0.13	1.97	7.61	0.48	13.49	31.07	4.70
Guespalapa	Guespalapa shield lava	**S12A**			109	20.13	133.95	199.34	28.19	7	13.03	1.49	0.94	284	0.70	0.18	22.98	0.15	2.06	7.96	0.50	14.26	32.40	5.14
Guespalapa	Guespalapa shield lava	**CH-07-9**			125	19.57	130.87	288.80	28.89		12.00	1.08	0.83	240	0.57	0.31	19.48	0.13	1.68	4.08	0.26	9.55	22.47	4.08
Guespalapa	Guespalapa shield lava	**CH-05-11**			104	17.80	123.74	132.66	27.52		12.20	1.59	0.87	274	0.70	0.15	21.57	0.16	2.08	8.82	0.53	14.32	34.08	4.81
Guespalapa	Guespalapa shield lava	**CH-05-14**			101	18.22	123.02	142.35	26.68		13.52	1.45	0.89	277	0.68	0.17	21.52	0.14	2.02	7.58	0.46	13.46	31.52	4.81
Guespalapa	Guespalapa shield lava	**CH-08-3**			137	13.86	101.30	204.49	21.41	9	20.03	1.50	1.42	464	1.18	0.30	35.68	0.22	3.76	7.26	0.46	19.19	39.88	7.49
Guespalapa	El Caballito bomb	**MCH-06-8**	1.3		118	19.40	134.21	168.81	30.58		11.44	1.64	0.90	272	0.75	0.13	22.40	0.21	2.28	13.56	0.82	16.39	37.91	4.52
Guespalapa	El Caballito bomb	**MCH-06-9**	1.8		115	19.78	136.58	166.74	30.51		13.01	1.67	0.90	280	0.75	0.10	21.90	0.21	2.28	13.61	0.82	16.55	37.96	4.45
Guespalapa	El Caballito bomb	**CH-05-10**	1.3		114	18.82	131.13	152.15	30.34		12.91	1.75	0.84	273	0.74	0.11	21.69	0.21	2.24	13.36	0.81	16.54	38.40	4.39
Guespalapa	El Caballito bomb	**CH-07-5**			96	17.13	118.70	144.09	27.54		13.21	1.62	0.83	260	0.69	0.13	21.16	0.17	2.03	10.76	0.67	15.14	34.12	4.46
Guespalapa	Manteca bomb	**CH-07-6**			110	17.48	120.09	217.87	26.21		12.75	1.16	0.84	264	0.61	0.83	19.88	0.12	1.78	4.40	0.27	10.29	24.04	4.44
Guespalapa	Manteca bomb	**CH-07-7**			102	17.17	117.30	174.60	24.94		13.27	1.15	0.84	260	0.62	0.12	20.10	0.12	1.79	4.44	0.28	10.42	23.83	4.39
Guespalapa	El Hoyo crater rim lava	**CH-07-11**			159	21.60	142.99	268.42	31.96		11.14	1.03	0.66	209	0.50	0.10	16.18	0.12	1.53	3.61	0.22	9.12	21.57	3.62
Guespalapa	El Hoyo crater rim lava	**CH-07-12**			164	21.64	142.23	310.74	32.56		11.24	1.03	0.66	214	0.49	0.10	16.22	0.12	1.52	3.61	0.22	9.08	21.69	3.65
Guespalapa	El Hoyo crater rim lava	**CH-07-13**			167	21.50	141.91	378.26	32.52		10.95	1.01	0.66	214	0.48	0.09	16.06	0.12	1.50	3.53	0.22	9.04	21.48	3.59
Guespalapa	El Hoyo crater rim lava	**CH-07-10**			159	21.50	139.52	284.48	31.98		10.84	1.05	0.67	210	0.49	0.10	16.14	0.12	1.53	3.65	0.22	9.07	21.34	3.68
Guespalapa	El Hoyo crater vesicular lava	**CH-07-14**			165	21.16	138.61	292.87	32.38		11.20	1.01	0.73	213	0.54	0.04	17.02	0.13	1.65	3.72	0.23	9.29	21.92	3.79

Suchiooc	Suchiooc Cone lava	**CH-05-2**	5.1		158	21.25	153.40	287.84	35.48		9.56	1.35	0.53	289	0.67	0.11	17.94	0.15	2.21	11.40	0.67	16.58	37.64	4.71
Suchiooc	Suchiooc Cone bomb	**CH-08-8**	6.1		169	20.64	148.96	307.26	36.26	4	8.70	1.33	0.48	272	0.64	0.08	15.53	0.14	2.14	10.32	0.61	15.91	35.62	4.14
Suchiooc	Suchiooc Cone bomb	**CH-08-7**	5.7		182	19.05	133.04	222.03	34.21	6	10.28	1.70	0.63	320.15	0.79	0.11	20.25	0.22	2.47	17.39	1.03	19.27	41.64	4.65
Suchiooc	Suchiooc distal lavas	**CH-08-11**	14.2		123	21.54	152.12	231.80	35.55	4	9.46	1.79	0.29	273.53	0.70	0.05	16.31	0.23	2.46	24.37	1.43	21.08	45.61	3.64
Suchiooc	Suchiooc distal lavas	**CH-08-10**	11.2		136	19.64	136.13	242.52	33.81	4	8.08	1.76	0.41	268.89	0.74	0.07	15.97	0.23	2.27	21.32	1.25	19.62	42.32	3.73
Suchiooc	Suchiooc distal lavas	**CH-08-9**	11.6		143	20.39	143.34	249.05	35.36	4	8.62	1.79	0.44	271.10	0.76	0.07	18.03	0.26	2.37	22.14	1.31	19.16	41.44	3.75
Suchiooc	Suchiooc distal lavas	**CH-08-12**	6.9		133	21.21	152.27	252.15	34.18	10	11.17	2.01	0.82	313.66	0.89	0.13	23.69	0.44	2.77	24.32	1.42	24.16	51.66	4.78
Suchiooc	Suchiooc proximal lava	**CH-08-6**[h]	7.5		207	20.74	148.26	263.69	37.86	5	9.87	1.36	0.50	288	0.67	0.14	16.10	0.16	2.22	11.13	0.67	16.26	36.40	4.30
Suchiooc	Suchiooc proximal lava	**CH-08-13**	6.0		185	20.43	146.65	287.56	37.06	4	9.40	1.37	0.53	281	0.65	0.92	16.61	0.15	2.15	10.81	0.65	16.06	35.91	4.21
Suchiooc	Suchiooc proximal lava	**CH-08-14**	6.7		210	20.62	147.69	320.04	38.11	5	8.76	1.39	0.54	295	0.67	0.14	18.77	0.16	2.20	11.08	0.66	16.31	36.45	4.28
Suchiooc	Suchiooc proximal lava	**CH-07-1**	4.2		197	20.29	140.28	294.60	36.23		9.46	1.43	0.49	270	0.65	0.13	16.82	0.17	2.13	12.92	0.75	16.34	35.06	4.26
Suchiooc	Suchiooc proximal lava	**CH-05-1**	6.2		133	20.14	140.13	184.42	33.14		10.84	1.74	0.63	339.42	0.84	0.31	20.29	0.20	2.67	18.92	1.10	22.05	48.74	5.83
Suchiooc	Suchiooc proximal lava	**CH-07-2**	10.1		149	20.34	141.65	252.40	35.03		9.97	1.76	0.45	287.87	0.74	0.07	18.03	0.23	2.35	21.81	1.29	20.22	43.94	3.78
Tuxtepec	Tuxtepec bomb	**S15**[i]	17.7	0.7	220	27.11	187.72	494.14	40.23	2	13.04	1.46	0.35	1095	1.12	0.05	18.50	0.12	4.70	3.82	0.22	43.99	101.83	10.08
Tuxtepec	Tuxtepec bomb	**S14**	18.4	0.7	249	27.05	182.01	498.27	41.70	2	11.77	1.22	0.42	1054	0.87	0.63	21.44	0.10	4.55	3.71	0.21	42.81	101.94	10.28
Yecahuazac	Yecahuazac Cone bomb	**CH-08-15**	6.6		168	20.94	154.49	280.84	32.06	4	10.18	1.61	0.68	637	1.35	0.17	24.66	0.16	4.03	8.44	0.48	31.16	68.96	7.34
Yecahuazac	Yecahuazac lava flow	**CH-08-16**	5.7		161	19.09	134.25	229.52	33.28	6	10.19	1.74	0.56	306.81	0.81	0.12	19.24	0.22	2.53	18.08	1.07	19.72	42.64	4.64
Yecahuazac	Yecahuazac lava flow?	**MCH-06-6**[j]	n.a.		141	19.17	135.32	221.16	31.63		10.94	1.62	0.62	309.90	0.78	0.11	20.36	0.19	2.43	16.77	0.97	18.76	40.93	4.99
Yecahuazac	Yecahuazac lava flow?	**CH-05-3**[j]	n.a.		142	19.59	135.85	213.48	32.27		10.79	1.68	0.61	308.88	0.80	0.22	20.14	0.20	2.44	18.83	1.10	19.63	42.96	4.73
C. Magdalena	Shallow intrusive	**CH-05-19**			7	7.90	55.12	12.75	6.97		15.75	1.48	0.17	540	0.67	0.25	17.31	0.13	4.45	5.64	0.42	17.90	36.64	11.27
Cuatzin	Cuatzin lava flow	**CH-09-16**			37	12.67	84.35	77.09	13.74		26.88	1.84	2.21	574	1.63	0.40	50.62	0.24	4.78	9.11	0.59	23.50	47.26	9.67
Ocusacayo	Ocusacayo distal lava	**CH-05-6**			34	11.85	82.59	84.06	12.11		24.23	1.60	2.08	600	1.76	0.46	46.14	0.19	5.17	6.61	0.44	24.51	49.62	11.00
Ocusacayo	Ocusacayo lava	**CH-05-7**			37	11.94	82.12	92.02	12.10		24.64	1.61	2.16	586	1.75	0.43	48.03	0.19	5.14	6.58	0.45	23.13	47.32	10.98
Pelado	Pelado Cone bomb	**CH-07-3**			80	14.02	93.44	171.72	18.30		19.83	1.63	1.52	537	1.25	0.29	41.23	0.25	4.17	10.78	0.66	22.31	46.49	8.39
Pelado	Pelado shield lava	**ASC42**			55	13.88	93.67	111.51	18.05	11	19.25	1.91	1.39	517	1.32	0.33	38.43	0.30	4.20	14.18	0.86	24.09	49.55	8.24
Pelado	Pelado shield lava	**CH-07-4**			67	15.45	100.51	159.30	19.58		18.32	1.65	1.41	496	1.17	0.27	37.27	0.23	3.77	11.51	0.71	20.85	43.72	7.62
Teuthli	V.Teuthli shield lava	**CH-05-9**			7	17.83	126.73	38.62	16.23		14.57	1.36	1.82	376	1.13	0.25	37.17	0.31	3.61	5.58	0.38	16.04	34.92	7.14
Teuthli	V.Teuthli shield lava	**CH-05-8**			11	25.67	174.58	111.17	22.38		13.12	1.27	1.23	239	0.71	0.18	26.91	0.36	2.54	5.44	0.35	12.76	30.13	4.95
Tlaloc	V.Tlaloc shield lava	**CH-05-4**			74	16.24	112.90	124.52	20.42		17.28	1.38	1.34	435	0.99	0.26	31.82	0.26	3.21	6.34	0.40	17.04	36.31	7.62
Tlaloc	V.Tlaloc shield lava	**CH-05-5**			67	16.53	115.05	113.93	20.04		16.75	1.40	1.30	430	0.96	0.25	30.89	0.26	3.17	6.41	0.40	17.02	36.40	7.43
Zempoala Ridge	Shallow intrusive	**CH-05-18**			17	14.51	99.25	46.19	13.77		17.64	1.41	1.04	448	1.14	0.27	38.70	0.12	3.28	5.34	0.35	15.66	31.62	7.81
Zempoala Ridge	Shallow intrusive	**CH-05-17**			16	11.84	77.16	55.11	10.49		18.12	1.52	1.13	493	1.28	0.33	42.37	0.14	4.09	5.58	0.37	17.69	35.01	13.44
flow near Tuxtepec	Lava flow	**S13**			47	24.51	174.93	158.58	29.94	20	16.15	1.70	1.31	311	0.88	0.20	25.73	0.36	2.69	13.25	0.77	19.89	43.97	5.29
Muneco	Lava flow	**S16**			52	15.92	110.98	112.75	17.71	9	18.16	1.47	1.52	429	1.07	0.28	33.67	0.16	3.09	6.54	0.43	18.18	35.66	7.25

(Continued)

Table 1. *Major and trace element data of calc-alkaline and high-Nb magmas of the central Mexican Volcanic Belt (Continued)*

Volcano	Stratigraphic Unit	Sample_ID	Pr	Mo	Sr	Nd	Sm	Hf	Zr	Eu	Sn	Sb	Gd	Tb	Dy	Y	Ho	Er	Yb	Lu	$^{87}Sr/^{86}Sr$	2 STEM	$^{143}Nd/^{144}Nd$	2 STEM	eNd
Popocatepetl	Ventorillo aphyric dacite	**CH-09-18**	5.67	1.17	486	23.52	4.98	4.79	190	1.33	1.42	0.23	4.33	0.63	3.45	19.87	0.67	1.86	1.74	0.27					
Popocatepetl	Dyke in upper B.Nexpayantla	**CH-09-19**	4.13	0.96	427	16.79	3.46	3.67	167	0.94	1.21	0.18	2.93	0.41	2.38	12.95	0.48	1.26	1.29	0.18	0.70458	9	0.51277	6	2.66
Popocatepetl	Volcaniclast in upper B.Nexpayantla	**CH-09-20**	4.27	0.88	482	17.45	3.77	4.14	167	1.15	1.15	0.17	3.47	0.52	2.95	17.27	0.59	1.65	1.57	0.24	0.70429	8	0.51286	5	4.41
Popocatepetl	Volcaniclast in upper B.Nexpayantla	**CH-09-21**	5.01	1.05	470	20.76	4.45	4.51	186	1.29	1.22	0.20	4.01	0.60	3.35	19.59	0.66	1.84	1.73	0.26					
Popocatepetl	Dyke in lower B. Nexpayantla	**CH-09-22**	4.39	0.69	455	19.05	4.37	4.03	165	1.31	1.02	0.07	4.15	0.63	3.56	20.57	0.71	1.98	1.86	0.28					
Popocatepetl	Nealtican flows	**Popo5**	5.04		468	20.35	4.46	4.50	185	1.29	1.40	0.18	4.16	0.63	3.58	20.51	0.71	1.98	1.89	0.29					
Cuatepel	Cuatepel lava flow	**CH-08-26**																							
Cuatepel	Cuatepetel Cone	**CH-08-25**																							
Guespalapa	Guespalapa shield lava	**CH-09-1**	4.08	0.82	477	17.64	4.13	4.06	182	1.37	1.11	0.10	4.09	0.63	3.64	20.73	0.72	1.98	1.85	0.28					
Guespalapa	Guespalapa shield lava	**CH-07-8**	3.11	0.87	374	14.06	3.55	3.26	131	1.20	0.94	0.10	3.55	0.56	3.22	18.50	0.65	1.78	1.66	0.25	0.70419	7	0.51293	6	5.65
Guespalapa	Guespalapa shield lava	**CH-09-2**	4.17	0.84	488	18.11	4.24	4.15	186	1.41	1.14	0.10	4.19	0.65	3.71	21.24	0.74	2.03	1.89	0.29	0.70381	9	0.51294	5	5.82
Guespalapa	Guespalapa shield lava	**S12A**	4.38		510	19.06	4.51	4.34	188	1.48	1.27	0.10	4.39	0.68	3.93	22.27	0.79	2.15	1.99	0.31	0.70383	9	0.51294	6	5.85
Guespalapa	Guespalapa shield lava	**CH-07-9**	3.13	0.85	373	14.11	3.49	3.28	131	1.17	0.89	0.10	3.59	0.56	3.25	18.69	0.66	1.80	1.64	0.25	0.70417	7	0.51292	6	5.54
Guespalapa	Guespalapa shield lava	**CH-05-11**	4.38	0.90	505	19.00	4.38	4.22	192	1.48	1.20	0.09	4.32	0.68	3.85	21.60	0.77	2.10	1.96	0.29	0.70376	7	0.51294	6	5.96
Guespalapa	Guespalapa shield lava	**CH-05-14**	4.12	0.86	484	18.24	4.18	4.07	185	1.42	1.15	0.09	4.19	0.65	3.74	20.82	0.75	2.03	1.91	0.28	0.70384	9	0.51294	5	5.85
Guespalapa	Guespalapa shield lava	**CH-08-3**	5.08	1.06	456	20.69	4.29	4.47	186	1.27	1.14	0.14	3.86	0.58	3.29	19.18	0.65	1.83	1.74	0.27	0.70442	7	0.51282	4	3.61
Guespalapa	El Caballito bomb	**MCH-06-8**	4.82	1.03	511	20.98	4.75	4.42	206	1.57	1.38	0.11	4.74	0.74	4.24	23.83	0.85	2.33	2.17	0.33	0.70363	8	0.51292	11	5.43
Guespalapa	El Caballito bomb	**MCH-06-9**	4.86	0.99	514	21.00	4.74	4.45	205	1.58	1.38	0.12	4.75	0.74	4.27	24.07	0.85	2.34	2.17	0.33	0.70363	6	0.51293	6	5.66
Guespalapa	El Caballito bomb	**CH-05-10**	4.88	1.02	529	20.89	4.75	4.48	209	1.58	1.34	0.10	4.69	0.73	4.20	23.71	0.84	2.32	2.16	0.32					
Guespalapa	El Caballito bomb	**CH-07-5**	4.53	1.45	511	19.54	4.50	4.29	197	1.49	1.23	0.11	4.44	0.69	3.95	22.51	0.79	2.16	1.99	0.30	0.70366	6	0.51295	12	6.06
Guespalapa	Manteca bomb	**CH-07-6**	3.31	0.89	401	14.76	3.62	3.48	140	1.22	0.96	0.09	3.63	0.57	3.24	18.44	0.65	1.77	1.64	0.25	0.70415	8	0.51293	6	5.60
Guespalapa	Manteca bomb	**CH-07-7**	3.32	0.94	403	14.73	3.56	3.52	140	1.20	0.87	0.09	3.65	0.57	3.23	18.47	0.64	1.73	1.61	0.25	0.70414	6	0.51292	9	5.59
Guespalapa	El Hoyo crater rim lava	**CH-07-11**	3.04	0.80	362	13.82	3.52	3.10	123	1.14	0.83	0.09	3.64	0.57	3.40	19.76	0.69	1.92	1.80	0.27					
Guespalapa	El Hoyo crater rim lava	**CH-07-12**	3.02	0.78	362	13.74	3.52	3.11	122	1.16	0.82	0.08	3.63	0.58	3.36	19.65	0.68	1.92	1.84	0.28	0.70411	9	0.51295	9	6.04
Guespalapa	El Hoyo crater rim lava	**CH-07-13**	3.01	0.79	351	13.70	3.46	3.07	121	1.14	0.82	0.09	3.59	0.57	3.35	19.60	0.68	1.91	1.81	0.28					
Guespalapa	El Hoyo crater rim lava	**CH-07-10**	3.00	0.78	360	13.92	3.57	3.12	122	1.18	0.82	0.08	3.62	0.57	3.30	19.39	0.68	1.94	1.83	0.28					
Guespalapa	El Hoyo crater vesicular lava	**CH-07-14**	3.06	0.72	349	13.70	3.43	3.11	124	1.13	0.83	0.10	3.58	0.56	3.33	19.17	0.67	1.84	1.73	0.27	0.70418	6	0.51292	7	5.42
Suchiooc	Suchiooc Cone lava	**CH-05-2**	5.05	1.27	547	21.63	4.94	3.91	171	1.52	1.13	0.07	4.69	0.73	4.20	24.18	0.84	2.32	2.15	0.32	0.70363	6	0.51297	5	6.42
Suchiooc	Suchiooc Cone bomb	**CH-08-8**	4.78	0.97	527	20.55	4.61	3.68	161	1.45	1.04	0.06	4.48	0.68	3.98	22.83	0.80	2.19	2.03	0.31	0.70365	7	0.51296	5	6.24
Suchiooc	Suchiooc Cone bomb	**CH-08-7**	5.40	1.35	537	22.78	5.07	4.44	206.44	1.63	1.28	0.07	4.95	0.76	4.39	25.12	0.88	2.40	2.23	0.34	0.70360	8	0.51295	7	6.01
Suchiooc	Suchiooc distal lavas	**CH-08-11**	5.80	1.20	548	24.51	5.49	4.78	229.52	1.80	1.44	0.06	5.51	0.86	5.02	28.39	1.00	2.73	2.51	0.38	0.70337	9	0.51294	6	5.93

Suchiooc	Suchiooc distal lavas	**CH-08-10**	5.41	1.44	535	22.84	5.10	4.43	212.01	1.67	1.31	0.07	5.10	0.79	4.61	26.40	0.93	2.54	2.34	0.35	0.70342	8	0.51294	6	5.89
Suchiooc	Suchiooc distal lavas	**CH-08-9**	5.34	1.53	537	22.56	5.05	4.62	219.80	1.67	1.42	0.08	5.04	0.79	4.60	26.08	0.92	2.52	2.33	0.35					
Suchiooc	Suchiooc distal lavas	**CH-08-12**	6.65	1.62	500	28.07	6.12	5.31	262.32	1.93	1.48	0.13	6.03	0.92	5.39	30.75	1.08	2.94	2.71	0.41					
Suchiooc	Suchiooc proximal lava	**CH-08-6**[h]	4.87	1.00	532	20.95	4.73	3.79	172	1.49	1.08	0.05	4.56	0.70	4.04	23.30	0.81	2.23	2.08	0.32	0.70360	8	0.51294	6	5.93
Suchiooc	Suchiooc proximal lava	**CH-08-13**	4.80	1.01	524	20.63	4.66	3.70	164	1.45	1.10	0.06	4.51	0.69	3.98	22.98	0.80	2.20	2.05	0.31					
Suchiooc	Suchiooc proximal lava	**CH-08-14**	4.89	1.01	529	20.94	4.71	3.78	166	1.49	1.07	0.06	4.55	0.70	4.05	23.11	0.81	2.23	2.07	0.31					
Suchiooc	Suchiooc proximal lava	**CH-07-1**	4.71	1.43	507	20.60	4.68	3.85	169	1.50	1.13	0.06	4.55	0.71	3.95	22.88	0.80	2.27	2.08	0.32	0.70360	8	0.51294	6	5.93
Suchiooc	Suchiooc proximal lava	**CH-05-1**	6.37	1.80	579	26.65	5.94	4.78	225.32	1.82	1.41	0.09	5.57	0.86	4.91	27.72	0.97	2.65	2.46	0.37	0.70357	7	0.51294	6	5.99
Suchiooc	Suchiooc proximal lava	**CH-07-2**	5.61	1.97	541	23.99	5.41	4.63	218.07	1.74	1.38	0.07	5.35	0.83	4.80	27.50	0.96	2.69	2.43	0.37	0.70341	6	0.51292	8	5.56
Tuxtepec	Tuxtepec bomb	**S15**[i]	13.97		938	61.26	12.32	3.94	160	3.15	1.01	0.04	9.19	1.17	5.36	28.56	0.98	2.56	2.15	0.32					6.87
Tuxtepec	Tuxtepec bomb	**S14**	13.62		954	59.76	12.08	3.81	155	3.06	0.94	0.05	8.87	1.14	5.36	27.67	0.96	2.51	2.09	0.31					
Yecahuazac	Yecahuazac Cone bomb	**CH-08-15**	9.20	1.19	766	38.35	7.65	4.61	199	2.17	1.01	0.06	6.03	0.82	4.23	22.97	0.79	2.12	1.88	0.28	0.70430	8	0.51293	7	5.78
Yecahuazac	Yecahuazac lava flow	**CH-08-16**	5.51	1.41	539	23.27	5.18	4.52	211.23	1.66	1.30	0.08	5.06	0.78	4.51	25.74	0.90	2.47	2.28	0.34					
Yecahuazac	Yecahuazac lava flow?	**MCH-06-6**[j]	5.42	1.64	536	22.49	5.11	4.41	200.82	1.59	1.28	0.08	4.87	0.76	4.42	25.08	0.88	2.40	2.22	0.34					
Yecahuazac	Yecahuazac lava flow?	**CH-05-3**[j]	5.58	1.80	537	23.27	5.28	4.54	212.59	1.65	1.35	0.08	5.06	0.79	4.60	25.95	0.91	2.51	2.30	0.35					
C.Magdalena	Shallow intrusive	**CH-05-19**	4.63	0.56	408	17.78	3.83	2.21	64	1.02	1.26	0.19	3.25	0.50	2.71	15.82	0.52	1.43	1.31	0.19					
Cuatzin	Cuatzin lava flow	**CH-09-16**	5.89	1.07	430	23.44	4.80	5.23	232	1.40	1.33	0.15	4.35	0.66	3.66	21.79	0.73	2.01	1.88	0.29	0.70450	7	0.51284	5	3.87
Ocusacayo	Ocusacayo distal lava	**CH-05-6**	6.58	1.54	499	25.13	5.26	5.12	209	1.40	1.26	0.20	4.22	0.64	3.46	19.20	0.66	1.79	1.66	0.25	0.70443	9	0.51290	5	5.07
Ocusacayo	Ocusacayo lava	**CH-05-7**	6.17	1.51	470	23.62	5.00	5.09	207	1.35	1.29	0.21	4.07	0.62	3.43	19.06	0.65	1.79	1.68	0.25					
Pelado	Pelado Cone bomb	**CH-07-3**	5.66	1.58	460	22.95	4.78	5.02	212	1.36	1.23	0.12	4.44	0.67	3.65	21.11	0.71	2.00	1.88	0.29					
Pelado	Pelado shield lava	**ASC42**	6.21	1.43	497	25.04	5.19	5.35	241	1.54	1.42	0.11	4.77	0.72	4.07	23.50	0.81	2.22	2.07	0.31					
Pelado	Pelado shield lava	**CH-07-4**	5.40	1.57	453	21.97	4.67	4.80	209	1.37	1.27	0.11	4.38	0.67	3.74	21.57	0.74	2.06	1.93	0.29					
Teuthli	V.Teuthli shield lava	**CH-05-9**	4.65	0.93	534	18.91	4.32	4.15	171	1.29	1.20	0.17	3.99	0.63	3.66	20.98	0.73	2.04	1.93	0.29					
Teuthli	V.Teuthli shield lava	**CH-05-8**	4.19	0.78	433	18.07	4.53	3.96	169	1.41	1.35	0.10	4.70	0.78	4.75	27.60	0.97	2.73	2.62	0.40					
Tlaloc	V.Tlaloc shield lava	**CH-05-4**	4.74	1.17	451	19.06	4.30	4.40	179	1.28	1.14	0.15	3.98	0.63	3.62	21.18	0.72	2.03	1.94	0.29					
Tlaloc	V.Tlaloc shield lava	**CH-05-5**	4.80	1.15	452	19.15	4.34	4.37	181	1.29	1.14	0.14	4.00	0.64	3.68	21.23	0.73	2.04	1.95	0.29					
Zempoala Ridge	Shallow intrusive	**CH-05-18**	4.29	1.15	422	16.86	3.88	4.31	170	1.12	1.07	0.13	3.49	0.55	3.16	18.26	0.61	1.70	1.59	0.24					
Zempoala Ridge	Shallow intrusive	**CH-05-17**	4.48	0.88	448	17.20	3.77	4.28	171	1.08	1.00	0.13	3.21	0.49	2.77	15.61	0.54	1.48	1.38	0.21	0.70422	9	0.51286	6	4.29
flow near Tuxtepec	Lava flow	**S13**	5.66		350	24.17	5.52	4.78	218	1.70	1.87	0.07	5.88	0.95	5.76	33.85	1.20	3.33	3.13	0.48					
Muneco	Lava flow	**S16**	4.71		482	19.51	4.25	4.37	180	1.29	1.04	0.12	3.96	0.60	3.36	20.42	0.68	1.89	1.76	0.27					

Major element oxide abundances in wt%. Trace element abundances in ppm. Total Fe given as Fe_2O_3.

Run error of isotope ratios given as 2 standard errors of the mean in ppm.

[a]Indicates laboratory were major and trace elements were obtained (Harvard, Langmuir Laboratory; WSU, Washington State University; CGEO, Centro de Geociencias).

[b]Andesitic (quartz-normative) and basaltic (olivine-normative). Inferred from petrograpy and stratigraphic context for samples without major element oxides.

[c]Division in calc-alkaline (<16 ppm Nb) and high-Nb arc magmas (>16 ppm Nb).

[d]Denotes sample for which 3He/4He of olivines have been obtained (see Straub *et al.* 2011).

[e]Selected data (SiO_2, FeO*, MgO, Ni, Sr, Nb, La, Gd, Y, Yb, Nd isotope ratios, normative Qtz, normative Oliv) for some samples have previously been reported by Straub *et al.* (2011).

[f]Bulk rock Mg#, calculated assuming 18% ferric Fe.

[g]CIPW norm: Qtz, quartz-normative; Oliv, olivine normative. Calculation with program created by K. Hollocher at the Union College of Schenectady (http://www.union.edu/PUBLIC/GEODEPT/COURSES/petrology/norms.htm)

[h]Sr and Nd isotope ratios of this sample are those of sample CH07-1 that was taken at the same location.

[i]Sr and Nd isotopes are those of sample pw115 (LaGatta), a sample from same location with identical major and trace elements.

[j]Origin from Yecahuazac Cone possible, but not certain.

western flank of Yecahuazac cones actually derives from V. Suchiooc, and not from Yecahuazac as stated by Wallace & Carmichael (1999) and LaGatta (2003).

Volcan Tuxtepec is a small scoria cone to the east of the village of Capulin (Wallace & Carmichael 1999). It erupts olivine- and plagioclase-phyric olivine-normative calc-alkaline basalts with the lowest $SiO_2 = 49.9–50.2$ wt% and Nb = 3.7–3.8 ppm, but the highest MgO = 9.7–10.0 wt% and Mg# = 71.6–72.3 reported from the Sierra Chichinautzin Volcanic Field.

V. Popocatepetl is a major calc-alkaline MVB composite volcano that is well known for the presence of olivine phenocrysts in basaltic andesitic and andesitic magmas (e.g. Straub & Martin-Del Pozzo 2001; LaGatta 2003; Schaaf *et al.* 2005). One quartz-normative basaltic andesite (SiO_2 = 56.7 wt%, LaGatta 2003) from the oldest exposed series in Barranco Nexpayantla provided sufficient olivine for He isotope analyses (Straub *et al.* 2011). The most recent dating suggests a maximum age of *c.* 600 ka for the oldest exposed Popo series (R. Espinasa-Pereña, work in progress).

Analytical methods

All new data are reported in Table 1. In addition, we report re-analyses of major element oxides by DCP of samples reported by Wallace & Carmichael (1999) for which LaGatta (2003) had analysed for trace element and isotope ratios (see Supplementary material).

Bulk rock powders were analysed for major elements (50 samples) either at the Geoanalytical Laboratory of Washington State University, USA (XRF method, 44 samples), or at the Department of Earth and Planetary Sciences (Langmuir Laboratory) at Harvard University, USA (DCP method). Trace elements were determined by ICP-MS methods at the Centro de Geociencias (59 samples), Universidad Nacional Autónoma de

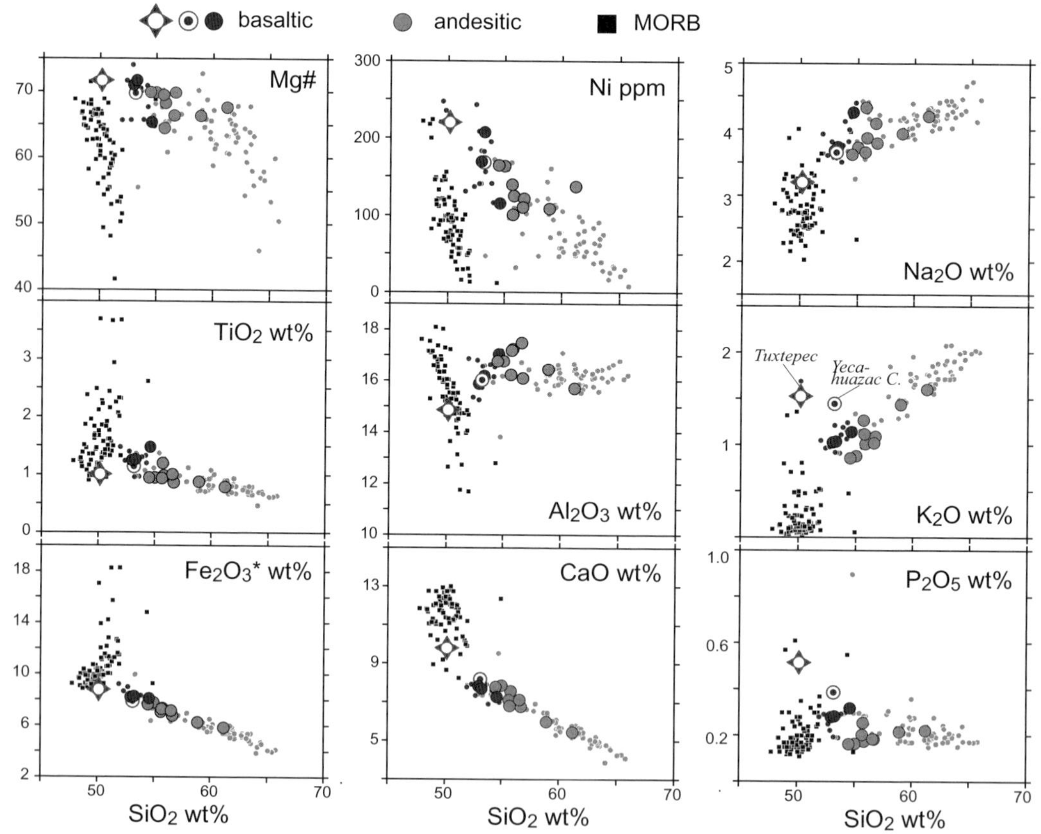

Fig. 3. Whole-rock Mg# numbers and major element oxides v. SiO_2 (in wt%) of central MVB basaltic to andesitic magmas compared to East Pacific Rise MORB (filled squares) from Niu & Batiza (1997). Mg# number (=molar ratio of $(Mg/(Mg + Fe^{2+}))$) is calculated using 18% ferric iron after Straub *et al.* (2008).

México, Querétaro, México or at Department of Earth and Planetary Sciences (Langmuir Laboratory) at Harvard University, USA. A selected subset of 29 samples was analysed for Sr and Nd isotope ratios by thermal ionization mass spectrometry at the Institute for Earth Sciences, Academia Sinica, Taipei, Taiwan. Details of sample preparation, analytical methods and assessment of interlaboratory comparability are given in the Supplementary material.

There is a small overlap with abundance data and $^{143}Nd/^{144}Nd$ ratios previously published by Straub *et al.* (2011). This is indicated in the footnote to Table 1. Olivine compositions and $^3He/^4He$ relevant to the discussion have been reported by Straub *et al.* (2008, 2011).

Results

Major elements

Major element oxides are presented in Figures 1 and 3. The primitive nature of the calc-alkaline magmas is evident in the high Ni and Mg# values that are comparable to mid-ocean ridge basalts (MORB), which best exemplify primitive melts from upper mantle peridotite. On the other hand, the central MVB magmas differ from MORB by the arc-typical enrichments in SiO_2, K_2O and Na_2O, and depletions of CaO, and Fe_2O_3*, MnO (not shown) and TiO_2, all of which are at the lower end, or below MORB. Only Al_2O_3, P_2O_5 and MgO (not shown) abundances compare with MORB. When plotted against SiO_2, the major element oxides follow strikingly linear trends away from the field of MORB. Such trends are typical for melt mixing rather than for fractional crystallization (e.g. Reubi & Blundy 2009). The linear trends are especially tight for major elements TiO_2, CaO, Fe_2O_3* and MgO that are commonly thought to be mantle-derived in arc magmas. There is more variability of K_2O, Na_2O and P_2O_5 that may be influenced by the slab flux (Straub *et al.* 2008, 2013). The basalts from Tuxtepec (S15) and Yecahuazac Cone (CH08–15), respectively, stand out because of their high K_2O and P_2O_5 abundances.

Incompatible trace elements

Multi-element diagrams of incompatible elements reveal the arc-typical relative enrichments of Cs, Ba, U, K, Sr and Li that are paired with relative depletions of Nb, Ta and Ti (Fig. 4; only samples high-$^3He/^4He$ olivines are shown for clarity). A notable feature is that the andesites are more depleted and more homogeneous in incompatible elements than the basalts. The Guespalapa andesites

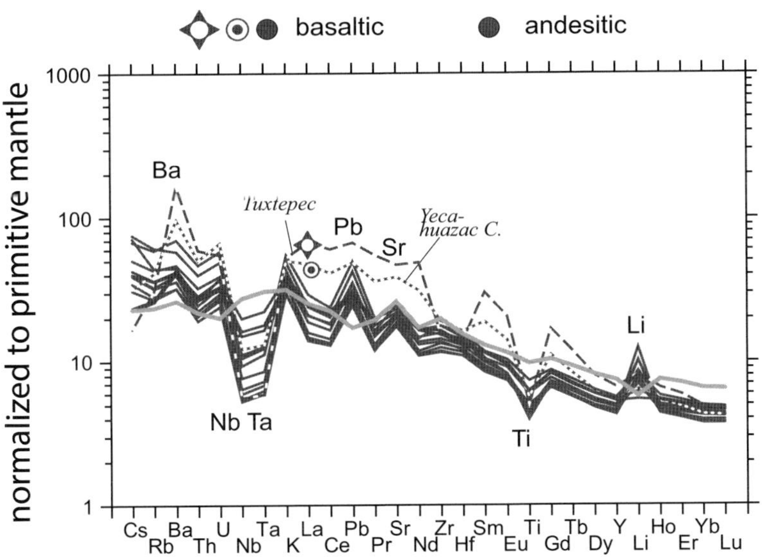

Fig. 4. Multi-element diagram of incompatible trace elements for whole-rock samples from the central MVB normalized to primitive mantle from McDonough & Sun (1995). For clarity, only samples with high-$^3He/^4He$ olivines are shown. Blue lines show basaltic (olivine-normative) melts, with enriched high-La basalts Tuxtepec and Yecahuazac Cone being denoted by separate blue–white symbols; red lines andesitic (quartz-normative) melts. Old Texcal Flow magma (green line) is a high-Nb arc basalt of the Sierra Chichinautzin that is only minimally affected by slab additions and best represents a melt from background mantle prior to subduction processing (Straub *et al.* 2013).

have the lowest abundances of light rare earth elements (LREE) and Nb, while the basalts and, in particular those of Tuxtepec and Yecahuazac Cone, tend to be more enriched. Note that the Tuxtepec and Yecahuazac Cone basalts were targeted for the $^{3}He/^{4}He$ work because of their unusual trace element patterns, but are otherwise rare in the Sierra Chichinautzin.

The unusual composition of the Tuxtepec and Yecahuazac Cone basalts also shows in trends of individual trace elements v. SiO_2 (Fig. 5). In general, the highly incompatible LILE (Cs, Ba, Rb, U, Pb, Th, Pb, Sr) increase with SiO_2, while LREE and HFSE are largely indifferent towards SiO_2. The Tuxtepec and Yecahuazac Cone basalts, however, are enriched in LREE and HFSE at low SiO_2, except for Nb and Ta, which are lower or comparable to the other calc-alkaline magmas. The heavy rare earth elements (HREE; Ho, Er, Yb and Lu) of all magmas always decrease slightly with SiO_2 in tight trends that resemble those of the major element oxides TiO_2, CaO, Fe_2O_3 and MgO.

Most central MVB magmas have fairly flat, HREE patterns with Dy/Yb ($= 1.95 \pm 0.14$, $n = 109$) that are only slightly higher than MORB (Dy/Yb $= 1.69 \pm 0.10$, $n = 80$) and lack a recognizable 'garnet signature' (Fig. 6). Again, the exceptions are the Tuxtepec and Yecahuazac Cone basalts that have high Dy/Yb of 2.5 and 2.3, respectively, which suggests a role of garnet in their genesis. Following Gomez-Tuena *et al.* (2007*a*), who report young MVB volcanic rocks with similar trace element signatures from the Valle de Bravo further to the west, Tuxtepec and Yecahuazac Cone basalts will be referred to as '*high-La basalts*'. All other calc-alkaline basalts and andesites will be referred to as '*low-La magmas*'.

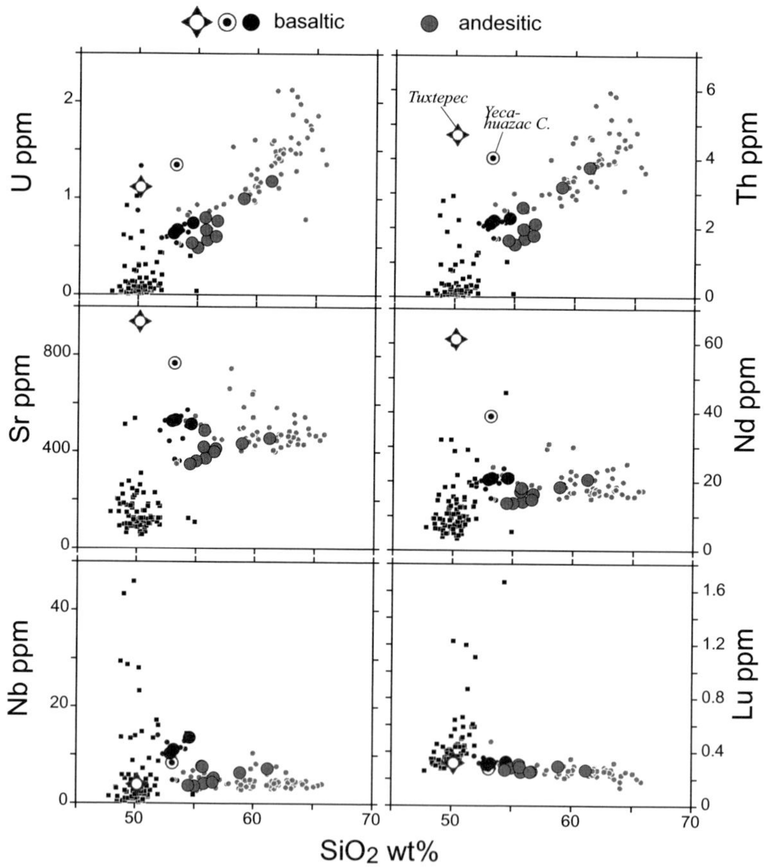

Fig. 5. Selected trace elements (in ppm) v. SiO_2 (wt%) of whole rocks of central MVB basaltic to andesitic magmas compared with East Pacific Rise MORB (filled squares) from Niu & Batiza (1997). Note diverging behaviour of U and Nb despite similar incompatibility during melting and crystallization.

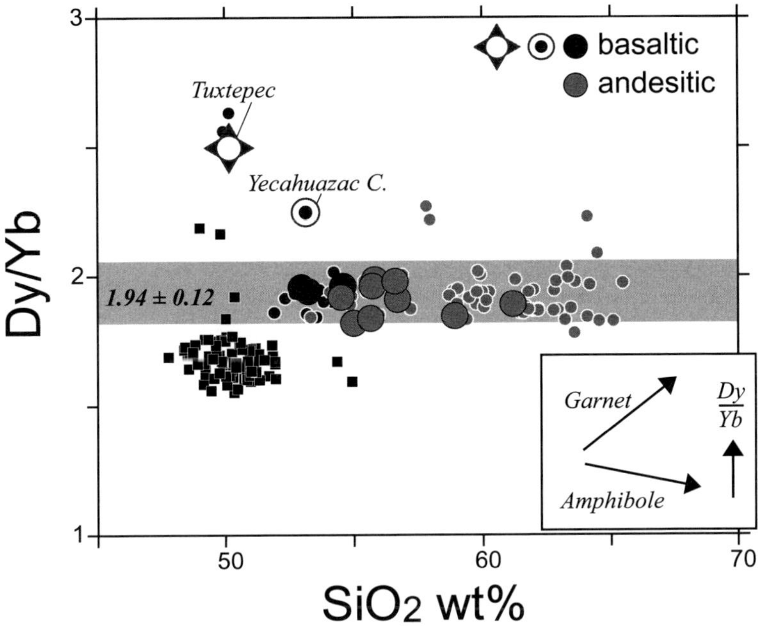

Fig. 6. Bulk rock Dy/Yb v. SiO_2 (wt%) in central MVB calc-alkaline basalts and andesites, compared with East Pacific Rise MORB from Niu & Batiza (1997) (filled squares). Grey band is average of low-La basalts and andesite. Inset after Davidson *et al.* (2007) indicates the trends induced by fractional crystallization of amphibole and garnet.

Sr and Nd isotope ratios

It is no surprise that our new $^{87}Sr/^{86}Sr$ and $^{143}Nd/^{144}Nd$ data are similar to those reported previously from the central MVB (e.g. Siebe *et al.* 2004*a*; Schaaf *et al.* 2005; Straub *et al.* 2008). This means that the high-$^{3}He/^{4}He$ samples have fairly unradiogenic $^{87}Sr/^{86}Sr$ (=0.70350–0.70441) and radiogenic $^{143}Nd/^{144}Nd$ (= 0.51282–0.51299), which are close to Pacific MORB and only marginally overlap with crustal compositions from within and around the MVB (Straub *et al.* 2008, 2013). $^{87}Sr/^{86}Sr$ and $^{143}Nd/^{144}Nd$ correlate broadly with SiO_2, and andesites have systematically higher $^{87}Sr/^{86}Sr$ and lower $^{143}Nd/^{144}Nd$ than the basalts (Fig. 7). This shows that the basaltic and andesitic series cannot be related by simple fractional crystallization despite the coherent trends of major elements v. SiO_2, and despite their proximity in time and space.

Discussion

The trace element and Sr–Nd isotope flux in arcs

At first view, the trace elements of the calc-alkaline magmas display features that agree with the pyroxenite model proposed by Straub *et al.* (2011). For example, this model predicts correlations between slab-derived elements and melt SiO_2, because high-SiO_2 initial mantle melts are produced from silica-excess pyroxenites that contain higher amounts of the slab component. Figure 5 shows such correlations for highly incompatible elements (e.g. U, Th) that are generally slab-controlled in arc magmas. For the same reason, and assuming the subducted sediment as source of the radiogenic Sr and unradiogenic Nd, the pyroxenite model predicts – as broadly observed – an increase in $^{87}Sr/^{86}Sr$ and a decrease in $^{143}Nd/^{144}Nd$ with increasing melt SiO_2 (Fig. 7). The pyroxenite model can also account for the difference in trace element signatures between the low-La calc-alkaline magmas and the high-La calc-alkaline basalts, as it proposes the flux of many individual slab components that may have very different trace element signatures and that infiltrate a heterogeneous subarc mantle (e.g. Straub *et al.* 2008; Tollstrup *et al.* 2010; Gomez-Tuena *et al.* 2011).

On the other hand, open questions remain. For example, the abundance data only point to a slab flux of the highly fluid-mobile LILE, but do not provide information about a slab flux of the middle rare earth elements (MREE) and HFSE. Moreover, it is not clear how the large amount of slab material required by the pyroxenite model fits

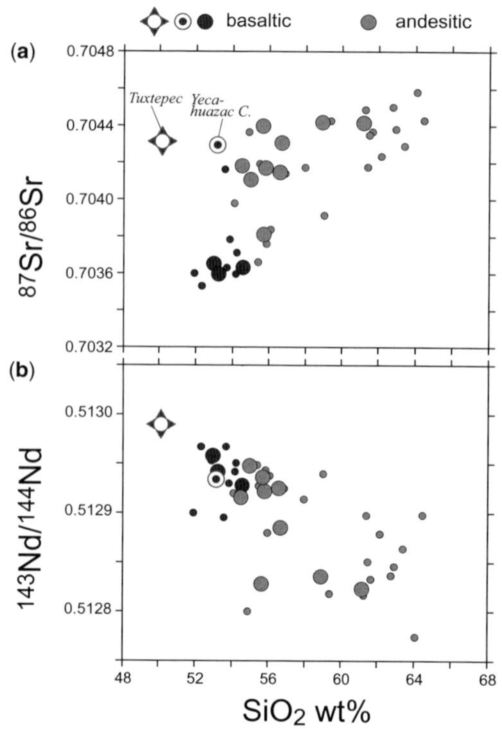

Fig. 7. Whole rock (**a**) $^{87}Sr/^{86}Sr$ and (**b**) $^{143}Nd/^{144}Nd$ v. melt SiO_2 (wt%). Note systematic difference in basaltic and andesitic series which excludes a fractional crystallization link between the two series and implies source diversity.

with the flat HREE pattern of the magmas. The minimum amount of slab-derived SiO_2 in the reaction pyroxenite is fixed by the stochiometric reaction that transforms mantle olivine to 'reaction orthopyroxene': $SiO_2 + (Mg, Fe)_2SiO_4 = (Mg, Fe)_2Si_2O_6$. Assuming an original background mantle with *c.* 50–60% olivine, the resulting reaction pyroxenite (which consists of *c.* 50–60% reaction orthopyroxene, plus the original mantle ortho- and clinopyroxene) must contain at least *c.* 15–18% slab-derived material. The formation of silica-excess pyroxenite requires even higher amounts (e.g. Straub *et al.* 2011). Given these amounts, why is there no garnet signature in the majority of the calc-alkaline magmas? In the following, we will address this question by forward and inverse modelling of the slab flux.

Modelling the flux of REE and Y. Forward modelling the slab-to-arc flux provides one way to evaluate the impact of recycled trace elements on arc magmas. However, such modelling is complicated, because it involves many poorly known variables (e.g. Straub & Zellmer 2012). The slab flux depends on slab composition, residual phases and partitioning behaviour of the individual elements, all of which vary strongly as the slab is processed at gradually increasing pressures and temperatures (e.g. Kessel *et al.* 2005; Klimm *et al.* 2008; Skora & Blundy 2010). In order to limit the uncertainties inherent in forward modelling, we focused on the REE and Y. These elements have the advantage of behaving coherently in petrogenetic processes, span a wide range of incompatible to moderately incompatible elements, and are sensitive to residual garnet. While the inherent uncertainties do not allow one to obtain exact numbers, they allow for testing whether the REE and Y pattern of the arc magmas can be consistent with >15% silicic slab component in the mantle source. Modelling parameters are compiled in Table 2; see also the Supplementary material.

The models calculate first the REE and Y of partial slab melts from the subducted sediment and basaltic igneous crust, respectively. (Note: We do not distinguish between a 'fluid-like' or 'melt-like' nature of the slab component, but prefer the term 'slab melt', because the mobilization of melt-mobile elements (LREE) is implied.) A composite of the two slab melts is then added to the 'background mantle', which is the subarc mantle prior to subduction modification. Finally, a partial melt of this metasomatized mantle is calculated, assuming source transformation from peridotite to a solid pyroxenite lithology prior to melting. Figure 8 compares the modelled melts with target compositions of representative calc-alkaline basalts and andesites of the central MVB. Details of the models are as follows:

(1) Primitive mantle from McDonough & Sun (1995), partially depleted by previous melting events, was chosen as background mantle. This choice is based on Straub *et al.* (2013), who identified the high-Nb basalts of the 'old Texcal Flow' (*c.* 17–19 ppm Nb), which erupts a few kilometres SE of Guespalapa, as the central MVB magma type least affected by slab metasomatism. These basalts are very similar to a *c.* 3–5% partial melt of primitive mantle, but are too enriched in TiO_2, LREE and HFSE to serve as source of the calc-alkaline magmas. However, residues of this mantle, which resemble depleted MORB-type mantle, have suitable abundances. While the subarc mantle could be inherently heterogeneous, we favour a melting link between enriched and depleted compositions because an overall compositional continuum exists between calc-alkaline and high-Nb magmas in the central MVB (LaGatta 2003; Gomez-Tuena *et al.* 2007*a*).

Table 2. *Summary of trace element modelling*

Sample_ID Volcano Eruptive unit		CH07-6 Guespalapa Manteca	CH07-12 Guespalapa El Hoyo	SPO60 Cuatepel Cone	MCH06-9 Guespalapa El Caballito	CH08-15 Yecahuazac Cone	S15 Tuxtepec Cone
Background mantle (mantle without subduction component) from McDonough & Sun (1995)		Residual mantle after extraction of 5% melt	Residual mantle after extraction of 5% melt	Residual mantle after extraction of 5% melt	Residual mantle after extraction of 5% melt	Residual mantle after extraction of 7% melt	Residual mantle after extraction of 7% melt
Modal composition of metasomatized mantle	GRT	–	–	–	0.5%	1%	0.5%
	OLIV	2%	10%	–	20%	14%	16
	OPX	90%	80%	90%	67%	80%	83
	CPX	8%	10%	10%	12%	5%	1
Ratio of sediment: igneous crust in composite slab component		7: 93	10: 90	20: 80	25: 75	7: 93	10: 90
Percentage of composite slab component in metasomatized mantle		30	25	25	15	20	30
Extent of melting of metasomatized mantle (%)		7	7	5	3	2	3
Least squares fit*		0.68	0.60	0.77	0.49	0.66	0.84
Percentages of elements added from slab	La	91	90	93	90	93	97
	Ce	83	82	86	79	86	94
	Nd	72	68	73	61	71	86
	Sm	61	56	60	45	53	76
	Eu	52	46	48	34	41	67
	Y	13	10	11	6	9	24
	Yb	5	4	4	2	4	11
	Lu	4	3	3	2	3	11

*The least squares fit assesses the fit between model and measured composition. For each element, the values of x was calculated as follows: $x = ((\text{measured values} - \text{modelled value})/10\% \text{ of measured values})^2$. If the average of x for all elements was <1, the fit was considered satisfactory.
For other reservoir compositions and partition coefficients, see Supplementary material.

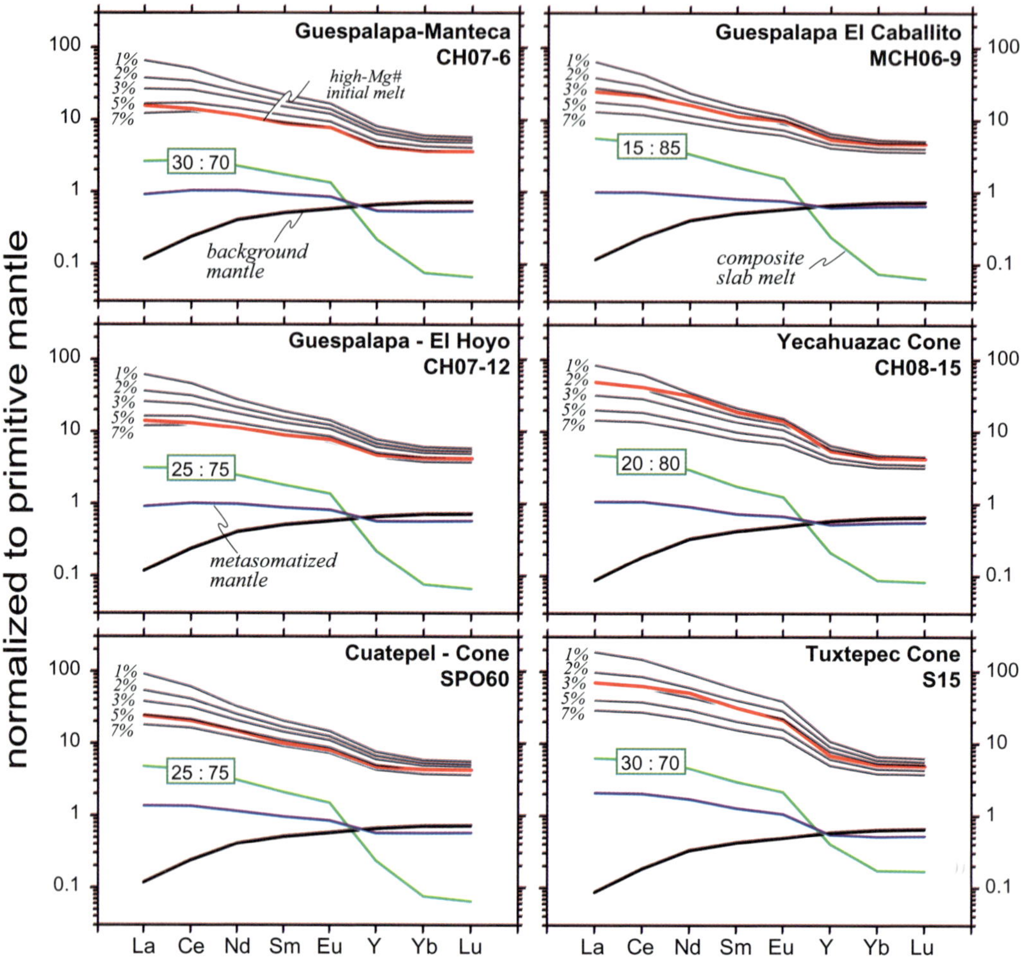

Fig. 8. Model calculations of REE and Y in selected low-La and high-La calk-alkaline andesites. Modelling and input parameters are given in Table 2 and in the Supplementary material. Data normalized to primitive mantle of McDonough & Sun (1995). Thick red line, composition of high-Mg# arc magma; thin black lines, forward models for 1, 2, 3, 5 and 7% extents of melting; thick green line, composite slab component of melts from subducted sediment melt and igneous crust, with the mixing ratio displayed in box; thick black line, background mantle; thick blue line, metasomatized mantle, which means background mantle and slab component. For discussion see text.

(2) We use the average terrigenous trench sediment from LaGatta (2003). For subducted basaltic igneous crust, we use the average recycled MORB from East Pacific Rise from Donnelly *et al.* (2004) (note that the N-type MORB from Sun & McDonough (1989) also produces acceptable results for most samples). No loss of REE and Y beneath the forearc was considered, which is consistent with the fluid-immobile nature of these elements. Partial slab melts were calculated with the batch melting equation after Langmuir *et al.* (1992) and the experimentally determined bulk partition coefficients (*D*) for elements La, Ce, Nd, Sm, Eu, Y, Yb and Lu from Skora & Blundy (2010) (sediment) and Klimm *et al.* (2008) (basaltic igneous crust), which were obtained in the same laboratory by similar techniques. All models use a 5% partial melt of either sediment or igneous crust.

(3) A composite of the partial melts from sediment and basaltic crust was calculated and added to the subarc mantle. The mixing ratio of the two slab components in the composite melt is arbitrary, and was adjusted to acquire the best results. Finally, we assumed an amount of *c.* 15–30% of composite slab

melt in the mantle source, which corresponds to the percentages required by the pyroxenite model.

(4) Melting of the metasomatized mantle was done by a batch melting model after Langmuir *et al.* (1992) with mineral-melt partition coefficients (K_d) from Donnelly *et al.* (2004). Olivine, orthopyroxene, clinopyroxene and sometimes traces of garnet were residual phases. The proportion of olivine and orthopyroxene varied in agreement with the reaction model, but had no perceptible effect on the REE and Y abundances in the melt, because of the very similar K_d values of olivine and orthopyroxene.

(5) The modelled melt compositions were compared with representative calc-alkaline basalts and andesites with high-$^3He/^4He$ olivines. The high values of Mg# >65.8–72.8 of these magmas suggest only minimum loss of mafic silicates from the initial melts, and hence they should closely resemble the initial mantle melts. Initial melt compositions were calculated by adding equilibrium olivine in 1% steps until the maximum Mg# = 72.8 observed (corresponds to olivine with $Fo_{89.8}$). The REE and Y were corrected accordingly, assuming perfect incompatibility in olivine. The agreement of the modelled melts with fractionation-corrected 'target magmas' was tested by a least squares fit calculated as follows for each element: $x =$ ((measured values − modelled value)/10% of measured values). An average of ≤ 1 for all values of x – which means that results are within 10% of observed data – was considered satisfactory. The principal results of the forward models are shown in Figure 8 and can be summarized as follows:

(a) The strong garnet signature of the slab melts is not transmitted to arc magmas, even if the mantle source contained as much as 15–40% of the composite melt. This is due to the fact that the high values of D of the slab (see Supplementary material) efficiently retain the HREE in the slab. Consequently, the HREE (*c.* Ho to Lu) of the background mantle are an order of magnitude higher than those of the slab melts, and the mantle HREE dominate the HREE budget of the partial melts. This results in the flat HREE observed. On the other hand, as suggested by many studies (e.g. Elliott *et al.* 1997, and later studies), the slab flux significantly increases the LREE in the subarc mantle, up to a point where the slab flux controls the LREE in the arc magmas, especially if the background mantle was depleted by previous melting.

(b) There is no problem of reproducing the REE and Y patterns of the low-La calc-alkaline basalts and andesites. In these models, values of the slab bulk D were always the same, and the proportions of sediment melt in the slab composite melt (7–25%) and of composite slab component in mantle (15–30%) were similar. Only one model (for sample MCH06-9) required residual garnet (0.5%) in the mantle source. The extent of mantle melting *c.* 3–7% is fairly low, but is probably underestimated because the model results depend here on the choice of bulk D of slab. The high values of D for the igneous crust (obtained at 2.5 GPa, Klimm *et al.* 2008) require lower extents of melting, but the lower D at the higher pressures of *c.* 3 GPa more appropriate for arc front depth allows for extents of mantle melting >5%.

(c) The steeper REE patterns of the high-La basalts from Tuxtepec and Yecahuazac Cone are more difficult to model. Notably, the problem is not so much matching their steeper HREE, but their fairly flat LREE. A different choice of bulk D for sediment melting from Skora & Blundy (2010) produces acceptable results, but requires low extents of mantle melting (2–3%) and a comparatively large proportion of sediment in the slab composite melt (10–25%), as well as a large proportion of composite slab melt (20–30%) in mantle source. Again, these latter three variables all depend on the bulk D values of the LREE which are highly variable and uncertain across the relevant P–T range. Unfortunately, only measured bulk D can be used (Kessel *et al.* 2005; Klimm *et al.* 2008; Skora & Blundy 2010), as partitioning data for clinopyroxene, a major residual phase in slab, are not available. Calculated bulk D, from the sum of the individual values of K_d, that includes the dependency on slab temperature, pressure and proportion of residual phases, may provide better results. Thus, despite the successful fit with still reasonable parameters, the numbers derived by the model must be considered as approximations, and not as a constraint.

In summary, the REE signature of the calc-alkaline MVB magmas can be reconciled with the large amounts of slab component in source required by the pyroxenite model. This result is a consequence of the high D values >1 for the two slab components and, in particular, of the high D values of the HREE which reach 100 and more (see Supplementary material). At such high D values, the composite slab component is effectively devoid of the HREE so that the HREE of the background mantle control the HREE in melts from metasomatized mantle. Such results cannot be achieved with lower slab D values <1 (e.g. Stolper & Newman 1994; Gomez-Tuena *et al.* 2007*a*). For example, the calculation of Gomez-Tuena *et al.* (2007*a*) used slab D values that are lower by an order of magnitude. At such low D values, the mantle wedge can absorb only up to 10% slab component without transmitting a garnet signature to a partial melts, and additional 'chromatographic' exchange between the advancing slab component and the surrounding mantle 'column' was required to obtain a more mantle-like composition (e.g. Stolper & Newman 1994). On the other hand, the forward calculations predict that >50–98% of the LREE in the central MVB magmas were slab-derived (Table 2). Thus, despite the high slab D values of Skora *et al.* (2010) and Klimm *et al.* (2008), the slab flux still strongly affects the composition of the subarc mantle.

The magnitude of the slab flux: the inverse approach. LREE are considered as 'melt-mobile', and hence it seems likely that other melt-mobile elements, such as the MREE and HFSE, were also added from slab in significant amounts (Elliott *et al.* 1997; Woodhead *et al.* 1998, 2001; Plank 2005; Gomez-Tuena *et al.* 2007*a*, 2011). Forward modelling, however, is too uncertain for deriving robust information on the mobility of the MREE and HFSE. Therefore, we test this hypothesis with the inverse approach of Pearce *et al.* (1995). This method assumes that the ratios of incompatible trace elements in arc magmas resemble those of the mantle source. Partial melts from 'background mantle' (which is the mantle without slab components) thus have 'smooth' patterns of (normalized) incompatible elements, whereas the mantle infiltrated by slab components produces arc magmas with strong relative enrichment of slab-derived elements. The different abundance of a given element between melts from the background and the metasomatized mantle, respectively, then provides a direct measure of the percentage of the element contributed from slab.

An important reference in this approach is the trace element composition of the melt from the unmetasomatized background mantle. Such melts commonly do not erupt in arcs, but their trace element patterns can be reconstructed from the erupted arc magmas on the basis of few elements that are principally mantle-derived in a given set of arc magmas. Pearce *et al.* (1995) used Nb and Yb to this purpose, assuming these elements to be entirely mantle-derived. While this works well in some arcs, a principal mantle origin (or 'conservative' behaviour) of Nb is doubtful in the central MVB. One immediate problem is the huge range in Nb of a factor of 10 (from 3.6 to 33.5 ppm Nb) in high-Mg# >60, high-$^3He/^4He$ magmas, that erupt within a few kilometres and a few thousand years from each other (Fig. 9) (Straub *et al.* 2013, and this study). This range of Nb cannot result from different extents of melting, fractional crystallization or crustal contamination, but must reflect mantle source heterogeneity. Moreover, slab signatures emerge in the magmas with the highest and lowest Nb, while the magmas least influenced by slab additions (the 'old Texcal Flow basalts') have intermediate Nb *c.* 17–19 ppm (Fig. 9). Notably, magmas with higher Nb >20 ppm also have higher ratios of Nb–Ta > 17 that have been related to the addition of high Nb–Ta components from slab (e.g. Stolz *et al.* 1996; Gomez-Tuena *et al.* 2011; Koenig & Schuth 2011).

On the other hand, TiO_2 abundances and variations are consistent with a principal mantle origin. The old Texcal Flow basalts have the highest $TiO_2 = 1.96$ wt% and high TiO_2–Lu = 4.6 in central Mexico (Fig. 9, Straub *et al.* 2013), while all other HNABs and calc-alkaline magmas have lower TiO_2 of <2–0.8 wt%, and TiO_2–Lu <4.6–2.5. These ranges are similar to those of MORB, which provides a reference for the variability of melts derived from an unmetasomatized upper mantle (Pearce *et al.* 1995; Pearce & Peate 1995). Therefore, we take TiO_2–Lu to approximate the MVB background mantle. The results are shown in Figures 10 and 11, where TiO_2–Lu is plotted against the Lu-normalized element investigated for slab addition. If the investigated element – which is the numerator of the ratio on the y-axis – is mantle-controlled like TiO_2 and Lu, the MVB data will overlap with mantle magmas. When the third element is added from slab, it shows a positive deviation from the range of the mantle melts that scales to the percentage of the slab-derived element in the arc magmas.

Calculation of the background mantle. We assume the background mantle to be primitive mantle based on the nearby erupting old Texcal Flow basalts which resemble a 3.5% melt from primitive mantle with minimal slab additions (Straub *et al.* 2013). Assuming a pyroxenite source

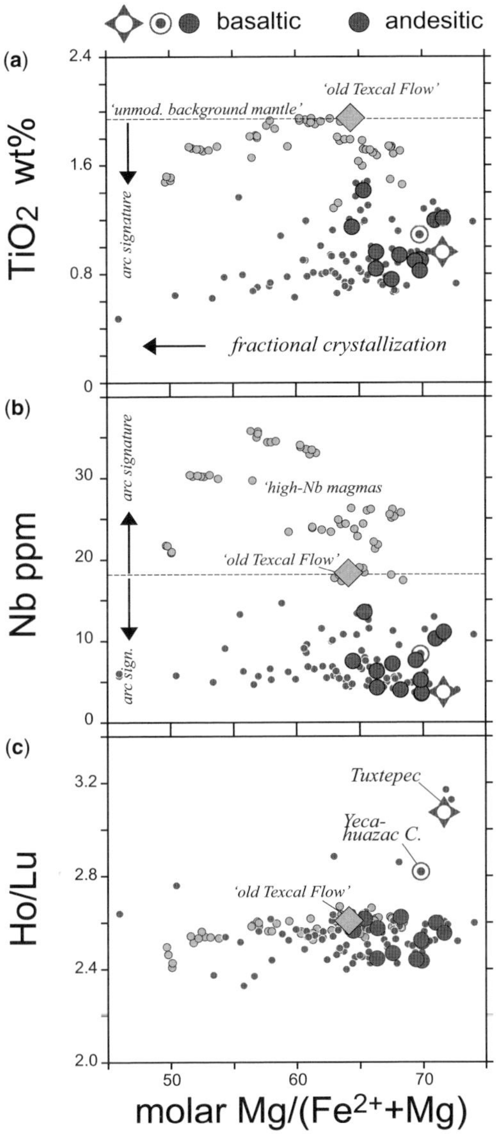

Fig. 9. **(a–c)** Nb (ppm), TiO_2 (wt%) and Ho/Lu v. Mg# of central MVB calc-alkaline basalts and andesites. High-Nb arc magmas (grey circles) of the central MVB after Straub *et al.* (2013), LaGatta (2003) and this study. Old Texcal Flow magma is a high-Nb arc basalt from Straub *et al.* (2013) that approximates best a melt from the central MVB mantle prior to subduction processing. Stippled horizontal lines abundance levels of melts from unmodified background mantle prior to subduction modification (which is similar to primitive mantle). In mantle-controlled elements, depletion by melting can only lower the abundance in later melts (e.g. TiO_2). The existence of the high-Nb magmas thus indicates Nb addition from slab, which is consistent with the 'arc signatures' of the magmas (e.g. Nb/La) decreasing above and below the 'background mantle line'.

lithology (orthopyroxene–clinopyroxene–garnet = 85:15:0.3) and the K_d values of Donnelly *et al.* (2004), we then calculated a range of melts from this primitive mantle with different extents of melting, ranging from 3.5 to 5, 10, 20 and 50%. Figure 10 shows that the lowest TiO_2/Lu observed is not reproduced by even an extent of melting of 50%, which is unrealistically high. However, the low TiO_2/Lu can be reproduced assuming re-melting of a residual mantle after several per cent melt extraction. Repetitive mantle melting, triggered by new hydrous slab additions, is a realistic model for the central MVB that is characterized by closely spaced, successive volcanic eruptions (e.g. Siebe *et al.* 2004*b*, 2005). The exact amount of the previously extracted melt and the number of melting events is not known, but it serves to assume 5% melt extraction in order to illustrate the resulting chemical variations. Melting curves for the original (primitive) mantle (solid lines), and for the depleted residual mantle (stippled) are shown in Figures 10 and 11 together with arc magmas with Mg# > 60.

Highly incompatible elements. Figure 10 illustrates the behaviour of highly incompatible elements (those to the left of Sr in multi-element diagrams) by the examples of Th, Nb and K_2O. Note that data are filtered for Mg# > 60, which minimized the influence of crustal contamination. The interpretation of the Lu-normalized ratios of these elements requires some caution for two reasons. First, these ratios increase at an extent of melting <10% during melting, but still to a far lesser extent than the increase caused by slab additions. Second, the Lu-normalized ratios differ significantly in melts from enriched and depleted mantle sources, respectively (Fig. 10). However, despite these limitations, it is obvious that these elements must be slab-controlled. Even relative to an enriched, primitive mantle – which illustrates the minimum slab addition – the slab-derived percentages in the magmas exceed >50% (except for Nb and Ta). If the mantle is depleted by previous melting, the slab-derived percentages commonly exceed >90% and approximate 100% in many cases. The case of Nb (and Ta) is different, as a primitive mantle cannot produce the low Nb/Lu of some of the calc-alkaline magmas (Fig. 10b). However, after the loss of only a few per cent of melts, the residual mantle is practically devoid of Nb, and very sensitive to leverage by the slab flux. Even a slab component with low Nb relative to the other highly incompatible elements may then add significant Nb to the mantle. While percentages of slab-derived Nb in arc magmas may not be as high as those of the fluid-mobile elements, slab-derived Nb may still account for >50% of the low-Nb arc magmas with only 3–4 ppm Nb (Fig. 10b). If Nb was mobilized to the same extent as the other highly incompatible elements, the slab flux of Nb must clearly control Nb abundances in arc magmas. In summary, while exact percentages of the slab-derived highly compatible elements are difficult to assess, the inverse approach suggests that slab flux of these elements dominates the arc outflux, and especially so if the mantle has been depleted by previous melting.

Moderately incompatible elements. For moderately incompatible elements (to the right of Sr in multi-element diagrams), which include the MREE and HREE, previous melt extraction is far less efficient in modifying the Lu-normalized ratios because their values of D are less different from D_{TiO2} during mantle melting (Fig. 11). Thus, the assessment against TiO_2/Lu provides fairly robust estimates of the slab-derived percentages of the elements in the arc magmas.

The inversion indicates significant slab additions for the middle REE down to Gd, Tb and Dy. In general, between *c.* 20 and 90% of these elements may be slab-derived in the low-La calc-alkaline magmas, whereby the slab-derived percentage decreases with decreasing incompatibility. The high-La basalts have high slab contributions of these elements, which suggest the presence of rare, unusually enriched slab components. Notably, the inversion indicates that as much as 50% of the Gd (placed just to the right of TiO_2 in multi-element diagrams) may be slab-derived low-La calc-alkaline magmas (and up to 70% in the high-La basalts). Thus, recycling of Eu (just to the right of TiO_2 in multi-element diagram) and Gd from slab, combined with TiO_2 retention in the slab, would cause the pronounced negative TiO_2 anomaly relative to Eu and Gd (Fig. 4). In other words, while the Eu and Gd abundances in arc magmas reflect both mantle and slab contributions, TiO_2 varies only by depletion through melting. Indeed, no alternative process can account for the relative TiO_2 depletion in the high-Mg# magmas. Fractional crystallization, different extents of partial melting and crustal contamination are all precluded by the significant range of TiO_2/Gd at comparable high Mg# values.

The comparative robustness of the Lu-normalized ratios of the moderately incompatible elements towards the background mantle allows for correlating their slab-derived percentages with melt

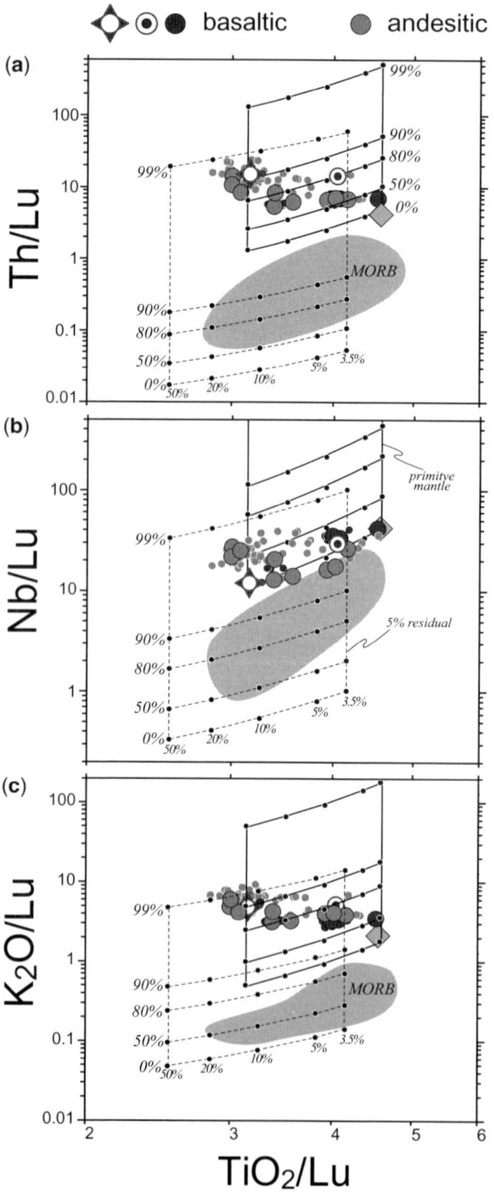

Fig. 10. (**a–c**) Lu-normalized ratios of Th, Nb and K_2O v. TiO_2/Lu for central MVB calc-alkaline basalts and andesites. All samples have Mg# > 60. Field of MORB with < 10 ppm Nb after Niu & Batiza (1997). Raster with solid lines *x*-axis indicates variations of 3.5, 5, 10, 20 and 50% melts (sub-horizontal lines) from primitive mantle after McDonough & Sun (1995); partition coefficient from Donnelly *et al.* (2004). Residual source: orthopyroxene–clinopyroxene–garnet, 85:15:0.3. Vertical lines along *y*-axis indicate percentage of element derived from subducted slab. Raster with stippled lines – as before, only assuming residual primitive mantle after 5% melt extraction as source with orthopyroxene–clinopyroxene–garnet, 95:5:0.3. Old Texcal Flow magma (large green diamond) is a high-Nb arc basalt of the Sierra Chichinautzin that is only minimally affected by slab additions and best represents a melt from background mantle prior to subduction processing (Straub *et al.* 2013).

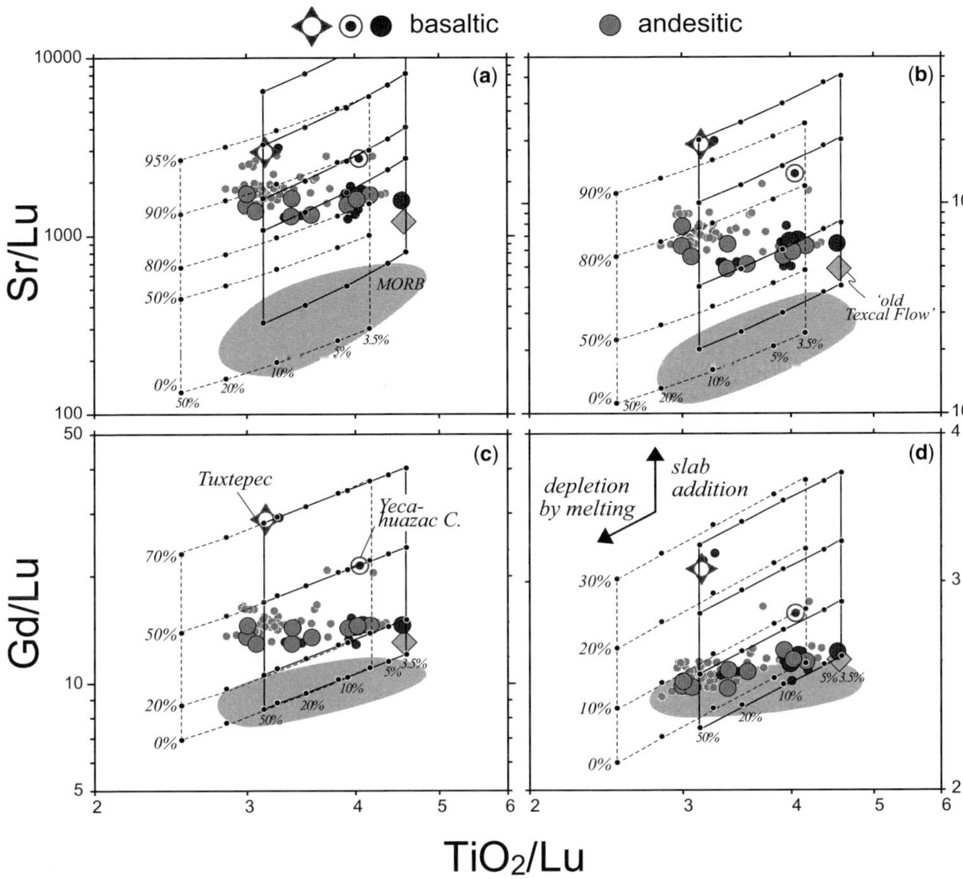

Fig. 11. (**a–d**) Lu-normalized ratios of Nd, Sr, Gd and Ho v. TiO_2/Lu for central MVB calc-alkaline basalts and andesites. All other variables as in Figure 10.

SiO_2. In the low-La calc-alkaline magmas, the slab-derived percentage increases with melt SiO_2 as predicted by forward models (Fig. 12). As expected, the slope of the increase v. melt SiO_2 becomes flatter, as the percentage of the slab-derived element decreases, and the subarc mantle increasingly buffers the lesser slab flux of the least incompatible elements. Notably, in agreement with the forward models, only the HREE to the right of Ho (Er and Yb, Lu by definition) lack correlation with melt SiO_2, which is consistent with their principal origin from background mantle and their efficient retention in the subducted slab by residual garnet.

In the high-La basalts from Tuxtepec and Yecahuazac Cone the slab flux increases independently from melt SiO_2 (Fig. 12). However, these excursions are accountable in the context of the pyroxenite model, when considering that the slab flux must not be homogenous. Such extreme enrichments may simply reflect rare slab components that are extremely enriched in REE and some other incompatible elements (e.g. Sr, P_2O_5), but depleted in the Nb and Ta. If the mantle was already depleted by melting, the full signature of this slab component must then show against the background, including the garnet signature. Moreover, in this case the high element abundances do not depend on multiple infiltrations of silicic slab components, and the addition of one such extremely enriched component may suffice to create silica-deficient pyroxenite that then produces enriched basaltic melts. Obviously, the rarity of the high-La basalts shows that these conditions of formation are not commonly met, in contrast to the more ubiquitous flux of slab components that produce the low-La calc-alkaline series.

In summary, the inverse approach agrees with the results of the forward models in that most incompatible elements are either completely or strongly controlled by the slab flux. Only a few elements,

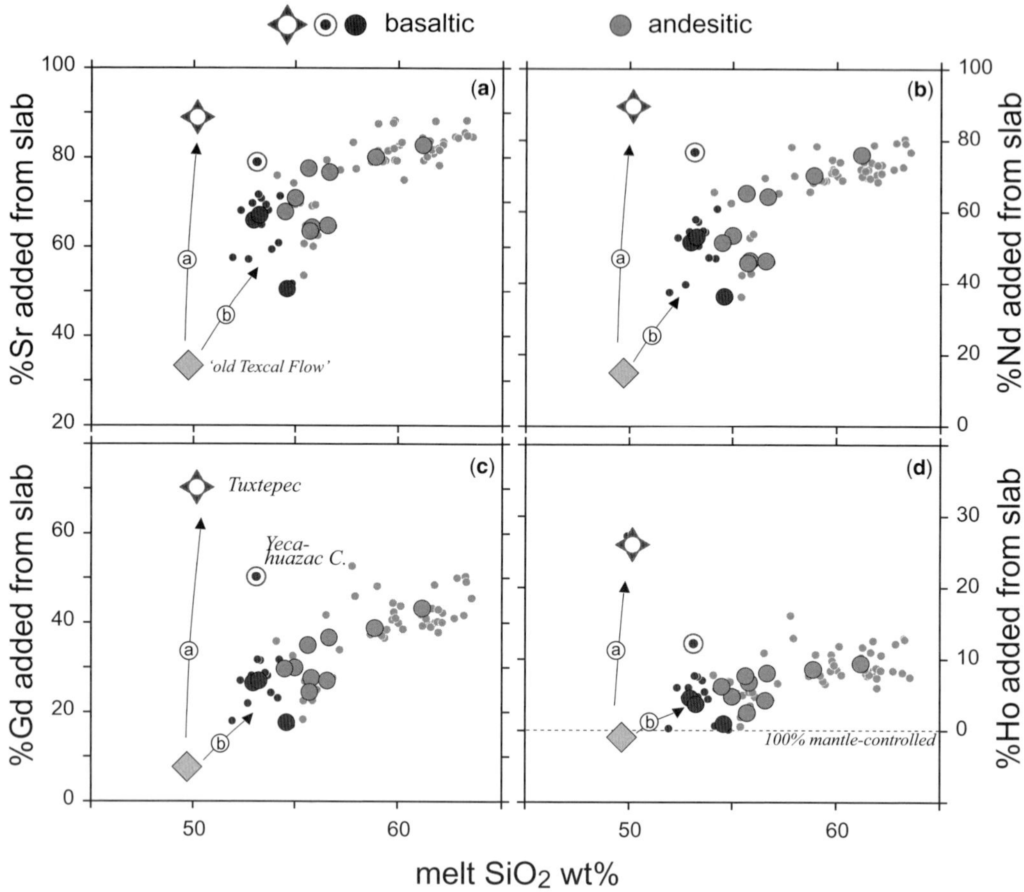

Fig. 12. (**a–d**) Percentage of moderately incompatible slab-derived elements Sr, Nd, Gd and Ho in arc magmas v. melt SiO_2 for samples with Mg# > 60. Old Texcal Flow magma (large green diamond) is a high-Nb arc basalt from Straub *et al.* (2013) that is only minimally affected by slab additions and best represents a melt from the central MVB background mantle prior to subduction processing. Trend labelled 'a' indicates addition of highly enriched slab melt to silica-deficient 'reaction pyroxenite' producing basaltic melts. Trend labelled 'b' indicates addition of multiple silicic slab components that create silica-excess pyroxenite. Note transitional character of Yecahuazac Cone basalt.

such as TiO_2 and the HREE Er, Yb and Lu principally derive from mantle, which accounts for the flat HREE patterns and the negative TiO_2 anomaly in the arc magmas. Additional variability can be attributed to rare, extremely enriched slab components that infiltrate a subarc mantle that has already been depleted by melting.

Sr and Nd isotope mixing trends. Lastly, we test these results in Sr–Nd isotope space. It has been suggested that the moderate deviations of $^{87}Sr/^{86}Sr$ and $^{143}Nd/^{144}Nd$ from MORB-type mantle magmas were caused by crustal contamination (e.g. Martinez-Serrano *et al.* 2004; Siebe *et al.* 2004*a*, 2005; Schaaf & Carrasco-Núñez 2010). Clearly, on the basis of the $^{87}Sr/^{86}Sr$ and $^{143}Nd/^{144}Nd$ data alone, it is not possible to distinguish conclusively between the models. However, the entirety of the data available for this sample set allows for considering a possible slab control, and justifies a detailed discussion.

Both forward and inverse models suggest that most of Sr (*c.* 50–80%) and Nd (*c.* 40–80%) in the low-La calc-alkaline magmas were slab-derived, with the slab addition being even higher in the high-La magmas (Figs 11 & 12). Consequently, the moderate deviations of $^{87}Sr/^{86}Sr$ and $^{143}Nd/^{144}Nd$ from upper mantle should largely be controlled by the slab flux (Fig. 13). However, if subducted sediment was the only source of slab-derived Sr and Nd in the arc, the $^{87}Sr/^{86}Sr$ and $^{143}Nd/^{144}Nd$ of arc should be much closer to the trench sediment

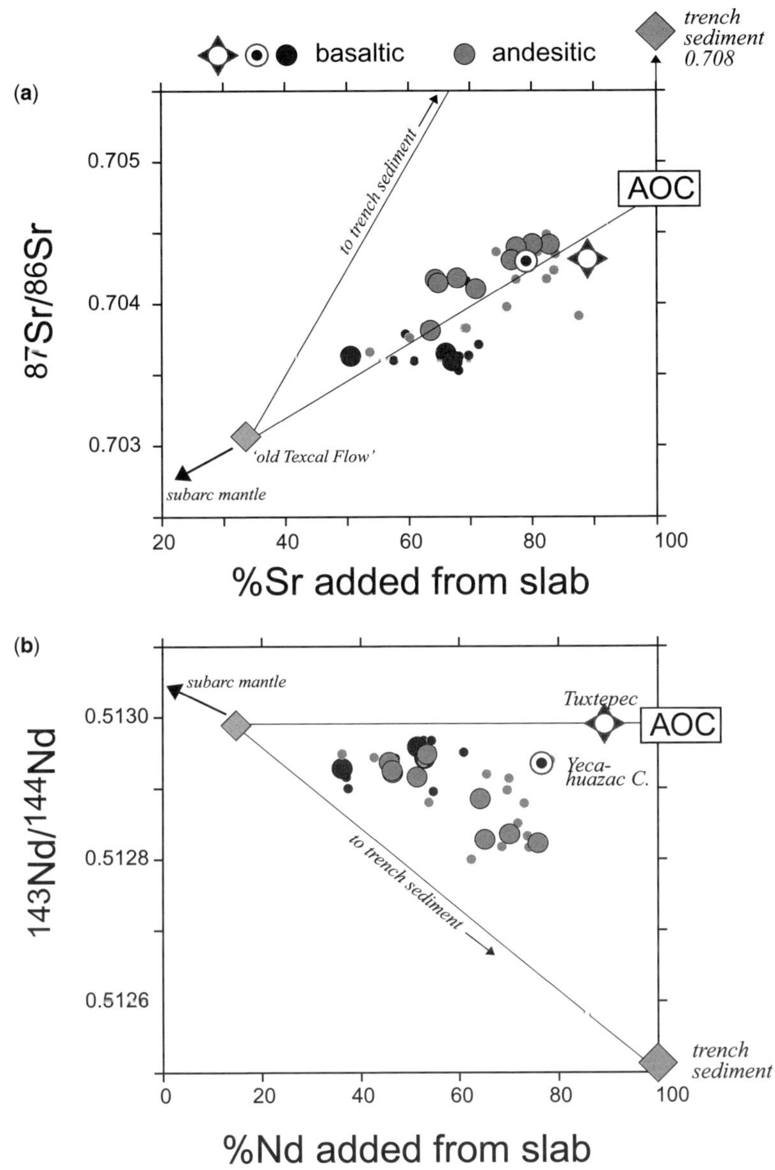

Fig. 13. (**a**, **b**) Percentage of slab-derived elements Sr and Nd in the arc magmas v. $^{87}Sr/^{86}Sr$ and. $^{143}Nd/^{144}Nd$, respectively. All samples have Mg# > 60. AOC, altered oceanic crust. Trench sediment after Gomez-Tuena *et al.* (2003). Old Texcal Flow magma (large green diamond) is a high-Nb arc basalt from Straub *et al.* (2013) that is only minimally affected by slab additions and best represents a melt from the central MVB background mantle prior to subduction processing.

than observed. On the other hand, the Sr and Nd flux from sediment can be buffered by the concomitant Sr and Nd flux from the far more voluminous subducted igneous crust, which results in fairly unradiogenic $^{87}Sr/^{86}Sr$ and radiogenic $^{143}Nd/^{144}Nd$, despite a high slab flux (e.g. Straub & Zellmer 2012). Figure 13 correlates the percentage of slab-derived Sr and Nd with their respective isotope ratios. In both diagrams, the arc data do not plot on a mixing line between mantle and trench sediment, but show deviations that agree with the addition of Nd and Sr from the subducted altered igneous crust (referred to as 'AOC' (altered oceanic crust) hereafter). In %Sr v. $^{87}Sr/^{86}Sr$ space, the

Sr-rich slab component has only moderately elevated $^{87}Sr/^{86}Sr$, *c.* 0.7048. As suggested previously for the MVB (Gomez-Tuena *et al.* 2007*a*), this points to the AOC ($^{87}Sr/^{86}Sr$ up to *c.* 0.705, e.g. Staudigel *et al.* 1995) as a principal source of the excess arc Sr, and not the subducted trench sediment. Because $^{143}Nd/^{144}Nd$ remains unchanged during the alteration of the oceanic crust, Nd flux of the trench sediment must be responsible for the lower $^{143}Nd/^{144}Nd$ of the arc magmas. However, there is no clear trend to the trench sediment either. All arc data – and especially the high-La basalts – require radiogenic Nd from slab which must derive from the AOC.

Importantly, if Sr and Nd were added from igneous subducted crust as well, mixing relationships in the Sr–Nd isotope space are complicated, because two of the source components (mantle and igneous crust) overlap in Sr and Nd isotope space. Quantification is further complicated by the fact that the arc data require the slab components to have different ratios of Sr/Nd. For example, the low-La calc-alkaline magmas have higher Sr/Nd = 22–27 than the high-La basalts (low Sr/Nd = 15–20). These ratios are all higher than the Sr/Nd of subarc mantle (Sr/Nd = 16), subducted igneous crust (Sr/Nd = 12.3) and trench sediment (Sr/Nd = 7), and thus agree with the idea of preferential Sr release from slab relative to the less mobile Nd. However, it is also clear that the extreme enrichment of the high-La basalts requires a slab component that is enriched both in Sr and Nd, but only minimally fractionated.

Because there are so many variables, there are only possible, not definite, solutions in the two-dimensional $^{87}Sr/^{86}Sr$ v. $^{143}Nd/^{144}Nd$ space. Figure 14 presents two possible solutions for low-La calc-alkaline magmas (panel a) and the high-La basalts (panel b). Model parameters are listed in Table 3. The modelling is done stepwise, but in reality the slab components may be generated and mix with the mantle wedge in any order. Both models first calculate a composite slab component from subducted sediment and igneous crust (mixing curve **a**). The curve is constrained to pass through the HNABs of central Mexico that define an end member in Sr–Nd isotope mixing space. In order to pass through the observed data, model curve **a** assumes some (*c.* 15%) loss of sedimentary Sr and preferential release (by a factor of 2) of Sr from igneous crust, which is in agreement with progressive slab dehydration (e.g. Kessel *et al.* 2005; Savov *et al.* 2005). In a second step, a composite of this slab fluid, which undergoes additional fractionation of Sr/Nd by preferential retention of Nd in slab, creates a slab component with high Sr/Nd

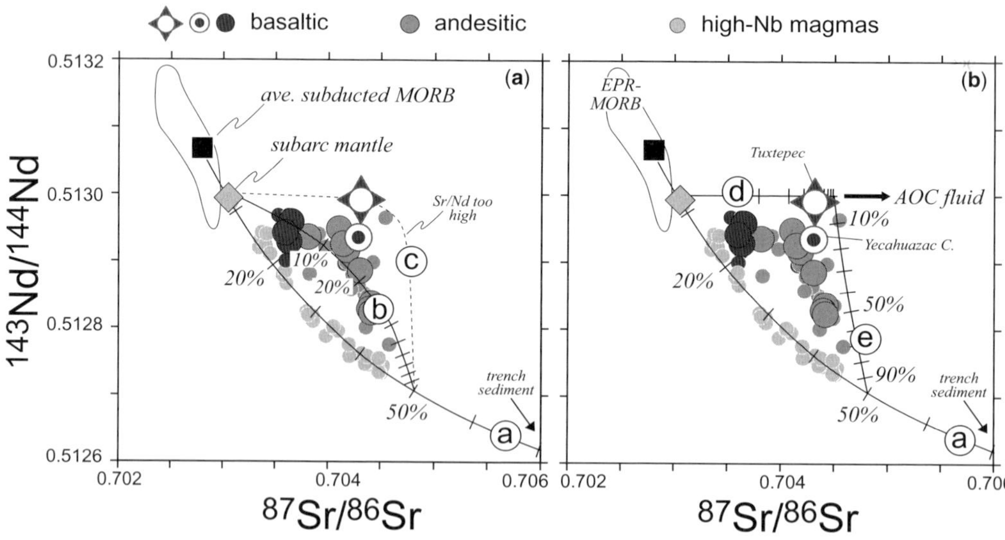

Fig. 14. $^{87}Sr/^{86}Sr$ v. $^{143}Nd/^{144}Nd$ mixing models for central MVB calc-alkaline basalts and andesites. Model parameters are compiled in Table 3. High-Nb arc magmas (grey circles) of the central MVB after Straub *et al.* (2013), LaGatta (2003) and this study. (**a**) Three-component mixing model for low-La calc-alkaline magmas, assuming Sr-rich fluid-like composite component from subducted sediment and igneous crust to infiltrate subarc mantle. (**b**) Four-component mixing model of high-La calc-alkaline basalts, assuming in addition an Sr-rich fluid with high $^{87}Sr/^{86}Sr$ from altered oceanic crust. Zero-age MORB from East Pacific Rise after PetDB (2011) Subducting igneous oceanic crust and terrigenous sediment from LaGatta (2003) and Gomez-Tuena *et al.* (2007*a*).

Table 3. *Summary of Sr and Nd isotope modelling*

	$^{87}Sr/^{86}Sr$	$^{143}Nd/^{144}Nd$	Nd	Sr	Sr/Nd
End member compositions					
Terrigenous trench sediment[1]	0.70846	0.51252	23.4	164.1	7.0
Subducting igneous crust[2]	0.70278	0.51307	12.28	123.8	10.1
Mantle wedge[3]	0.70305	0.51300	1.25	19.9	15.9
*Curve **a***					
Mixing of subducted sediment and igneous crust					
terrigenous trench sediment[4]	0.70846	0.51252	23.40	140[4]	6.0
subducting igneous crust[5]	0.70278	0.51307	12.28	250[5]	20.4
*Curve **b***					
Mixing of mantle and composite slab component					
mantle wedge[3]	0.70305	0.51307	1.25	19.9	10.1
Composite slab component	0.70482	0.51271	4[6]	195	48.8
*Curve **c***					
Mixing of mantle and composite slab component					
mantle wedge[3]	0.70305	0.51307	1.25	19.9	10.1
Composite slab component	0.70482	0.51271	0.2[7]	195	975
*Curve **d***					
Mixing of mantle and Sr-rich fluid from subducted igneous crust					
Mantle wedge	0.70305	0.51300	1.25	19.9	10.1
Sr-rich fluid from igneous crust[8]	0.70450	0.51300[8]	12.28	200[8]	16.3
*Curve **e***					
Mixing of impregnated mantle and composite slab component					
Pre-impregnated mantle wedge	0.70450	0.51300[8]	12.28	200[8]	16.3
Composite slab component[9]	0.70482	0.51271	17.84	195	10.9

[1]Terrigenous trench sediment from DSDP Site 467 after LaGatta (2003).
[2]Isotope ratios estimated from average zero-age East Pacific Rise MORB (PetDB 2011); Sr and Nd abundances are for average EPR (Donnelly *et al.* 2004).
[3]Isotope ratios based on 'old Texcal Flow' from Straub *et al.* (2013), Sr and Nd abundances for primitive mantle from McDonough & Sun(1995).
[4]Sr reduced by 15%, assuming early loss of sedimentary Sr beneath forearc.
[5]Sr increased by a factor of *c.* 2, assuming preferential mobilization of Sr from igneous crust.
[6]Nd reduced by 78%, assuming retention of Nd in slab, thus creating a slab component with high Sr/Nd.
[7]Nd reduced by 98%, assuming near complete Nd retention in slab, thus creating a slab component with very high Sr/Nd.
[8]Assuming $^{87}Sr/^{86}Sr = 0.7045$ and Sr = 200 ppm for fluid from altered oceanic crust.
[9]Assuming no additional fractionation of Sr/Nd in composite slab component

(*c.* 49) that mixes with the low Sr/Nd mantle (*c.* 16) to produce arc magmas with Sr/Nd *c.* 20–30 (curve **b**). A still higher Sr/Nd of the slab component would reproduce the extreme $^{87}Sr/^{86}Sr$ and $^{143}Nd/^{144}Nd$ of the high-La basalts (stippled curve **c**), but only at unrealistically high (not observed) Sr/Nd, *c.* 100, of the arc magmas. On the other hand (Fig. 14b), the low Sr/Nd, *c.* 15–20, of the high-La Tuxtepec and Yecahuazac Cone basalts can be generated, assuming the existence of a Sr-rich slab fluid with Sr/Nd *c.* 16 and high $^{87}Sr/^{86}Sr$, *c.* 0.7045, from altered igneous crust (curve **d** in panel b). Mixing of a composite slab fluid with low, unfractionated Sr/Nd, *c.* 11, may then create source with high $^{87}Sr/^{86}Sr$ at $^{143}Nd/^{144}Nd$, and the low Sr/Nd of the high-La basalts (curve **e**).

In summary, mixing models show that – despite the deceptively limited range in $^{87}Sr/^{86}Sr$ and $^{143}Nd/^{144}Nd$ – the calc-alkaline magmas are a complex mixture of source components with similar $^{87}Sr/^{86}Sr$ and $^{143}Nd/^{144}Nd$ and different Sr/Nd. No single solution exists, and it is possible that each magma batch, regardless of whether it is erupted from the same volcano or from a different volcano, represents a slightly different mixture of source components.

Implications for arc fluxes and arc crustal growth

Overall, the results of our study show that there is nothing in the incompatible trace elements and $^{87}Sr/^{86}Sr$ and $^{143}Nd/^{144}Nd$ of the calc-alkaline magmas that would invalidate the pyroxenite model. Moreover, the results are in line with models that propose a strong causal link between andesite melt formation and plate subduction (Ringwood 1974; Defant *et al.* 1991; Kelemen 1995; Yogodzinski *et al.* 1995). There seems to be no need for a causal role of crustal differentiation. Clearly, as such the pyroxenite model of andesite genesis has far-reaching implications for connectivity between trench and arc, the magnitude of the arc outflux and arc crustal growth. These will be shortly discussed in the following.

The link between andesite genesis and plate subduction. The strength of the link between andesite formation and plate subduction is best seen in the major element oxides that form >99% of the mass of the erupted arc magmas. If arc magmas were principally partial melts of 'reaction pyroxenites', then the increased melt SiO_2 of the arc magmas reflects their origin of as hybrid of mantle and silicic slab components. Such connections should also affect the variations of other major elements. Major element oxides TiO_2, FeO, MnO, MgO and CaO are not mobilized in the silicic slab component and must originate from the subarc mantle (e.g. Beard & Lofgren 1991; Rapp & Watson 1995; Kessel *et al.* 2005; Klimm *et al.* 2008; Skora & Blundy 2010). Thus, they should – as observed – correlate inversely with melt SiO_2 (Fig. 2). In contrast, major element oxides that co-correlate with SiO_2 in abundance space, such as Na_2O and K_2O, should be slab derived. Moreover, P_2O_5 behaves similar to Nd, and may also be partially recycled. As incompatible elements, K_2O, Na_2O and P_2O_5 can be assessed by the inverse approach (Figs 10 & 15). While the slab control for K_2O is obvious, inversion reveals significant slab contributions to Na_2O and P_2O_5 (Fig. 15). The case of Al_2O_3, however, is more difficult, because it cannot be

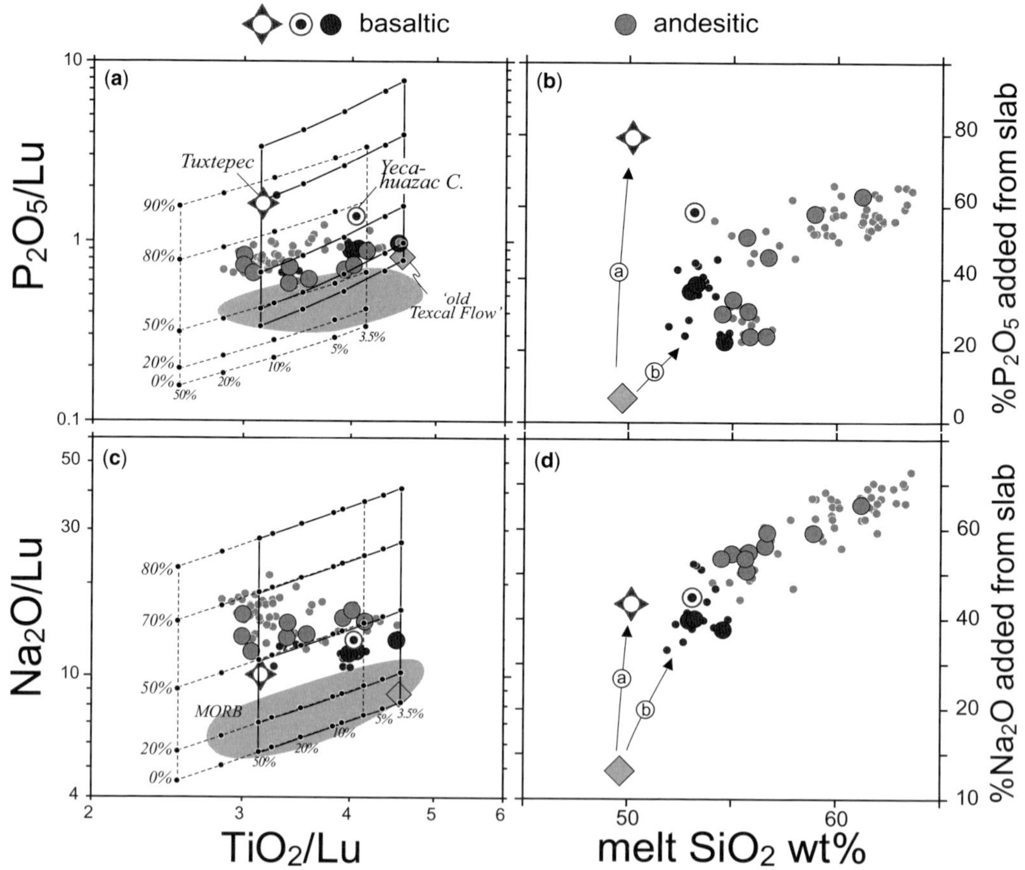

Fig. 15. (**a**, **b**) Na_2O/Lu and P_2O_5/Lu v. TiO_2/Lu, and (**c**, **d**) percentage of slab-derived elements Na and P_2O_5 in the arc magmas v. in central MVB calc-alkaline basalts and andesites. All samples have Mg# > 60. The bulk partition coefficient of Na (D_{Na}) is assumed to be that of D_{TiO2} during mantle melting. All symbols and data sources as in Figures 10 and 12.

assessed by the inverse approach. There is nothing in abundance space that would suggest Al_2O_3 recycling, but this is inconclusive. Al_2O_3 is a major component in alumosilicate slab melts (Beard & Lofgren 1991; Rapp & Watson 1995; Kessel *et al.* 2005), but also a significant constituent of the upper mantle. Thus, mantle melting may buffer the slab flux and obscure easy evidence of slab contributions of Al_2O_3 in arc magmas. In summary, despite some uncertainties, our extended dataset fully agrees with our earlier results that the major element oxides of the central MVB andesites were blends of slab and mantle materials, and did not result from shallow crustal processing of partial peridotite melts (Straub *et al.* 2008, 2013).

The effects of crustal differentiation. While the pyroxenite model does not require the overlying crust in order to create andesitic magmas, it does not imply the lack of subsequent intra-crustal differentiation. Without crustal differentiation, all central MVB magmas had high Mg# > 70 numbers (Straub *et al.* 2011). However, the fact that most erupted magmas have lower values of Mg# requires additional differentiation that should take place in the arc crust, which poses a physical and thermal barrier to the ascending mantle melts.

Because initial andesitic melts already have low melt MgO and FeO, moderate loss of mafic silicates (olivine, pyroxenes) suffices to produce the low-Mg# melts observed. Loss of mafic silicates will mostly affect melt MgO, which is rapidly lowered. In contrast, the abundance levels of melt SiO_2 and FeO – with bulk D close to unity – are largely maintained (Straub *et al.* 2011). Olivine, the ubiquitous and often only phenocrysts of the high-Mg# magmas, clearly plays a major role in lowering melt Mg#. However, results from melting experiments provide an interesting alternative as they demonstrate that, at the higher pressures in the lower crust, pyroxenes, not olivines, are on the liquidus of the andesitic melts (Weaver *et al.* 2011; Weber *et al.* 2011). While the presence of high-Mg# pyroxenes in our sample set remains to be confirmed, there are revealing trends of olivine in Fo v. Ni space, where olivines from various lava flows are aligned in conspicuous sub-parallel trends. This is true for olivines from different volcanoes, but also for olivines of different eruptive series of single volcanic centres (exemplified for Guespalapa volcanics in Fig. 16a). Excepting rare cases where complex crustal intracrustal differentiation may play a role (Straub *et al.* 2011), there are two ways to create such trends: (1) the initial melts have different values of Mg#, either by inherent source heterogeneity or by progressive depletion through serial melting; or (2) the values of Mg# of initial melts are variably lowered by early pyroxene fractionation in the lower crust, or even in the uppermost lithospheric mantle (Fig. 16b). This is because crystallizing pyroxenes also lowers melt Mg# (if at a lesser rate than olivine),

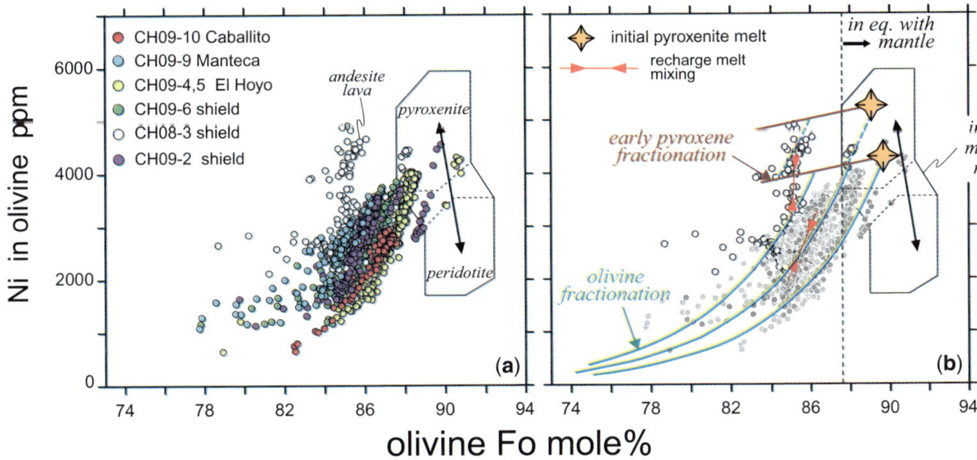

Fig. 16. (**a**) Olivine Ni (ppm) v. Fo (mol%) for samples from Guespalapa Volcanic Complex. All olivine data are from Straub *et al.* (2011). Fields labelled 'pyroxenite' and 'peridotite' indicate olivine composition in equilibrium with initial partial melts of pyroxenite and peridotite after Straub *et al.* (2011). Large black arrow indicates mixing between pyroxenite and peridotite initial component melts. (**b**) Possible interpretation of olivine trends after Straub *et al.* (2011, 2013). Few olivine are in equilibrium with initial mantle melts (white field with hybrids of pyroxenite and peridotite melts), and most olivine crystallize from derivate melts. Thick brown lines delineate early pyroxene fractionation; thick green lines delineate olivine fractionation trends. Red lines with arrows indicate recharge mixing.

but barely affect melt Ni given the much lower-K_d Ni in pyroxenes (e.g. Beattie *et al.* 1991). In particular, the higher the SiO_2 of the initial melt, the more likely it is that pyroxenes will be on the liquidus. Such early pyroxenes may be left behind as residues at the walls of conduits walls through which the magmas ascend to surface.

Evidence for crustal melt processing is ubiquitous in the complex zoning patterns of olivines that are recorded in all calc-alkaline magmas mixing between two or more component melts (Straub *et al.* 2011). Given their low to moderate H_2O abundances of the central MVB magmas (Cervantes & Wallace 2003*a*, *b*; Johnson *et al.* 2009; Roberge *et al.* 2009), olivine will only crystallize in the upper crust. Thus, most melt mixing must happen in the upper crust, and probably occurs in the shallow magma plumbing system just prior to and during eruption. Much of this mixing is probably 'recharge mixing', where newly ascending melt batches charge into slower ascending or stagnant melts at higher level and ultimately trigger the eruptions (Kent *et al.* 2010). Clearly, early loss of mafic silicates and late-stage melt mixing will obscure the spectrum of initial melt that make up the central MVB calc-alkaline andesites (Straub *et al.* 2011). However, neither process seems efficient in entirely eliminating the characteristics of the initial mantle melts.

Magnitude of the arc outflux and connectivity to slab. The pyroxenite model also has consequences for the magnitude of the arc outflux. Models that assume a basaltic flux to the Moho calculate the arc outflux for the lower elemental concentration appropriate to basaltic melts (e.g. Plank & Langmuir 1993; Straub & Layne 2003). This limitation, however, does not exist for initial arc magmas produced by reaction pyroxenites that are more silicic and richer in many incompatible elements. Because the element concentrations of the initial arc magmas relate linearly to the magnitude of the arc outflux, the arc outflux of incompatible elements increases accordingly. This aspect is particularly intriguing for slab-controlled, recycled elements in arc magmas that tend to be concentrated in the silicic melts released from slab (e.g. highly incompatible mobile element like Cs, Ba, Rb, K, U, Th, Pb, Sr), but also for the recycled, climatically active volatile elements (e.g. H_2O, CO_2, halogens). Moreover, being transported to surface in the SiO_2-rich mantle melts obviates the need for extensive and laborious crustal differentiation, and ensures better connectivity between slab input and arc output. Revised estimates, based on the pyroxenite model, may significantly factor into models of the feedback between subduction cycling and the mechanisms of the ocean–atmosphere system.

Implications for arc crustal growth. The pyroxenite model has implications for models of arc crustal growth, as it offers the possibility that the arc crust may grow through accretion of a broad spectrum of basaltic, andesitic and dacitic mantle melts. Depending on the ratio of basaltic to dacitic melts (an unknown variable), the newly accreted crust may be andesitic on average. This would much reduce, and possibly even forgo, the periodic delamination of mafic crustal residues that is needed in order to create andesitic crust from initial basaltic melts (e.g. Tatsumi 2000). Furthermore, the pyroxenite model also affects the rates of arc crustal growth. One method to calculate arc crustal growth is to estimate the composition and volume of the extrusive magma series, and infer the volume of the mafic intrusives that must be left behind if the flux to the Moho was basaltic (Crisp 1984; White *et al.* 2006). However, this approach fails if the subarc mantle produces a broad range of mafic and silicic compositions, as in this case the eruptive volumes do not permit inferences on the volumes of the intrusive rocks. Rather, the intrusive volume is an unknown quantity, as there is no knowledge on how much of the respective basaltic, andesitic and dacitic mantle melts will erupt, and whether the erupted melt volumes will at least reflect the correct proportions of the initial melts. Kent *et al.* (2010) recently proposed that mafic melts (by their higher density) and dacitic–rhyolitic melts (by their viscosity) may preferentially become sequestered in the crust, while intermediate andesites preferentially erupt. There is also the possibility that the proportions of mafic v. silicic mantle melts change among different arcs depending on the subduction parameters, or during the life-time of arcs that typically comprise tens of millions of years (Price *et al.* 2005; Condie 2007). Such concepts render it impossible to use the erupted volcanic volumes as a proxy for the total magmatic volume produced per increment of time. Alternative approaches are thus required to deduce the volume and accretion rates of intrusive bodies directly. Geophysical constraints on geothermal gradients in continental magmatic arcs (Rothstein & Manning 2003) and on how arc surface heat flux changes in tandem with subduction parameters (e.g. convergence rate, Zellmer 2008, 2009) may help to provide some valuable insights in the future.

Concluding remarks

The major results of this study can be summarized as follows:

(1) The systematics of incompatible trace elements and $^{87}Sr/^{86}Sr$ and $^{143}Nd/^{144}Nd$ ratios of the central MVB calc-alkaline basalts

to andesites agree with their proposed origin as partial melts from 'reaction pyroxenite' in the mantle wedge.

(2) Andesite formation in the central MVB is primarily a consequence of reactions between silicic slab components and the peridotite mantle, and not of intra-crustal melt differentiation.

(3) The pyroxenite model implies strong causal links between andesite genesis and plate subduction with implications for models of arc crustal growth, the efficiency of slab recycling and the connectivity between the arc magmatic fluxes and the Earth's exosphere.

Special thanks to V. L. Becerra, J. V. Gonzalez, H. S. H. Fernandez and S. N. Licona for a memorable sampling trip to Popo. R. Conrey, W.-Y. Hsu, D. Johnson and O. Pérez-Arvizu are thanked for laboratory support. G. Mesko helped with sample preparation, and J. Jweda, L. Mori, P. Sasnett and J. van Tongeren assisted in the field. Constructive reviews of two anonymous reviewers and the editorial support of Phil Leat are gratefully acknowledged. This study was financially supported by the US National Science Foundation (grant EAR-07-38707 to S.M.S.) and the National Science Council of Taiwan (grants 96-2811-M-001-023 and 98-2811-M-001-052 to G.F.Z.).

References

Agustín-Flores, J., Siebe, C. & Guilbaud, M. N. 2011. Geology and geochemistry of Pelagatos, Cerro del Agua, and Dos Cerros monogenetic volcanoes in the Sierra Chichinautzin Volcanic Field, south of México City. *Journal of Volcanology and Geothermal Research*, **201**, 143–162, http://dx.doi.org/10.1016/j.jvolgeores.2010.08.010

Baker, M. B., Grove, T. L. & Price, R. 1994. Primitive basalts and andesites from the Mt. Shasta region, N. California: products of varying melt fraction and water content. *Contributions to Mineralogy and Petrology*, **118**, 111–129.

Ballentine, C. J. & Burnard, P. G. 2002. Production, release and transport of noble gases in the continental crust. *Reviews in Mineralogy and Geochemistry*, **47**, 481–538.

Beard, J. S. 1995. Experimental, geological and geochemical constraints on the origins of low-Ks silicic magmas in oceanic arcs. *Journal of Geophysical Research*, **100**, 15593–15600.

Beard, J. S. & Lofgren, G. E. 1991. Dehydration melting and water-saturated melting of basaltic and andesitic greenstones and amphibolites at 1, 3 and 6.9 kbar. *Journal of Petrology*, **32**, 365–401.

Beattie, P., Ford, C. & Russell, D. 1991. Partition coefficients for olivine-melt and orthopyroxene-melt systems. *Contributions to Mineralogy and Petrology*, **109**, 212–224.

Cervantes, P. & Wallace, P. J. 2003*a*. The role of H_2O in subduction zone magmatism: new insights from melt inclusions in high-Mg basalts from central Mexico. *Geology*, **31**, 235–238.

Cervantes, P. & Wallace, P. J. 2003*b*. Magma degassing and basaltic eruption styles: a case study of V2000 year BP Xitle volcano in central Mexico. *Journal of Volcanology and Geothermal Research*, **120**, 249–270.

Condie, K. C. 2007. Accretionary orogens in space and time. *In*: Hatcher, R. D., Jr., Carlson, M. P., McBride, J. H. & Martínez Catalán, J. R. (eds) *4-D Framework of Continental Crust*. Geological Society of America, Memoirs, Boulder, **200**, 145–158.

Crisp, J. A. 1984. Rates of magma emplacement and volcanic output. *Journal of Volcanology and Geothermal Research*, **20**, 177–211.

Davidson, J., Turner, S., Handley, H., Macpherson, C. & Dosseto, A. 2007. Amphibole 'sponge' in arc crust? *Geology*, **35**, 787–790.

Defant, M. & Drummond, M. 1990. Derivation of some modern arc magmas by melting of young subducted lithosphere. *Nature*, **347**, 662–665.

Defant, M. J., Richerson, P. M. *et al.* 1991. Dacite genesis via both slab melting and differentiation: petrogenesis of La Yeguada volcanic complex, Panama. *Journal of Petrology*, **32**, 1101–1142.

Donnelly, K., Goldstein, S. L., Langmuir, C. H. & Spiegelman, M. 2004. Origin of Enriched Ocean Ridge Basalt and Implications for mantle dynamics. *Earth and Planetary Science Letters*, **226**, 347–366.

Eichelberger, J. C. 1978. Andesitic volcanism and crustal evolution. *Nature*, **275**, 21–27.

Elliott, T., Plank, T., Zindler, A., White, W. & Bourdon, B. 1997. Element transport from subducted slab to juvenile crust at the Mariana arc. *Journal of Geophysical Research*, **102**, 14991–15019.

Espinasa-Perena, R. 2006. Lava tubes of the Suchiooc Volcano, Mexico. *Bulletin of the Association for Mexican Cave Studies*, **17**, 80.

Ferrari, L., Orozco-Esquivel, M. T., Manea, V. C. & Manea, M. 2011. The dynamic history of the Trans-Mexican Volcanic Belt and the Mexico subduction zone. *Tectonophysics*, **522–523**, 122–149, http://dx.doi.org/10.1016/j.tecto.2011.09.018

Gill, J. 1981. *Orogenic Andesites and Plate Tectonics*. Springer, Berlin.

Gómez-Tuena, A., LaGatta, A., Langmuir, C. H., Goldstein, S. L., Ortega-Gutiérrez, F. & Carrasco-Nunez, G. 2003. Temporal control of subduction magmatism in the eastern Trans-Mexican Volcanic Belt: mantle sources, slab contributions, and crustal contamination. *Geochemistry Geophysics Geosystems*, **8**, 8913.

Gómez-Tuena, A., Langmuir, C. H., Goldstein, S. L., Straub, S. M. & Ortega-Gutierrez, F. 2007*a*. Geochemical evidence for slab melting in the Trans-Mexican Volcanic Belt. *Journal of Petrology*, **48**, 537–562, http://dx.doi.org/10.1093/petrology/egl071

Gómez-Tuena, A., Orozco-Esquivel, M. T. & Ferrari, L. 2007*b*. Igneous petrogenesis of the Trans-Mexican Volcanic Belt. *In*: Alaniz-Álvarez, S. A. & Nieto-Samaniego, Á. F. (eds) *Geology of México: Celebrating the Centenary of the Geological Society of México*.

Geological Society of America Special Papers, **422**, 129–181, http://dx.doi.org/10.1130/2007.2422(05)

Gómez-Tuena, A., Laura Mori, L., Goldstein, S. L. & Perez-Arvizu, O. 2011. Magmatic diversity of western Mexico as a function of metamorphic transformations in the subducted oceanic plate. *Geochimica et Cosmochimica Acta*, **75**, 213–241, http://dx.doi.org/10.1016/j.gca.2010.09.029

Hildreth, W. & Moorbath, S. 1988. Crustal contributions to arc magmatism in the Andes of Central Chile. *Contributions to Mineralogy and Petrology*, **98**, 455–489.

Johnson, E. R., Wallace, P. J., Delgado-Granados, H., Manea, V. C., Kent, A. J. R., Bindeman, I. N. & Donegan, C. S. 2009. Subduction-related volatile recycling and magma generation beneath central Mexico: insights from melt inclusions, oxygen isotopes and geodynamic models. *Journal of Petrology*, **50**, 1729–1764, http://dx.doi.org/10.1093/petrology/egp051

Kay, R. W. 1978. Aleutian magnesian andesites: melts from Subducted Pacific Ocean crust. *Journal of Volcanology and Geothermal Research*, **4**, 117–132.

Kelemen, P. B. 1995. Genesis of high Mg# andesites and the continental crust. *Contributions to Mineralogy and Petrology*, **120**, 1–19.

Kent, A. J. R., Darr, C., Koleszar, A. M., Salisbury, M. J. & Cooper, K. M. 2010. Preferential eruption of andesitic magmas through recharge filtering. *Nature Geoscience*, **3**, 631–636, http://dx.doi.org/10.1038/ngeo924

Kessel, R., Schmidt, M. W., Ulmer, P. & Pettke, T. 2005. Trace element signature of subduction-zone fluids, melts and supercritical liquids at 120–180 km depth. *Nature*, **437**, 724–727, http://dx.doi.org/10.1038/nature03971

Kim, Y., Clayton, R. W. & Jackson, J. M. 2010. Geometry and seismic properties of the subducting Cocos plate in central Mexico. *Journal of Geophysical Research*, **115**, http://dx.doi.org/10.1029/2009JB006942

Klimm, K., Blundy, J. D. & Green, T. H. 2008. Trace element partitioning and accessory phase saturation during H_2O-saturated melting of basalt with implications for subduction zone chemical fluxes. *Journal of Petrology*, **49**, 523–553, http://dx.doi.org/10.1093/petrology/egn001

Koenig, S. & Schuth, S. 2011. Deep melting of old subducted oceanic crust recorded by superchondritic Nb/Ta in modern island arc lavas. *Earth and Planetary Science Letters*, **301**, 265–274, http://dx.doi.org/10.1016/j.epsl.2010.11.007

LaGatta, A. B. 2003. *Arc magma genesis in the Eastern Mexican Volcanic Belt*. PhD thesis, Columbia University, New York.

Langmuir, C. H., Klein, E. M. & Plank, T. 1992. Petrological systematics of mid-ocean ridge basalts: constraints on melt generation beneath ocean ridges. *In*: Morgan, J. P., Blackman, D. K. & Sinton, J. M. (eds) *Mantle Flow and Melt Generation at Mid-Ocean Ridges*. Geophysical Monograph. American Geophysical Union, Washington, DC, 183–280.

Martinez-Serrano, R. G., Schaaf, P., Solids-Pichardo, G., Hernandez-Bernal, M. S., Hernandez-Trevino, T., Morales-Contreras, J. J. & Macias, J. L. 2004. Sr, Nd and Pb isotope and geochemical data from the Quaternary Nevado the Toluca volcano, a source of recent adakite magmatism, and the Tenango Volcanic Field. *Journal of Volcanology and Geothermal Research*, **138**, 77–110.

McDonough, W. F. & Sun, S. S. 1995. The composition of the Earth. *Chemical Geology*, **120**, 223–253.

Meriggi, L., Macías, J. L., Tommasini, S., Capra, L. & Conticelli, S. 2008. Heterogeneous magmas of the Quaternary Sierra Chichinautzin volcanic field (central Mexico): the role of an amphibole-bearing mantle and magmatic evolution processes. *Revista Mexicana de Ciencias Geológicas*, **25**, 197–216.

Morris, J. D., Leeman, W. P. & Tera, F. 1990. The subducted component in island arc lavas: constraints from Be isotopes and B-Be systematics. *Nature*, **344**, 31–36.

Niu, Y. & Batiza, R. 1997. Trace element evidence from seamounts for recycled oceanic crust in the Eastern Pacific mantle. *Earth and Planetary Science Letters*, **148**, 471–483.

Ortega-Gutiérrez, F., Elías-Herrera, M., Gómez-Tuena, A., Mori, L., Reyes-Salas, M., Macías-Romo, C. & Solari, L. A. 2012. Petrology of high-grade crustal xenoliths in the Chalcatzingo Miocene subvolcanic field, southern Mexico: buried basement of the Guerrero-Morelos platform and tectonostratigraphic implications. *International Geology Review*, **54**, 1597–1634, http://dx.doi.org/10.1080/00206814.2011.649956

Pardo, M. & Suarez, G. 1995. Shape of the subducted Rivera and Cocos plate in southern Mexico: seismic and tectonic implications. *Journal of Geophysical Research*, **100**, 12357–12373.

Pearce, J. A. & Peate, D. W. 1995. Tectonic implications of the compositions of volcanic arc magmas. *Annual Review of Earth and Planetary Sciences*, **23**, 251–285.

Pearce, J. A., Baker, P. E., Harvey, P. K. & Luff, I. W. 1995. Geochemical evidence for subduction fluxes, mantle melting and fractional crystallization beneath the South Sandwich Island Arc. *Journal of Petrology*, **36**, 1073–1109.

Perez-Campos, X., Kim, Y. H. et al. 2008. Horizontal subduction and truncation of the Cocos Plate beneath central Mexico. *Geophysical Research Letters*, **35**, 18303, http://dx.doi.org/10.1029/2008GL035127

Pertermann, M. & Hirschmann, M. M. 2003. Anhydrous partial melting experiments on MORB-like eclogite: phase relations, phase compositions and mineral-melt partitioning of major elements at 2–3 GPa. *Journal of Petrology*, **44**, 2173–2201.

PETDB 2011. *Information System for Geochemical Data of Igneous and Metamorphic Rocks from the Ocean Floor*, http://www.earthchem.org/petdb

Plank, T. 2005. Constraints from Thorium/Lanthanum on sediment recycling at subduction zones and the evolution of the continents. *Journal of Petrology*, **46**, 921–944, http://dx.doi.org/10.1093/petrology/egi005

Plank, T. & Langmuir, C. H. 1988. An evaluation of the global variations in the major element chemistry of arc basalts. *Earth and Planetary Science Letters*, **90**, 349–370.

PLANK, T. & LANGMUIR, C. H. 1993. Tracing trace elements from sediment input to volcanic output at subduction zones. *Nature*, **362**, 739–743.

PRICE, R. C., GAMBLE, J. A., SMITH, I. E. M., STEWART, R. B., EGGINS, S. & WRIGHT, I. C. 2005. An integrated model for the temporal evolution of andesites and rhyolites and crustal development in New Zealand's North Island. *Journal of Volcanology and Geothermal Research*, **140**, 1–24, http://dx.doi.org/10.1016/j.jvolgeores.2004.07.013

RAPP, R. P. & WATSON, E. B. 1995. Dehydration melting of metabasalt at 8–32 kbar: implications for continental growth and crust-mantle recycling. *Journal of Petrology*, **36**, 891–931.

REUBI, O. & BLUNDY, J. 2008. Assimilation of plutonic roots, formation of high-K 'Exotic' Melt inclusions and genesis of andesitic magmas at Volcan De Colima, Mexico. *Journal of Petrology*, **49**, 2221–2243, http://dx.doi.org/10.1093/petrology/egn066

REUBI, O. & BLUNDY, J. 2009. A dearth of intermediate melts at subduction zone volcanoes and the petrogenesis of arc andesites. *Nature*, **461**, 1269–1273, http://dx.doi.org/10.1038/nature08510

RINGWOOD, A. E. 1974. The petrological evolution of island arc systems. *Journal of the Geological Society, London*, **130**, 183–204.

ROBERGE, J., DELGADO-GRANADOS, H. & WALLACE, P. J. 2009. Mafic magma recharge supplies high CO_2 and SO_2 gas fluxes from Popocatépetl volcano, Mexico. *Geology*, **37**, 107–110.

ROTHSTEIN, D. A. & MANNING, C. E. 2003. Geothermal gradients in continental magmatic arcs: constraints from the eastern Peninsular Ranges batholith, Baja California, México. *In*: JOHNSON, S. E., PATERSON, S. R., FLETCHER, J. M., GIRTY, G. H., KIMBROUGH, D. L. & MARTÍN-BARAJAS, A. (eds) *Tectonic Evolution of North- Western México and the Southwestern USA*. Geological Society of America Special Papers, **374**, 337–354.

RUDNICK, R. 1995. Making continental crust. *Nature*, **378**, 571–578.

RUSCITTO, D. M., WALLACE, P. J., COOPER, L. B. & PLANK, T. 2012. Global variations in H_2O/Ce: 2. Relationships to arc magma geochemistry and volatile fluxes. *Geochemistry Geophysics Geosystems*, **13**, Q03025, http://dx.doi.org/10.1029/2011GC003887

SAVOV, I. P., RYAN, J. G., D'ANTONIO, M., KELLEY, K. & MATTIE, P. 2005. Geochemistry of serpentinized peridotites from the Mariana Forearc Conical Seamount, ODP Leg 125: implications for the elemental reycycling in subduction zones. *Geochemistry Geophysics Geosystems*, **6**, http://dx.doi.org/10.1029/2004GC000777

SCHAAF, P. & CARRASCO-NÚÑEZ, G. 2010. Geochemical and isotopic profile of Pico de Orizaba (Citlaltépetl) volcano, Mexico: insights for magma generation processes. *Journal of Volcanology and Geothermal Research*, **197**, 108–122, http://dx.doi.org/10.1016/j.jvolgeores.2010.02.016

SCHAAF, P., STIMAC, J., SIEBE, C. & MACIAS, J. L. 2005. Geochemical evidence for mantle origin and crustal processes in volcanic rocks from Popocatepetl and surrounding monogenetic volcanoes, central Mexico. *Journal of Petrology*, **46**, 1243–1282.

SIEBE, C., RODRIGUEZ-LARA, V., SCHAAF, P. & ABRAMS, M. 2004*a*. Geochemistry, Sr–Nd isotope composition, and tectonic setting of Holocene Pelado, Guespalapa and Chichinautzin scoria cones, south of Mexico City. *Journal of Volcanology and Geothermal Research*, **130**, 197–226.

SIEBE, C., RODRIGUEZ-LARA, V., SCHAAF, P. & ABRAMS, M. 2004*b*. Radiocarbon ages of Holocene Pelado, Guespalapa, and Chichinautzin scoria cones, south of Mexico City: implications for archeology and future hazards. *Bulletin of Volcanology*, **66**, 203–225, http://dx.doi.org/10.1007/s00445-003-0304-z

SIEBE, C., ARANA-SALINAS, L. & ABRAMS, M. 2005. Geology and radiocarbon ages of Tlaloc, Tlacotenco, Cuauhtzin, Hijo del Cuauhtzin, Teuhtli, and Ocusacayo monogenetic volcanoes in the central part of the Sierra Chichinautzin, Mexico. *Journal of Volcanology and Geothermal Research*, **141**, 225–243.

SKORA, S. & BLUNDY, J. 2010. High-pressure hydrous phase relations of radiolarian clay and implications for the involvement of subducted sediment in arc magmatism. *Journal of Petrology*, **51**, 2211–2243, http://dx.doi.org/10.1093/petrology/egq054

STAUDIGEL, H., DAVIES, G. R., HART, S. R., MARCHANT, K. M. & SMITH, B. M. 1995. Large scale isotopic Sr, Nd and O isotopic anomaly of altered oceanic crust: DSDP/ODP sites 417/418. *Earth and Planetary Science Letters*, **130**, 169–185.

STOLPER, E. & NEWMAN, S. 1994. The role of water in the petrogenesis of Mariana trough magmas. *Earth and Planetary Science Letters*, **121**, 293–325.

STOLZ, A. J., JOCHUM, K. P., SPETTEL, B. & HOFMANN, A. W. 1996. Fluid- and melt-related enrichment in the subarc mantle: evidence from Nb/Ta variations in island-arc basalts. *Geology*, **24**, 587–590.

STRAUB, S. M. & LAYNE, G. D. 2003. The systematics of chlorine, fluorine and water in Izu arc front volcanic rocks: implication for volatile recycling in subduction zones. *Geochimica et Cosmochimica Acta*, **67**, 4179–4203.

STRAUB, S. M. & MARTIN-DEL POZZO, A. L. 2001. The significance of phenocryst diversity in tephra from recent eruptions at Popocatepetl volcano (Central Mexico). *Contributions to Mineralogy and Petrology*, **140**, 487–510.

STRAUB, S. M. & ZELLMER, G. F. 2012. Volcanic arcs as archives of plate tectonic change. *Gondwana Research*, **21**, 495–516, http://dx.doi.org/10.1016/j.gr.2011.10.006

STRAUB, S. M., LAGATTA, A. B., MARTIN-DEL POZZO, A. L. & LANGMUIR, C. H. 2008. Evidence from high Ni olivines for a hybridized peridotite/pyroxenite source for orogenic andesites from the central Mexican Volcanic Belt. *Geochemistry Geophysics Geosystems*, **9**, Q03007, http://dx.doi.org/10.1029/2007GC001583

STRAUB, S. M., GOMEZ-TUENA, A., STUART, F. M., ZELLMER, G. F., ESPINASA-PERENA, R., CAI, M. Y. & IIZUKA, Y. 2011. Formation of hybrid arc andesites beneath thick continental crust. *Earth and Planetary Science Letters*, **303**, 337–347, http://dx.doi.org/10.1016/j.epsl.2011.01.013

STRAUB, S. M., GOMEZ-TUENA, A. ET AL. 2013. The processes of melt differentiation in arc volcanic rocks: insights from OIB-type arc magmas in the central Mexican Volcanic Belt. *Journal of Petrology*, **54**, 665–701, http://dx.doi.org/10.1093/petrology/egs081

SUN, S. S. & MCDONOUGH, W. F. 1989. Chemical and isotopic systematics of oceanic basalts: implications for mantle composition and processes. *In*: SAUNDERS, A. D. & NORRY, M. J. (eds) *Magmatism in the Ocean Basins*. Geology Society, London, Special Publications, **42**, 313–345.

TAMURA, Y. & TATSUMI, Y. 2002. Remelting of an andesitic crust as a possible origin for rhyolitic magma in oceanic arcs: an example from the Izu–Bonin Arc. *Journal of Petrology*, **43**, 1029–1047.

TATSUMI, Y. 2000. Continental crust formation by crustal delamination in subduction zones and complementary accumulation of the enriched mantle I component in the mantle. *Geochemistry Geophysics Geosystems*, **1**, http://dx.doi.org/10.1029/2000GC000094

TOLLSTRUP, D., GILL, J. B., KENT, A. J. R., PRINKEY, D., WILLIAMS, R., TAMURA, Y. & ISHIZUKA, O. 2010. Across-arc geochemical trends in the Izu–Bonin arc: contributions from the subducting slab, revisited. *Geochemistry Geophysics Geosystems*, **11**, http://dx.doi.org/10.1029/2009GC002847

URRUTIA-FUCUGAUCHI, J. & MARTIN DEL POZZO, A. L. 1993. Implicaciones de los datos paleomagneticos sobra la edad de la sierra Chichnautzin, Cuenca de Mexico. *Geofisica Internationale*, **33**, 523–533.

VERMA, S. P. 1999. Geochemistry of evolved magmas and their relationship to subduction-unrelated mafic volcanism at the volcanic front of the central Mexican Volcanic Belts. *Journal of Volcanology and Geothermal Research* **93**, 151–171.

WALLACE, P. J. & CARMICHAEL, I. S. E. 1999. Quaternary volcanism near the Valley of Mexico: implications for subduction zone magmatism and the effects of crustal thickness variations on primitive magma compositions. *Contributions to Mineralogy and Petrology*, **135**, 291–314.

WEAVER, S. L., WALLACE, P. J. & JOHNSTON, D. A. 2011. A comparative study of continental v. intraoceanic arc mantle melting:Experimentally determined phase relations of hydrous primitive melts. *Earth and Planetary Science Letters*, **308**, 97–106, http://dx.doi.org/10.1016/j.epsl.2011.05.040

WEBER, R. M., WALLACE, P. J. & JOHNSTON, A. D. 2011. Experimental insights into the formation of high-Mg basaltic andesites in the trans-Mexican volcanic belt. *Contributions to Mineralogy and Petrology*, **163**, 825–840, http://dx.doi.org/10.1007/s00410-011-0701-9

WHITE, S. M., CRISP, J. A. & SPERA, F. J. 2006. Long-term volumetric eruption rates and magma budgets. *Geochemistry Geophysics Geosystems*, **7**, 1–20.

WOODHEAD, J. D., EGGINS, S. M. & JOHNSON, R. W. 1998. Magma genesis in the New Britain Island arc: further insights into melting and mass transfer processes. *Journal of Petrology*, **39**, 1641–1668.

WOODHEAD, J. D., HERGT, J. M., DAVIDSON, J. P. & EGGINS, S. M. 2001. Hafnium isotope evidence for 'conservative' mobility during subduction zone processes. *Earth and Planetary Science Letters*, **192**, 331–346.

YOGODZINSKI, G. M., KAY, R. W., VOLYNETS, O. N., KOLOSKOV, A. V. & KAY, S. M. 1995. Magnesian andesite in the western Aleutian Komandorsly region: implications for slab melting and processes in the mantle wedge. *Geological Society of America Bulletin*, **107**, 505–519.

YOGODZINKSI, G., LEES, J. M., CHURIKOVA, T. G., DORENDORF, F., WOERNER, G. & VOLYNETS, O. N. 2001. Geochemical evidence for the melting of the subducting oceanic lithosphere at the plate edges. *Nature*, **409**, 500–504.

ZELLMER, G. F. 2008. Some first-order observations on magma transfer from mantle wedge to upper crust at volcanic arcs. *In*: ANNEN, C. & ZELLMER, G. F. (eds) *Dynamics of Crustal Magma Transfer, Storage and Differentiation*. Geological Society, London, Special Publications, **304**, 15–31, http://dx.doi.org/10.1144/SP304.2

ZELLMER, G. F. 2009. Petrogenesis of Sr-rich adakitic rocks at volcanic arcs: insights from global variations of eruptive style with plate convergence rates and surface heat flux. *Journal of the Geological Society, London*, **166**, 725–734.

Andesite petrogenesis by slab-derived plume pollution of a continental rift

ARTURO GÓMEZ-TUENA[1]*, BEATRIZ DÍAZ-BRAVO[1], ALMA VÁZQUEZ-DUARTE[1], OFELIA PÉREZ-ARVIZU[1] & LAURA MORI[2]

[1]*Centro de Geociencias, Universidad Nacional Autónoma de México, Querétaro 76230, México*

[2]*Instituto de Geología, Universidad Nacional Autónoma de México, México D.F. 04510, México*

**Corresponding author (e-mail: tuena@geociencias.unam.mx)*

Abstract: The western Mexican subduction zone is characterized by steep subduction of the Rivera plate, and by the existence of a continental rift at the rear arc under which the slab rests at >300 km deep. Mafic magmatism at the volcanic front is potassic lamprophyric, interpreted to be influenced by deep and hot slab melts or supercritical fluids. In contrast, mafic rocks at the rear arc are intraplate-like basalts that derive from low extents of melting of a dryer mantle source. Although a transition from a volcanic arc front to an extensional rear arc is apparent, calc-alkaline andesitic stratovolcanoes with trace element characteristics that suggest a key role of residual amphibole have been constructed at the rear arc during the past ~200 ka. Crystal fractionation of basalts and partial melting of crustal amphibolites are not viable mechanisms for andesites, whereas melting of slab amphibolites beneath the rear arc is also problematic because the oceanic plate rests too deep. We thus suggest that andesites are partial melts of rising diapirs made by mixtures of hydrous mantle, sediments, and possibly eroded crustal blocks, which detach buoyantly from the downgoing slab as discrete plumes that 'pollute' the upwelling regime of a continental rift.

Andesites are the most abundant volcanic rock-type in convergent margins, and have played a key role in the construction of Earth's continents (Rudnick & Gao 2003; Brown & Rushmer 2006). Notwithstanding their importance and ubiquity, fundamental questions about their origin, evolution and specific geodynamic setting remain the subject of vigorous debate (Gómez-Tuena *et al.* this volume, in prep.). Central to the discussion is whether andesites are secondary products of basalt differentiation, mixtures between derivative felsic liquids and new mafic inputs (Kent *et al.* 2010), or direct mantle extracts closely associated to the hybridization of mantle and crust materials inherent to convergent margins (Taylor 1967). If the primary mantle contribution to arcs is basaltic in nature, one caveat is that mafic and dense residues left after basalt differentiation must be periodically recycled back into the mantle through delamination or lithospheric foundering, because otherwise the composition of continents as a whole would have to remain basaltic (Ringwood & Green 1966). Mantle hybridization by slab-derived siliceous melts may create andesites without the need for a basaltic precursor (Kelemen 1998; Straub *et al.* 2011), but the specific transport pathways followed by the subduction fluxes and their relation to the thermal state of a convergent margin still remain controversial. A nearly vertical reactive transport from the slab surface to the location of arc volcanoes appears consistent with evidence from newly developed geothermometers (Cooper *et al.* 2012), whereas petrological evidence from ultra-high pressure (UHP) terranes and numerical simulations favour diapiric exhumation of deeply subducted lithologies that may instead melt within the hot core of the mantle wedge (Gerya & Yuen 2003; Behn *et al.* 2011). Other numerical parameterizations also suggest that the locations of volcanic arcs are controlled by mantle wedge dynamics, and not by the local release of fluids at the slab surface (England & Katz 2010). Disentangling the mechanisms of geochemical recycling in convergent margins is thus essential to understand the origin of andesites and their role in the construction of continents.

In this contribution we investigate andesite petrogenesis in the context of the western Trans-Mexican Volcanic Belt (TMVB), a peculiar arc segment of the Mexican subduction zone in which slab convergence and extensional tectonics have been instrumental in the construction of one of the most diverse magmatic arcs on the planet (Gómez-Tuena *et al.* 2007*b*). We report new whole-rock geochemical data from two representative andesitic volcanoes in western Mexico, Tequila and Sangangüey, and compare their compositions to those of the most primitive mafic magmas emplaced in the region. Using this dataset, and taking into

From: GÓMEZ-TUENA, A., STRAUB, S. M. & ZELLMER, G. F. (eds) 2014. *Orogenic Andesites and Crustal Growth*. Geological Society, London, Special Publications, **385**, 65–101.
First published online June 11, 2013, http://dx.doi.org/10.1144/SP385.4

consideration the regional geological and tectonic context, we elucidate the mechanisms of element recycling involved in the compositional diversity. We critically evaluate key models for andesite petrogenesis, identifying inconsistencies and unresolved observations, and advance a new explanation that may help to reconcile conflicting hypothesis for Mexico, and for arcs in general.

Tectonic and geological framework

The Mexican subduction zone comprises two independent tectonic plates, Cocos and Rivera, with different ages, convergence rates and relative motions (Mammerickx & Klitgord 1982; Pardo & Suárez 1995; DeMets & Traylen 2000; Yang *et al.* 2009). The lack of deep earthquakes precludes a precise definition of the subduction geometry (Pardo & Suárez 1993), but recent seismic experiments have shown that subduction of the young Rivera plate is very steep (>50°), whereas subduction of Cocos becomes flat to the east (Pérez-Campos *et al.* 2008; Yang *et al.* 2009), displacing the volcanic arc front more than 300 km inland (Fig. 1). The boundary between the two plates intersects the continent at the southern edge of the Colima rift, it extends below Colima volcano, and then becomes gradually divergent to the north with the opening of a slab window (Yang *et al.* 2009). And yet Colima volcano has been constructed relatively close to the trench and lies ~90 km above the boundary between Rivera and Cocos (Fig. 1), whereas the monogenetic volcanoes that define the active volcanic front are mainly distributed along a NW–SE-trending belt under which Rivera rests at 140–150 km depth (Fig. 2). Interestingly, further to the north, within the so-called Tepic–Zacoalco rift

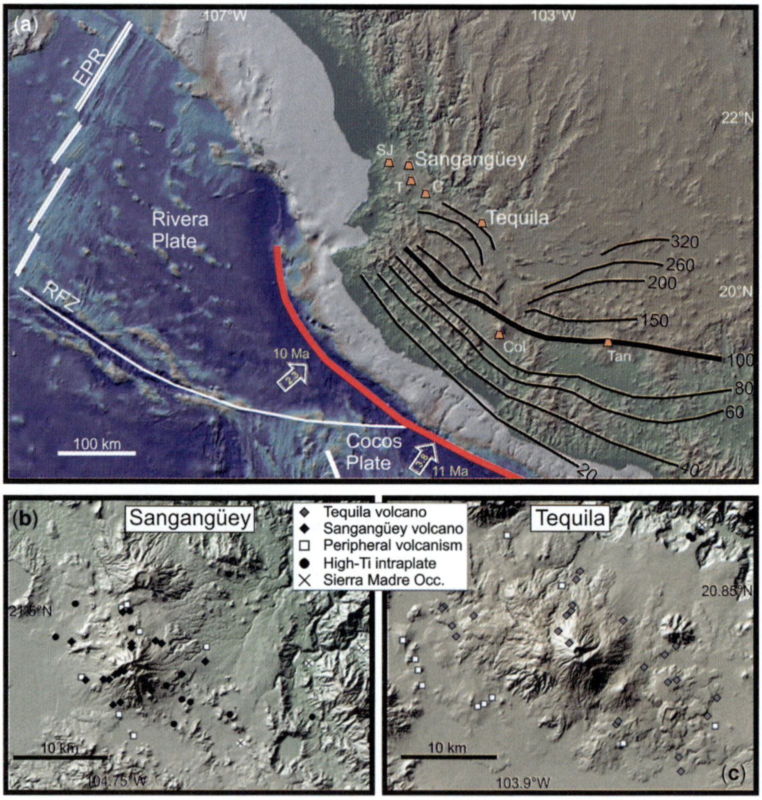

Fig. 1. (**a**) Simplified tectonic map of the western Mexican Subduction Zone, showing slab depth contours and the location of andesitic stratovolcanoes: Sangangüey, Tequila, San Juan (SJ), Tepetiltic (T), Ceboruco (C), Colima (Col) and Tancítaro (Tan). Slab ages (Ma) and convergence rates (cm a^{-1}) are from DeMets & Traylen (2000). Slab depth contours (km) are based on seismic tomographies (Yang *et al.* 2009; Pérez-Campos *et al.* 2008). Seafloor features: East Pacific Rise (EPR), Rivera Fracture Zone (RFZ) General map modified from Ferrari *et al.* (2012). (**b**) Samples locations for Sangangüey volcano. (**c**) Sample locations for Tequila volcano. Maps were constructed using GeoMapApp (http://www.geomapapp.org).

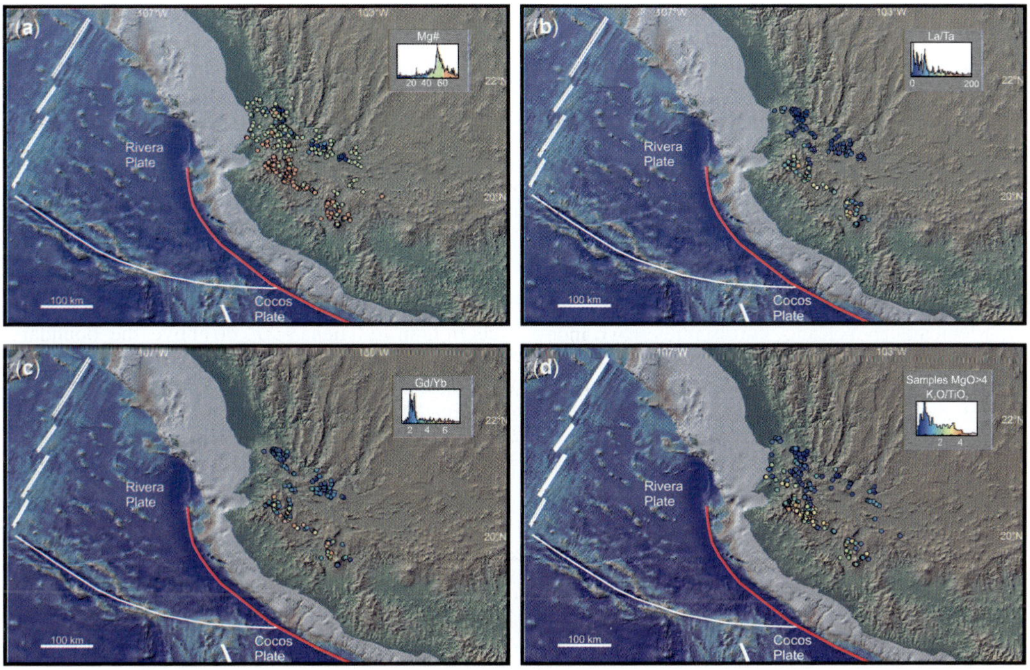

Fig. 2. Pliocene to Recent magmatic diversity of the western Trans-Mexican Volcanic Belt. (**a**) Mg#, (**b**) La/Ta, (**c**) Gd/Yb, (**d**) K_2O/TiO_2 for samples with MgO contents >4 wt%. Geochemical data of volcanic rocks compiled from Gómez-Tuena *et al.* (2007*b*), the GEOROC database (http://georoc.mpch-mainz.gwdg.de/georoc/) and the present study. Colour-coded histograms represent the relative abundance of analysed samples and not necessarily the relative volume of different rock compositions. Maps were constructed using GeoMapApp (http://www.geomapapp.org).

(TZR), andesitic stratovolcanoes like Tequila, Ceboruco, Sangangüey, Tepetiltic and San Juan have been constructed over a subducted slab that lies deeper than 250–300 km (Fig. 1), or may be even absent (Yang *et al.* 2009).

The magmatic history of the western TMVB is different to that of other sectors of the Mexican arc, and unusual when compared to most magmatic arcs on the planet (Gómez-Tuena *et al.* 2007*b*). This is because extensional tectonics and volcanism have evolved hand in hand since the late Miocene (Ferrari & Rosas-Elguera 2000), presumably as a direct response to the complex evolution of the convergent margin (DeMets & Traylen 2000). The oldest exposed rocks belong to the Jalisco block, a fault-bounded Cretaceous batholith (*c.* 100–75 Ma) that experienced significant uplift before the Pliocene that may continue to the present times (Righter *et al.* 2010; Ramírez-Herrera *et al.* 2011). The following magmatic episode in the region belongs to the Oligocene–Early Miocene Sierra Madre Occidental (SMO), one of the largest silicic igneous provinces on Earth (Bryan & Ernst 2008). Ignimbrites of the SMO directly overlie Cretaceous rocks within the TZR (Frey *et al.* 2007), but they have never been found in the Jalisco Block further to the south, either because they were eroded away during uplift, or because they were never emplaced above it. After a *c.* 10 Ma magmatic hiatus at the end of the Miocene, volcanism resumed in the area with the emplacement of a voluminous sequence of tholeiitic and calc-alkaline basalts (11–8 Ma) that has been interpreted as the magmatic manifestation of a mantle plume (Moore *et al.* 1994), a slab detachment event (Ferrari 2004), or a lithospheric foundering episode (Mori *et al.* 2009). Volcanism changed during the Pliocene (5.5–3 Ma), becoming bimodal towards the rear arc, and potassic lamprophyric to calc-alkaline within the Jalisco Block (Lange & Carmichael 1991; Frey *et al.* 2007; Ownby *et al.* 2008). Voluminous rhyolites and ignimbrites were emplaced during this period along the main axis of the TZR, in close association with high-TiO_2(Nb) intraplate basalts (Frey *et al.* 2007). In contrast, potassic lamprophyres started to erupt at 4.6 Ma in the western flank of the Colima rift and gradually migrated to the NW along the so-called Central Jalisco Volcanic Lineament (Bandy *et al.* 2001; Ownby *et al.* 2008), a trench-parallel belt of monogenetic fields locally emplaced inside small graben (Tapalpa, Ayutla, Los Volcanes, Mascota and San

Sebastian). Less voluminous bimodal volcanism continues to the present times within the TZR, whereas lamprophyric monogenetic volcanoes similar to those found within the Jalisco Block have erupted very recently on the NW flank of Colima volcano (Carmichael *et al.* 2006). Surprisingly, despite the widespread distribution of fissural lava flows and monogenetic volcanoes along the entire area, large andesitic stratovolcanoes, that are the quintessential feature of volcanic arcs world-wide, only started to appear in western Mexico during the Pleistocene. The oldest dated rock attributed to a composite volcano corresponds to Cántaro volcano ($\sim$1.5 Ma, Allan 1986), a highly eroded precursor stratovolcano emplaced at the northern end of the Colima Volcanic Complex. Volcanic activity gradually migrated southwards and is currently localized at Fuego de Colima volcano, North America's most active and productive central volcano (Savov *et al.* 2008). The youngest volcanic manifestations within the TZR are concentrated around Ceboruco volcano, a late Quaternary andesitic stratovolcano that last erupted in 1870 (Sieron & Siebe 2008). No other volcano within the TZR is considered active, but stratovolcanoes like San Juan, Sanganguey and Tequila were all constructed over the last 200 ka (Nelson & Livieres 1986; Nelson & Hegre 1990; Luhr 2000; Lewis-Kenedi *et al.* 2005), with reported ages that can be as young as *c.* 60 ka to *c.* 15 ka (Luhr 2000; Lewis-Kenedi *et al.* 2005).

Samples and methods

In this contribution we report new comprehensive geochemical data (Tables 1 & 2) and petrographic observations of volcanic rocks from Tequila and Sangangüey volcanoes, including most of their peripheral monogenetic cones (Fig. 1b, c). We also include data of post-Pliocene intraplate volcanic rocks emplaced within the TZR, with special emphasis in the areas surrounding Tequila and Sangangüey volcanoes. Some selected basement rock samples belonging to the Jalisco Block and the SMO province are also reported. For the purpose of comparison, some plots include previously published geochemical data of the most primitive rocks collected from the Mascota and Colima volcanic fields, representative of the compositional diversity found at the current volcanic front (Gómez-Tuena *et al.* 2011).

Major element compositions were obtained by X-ray fluorescence on a Siemens SRS-3000 instrument at the Instituto de Geología of the Universidad Nacional Autónoma de México (UNAM), following procedures described elsewhere (Lozano & Bernal 2005). Trace element concentrations, as well as some major elements on selected samples, were determined by inductively coupled plasma mass spectrometry (ICP-MS) using a Thermo Series X^{II} at the Centro de Geociencias (CGEO), UNAM, following sample preparation and measurement procedures described in Mori *et al.* (2007). The long-term reproducibility of the trace element data at CGEO is given by the average concentrations and standard deviations of multiple digestions of the US Geological Survey rock standards AGV-2, BHVO-2, BCR-2, and the Geological Survey of Japan JB-2, and has been reported in several previous publications (Mori *et al.* 2007, 2009; Gómez-Tuena *et al.* 2011).

Sample preparations for Sr, Nd and Pb isotopic analyses were performed at the clean lab facilities of the CGEO following previously established chemical procedures (Gómez-Tuena *et al.* 2003, 2011). Sr, Nd and Pb isotope ratios were measured using a Thermo Neptune *Plus* multicollector ICP-MS at CGEO. Nd and Pb isotopic compositions of samples and standards were measured in freshly prepared *c.* 200 ppb solutions, whereas Sr isotopes were determined using solutions diluted to *c.* 300 ppb concentrations. Total processing blanks (including column chemistry) were $<$200 pg for Sr and $<$60 pg for Pb, which are negligible in comparison to concentrations in the analysed samples. Analyses of samples and standards consist of 70–100 static measurement cycles, with each cycle measured in *c.* 4 s integrations. Beam intensities of 50 V/ppm, 17 V/ppm and 100 V/ppm were routinely achieved for ^{88}Sr, ^{142}Nd and ^{208}Pb, respectively, using wet plasma and a 100 μl/min free aspirating nebulizer.

For Sr isotope ratio measurements, the collector array was configured to statically monitor all isotopes of interest, as well as the most critical isobaric interferences of Rb and Kr. Rb and the rest of the matrix elements are effectively removed from Sr during column chromatography, whereas Kr is always present as an impurity in the Ar carrier gas, although at extremely low concentrations (^{83}Kr $<$0.5 mV) that remained fairly constant throughout the course of this study. The measured ^{87}Sr, ^{86}Sr and ^{84}Sr intensities were none the less corrected for any Rb and Kr contribution using the within-run measurements of ^{85}Rb and ^{83}Kr. The interference-corrected $^{87}Sr/^{86}Sr$ ratios were exponentially normalized for mass bias to $^{86}Sr/^{88}Sr = 0.1194$ and corrected to a NBS-987 standard ratio of $^{87}Sr/^{86}Sr = 0.710230$. During the course of this study, the average measured value for NIST SRM 987 standard was 0.710264 ± 0.000014 (2σ, $n = 60$).

Nd isotopic ratios were measured statically using the entire collector array in order to monitor all relevant isotopic species and the potentially critical isobaric interference of Sm. Although Sm was effectively removed along with the rest of the rare earth

Table 1. *Major and trace elements of the studied rock suites*[a]

Suite	Sanga	Sanga	Sanga	Sanga	Sanga	Sanga	Sanga	Sanga	Sanga	Sanga	Sanga	Sanga	Sanga	Sanga	Sanga	Sanga
Sample	TPZ-10-05	TPZ-10-27	TPZ-10-28	TPZ-10-04	TPZ-10-10	SAN-10-07	SAN-10-10	SAN-10-11	SAN-10-14	SAN-10-15	SAN-10-16	SAN-10-17	SAN-10-18	SAN-10-19	SAN-10-19B	RTZ-11-18
Long. W	−104.762	−104.709	−104.703	−104.757	−104.713	−104.682	−104.746	−104.752	−104.750	−104.746	−104.746	−104.745	−104.745	−104.749	−104.781	−104.795
Lat. N	21.434	21.471	21.469	21.434	21.424	21.439	21.412	21.410	21.439	21.440	21.440	21.440	21.440	21.439	21.434	21.471
Major elements (wt%)																
	XRF	XRF	XRF	XRF	XRF	XRF	XRF	ICP-MS	XRF	XRF	XRF	XRF	XRF	XRF	ICP-MS	ICP-MS
SiO_2	60.9	61.5	60.4	59.2	59.7	60.7	60.6		61.4	62.4	61.0	60.2	61.7	61.5		
TiO_2	0.73	0.76	0.76	0.77	0.76	0.77	0.76	0.74	0.75	0.77	0.75	0.74	0.76	0.76	0.79	0.77
Al_2O_3	17.5	17.0	18.0	18.0	17.8	17.7	18.1		17.5	17.4	17.4	17.9	17.4	17.9		
$Fe_2O_3^{tot}$	5.5	5.6	5.6	5.8	5.5	5.5	5.3		5.3	5.5	5.4	5.4	5.4	5.4		
MnO	0.083	0.090	0.090	0.085	0.081	0.091	0.098	0.083	0.091	0.089	0.089	0.088	0.083	0.088	0.088	0.093
MgO	2.8	2.9	2.8	3.2	2.9	2.9	2.9	2.9	2.6	2.5	2.7	2.8	2.8	2.6	2.6	2.7
CaO	5.2	5.4	5.2	5.7	5.4	5.5	5.4	5.8	5.0	4.9	5.1	5.2	5.5	5.5	4.7	5.0
Na_2O	4.0	4.3	4.2	3.8	3.8	3.9	3.0		4.1	4.1	3.9	4.0	3.6	3.6		
K_2O	2.2	2.0	2.1	2.1	2.1	2.1	2.2	2.2	2.1	2.1	2.3	2.1	2.0	2.0	2.2	2.4
P_2O_5	0.28	0.001	0.16	0.34	0.18	0.18	0.18	0.20	0.22	0.22	0.20	0.19	0.22	0.18	0.22	0.23
LOI	0.62	0.24	0.60	0.89	1.48	0.80	0.98		0.45	0.36	0.89	0.83	0.70	0.26		
Total	99.9	99.7	99.9	99.8	99.6	100.1	99.7		99.5	100.4	99.7	99.4	100.2	100.0		
Trace elements (ppm)																
	ICP-MS	ICP-MS	ICP-MS	ICP-MS	ICP-MS	ICP-MS	ICP-MS	ICP-MS	ICP-MS	ICP-MS	ICP-MS	ICP-MS	ICP-MS	ICP-MS	ICP-MS	ICP-MS
Sc	13.1	11.8	11.4	13.1	12.9	14.1	13.6	12.6	13.3	12.8	12.0	12.2	13.8	14.4	13.2	12.0
V	98.2	101	102	116	112	115	116	111	98.6	105	103	103	109	109	88.0	86.5
Cr	25.5	25.2	25.8	26.9	27.3	40.7	33.4	26.2	27.7	33.2	43.4	27.2	33.0	30.2	22.8	21.8
Co	13.5	13.8	13.9	15.5	14.6	15.5	15.8	14.5	13.5	14.6	14.0	14.1	14.7	14.6	11.7	11.8
Ni	17.6	17.4	17.5	20.0	19.5	22.7	24.8	16.8	19.6	44.4	24.7	17.0	19.7	19.6	17.5	16.7
Cu	21	17	14	22	23	21	22	20	20	21	17	18	20	21	19	19
Zn	58	64	65	61	59	64	64	60	70	65	65	65	65	64	65	67
Ga	18.8	19.5	19.8	19.6	19.5	20.0	20.0	19.6	19.9	20.0	19.7	19.8	19.8	19.[illegible]	19.9	19.5
Li	14.2	14.9	14.5	11.9	13.8	13.6	14.5	14.0	10.9	9.4	16.7	16.9	14.8	12.3	12.2	14.0
Be	1.59	1.52	1.53	1.40	1.49	1.46	1.54	1.52	1.80	1.71	1.74	1.59	1.56	1.43	1.62	1.55
Rb	34	32	31	30	31	32	33	33	35	35	33	32	34	36	32	33
Sr	583	613	605	680	629	629	623	618	595	578	602	609	611	652	579	616
Y	18.1	17.9	17.8	17.8	21.2	21.3	22.5	24.1	19.1	18.0	21.2	19.5	18.3	17.6	28.8	18.8
Zr	152	157	160	127	137	140	141	139	165	163	159	159	154	149	198	199
Nb	7.60	7.81	8.41	6.21	6.66	7.05	6.96	6.84	9.17	9.18	9.05	8.63	7.92	7.51	9.18	9.24
Mo	1.58	1.52	1.65	1.43	1.48	2.51	2.13	1.40	2.00	3.53	3.82	1.47	1.94	1.34	1.30	1.68
Sn	1.02	0.96	1.02	0.98	1.04	0.93	0.92	0.89	0.98	1.01	0.99	0.97	0.95	0.71	1.03	1.04
Sb	0.20	0.16	0.16	0.18	0.20	0.14	0.14	0.13	0.13	0.12	0.15	0.14	0.14	0.10	0.16	0.19
Cs	0.72	0.68	0.68	0.71	0.71	0.70	0.74	0.74	0.41	0.55	0.75	0.71	0.72	0.73	0.57	0.76
Ba	839	846	889	757	830	879	863	846	889	897	855	892	852	879	998	881
La	19.3	18.1	18.9	17.6	23.2	20.8	25.2	25.3	20.2	20.6	22.2	21.2	18.3	17.9	28.2	19.2
Ce	36.0	36.7	37.2	32.2	37.2	36.7	35.6	35.5	38.6	38.2	37.4	38.8	36.8	36.9	39.7	38.6
Pr	4.53	4.63	4.76	4.25	5.01	5.05	5.79	5.58	5.00	5.10	5.34	5.07	4.61	4.76	6.41	4.82
Nd	18.5	18.2	18.6	17.1	19.9	20.0	22.6	22.2	19.7	19.8	21.5	19.8	18.2	16.4	25.1	18.8

(*Continued*)

Table 1. *Major and trace elements of the studied rock suites (Continued)*

Suite	Sanga	Sanga	Sanga	Sanga	Sanga	Sanga	Sanga	Sanga	Sanga	Sanga	Sanga	Sanga	Sanga	Sanga	Sanga	Sanga
Sample	TPZ-10-05	TPZ-10-27	TPZ-10-28	TPZ-10-04	TPZ-10-10	SAN-10-07	SAN-10-10	SAN-10-11	SAN-10-14	SAN-10-15	SAN-10-16	SAN-10-17	SAN-10-18	SAN-10-19	SAN-10-19B	RTZ-11-18
Long. W	−104.762	−104.709	−104.703	−104.757	−104.713	−104.682	−104.746	−104.752	−104.750	−104.746	−104.746	−104.745	−104.745	−104.749	−104.781	−104.795
Lat. N	21.434	21.471	21.469	21.434	21.424	21.439	21.412	21.410	21.439	21.440	21.440	21.440	21.440	21.439	21.434	21.471
	ICP-MS	ICP-MS	ICP-MS	ICP-MS	ICP-MS	ICP-MS	ICP-MS	ICP-MS	ICP-MS	ICP-MS	ICP-MS	ICP-MS	ICP-MS	ICP-MS	ICP-MS	ICP-MS
Sm	3.72	3.84	3.90	3.60	4.05	4.16	4.49	4.48	4.04	4.04	4.35	4.04	3.82	3.49	5.04	3.94
Eu	1.09	1.11	1.15	1.09	1.20	1.21	1.29	1.27	1.18	1.19	1.20	1.16	1.10	1.14	1.48	1.17
Gd	3.57	3.66	3.66	3.43	4.01	4.02	4.32	4.47	3.71	3.68	4.12	3.76	3.52	3.27	4.96	3.57
Tb	0.541	0.534	0.535	0.514	0.591	0.605	0.649	0.664	0.561	0.550	0.609	0.566	0.540	0.555	0.728	0.539
Dy	3.16	3.14	3.15	2.93	3.41	3.54	3.76	3.90	3.29	3.16	3.51	3.29	3.15	3.22	4.31	3.17
Ho	0.63	0.64	0.63	0.59	0.70	0.72	0.76	0.80	0.66	0.63	0.72	0.66	0.64	0.65	0.87	0.63
Er	1.79	1.75	1.74	1.63	1.92	1.99	2.06	2.19	1.83	1.76	1.96	1.82	1.76	1.79	2.40	1.77
Yb	1.73	1.74	1.73	1.57	1.83	1.94	1.99	2.06	1.82	1.74	1.92	1.80	1.74	1.78	2.19	1.75
Lu	0.264	0.260	0.260	0.240	0.280	0.295	0.303	0.318	0.281	0.262	0.296	0.275	0.267	0.265	0.336	0.268
Hf	3.75	3.95	4.01	3.14	3.36	3.45	3.49	3.42	3.92	3.91	3.79	3.81	3.71	3.63	3.75	3.87
Ta	0.53	0.59	0.64	0.43	0.47	0.49	0.49	0.48	0.64	0.63	0.63	0.59	0.55	0.52	0.63	0.64
Tl	0.148	0.151	0.153	0.129	0.136	0.031	0.030	0.030	0.028	0.027	0.028	0.028	0.028	0.106	0.141	0.153
Pb	9.6	9.8	9.9	8.6	9.3	9.6	9.8	9.7	10.2	9.9	9.8	9.9	9.9	9.5	9.2	9.8
Th	2.61	2.61	2.77	2.24	2.44	2.51	2.58	2.52	2.82	2.89	2.79	2.72	2.64	2.56	2.63	2.76
U	0.796	0.795	0.830	0.690	0.739	0.748	0.773	0.760	0.879	0.877	0.865	0.834	0.806	0.774	0.818	0.882

Suite	Sanga	Sanga	Sanga	Sanga	Per-Sanga	Per-Sanga	Per-Sanga	Per-Sanga	Per-Sanga	Per-Sanga	Per-Sanga	HiTi-Sanga	HiTi-Sanga	HiTi-Sanga	HiTi-Sanga	HiTi-Sanga	HiTi-Sanga
Sample	RTZ-11-23	RTZ-11-24A	RTZ-11-25	RTZ-11-29	TPZ-10-25	TPZ-10-26	RTZ-11-28	TPZ-10-06	SAN-10-12	RTZ-11-21	RTZ-11-07	TPZ-10-11	TPZ-10-21	TPZ-10-23	TPZ-10-14	SAN-10-01A	SAN-10-06
Long. W	−104.734	−104.742	−104.734	−104.658	−104.743	−104.739	−104.656	−104.747	−104.786	−104.726	−104.732	−104.674	−104.811	−104.793	−104.633	−104.691	−104.697
Lat. N	21.469	21.499	21.466	21.452	21.503	21.505	21.465	21.400	21.437	21.480	21.381	21.413	21.475	21.506	21.401	21.393	21.429
Major elements (wt%)																	
	ICP-MS	ICP-MS	ICP-MS	ICP-MS	XRF	XRF	ICP-MS	XRF	XRF	XRF	XRF	XRF	XRF	XRF	XRF	XRF	XRF
SiO_2				55.6	55.1			56.7	54.0	57.3	52.9	48.5	48.2	48.9	48.5	47.8	48.4
TiO_2	0.69	0.82	0.78	0.70	1.12	1.20	0.57	1.03	1.48	1.14	1.41	2.99	2.01	2.17	2.65	1.81	2.08
Al_2O_3					18.6	18.7		17.9	18.9	18.7	18.8	16.7	15.8	18.2	16.4	17.6	17.7
$Fe_2O_3^{tot}$					7.3	7.8		7.3	8.6	7.3	8.5	13.3	9.7	9.5	14.2	11.5	9.1
MnO	0.083	0.088	0.083	0.082	0.124	0.124	0.081	0.124	0.148	0.115	0.138	0.212	0.159	0.157	0.260	0.189	0.146
MgO	2.8	2.9	2.9	2.5	3.9	4.1	0.75	3.4	3.8	3.3	4.9	4.1	7.2	6.1	3.8	6.8	8.3
CaO	5.4	5.4	5.3	5.1	6.5	7.0	2.6	5.9	6.8	6.4	7.7	6.9	8.5	8.2	7.0	10.5	9.2
Na_2O					4.4	4.2		4.5	3.9	3.9	3.0	4.2	3.7	4.0	4.3	3.2	3.5
K_2O	2.4	2.5	1.9	2.1	1.4	1.4	2.6	1.5	1.2	1.5	1.2	1.7	1.6	2.0	1.8	0.6	1.6
P_2O_5	0.18	0.20	0.19	0.22	0.29	0.15	0.22	0.38	0.46	0.34	0.41	0.97	0.57	0.74	1.64	0.27	0.47
LOI					−0.04	0.17		0.71	0.32	0.41	0.72	0.40	-0.03	0.08	−0.31	−0.48	0.16
Total					99.1	99.8		99.4	99.7	100.4	99.7	99.9	97.4	100.1	100.3	100.0	100.7

Trace elements (ppm)	ICP-MS	ICP-MS	ICP-MS	ICP-MS	ICP-MS	ICP-MS	ICP-MS	ICP-MS	ICP-MS	ICP-MS	ICP-MS	ICP-MS	ICP-MS	ICP-MS	ICP-MS	ICP-MS	ICP-MS
Sc	11.9	13.1	13.2	12.4	14.6	16.0	7.3	14.4	19.1	14.2	20.5	23.0	24.4	23.8	24.0	30.1	25.7
V	105	99.3	97.9	101	136	150	17.8	122	169	110	164	233	187	198	143	287	193
Cr	29.5	25.4	23.5	25.5	33.2	38.0	2.04	45.6	47.8	26.1	127	1.52	226	153	1.42	23.6	287
Co	14.3	12.9	13.3	13.5	18.8	20.6	2.20	16.9	22.9	14.7	25.6	26.2	34.6	31.6	23.9	42.5	36.8
Ni	21.0	20.2	19.1	16.7	20.7	26.4	1.33	24.1	26.9	18.3	62.2	1.64	111.3	79.9	1.29	33.4	145
Cu	21	21	22	20	38	32	4.1	22	30	21	33	20	46	48	17	20	51
Zn	59	63	62	63	82	80	75	86	93	74	82	128	76	74	145	91	72
Ga	17.9	19.7	19.3	18.7	21.2	21.0	19.6	21.6	22.1	20.4	20.3	24.5	20.8	22.5	25.1	20.7	19.1
Li	12.4	13.6	12.0	13.0	9.0	7.9	11.6	12.4	10.6	9.5	6.2	9.7	6.8	8.0	10.6	5.6	6.2
Be	1.36	1.42	1.31	1.69	1.52	1.42	2.41	1.61	1.63	1.46	1.19	2.21	2.08	2.31	2.41	1.15	1.99
Rb	33	33	29	35	13	14	43	15	11	16	12	17	18	32	22	6.9	21
Sr	578	645	673	568	739	759	439	705	700	698	613	480	476	543	526	564	498
Y	16.6	18.5	16.5	20.7	18.7	19.8	25.7	21.0	27.4	20.2	26.9	50.4	31.9	34.5	54.9	26.3	28.2
Zr	131	147	129	156	138	132	241	163	147	143	138	290	252	303	299	165	218
Nb	6.40	7.16	6.32	8.36	10.9	10.2	20.3	10.8	12.3	11.3	13.0	52.4	43.0	64.5	48.6	21.9	49.8
Mo	1.43	1.49	1.33	1.41	1.19	1.15	2.15	1.41	0.70	1.20	1.03	2.46	2.06	3.65	3.34	3.30	2.55
Sn	0.83	1.00	0.86	0.92	0.89	0.83	1.5	0.97	0.92	0.88	0.98	2.2	1.7	2.1	2.2	1.1	1.4
Sb	0.19	0.19	0.15	0.17	0.10	0.10	0.16	0.14	0.08	0.10	0.10	0.04	0.05	0.10	0.06	0.03	0.04
Cs	0.73	0.78	0.74	0.76	0.20	0.27	1.03	0.35	0.15	0.29	0.20	0.05	0.15	0.31	0.18	0.12	0.20
Ba	820	877	792	936	597	594	933	771	689	681	713	556	213	353	580	191	285
La	16.5	17.9	16.2	22.3	16.1	15.8	29.5	19.2	22.4	18.5	18.8	39.3	28.1	41.0	41.5	16.2	30.4
Ce	33.2	35.5	32.3	39.3	33.4	32.8	53.7	41.0	39.9	34.0	35.3	73.9	55.3	75.4	87.6	35.5	59.3
Pr	4.10	4.46	4.18	5.40	4.50	4.41	7.10	5.43	6.16	4.82	5.31	10.1	6.76	8.85	12.0	4.86	7.01
Nd	16.3	17.6	16.8	21.2	18.8	18.8	27.3	22.3	26.2	19.9	23.4	41.7	26.7	33.2	51.2	21.3	27.0
Sm	3.49	3.79	3.66	4.36	4.17	4.22	5.41	4.77	5.79	4.36	5.24	9.53	5.99	6.90	12.0	5.16	5.67
Eu	1.01	1.10	1.10	1.22	1.46	1.49	1.51	1.59	1.90	1.51	1.75	3.03	1.96	2.20	3.82	1.72	1.79
Gd	3.29	3.55	3.39	4.07	4.04	4.17	4.75	4.52	5.62	4.13	5.22	9.25	6.21	6.99	11.4	5.19	5.63
Tb	0.499	0.534	0.509	0.605	0.584	0.603	0.715	0.653	0.834	0.607	0.771	1.38	0.941	1.02	1.676	0.792	0.869
Dy	2.95	3.17	2.96	3.58	3.37	3.43	4.19	3.79	4.86	3.51	4.47	8.32	5.73	6.15	9.95	4.66	5.19
Ho	0.60	0.63	0.59	0.73	0.67	0.69	0.87	0.75	0.97	0.69	0.88	1.60	1.13	1.21	1.87	0.92	1.03
Er	1.64	1.75	1.60	2.01	1.80	1.84	2.49	2.06	2.60	1.86	2.35	4.35	3.10	3.32	4.99	2.43	2.80
Yb	1.58	1.70	1.55	1.92	1.67	1.67	2.50	1.97	2.40	1.70	2.02	3.93	2.92	3.16	4.39	2.16	2.56
Lu	0.241	0.258	0.236	0.293	0.248	0.249	0.386	0.294	0.360	0.254	0.305	0.591	0.434	0.477	0.653	0.319	0.382
Hf	3.33	3.52	3.15	3.79	3.33	3.14	5.61	3.89	3.42	3.10	3.25	6.04	5.07	6.14	6.33	3.42	4.45
Ta	0.46	0.50	0.45	0.59	0.74	0.68	1.37	0.72	0.73	0.72	0.80	3.12	2.75	4.14	2.84	1.31	2.88
Tl	0.141	0.148	0.135	0.165	0.059	0.074	0.160	0.068	0.020	0.047	0.045	0.025	0.020	0.074	0.064	0.020	0.021
Pb	9.2	9.7	8.5	9.4	5.9	5.6	10.7	7.6	5.7	5.6	5.5	2.8	2.0	2.9	3.9	1.4	2.1
Th	2.40	2.57	2.31	2.69	1.75	1.64	4.26	1.50	1.02	1.78	1.14	3.54	3.13	5.38	3.28	1.47	3.65
U	0.734	0.771	0.717	0.827	0.563	0.532	1.382	0.520	0.375	0.581	0.370	0.643	1.03	1.63	1.12	0.509	1.07

(Continued)

Table 1. *Major and trace elements of the studied rock suites (Continued)*

Suite	HiTi-Sanga	HiTi-Sanga	HiTi-Sanga	HiTi-Sanga	Tequila	Tequila	Tequila	Tequila	Tequila	Tequila	Tequila	Tequila	Tequila	Tequila	Tequila	Tequila	Tequila
Sample	SAN-10-09	RTZ-11-22	RTZ-11-27	RTZ-11-12	TZ-09-02	TZ-09-03	TZ-09-05	TZ-09-15	TZ-09-16	TZ-09-17	TZ-09-20	TZ-09-27	TZ-09-29	TZ-09-18	TZ-09-19	TZ-09-22	TZ-09-28
Long. W	−104.682	−104.733	−104.735	−104.549	−103.745	−103.799	−103.846	−103.842	−103.850	−103.851	−103.853	−103.803	−103.805	−103.856	−103.864	−103.982	−103.805
Lat. N	21.416	21.483	21.500	21.399	20.798	20.821	20.861	20.868	20.835	20.831	20.800	20.725	20.722	20.827	20.812	20.834	20.722
Major elements (wt%)																	
	XRF	XRF	ICP-MS	XRF	XRF	XRF	XRF	XRF	XRF	XRF	XRF	XRF	XRF	XRF	XRF	XRF	ICP-MS
SiO_2	48.4	49.2		47.6	60.1	60.3	61.8	61.0	62.2	62.5	62.0	58.1	60.5	62.1	61.5	57.4	
TiO_2	2.14	2.34	2.30	3.39	0.85	0.81	0.71	0.77	0.72	0.72	0.70	0.87	0.82	0.71	0.71	1.04	0.90
Al_2O_3	18.2	18.6		16.2	17.0	16.6	16.5	16.5	16.6	16.5	16.4	16.9	16.5	16.4	16.3	16.8	
$Fe_2O_3^{tot}$	9.3	10.1		14.5	6.0	5.7	5.0	5.6	5.1	5.1	4.9	6.4	5.7	5.1	5.1	7.0	
MnO	0.166	0.185	0.169	0.243	0.094	0.086	0.074	0.082	0.075	0.075	0.074	0.104	0.089	0.079	0.077	0.108	
MgO	7.1	5.1	5.4	4.8	3.5	3.1	2.5	2.9	2.6	2.7	2.5	3.5	3.1	2.6	2.6	3.5	3.6
CaO	8.7	7.0	8.0	8.1	5.8	5.7	5.1	5.5	5.1	5.2	5.0	6.1	5.5	5.1	5.1	5.7	6.5
Na_2O	3.1	3.4		3.3	3.7	3.7	3.7	3.6	3.5	3.5	3.8	3.9	4.0	3.7	3.8	4.0	
K_2O	1.7	2.0	2.1	1.2	2.3	2.2	2.6	2.6	2.7	2.7	2.3	1.9	2.2	2.6	2.3	2.1	2.1
P_2O_5	0.56	0.67	0.71	0.82	0.20	0.18	0.14	0.17	0.15	0.14	0.14	0.29	0.21	0.16	0.15	0.31	0.35
LOI	0.06	0.99		−0.58	0.81	0.86	1.16	0.87	1.15	1.18	2.29	1.20	0.61	0.83	2.35	2.51	
Total	99.6	99.6		99.6	100.3	99.2	99.3	99.6	99.8	100.3	100.2	99.3	99.3	99.3	100.0	100.5	
Trace elements (ppm)																	
	ICP-MS	ICP-MS	ICP-MS	ICP-MS	ICP-MS	ICP-MS	ICP-MS	ICP-MS	ICP-MS	ICP-MS	ICP-MS	ICP-MS	ICP-MS	ICP-MS	ICP-MS	ICP-MS	ICP-MS
Sc	25.1	22.4	21.7	30.3	15.3	14.4	12.3	14.0	12.6	12.5	12.3	14.4	14.1	12.3	13.4	16.7	15.0
V	195	175	181	333	120	116	104	111	104	105	104	116	115	105	107	129	120
Cr	210	74.5	85.8	9.29	60.7	40.6	32.5	40.6	33.1	33.7	27.3	49.6	51.2	34.5	51.4	60.7	56.2
Co	34.9	28.6	29.2	35.0	18.0	16.6	14.2	15.9	14.6	14.4	14.3	18.3	16.9	14.4	15.6	20.8	18.6
Ni	112	46.8	52.1	10.1	40.6	33.2	27.0	31.3	28.0	27.8	26.8	35.7	35.7	27.8	34.0	47.6	41.2
Cu	52	45	39	27	32	31	29	21	29	24	28	32	30	28	24	33	33
Zn	76	91	81	138	62	59	55	58	60	54	54	70	66	54	56	71	71
Ga	20.2	22.8	20.9	24.4	18.5	18.2	18.0	18.1	18.3	18.1	18.1	20.0	18.5	18.3	18.2	19.8	19.8
Li	6.8	8.6	7.6	8.7	11.4	11.1	11.1	10.7	12.6	13.0	16.4	10.4	13.6	12.4	13.1	11.8	10.7
Be	2.13	2.60	2.61	1.57	1.57	1.48	1.53	1.48	1.54	1.51	1.56	1.78	1.68	1.58	1.61	1.97	1.66
Rb	22	20	26	16	46	48	57	51	55	55	56	33	46	56	53	37	35
Sr	506	454	485	508	533	530	507	521	488	514	491	799	515	509	500	574	820
Y	30.7	41.1	35.0	47.2	21.8	17.8	17.8	17.8	17.7	17.2	17.3	31.5	19.8	17.4	17.6	21.7	19.7
Zr	243	422	302	265	194	170	156	168	151	156	151	211	203	159	164	256	209
Nb	53.0	51.6	61.3	40.6	9.67	8.47	7.46	8.24	7.35	7.43	7.73	12.36	9.98	7.43	7.46	16.0	12.6
Mo	2.21	1.87	4.15	2.24	2.10	1.92	1.90	1.89	1.75	1.87	1.98	1.86	2.13	1.85	1.84	2.31	2.69
Sn	1.6	2.2	1.9	1.7	1.2	1.1	1.1	1.1	1.1	1.1	1.1	1.1	1.2	1.1	1.1	1.3	0.77
Sb	0.05	0.06	0.10	0.04	0.16	0.18	0.22	0.18	0.21	0.20	0.21	0.10	0.18	0.19	0.17	0.45	0.05
Cs	0.20	0.08	0.23	0.02	0.78	0.86	1.10	0.90	1.13	1.12	1.14	0.53	0.81	1.14	1.06	0.51	0.54
Ba	272	226	331	390	794	765	810	774	823	800	825	628	826	815	814	819	659
La	32.5	32.2	36.9	30.2	24.7	20.9	21.7	21.1	21.6	21.0	21.5	38.9	23.7	21.2	21.4	28.3	25.2
Ce	63.4	65.5	72.6	64.7	44.2	41.2	40.7	41.1	41.6	40.4	40.8	52.1	45.8	42.8	43.4	53.6	52.0
Pr	7.53	8.06	8.51	8.80	5.73	4.97	5.01	5.00	4.85	4.89	4.93	8.79	5.62	4.89	4.98	6.76	6.61
Nd	29.1	31.8	32.9	37.5	21.8	18.8	18.8	18.9	18.5	18.3	18.2	33.2	21.2	18.5	18.8	25.6	23.1
Sm	6.16	7.19	7.07	8.94	4.44	3.84	3.79	3.84	3.65	3.68	3.71	6.35	4.26	3.72	3.77	5.11	4.59

Eu	1.92	2.19	2.12	2.79	1.25	1.10	1.03	1.09	1.02	1.02	1.02	1.63	1.19	1.06	1.07	1.51	1.48
Gd	6.13	7.26	6.99	8.92	4.13	3.54	3.47	3.51	3.39	3.36	3.36	5.91	3.90	3.41	3.46	4.57	4.06
Tb	0.946	1.14	1.08	1.33	0.616	0.532	0.522	0.530	0.509	0.506	0.516	0.889	0.586	0.523	0.529	0.673	0.647
Dy	5.63	6.94	6.46	7.90	3.61	3.10	3.05	3.09	3.01	2.97	2.96	5.01	3.42	2.96	3.01	3.84	3.63
Ho	1.12	1.36	1.28	1.51	0.73	0.62	0.62	0.62	0.61	0.60	0.60	1.02	0.69	0.60	0.61	0.76	0.72
Er	3.04	3.76	3.51	4.05	2.03	1.72	1.71	1.71	1.67	1.66	1.66	2.81	1.90	1.66	1.70	2.08	1.99
Yb	2.80	3.51	3.26	3.52	1.93	1.67	1.68	1.69	1.65	1.64	1.66	2.62	1.85	1.64	1.70	1.98	1.89
Lu	0.415	0.517	0.484	0.511	0.297	0.254	0.258	0.256	0.251	0.251	0.256	0.389	0.281	0.253	0.260	0.300	0.279
Hf	4.90	6.17	5.93	5.14	4.31	3.90	3.72	3.89	3.71	3.71	3.68	4.64	4.53	3.81	3.89	5.26	4.61
Ta	3.07	3.16	3.79	2.49	0.68	0.62	0.58	0.62	0.58	0.58	0.61	0.79	0.71	0.57	0.57	1.03	0.79
Tl	0.021	0.035	0.065	0.030	0.214	0.203	0.258	0.222	0.234	0.257	0.235	0.147	0.157	0.240	0.230	0.121	0.096
Pb	2.2	2.6	2.6	2.8	9.2	9.0	10.2	9.2	10.3	9.6	10.8	7.8	9.5	9.6	9.6	8.7	7.9
Th	3.81	3.44	4.38	2.49	3.98	4.21	4.90	4.37	4.91	4.88	5.28	3.21	4.18	4.96	4.85	3.44	3.26
U	1.14	1.20	1.38	0.670	1.18	1.20	1.35	1.25	1.32	1.34	1.42	1.02	1.25	1.35	1.34	1.15	1.01

Suite	Tequila	Tequila	Tequila	Tequila	Tequila	Tequila	Tequila	Tequila	Tequila	Tequila	Tequila	Tequila	Tequila	Per-Teq	Per-Teq	Per-Teq	Per-Teq
Sample	TQ-10-02A	TQ-10-03	TQ-10-04	TQ-10-06	TQ-10-07	TQ-10-09	TQ-10-11	TQ-10-12	TQ-10-14	RTZ-11-46	RTZ-11-47	RTZ-11-49	RTZ-11-50	TZ-09-09	TZ-09-21A	TQ-10-08	TZ-09-04
Long. W	−103.703	−103.757	−103.777	−103.745	−103.754	−103.981	−103.972	−103.950	−103.968	−103.705	−103.713	−103.740	−103.742	−103.731	−103.916	−103.982	−103.861
Lat. N	20.748	20.789	20.777	20.765	20.762	20.835	20.816	20.826	20.807	20.720	20.730	20.678	20.690	20.901	20.903	20.878	20.855
Major elements (wt%)																	
	XRF	XRF	XRF	XRF	XRF	XRF	XRF	ICP-MS	XRF	ICP-MS	XRF	ICP-MS	ICP-MS	ICP-MS	XRF	XRF	XRF
SiO_2	61.0	60.3	59.8	60.4	61.4	53.6	58.4		58.3		63.6				74.6	76.6	74.4
TiO_2	0.75	0.87	0.79	0.85	0.75	1.05	1.02	1.13	1.09	1.38	0.58	0.57	0.85	0.14	0.12	0.14	0.17
Al_2O_3	16.9	17.0	17.8	17.5	16.9	17.3	16.8		17.3		16.5				13.1	10.7	13.8
$Fe_2O_3^{tot}$	5.6	5.8	6.0	5.7	5.6	7.0	6.8		6.9		4.8				1.6	3.0	1.9
MnO	0.091	0.098	0.098	0.099	0.095	0.115	0.111		0.111	0.119	0.078	0.078	0.097	0.036	0.060	0.079	0.051
MgO	2.8	3.4	3.0	2.8	2.8	3.4	3.8	3.6	3.6	3.3	2.7	2.3	3.3	0.19	0.21	0.14	0.18
CaO	5.4	5.9	6.1	5.6	5.4	5.6	6.2	6.5	6.3	8.0	4.9	4.8	5.8	0.27	0.41	0.18	0.33
Na_2O	3.6	3.5	3.3	3.3	3.7	3.9	3.9		3.9		3.1				4.4	4.8	4.3
K_2O	2.5	2.5	1.9	2.2	2.3	2.2	2.2	2.5	2.4	1.5	2.7	2.7	2.2	5.4	4.9	4.4	5.1
P_2O_5	30.24	0.22	0.26	0.23	0.23	0.33	0.31	0.37	0.34	0.40	0.14	0.18	0.27	0.03	0.02	0.02	0.03
LOI	0.74	0.88	1.53	0.88	1.06	0.66	0.36		0.36		1.06				0.46	0.42	0.52
Total	99.6	100.5	100.6	99.7	100.3	100.3	100.0		100.7		100.1				99.8	100.5	100.7
Trace elements (ppm)																	
	ICP-MS	ICP-MS	ICP-MS	ICP-MS	ICP-MS	ICP-MS	ICP-MS	ICP-MS	ICP-MS	ICP-MS	ICP-MS	ICP-MS	ICP-MS	ICP-MS	ICP-MS	ICP-MS	ICP-MS
Sc	13.7	15.1	14.4	14.6	13.9	16.8	16.8	16.4	17.0	17.5	11.7	11.7	13.6	1.4	3.9	4.2	3.8
V	106	122	116	111	106	137	139	144	144	146	87.8	93.4	96.7	3.88	4.48	0.59	3.55
Cr	44.1	61.8	40.9	36.9	44.4	57.0	63.2	43.8	43.5	36.9	46.3	41.6	46.0	1.95	3.38	2.87	1.66
Co	15.8	18.8	17.7	16.0	15.9	21.3	21.4	20.8	20.5	21.3	13.5	12.8	16.9	0.88	0.65	0.22	0.62

(Continued)

Table 1. *Major and trace elements of the studied rock suites* (*Continued*)

Suite	Tequila	Tequila	Tequila	Tequila	Tequila	Tequila	Tequila	Tequila	Tequila	Tequila	Tequila	Tequila	Tequila	Per-Teq	Per-Teq	Per-Teq	Per-Teq
Sample	TQ-10-02A	TQ-10-03	TQ-10-04	TQ-10-06	TQ-10-07	TQ-10-09	TQ-10-11	TQ-10-12	TQ-10-14	RTZ-11-46	RTZ-11-47	RTZ-11-49	RTZ-11-50	TZ-09-09	TZ-09-21A	TQ-10-08	TZ-09-04
Long. W	−103.703	−103.757	−103.777	−103.745	−103.754	−103.981	−103.972	−103.950	−103.968	−103.705	−103.713	−103.740	−103.742	−103.731	−103.916	−103.982	−103.861
Lat. N	20.748	20.789	20.777	20.765	20.762	20.835	20.816	20.826	20.807	20.720	20.730	20.678	20.690	20.901	20.903	20.878	20.855
	ICP-MS	ICP-MS	ICP-MS	ICP-MS	ICP-MS	ICP-MS	ICP-MS	ICP-MS	ICP-MS	ICP-MS	ICP-MS	ICP-MS	ICP-MS	ICP-MS	ICP-MS	ICP-MS	ICP-MS
Ni	30.0	42.3	33.6	27.6	29.9	45.0	48.2	38.1	36.8	21.8	28.9	25.7	36.8	1.56	1.54	1.54	0.38
Cu	29	32	29	13	29	37	34	27	29	40	23	16	30	7.0	5.1	3.5	3.4
Zn	66	63	67	67	65	74	73	74	74	76	53	56	65	131	59	122	52
Ga	19.3	19.4	20.7	19.3	19.7	21.1	19.8	20.4	20.1	19.3	17.0	17.5	19.3	29.6	21.2	28.2	22.4
Li	14.0	13.0	11.7	13.4	13.1	11.5	10.6	12.2	11.1	9.2	14.8	14.0	10.6	27.5	35.3	43.1	11.8
Be	1.70	1.61	1.50	1.82	1.58	1.99	1.76	1.82	1.79	1.59	1.36	1.86	1.49	9.95	6.09	7.34	4.90
Rb	44	44	33	41	43	39	38	36	37	22	58	58	32	202	152	129	137
Sr	602	515	684	630	590	535	584	595	599	629	436	458	837	12.3	9.07	1.63	21.7
Y	18.6	20.8	22.3	24.1	19.2	25.2	21.5	23.0	22.0	42.1	16.6	20.2	18.0	95.7	44.3	74.4	68.0
Zr	203	194	184	192	206	249	245	250	251	201	147	193	203	618	252	756	286
Nb	10.8	9.76	10.0	10.2	10.9	16.4	15.5	16.5	15.9	15.7	6.73	8.32	10.6	83.8	46.5	71.3	44.7
Mo	2.12	1.85	2.09	1.96	2.18	2.11	2.12	2.69	2.16	1.74	1.61	2.01	1.70	2.85	6.49	9.34	2.51
Sn	0.80	0.87	0.79	0.75	0.85	1.02	0.91	0.97	0.92	1.02	0.96	0.96	0.97	7.9	4.3	5.0	4.0
Sb	0.07	0.10	0.06	0.06	0.07	0.08	0.09	0.09	0.09	0.09	0.16	0.15	0.11	0.88	0.92	0.78	0.69
Cs	0.75	0.75	0.60	0.73	0.75	0.54	0.63	0.58	0.60	0.33	1.32	1.17	0.56	5.22	3.79	3.71	1.91
Ba	764	774	643	746	747	830	790	820	802	781	867	854	648	30.2	69.9	1.23	377
La	23.7	23.6	25.6	29.9	23.7	32.5	26.1	28.1	26.7	29.4	21.6	29.1	23.8	53.0	51.9	65.1	85.8
Ce	47.7	43.5	44.0	46.1	47.2	54.1	52.4	56.1	53.5	46.2	41.1	45.2	48.3	108.2	102.2	127.9	84.7
Pr	5.84	5.37	5.93	7.20	5.79	7.24	6.46	6.89	6.65	6.84	4.77	6.29	6.03	15.2	11.5	15.6	18.8
Nd	20.0	21.1	23.1	24.0	20.8	28.1	23.6	25.5	24.0	27.7	17.9	23.0	23.0	56.9	42.1	63.7	65.9
Sm	4.05	4.35	4.64	4.79	4.22	5.64	4.80	5.22	4.88	5.90	3.51	4.49	4.51	14.4	8.06	13.4	12.5
Eu	1.27	1.19	1.33	1.31	1.26	1.56	1.48	1.57	1.52	1.68	0.957	1.12	1.34	0.27	0.195	0.164	0.694
Gd	3.65	4.01	4.28	4.46	3.77	5.10	4.32	4.69	4.40	5.98	3.20	4.07	3.93	14.0	7.32	12.5	10.8
Tb	0.590	0.587	0.615	0.745	0.583	0.728	0.671	0.717	0.685	0.892	0.487	0.617	0.574	2.25	1.22	1.86	1.73
Dy	3.37	3.39	3.51	4.31	3.34	4.11	3.84	4.11	3.90	5.35	2.83	3.58	3.24	16.1	7.50	11.5	10.0
Ho	0.67	0.69	0.71	0.86	0.67	0.81	0.75	0.81	0.77	1.10	0.57	0.71	0.63	2.94	1.53	2.35	2.06
Er	1.87	1.89	1.95	2.38	1.85	2.25	2.08	2.22	2.14	3.02	1.60	1.97	1.74	8.53	4.55	6.85	5.86
Yb	1.85	1.82	1.85	2.27	1.82	2.10	2.00	2.10	2.03	2.49	1.61	2.53	1.66	9.23	4.75	6.90	5.68
Lu	0.275	0.270	0.276	0.331	0.273	0.317	0.300	0.321	0.305	0.371	0.248	0.294	0.252	1.215	0.721	1.053	0.817
Hf	4.51	4.22	3.97	4.34	4.48	5.20	5.08	5.34	5.19	4.13	3.64	4.52	4.39	17.30	8.40	17.17	7.49
Ta	0.73	0.65	0.63	0.68	0.72	1.00	0.98	1.04	0.99	0.93	0.55	0.65	0.69	5.88	3.14	4.13	2.88
Tl	0.142	0.142	0.134	0.176	0.129	0.108	0.102	0.107	0.093	0.143	0.269	0.297	0.145	1.036	0.654	0.550	0.585
Pb	9.0	9.1	7.8	8.5	9.0	9.0	8.5	8.7	8.5	5.7	10.3	10.2	7.8	31.7	20.3	24.3	18.5
Th	4.05	4.07	3.32	3.81	4.00	3.56	3.37	3.51	3.34	2.39	5.47	5.18	3.27	20.11	15.41	16.85	13.79
U	1.19	1.16	0.957	1.13	1.17	1.18	1.11	1.12	1.12	0.775	1.53	1.48	1.03	6.67	5.45	6.10	4.50

Suite	Per-Teq	Per-Teq	Per-Teq	Per-Teq	Per-Teq	Per-Teq	Per-Teq	Per-Teq	HighTi-Teq	HighTi-Teq	HighTi-Teq	HighTi-Teq	Per-Cebo	Per-Cebo	Per-Cebo	SMOcc
Sample	TZ-09-25	TZ-09-31D	TZ-09-43	TZ-09-32	TQ-10-15	TQ-10-16	TQ-10-17	TQ-10-18	TZ-09-07	TZ-09-08	RTZ-11-35	TZ-09-11	TZ-10-03	RTZ-11-04	RTZ-11-05B	TZ-09-13
Long. W	−103.799	−103.947	−104.022	−103.940	−104.010	−104.006	−104.002	−103.931	−103.729	−103.731	−103.983	−103.719	−104.572	−104.559	−104.554	−103.719
Lat. N	20.704	20.739	20.803	20.742	20.784	20.774	20.760	20.748	20.901	20.905	21.028	20.906	21.211	21.213	21.203	20.912
Major elements (wt%)																
	XRF	XRF	XRF	XRF	XRF	XRF	XRF	XRF	XRF	XRF	XRF	XRF	XRF	XRF	XRF	XRF
SiO_2	58.6	59.7	63.3	63.6	69.0	59.2	64.8	65.0	47.6	49.2	48.8	51.3	52.8	51.4	52.4	75.6
TiO_2	1.24	1.41	1.23	1.05	0.47	1.28	1.07	1.01	2.46	2.83	1.95	2.09	1.22	1.52	1.22	0.31
Al_2O_3	16.4	16.1	16.1	15.7	15.3	16.8	15.6	16.0	15.9	15.4	17.2	16.3	18.6	18.5	19.4	12.7
$Fe_2O_3^{tot}$	7.0	7.2	6.1	5.4	3.0	7.3	5.4	5.1	12.6	12.5	12.0	11.1	8.2	9.2	8.2	1.7
MnO	0.112	0.116	0.108	0.098	0.078	0.112	0.094	0.097	0.191	0.161	0.177	0.163	0.125	0.146	0.146	0.046
MgO	3.1	2.5	2.0	1.6	0.57	3.1	1.8	1.5	5.9	4.3	5.8	3.9	4.8	5.3	4.5	0.40
CaO	5.8	4.9	4.2	3.6	1.4	6.0	3.7	3.4	8.8	6.9	9.1	7.0	7.6	7.9	7.8	1.1
Na_2O	4.2	4.6	4.4	4.8	4.7	4.1	4.2	4.7	3.5	3.8	2.9	4.0	3.9	3.0	3.4	3.4
K_2O	2.2	2.5	2.9	3.1	4.8	2.1	3.1	3.1	1.1	1.7	0.8	2.0	1.3	1.3	1.2	4.4
P_2O_5	30.40	0.52	0.44	0.37	0.11	0.42	0.35	0.34	0.52	0.63	0.29	0.55	0.68	0.38	0.33	0.08
LOI	0.43	0.02	−0.01	0.24	1.07	0.16	0.10	0.00	1.23	2.48	0.08	0.98	-0.01	0.80	0.33	0.75
Total	99.5	99.4	100.7	99.5	100.6	100.5	100.4	100.3	99.9	99.8	99.1	99.5	99.3	99.4	98.9	100.5
Trace elements (ppm)																
	ICP-MS	ICP-MS	ICP-MS	ICP-MS	ICP-MS	ICP-MS	ICP-MS	ICP-MS	ICP-MS	ICP-MS	ICP-MS	ICP-MS	ICP-MS	ICP-MS	ICP-MS	ICP-MS
Sc	17.8	15.7	14.2	11.0	11.4	17.4	13.6	12.9	32.0	24.3	28.5	20.4	19.3	22.7	18.2	5.78
V	140	133	104	73.4	4.94	157	92.9	62.6	277	279	257	229	183	181	179	29.5
Cr	27.3	2.33	8.90	2.11	1.25	18.9	3.79	1.07	36.4	8.73	36.9	1.80	50.6	93.7	29.8	2.39
Co	17.5	14.4	11.1	8.21	1.12	20.3	10.5	7.42	40.7	33.3	42.4	28.8	26.2	31.1	25.5	2.35
Ni	18.3	3.05	4.32	1.43	0.18	21.6	4.21	0.14	39.2	19.3	52.0	11.0	38.3	56.3	32.7	1.31
Cu	34	22	18	12	3.2	39	17	8.9	49	41	51	45	39	41	39	5.8
Zn	78	80	72	74	66	76	67	68	101	122	103	119	80	86	86	33
Ga	20.0	20.4	19.9	19.7	20.9	21.2	21.0	21.3	21.1	22.9	21.9	22.1	22.3	20.7	21.9	14.7
Li	13.9	15.8	19.1	18.8	17.6	12.4	22.0	21.7	7.3	10.6	6.4	12.4	8.5	8.1	9.0	19.8
Be	1.83	2.05	2.45	2.49	3.58	1.83	2.54	2.46	1.66	2.11	1.33	1.97	1.36	1.41	1.41	2.13
Rb	42	46	61	60	85	39	64	60	13	24	11	24	15	15	15	126
Sr	512	500	401	400	167	512	358	382	495	548	532	550	1038	852	1198	138
Y	25.5	27.8	27.9	27.6	41.4	25.7	27.9	29.8	33.8	35.5	41.7	31.3	18.9	22.4	19.2	24.0
Zr	211	242	253	258	528	205	242	261	216	266	202	213	142	164	140	81.1
Nb	16.3	19.8	21.7	19.6	30.7	15.5	21.0	21.3	28.7	27.0	18.1	19.1	9.19	15.1	9.64	8.80
Mo	2.20	2.71	3.44	3.14	4.21	2.02	3.58	3.17	1.71	2.24	1.24	1.79	0.94	0.99	1.22	1.86
Sn	1.4	1.6	1.8	1.7	2.1	1.1	1.5	1.5	1.7	1.8	1.2	1.5	0.93	1.08	0.92	1.08
Sb	0.17	0.19	0.26	0.23	0.42	0.09	0.18	0.13	0.07	0.12	0.05	0.16	0.07	0.14	0.10	1.05
Cs	0.85	0.91	1.41	1.31	1.96	0.82	1.60	1.35	0.05	0.15	0.06	0.38	0.23	0.30	0.22	2.44
Ba	739	830	866	916	1220	741	842	928	371	676	286	765	550	523	585	1059

(Continued)

Table 1. *Major and trace elements of the studied rock suites* (*Continued*)

Suite	Per-Teq	Per-Teq	Per-Teq	Per-Teq	Per-Teq	Per-Teq	Per-Teq	Per-Teq	HighTi-Teq	HighTi-Teq	HighTi-Teq	HighTi-Teq	Per-Cebo	Per-Cebo	Per-Cebo	SMOcc
Sample	TZ-09-25	TZ-09-31D	TZ-09-43	TZ-09-32	TQ-10-15	TQ-10-16	TQ-10-17	TQ-10-18	TZ-09-07	TZ-09-08	RTZ-11-35	TZ-09-11	TPZ-10-03	RTZ-11-04	RTZ-11-05B	TZ-09-13
Long. W	−103.799	−103.947	−104.022	−103.940	−104.010	−104.006	−104.002	−103.931	−103.729	−103.731	−103.983	−103.719	−104.572	−104.559	−104.554	−103.719
Lat. N	20.704	20.739	20.803	20.742	20.784	20.774	20.760	20.748	20.901	20.905	21.028	20.906	21.211	21.213	21.203	20.912
	ICP-MS	ICP-MS	ICP-MS	ICP-MS	ICP-MS	ICP-MS	ICP-MS	ICP-MS	ICP-MS	ICP-MS	ICP-MS	ICP-MS	ICP-MS	ICP-MS	ICP-MS	ICP-MS
La	26.1	29.3	32.0	31.9	47.8	25.7	31.4	32.9	24.8	28.9	20.3	25.1	18.8	20.8	20.4	25.6
Ce	52.9	61.1	64.6	63.9	81.3	51.5	61.4	64.5	53.7	63.0	34.0	54.7	40.9	45.4	44.2	50.2
Pr	6.88	7.76	8.03	7.97	11.4	6.58	7.60	7.96	6.89	8.33	5.99	7.20	5.49	5.95	5.94	6.18
Nd	27.4	31.1	31.0	30.6	41.7	27.1	29.4	31.6	28.9	34.3	26.4	30.1	22.6	25.0	24.3	22.4
Sm	5.86	6.52	6.43	6.35	8.18	5.83	6.06	6.58	6.62	7.80	6.64	6.78	4.73	5.34	4.98	4.53
Eu	1.67	1.84	1.73	1.70	1.77	1.62	1.56	1.69	2.16	2.37	2.15	2.06	1.56	1.69	1.59	0.831
Gd	5.37	5.94	5.85	5.71	7.40	5.33	5.49	5.88	6.64	7.41	7.06	6.52	4.41	4.92	4.35	4.04
Tb	0.804	0.879	0.864	0.851	1.14	0.775	0.806	0.853	0.997	1.110	1.078	0.979	0.623	0.728	0.628	0.642
Dy	4.56	5.07	5.01	4.91	6.69	4.37	4.64	4.93	5.88	6.47	6.57	5.71	3.47	4.16	3.53	3.94
Ho	0.90	0.99	0.98	0.97	1.34	0.86	0.92	0.97	1.15	1.26	1.28	1.12	0.69	0.81	0.68	0.81
Er	2.45	2.71	2.71	2.67	3.84	2.34	2.55	2.70	3.08	3.43	3.42	3.09	1.85	2.20	1.84	2.35
Yb	2.27	2.51	2.59	2.54	3.82	2.15	2.47	2.59	2.73	3.15	3.04	2.83	1.67	1.96	1.68	2.45
Lu	0.340	0.372	0.382	0.381	0.587	0.317	0.364	0.385	0.404	0.457	0.440	0.421	0.249	0.293	0.253	0.364
Hf	4.77	5.37	5.73	5.92	11.17	4.53	5.58	6.07	4.61	5.63	3.52	4.85	3.51	3.76	3.35	2.73
Ta	1.11	1.32	1.53	1.38	1.85	0.97	1.43	1.42	1.76	1.65	1.13	1.21	0.60	0.94	0.57	0.71
Tl	0.163	0.170	0.225	0.219	0.341	0.111	0.189	0.190	0.025	0.066	0.025	0.074	0.070	0.043	0.073	0.452
Pb	9.3	9.2	10.8	11.0	15.3	8.3	11.3	11.3	2.9	6.3	2.0	8.4	6.0	5.2	6.3	17.3
Th	4.35	4.71	6.57	6.21	8.38	4.03	7.42	6.65	1.70	2.31	1.32	1.86	1.76	1.89	1.85	10.42
U	1.36	1.59	2.22	2.04	2.99	1.26	2.40	2.11	0.568	0.838	0.421	0.676	0.582	0.633	0.600	3.12

Suite	SMOcc	SMOcc	SMOcc	SMOcc	SMOcc	SMOcc	SMOcc	SMOcc	BJ	BJ	BJ	BJ	BJ	BJ	BJ	BJ
Sample	TZ-09-38	RTZ-11-38	TZ-09-40	TPZ-10-02	TPZ-10-01	TPZ-10-16	RTZ-11-16	RTZ-11-14B	RTZ-11-30	RTZ-11-31	RTZ-11-32	JAL-05-04	JAL-05-14	JAL-05-32	TEC-06-08	JAL-07-19
Long. W	−104.074	−103.935	−104.216	−104.198	−104.172	−104.622	−104.499	−104.553	−104.893	−104.818	−104.831	−104.465	−104.514	104.933	−104.110	−103.759
Lat. N	20.747	21.037	20.788	21.017	21.009	21.374	21.465	21.451	21.145	21.110	21.132	20.815	20.637	20.554	20.208	19.907
Major elements (wt%)																
	XRF	ICP-MS	XRF	XRF	XRF	XRF	ICP-MS	XRF	XRF	ICP-MS	ICP-MS	XRF	XRF	XRF	XRF	XRF
SiO_2	77.2		77.8	76.6	63.6	74.6		71.9	77.6			68.3	59.0	70.9	61.1	52.4
TiO_2	0.14	0.42	0.14	0.17	0.76	0.34	0.14	0.42	0.10	0.11	0.52	0.46	0.75	0.32	0.69	0.81
Al_2O_3	13.3		12.1	11.4	16.9	13.0		13.4	13.1			16.4	17.3	14.9	15.9	20.0
$Fe_2O_3^{tot}$	0.3		1.3	1.6	4.7	2.3		2.2	0.6			3.0	6.8	2.9	6.6	7.5
MnO	0.013	0.052	0.050	0.047	0.106	0.028	0.045	0.051	0.015	0.009	0.076	0.005	0.136	0.073	0.101	0.107
MgO	0.76	0.45	0.32	0.38	1.8	0.66	0.20	0.69	0.40	0.35	2.7	0.91	3.2	0.73	3.1	2.1
CaO	0.1	1.7	0.17	0.28	3.6	0.19	0.66	1.6	0.10	0.28	5.3	3.2	5.2	2.1	5.8	7.0
Na_2O	1.7		2.5	2.3	4.5	0.86		2.2	1.6			4.1	3.4	4.5	2.8	4.8
K_2O	5.0	4.0	4.5	5.9	3.2	5.5	4.5	4.8	4.8	3.7	1.8	3.2	2.7	3.1	2.2	1.6
P_2O_5	0.02	0.12	0.03	0.15	0.26	0.20	0.05	0.06	0.00	0.03	0.14	0.15	0.17	0.11	0.15	0.17
LOI	1.77		1.62	0.73	0.66	2.51		2.90	1.43			0.57	1.84	0.78	1.39	2.81
Total	100.3		100.5	99.6	100.0	100.2		100.3	99.9			100.3	100.5	100.4	99.9	99.5

Trace elements (ppm)																
	ICP-MS	ICP-MS	ICP-MS	ICP-MS	ICP-MS	ICP-MS	ICP-MS	ICP-MS	ICP-MS	ICP-MS	ICP-MS	ICP-MS	ICP-MS	ICP-MS	ICP-MS	ICP-MS
Sc	2.69	8.32	7.78	3.29	10.5	4.08	4.23	6.23	4.08	2.26	10.9	5.44	18.5	4.66	21.7	24.5
V	10.9	37.2	55.7	8.87	75.0	23.7	8.65	16.8	3.35	8.94	88.7	35.7	143	25.8	161	242
Cr	2.17	2.59	18.9	2.25	7.65	5.32	2.16	3.48	3.32	2.25	24.7	3.57	6.24	2.64	23.7	11.5
Co	0.53	1.85	7.62	1.18	8.04	2.44	1.25	1.18	0.65	1.22	13.0	4.10	17.2	3.00	16.6	17.0
Ni	0.60	1.67	14.4	1.17	3.63	2.15	1.30	1.28	1.01	0.90	19.6	1.50	6.47	1.13	8.49	7.88
Cu	3.1	4.0	14	6.1	14	5.5	2.0	4.4	4.1	1.6	14	2.9	40	2.1	27	116
Zn	25	67	59	45	158	43	38	62	11	45	53	72	79	52	62	68
Ga	17.9	18.5	19.5	13.1	20.1	15.9	16.1	16.0	17.5	14.3	17.8	21.6	17.9	18.1	16.7	19.4
Li	9.0	19.3	12.8	16.6	15.2	58.0	23.8	28.7	8.5	10.9	9.5	16.5	29.8	31.9	15.3	16.3
Be	2.16	2.58	2.35	2.31	2.28	2.77	3.14	2.43	1.41	1.89	1.26	2.09	1.27	2.25	1.04	0.81
Rb	141	111	95	143	93	187	158	118	117	143	28	84	114	113	64	35
Sr	86.9	254	273	140	395	86.2	69.9	126	95.8	84.6	666	494	414	291	372	629
Y	23.6	30.2	26.0	26.0	23.5	24.6	20.3	36.7	19.4	19.1	12.8	12.4	24.3	20.9	19.9	22.6
Zr	76.7	92.1	123	121	14.9	101	5.52	116	139	85.4	89	4.35	35.6	47.3	14.4	70.0
Nb	16.1	15.3	13.9	15.8	12.5	12.6	13.1	17.0	8.55	8.64	3.87	8.88	6.73	8.78	4.97	3.33
Mo	0.27	0.92	1.21	0.28	2.24	0.59	1.28	0.80	0.25	0.70	1.40					1.23
Sn	1.3	0.92	1.6	1.9	1.3	1.8	1.8	0.86	1.7	2.5	0.54	1.6	1.2	2.4	0.94	0.71
Sb	0.63	1.08	0.41	1.09	0.85	14.60	0.85	1.48	1.00	0.75	0.20	0.07	4.17	1.08	0.44	0.42
Cs	1.42	4.41	1.85	1.55	1.75	11.0	2.89	6.04	3.16	4.11	0.66	1.47	4.14	3.74	3.01	0.30
Ba	811	1365	604	1190	1206	1776	626	1265	838	858	980	1165	730	987	759	621
La	27.5	38.8	29.9	26.3	38.7	38.3	28.2	43.4	28.1	27.9	14.5	30.4	17.2	26.4	13.7	9.23
Ce	58.8	82.9	57.3	55.1	76.5	60.9	54.6	85.2	55.7	52.3	25.9	60.5	38.2	50.9	28.2	20.2
Pr	7.19	10.1	7.08	6.58	9.26	8.61	6.37	9.73	6.22	6.05	3.35	7.30	4.96	6.06	3.62	2.77
Nd	23.8	38.4	25.5	23.5	33.5	33.3	22.1	36.4	20.7	20.7	13.3	26.8	19.9	21.8	14.6	12.4
Sm	4.10	7.70	5.14	4.87	6.22	6.78	4.19	7.31	4.19	3.82	2.71	4.87	4.52	4.11	3.34	3.19
Eu	0.470	1.89	0.853	0.577	1.55	1.64	0.475	1.52	0.575	0.501	0.873	1.18	1.01	0.894	0.980	0.943
Gd	3.40	6.53	4.52	4.52	5.25	5.60	3.48	6.26	3.09	3.15	2.51	3.63	4.36	3.63	3.41	3.44
Tb	0.549	0.975	0.702	0.678	0.754	0.846	0.560	0.952	0.410	0.502	0.377	0.494	0.690	0.567	0.546	0.546
Dy	3.45	5.57	4.03	4.22	4.25	4.73	3.29	5.53	2.31	2.94	2.16	2.38	4.17	3.35	3.35	3.42
Ho	0.74	1.08	0.82	0.86	0.83	0.90	0.67	1.07	0.52	0.60	0.44	0.42	0.84	0.69	0.70	0.73
Er	2.32	2.96	2.32	2.56	2.27	2.54	1.96	2.94	1.69	1.75	1.20	1.07	2.36	1.97	1.95	2.05
Yb	2.71	2.83	2.34	2.92	2.09	2.59	2.10	2.86	2.27	1.93	1.17	.89	2.32	2.00	1.90	2.01
Lu	0.415	0.404	0.339	0.426	0.290	0.370	0.309	0.391	0.358	0.296	0.180	0.123	0.343	0.301	0.288	0.298
Hf	3.21	3.10	3.58	4.40	0.51	2.65	0.28	2.71	3.45	2.77	2.42	0.11	1.16	1.68	0.66	2.02
Ta	1.15	0.96	0.92	1.18	0.84	0.94	1.02	1.06	0.75	0.89	0.32	0.67	0.55	0.73	0.35	0.22
Tl	0.807	0.345	0.190	1.155	0.539	1.398	0.952	0.639	0.745	0.883	0.174	0.369	0.480	0.533	0.283	0.032
Pb	2.8	18.4	9.2	11.9	25.0	17.5	16.4	19.4	5.3	14.7	10.0	16.6	18.3	15.8	9.4	8.1
Th	9.91	9.61	7.89	9.76	3.71	11.11	9.67	11.30	15.01	15.81	2.11	7.15	11.52	10.81	4.15	2.33
U	3.40	2.42	1.73	2.91	2.44	3.18	2.86	2.22	3.63	4.54	0.704	1.41	3.59	3.87	1.37	0.837

[a]Rock suite notation: Sangangüey volcano (Sanga), peripheral monogenetic volcanism to Sangangüey volcano (Per-Sanga), high TiO_2 volcanism from Sangangüey area (HiTi-Sanga), Tequila volcano (Tequila), peripheral monogenetic volcanism to Tequila (Per-Teq), high-TiO_2 volcanism from Tequila area (HiTi-Tequila), peripheral volcanism to Ceboruco volcano (Per-Cebo), Sierra Madre Occidental (SMOcc), Bloque Jalisco (BJ).

Table 2. *Sr, Nd and Pb isotopic compositions for the studied rock suites*

Sample	$^{87}Sr/^{86}Sr$	2σ	$^{206}Pb/^{204}Pb$	2σ	$^{207}Pb/^{204}Pb$	2σ	$^{208}Pb/^{204}Pb$	2σ	$^{143}Nd/^{144}Nd$	2σ
Sangangüey volcano										
TPZ-10-04	0.704055	10	18.7154	6	15.6016	6	38.5115	16	0.512781	10
SAN-10-10	0.704122	13	18.7265	7	15.6064	7	38.5375	17	0.512759	9
SAN-10-16	0.704113	10	18.7311	7	15.6047	6	38.5329	15	0.512765	11
SAN-10-17	0.704127	14	18.7331	7	15.6063	6	38.5394	16	0.512759	10
SAN-10-18	0.704131	12	18.7323	6	15.6063	5	38.5393	14	0.512767	9
SAN-10-19	0.704135	14	18.7305	7	15.6052	6	38.5362	16	0.512753	17
Peripheral volcanism to Sangangüey volcano										
TPZ-10-26	0.703962	8	18.6926	7	15.5824	6	38.4062	18	0.512852	7
SAN-10-12	0.704035	8	18.6961	6	15.5896	5	38.4493	15	0.512819	5
High-TiO_2 Basalts (Sangangüey area)										
TPZ-10-23	0.702977	8	18.9053	7	15.5793	6	38.4718	16	0.512980	7
TPZ-10-14	0.703908	6	18.9216	5	15.6070	5	38.5693	15	0.512867	6
SAN-10-01A	0.703736	6	19.0680	10	15.5980	9	38.6793	23	0.512900	7
SAN-10-06	0.702917	11	18.9251	11	15.5728	9	38.4843	25	0.512972	7
SAN-10-09	0.702892	9	18.8891	8	15.5691	7	38.4345	19	0.512993	7
RTZ-11-22	0.702894	8	18.7517	7	15.5488	7	38.2695	18	0.513028	8
RTZ-11-12	0.703805	8	18.9352	8	15.5989	7	38.5390	18	0.512882	6
Tequila volcano										
TZ-09-03	0.704101	6	18.7148	8	15.5995	7	38.5091	20	0.512788	9
TZ-09-05	0.703956	6	18.7101	6	15.5991	6	38.5102	15	0.512784	8
TZ-09-20	0.703966	8	18.7096	6	15.5990	6	38.5123	15	0.512791	10
TZ-09-27	0.703415	7	18.7030	7	15.5925	7	38.4577	19	0.512880	7
TZ-09-19	0.703977	8	18.7121	6	15.6011	6	38.5155	15	0.512783	7
TQ-10-02A	0.703765	7	18.7072	7	15.5911	7	38.4682	16	0.512836	7
TQ-10-03	0.703968	10	18.7187	5	15.5974	5	38.5042	14	0.512795	7
TQ-10-04	0.703632	8	18.7019	6	15.5910	5	38.4631	14	0.512859	8
TQ-10-06	0.703708	8	18.7078	7	15.5911	7	38.4646	18	0.512835	7
TQ-10-09	0.703793	7	18.6958	6	15.5842	6	38.4325	16		
TQ-10-14			18.6975	6	15.5864	5	38.4405	15	0.512874	11
Peripheral volcanism to Tequila volcano										
TZ-09-21A			18.7300	6	15.5875	5	38.4521	14	0.512900	6
TZ-09-04			18.7332	6	15.5915	6	38.4674	15	0.512888	7
TZ-09-25	0.703860	6	18.7110	7	15.5936	6	38.4751	15	0.512859	7
TZ-09-43	0.703801	10	18.7101	6	15.5901	5	38.4556	15	0.512876	8
TZ-09-32	0.703873	8	18.7130	6	15.5939	5	38.4752	14	0.512863	7
TQ-10-16	0.703897	7	18.7116	7	15.5936	6	38.4781	16	0.512861	8
TQ-10-17	0.703853	10	18.7130	7	15.5912	6	38.4630	16	0.512865	10
High-TiO_2 Basalts (Tequila area)										
TZ-09-07	0.703392	6	18.7716	10	15.5775	9	38.3898	22	0.512961	7
TZ-09-08	0.703808	7	18.7180	7	15.5889	6	38.4431	18	0.512892	7
RTZ-11-35	0.703690	9								
TZ-09-11	0.703947	6	18.7096	5	15.5928	5	38.4623	14	0.512872	7
Peripheral volcanism to Ceboruco volcano										
RTZ-11-04	0.703653	7	18.6941	6	15.5859	5	38.4209	14	0.512890	7
RTZ-11-05B	0.703717	10	18.6710	6	15.5827	6	38.3998	16	0.512874	6
Sierra Madre Occidental										
TZ-09-13	0.710306	10	18.7320	4	15.5951	4	38.4904	12	0.512806	12
RTZ-11-38	0.704997	9	18.8042	4	15.6193	4	38.6439	12	0.512700	7
TZ-09-40	0.710836	20	18.8390	5	15.6152	4	38.6339	12		
TPZ-10-01	0.704818	10	18.7905	5	15.6174	5	38.6207	13	0.512707	7
RTZ-11-14B	0.705708	10	18.8168	6	15.6175	5	38.6463	15	0.512773	30

(*Continued*)

Table 2. *Continued*

Sample	$^{87}Sr/^{86}Sr$	2σ	$^{206}Pb/^{204}Pb$	2σ	$^{207}Pb/^{204}Pb$	2σ	$^{208}Pb/^{204}Pb$	2σ	$^{143}Nd/^{144}Nd$	2σ
Jalisco Block										
RTZ-11-30	0.708179	22	19.0984	7	15.6360	7	38.9190	15		
JAL-05-14	0.704885	9	18.8266	6	15.6204	5	38.6584	13	0.512698	6
TEC-06-08	0.704952	12	18.5484	4	15.5900	4	38.3721	11		

Reported values are not age-corrected. 2σ errors for individual measurements are multiplied by 10^6 for Sr and Nd, and by 10^4 for Pb isotopic ratios.

elements (REE) during chromatographic purification, its very small contribution was corrected by measuring the intensity of the non-interfered isotope ^{147}Sm. The calculated $^{143}Nd/^{144}Nd$ ratios were exponentially normalized for mass bias to $^{146}Nd/^{144}Nd = 0.72190$, and further corrected to a JNdi standard value of $^{143}Nd/^{144}Nd = 0.512115$ (Tanaka *et al.* 2000), repeatedly measured during the same analytical session. Over the course of this study, the measured $^{143}Nd/^{144}Nd$ ratios of the JNdi standard were 0.512089 ± 0.000011 (2σ, $n = 69$). Using these corrections, the $^{143}Nd/^{144}Nd$ ratios for the La Jolla standard as an external monitor were 0.511860 ± 0.000009 (2σ, $n = 25$).

Pb isotopic compositions were performed by static multicollection on masses 202 to 208. A NIST SRM 997 Tl solution with a reference $^{205}Tl/^{203}Tl$ value of 2.3871 was used to monitor instrumental mass fractionation, and ^{202}Hg was monitored and used to correct the isobaric interference of ^{204}Hg on ^{204}Pb using the Hg natural abundances. The measured beam intensities for ^{202}Hg were *c.* 0.05 mV making the interference correction on ^{204}Pb almost negligible. To improve reproducibility and accuracy, samples and standards were freshly prepared for each analytical session, in order to avoid possible photoxidation of the Tl solution (Kamenov *et al.* 2004), and analysed at the same Pb/Tl ratio of *c.* 4 to ensure that standards and samples are effectively matrix-matched. The Tl fractionation-corrected values in the samples were further adjusted to the NIST SRM-981 standard values of $^{206}Pb/^{204}Pb = 16.9356$, $^{207}Pb/^{204}Pb = 15.4891$, and $^{208}Pb/^{204}Pb = 36.7006$ (Todt *et al.* 1996). Over the course of this study, the Tl fractionation-corrected Pb isotopic compositions of the NIST SRM-981 standard were $^{206}Pb/^{204}Pb = 16.9308$, $^{207}Pb/^{204}Pb = 15.4841$, $^{208}Pb/^{204}Pb = 36.6745$ (2σ of 63, 67, 62 ppm, respectively, $n = 69$).

Results

Petrography

Studied rocks from Sangangüey and Tequila stratovolcanoes are porphyritic with a phenocryst assemblage dominated by plagioclase with minor amounts of orthopyroxene, clinopyroxene and, in some cases, amphibole. In these rocks, fine- (<1 mm) to medium-grained (up to *c.* 4 mm) plagioclase phenocrysts commonly exhibit disequilibrium features, such as complex zoning, sieve textures with subsequent overgrowths, or corroded shapes; other crystals show wavy extinction or wedge-shaped twinning, indicative of mechanical deformation. Pyroxene phenocrysts are fine- to medium-grained (up to *c.* 2 mm); orthopyroxene exhibits typical pleochroism in pinkish-pale green tones, whereas clinopyroxene often shows evidence of deformation, such as mechanical twinning or wavy extinction. Fine- to medium-grained (up to *c.* 4 mm) amphibole phenocrysts are pleochroic in light brown-reddish tones; they generally display opacitic rims due to devolatilization and some are completely oxidized. As commonly observed in many stratovolcanoes (Reubi & Blundy 2008), plagioclase and pyroxene phenocrysts are often grouped into glomeroporphyritic aggregates, or gabbroic clots, that occasionally include amphibole. The hypocrystalline groundmass is made of plagioclase laths that occasionally display hyalopilitic textures, pyroxene, and opaque minerals, with partially oxidized amphibole only seen in the most differentiated samples. No olivine has been found in the studied thin sections of Tequila and Sangangüey, although its presence has been reported in similar lavas from Cebrouco (Nelson 1980) and Tepetiltic (Petrone 2010).

Monogenetic cones peripheral to Sanganguey volcano are also porphyritic, with variable phenocryst associations in relation to their mafic or intermediate character: plagioclase, olivine and clinopyroxene in the most mafic rocks; and plagioclase, orthopyroxene and amphibole in more evolved ones. As noticed in previous studies (Frey & Lange 2011), monogenetic volcanism peripheral to Tequila stratovolcano is scarcely porphyritic, with a phenocryst assemblage made of plagioclase and minor clinopyroxene; with a few samples also containing minor amounts of orthopyroxene and amphibole. In all samples, fine- to medium-grained (up to *c.* 3.5 mm) plagioclase phenocrysts often show rounded or embayed shapes, sieve textures, as well as concentric or more complex zoning; other crystals exhibit wavy extinction or wedge-shaped twinning. Olivine phenocrysts have fine- to medium-grained dimensions (up to 3.5 mm); they generally display fresh surfaces, but some crystals show partial iddingsitization along their rims and fractures. Clinopyroxene and orthopyroxene phenocrysts are very scarce, and are usually grouped into glomeroporphyritic aggregates; the former have fine- to medium-grained dimensions (up to 1.5 mm), and occasionally show mechanical twinning or wavy extinction; the latter are mainly fine-grained, with typical pleochroism in pale green-pinkish tones. Fine-grained amphibole phenocrysts are pleochroic in light brown-reddish or greenish tones and generally display opacitic rims. In the most mafic samples, the phenocrysts are embedded in a hypocrystalline groundmass of plagioclase laths sometimes forming hyalopilitic textures, variably iddingsitized olivine crystals, clinopyroxene, and opaque minerals. The groundmass of amphibole-bearing rocks is composed of plagioclase laths, pyroxene and oxides. Dacitic to rhyolitic domes peripheral to Tequila have very low porphyricity indices, with fine- to medium-grained (up to 2 mm) phenocrysts of feldspar and plagioclase, embedded in a glassy cryptocrystalline groundmass made of quartz, feldspar and opaque minerals.

Intraplate-like mafic rocks generally display porphyritic textures with a phenocryst assemblage made of olivine and plagioclase, with a few samples also containing scarce clinopyroxene phenocrysts. Fine-grained olivine phenocrysts occasionally show partial iddingsitization along their rims and fractures; some crystals have rounded or embayed shapes. Plagioclase phenocrysts are fine-grained to very large (up to 10 mm); the larger crystals typically exhibit sieve textures, complex zoning, and corroded shapes, as well as wedge-shaped twinning or wavy extinction. In some cases, plagioclase phenocrysts form glomeroporphyritic aggregates with fine-grained phenocrysts and microphenocrysts of olivine. Clinopyroxene phenocrysts have medium-grained dimensions (*c.* 2 mm), and typically display rounded shapes or wavy extinction. The phenocrysts are embedded in a hypocrystalline groundmass of plagioclase occasionally showing hyalopilitic textures, variably iddingsitized olivine, clinopyroxene and opaque minerals.

Major and trace element compositions

The complex tectonic evolution of western Mexico has been instrumental in the construction of a very peculiar magmatic arc in which the chemical compositions of volcanic rocks differ strongly geographically (Fig. 2). Despite some scatter in the data, rocks at the volcanic front tend to be more mafic when MgO or Mg number (Mg#) are taken as index parameters for evolution (Fig. 2a). Rhyolites are abundant at the rear arc, and usually form a bimodal association with basalts, but they are notably absent at the volcanic front. Interestingly, and in contrast to what is observed in most arc volcanoes that increase their K_2O and incompatible trace element contents with distance from the trench (i.e. the K–h relationship; Dickinson 1975), rocks at the volcanic front in western Mexico are richer in K_2O and in most incompatible trace elements, and they usually display stronger subduction signatures than rocks at the rear arc (Lange & Carmichael 1991; Gómez-Tuena *et al.* 2011). These can be observed by comparing the ratios of large ion lithophile elements (LILE) to high field strength elements (HFSE) like Rb/Ta, but differences are better portrayed in the relative enrichment of the light rare earth elements (LREE) over the HFSE (Fig. 2b). The middle and heavy rare earth elements (MREE, HREE) can also be extremely fractionated in rocks from the volcanic front (Fig. 2c), suggesting important differences in the mechanisms of melt extraction for both regions.

Furthermore, if only the more mafic rocks are considered (MgO $> 4\%$), there is a clear transition from a high-K_2O volcanic front to a high-TiO_2 rear arc (Fig. 2d). Most mafic rocks at the volcanic front are potassium-alkaline (Fig. 3a) with lamprophyric mineral associations (minettes, hornblende-lamprophyres, absarokites and spessartites) and typical arc-related trace element features, such as prominent negative Nb–Ta anomalies and positive Pb spikes (Fig. 4a). In contrast, most mafic rocks at the rear arc are high-TiO_2 alkaline basalts, without hydrous minerals, that follow a near-tholeiitic enrichment in FeO during magmatic evolution (Fig. 3). Most of these rocks have much smoother trace element patterns, and can even display positive Nb–Ta anomalies (Fig. 4a), although more evolved rocks tend to display transitional features, lower TiO_2 contents, and incipient Nb–Ta troughs (see Petrone (2010) for an extended description of the compositional variability). Thus, in their most primitive form, these rocks have compositions that are very similar to those found in intraplate tectonic environments, and have been traditionally named as 'high-TiO_2(Nb) intraplate' or 'Ocean Island Basalt (OIB)-like' volcanic successions (Gómez-Tuena *et al.* 2007*b*).

At least five calc-alkaline andesitic stratovolcanoes with typical arc-related trace element patterns have been erupting at the rear arc for the past *c.* 200 ka (Figs 1–4). Their stratigraphic, mineralogical and chemical characteristics have been described in a number of publications (Nelson &

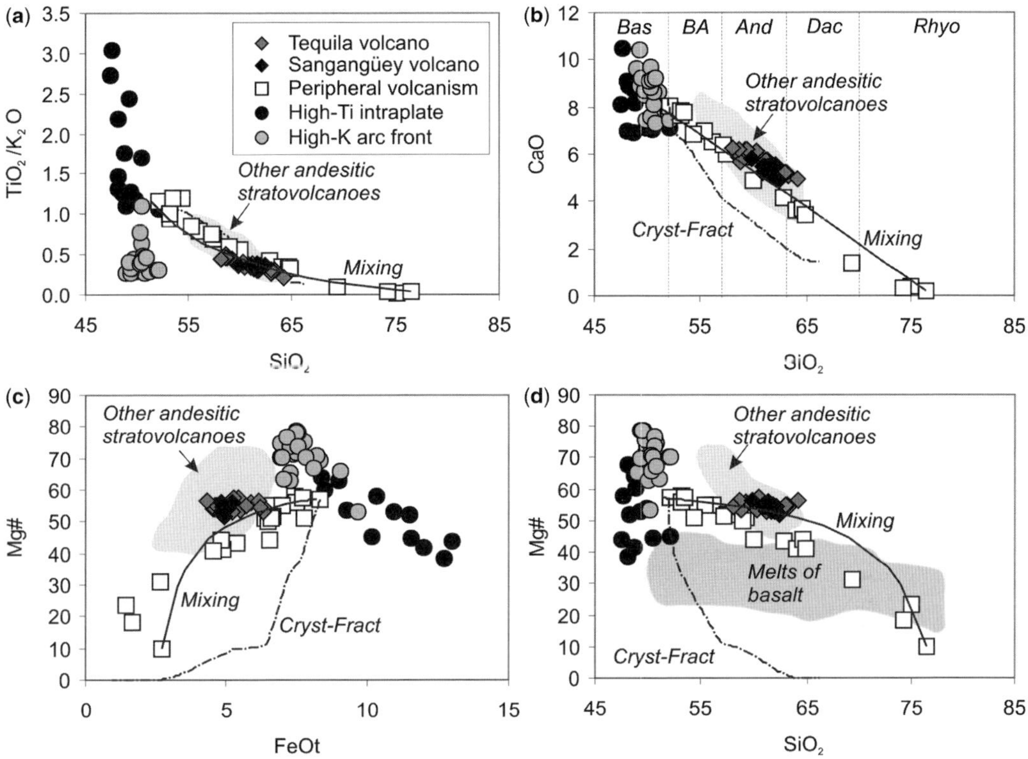

Fig. 3. Summary of major element compositions of the studied rock suites. Abundances of oxides are normalized to 100% volatile free. (**a**) SiO_2 v. TiO_2/K_2O, (**b**) SiO_2 v. CaO, showing classificatory compositional fields for basalts (Bas), basaltic andesites (BA), andesites (And), dacites (Dac) and rhyolites (Rhyol). (**c**) FeO* v. Mg#, (**d**) SiO_2 v. Mg#. Calculated molar Mg# = $Mg/(Mg + 0.85Fe^{2+}_{tot})$. Solid line is a simple mixing model between the most primitive calc-alkaline basaltic andesite in TZR (RTZ-11-04) with a very evolved rhyolite (TQ-10-08). Dash–dot line is an isobaric fractional crystallization MELTS model (Ghiorso & Sack 1995) of the same basaltic andesite at 1 GPa, starting at a liquid temperature of 1300 °C, with a fixed oxygen fugacity ($Fe_3/Fe_{total} = 0.1$). The rapid drop in Mg# at nearly constant SiO_2 is mainly caused by fractionation of olivine and pyroxene. Dark grey field shows the compositions of experimental melts of natural hydrous basalts at 1–4 GPa (Sen & Dunn 1994; Rapp 1995; Rapp & Watson 1995; Rapp *et al.* 1999). Light grey field shows the compositional variation of main edifice lavas from other stratovolcanoes in the region: Colima (Luhr & Carmichael 1990; Luhr 2002; Valdez-Moreno *et al.* 2006), Ceboruco and Tepetiltic (Petrone 2010), San Juan (Luhr 2000). Pyroclastic rocks can be more felsic than lavas but were omitted from the plots because their sources may be ambiguous, are often weathered, and their compositions scatter.

Livieres 1986; Verma & Nelson 1989; Wallace & Carmichael 1994; Lewis-Kenedi *et al.* 2005; Frey & Lange 2011), and even if some local variations exist among them, they all share remarkably similar compositional features and volcanological characteristics. Furthermore, stratovolcanoes at the rear arc are very similar to Colima volcano in terms of their major and trace element compositions (Figs 3 & 4) (Luhr 2002; Reubi & Blundy 2008; Savov *et al.* 2008), even though the latter is emplaced much closer to the trench and in a different tectonic setting (Fig. 1). Colima may be more voluminous, hydrous and explosive than Ceboruco (Luhr 1992), but rocks from Tequila and San Juan can have higher water contents than Colima and are also known to develop explosive activity (Wallace & Carmichael 1994; Luhr 2000).

Stratovolcanoes in western Mexico are mainly constituted by a monotonous set of andesitic rocks, although some basaltic-andesites and dacites are not uncommon, and pyroclastic deposits can extend to rhyolitic compositions (Luhr 2000; Chertkoff & Gardner 2004). Major element contents of rocks from these volcanoes mostly overlap with each other and follow linear trends in major element variation diagrams (Fig. 3), thus classifying as calc-alkaline, medium-K, metaluminous, high-Mg#

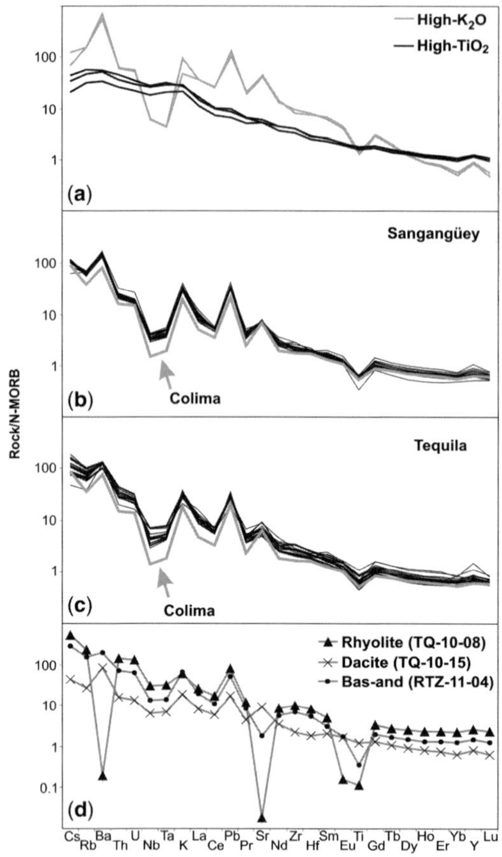

Fig. 4. N-MORB normalized trace element patterns of the studied rock suites: (**a**) compares the patterns of some of the most primitive end-members found in the western TMVB. The high-K_2O samples are from primitive minettes reported in (Gómez-Tuena *et al.* 2011). The high-TiO_2 end-member is represented by samples TPZ-10-21, TPZ-10-23 and RTZ-11-27 from this study. (**b**) Tequila volcano. (**c**) Sangangüey volcano. The average composition of Colima volcano (Luhr 2002) is also included for comparison. (**d**) Trace element compositions of representative peripheral volcanoes.

andesites. Peripheral volcanism to stratovolcanoes is represented by scoria cones and domes of variable compositions that can be either more mafic, or extend to dacitic–rhyolitic compositions (Fig. 3). Interestingly, and in spite of the extensive sampling efforts of numerous research groups and the abundance of mafic alkaline rocks in the area, true calc-alkaline basalts that can be parental to andesites are extremely rare in western Mexico. To our knowledge, the most primitive calc-alkaline sample that has been identified within TZR is a basaltic andesite that erupted from a scoria cone *c.* 10 km to the NW of Ceboruco volcano (Petrone 2010) (our sample RTZ-11-04, Fig. 3). The scenario is not much different at the volcanic front: the only true calc-alkaline basalt that has been documented belongs to the Tezontal or Tezontle scoria cone, emplaced at *c.* 25 km NE of Colima volcano (Luhr & Carmichael 1981; Gómez-Tuena *et al.* 2011). Trace element patterns of these basalts are similar to those of the andesitic stratovolcanoes, albeit with slightly higher HFSE and REE contents and thus less pronounced subduction signatures (Fig. 4d). In contrast, peripheral dacitic–rhyolitic domes are often glassy, weakly peraluminous, with prominent negative anomalies of Ba, Sr and Eu in their trace element patterns (Fig. 4d).

Isotope geochemistry

Sr–Nd isotopic compositions of volcanic rocks from western Mexico broadly overlap with each other, plotting between the East Pacific Rise–mid-ocean ridge basalts (EPR–MORB) and an upper crust-like component similar in composition to the Pacific sediments or the local basement (Fig. 5a). Yet when the isotopic variations are seen in more detail–for instance, by considering only those rocks that erupted in the context of the TZR (Fig. 5b)–andesites erupting from stratovolcanoes define a subparallel trend that is independent to that of more primitive intraplate basalts and other peripheral volcanism, including the most primitive calc-alkaline basaltic andesite. Despite the existence of a whole spectrum of rock compositions in the area, from basalts to rhyolites, the isotopic variations are also largely independent from silica (Fig. 5c). Thus andesites erupted from stratovolcanoes appear as isolated isotopic entities, with small influence from the sources involved in the formation of surrounding volcanism.

Pb isotopic compositions of rocks having positive Pb anomalies in their trace element diagrams (Fig. 4) also plot along a simple mixing line bracketed between the EPR–MORB and the crust; while the most primitive high-TiO_2 intraplate rocks, having negligible Pb spikes, form an independent correlation that is displaced towards higher $^{206}Pb/^{204}Pb$ ratios (Fig. 5d). The compositions of these intraplate rocks overlap a mixing trend defined by the Rivera plate MORB, in the depleted end, and a more enriched high μ component (HIMU; $\mu = {}^{238}U/^{204}Pb$) that has been observed in some Mexican Pacific islands, such as Isabel (Housh *et al.* 2010), a small volcanic island off the coast of Nayarit state constructed over the Rivera plate; and Socorro (Bohrson & Reid 1995), which is sitting on the Pacific plate and thus far from any subduction influence.

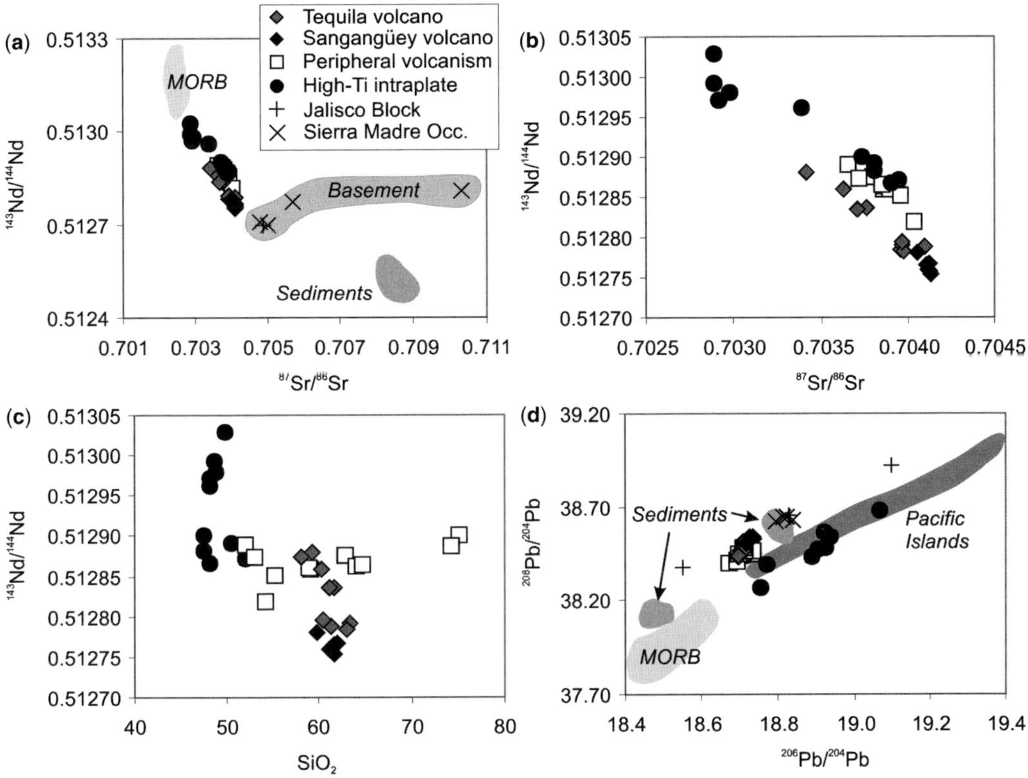

Fig. 5. Sr–Nd–Pb isotope variation diagrams for the studied rock suites and possible end-members. (**a**) $^{143}Nd/^{144}Nd$ v. $^{87}Sr/^{86}Sr$. (**b**) A more detailed portrait of the $^{143}Nd/^{144}Nd$ v. $^{87}Sr/^{86}Sr$ isotopic variations showing sub-parallel arrays within the suites. (**c**) $^{143}Nd/^{144}Nd$ isotopic variations are independent from SiO_2. (**d**) Pb isotopic variations. Also shown: data field of the 20°–24°N EPR–MORB, corresponding to Rivera Plate (Lehnert *et al.* 2000); http://www.petdb.org/); the isotopic composition of Cocos plate sediments collected at DSDP Site 487 (LaGatta 2003); and data field for the Mexican Pacific Islands of Socorro and Isabel (Bohrson & Reid 1997; Housh *et al.* 2010).

Discussion

Pliocene to Recent volcanism in the region is strongly diverse (Fig. 2). The variable isotopic composition indicates that magma sources must differ within and among the suites (Fig. 5), although different degrees of crystal fractionation must also contribute to this diversity (Petrone 2010). Nevertheless, and despite this variability, most mafic rocks in western Mexico depict a clear transition from a potassic volcanic front to a high-TiO_2 intraplate-like rear arc for which a simple explanation can be advanced: slab-derived fluxes concentrate below the volcanic front and are responsible for arc magmatism, whereas slab contributions become less influential farther away from the trench, where decompression melts of a dry upper mantle become dominant (Luhr 1997; Gómez-Tuena *et al.* 2011; Ferrari *et al.* 2012). Yet such a simple scenario becomes a lot more complicated since at least for the past *c.* 200 ka, calc-alkaline andesitic stratovolcanoes like Colima volcano have erupted close to the trench, but also at the rear arc, such as Tequila and Sangangüey. The compositional similarities among these volcanoes strongly speak for a common origin, but where exactly do these magmas come from, and what is their relation to the peculiar tectonic environment, are questions that still remain unanswered. In this section we critically evaluate the various competing hypotheses that have been proposed to explain the origin of andesitic volcanism in western Mexico, and provide a new model that may help to reconcile conflicting observations. To this end, we compare the chemical compositions of andesitic stratovolcanoes to that of monogenetic volcanism in the region, but we do not go into the details about their intricate differentiation pathways, as they have been thoroughly documented in previous studies (Nelson & Livieres 1986; Petrone 2010; Frey & Lange 2011).

Mixing, fractionation and crustal contamination

Ever since the seminal work of Nixon (1988) on Iztaccíhuatl, it has long been recognized that andesitic volcanoes in Mexico record some sort of mixing or mingling process between magmatic components that reside and fractionate in long-lived magma chambers. Mixing allows for homogenization and the creation of monotonous andesitic compositions, can explain disequilibrium mineral textures, as well as the linear arrays observed in most major element variation diagrams. Mixing is thus largely accepted as a fundamental petrologic mechanism, but the specific compositions of the end-members involved, as well as their primary origins, remain controversial and the subject of much debate.

Most workers agree that the Mexican andesites must include a strong mantle-derived contribution but no consensus has been reached regarding its composition (Nixon 1988; Wallace & Carmichael 1994; Márquez *et al.* 1999; Straub & Martin-Del Pozzo 2001; Petrone *et al.* 2003; Schaaf *et al.* 2005). Calc-alkaline basalts are surprisingly scarce in the western Mexican arc, and even if those few basalts that exist have been regarded as potential precursors to andesites (Luhr & Carmichael 1980, 1981; Petrone & Ferrari 2008), certain chemical characteristics of andesites, when seen in detail, are difficult to explain by basalt fractionation alone. Indeed, Figure 3c shows that most andesitic rocks erupting from stratovolcanoes retain Mg# that are too high to be derivative of any known basalt by fractional crystallization involving olivine, pyroxene and plagioclase, the most common mineral assemblage observed in basaltic thin sections, and predicted by simple fractionation models (Fig. 3c). Although retention of high Mg# with increasing SiO_2 may be explained by early fractionation of significant amounts of Fe–Ti oxides (Shellnutt & Zellmer 2010), this may only occur at much higher oxygen fugacities than what has been determined in Mexico and in other magmatic arcs (Blatter & Carmichael 2001).

Andesitic rocks from western Mexico also have a tendency to decrease HFSE and REE contents as well as Nb/Ta and Dy/Yb ratios with increasing SiO_2 (Fig. 6), a puzzling characteristic that has been often observed in most intermediate rocks all across the Mexican arc (Luhr & Carmichael 1985; Verma 1999; Siebe *et al.* 2004; Gómez-Tuena *et al.* 2007*b*). As previously shown, such variations are difficult to explain by fractional crystallization of a low pressure mineral assemblage dominated by olivine, pyroxene and plagioclase, because all these elements must remain incompatible. Yet it has been increasingly recognized that minerals like amphibole and garnet can join the fractionation assemblage in hydrous magmas at conditions relevant for the lower and middle crust, and thus imprint unusual trace element characteristics to the derivative liquids (Müntener *et al.* 2001; MacPherson *et al.* 2006; Davidson *et al.* 2007; Petrone & Ferrari 2008; Alonso-Perez *et al.* 2009; Zellmer *et al.* 2012).

Amphibole and garnet are rarely present as phenocrysts in the eruptive sequences, but the well-documented differential partitioning of the REE and the HFSE between garnet (Johnson 1998; Green *et al.* 2000) and amphibole (Bottazzi *et al.* 1999; Tiepolo *et al.* 2000*a*, *b*) allows their recognition as potential cryptic fractionation phases (Davidson *et al.* 2007). Indeed, Figures 6c, d show that the low Nb/Ta and Dy/Yb ratios of the Mexican andesites can be reproduced by simple crystal fractionation models, but only if amphibole completely overwhelms the fractionating assemblage and no garnet is involved. Some experiments have shown that an amphibole-dominated fractionation assemblage is possible at moderate pressures (*c.* 0.8 GPa) and water contents (*c.* 4 wt%) (Alonso-Perez *et al.* 2009), but crystallization of a significant amount of amphibole will also induce a decrease in Mg#, in a similar way as olivine and pyroxene (Fig. 3c), as well as the formation of peraluminous derivative liquids (Alonso-Perez *et al.* 2009). Major and trace element compositions consistent with moderate amounts of amphibole fractionation are observed in some peripheral volcanoes, including slightly peraluminous rhyolites that also display low Dy/Yb ratios. In contrast, andesites erupting from large stratovolcanoes have a similar or even stronger amphibole signature (lower Nb/Ta and Dy/Yb ratios), but they always remain metaluminous at constantly high Mg#. Concurrent fractionation of plagioclase and garnet may modulate Al_2O_3 contents and hinder peraluminosity, and it has also been suggested that cryptic fractionation of almandine-rich garnets (with Fe $>$ Mg) may keep andesites metaluminous and help retaining their elevated Mg# during crystallization (Alonso-Perez *et al.* 2009; Zellmer *et al.* 2012). Strongly negative Eu anomalies in rhyolites (Fig. 4d) are consistent with crystallization of abundant feldspar, but the precipitation of garnet is not supported in any of these rock sequences because even a small amount of it would imprint high Dy/Yb ratios to the derivative liquids (Fig. 6).

We thus conclude that fractionation of amphibole and feldspar was involved in the formation of evolved monogenetic volcanism in the area. Given that their Sr–Nd isotopic compositions overlap with those of basalts (Fig. 5b), it seems reasonable to suggest that the evolved peripheral products may indeed represent their derivative liquids. Yet

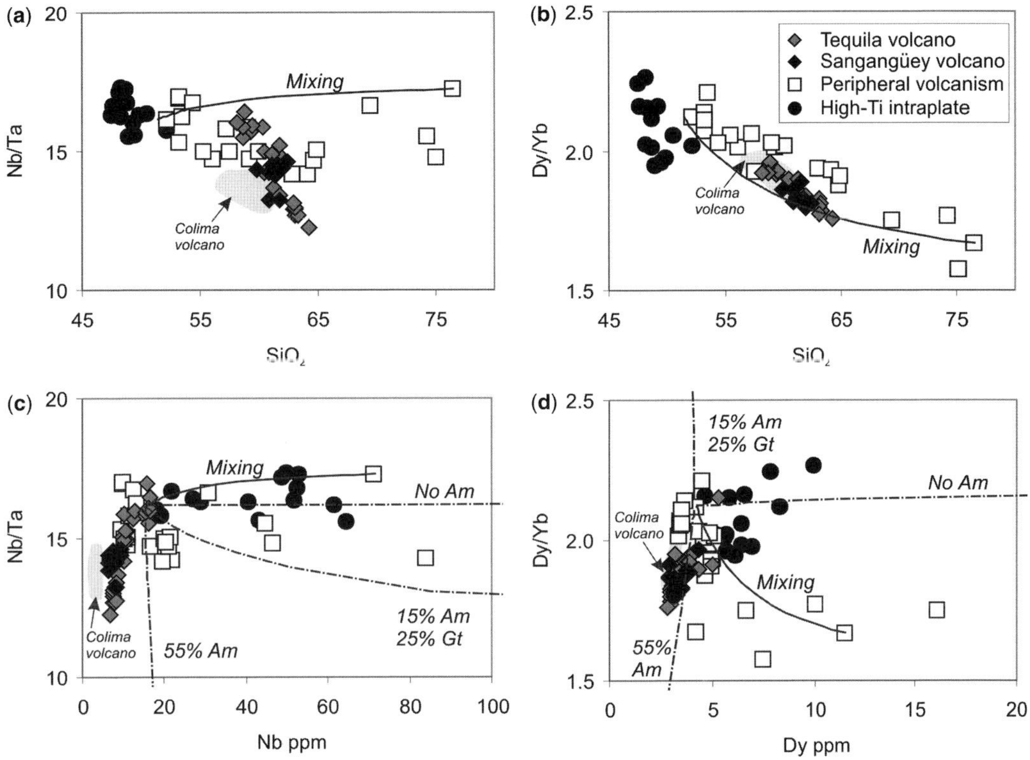

Fig. 6. Andesite rocks have a tendency to decrease (**a**) Nb/Ta, (**b**) Dy/Yb ratios, as well as (**c**) Nb and (**d**) Dy contents, with increasing SiO_2. Simple mixing (solid line) between mafic and felsic magmas cannot explain these characteristics, and are difficult to form by crystal fractionation unless amphibole dominates the fractionation assemblage. Fractional crystallization models (dash–dot lines) are shown for different proportions of crystallizing amphibole and garnet. Models use most the primitive calc-alkaline basaltic andesite found in the TZR (sample RTZ-11-04) as initial magma. A typical low pressure fractional crystallization model without amphibole (no Am) uses bulk $D_{Nb} = 0.0142$, $D_{Ta} = 0.0157$, $D_{Dy} = 0.1672$, and $D_{Yb} = 0.1769$ (assuming fractionation of 35% Ol, 34% Cpx, 20% Opx, 10% Pl, 1% Fe–Ti oxides). Fractional crystallization models involving amphibole and/or garnet use bulk $D_{Nb} = 0.2517$, $D_{Ta} = 0.1591$, $D_{Dy} = 1.112$, and $D_{Yb} = 2.221$ (assuming 45% Cpx, 15% Opx, 15% Am, 25% Grt, based on the experimental basaltic melt association at 1.2 GPa from Müntener *et al.* 2001), $D_{Nb} = 0.9159$, $D_{Ta} = 0.5694$, $D_{Dy} = 1.496$, and $D_{Yb} = 0.9681$ (55% Am, 43% Plag, 1% Fe–Ti oxides; based on the experimental basaltic melt association at 0.8 GPa from Alonso *et al.* 2009). Individual mineral Kd values from Hart & Dunn (1993), Kelemen *et al.* (1993); Johnson (1994); Bottazzi *et al.* (1999); Tiepolo *et al.* (2000*a*); Foley *et al.* (2002); Donnelly (2002) and Salters & Longhi (1999). Data field for Colima volcano (light grey) is taken from Luhr (2002).

the evidence also indicates that no combination of observed, or even cryptic, fractionation assemblages can explain the overall major and trace element compositions of andesites erupting from large stratovolcanoes.

If crystal fractionation of basalt is not the main mechanism involved in the formation of andesitic stratovolcanoes, a simple mixing process between basaltic and rhyolitic melts, or the more recently proposed scenario of bulk assimilation of a pre-solidified basalt at depth by ascending evolved magmas (Reubi & Blundy 2009), may provide the simplest explanation. Both processes, albeit conceptually distinct, will produce similar compositional results in the whole rocks, as they both create linear arrays in major element diagrams and roughly constant Mg# at intermediate SiO_2 contents (Fig. 3c). Yet, and even if mixing between contemporaneous basalt and rhyolite will be a sufficient explanation for some minor andesitic monogenetic cones, it will fail to explain the low incompatible trace element contents of andesitic stratovolcanoes, as well as their independent trends in most isotopic and trace element diagrams (Figs 5 & 7). Contemporaneous peripheral rhyolites are thus not representative of the felsic component involved in the mixing mechanism that formed andesitic stratovolcanoes.

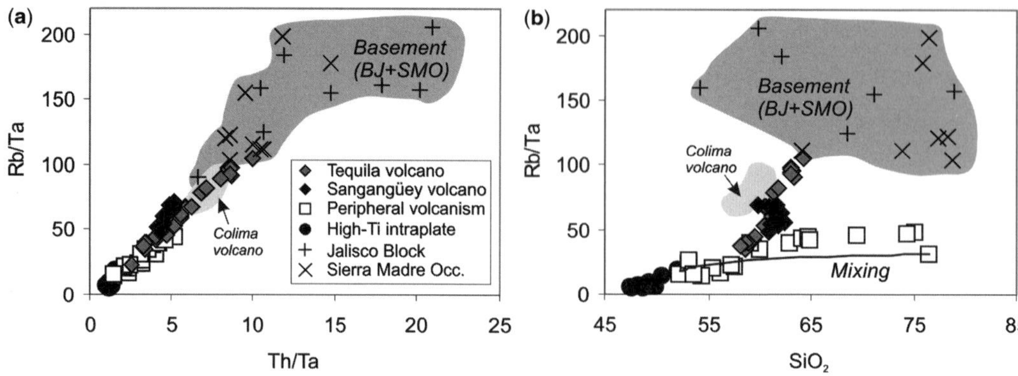

Fig. 7. Andesites extend to higher (**a**) Rb/Ta and Th/Ta ratios at increasing (**b**) SiO_2 contents, indicating a mixing relationship that cannot be explained by local rhyolites as the felsic end-member. The data thus suggest recycling of pre-existent crustal rocks from the Jalisco Block and/or the Sierra Madre Occidental. Data field for Colima volcano (light grey) is taken from Luhr (2002).

The isotopic composition of basaltic magmas may be modified by assimilation of pre-existent crust, and contamination has always been regarded as an important mechanism to explain the enriched isotopic compositions of volcanic rocks in thick crusted continental arcs as the TMVB (Wilcox 1954; Chesley *et al.* 2002). Nevertheless, and even though the volcanic successions emplaced within the TZR display a broad range of isotopic compositions that point to crustal contributions (Fig. 5a), their isotopic variations do not show a simple correlation with indices of fractionation (Fig. 5c), perhaps the most robust geochemical tests for an assimilation-fractional crystallization processes (AFC; DePaolo 1981). Contamination effects may be recorded in some of the most evolved intraplate magmas which tend to be slightly more enriched in Sr–Nd isotopes than most primitive ones, and which display relatively low $^{206}Pb/^{204}Pb$ isotopic ratios that are more akin to the local continental crust (Fig. 5d). Yet it is interesting to note that andesites do not plot as an evolved extension of these basalts: their entire isotopic range is almost as variable as that of basalts, but it follows a sub-parallel array that varies independently from silica (Fig. 5c) and at nearly constant Mg# (Fig. 3). Remarkable as well is the lack of significant isotopic variation in Sr–Nd space displayed by peripheral volcanism, despite extending to rhyolitic compositions. We thus conclude that crustal contamination may have occurred during crystallization of some high-TiO_2 intraplate rocks and peripheral volcanoes; and even if an open-system evolution may be unavoidable during the construction of large stratovolcanoes, the isotopic and major element data indicate that an AFC process cannot be the single most important mechanism governing their origin.

Andesites by partial melting of a pre-existent arc crust

Continuous underplating of mantle-derived magmas induces heat advection through the crust that can lead to anatexis of ancestral or newly emplaced magmatic rocks, a mechanism that has been suggested to generate felsic magmas in continental rifts (Chayes 1963). Intermediate rocks are uncommon in these environments, as they usually display bimodal magmatic associations. Furthermore most natural migmatites (Moyen 2009; Reichardt & Weinberg 2012) and experimental anatectic melts of basalt (Sen & Dunn 1994; Rapp 1995; Rapp & Watson 1995; Rapp *et al.* 1999) have much lower Ni, Cr and Mg# than those observed in the Mexican andesites (Fig. 3c), indicating that they cannot represent pristine crustal melts, and that they require, in any case, a large mantle-derived contribution.

As shown above, mixing between contemporaneous basalts and rhyolites is inconsistent with the trace element and isotopic data, but local basement rocks from the SMO or the Jalisco Block display much higher LILE/HFSE ratios and Sr–Nd isotopes (Figs 5 & 7), and thus could act as the source of the required siliceous end-member if undergoing partial melting. None the less, exposed basement rocks are usually feldspar-rich plutons and ignimbrites, and will generate anatectic melts with negative Eu, Sr and Ba anomalies that are not observed in andesites; indicating that melting could not have occurred at upper crustal levels. Melting of a deep garnet-bearing crust is also precluded because of the low Dy/Yb ratios of andesites (Fig. 6), but the presence of residual amphibole is supported by the relative MREE depletion and low Nb/Ta ratios of andesites displaying the highest SiO_2 contents (Fig. 6). In this sense, melting of an ancestral middle-lower crustal arc rock, in which abundant amphibole remains in the residue or is later created as a secondary anatectic reaction product (Gardien *et al.* 2000; Reichardt & Weinberg 2012), might in principle generate a siliceous melt that satisfies the mixing relationships of andesites erupted from large stratovolcanoes. In detail, however, several lines of evidence deem this process unlikely.

Dehydration melting of a pre-existent amphibolite requires a substantial amount of heat (>950 °C) that may come from the protracted intrusion of mantle-derived magmas (*c.* 1200 °C), but the amount of water in the source is limited to that bound in hydrous minerals (Annen *et al.* 2006). Yet amphibole is not particularly a water-rich phase: a hypothetical rock with *c.* 50% amphibole would not contain more than 1 wt% H_2O. If the partitioning of water during partial melting of amphibole is considered to be similar to that of Ce with a $K_d = c.\ 0.5$ (Hauri *et al.* 2006), and water behaves as perfectly incompatible in any other mineral, the maximum amount of water that can be released from a source with 50% amphibole is *c.* 4 wt%, at extremely low extents of melting. Some andesites from Tequila volcano, for instance, had a minimum of 7.5 wt% dissolved water (Wallace & Carmichael 1994), whereas 6–16 wt% pre-eruptive water contents have been considered for Mexican andesites by experimental phase equilibria and thermodynamic considerations (Carmichael 2002). Clearly, the Mexican andesites require an additional influx of water to what can be extracted by dehydration melting of amphibole alone. External influence of H_2O-rich fluids is also needed if amphibole has to remain stable after melt formation and extraction, either as a residual phase or as a secondary peritectic melt reaction product (Gardien *et al.* 2000; Reichardt & Weinberg 2012).

Availability of large amounts of free fluids in the deep crust is difficult to envisage, at least in the context of the TZR, because arc magmatism is concentrated at the volcanic front and the most mafic post-Pliocene volcanism at the rear arc is intraplate-like and thus essentially dry (Figs 2–4). Deep crystallization of intraplate basalts can, in principle, provide some additional volatiles, but it remains uncertain how effectively they can separate from their residual melts when still residing at deep crustal levels, since evolved siliceous liquids like these become water-saturated only closer to the surface (<0.5 GPa) or during eruption (Annen *et al.* 2006).

But perhaps the strongest arguments against a crustal origin for the siliceous component are geological, rather than petrological or geochemical. About 10 Ma after the waning of the SMO, vigorous tholeiitic mafic magmatism was established in the area between *c.* 11 and 8 Ma (Ferrari *et al.* 2000; Mori *et al.* 2009), and was later followed by a period of intraplate mafic magmatism that likely triggered a small-scale ignimbritic flare-up event at 5–3 Ma (Frey *et al.* 2007). Interestingly, andesitic stratovolcanoes like Tequila and Sangangüey only started to appear within the TZR over the last 200 ka, a period during which bimodal magmatism was still being emplaced but at a smaller volume (Lewis-Kenedi *et al.* 2005). And herein lies a conundrum: if a fertile continental arc crust has been residing in the area at least since the late Cretaceous (i.e. the Jalisco Block), and has been the subject of an almost continuous mafic influx for at least 11 Ma, why is it that this particular kind of amphibole-rich crust starts to melt at an almost instantaneous geological moment during the past 200 ka, in order to form similar volcanoes at different geographical locations? Basaltic intrusion rates through the Moho are difficult to determine from the eruptive record, but calculated incubation times between the first intruded basalt and the beginning of melting of a hypothetical amphibolite layer located at 20–30 km depth in the crust are on the order of 3 Ma or less for a wide range of geologically relevant intrusion rates (Annen *et al.* 2006). Under these circumstances, andesitic volcanoes, or amphibolite melts, should have been more common in Miocene times, when the basaltic influx was stronger, or in any case be more erratically distributed in time and space. Also puzzling is that despite the abundance of rhyolites in the area (which may indeed represent crustal melts), this hypothetical amphibolite magma has never erupted pristinely, it always has to mix with basalt, even if it is less dense and viscous than nearby rhyolites due to its much higher water contents. But perhaps the more troubling implication of the crustal melt model is that, at least locally, no subduction influence will be required for the formation of these andesites, since they will only be recycling ancient arc lithologies that already reside in the crust, and which mix with basaltic magmas generated in the context of a continental rift. Such an idea is not new for the Mexican arc (Márquez *et al.* 1999; Verma 2002), and even if it may find some support when only the geochemical data are considered, it becomes significantly weakened in the light of the much larger geological and petrological context (Ferrari *et al.* 1999; Blatter *et al.* 2003; Torres-Alvarado & Verma 2003; Siebe *et al.* 2004).

Subduction zone recycling and the formation of andesites

If the Mexican andesites are not differentiates of basalt, nor mixtures between basalts and direct melts of the upper plate, a possibility exists that they can be direct mantle extracts. Several lines of evidence appear consistent with this view (Blatter & Carmichael 1998, 2001; Martínez-Serrano *et al.* 2004; Gómez-Tuena *et al.* 2007*a*, 2008; Mori *et al.* 2007; Straub *et al.* 2011), but the specific process by which a silica-rich melt can be generated from an ultramafic source is still a matter of debate.

Some authors have proposed that melting of peridotites at conditions close to water saturation could create silica-rich melts directly from the mantle (Blatter & Carmichael 2001; Carmichael 2002; Weber *et al.* 2012), in accordance with some experimental evidence (Tatsumi 1982; Grove *et al.* 2002; Parman & Grove 2004). Yet, to our knowledge, no hydrous melting experiment has been able to produce liquids with high enough SiO_2 contents that could match those of the andesitic rocks studied here. Furthermore, experimental data and partitioning modelling indicate that a very narrow window of conditions must be met in order to form liquids that approach andesitic compositions (Parman & Grove 2004): not only must water contents be kept at near-saturation levels (>7 wt%), but also temperatures and pressures of the last equilibration should be low (<1000 °C and 1 GPa), thus restricting primitive andesite formation to near Moho depths for most arcs. Since subtle variations in these parameters, or in the fertility of the source, will inevitably render the melts basaltic, it seems highly unlikely that all those circumstances can be universally met at the widely diverse tectonic conditions in which homogeneous andesites erupt, not only in Mexico but world-wide.

Other authors have shown that at least some Mexican andesites can be the result of melt–rock interactions between slab-derived siliceous melts and the mantle wedge (Gómez-Tuena *et al.* 2003, 2008; Martínez-Serrano *et al.* 2004; Mori *et al.* 2007). This idea has been further supported by the discovery of high-Ni olivines from some Mexican andesites, which are interpreted to be the result of reactive formation of secondary mantle pyroxenites during slab-derived silica influx of the upper mantle (Straub *et al.* 2011, 2013). None the less, and even if a strong slab-melt signature has been observed in some volcanic sequences, it also true that not all andesites in Mexico have the high Sr/Y ratios that are indicative of melting in the presence of eclogite residuals (Defant & Drummond 1990). In fact, andesites studied here are remarkable in this regard, since they display some of the lowest Sr/Y ratios observed in Mexico (<40), even if they have erupted above one of the warmest subduction zones on Earth (Syracuse *et al.* 2010; Ferrari *et al.* 2012). The evidence from Mexico thus makes clear that the mechanisms of element recycling involved in the formation of primitive andesites cannot be simply the result of a specific slab geotherm.

The geochemical evidence indicates that continental crust is being recycled during the formation of the Mexican andesites: their trace element and isotopic compositions show clear mixing systematics, in which the high-SiO_2 component has relatively higher Rb/Ta and Th/Ta ratios that can only be derived from a crustal source (Fig. 7). If, as shown above, these features are difficult to explain by crustal contamination or direct anatexis of the upper plate, they must have been acquired by the introduction of enriched crustal materials into the subduction zone. This may be in the form of subducted pelagic sediments, but most likely is of terrigenous and continental origin, since the Rivera plate is too young to have a thick pelagic cover (Manea *et al.* 2003), and because the mixing trends displayed by andesites extend to the compositions of the local basement: the Jalisco Block and SMO (Fig. 7). This evolved igneous crust has been uplifting (Righter *et al.* 2010; Ramírez-Herrera *et al.* 2011), eroding, and experimenting subduction erosion (Clift & Vannucchi 2004) at least since the Oligocene, and maybe as far back as the late Cretaceous, and therefore represents the only important continental influence to the mantle source in the region. But how exactly this crust is incorporated into the source of arc magmas requires a more careful evaluation of the trace element systematics.

Figure 8 shows that high-TiO_2 intraplate basalts have similar Nb/Ta and Rb/Ta ratios than MORB at slightly higher Dy/Yb values, geochemical characteristics that are consistent with their formation by low extents of melting of a dry garnet peridotite source that has been negligibly affected by the subduction agents (Luhr 1997; Petrone *et al.* 2003; Gómez-Tuena *et al.* 2011). In contrast, potassic lamprophyres from the volcanic front, as well as andesitic stratovolcanoes, display much higher LILE/HFSE ratios, such as Rb/Ta, that are indicative of a strong subduction-induced crustal influence. Interestingly, while both suites display a near-equivalent relative enrichment of the LILE, potassic rocks extend to much higher Nb/Ta, La/Ta and Dy/Yb ratios than andesitic stratovolcanoes, which generally follow a diverging trend in these diagrams. Since correlations like these cannot be formed by different extents of melting of a common source, they must imply variations in the mechanisms of element recycling in the subduction zone.

In a recent study, Gómez-Tuena *et al.* (2011) interpreted the geochemical variability among magmatic sequences of western Mexico as a function of the prograde metamorphic transformations that affect the subducted slab at increasing depths. Based on partitioning experimental data describing

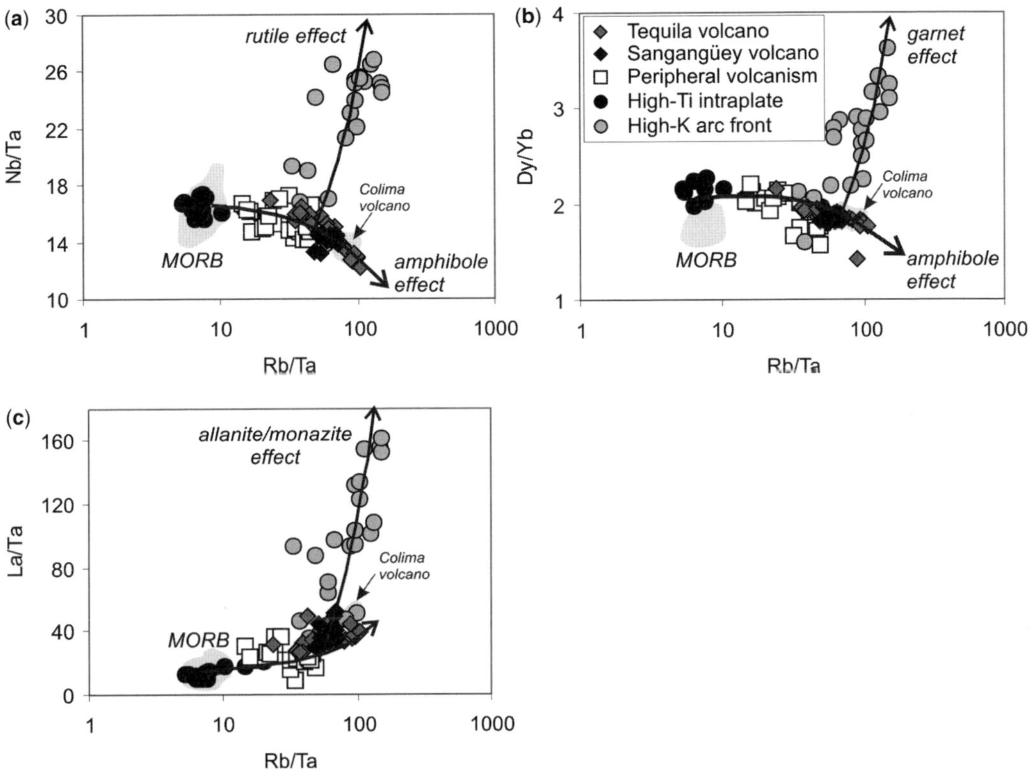

Fig. 8. Process of element recycling in the Mexican Subduction zone. (**a**) Rb/Ta ratios correlate positively with Nb/Ta, (**b**) La/Ta ratios and (**c**) Dy/Yb in potassic volcanic rocks from the volcanic front, whereas andesitic rocks follow opposite trends. Potassic rocks require a high temperature subduction flux extracted from eclogite, while andesitic stratovolcanoes necessitate a flux derived from an amphibole-rich lithology. Data field for Colima volcano (light grey) is taken from Luhr (2002).

trace element mobility at different temperatures (Kessel *et al.* 2005), these authors concluded that calc-alkaline magmas at the volcanic front of western Mexico were affected by low-temperature subduction fluxes, whereas potassium-rich rocks had stronger contributions from a deeper and hotter melt or supercritical fluid related to phengite and allanite/monazite disintegration. Since no detailed analysis was performed on andesitic rocks during their study, Gómez-Tuena *et al.* (2011) adhered to the traditional interpretation that the magmatic calc-alkaline flux had to be basaltic in nature (Luhr & Carmichael 1981), and that andesites erupted from stratovolcanoes represented crustal differentiates that preserved the trace element and isotopic compositions of their parental basalts (Luhr & Carmichael 1980, 1981; Lassiter & Luhr 2001). None the less, given that crystal fractionation does not appear to be a viable mechanism for andesite formation as shown above, and that their trace element variations correlate with silica enrichment (Fig. 7), we conclude that the low-temperature subduction flux must also be a siliceous melt rather than an aqueous fluid.

As a result, the correlations observed in Figure 8 can be interpreted in terms of the different mineral assemblages that remain in the residue after melt extraction. Partial melts leaving a rutile-bearing eclogite residue will display high Nb/Ta and Dy/Yb ratios due to the presence of rutile (Brenan *et al.* 1994; Xiong *et al.* 2005) and garnet (Johnson 1998; Green *et al.* 2000), whereas partial melts extracted from an amphibole-rich lithology will display the opposite effect due to the differential partitioning between HFSE and REE in low Mg# amphiboles (Bottazzi *et al.* 1999; Tiepolo *et al.* 2000*b*; Foley *et al.* 2002). A K- and LREE-rich metasomatic agent will be released if minerals like phengite, allanite and monazite are significantly depleted, or even exhausted during melt extraction, but these elements will be buffered at relatively low concentrations if these phases remain stable in the residual mineral assemblage. In other words, the recycled component involved in the formation of calc-alkaline andesitic volcanoes and high-K_2O rocks may be essentially the same in terms of bulk composition; what differs is the mineral assemblage that remains in the melting residue at different P–T conditions. For andesitic stratovolcanoes, melting and recycling must have occurred at relatively shallow depths and lower temperatures, within the stability field of amphibole, whereas for high-potassium magmas it must have occurred much deeper, at eclogite metamorphic facies.

Insights from the H_2O/Ce geothermometer

The overall geochemical evidence thus indicates that potassic rocks and andesites can all be mantle derived; their contrasting compositions governed by the chemical nature of the subduction flux. Since these parameters are controlled by the thermal state of the convergent margin, the geochemical characteristics of the volcanic rocks may in principle be used to constrain the temperature at which the different metasomatic agents have been extracted from the subducted lithologies. The traditional approach is to use incompatible trace element ratios and compare them to experimental solid/fluid partition coefficients at different temperatures (Kessel *et al.* 2005; Gómez-Tuena *et al.* 2011), but more recent studies have recognized that some specific geochemical tracers can effectively constrain temperatures of fluid or melt extraction, since they are a simple function of the temperature-dependent solubilities of key mineral species like phengite (Hermann & Spandler 2008), rutile (Klimm *et al.* 2008) or monazite/allanite (Plank *et al.* 2009; Cooper *et al.* 2012).

In order to gain additional insights into the thermal conditions of element recycling in western Mexico we applied the newly developed H_2O/Ce geothermometer (Plank *et al.* 2009; Cooper *et al.* 2012). To this end, we compiled published water contents directly measured in olivine-hosted melt inclusions for a number of well-studied scoria cones emplaced at the volcanic front of western Mexico (Maria & Luhr 2008; Vigouroux *et al.* 2008). Direct measurements of water are not available for andesitic stratovolcanoes where olivine melt inclusions are scarce or non-existent, but estimates of minimum pre-eruptive water contents have been reported using plagioclase-liquid hygrometers for Ceboruco, Tequila, Colima and San Juan volcanoes (Luhr 1992, 2000; Wallace & Carmichael 1994). For the purpose of this study, we considered only the samples with the highest water contents from each volcano. Ce and Nb concentrations used in the calculations come from high-precision ICP–MS whole-rock analyses, either reported in same studies quoting water contents, or obtained by our group in equivalent rock samples (Table 3). The measured H_2O/Ce ratios were further corrected for mantle contributions using the whole-rock Nb/Ce ratios following previously described methods, and using an E-MORB background mantle value as assumed for Mexico (Cooper *et al.* 2012). Since the H_2O/Ce ratios of volcanic rocks may be affected to some extent by crystal fractionation or decompression-induced degassing, the calculated fluid temperatures must always be considered as maxima (Plank *et al.* 2009; Cooper *et al.* 2012).

Measured or calculated water contents are highly variable in the western Mexican arc, spanning from 7.5 wt% in calc-alkaline andesites from Tequila volcano, to 0.5 wt% in highly enriched potassic lamprophyres erupted at the volcanic front (Table 3). As a result, the calculated fluid temperatures for the whole western Mexican arc fluctuate between 806 °C and 1208 °C (Fig. 9). Temperatures higher than 1000 °C are above the stability field of allanite and monazite, and thus fall outside the parameterization potential of the H_2O/Ce geothermometer (Plank *et al.* 2009); yet they are qualitatively consistent with the interpretation that potassic rocks from western Mexico were affected by a high-temperature flux in which allanite, monazite and phengite could have been entirely exhausted from their source (Gómez-Tuena *et al.* 2011). Therefore, even if an accurate temperature cannot be established for these cases, it is none the less clear that the metasomatic agent involved in the genesis of potassic magmas must have reached at least 1000 °C, the highest temperatures registered so far for any arc on Earth (Cooper *et al.* 2012). But, perhaps more interesting is that, despite the extremely high temperatures observed along the volcanic front, and with the exception of the slightly less hydrous Ceboruco volcano (Luhr 1992), most calc-alkaline volcanoes record fluid temperatures below 870 °C, regardless of whether they are located closer to a trench, like Colima volcano, or *c.* 120 km inland, like Tequila and San Juan. For the case of western Mexico, this geographical difference is not trivial, since it translates into a *c.* 200 km difference in slab depth due to the steep subduction of the Rivera plate (Fig. 1).

Recent numerical models have predicted slab surface temperatures of *c.* 800 °C underneath the volcanic front of western Mexico, and up to 950 °C below the rear arc (Ferrari *et al.* 2012). These temperatures are well above the wet solidus of subducted basalts and sediments (Poli & Schmidt 2002), and within the range of those calculated for stratovolcanoes using the H_2O/Ce thermometer. Yet the predicted temperatures

Table 3. *Slab fluid compositions and temperatures*

Volcano	Sample	H_2O* wt%	Ce[†] ppm	Nb[†] ppm	H_2O/Ce ppm/ppm	Nb/Ce ppm/ppm	H_2O/Ce (corr)[‡] ppm/ppm	Temperature[§] °C	Data[‖] Sources
Stratovolcanoes									
Tequila	Q-39	7.5	54	11	1389	0.20	1928	848	1, 2
San Juan	1004-1	5.6	37	8.0	1501	0.21	2172	837	3
Ceboruco	983-3	2.7	58	5.8	468	0.10	504	971	4, 5
Colima	Col-15	3.4	26	3.3	1313	0.13	1546	869	4
Peripheral cinder cones to Colima volcano									
Tezontle	SAY-22F-L	0.9	25	3.9	366	0.16	414	989	6, 7
Apaxtepec	VF99-01B-A	1.6	69	13	235	0.19	250	1036	6, 7
San Isidro	VF99-07A-Ja	2.0	149	13	136	0.09	129	1097	6, 7
Telcampana	VF99-09A-Bb	1.6	91	7.6	175	0.08	172	1070	6, 7
La Erita	VF9908-F.A	7.3	102	9.6	709	0.09	768	933	6, 7
Novillero	MAS02-1B-J	1.2	118	13	98	0.11	81	1139	6, 7
Molcajete Galope	MAS02-2B-E	5.0	71	6.0	704	0.08	751	935	6
La Esperanza	MAS02-3B-J	3.0	29	6.0	1036	0.21	1440	875	6
Molcajete	MAS02-6A-B	0.5	110	6.0	43	0.05	38	1208	6
La Erita	LE-02-04	4.5	102	9.6	440	0.09	468	978	7, 8
Carpientero Norte	CN-18	1.6	192	14	83	0.07	75	1146	7, 8
Apaxtepex	AP-02-04	4.4	69	13	642	0.19	833	925	7, 8
Apaxtepex	AP-03-13	6.2	69	13	905	0.19	1210	891	7, 8
Cerro Usmajac	CU-02-02	5.5	54	22	1020	0.41	3055	806	7, 8

*Maximum amount of water for each volcano as measured in olivine melt inclusions (cinder cones) or using plagioclase hygrometers (stratovolcanoes).
[†]Ce and Nb from average whole-rock compositions, equivalent whole-rock samples, or from melt inclusions.
[‡]Mantle-corrected H_2O/Ce compositions of slab fluids assuming an E-MORB mantle source with $H_2O/Ce = 200$ and $Nb/Ce = 0.55$ (Cooper *et al.* 2012).
[§]Slab fluid temperatures at 4 GPa based on mantle-corrected H_2O/Ce ratios using equation (1) from Plank *et al.* (2009).
[‖]Data sources: (1) Wallace & Carmichael (1994); (2) this work; (3) Luhr (2000); (4) Luhr (1992); (5) Petrone *et al.* (2003); (6) Maria & Luhr (2008); (7) Gómez-Tuena *et al.* (2011); (8) Vigouroux *et al.* (2008).

underneath the arc front are still below the dehydration melting temperature of phengite and allanite/monazite, and are thus unable to explain the temperatures and compositions recorded by the highly enriched potassic lamprophyres of the volcanic front. But even more important is that the predicted thermal structure will not allow the transfer of a low-temperature fluid to the rear arc, where the slab is at more than 300 km depth. Applying a pressure correction to the H_2O/Ce thermometer for the rear arc will increase fluid temperatures to *c.* 1300 °C below Tequila and San Juan stratovolcanoes, values that are unreasonably high for any known slab geotherm at those depths (Syracuse *et al.* 2010). Moreover, these temperatures and pressures will also be entirely inconsistent with the trace element constraints discussed above, that envisage residual amphibole during hydrous melting of a recycled crustal lithology. So either the numerical thermal models for the Mexican subduction zone are incorrect, or the H_2O/Ce ratios are not recording temperatures of a fluid or melt that has been extracted from the top of the subducted slab.

A case for slab plumes?

The transfer of fluids or melts from the slab to the mantle wedge has often been considered to follow a vertical path, and the broad correlations observed between modelled sub-arc slab surface temperatures (Syracuse *et al.* 2010) and those calculated by the H_2O/Ce thermometer (Cooper *et al.* 2012) appear to be consistent with this view for the global spectrum of arcs. Yet these correlations consider average temperatures along the entire length of a convergent margin, and thus obscure most of the complexities involved in the construction of shorter segments, not to mention the significant variations observed among individual volcanic centres. As shown above, the western Mexican subduction zone is noteworthy in this regard

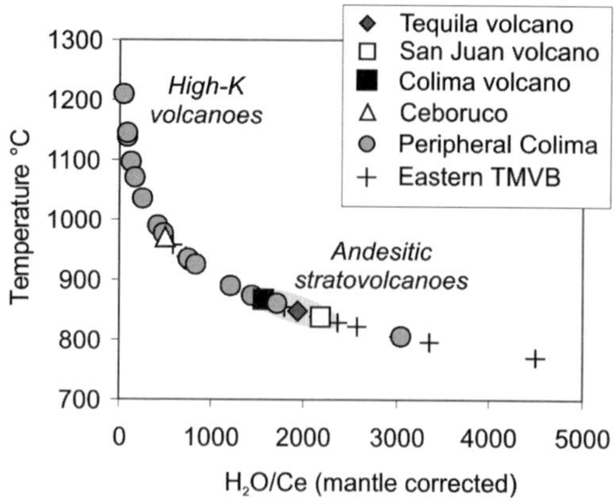

Fig. 9. Mantle-corrected H_2O/Ce estimates of slab fluid temperatures following Plank *et al.* (2009) and Cooper *et al.* (2012). Highest temperatures are observed in potassic lamprophyres from the volcanic front. With the exception of Ceboruco volcano, stratovolcanoes in western Mexico show the lowest temperatures regardless of their position above the subducted slab. Calculated fluid temperatures for the eastern TMVB are shown for comparison (Cooper *et al.* 2012).

because no obvious relationship exists between the predicted thermal structure and rock compositions: that is higher fluid temperatures are recorded at the volcanic front than at the rear arc. So unless andesitic stratovolcanoes have no relation to the active convergent margin (with the implications discussed above), then a mechanism to channelize a relatively cooler subduction flux farther away from the trench must be envisaged.

Andesitic stratovolcanoes in western Mexico have been emplaced in close association with potassic-alkaline magmas at the volcanic front, or aside high-TiO_2 intraplate volcanoes towards the rear arc (Figs 1 & 2), but no interaction among the different magmatic suites is apparent from the geochemical data. If these sequences are all mantle derived, then the upper mantle below western Mexico has to be not only extremely heterogeneous, but also one in which very sharp boundaries must exist: andesitic stratovolcanoes should be fed by a source that is compositionally distinct from that supplying the adjacent scoria cones, as can be interpreted from the contrasting Pb isotopic arrays (Fig. 5d). Therefore the recycled crustal components involved in andesite petrogenesis do not pervasively permeate the entire upper mantle below the arc, but rather form discrete and compositionally distinct parcels, that can even be entirely surrounded by a dry peridotite. In addition, the geochemical evidence indicates that the recycled crust must have experienced flux melting, because otherwise amphibole will be inevitably consumed, and the magmas will hardly be water-rich (Gardien *et al.* 2000). The question then is to decipher the geological mechanism that allows the transfer of discrete parcels of recycled hydrous crustal components at distinct locations away from the slab, in order to form isolated homogeneous andesitic volcanoes upon partial melting, in a context of genetically unrelated mafic magmatism.

Although counterintuitive at first sight, it has become increasingly recognized that most subducted lithologies are actually quite buoyant with respect to mantle peridotites, even if they have been buried to UHP conditions (Currie *et al.* 2007; Behn *et al.* 2011; Hacker *et al.* 2011). Thermal models have shown that subducted sediments and eroded quartz-feldspathic continental rocks are particularly difficult to subduct, and thus they may easily detach from the downgoing slab as buoyant diapirs (Currie *et al.* 2007; Hacker *et al.* 2011). Subducted oceanic basalt and eroded continental mafic lithologies become much denser during prograde metamorphism, and will likely sink into the mantle unless they become fragmented and surrounded by other less dense felsic material (Hacker *et al.* 2011; Little *et al.* 2011; Martens *et al.* 2012). Other numerical simulations have shown that mixtures of hydrous mantle and partially molten basalts and sediments can rise buoyantly through a network of conduits and diapirs that develop spontaneously at the slab–mantle interface as Rayleigh–Taylor thermochemical instabilities (Gerya & Yuen 2003; Gerya *et al.* 2004; Gerya 2011). These diapiric mélanges can be colder than the surrounding mantle wedge, but they are able to propagate upwards by being hydrous and possibly partially molten, with calculated upward velocities than may exceed 1 m a^{-1} (Gerya *et al.* 2004). Diapiric exhumation of deeply subducted lithologies can thus have important implications for arc magmatism and the construction of continents because it brings all the necessary ingredients to form primitive andesitic magmas upon partial melting.

The petrologic and geochemical evidence discussed above indicates that the source feeding andesitic stratovolcanoes is likely a hybrid mechanical mixture of continental crustal lithologies, either in the form of subducted continental sediments and/or eroded crustal fragments, and possibly fragmented amphibolite-facies MORB, that are intermingled or surrounded with portions of highly hydrated mantle. Given the similarity between andesites at the front (e.g. Colima volcano), from those at the rear arc (e.g. Tequila and Sangangüey), these mélanges should be able to detach from the slab and be channelized through a well-connected network of conduits and diapirs, like those predicted by numerical models (Gerya & Yuen 2003; Gerya *et al.* 2004; Gerya 2011). Once separated from the slab, these diapirs will be incorporated into the hot core of the mantle wedge where they will gradually heat up by diffusion and, given the relatively low solidus temperature of most of its constituents, almost inevitable undergo extensive partial melting (Katz & Rudge 2011). Although the complex interplay between dehydration and melting reactions are difficult to constrain with the information available, we speculate that the additional source of water required to allow flux melting of the crustal lithologies may come from the temperature-dependent decomposition of minerals like serpentine (Fumagalli & Poli 2005), which likely comprise a significant volume of the rising diapirs (Gerya *et al.* 2004). Thus, while most of the water in the magmas may come from serpentine dehydration, their trace element and isotopic signatures must derive from melting of recycled sediments or eroded crustal fragments, which at least in the case of western Mexico should have the same compositions as the SMO and the Jalisco Block.

The process of element recycling for highly potassic igneous rocks differs from that of andesitic volcanism by being relatively deeper and hotter, presumably due to partial melting of the same subducted lithologies, but at higher metamorphic stages (Gómez-Tuena *et al.* 2011). In contrast to what is observed in andesitic stratovolcanoes, the low water contents of potassic lamprophyres, as well as their highly enriched trace element contents (i.e. low H_2O/Ce), indicate that melting occurred by dehydration melting of phengite, at temperatures that may exceed 1000 °C. Although the western subduction zone is regarded as one of the warmest on the planet (Syracuse *et al.* 2010), these fluid temperatures are at least 200 °C higher than the slab surface temperatures calculated below the volcanic front by the most recent thermal models (Ferrari *et al.* 2012). So either the subducted Rivera plate has a steeper geotherm due to sinking at a nearly vertical angle, allowing slab melts or supercritical fluids to ascend from *c.* 300 km depth; or the melting process involved in formation of the potassic front did not occur at the slab surface, but also within the mantle wedge where temperatures as high as 1300 °C can easily be reached. Although such an interpretation would help in reconciling the apparent contradiction between numerical models and the geochemical evidence, it would imply, none the less, that the mantle wedge below western Mexico has been under a very complex internal motion. Can plumes be detached from the slab at different depths? Are there more than one set of slab plumes formed at different times? If true, then why is it that some diapirs are able to travel laterally, while others experience only vertical transport? Although answering these questions would require different kinds of studies, in the next section we will evaluate some tectonic implications that may be worth exploring in future research.

Tectonic considerations

It has been well documented that, starting at *c.* 10 Ma, a gradual decrease in convergence rate between Rivera and North America induced slab rollback and promoted the establishment of an extensional triple junction in the overriding plate (Demant 1981; DeMets & Traylen 2000; Ferrari & Rosas-Elguera 2000). Within the context of the TZR, decompression melting should be at least partially responsible for the generation of intraplate magmatism since the Pliocene (Luhr 1997; Ferrari *et al.* 2001; Gómez-Tuena *et al.* 2011), not only because most of these magmas have negligible subduction signatures, but also because the locus of arc magmatism started to appear not much later farther south, close to the location of the current volcanic front. The oldest potassic lamprophyres appeared in the vicinity of Tapalpa at around 4.69 Ma (Allan 1986), and gradually migrated to the NW until reaching the San Sebastian area at 1.52 Ma (Lange & Carmichael 1991). It is during this period of time that the largest volumes of intraplate and rhyolitic magmas were erupted within the TZR (Frey *et al.* 2007). Since the construction of the potassic volcanic front is, for the most part, contemporaneous with the emplacement of intraplate magmas at the rear arc, then the subducted slab must have retreated rather rapidly, and acquired a similar configuration as the one it has today, at least by the Early Pliocene.

Interestingly, during the period between 2.6 and 1 Ma, convergence between Rivera and North America gradually rotated to a direction parallel to the trench, while diminishing its velocity to almost complete stagnation (DeMets & Traylen 2000). Thus, even if only right lateral strike-slip motion occurred at this plate boundary for *c.* 1.6 Ma, both potassic magmatism at the front and intraplate bimodal volcanism at the rear arc were still being generated, indicating that the mantle wedge must have remained in continuous motion despite the stagnation of plate convergence. As speculated by DeMets & Traylen (2000), it is likely that Rivera continued to sink vertically during this period of time, modifying its pressure–temperature path, while influencing extension-induced mantle upwelling at the rear arc. Melting of a steeply dipping slab (Gómez-Tuena *et al.* 2011), or the possible detachment of small fragments of buoyant materials at increasing depth during slab rollback, seems a viable mechanism for the generation of potassic magmas at the volcanic front starting in the Early Pliocene.

Orthogonal convergence to the trench resumed and increased in velocity at about 1 Ma, presumably as a response to a clockwise rotation of the Pacific–Rivera spreading ridge (DeMets & Traylen 2000). Potassic volcanism continued to erupt during that time but became more diffuse, reappearing in the vicinity of Colima volcano; whereas intraplate and rhyolitic volcanism at the rear arc was significantly reduced (Frey *et al.* 2007). Interestingly, it is precisely around this period of resumed convergence that andesitic stratovolcanoes started to appear throughout the region. If these andesites are related to melting of thermochemical slab plumes, as suggested above, then it is likely that the re-establishment of convergence was intimately linked to plume formation. Plumes could have been formed before, of course, but recent numerical simulations have shown that slab rollback is an unfavourable tectonic condition for large diapir development, albeit it can enhance back-arc spreading and decompression melting (Vogt *et al.* 2012), the mechanisms responsible for the formation of intraplate magmatism at the rear arc. Furthermore, if large plumes developed as a response to slab rollback, then there is no reason for them not to be melting and forming andesitic stratovolcanoes during the entire Pliocene period, since slab retreat was well underway by then if not before.

Tectonic compression appears to be the most favourable condition for plume formation (Vogt *et al.* 2012), but no structural indication exists for compression in western Mexico during the past 1 Ma. Yet evidence from bedrock incision rates and emerging palaeoshorelines indicates that the Jalisco Block has been experiencing uplift since the Pliocene, at an increasing rate during the Quaternary (Righter *et al.* 2010; Ramírez-Herrera *et al.* 2011). We thus speculate that the sudden restoration of convergence between Rivera and North America at *c.* 1 Ma could have played a role in the uplift of the Jalisco Block, but also may have facilitated, or even triggered, slab plume formation. That would imply, however, that it will take no more than *c.* 800 ka for slab plumes to be detached from the slab, undergo partial fusion, and erupt to the surface as primitive andesites. Unfortunately, it is not possible to know the exact depth at which these plumes may have detached from the slab; and even if we can be confident that they underwent melting at less than 70 km depth because of the presence of residual amphibole, it is difficult to know its precise melting location. None the less, assuming that plumes ascended from *c.* 70 km depth, then exhumation rates of at least 5 cm a^{-1} will be needed for an ascending diapir to reach the base of the local continental crust (*c.* 40 km) in *c.* 800 ka; while its partial melts should have ascended even more rapidly. Although this minimum velocity is almost twice as fast as the current convergent rate between Rivera and North America, an exhumation rate 5–6 cm a^{-1} is still within the range of most tectonic processes, and similar to those calculated for natural UHP terranes (Ellis *et al.* 2011).

Another important tectonic aspect that deserves attention is that stratovolcanoes in western Mexico have all been constructed in the context of continental extension: they are along the trace of the trench-parallel TZR at the rear arc, or at the centre of the trench-normal Colima rift at the volcanic front (Fig. 1). Importantly, while stratovolcanoes are all relatively recent features, rifting in these areas has been active at least since the early Pliocene (Allan 1986; Ferrari & Rosas-Elguera 2000). Andesitic stratovolcanoes are thus occupying or 'polluting' zones of mantle upwelling and crustal weakness, in which an attenuated and faulted crust may have provided preferential ascent pathways. This is not to say that andesites are the result of extensional tectonics, but that rifting, upwelling and exhumation of deeply seated slab-plumes may all positively reinforce each other, promoting the construction of andesitic stratovolcanoes within a tectonic context that otherwise will only be experiencing decompression-induced mantle melting. In other words, thermo-chemical slab plumes may form at any location above the slab, but they are able to ascend more effectively, and thus melt to a higher extent, if they become trapped into the upwelling regime of a continental rift. Perhaps very few of these plumes meet the necessary conditions to form stratovolcanoes, some may stagnate and underplate the base of the crust if it happens to be too thick, strongly compressed, or insufficiently weak. Likewise, evidence from potassic volcanoes also indicates that not all subducted lithologies become buoyant at relatively shallow pressure, some of them can also be down-dragged to much greater depths, and either melt close to the slab surface, or likely detach from it at a later stage in order to melt in the hot core of the mantle wedge. Although other possibilities can be envisaged, it is clear that the petrological and tectonic complexity of the western Mexican subduction zone are intimately linked, offering the possibility to test emerging hypotheses on arc magmatism.

Concluding remarks

The geochemical evidence has a certain level of ambiguity regarding the origin of andesites, especially when examined in its own exclusive context. Processes like fractional crystallization of a basaltic melt, mixing of basalt and rhyolite, crustal contamination, or direct anatexis of pre-existent arc crust, are all able to satisfy certain observations, and even if they may provide explanations for some of the studied rock suites, they can also be largely inadequate for others. For instance, intraplate basalts in western Mexico do seem to differentiate and contaminate at crustal levels, and are potential precursors to contemporaneous rhyolites, either through re-melting or crystal fractionation. Mixing between these two magmatic end-members may have formed at least some andesites erupted from peripheral volcanoes (Figs 3 & 7). Yet these seemingly ubiquitous processes and components cannot account for the generation of large andesitic stratovolcanoes, which at least require constituents of a different kind.

Subduction of crustal materials and the consequent hybridization of the mantle are inherent to all convergent margins; and, despite variations in their thermal state, geological setting and local tectonic conditions, almost without exception, global arcs produce andesitic volcanoes that are surprisingly homogeneous in terms of bulk compositions. Nevertheless, each volcano shows subtle but important differences in trace element patterns and isotopic compositions, that attest to different contributions of mantle and crustal lithologies and the metamorphic conditions under which partial melting occurs. In this sense, mechanical mixing of subducted crustal materials within the mantle, and either melting of the entire hybrid mélange (Castro *et al.* 2010), or the eventual reaction of siliceous crustal melts with peridotite (Straub *et al.* 2011), have the potential to form andesites without the need for a basaltic precursor.

Melting and recycling may occur directly at the slab surface below most arcs, but evidence from western Mexico makes clear that the subduction fluxes do not always follow a vertical path, and that the compositions of volcanic rocks at the surface may not bear a direct relationship to a specific slab geotherm. Thus, numerical thermal models of a subduction zone and thermochemical tracers like H_2O/Ce ratios are not necessarily a measure of the same phenomenon. Although other processes can be foreseen, and the complex interplay among evolving subduction parameters may have an influence in melt formation (Vogt *et al.* 2012), the existence of thermochemical slab plumes can help to reconcile this apparent discrepancy. Indeed, if subduction mélanges can detach from the down-going slab, their transport pathways will be mainly governed by buoyancy, but they will also be at the mercy of local mantle convection: they can be down-dragged to deeper levels, or be rapidly transported to other locations if they become trapped in convecting cells that follow an opposite direction. Once detached from the slab, the temperature at which they undergo melting could be very different to that at the slab surface (Hacker *et al.* 2011). It is perhaps in this context that large buoyant diapirs can melt at shallow depth and at lower temperatures, creating andesitic stratovolcanoes, whereas smaller plumes can melt at higher depth and participate in the formation of highly potassic cinder cones. Diapiric exhumation of subducted mélanges can thus be an important phenomenon involved in the generation of arc magmas at a global scale, but it is in the peculiar tectonic setting of western Mexico where they can be effectively sorted and isolated at seemingly unexpected locations.

Our most sincere thanks to R. Leija, M. Albarrán, E. Salazar, J. Vázquez, C. Ortega, P. Girón and R. Lozano during sample preparation and analyses. Stimulating discussions with F. Ortega, V. Manea, L. Capra and S. Straub are highly appreciated. The manuscript benefited greatly from careful reviews by C. Petrone and L. Ferrari. Editorial handling by G. Zellmer is also appreciated. B. Díaz-Bravo and A. Vázquez-Duarte were supported by CONACyT graduate studies scholarships. This study was supported by PAPIIT-UNAM project IN107810 to A. Gómez-Tuena.

References

ALLAN, J. F. 1986. Geology of the Colima and Zacoalco grabens, SW Mexico: Late Cenozoic rifting in the Mexican Volcanic Belt. *Geological Society of America Bulletin*, **97**, 473–485.

ALONSO-PEREZ, R., MÜNTENER, O. & ULMER, P. 2009. Igneous garnet and amphibole fractionation in the roots of island arcs: experimental constraints on andesitic liquids. *Contributions to Mineralogy and Petrology*, **157**, 541–558, doi: http://dx.doi.org/10.1007/s00410-008-0351-8

ANNEN, C., BLUNDY, J. D. & SPARKS, R. S. J. 2006. The genesis of intermediate and silicic magmas in deep crustal hot zones. *Journal of Petrology*, **47**, 505–539, doi: http://dx.doi.org/10.1093/petrology/egi084

BANDY, W., URRUTIA-FUCUGAUCHI, J., MCDOWELL, F. & MORTON-BERMEA, O. 2001. K–Ar ages of four mafic lavas from the Central Jalisco Volcanic Lineament: supporting evidence for a NW migration of volcanism within the Jalisco block, western Mexico. *Geofísica Internacional*, **40**, 259–269.

BEHN, M. D., KELEMEN, P. B., HIRTH, G., HACKER, B. R. & MASSONNE, H.-J. 2011. Diapirs as the source of the sediment signature in arc lavas. *Nature Geoscience*, **4**, 641–646, doi: http://dx.doi.org/10.1038/ngeo1214

BLATTER, D. L. & CARMICHAEL, I. S. E. 1998. Hornblende peridotite xenoliths from central Mexico reveal the highly oxidized nature of subarc upper mantle. *Geology*, **26**, 1035–1038.

BLATTER, D. L. & CARMICHAEL, I. S. E. 2001. Hydrous phase equilibria of a Mexican high-silica andesite: a candidate for mantle origin?. *Geochimica et Cosmochimica Acta*, **65**, 4043–4065.

BLATTER, D. L., CARMICHAEL, I. S. E., DEINO, A. L. & RENNE, P. 2003. Discussion and reply: neogene volcanism at the front of the central Mexican Volcanic Belt: basaltic andesites to dacites, with contemporaneous shoshonites and high-TiO2 lava – discussion. *Geological Society of America Bulletin*, **115**, 1021–1024, doi: http://dx.doi.org/10.1130// B25353R.1

BOHRSON, W. A. & REID, M. R. 1995. Petrogenesis of alkaline basalts from Socorro Island, Mexico: trace element evidence for contamination of ocean island basalt in the shallow ocean crust (vol 24, pg 555, 1995). *Journal of Geophysical Research: Solid Earth*, **100**, 24 555–24 576.

BOHRSON, W. A. & REID, M. R. 1997. Genesis of silicic peralkaline volcanic rocks in an Ocean Island setting by crustal melting and open-system processes: Socorro Island, Mexico. *Journal of Petrology*, **38**, 1137–1166, doi: http://dx.doi.org/10.1093/petroj/ 38.9.1137

BOTTAZZI, P., TIEPOLO, M., VANNUCCI, R., ZANETTI, A., BRUMM, R., FOLEY, S. & OBERTI, R. 1999. Distinct site preferences for heavy and light REE in amphibole and the prediction of Amph/LDREE. *Contributions to Mineralogy and Petrology*, **137**, 36–45.

BRENAN, J., SHAW, H., PHINNEY, D. & RYERSON, F. 1994. Rutile-aqueous fluid partitioning of Nb, Ta, Hf, Zr, U and Th: U and Th: implications for high field strength element depletions in island-arc basalts. *Earth and Planetary Science Letters*, **128**, 327–339.

BROWN, M. & RUSHMER, T. 2006. *Evolution and Differentiation of the Continental Crust*. Cambridge University Press, Cambridge.

BRYAN, S. E. & ERNST, R. E. 2008. Revised definition of Large Igneous Provinces (LIPs). *Earth-Science Reviews*, **86**, 175–202, doi: http://dx.doi.org/10. 1016/j.earscirev.2007.08.008

CARMICHAEL, I. S. E. 2002. The andesite aqueduct: perspectives on the evolution of intermediate magmatism in west-central (105°–99°W) Mexico. *Contributions to Mineralogy and Petrology*, **143**, 641–663.

CARMICHAEL, I. S. E., FREY, H. M. & LANGE, R. A. 2006. The Pleistocene cinder cones surrounding Volcán Colima, Mexico re-visited: eruption ages and volumes, oxidation states, and sulfur content. *Bulletin of Volcanology*, **68**, 407–419.

CASTRO, A., GERYA, T., GARCÍA-CASCO, A., FERNÁNDEZ, C., DÍAZ-ALVARADO, J., MORENO-VENTAS, I. & LÖW, I. 2010. Melting relations of MORB–sediment mélanges in underplated mantle wedge plumes; implications for the origin of cordilleran-type batholiths. *Journal of Petrology*, **51**, 1267–1295, doi: http://dx. doi.org/10.1093/petrology/egq019

CHAYES, F. 1963. Relative abundance of intermediate members of the oceanic basalt–trachyte association. *Journal of Geophysical Research*, **68**, 1519–1534.

CHERTKOFF, D. G. & GARDNER, J. E. 2004. Nature and timing of magma interactions before, during, and after the caldera-forming eruption of Volcán Ceboruco, Mexico. *Contributions to Mineralogy and Petrology*, **146**, 715–735, doi: http://dx.doi.org/10.1007/ s00410-003-0530-6

CHESLEY, J., RUIZ, J., RIGHTER, K., FERRARI, L. & GÓMEZ-TUENA, A. 2002. Source contamination versus assimilation: an example from the Trans-Mexican Volcanic Arc. *Earth and Planetary Science Letters*, **195**, 211–221.

CLIFT, P. & VANNUCCHI, P. 2004. Controls on tectonic accretion versus erosion in subduction zones: implications for the origin and recycling of the continental crust. *Reviews of Geophysics*, **42**, RG2001, doi: http://dx.doi.org/10.1029/2003RG000127

COOPER, L. B., RUSCITTO, D. M., PLANK, T., WALLACE, P. J., SYRACUSE, E. M. & MANNING, C. E. 2012. Global variations in H_2O/Ce: 1. Slab surface temperatures beneath volcanic arcs. *Geochemistry, Geophysics, Geosystems*, **13**, 27, doi: http://dx.doi.org/10. 1029/2011GC003902

CURRIE, C. A., BEAUMONT, C. & HUISMANS, R. S. 2007. The fate of subducted sediments: a case for backarc intrusion and underplating. *Geology*, **35**, 1111–1114, doi: http://dx.doi.org/10.1130/G24098A.1

DAVIDSON, J., TURNER, S., HANDLEY, H., MACPHERSON, C. & DOSSETO, A. 2007. Amphibole 'sponge' in arc crust? *Geology*, **35**, 787–790, doi: http://dx.doi.org/10. 1130/G23637A.1

DEFANT, M. & DRUMMOND, M. 1990. Derivation of some modern arc magmas by melting of young subducted lithosphere. *Nature*, **347**, 662–665.

DEMANT, A. 1981. Interpretación geodinámica del volcanismo del Eje Neovolcánico Transmexicano. *Revista Instituto de Geología*, **5**, 217–222.

DEMETS, C. & TRAYLEN, S. 2000. Motion of the Rivera plate since 10 Ma relative to the Pacific and North American and the mantle. *Tectonophysics*, **318**, 119–159.

DEPAOLO, D. 1981. Trace element and isotopic effects of combined wallrock assimilation and fractional crystallization. *Earth and Planetary Science Letters*, **53**, 189–202.

DICKINSON, W. 1975. Potash-depth (k–h) relations in continental margin and intra-oceanic magmatic arcs. *Geology*, **3**, 53–56.

DONNELLY, K. 2002. *The Genesis of E-MORB: Extensions and Limitations of the Hot Spot Model*. Columbia University, Palisades, NY.

ELLIS, S. M., LITTLE, T. A., WALLACE, L. M., HACKER, B. R. & BUITER, S. J. H. 2011. Feedback between rifting and diapirism can exhume ultrahigh-pressure rocks. *Earth and Planetary Science Letters*, **311**, 427–438.

ENGLAND, P. C. & KATZ, R. F. 2010. Melting above the anhydrous solidus controls the location of volcanic arcs. *Nature*, **467**, 700–703, doi: http://dx.doi.org/ 10.1038/nature09417

FERRARI, L. 2004. Slab detachment control on mafic volcanic pulse and mantle heterogeneity in central Mexico. *Geology*, **32**, 77–80.

FERRARI, L. & ROSAS-ELGUERA, J. 2000. Late Miocene to Quaternary extension at the northern boundary of the Jalisco block, western Mexico: the Tepic-Zacoalco rift revised. *In*: AGUIRRE-DÍAZ, G., DELGADO-GRANADOS, H. & STOCK, J. M. (eds) *Cenozoic Tectonics and Volcanism of Mexico*. Geological Society of America Special Papers, **334**, 42–64.

FERRARI, L., ROSAS-ELGUERA, J., MÁRQUEZ, A., OYARZUN, R., DOBLAS, M. & VERMA, S. P. 1999. Alkalic (ocean-island basalt type) and calc-alkalic volcanism in the Mexican volcanic belt: a case for plume-related magmatism and propagating rifting at an active margin?: comment and reply. *Geology*, **27**, 1055–1056, doi: http://dx.doi.org/10.1130/0091-7613(1999)027<1055:AOIBTA>2.3.CO;2

Ferrari, L., Vaggelli, G., Petrone, C. M., Manetti, P. & Conticelli, S. 2000. Late Miocene volcanism and intra-arc tectonics during the early development of the Trans-Mexican Volcanic Belt. *Tectonophysics*, **318**, 161–185.

Ferrari, L., Petrone, C. M. & Francalanci, L. 2001. Generation of oceanic-island basalt-type volcanism in the western Trans-Mexican volcanic belt by slab rollback, asthenosphere infiltration, and variable flux melting. *Geology*, **20**, 507–510.

Ferrari, L., Orozco-Esquivel, T., Manea, V. & Manea, M. 2012. The dynamic history of the Trans-Mexican Volcanic Belt and the Mexico subduction zone. *Tectonophysics*, **522–523**, 122–149, doi: http://dx.doi.org/10.1016/j.tecto.2011.09.018

Foley, S., Tiepolo, M. & Vannucci, R. 2002. Growth of early continental crust controlled by melting of amphibolite in subduction zones. *Nature*, **417**, 837–840, doi: http://dx.doi.org/10.1038/nature00799

Frey, H. & Lange, R. 2011. Phenocryst complexity in andesites and dacites from the Tequila volcanic field, Mexico: resolving the effects of degassing vs. magma mixing. *Contributions to Mineralogy and Petrology*, **162**, 415–445, doi: http://dx.doi.org/10.1007/s00410-010-0604-1

Frey, H. M., Lange, R. A., Hall, C. M., Delgado-Granados, H. & Carmichael, I. S. E. 2007. A Pliocene ignimbrite flare-up along the Tepic–Zacoalco rift: Evidence for the initial stages of rifting between the Jalisco block (Mexico) and North America. *Geological Society of America Bulletin*, **119**, 49–64, doi: http://dx.doi.org/10.1130/B25950.1

Fumagalli, P. & Poli, S. 2005. Experimentally determined phase relations in hydrous peridotites to 6·5 GPa and their consequences on the dynamics of subduction zones. *Journal of Petrology*, **46**, 555–578, doi: http://dx.doi.org/10.1093/petrology/egh088

Gardien, V., Thompson, A. B. & Ulmer, P. 2000. Melting of biotite + plagioclase + quartz gneisses: the role of H_2O in the stability of amphibole. *Journal of Petrology*, **41**, 651–666, doi: http://dx.doi.org/10.1093/petrology/41.5.651

Gerya, T. 2011. Future directions in subduction modeling. *Journal of Geodynamics*, **52**, 344–378, doi: http://dx.doi.org/10.1016/j.jog.2011.06.005

Gerya, T. V. & Yuen, D. A. 2003. Rayleigh–Taylor instabilities from hydration and melting propel ‘cold plumes’ at subduction zones. *Earth and Planetary Science Letters*, **212**, 47–62, doi: http://dx.doi.org/10.1016/S0012-821X(03)00265-6

Gerya, T. V., Yuen, D. A. & Sevre, E. O. D. 2004. Dynamical causes for incipient magma chambers above slabs. *Geology*, **32**, 89–92, doi: http://dx.doi.org/10.1130/G20018.1

Ghiorso, M. S. & Sack, R. O. 1995. Chemical mass transfer in magmatic processes IV. A revised and internally consistent thermodynamic model for the interpolation and extrapolation of liquid-solid equilibria in magmatic systems at elevated temperatures and pressures. *Contributions to Mineralogy and Petrology*, **119**, 197–212, doi: http://dx.doi.org/10.1007/BF00307281

Gómez-Tuena, A., LaGatta, A., Langmuir, C. H., Goldstein, S. L., Ortega-Gutiérrez, F. & Carrasco-Núñez, G. 2003. Temporal control of subduction magmatism in the Eastern Trans-Mexican Volcanic Belt: mantle sources, slab contributions and crustal contamination. *Geochemistry, Geophysics, Geosystems*, **4**, doi: http://dx.doi.org/10.1029/2003GC000524

Gómez-Tuena, A., Langmuir, C. H., Goldstein, S. L., Straub, S. & Ortega-Gutiérrez, F. 2007*a*. Geochemical evidence for slab melting in the Trans-Mexican Volcanic Belt. *Journal of Petrology*, **48**, 537–562.

Gómez-Tuena, A., Orozco-Esquivel, M. T. & Ferrari, L. 2007*b*. Igneous petrogenesis of the transmexican volcanic belt. *In*: Alaniz-Álvarez, S. A. & Nieto-Samaniego, A. F. (eds) *Geology of México: Celebrating the Centenary of the Geological Society of México*. Geologial Society of America Special Papers, **422**, 129–181.

Gómez-Tuena, A., Mori, L., Rincón-Herrera, N. E., Ortega-Guitérrez, F., Solé, J. & Iriondo, A. 2008. The origin of a primitive trondhjemite from the Trans-Mexican Volcanic Belt and its implications for the construction of a modern continental arc. *Geology*, **36**, 471–474.

Gómez-Tuena, A., Mori, L., Goldstein, S. L. & Pérez-Arvizu, O. 2011. Magmatic diversity of western Mexico as a function of metamorphic transformations in the subducted oceanic plate. *Geochimica et Cosmochimica Acta*, **75**, 213–241, doi: http://dx.doi.org/10.1016/j.gca.2010.09.029

Gómez-Tuena, A., Straub, S. M. & Zellmer, G. In prep. Orogenic andesites and crustal growth. *In*: Gómez-Tuena, A., Straub, S. M. & Zellmer, G. F. (eds) *Orogenic Andesites and Crustal Growth*. Geological Society, London, Special Publications, **385**.

Green, T., Blundy, J., Adam, J. & Yaxley, G. 2000. SIMS determination of trace element partition coefficients between garnet, clinopyroxene and hydrous basaltic liquids at 2–7.5 GPa and 1080–12008C. *Lithos*, **53**, 165–187.

Grove, T., Parman, S., Bowring, S., Price, R. & Baker, M. 2002. The role of an H_2O-rich fluid component in the generation of primitive basaltic andesites and andesites from the Mt. Shasta region, N California. *Contributions to Mineralogy and Petrology*, **142**, 375–396, doi: http://dx.doi.org/10.1007/s00410010 0299

Hacker, B. R., Kelemen, P. B. & Behn, M. D. 2011. Differentiation of the continental crust by relamination. *Earth and Planetary Science Letters*, **307**, 501–516, doi: http://dx.doi.org/10.1016/j.epsl.2011.05.024

Hart, S. & Dunn, T. 1993. Experimental cpx/melt partitioning of 24 trace elements. *Contributions to Mineralogy and Petrology*, **113**, 1–8.

Hauri, E., Gaetani, G. & Green, T. 2006. Partitioning of water during melting of the Earth's upper mantle at H_2O-undersaturated conditions. *Earth and Planetary Science Letters*, **248**, 715–734.

Hermann, J. & Spandler, C. 2008. Sediment melts at sub-arc depths: an experimental study. *Journal of Petrology*, **49**, 717–740.

Housh, T. B., Aranda-Gómez, J. J. & Luhr, J. F. 2010. Isla Isabel (Nayarit, México): Quaternary alkalic basalts with mantle xenoliths erupted in the mouth of the Gulf of California. *Journal of Volcanology and Geothermal Research*, **197**, 85–107, doi: http://dx.doi.org/10.1016/j.jvolgeores.2009.06.011

Johnson, K. 1994. Experimental cpx/ and garnet/melt partitioning of REE and other trace elements at high pressures; petrogenetic implications. *Mineralogical Magazine*, **58A**, 454–455.

Johnson, T. 1998. Experimental determination of partition coefficients for rare earth and high-field-strength elements between clinopyroxene, garnet, and basaltic melt at high pressures. *Contributions to Mineralogy and Petrology*, **133**, 60–68.

Kamenov, G. D., Mueller, P. A. & Perfit, M. R. 2004. Optimization of mixed Pb/Tl solutions for high precision isotopic analyses by MC-ICP-MS. *Journal of Analytical Atomic Spectrometry*, **19**, 1262, doi: http://dx.doi.org/10.1039/b403222e

Katz, R. F. & Rudge, J. F. 2011. The energetics of melting fertile heterogeneities within the depleted mantle. *Geochemistry, Geophysics, Geosystems*, **12**, 22, doi: http://dx.doi.org/201110.1029/2011GC003834

Kelemen, P. B. 1998. Silica enrichment in the continental upper mantle via melt/rock reaction. *Earth and Planetary Science Letters*, **164**, 387–406.

Kelemen, P. B., Shimizu, N. & Dunn, T. 1993. Relative depletion of niobium in some arc magmas and the continental crust: partitioning of K, Nb, La and Ce during melt/rock interaction in the upper mantle. *Earth and Planetary Science Letters*, **120**, 111–134.

Kent, A. J. R., Darr, C., Koleszar, A. M., Salisbury, M. J. & Cooper, K. M. 2010. Preferential eruption of andesitic magmas through recharge filtering. *Nature Geoscience*, **3**, 631–636, doi: http://dx.doi.org/10.1038/ngeo924

Kessel, R., Schmidt, M., Ulmer, P. & Pettke, T. 2005. Trace element signature of subduction-zone fluids, melts and supercritical liquids at 120–180 km depth. *Nature*, **437**, 724–727.

Klimm, K., Blundy, J. & Green, T. 2008. Trace element partitioning and accessory phase saturation during H_2O-saturated melting of basalt with implications for subduction zone chemical fluxes. *Journal of Petrology*, **49**, 523–553.

LaGatta, A. 2003. *Arc magma genesis in the eastern Mexican volcanic belt*. PhD thesis, Columbia University.

Lange, R. A. & Carmichael, I. S. E. 1991. A potassic volcanic front in western Mexico: the lamprophyric and related lavas of San Sebastian. *Geological Society of America Bulletin*, **103**, 928–940.

Lassiter, J. L. & Luhr, J. F. 2001. Osmium abundance and isotope variations in mafic Mexican volcanic rocks: evidence for crustal contamination and constraints on the geochemical behavior of osmium during partial melting and fractional crystallization. *Geochemistry, Geophysics, Geosystems*, **2**, doi: http://dx.doi.org/10.1029/2000GC000116

Lehnert, K., Su, Y., Langmuir, C., Sarbas, B. & Nohl, U. 2000. A global geochemical database structure for rocks. *Geochemistry, Geophysics, Geosystems*, **1**, doi: http://dx.doi.org/10.1029/1999GC000026

Lewis-Kenedi, C. B., Lange, R. A., Hall, C. M. & Delgado-Granados, H. 2005. The eruptive history of the Tequila volcanic field, western Mexico: ages, volumes, and relative proportions of lava types. *Bulletin of Volcanology*, **67**, 391–414.

Little, T. A., Hacker, B. R., Gordon, S. M., Baldwin, S. L., Fitzgerald, P. G., Ellis, S. & Korchinski, M. 2011. Diapiric exhumation of Earth's youngest (UHP) eclogites in the gneiss domes of the D'Entrecasteaux Islands, Papua New Guinea. *Tectonophysics*, **510**, 39–68, doi: http://dx.doi.org/10.1016/j.tecto.2011.06.006

Lozano, R. & Bernal, J. P. 2005. Characterization of a new set of eight geochemical reference materials for XRF major and trace element analysis. *Revista Mexicana de Ciencias Geológicas*, **22**, 329–344.

Luhr, J. F. 1992. Slab-derived fluids and partial melting in subduction zones: insights from two contrasting Mexican volcanoes (Colima and Ceboruco). *Journal of Volcanology and Geothermal Research*, **54**, 1–18, doi: http://dx.doi.org/10.1016/0377-0273(92)90111-P

Luhr, J. F. 1997. Extensional tectonics and the diverse primitive volcanic rocks in the western Mexican Volcanic Belt. *Canadian Mineralogist*, **35**, 473–500.

Luhr, J. F. 2000. The geology and petrology of Volcan San Juan Nayarit, Mexico and the compositionally zoned Tepic Pumice. *Journal of Volcanology and Geothermal Research*, **95**, 109–156.

Luhr, J. F. 2002. Petrology and geochemistry of the 1991 and 1998–1999 lava flows from Volcán de Colima, México: implications for the end of the current eruptive cycle. *Journal of Volcanology and Geothermal Research*, **117**, 169–194.

Luhr, J. F. & Carmichael, I. S. E. 1980. The Colima Volcanic Complex, Mexico. I. Post-caldera andesites from Volcán Colima. *Contributions to Mineralogy and Petrology*, **71**, 343–372.

Luhr, J. F. & Carmichael, I. S. E. 1981. The Colima Volcanic Complex, Mexico. II. Late-Quaternary cinder cones. *Contributions to Mineralogy and Petrology*, **76**, 127–147.

Luhr, J. F. & Carmichael, I. S. E. 1985. Jorullo Volcano, Michoacán, Mexico (1759–1774): the earliest stages of fractionation in calc-alkaline magmas. *Contributions to Mineralogy and Petrology*, **90**, 142–161.

Luhr, J. F. & Carmichael, I. S. E. 1990. Petrological monitoring of cyclical eruptive activity at Volcán Colima, México. *Journal of Volcanology and Geothermal Research*, **42**, 235–260.

MacPherson, C., Dreher, S. & Thirlwall, M. F. 2006. Adakites without slab melting: high pressure differentiation of island arc magma, Mindanao, the Philippines. *Earth and Planetary Science Letters*, **243**, 581–593.

Mammerickx, J. & Klitgord, K. 1982. North East Pacific Rise: evolution from 25 m.y.B.P. to the present. *Journal of Geophysical Research*, **87**, 6751–6759.

Manea, M., Manea, V. C. & Kostoglodov, V. 2003. Sediment fill in the Middle America Trench inferred from gravity anomalies. *Geofísica Internacional*, **42**, 603–612.

MARIA, A. H. & LUHR, J. F. 2008. Lamprophyres, basanites, and basalts of the Western Mexican volcanic belt: volatile contents and a vein–wallrock melting relationship. *Journal of Petrology*, **49**, 2123–2156, doi: http://dx.doi.org/10.1093/petrology/egn060

MÁRQUEZ, A., OYARZÚN, R., DOBLAS, M. & VERMA, S. P. 1999. Alkalic (oceanic-island basalt type) and calc-alkalic volcanism in the Mexican volcanic belt: a case for plume-related magmatism and propagating rifting at an active margin? *Geology*, **27**, 51–54.

MARTENS, U. C., BRUECKNER, H. K., MATTINSON, C. G., LIOU, J. G. & WOODEN, J. L. 2012. Timing of eclogite-facies metamorphism of the Chuacús complex, Central Guatemala: record of Late Cretaceous continental subduction of North America's sialic basement. *Lithos*, **146–147**, 1–10, doi: http://dx.doi.org/10.1016/j.lithos.2012.04.021

MARTÍNEZ-SERRANO, R., SCHAAF, P., SOLÍS-PICHARDO, G., HERNÁNDEZ-BERNAL, M. S., HERNÁNDEZ-TREVIÑO, T., MORALES-CONTRERAS, J. J. & MACÍAS, J. L. 2004. Sr, Nd and Pb isotope and geochemical data from the Quaternary Nevado de Toluca volcano, a source of recent adakitic magmatism, and the Tenango Volcanic Field, Mexico. *Journal of Volcanology and Geothermal Research*, **138**, 77–110.

MORI, L., GÓMEZ-TUENA, A., CAI, Y. & GOLDSTEIN, S. L. 2007. Effects of prolonged flat subduction on the Miocene magmatic record of the central Trans-Mexican Volcanic Belt. *Chemical Geology*, **244**, 452–473.

MORI, L., GÓMEZ-TUENA, A., SCHAAF, P., GOLDSTEIN, S. L., PÉREZ-ARVIZU, O. & SOLÍS-PICHARDO, G. 2009. Lithospheric removal as a trigger for flood basalt magmatism in the Trans-Mexican volcanic belt. *Journal of Petrology*, **50**, 2157–2186, doi: http://dx.doi.org/10.1093/petrology/egp072

MOORE, G., MARONE, C., CARMICHAEL, I. S. E. & RENNE, P. 1994. Basaltic volcanism and extension near the intersection of the Sierra Madre volcanic province and the Mexican Volcanic Belt. *Geological Society of America Bulletin*, **106**, 383–394.

MOYEN, J.-F. 2009. High Sr/Y and La/Yb ratios: the meaning of the 'adakitic signature'. *Lithos*, **112**, 556–574, doi: http://dx.doi.org/10.1016/j.lithos.2009.04.001

MÜNTENER, O., KELEMEN, P. & GROVE, T. 2001. The role of H_2O during crystallization of primitive arc magmas under uppermost mantle conditions and genesis of igneous pyroxenites: an experimental study. *Contributions to Mineralogy and Petrology*, **141**, 643–658.

NELSON, S. A. 1980. Geology and petrology of Volcán Ceboruco, Nayarit, México. *Geological Society of America Bulletin*, **91**, 2290–2431.

NELSON, S. A. & HEGRE, J. 1990. Volcán Las Navajas, a Pliocene–Pleistocene trachyte/peralkaline rhyolite volcano in the northwestern Mexican volcanic belt. *Bulletin of Volcanology*, **52**, 186–204.

NELSON, S. A. & LIVIERES, R. A. 1986. Contemporaneous calc-alkaline and alkaline volcanism at Sanganguey Volcano, Nayarit, Mexico. *Geological Society of America Bulletin*, **97**, 798–808.

NIXON, G. T. 1988. Petrology of the younger andesites and dacites of Iztaccíhuatl volcano, México: II. Chemical stratigraphy, magma mixing, and the composition of basaltic magma influx. *Journal of Petrology*, **29**, 265–303.

OWNBY, S. E., LANGE, R. A. & HALL, C. M. 2008. The eruptive history of the Mascota volcanic field, western Mexico: age and volume constraints on the origin of andesite among a diverse suite of lamprophyric and calc-alkaline lavas. *Journal of Volcanology and Geothermal Research*, **177**, 1077–1091, doi: http://dx.doi.org/10.1016/j.jvolgeores.2008.08.002

PARDO, M. & SUÁREZ, G. 1993. Steep subduction geometry of the Rivera plate beneath the Jalisco Block in western Mexico. *Geophysical Research Letters*, **20**, 2391–2394.

PARDO, M. & SUÁREZ, G. 1995. Shape of the subducted Rivera and Cocos plate in southern Mexico: seismic and tectonic implications. *Journal of Geophysical Research*, **100**, 12357–12373.

PARMAN, S. & GROVE, T. 2004. Harzburgite melting with and without H_2O: experimental data and predictive modeling. *Journal of Geophysical Research*, **109**, doi: http://dx.doi.org/10.1029/2003JB002566

PÉREZ-CAMPOS, X., KIM, Y. ET AL. 2008. Horizontal subduction and truncation of the Cocos Plate beneath central Mexico. *Geophysical Research Letters*, **35**, doi: http://dx.doi.org/10.1029/2008GL035127

PETRONE, C. M. 2010. Relationship between monogenetic magmatism and stratovolcanoes in western Mexico: the role of low-pressure magmatic processes. *Lithos*, **119**, 585–606.

PETRONE, C. & FERRARI, L. 2008. Quaternary adakite—Nb-enriched basalt association in the western Trans-Mexican Volcanic Belt: is there any slab melt evidence? *Contributions to Mineralogy and Petrology*, **156**, 73–86, doi: http://dx.doi.org/10.1007/s00410-007-0274-9

PETRONE, C. M., FRANCALANCI, L., CARLSON, R. W., FERRARI, L. & CONTICELLI, S. 2003. Unusual coexistence of subduction-related and intraplate-type magmatism: Sr, Nd and Pb isotope and trace element data from the magmatism of the San Pedro–Ceboruco graben (Nayarit, Mexico). *Chemical Geology*, **193**, 1–24.

PLANK, T., COOPER, L. B. & MANNING, C. E. 2009. Emerging geothermometers for estimating slab surface temperatures. *Nature Geoscience*, **2**, 611–615, doi: http://dx.doi.org/10.1038/ngeo614

POLI, S. & SCHMIDT, M. 2002. Petrology of subducted slabs. *Annual Review of Earth and Planetary Sciences*, **30**, 207–235.

RAMÍREZ-HERRERA, M. T., KOSTOGLODOV, V. & URRUTIA-FUCUGAUCHI, J. 2011. Overview of recent coastal tectonic deformation in the Mexican Subduction Zone. *Pure and Applied Geophysics*, **168**, 1415–1433, doi: http://dx.doi.org/10.1007/s00024-010-0205-y

RAPP, R. 1995. The amphibole-out phase boundary in partially melted metabasalt, and its control over melt fraction and composition, and source permeability. *Journal of Geophysical Research*, **100**, 15 601–15 610.

RAPP, R. & WATSON, E. 1995. Dehydration melting of metabasalt at 8–32 kb: implications for continental growth and crust-mantle recycling. *Journal of Petrology*, **36**, 891–931.

RAPP, R., SHIMIZU, N., NORMAN, M. & APPLEGATE, G. 1999. Reaction between slab-derived melts and peridotite in the mantle wedge: experimental constraints at 3.8 GPA. *Chemical Geology*, **160**, 335–256.

REICHARDT, H. & WEINBERG, R. F. 2012. Hornblende chemistry in meta- and diatexites and its retention in the source of leucogranites: an example from the Karakoram Shear Zone, NW India. *Journal of Petrology*, **53**, 1287–1318, doi: http://dx.doi.org/10.1093/petrology/egs017

REUBI, O. & BLUNDY, J. 2008. Assimilation of plutonic roots, formation of high-K 'exotic' melt inclusions and genesis of andesitic magmas at volcán Dd colima, Mexico. *Journal of Petrology*, **49**, 2221–2243, doi: http://dx.doi.org/10.1093/petrology/egn066

REUBI, O. & BLUNDY, J. 2009. A dearth of intermediate melts at subduction zone volcanoes and the petrogenesis of arc andesites. *Nature*, **461**, 1269–1273, doi: http://dx.doi.org/10.1038/nature08510

RIGHTER, K., CAFFEE, M., ROSAS-ELGUERA, J. & VALENCIA, V. 2010. Channel incision in the Rio Atenguillo, Jalisco, Mexico, defined by 36Cl measurements of bedrock. *Geomorphology*, **120**, 279–292, doi: http://dx.doi.org/10.1016/j.geomorph.2010.04.001

RINGWOOD, A. E. & GREEN, D. H. 1966. An experimental investigation of the Gabbro-Eclogite transformation and some geophysical implications. *Tectonophysics*, **3**, 383–427, doi: http://dx.doi.org/10.1016/0040-1951(66)90009-6

RUDNICK, R. & GAO, S. 2003. Composition of the continental crust. *In*: RUDNICK, R. (ed.) *Treatise on Geochemistry*. Elsevier, Oxford, 1–64.

SALTERS, V. J. M. & LONGHI, J. 1999. Trace element partitioning during the initial stages of melting beneath mid-ocean ridges. *Earth and Planetary Science Letters*, **166**, 15–30, doi: http://dx.doi.org/10.1016/S0012-821X(98)00271-4

SAVOV, I. P., LUHR, J. F. & NAVARRO-OCHOA, C. 2008. Petrology and geochemistry of lava and ash erupted from Volcán Colima, Mexico, during 1998–2005. *Journal of Volcanology and Geothermal Research*, **174**, 241–256, doi: http://dx.doi.org/10.1016/j.jvolgeores.2008.02.007

SCHAAF, P., STIMAC, J., SIEBE, C. & MACÍAS, J. L. 2005. Geochemical evidence for mantle origin and crustal processes in volcanic rocks from Popocatépetl and surrounding monogenetic volcanoes, Central Mexico. *Journal of Petrology*, **46**, 1243–1282.

SEN, C. & DUNN, T. 1994. Dehydration melting of a basaltic composition amphibolite at 1.5 and 2.0 GPa: implications for the origin of adakites. *Contributions to Mineralogy and Petrology*, **117**, 394–409.

SHELLNUTT, J. G. & ZELLMER, G. F. 2010. High-Mg andesite genesis by upper crustal differentiation. *Journal of the Geological Society, Lonndon*, **167**, 1081–1088, doi: http://dx.doi.org/10.1144/0016-76492010-070

SIEBE, C., RODRÍGUEZ-LARA, V., SCHAAF, P. & ABRAMS, M. 2004. Geochemistry, Sr–Nd isotope composition, and tectonic setting of Holocene Pelado, Guespalapa and Chichinautzin scoria cones, south of Mexico City. *Journal of Volcanology and Geothermal Research*, **130**, 197–226.

SIERON, K. & SIEBE, C. 2008. Revised stratigraphy and eruption rates of Ceboruco stratovolcano and surrounding monogenetic vents (Nayarit, Mexico) from historical documents and new radiocarbon dates. *Journal of Volcanology and Geothermal Research*, **176**, 241–264, doi: http://dx.doi.org/10.1016/j.jvolgeores.2008.04.006

STRAUB, S. & MARTIN-DEL POZZO, A. L. 2001. The significance of phenocryst diversity in tephra from recent eruptions at Popocatépetl Stratovolcano (central Mexico). *Contributions to Mineralogy and Petrology*, **140**, 487–510.

STRAUB, S. M., GOMEZ-TUENA, A., STUART, F. M., ZELLMER, G. F., ESPINASA-PERENA, R., CAI, Y. & IIZUKA, Y. 2011. Formation of hybrid arc andesites beneath thick continental crust. *Earth and Planetary Science Letters*, **303**, 337–347, doi: http://dx.doi.org/10.1016/j.epsl.2011.01.013

STRAUB, S. M., GÓMEZ-TUENA, A. *ET AL.* 2013. The processes of melt differentiation in arc volcanic rocks: insights from OIB-type arc magmas in the Central Mexican Volcanic Belt. *Journal of Petrology*, **54**, 665–701, doi: http://dx.doi.org/10.1093/petrology/egs081

SYRACUSE, E. M., VAN KEKEN, P. E. & ABERS, G. A. 2010. The global range of subduction zone thermal models. *Physics of the Earth and Planetary Interiors*, **183**, 73–90, doi: http://dx.doi.org/10.1016/j.pepi.2010.02.004

TANAKA, T., TOGASHI, S. *ET AL.* 2000. JNdi-1: a neodymium isotopic reference in consistency with LaJolla neodymium. *Chemical Geology*, **168**, 279–281, http://dx.doi.org/10.1016/S0009-2541(00)00198-4

TATSUMI, Y. 1982. Origin of high-magnesian andesites in the Setouchi volcanic belt, southwest Japan, II. Melting phase relations at high pressures. *Earth and Planetary Science Letters*, **60**, 305–317, doi: http://dx.doi.org/10.1016/0012-821X(82)90009-7

TAYLOR, S. R. 1967. The origin and growth of continents. *Tectonophysics*, **4**, 17–34, doi: http://dx.doi.org/10.1016/0040-1951(67)90056-X

TIEPOLO, M., VANNUCCI, R., BOTTAZZI, P. R. O., ZANETTI, A. & FOLEY, S. 2000*a*. Partitioning of rare earth elements, Y, Th, U, and Pb between pargasite, kaersutite, and basanite to trachyte melts: implications for percolated and veined mantle. *Geochemistry Geophysics Geosystems*, **1**, 1039–32, doi: http://dx.doi.org/10.1029/2000GC000064

TIEPOLO, M., VANNUCCI, R., OBERTI, R., FOLEY, S., BOTTAZZI, P. & ZANETTI, A. 2000*b*. Nb and Ta incorporation and fractionation in titanian pargasite and kaersutite: crystal-chemical constraints and implications for natural systems. *Earth and Planetary Science Letters*, **176**, 185–201.

TODT, W., CLIFF, R., HANSER, A. & HOFMANN, A. W. 1996. Evaluation of a 202Pb–205Pb double spike for high-precision lead isotope analysis. *In*: BASU, A. & HART, S. (eds) *Earth Processes: Reading the Isotopic Code*. American Geophysical Union, Washington, DC, Geophysical Monograph Series, **95**, 429–437, doi: http://dx.doi.org/10.1029/GM095

Torres-Alvarado, I. S. & Verma, S. P. 2003. Discussion and reply: Neogene volcanism at the front of the central Mexican Volcanic Belt: basaltic andesites to dacites, with contemporaneous shoshonites and high-TiO2 lava—discussion. *Geological Society of America Bulletin*, **115**, 1020–1021, doi: http://dx.doi.org/10.1130//B25276D.1

Valdez-Moreno, G., Schaaf, P., Macías, J. L. & Kusakabe, M. 2006. New Sr–Nd–Pb–O isotope data for Colima volcano and evidence for the nature of the local basement. *Geological Society of America Special Papers*, **402**, 45–63, doi: http://dx.doi.org/10.1130/2006.2402(02)

Verma, S. P. 1999. Geochemistry of evolved magmas and their relationship to subduction-unrelated mafic volcanism at the volcanic front of the central Mexican Volcanic Belt. *Journal of Volcanology and Geothermal Research*, **93**, 151–171.

Verma, S. P. 2002. Absence of Cocos plate subduction-related basic volcanism in southern Mexico: a unique case on Earth? *Geology*, **30**, 1095–1098.

Verma, S. P. & Nelson, S. A. 1989. Isotopic and trace element constraints on the origin and evolution of alkaline and calc-alkaline magmas in the northwestern Mexican volcanic belt. *Journal of Geophysical Research*, **94**, 4531–4544.

Vigouroux, N., Wallace, P. J. & Kent, A. J. R. 2008. Volatiles in high-K magmas from the Western Trans-Mexican Volcanic Belt: evidence for fluid fluxing and extreme enrichment of the mantle wedge by subduction processes. *Journal of Petrology*, **49**, 1589–1618, doi: http://dx.doi.org/10.1093/petrology/egn039

Vogt, K., Gerya, T. V. & Castro, A. 2012. Crustal growth at active continental margins: numerical modeling. *Physics of the Earth and Planetary Interiors*, **192–193**, 1–20.

Wallace, P. J. & Carmichael, I. S. E. 1994. Petrology of Volcán Tequila, Jalisco, México: disequilibrim phenocryst assemblages and evolution of the subvolcanic magma system. *Contributions to Mineralogy and Petrology*, **117**, 345–361.

Weber, R. M., Wallace, P. J. & Dana Johnston, A. 2012. Experimental insights into the formation of high-Mg basaltic andesites in the trans-Mexican volcanic belt. *Contributions to Mineralogy and Petrology*, **163**, 825–840, doi: http://dx.doi.org/10.1007/s00410-011-0701-9

Wilcox, R. E. 1954. Petrology of Paricutin Volcano, Mexico. *United States Geological Survey Bulletin*, **965-C**, 281–354.

Xiong, X., Adam, J. & Green, T. 2005. Rutile stability and rutile/melt HFSE partitioning during partial melting of hydrous basalt: implications for TTG genesis. *Chemical Geology*, **218**, 339–359.

Yang, T., Grand, S. P., Wilson, D., Guzmán-Speziale, M., Gómez-González, J. M., Domínguez-Reyes, T. & Ni, J. 2009. Seismic structure beneath the Rivera subduction zone from finite-frequency seismic tomography. *Journal of Geophysical Research*, **114**, doi: http://dx.doi.org/10.1029/2008JB005830

Zellmer, G. F., Iizuka, Y., Miyoshi, M., Tamura, Y. & Tatsumi, Y. 2012. Lower crustal H_2O controls on the formation of adakitic melts. *Geology*, **40**, 487–490, doi: http://dx.doi.org/10.1130/G32912.1

Reappraisal of uranium-series isotope data in Kamchatka lavas: implications for continental arc magma genesis

ANTHONY DOSSETO[1]* & SIMON TURNER[2]

[1]*Wollongong Isotope Geochronology Lab, GeoQuEST Research Centre, School of Earth and Environmental Sciences, University of Wollongong, Northfields Avenue, Wollongong NSW 2522, Australia*

[2]*Department of Earth and Planetary Sciences, GEMOC Centre, Macquarie University, North Ryde NSW 2109, Australia*

**Corresponding author (e-mail: tonyd@uow.edu.au)*

Abstract: Uranium-series isotopes can be used to determine constraints on the timescale of slab dehydration and melt production at subduction zones. However, interpretations of U–Th–Ra data suggest very different timescales of slab dehydration. Here, we present new U–Th–Ra data from Kamchatka along with a number of alternative models for production of radioactive disequilibrium. Variations in (^{226}Ra/^{230}Th) and (^{231}Pa/^{235}U) activity ratios are best explained by crystal fractionation with host rock assimilation for a duration of less than *c.* 6000 years. The association of the largest ^{226}Ra excesses with high Sr/Th in the most primitive lavas suggests that Ra–Th fractionation is controlled by slab dehydration less than 10 ka ago. We show that U–Th data can be explained by dynamic melting of a recently (<10 ka) metasomatized mantle wedge. Dynamic melting of an oxidized source metasomatized several hundreds of thousands of years ago cannot produce significant ^{231}Pa excess. Because ^{238}U–^{230}Th disequilibrium is inferred to be controlled by partial melting, there is no requirement for multi-stage slab dehydration commencing ~150 ka. We suggest that Ra–Th disequilibria constrain the timing of slab dehydration, whilst U–Th fractionation is dominated by partial melting, at least at the Kamchatka arc.

Over the past twenty years, uranium-series isotopes have been used to quantify the timescale of magmatic processes (e.g. Bourdon *et al.* 2003*a*, *b*; Dosseto *et al.* 2010). This is because isotopes of the uranium decay chains fractionate during magma production and differentiation, and their abundance is time-dependent. Thus, radioactive disequilibrium in the ^{238}U–^{230}Th–^{226}Ra and ^{235}U–^{231}Pa systems measured in volcanic rocks that have undergone little differentiation (or for which differentiation can be accounted for) can be modelled to constrain conditions of partial melting (e.g. melting rate, porosity; McKenzie 1985; Rubin & Macdougall 1988; Spiegelman & Elliott 1993; Sims *et al.* 1995, 1999, 2002, 2008*a*, *b*; Asmerom *et al.* 2000; Spiegelman 2000; Zou & Zindler 2000; Landwehr *et al.* 2001; Bourdon *et al.* 2003*a*, *b*; Dosseto *et al.* 2003; Rubin *et al.* 2005; Turner *et al.* 2005; Stracke *et al.* 2006; Petford *et al.* 2008; Bourdon & Elliott 2010). Radioactive disequilibrium in co-genetic volcanic rocks can also be used to model rates of crystallization, since disequilibrium generally decreases over timescales similar to that of magma differentiation. Previous studies have inferred crystallization rates ranging from 10^{-5} to 6×10^{-4} a^{-1} (Sigmarsson 1996; Turner & Hawkesworth 1997; Turner *et al.* 1999, 2001; Cooper *et al.* 2001; Cooper & Reid 2003; Rogers *et al.* 2004; Blake & Rogers 2005; Peate 2005; Sigmarsson *et al.* 2005; Price *et al.* 2007; Touboul *et al.* 2007; Dosseto *et al.* 2008; Dosseto & Turner 2010). Moreover, in subduction zones, U-series isotope disequilibrium can be used to constrain the timing of slab dehydration under the volcanic arc (e.g. Elliott *et al.* 1997; Hawkesworth *et al.* 1997; Turner & Hawkesworth 1997; Turner *et al.* 1999, 2001; Thomas *et al.* 2002; Bourdon *et al.* 2003*a*, *b*; Dosseto *et al.* 2003; Reagan *et al.* 2006; Tepley III *et al.* 2006; Huang *et al.* 2008; DuFrane *et al.* 2009): previous studies have argued that excess of uranium-238 (^{238}U) over thorium-230 (^{230}Th) in arc lavas requires that slab dehydration occurs over timescales of tens to hundreds of thousands of years (e.g. Turner & Hawkesworth 1997). However, the observation of radium-226 (^{226}Ra) excesses over thorium-230 (^{230}Th) suggests that slab-derived fluids have been added less than 10 000 years before eruption, implying fast melting rates and rapid magma transport towards the surface (tens of m a^{-1}; Turner *et al.* 2001). In one model, the apparent discrepancy in ages of slab dehydration from ^{230}Th–^{238}U and

From: Gómez-Tuena, A., Straub, S. M. & Zellmer, G. F. (eds) 2014. *Orogenic Andesites and Crustal Growth*. Geological Society, London, Special Publications, **385**, 103–116.
First published online June 11, 2013, http://dx.doi.org/10.1144/SP385.5

$^{226}Ra-^{230}Th$ disequilibria was resolved by multi-stage slab dehydration, where $^{226}Ra-^{230}Th$ disequilibrium reflects the last increments of fluid addition (Turner *et al.* 2000).

In the Kamchatka arc, Dosseto *et al.* (2003) suggested that lava compositions are best explained by mixing between a melt derived from slab melting, which has a ^{230}Th excess, and a melt produced during partial melting of the metasomatized mantle, which has ^{238}U and ^{226}Ra excess. Based on the inferred end-member composition of the mantle-derived melt, they proposed that the mantle was hydrated during two episodes of slab dehydration at 150 ka and <10 ka. However, a subsequent study by Turner *et al.* (2007) did not observe ^{230}Th excesses in the same Kamchatka lavas. Although the potential absence of ^{230}Th excess would not affect conclusions on the timing of slab dehydration and production of melts in the mantle, this could preclude the existence of a slab-derived melt and this has motivated us to re-visit the samples analysed by Dosseto *et al.* (2003). The new U–Th–Ra data are discussed from the perspective of the ongoing debate on the apparent discrepancy between the slab dehydration timescales derived from $^{238}U-^{230}Th$ and $^{226}Ra-^{230}Th$ disequilibria (e.g. Turner *et al.* 2007; DuFrane *et al.* 2009; Avanzinelli *et al.* 2012).

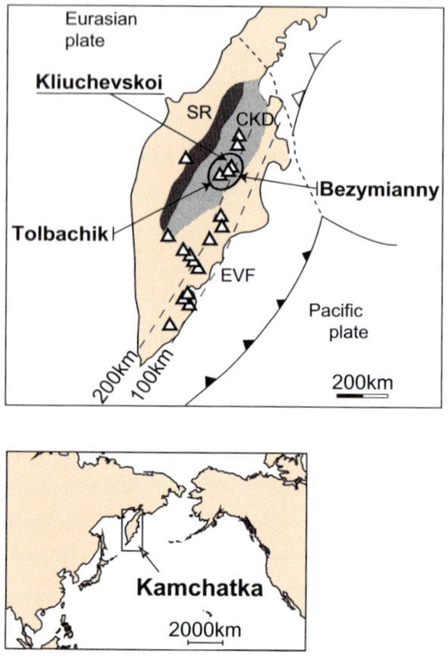

Fig. 1. Map of the Kamchatka peninsula with the three main volcanic regions: the Central Kamchatka Depression (CKD), the Sredinny Ridge (SR) and the Eastern Volcanic Front (EVF). Triangles show the location of the main active volcanoes. Dashed curves show the depth of the subducting plate surface. Modified from (Dosseto *et al.* 2003).

Study area and geochemistry of samples

The Kamchatka Peninsula is located in the north-western corner of the Pacific Ocean, where the Pacific plate, with an age of 80–90 Ma, subducts under the Eurasian plate at rate of 9 cm a^{-1} and with a dip angle of 49° (Fig. 1; DeMets *et al.* 1990). The Kamchatka arc is one of the most active volcanic arcs in the world, with the Kliuchevskoi volcano alone erupting *c.* 55 × 10^6 tons of magma per year (Melekestsev *et al.* 1991). Both tholeiitic and calc-alkaline lavas are produced, with lithologies varying from high-magnesian basalts to dacites and rhyolites. Radiogenic isotope compositions preclude any significant involvement of sediments in the source of the magmas (Kersting & Arculus 1995) but a significant contribution from a slab-derived fluid is inferred. This, along with the observation of enrichment in large ion lithophile elements (LILEs), suggests that the magmas are derived from partial melting of hydrated peridotite.

The samples analysed here are the same as those presented in Dosseto *et al.* (2003) and were collected within a 50 km diameter area from the Kliuchevskoi, Bezymianny and Tolbachik volcanoes (Fig. 1). These volcanoes are located in the Central Kamchatka Depression (CKD), one of the three volcanic regions of the peninsula. Samples analysed are two basalts from Kliuchevskoi, three from Tolbachik, one basic and two acid andesites from Bezymianny. Two of the basalts from Kliuchevskoi studied in Dosseto *et al.* (2003) were not available for analysis. The major, trace element and radiogenic isotope compositions were previously discussed in Dosseto *et al.* (2003). Sr and Nd isotope compositions suggest that studied rocks are derived from the same source (Dosseto *et al.* 2003). All samples are aphyric to subaphyric, except for sample PT30-4/1976 which is vesicular and has glomerocrysts of plagioclase and olivine (<1 cm). This sample also differs from the rest because it belongs to the high-K calc-alkaline series whereas other samples are calc-alkaline (following Miyashiro (1974) classification). It also has a different trace element pattern shifted to higher incompatible element concentrations. Oxygen isotope data indicate the samples are devoid of any alteration by meteoric water (Pineau *et al.* 1999).

Analytical techniques

Analyses were performed on the same powders as in Dosseto *et al.* (2003) (samples were not re-crushed). About 100 mg of powder was spiked with $^{236}U-^{229}Th$ and ^{228}Ra tracers. Samples were digested in a mixture of $HClO_4$, HCl, HNO_3 and HF. After driving off fluorides by evaporation at 100 °C, residual $HClO_4$ was dried down by step heating at 150, 170 and 200 °C. U, Th and Ra separation was performed using ionic chromatography techniques described in Dosseto *et al.* (2006) and Sims *et al.* (2008*a*, *b*).

U and Th concentrations and isotopic compositions were measured by multi-collector inductively-coupled plasma mass spectrometry (MC ICP-MS) on a Nu Instrument Nu Plasma at Macquarie University. Acquisition methods are described in Sims *et al.* (2008*a*, *b*). MC ICP-MS yields beams of several volts for ^{232}Th and >1000 counts per second (cps) for ^{230}Th, compared to only a few tens of mV for ^{232}Th and <100 cps for ^{230}Th by thermal ionization mass spectrometry (TIMS) using tungsten filaments in Dosseto *et al.* (2003). Ra concentration was determined by TIMS on a Thermo Triton at Macquarie University. ^{226}Ra and ^{228}Ra were measured dynamically on a secondary electron multiplier.

Accuracy and reproducibility of measurements were checked by analysis of the rock standard TML-3 and replicate analysis of two samples. For TML-3, (^{234}U/^{238}U) = 1.003 ± 0.003 (2σ; $n = 10$), (^{230}Th/^{238}U) = 0.994 ± 0.006 (2σ; $n = 10$), and the ^{230}Th/^{232}Th ratio deviates from the recommended value (5.816×10^{-6}; (Sims *et al.* 2008*a*, *b*)) by 0.5% ($n = 10$). Replicate analysis of two samples is given in Table 1.

Results

When compared with the original data (Dosseto *et al.* 2003), significant discrepancies are evident (Table 1). Note that reported relative 2-standard errors are calculated by taking the worst replicate analysis for each parameter (for instance PT33-1/82 for (^{226}Ra/^{230}Th) but AXO3/84 for (^{234}U/^{238}U)). The new results generally show a similar range of variation in U, Th and ^{226}Ra concentrations to those reported in Dosseto *et al.* (2003): between 0.333 and 1.035 ppm, 0.534 to 1.687 ppm and 188.6 to 676.8 fg g^{-1}, respectively. However, differences in U or Th concentrations up to 50% were observed for some samples. It is possible that in the initial study, the sample-spike homogenization was not complete and/or samples were not thoroughly powdered. ^{231}Pa concentrations were not determined in this study; however, (^{231}Pa/^{235}U) activity ratios were re-calculated using the new U concentrations and the natural ^{238}U/^{235}U ratio of 137.88.

All samples are in secular equilibrium for ^{234}U–^{238}U, confirming that they have not undergone significant low-temperature alteration, except for AXO3/84 which shows a slight depletion of ^{234}U relative to ^{238}U. (^{226}Ra/^{230}Th) activity ratios are also similar to those in Dosseto *et al.* (2003), with values between 0.98 and 3.17 (Fig. 2). Re-calculated (^{231}Pa/^{235}U) ratios based on new U concentrations but old ^{231}Pa abundances also show significant ^{231}Pa excess, as previously observed (Fig. 2). While we were not able to re-measure ^{231}Pa concentrations, the quality of the ^{231}Pa data is not a concern because the method employed to determine ^{231}Pa concentrations is similar to that used for ^{226}Ra concentrations, and old and new ^{226}Ra concentrations show a relatively good agreement. Like ^{226}Ra abundances, ^{231}Pa is determined by isotope dilution with ^{233}Pa, and although beam sizes at masses 263 ($^{231}PaO_2$) and 265 ($^{233}PaO_2$) were small (a few 100s of cps), the measured 263/265 ratios were close to 1, ensuring good precision and accuracy. The most striking difference with the previous study is in the (^{230}Th/^{238}U) ratios, where the new data indicate only values less than 1 (i.e. ^{238}U excess). Although sample heterogeneity could explain part of these discrepancies, this cannot explain the shift from ^{230}Th to ^{238}U excess. This difference could be derived from the difficulties encountered by Dosseto *et al.* (2003) in measuring ^{230}Th/^{232}Th ratios by TIMS using tungsten filaments on a Finnigan MAT262: less than 100 cps of ^{230}Th and only a few mV of ^{232}Th were obtained during analysis, and satisfying precision and accuracy could be difficult to achieve considering the large ^{232}Th/^{230}Th ratio measured (compared to isotopic dilution, e.g. ^{228}Ra/^{226}Ra or $^{233}PaO_2$/$^{231}PaO_2$, where ratios are close to unity).

Discussion

Reappraisal of the origin of CKD magmas

As shown above, although Ra–Th data do not show major discrepancies between the new and old measurements, U–Th data are significantly different and in this section we revisit the interpretations of Dosseto *et al.* (2003) in the light of the new data.

^{230}Th excesses were previously observed and this was interpreted as indicating the involvement of a slab-derived melt. The new data show no evidence for ^{230}Th excess and hence there is no requirement for such a slab-derived melt. The range in observed U–Th–Ra–Pa compositions was previously explained by mixing between a slab-derived melt and a mantle-derived melt, based on correlations between daughter–parent activity ratios. Although the new measurements show no correlation between (^{230}Th/^{238}U) and other daughter–parent activity ratios, the positive correlation between (^{226}Ra/^{230}Th) and (^{231}Pa/^{235}U) is still observed (Fig. 2b; (^{231}Pa/^{235}U) ratios were

Table 1. *Geochemical data for Kamchatka samples*

Sample name	Eruption date	Th (ppm)	U (ppm)	^{226}Ra (fg g^{-1})	Sr (ppm)	(^{234}U/^{238}U)	(^{230}Th/^{238}U)	(^{226}Ra/^{230}Th)	(^{231}Pa/^{235}U)	(^{238}U/^{232}Th)	(^{230}Th/^{232}Th)	Sr/Th
Tolbachik												
PT30-4/1976	August 1976	1.687	1.035	676.8	358	0.992	0.982	1.97	1.69	1.861	1.829	212
PT33-1/82	Holocene	0.884	0.557	188.6	285	0.996	0.986	1.02	1.13	1.914	1.887	322
PT33-1/82-R		0.834	0.524	190.1		0.998	0.989	1.09	1.20	1.908	1.887	
T1475/1975	30 July 1975	0.534	0.333	346.0	461	0.995	0.970	3.17	1.69	1.892	1.835	863
Bezymianny												
M-1564-1	post-1956 cycle	1.298	0.917	285.7	323	0.993	0.939	0.98	1.00	2.143	2.013	249
B82	1982	1.126	0.794	342.9	353	0.998	0.935	1.37	1.42	2.141	2.003	313
B8/4	1986	0.735	0.517	326.1	323	0.990	0.938	1.99	2.08	2.136	2.003	440
Kliuchevskoi												
AXO3/84	1984	0.736	0.481	261.5	376	0.982	0.940	1.71	1.26	1.983	1.864	511
AXO3/84-R		0.725	0.477	254.1		0.975	0.937	1.68	1.27	1.997	1.871	
K56P66	1966	0.722	0.473	264.5	362	0.999	0.962	1.72	1.37	1.991	1.916	502
Gravimetric standards												
BCR-2 ($n = 4$)		5.628	1.612			0.991	0.999			1.001	0.868	
TML-3 ($n = 11$ for U–Th; $n = 1$ for Ra)		29.69	10.59	3550		1.003	0.994	0.996		1.006	1.067	
Relative 2-standard error (%)		*0.0*	*0.0*	*0.0*		*0.0*	*0.0*	*0.0*	*0.0*	*0.0*	*0.0*	

Relative 2-standard errors are calculated using the worst sample replicate analysis (PT33-1/82 or AXO3/84).

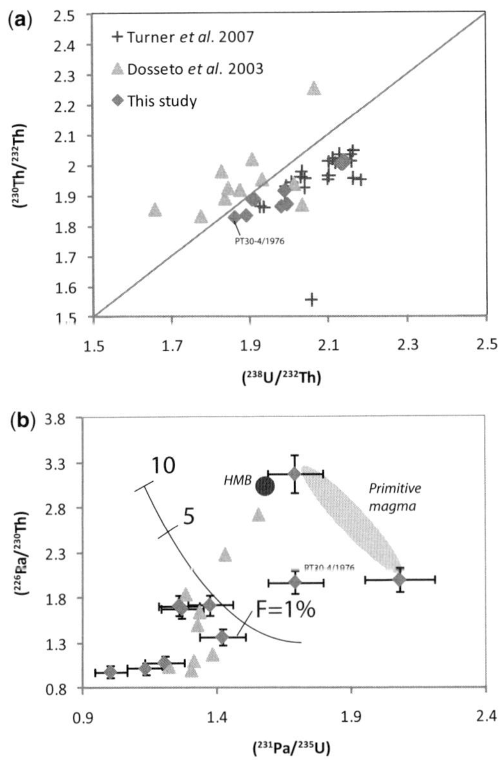

Fig. 2. U-series isotope compositions of Kamchatka lavas. Data from this study (diamonds) and Dosseto *et al.* (2003) (triangles) were obtained on the same samples. Crosses: Turner *et al.* (2007). (**a**) Significant differences are observed in a (^{230}Th/^{232}Th) v. (^{238}U/^{232}Th) equiline diagram between this study and Dosseto *et al.* (2003). This illustrates that this discrepancy is mostly the result of the poor quality of the Th data previously published. (**b**) The positive correlation between (^{226}Ra/^{230}Th) and (^{231}Pa/^{235}U) is still observed with the new measurements. Unlike in Dosseto *et al.* (2003) where the correlation is believed to represent mixing between melts derived from slab melting on the one hand, and mantle melting on the other hand, here it is proposed that magma differentiation produces this correlation, with a decrease in radioactive disequilibrium as differentiation takes place (see text for details). Compositions with the largest (^{226}Ra/^{230}Th) and (^{231}Pa/^{235}U) disequilibria are believed to represent those of a primitive melt produced from a metasomatized mantle: termed *HMB* in Dosseto *et al.* (2003) (dark grey dot) or *primitive magma* (light grey area). The curve shows calculated compositions using a continuous flux melting model as in Dosseto *et al.* (2003). It is clear that following this model, the production of ^{226}Ra excess requires a decrease in the (^{231}Pa/^{235}U) which is not observed here. Errors are not shown for Dosseto *et al.* (2003) and Turner *et al.* (2007) data.

re-calculated using new U concentrations, but samples were not re-analysed for ^{231}Pa). Accordingly, we suggest that these variations no longer reflect mixing between two primitive magmas with different origins (slab and mantle), but instead magma chamber processes whereby the most primitive magma is characterized by the largest radioactive disequilibria (see discussion below). Thus, the end-member composition of the mantle-derived melt in Dosseto *et al.* (2003), named HMB for 'high-magnesia basalt' and characterized by the largest ^{226}Ra excess, represents the composition of the primitive magma (Fig. 2b) and still constrains the origin of CKD magmas. Similarly to the conclusions in Dosseto *et al.* (2003), it is apparent on Figure 2b that flux melting cannot adequately explain the data as this will produce a decrease in (^{231}Pa/^{235}U) with increasing ^{226}Ra excess, whereas the inferred primitive magma is characterized by high (^{226}Ra/^{230}Th) and (^{231}Pa/^{235}U) ratios.

Rapid magma differentiation beneath the Central Kamchatka Depression

^{226}Ra–^{230}Th and ^{231}Pa–^{235}U disequilibria shows significant variations with Th concentrations (Fig. 3), used here as an index of magma differentiation. The broad negative correlations between the magnitude of these disequilibria and Th concentrations is interpreted as reflecting decay of radionuclides during differentiation along with fractionation and possible assimilation of host rock. This relationship can be used to place time constraints on the rate of magma differentiation in the crust. We use the model of assimilation and crystal fractionation presented in Dosseto *et al.* (2008). In this model, the change in activity ratios during differentiation is mostly controlled by the rate of crystallization (in g g^{-1} a^{-1}), and the assimilation/crystallization ratio. Parameters used in the model are shown in Table 2. The maximum residual magma fraction allowed is 0.4 as below this value, magma viscosity may become too high to allow physical separation of newly formed crystals. The U, Th, Sr concentrations and (^{226}Ra/^{230}Th) and (^{231}Pa/^{235}U) activity ratios in the parental magma were chosen to best reproduce observed compositions. Note that the model is relatively insensitive to the values taken for U, Th, Pa and Ra bulk partition coefficients (assuming realistic values), even in the case of crystallization of mineral phases such as amphibole, which is known to preferentially incorporate Ra over Th.

In the case where no assimilation occurs, it is not possible to produce Th concentrations higher than 1 ppm. Hence, we consider the scenario where the host rock is partially melted and assimilated into the differentiating magma. We assume that the assimilant (crustal melt) has the following

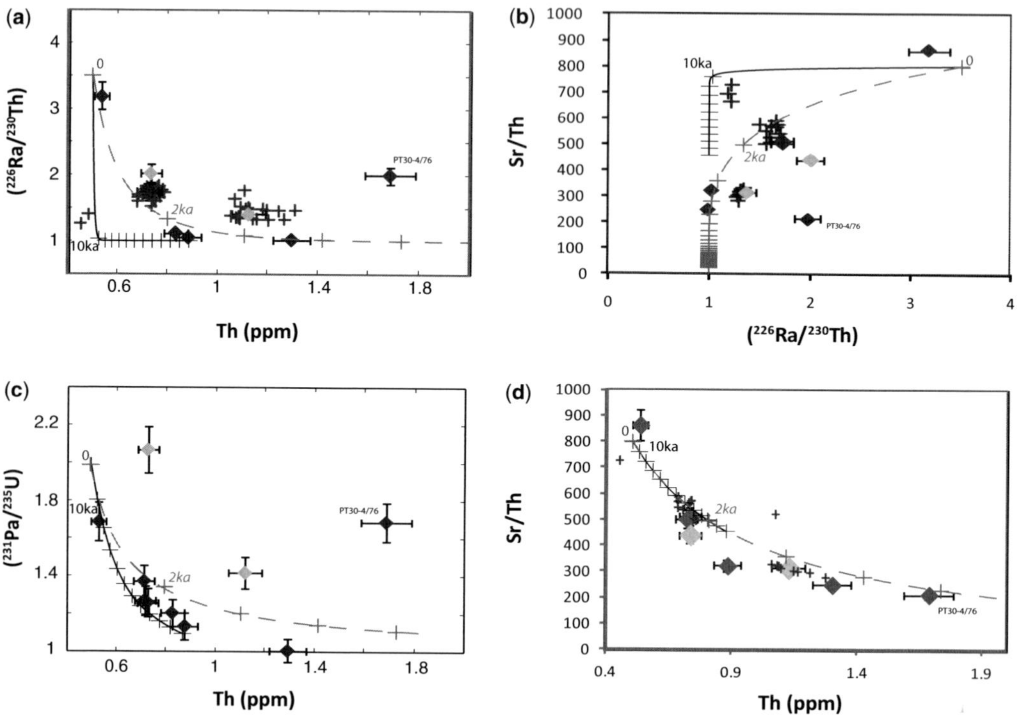

Fig. 3. Differentiation model. It is possible to account for observed compositions using an assimilation/crystallization model. However, only minimum crystallization rates and maximum durations of differentiation can be inferred. Compositions can be explained by continuous crystallization at a minimum rate of 5×10^{-6} g g^{-1} a^{-1} over a maximum duration of *c.* 6000 years. To account for high Th concentrations, assimilation is required and the ratio of assimilation to crystallization rates needs to be between 0.1 and 0.3. Moreover, high U and Th concentrations in the assimilant are required in order to reproduce Sr/Th and ($^{231}Pa/^{235}U$) compositions simultaneously. In these diagrams, the dashed model curve is shown as a possible solution where the most evolved compositions are produced after up to 6000 years of differentiation. The model curves are calculated with the following parameters: crystallization rate = 5×10^{-6} g g^{-1} a^{-1}, assimilation/crystallization rate ratio=0 (plain curve) and 0.3 (dashed curve); parental magma composition: U = 0.1 ppm, Th = 0.5 ppm, Sr = 400 ppm, ($^{226}Ra/^{230}Th$) = 3.5 and ($^{231}Pa/^{235}U$) = 1.7; assimilant composition: U = 60 ppm, Th = 100 ppm, Sr = 100 ppm, ($^{226}Ra/^{230}Th$) = ($^{231}Pa/^{235}U$) = 1. Bulk partition coefficients are: $D_{Th} = 0.01$, $D_U = 0.05$, $D_{Ra} = 0.1$, $D_{Pa} = 0.0001$ and $D_{Sr} = 0.8$ (a high coefficient is taken for Ra in order to maximize the effect of crystal fractionation; however, it has no effect on the calculated curves). Tick marks every 10 000 (plain curve) and 2000 years (dashed curve). Dark grey diamonds: high-magnesia and high-alumina basalts; light grey diamonds: andesites. Errors are not shown for Turner *et al.* (2007) data.

composition: Th, U and Sr concentrations are 100, 60 and 100 ppm, respectively, and ($^{226}Ra/^{230}Th$) = ($^{231}Pa/^{235}U$) = 1. High U and Th concentrations in the assimilant are required such that both Sr/Th and ($^{231}Pa/^{235}U$) data can be reproduced simultaneously. In order to keep the residual magma fraction at realistic values (≥ 0.4), a minimum assimilation/crystallization ratio, *r*, of 0.1 is required to produce observed Th concentrations. The maximum *r* value considered is 0.3 (Tegner *et al.* 2005). Modelled curves calculated with *r* values between 0.1 and 0.3 yield similar fits to the data. Because the model is under-constrained, only a minimum crystallization rate of 5×10^{-6} g g^{-1} a^{-1} can be determined. If the crystallization rate was lower than this value, ($^{226}Ra/^{230}Th$) ratios would decrease much faster than observed. This is used to infer a maximum differentiation timescale of 6000 years to evolve from the most primitive to the most evolved compositions observed. Figure 3 shows an example of differentiation model curves where the crystallization rate is set to 5×10^{-6} g g^{-1} a^{-1} and $r = 0$ or 0.3. In this case, the differentiation timescale is less than 6000 years (PT30-4/1976 is ruled out of the modelling as it is shown above that this sample is an outlier and is probably not genetically related to the other samples). Note that andesites deviate from the

Table 2. *Parameters used in the AFC model*

	Parental magma	Assimilant
Th (ppm)	0.5	100
U (ppm)	0.1	60
Sr (ppm)	400	100
(^{230}Th/^{238}U)	0.8	1
(^{231}Pa/^{235}U)	2	1
(^{226}Ra/^{230}Th)	3.5	1
Bulk partition coefficients		
Th	0.01	
U	0.05	
Sr	0.8	
Pa	10^{-4}	
Ra	0.1	
Maximum residual magma fraction		
0.4		

relationship formed by basalts in (^{231}Pa/^{235}U) v. Th concentration, while both andesites and basalts form a unique correlation in other diagrams. This deviation is not well understood and could indicate that andesites and basalts are not genetically related, where the source of the former would be characterized by a higher (^{231}Pa/^{235}U) (possibly reflecting a lower melting rate).

It is shown here that crustal assimilation is required to account for the geochemical characteristics of Kamchatka lavas. Crustal assimilation has not been considered previously because of the lack of variation between Sr isotopes and indices of differentiation. However, because the crust assimilated is young its Sr isotopic composition may be similar to that of the newly emplaced magmas and the lack of Sr isotope variation does not justify ruling out crustal assimilation.

Note that the ^{230}Th–^{238}U disequilibrium does not show any systematic variations with indices of differentiation. A correlation is expected since this radioactive system operates on the same timescales as the ^{231}Pa–^{235}U system and is equally sensitive to crystal fractionation and assimilation. This could be the result of (i) smaller variations in (^{230}Th/^{238}U) compared to (^{226}Ra/^{230}Th) and (^{231}Pa/^{235}U), such that the effect of assimilation is less clear than for the other radioactive systems; and (ii) conditions of partial melting of the host rock are such that the (^{226}Ra/^{230}Th) and (^{231}Pa/^{235}U) ratios of the crustal melt are significantly different from those of the primary magma (i.e. in secular equilibrium), whilst this would not be the case for (^{230}Th/^{238}U) ratios. This could be possible, for instance, if the porosity during partial melting of the crust was significant (e.g. >1%) such that no or little ^{226}Ra–^{230}Th and ^{231}Pa–^{235}U disequilibria could be produced, while generating ^{230}Th– ^{238}U disequilibrium.

Fluid origin of the ^{226}Ra–^{230}Th disequilibrium

High Ba/Th ratios are commonly used as an indicator of slab fluid addition. However, a positive correlation between (^{226}Ra/^{230}Th) and Ba/Th does not unequivocally require a fluid origin for ^{226}Ra excesses because Ba is more incompatible than Th and, under certain conditions, decreasing degrees of partial melting could also lead to increases in Ba/Th ratios. Contrastingly, because Sr is less compatible but more fluid mobile than Th, an increase in Sr/Th ratios can be explained only by fluid addition and a positive correlation with (^{226}Ra/^{230}Th) would imply that (i) ^{226}Ra excesses are fluid-derived and (ii) less than 10 ka elapsed between slab dehydration and eruption. Such a correlation was observed by Dosseto *et al.* (2003) and used to argue for a fluid origin of the ^{226}Ra–^{230}Th disequilibrium. Our new measurements confirm the positive correlation (Fig. 3b). However, as shown in the previous section, the range of (^{226}Ra/^{230}Th) and Sr/Th values is explained by magma differentiation. It is important to note that whilst the positive correlation in Figure 3b is accounted for by differentiation (cf. Huang *et al.* 2008), metasomatization of the magma source is still required to produce the high (^{226}Ra/^{230}Th) and Sr/Th ratios in the most primitive magmas. Huang *et al.* (2011) invoked partial melting to account for ^{226}Ra–^{230}Th disequilibrium in Kick'Em Jenny lavas (Lesser Antilles). However, if high (^{226}Ra/^{230}Th) ratios were produced during partial melting, they should be associated with low Sr/Th ratios (because the incompatibility order of these different elements is Ra $\gg$ Th > Sr). Thus, the primary ^{226}Ra–^{230}Th disequilibrium at the Kamchatka arc, and in arc lavas in general, is most likely derived from magma source metosomatism due to slab dehydration. The preservation of this ^{226}Ra–^{230}Th disequilibrium implies rapid magma generation and fast transport to the surface (Turner *et al.* 2001).

Magma genesis at subduction zones: an alternative hypothesis to multi-stage fluid addition

It is widely accepted that (^{238}U/^{230}Th) > 1 in arc lavas reflects metasomatism of the mantle wedge by slab-derived fluids (e.g. Turner *et al.* 2003). This contrasts with (^{238}U/^{230}Th) ratios <1 produced during decompression melting (e.g. mid-ocean ridge basalt (MORB) and ocean island basalt (OIB)) due to the greater incompatibility of Th relative to U. The timescales of fluid addition derived from (^{238}U/^{230}Th) > 1 are typically of the order of several tens of thousands of years (150 ka in the case of Kamchatka; Turner *et al.* 1998). However, $^{226}Ra-^{230}Th$ disequilibrium is also observed in most arc lavas. Because any $^{226}Ra-^{230}Th$ disequilibrium would decay after 10 ka, and this disequilibrium is attributed to slab dehydration, the combined observations from $^{226}Ra-^{230}Th$ and $^{238}U-^{230}Th$ systems have been interpreted in some models to reflect different stages of slab dehydration. Although it is anticipated that there might be more than one episode of slab dehydration (because of the numerous hydrous phases being subducted) it has proved harder to identify which hydrous phases are linked to specific disequilibria.

Several models have been developed to explain how magmas are produced in the mantle wedge (see Bourdon *et al.* 2003*a*, *b* for a review). For instance, Thomas *et al.* (2002) have proposed a *flux melting* model where melts are continuously produced while slab-derived fluids are added to the mantle wedge. However, as shown above and in Dosseto *et al.* (2003), this model cannot account for observed $^{226}Ra-^{230}Th$ and $^{231}Pa-^{235}U$ disequilibria (Fig. 2b). In this section, we investigate whether dynamic melting following recent fluid addition to the mantle wedge (<10 ka ago) could account for the observed $^{238}U-^{230}Th$ isotope compositions in Kamchatka. Dynamic melting involves a non-static source where matrix upwelling occurs (McKenzie 1985). In the case of subduction zones, we can consider that the source is advecting either as a result of fluid addition to the mantle wedge, creating a source that is more buoyant than the surrounding dry mantle, or as a result of downward convection along the slab (Dosseto *et al.* 2003). In our model, slab dehydration supplies U to magma sources shortly before melting (<10 ka ago), yielding a range in U/Th ratios at a constant ($^{230}Th/^{232}Th$). Assuming an upwelling or downwelling rate of 3 cm a^{-1}, a porosity of 1‰ and a final degree of partial melting of 20% (see Fig. 4 caption for other parameters), it is possible to reproduce the positive array observed on the ($^{230}Th/^{232}Th$) v. ($^{238}U/^{232}Th$) diagram (Fig. 4a). In this model, there is no demand for a multi-stage fluid addition to account for the different timescales of fluid addition inferred from $^{238}U-^{230}Th$ and $^{226}Ra-^{230}Th$. Melts produced have ($^{231}Pa/^{235}U$) activity ratio between 1.49 and 1.58 (Fig. 4b), which is comparable to (but somewhat lower than) the composition of the most primitive sample. To reproduce the highest ($^{231}Pa/^{235}U$) observed, one could consider that the source starts with a higher ($^{231}Pa/^{235}U$) as a result of the presence of Pa in the fluid (Avanzinelli *et al.* 2012). However, this would require the source to have ($^{231}Pa/^{235}U$) ratios between 1.24 and 1.44. This would suggest that more Pa than U is carried in the fluid, which is unrealistic (e.g. Bourdon *et al.* 2003*a*, *b*). Assuming that the metasomatized source has a ($^{226}Ra/^{230}Th$) of 3 (similar to the highest value observed in our samples), melting has little effect on this activity ratio, resulting in a slight increase to value of 3.2.

The model proposed above can explain U-series isotope compositions at other continental arcs. For instance, U–Th data of low-Ti lavas from

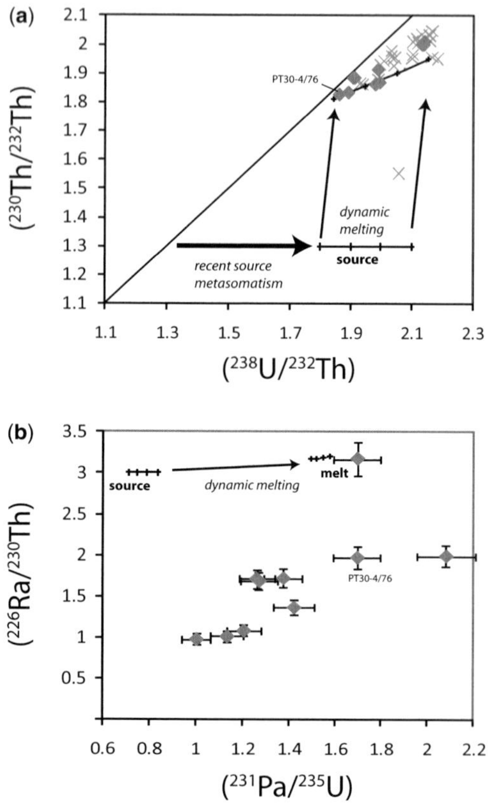

Fig. 4. Dynamic melting of a source that has undergone recent gain of U and Ra (<10 ka ago) via slab-derived fluid addition. Prior to melting, the metasomatized source has a ($^{230}Th/^{232}Th$) ratio of 1.3 (Turner *et al.* 1998) and ($^{238}U/^{232}Th$) ratios range between 1.8 and 2.1, reflecting the variable addition of fluids carrying U and Ra but no Th or Pa. This range of values is chosen such that the best fit to the data is produced. Bulk partition coefficients considered are: $D_{Th} = 0.00207$, $D_U = 0.00465$, $D_{Pa} = 0.000048$ and $D_{Ra} = 0.000001$ (based on a source with 52% olivine, 20% orthopyroxene, 20% clinopyroxene and 8% garnet). Final degree of partial melting: $F = 0.2$. Porosity: $\varphi = 0.001$. Melting rate $= 2.83 \times 10^{-4}$ g g^{-1} a^{-1} and up/downwelling rate = 3 cm a^{-1}. In (**a**), diamonds: this study, crosses: data from Turner *et al.* (2007). In (**b**), diamonds: this study. Errors are not shown for Turner *et al.* (2007) data.

Nicaragua (Reagan *et al.* 1994) can be reproduced using an up/downwelling rate of 6–7 cm a^{-1}, a range of ($^{238}U/^{232}Th$) values for the source prior to melting between 1.8 and 2.7, a final degree of partial melting of 8% and a source mineralogy composed of 52% olivine, 20% orthopyroxene, 20% clinopyroxene and 8% garnet (Fig. 5a). This model also yields ($^{231}Pa/^{235}U$) values between 1.48 and 1.70, which is comparable to the range measured by Thomas *et al.* (2002) for Nicaraguan lavas (1.29–1.74). Similarly, U-series data from the Parinacota volcano (Bourdon *et al.* 2000) can be re-interpreted with this model: in Figure 5b, ($^{230}Th/^{232}Th$) and ($^{238}U/^{232}Th$) ratios can be explained by dynamic melting of a recently metasomatized source to which sediments have been previously added in order to lower the source ($^{230}Th/^{232}Th$) ratio to 0.3. Melting parameters are similar to those used for the Nicaragua data, except for the up/downwelling rate (5 cm a^{-1}). Note that in all cases the up/downwelling rates inferred are lower than plate convergence rates. If these up/downwelling rates reflect convection in the mantle wedge, this implies an incomplete

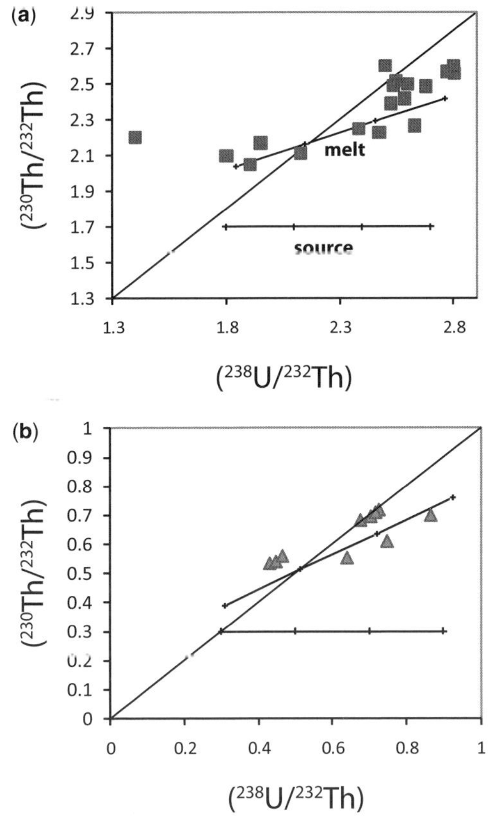

Fig. 5. Dynamic melting of a recently metasomatized source used to account for (a) Nicaragua and (b) Parinacota U–Th data. (**a**) ($^{230}Th/^{232}Th$) of the source = 1.5 and ($^{238}U/^{232}Th$) ranges between 1.5 and 2.7. $D_{Th} = 0.00207$, $D_U = 0.00465$ (based on 52% ol, 20% opx, 20% cpx and 8% gt), $F = 0.08$, $\varphi = 0.001$, melting rate = 2.64×10^{-4} g g^{-1} a^{-1} and up/downwelling rate = 7 cm a^{-1}. (**b**) ($^{230}Th/^{232}Th$) of the source = 0.3, following sediment addition, and ($^{238}U/^{232}Th$) ranges between 0.3 and 0.9. $D_{Th} = 0.002069$, $D_U = 0.004651$ (based on 52% ol, 20% opx, 20% cpx and 8% gt), $F = 0.08$, $\varphi = 0.001$, melting rate = 1.89×10^{-4} g g^{-1} $a^{-1}$and up/downwelling rate = 5 cm a^{-1}. Errors are not shown. Data are from Reagan *et al.* (1994) for Nicaragua and Bourdon *et al.* (2000) for Parinacota.

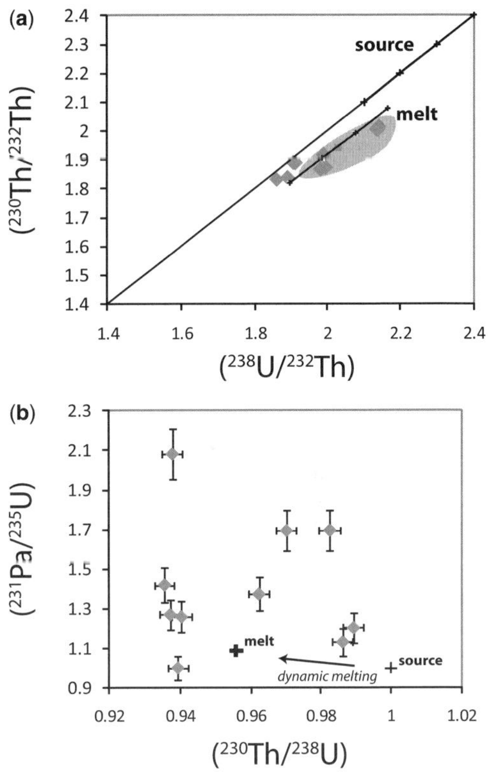

Fig. 6. Dynamic melting under high f_{O2}. The source has undergone addition of U during an old (>350 ka ago) fluid addition event, resulting in a range of ($^{230}Th/^{232}Th$) values prior to melting between 2.1 and 2.4. During melting, as a consequence of high oxygen fugacity, uranium is more incompatible than thorium ($D_U = D_{Th}/10$; see text for details). Other parameters are similar to previous models: $D_{Th} = 0.00207$ and $D_{Pa} = 0.000048$ (based on 52% ol, 20% opx, 20% cpx and 8% gt), $F = 0.15$, $\varphi = 0.001$, melting rate = 3.54×10^{-4} g g^{-1} a^{-1} and up/downwelling rate = 5 cm a^{-1}. Shaded area in (**a**): data from Turner *et al.* (2007). Errors are not shown for Turner *et al.* (2007) data.

coupling between the subducting slab and the overlying mantle.

It has recently been proposed that high-Mg basalts and andesites can be produced by partial melting of subarc peridotite and pyroxenite veins (Straub *et al.* 2011). Thus, an alternative model would be to consider a pyroxenite source. In this scenario, considering a source with 15% olivine, 80% orthopyroxene and 5% clinopyroxene (England & Davies 1973), it is possible to reproduce the ($^{238}U/^{232}Th$)–($^{230}Th/^{232}Th$) composition of Kamchatka rocks by simply considering a faster up/downwelling rate (4.5 cm a^{-1}, which is still realistic considering the convergence rate is 9 cm a^{-1}).

Another alternative model has been recently proposed where partial melting occurs under high oxygen fugacity (Beier *et al.* 2010). In this case, uranium becomes more incompatible than thorium (Lundstrom *et al.* 1994). Thus, the U-series isotope composition of lavas from the Manus Basin were explained by dynamic melting in the back-arc of a mantle previously metasomatized and oxidized by slab-derived fluids. In principle, this model can be used to reproduce the U–Th isotope composition of Kamchatka lavas. Because fluid addition occurs >350 ka ago, the metasomatized source has a high and variable ($^{230}Th/^{232}Th$) prior to melting (Fig. 6a). However, although significant ^{226}Ra–^{230}Th disequilibrium can be produced in this model (($^{226}Ra/^{230}Th$) ~ 3), no ^{231}Pa–^{235}U disequilibrium is produced (because D_U/D_{Pa} is only *c.* 4, v. *c.* 100 in other models) (Fig. 6b). This is at odds with the observation of significant ^{231}Pa excesses in the most primitive Kamchatka lavas (note that there are presently no ^{231}Pa data with which to test this model for the Manus Basin).

Conclusions

In the light of recent results by Turner *et al.* (2007), the U–Th–Ra isotope composition of volcanic rocks from the Central Kamchatka Depresssion was re-visited. We conclude that:

(i) New measurements show no ^{230}Th excess relative to ^{238}U, suggesting that there is no longer any evidence for a slab-derived melt.

(ii) Variations in U–Pa and Th–Ra isotope compositions are best explained by magma differentiation with crustal assimilation, rather than mixing between slab- and mantle-derived melts as initially suggested (Dosseto *et al.* 2003). A maximum duration of magma differentiation of *c.* 6000 years can be inferred, with a minimum crystallization rate of 5×10^{-6} g g^{-1} a^{-1}. Although not suggested by long-lived radiogenic isotopes (because the assimilant may have the same isotopic composition as the magma differentiating), assimilation is required and occurs at a minimum rate of 0.5×10^{-6} g g^{-1} a^{-1}.

(iii) High ($^{226}Ra/^{230}Th$) and Sr/Th ratios in the most primitive lavas emphasize that ^{226}Ra–^{230}Th disequilibrium in arc lavas is derived from dehydration of the subducted slab.

(iv) Multiple fluid additions to the mantle wedge are no longer required to account for U–Th–Pa–Ra isotope data. Ra–Th suggests recent fluid addition (<10 ka ago) whereas U–Th data can be reproduced by dynamic melting of a metasomatized source, enriched in U over Th prior to melting. This is in agreement with the conclusions of Dosseto *et al.* (2003) who proposed that the primitive melt is produced by equilibrium porous flow or dynamic melting of a recently metasomatized mantle. Thus, U–Th–Pa–Ra data at the Kamchatka arc can be explained by a simple model where a single episode of fluid addition occurs shortly (<few thousand years) before dynamic melting of the upwelling or downwelling source. This model can also account for U-series isotope data at other continental arcs (Andes, Nicaragua). In addition, the observation of high ($^{226}Ra/^{230}Th$) associated with high ($^{231}Pa/^{235}U$) ratios confirm that flux melting is not adequate to model magma genesis beneath the Kamchatka arc.

(v) An alternative model of arc magma production where the source is metasomatized more than 300 000 years ago and melting occurs under high oxygen fugacity cannot account for the observed ^{231}Pa–^{235}U disequilibrium in primitive magmas.

We would like to thank the editors for putting this Special Publication together, Georg Zellmer for editorial handling and Riccardo Avanzinelli and Ian Smith for their insightful reviews. We would also like to thank Tracy Rushmer, Hugh O'Neill, Gerhard Wörner, Bernard Bourdon and Bruce Schaefer for helpful discussions. The work was funded by an Australian Research Council Discovery Grant DP0988658. Dosseto acknowledges an Australian Research Council Future Fellowship FT0990447. Turner acknowledges an Australian Research Council Professorial Fellowship.

Appendix I

Assimilation–fractional crystallization model for U-series isotopes

The U-series isotope composition of co-genetic volcanic rocks can be modelled using an assimilation–fractional crystallization (AFC) model, similar to that presented in Dosseto *et al.* (2008).

In this model, the mass of differentiating magma evolves with time as follows:

$$\frac{dM}{dt} = -f \cdot M + Q_{assimil} \quad (A1)$$

where f is the crystallization rate (in a^{-1}) and $Q_{assimil}$ is the rate of wall rock assimilation (in g a^{-1}). We use the non-dimensional parameter, r, defined by DePaolo (1981): $r = Q_{assimil}/f.M$ so that equation (A1) becomes:

$$\frac{dM}{dt} = f \cdot M \cdot (r - 1) \quad (A2)$$

The variation with time of the mass of a nuclide in the magma, M_i, is:

$$\begin{aligned}\frac{dM_i}{dt} &= C_i \cdot \frac{dM}{dt} + M \cdot \frac{dC_i}{dt} \\ &= r \cdot f \cdot M \cdot C_{i,assimil} - f \cdot D_i \cdot M \cdot C_i \\ &\quad + \lambda_{i-1} \cdot M \cdot C_{i-1} - \lambda_i \cdot M \cdot C_i \quad (A3)\end{aligned}$$

where C_i and $C_{i\text{-}1}$ are, respectively, the concentrations of the nuclide and its parent in the magma (in g g^{-1}), $C_{i,assimil}$ is the concentration of the nuclide in the assimilated wall rock (in g g^{-1}), D_i is the partition coefficient of the nuclide and λ_i and λ_{i-1} are, respectively, the decay constants of the nuclide and its parent (in a^{-1}).

We can re-write (A3) so:

$$\begin{aligned}\frac{dC_i}{dt} &= r \cdot f \cdot C_{i,assimil} - C_i \cdot f(r + D_i - 1) \\ &\quad + \lambda_{i-1} \cdot C_{i-1} - \lambda_i \cdot C_i \quad (A4)\end{aligned}$$

This equation can be written for each nuclide (^{238}U, ^{230}Th, ^{226}Ra and ^{232}Th) and integrated so that the evolution of activity ratios with time can be calculated. The equation can also be solved for Sr concentrations, simply by removing the decay terms.

In the magma at the onset of differentiation, the U-series isotope composition of the magma is assumed to be similar to the most primitive composition observed: $(^{226}Ra/^{230}Th)_i = 3.5$ and $(^{231}Pa/^{235}U)_i = 2$. The thorium concentration, Th_i, is taken to be 0.5 ppm, the uranium concentration, U_i, 0.1 ppm, and the Sr concentration, Sr_i, 400 ppm, similar to concentrations observed in primitive lavas. The wall rock assimilated is assumed to be in secular equilibrium, as suggested previously (Berlo *et al.* 2004). The concentrations in the melted assimilant are taken to be $Th_{crust} = 100$ ppm, $U_{crust} = 60$ ppm and $Sr_{crust} = 100$ ppm. Bulk partition coefficients used are 0.8 for Sr, 10^{-5} for Pa (Blundy & Wood 2003), 10^{-2} for Th and 5×10^{-2} for U (for a mineralogical assemblage containing 10% clinopyroxene, which controls Th and U partitioning; Bacon & Druitt 1988). In order to maximize the effect of crystal fractionation, the bulk partition coefficient for Ra is taken to be 0.1 (as in presence of amphibole), although any values lower than this have no effect on the calculated compositions. U-series isotope compositions observed can be reproduced with a minimum crystallization rate of 5×10^{-6} g g^{-1} a^{-1}, an r value between 0.1 and 0.3 and a final residual magma fraction of 0.4.

Dynamic melting model

The dynamic melting model used in this study follows that described in (McKenzie 1985; Williams & Gill 1989). The activities for the different nuclides are calculated using the following equations (parameters defined in Table A1):

$$(^{238}U) = \frac{\beta(^{238}U)_i(1 - e^{-(\alpha_U + \beta)t})}{X \quad \alpha_U + \beta} \quad (A5)$$

$$(^{232}Th) = \frac{\beta(^{232}Th)_i(1 - e^{-(\alpha_{Th} + \beta)t})}{X \quad \alpha_{Th} + \beta} \quad (A6)$$

$$\begin{aligned}(^{230}Th) &= \frac{\beta(^{230}Th)_i - C_{Th}(^{238}U)_i}{X \quad \lambda_{230} + \alpha_{Th} + \beta}(1 - e^{-(\lambda_{230} + \alpha_{Th} + \beta)t}) \\ &\quad + \frac{C_{Th}(^{238}U)_i}{\alpha_U + \beta}(1 - e^{-(\alpha_U + \beta)t}) \quad (A7)\end{aligned}$$

$$\begin{aligned}(^{231}Pa) &= \frac{\beta(^{231}Pa)_i - C_{Pa}(^{235}U)_i}{X \quad \lambda_{231} + \alpha_{Pa} + \beta}(1 - e^{-(\lambda_{231} + \alpha_{Pa} + \beta)t}) \\ &\quad + \frac{C_{Pa}(^{235}U)_i}{\alpha_U + \beta}(1 - e^{-(\alpha_U + \beta)t}) \quad (A8)\end{aligned}$$

Table A1. *Definitions of parameters used in the dynamic melting model*

Parameter	Definition
X	Final degree of partial melting (unitless)
$(^{238}U)_i$, $(^{232}Th)_i$, $(^{230}Th)_i$, $(^{231}Pa)_i$ and $(^{226}Ra)_i$	Initial activities for ^{238}U, ^{232}Th, ^{230}Th, ^{231}Pa and ^{226}Ra, respectively (D_{Pa} g^{-1})
λ_{230}, λ_{231} and λ_{226}	Decay constants for ^{230}Th, ^{231}Pa and ^{226}Ra, respectively (a^{-1})
$\dot{M}$	Melting rate (g g^{-1} a^{-1})
ρ_f and ρ_s	Melt and source rock densities, respectively (g m^{-3})
ϕ	Porosity (unitless)
D_U, D_{Th}, D_{Pa} and D_{Ra}	Partition coefficients for U, Th, Pa and Ra, respectively (unitless)

$$(^{226}\text{Ra}) = \frac{\beta}{X}\left((^{226}\text{Ra})_i - \frac{C_{\text{Ra1}} - C_{\text{Ra2}}}{\lambda_{226} + \alpha_{\text{Ra}} - \lambda_{230} - \alpha_{\text{Th}}} - \frac{C_{\text{Ra2}}}{\lambda_{226} + \alpha_{\text{Ra}} - \alpha_U}\right) \times \frac{1 - e^{-(\lambda_{226} + \alpha_{\text{Ra}} + \beta)t}}{\lambda_{226} + \alpha_{\text{Ra}} + \beta} + (C_{\text{Ra1}} - C_{\text{Ra2}}) \times \frac{1 - e^{-(\lambda_{230} + \alpha_{\text{Ra}} + \beta)t}}{(\lambda_{226} + \alpha_{\text{Ra}} - \lambda_{230} - \alpha_{\text{Th}})(\lambda_{230} + \alpha_{\text{Th}} + \beta)} + C_{\text{Ra2}} \frac{1 - e^{-(\alpha_U + \beta)t}}{(\lambda_{226} + \alpha_{\text{Ra}} - \alpha_U)(\alpha_U + \beta)} \quad \text{(A9)}$$

with the following variables:

$$\beta = \frac{\dot{M}}{\rho_f \phi + \rho_s(1-\phi)}$$

$$F_U = \frac{1}{1 + D_U(\rho_s(1-\phi)/\rho_f\phi)},$$

$$F_{\text{Th}} = \frac{1}{1 + D_{\text{Th}}(\rho_s(1-\phi)/\rho_f\phi)},$$

$$F_{\text{Pa}} = \frac{1}{1 + D_{\text{Pa}}(\rho_s(1-\phi)/\rho_f\phi)},$$

$$F_{\text{Ra}} = \frac{1}{1 + D_{\text{Ra}}(\rho_s(1-\varphi)/\rho_f\varphi)}$$

$$\alpha_U = F_U \frac{(1-D_U)\dot{M}}{\rho_f\phi}, \quad \alpha_{\text{Th}} = F_{\text{Th}} \frac{(1-D_{\text{Th}})\dot{M}}{\rho_f\phi},$$

$$\alpha_{\text{Pa}} = F_{\text{Pa}} \frac{(1-D_{\text{Pa}})\dot{M}}{\rho_f\phi}, \quad \alpha_{\text{Ra}} = F_{\text{Ra}} \frac{(1-D_{\text{Ra}})\dot{M}}{\rho_f\phi}$$

$$C_{\text{Th}} = \frac{F_{\text{Th}}\lambda_{230}}{F_U(\lambda_{230}\alpha_{\text{Th}} - \alpha_U)}, \quad C_{\text{Pa}} = \frac{F_{\text{Pa}}\lambda_{231}}{F_U(\lambda_{231}\alpha_{\text{Pa}} - \alpha_U)},$$

$$C_{\text{Ra1}} = \frac{F_{\text{Ra}}\lambda_{226}(^{230}\text{Th})_i}{F_{\text{Th}}},$$

$$C_{\text{Ra2}} = \frac{F_{\text{Ra}}\lambda_{226}\lambda_{230}(^{238}\text{U})_i}{F_U(\lambda_{230} + \alpha_{\text{Th}} - \alpha_U)}$$

References

Asmerom, Y., Cheng, H., Thomas, R., Hirschmann, M. & Edwards, R. L. 2000. Melting of the Earth's lithospheric mantle inferred from protactinium–thorium–uranium isotopic data. *Nature*, **406**, 293–296.

Avanzinelli, R., Prytulak, J., Skora, S., Heumann, A., Koetsier, G. & Elliott, T. 2012. Combined ^{238}U–^{230}Th and ^{235}U–^{231}Pa constraints on the transport of slab-derived material beneath the Mariana Islands. *Geochimica et Cosmochimica Acta*, **92**, 308–328.

Bacon, C. R. & Druitt, T. H. 1988. Compositional evolution of the zoned calcalkaline magma chamber of Mount-Mazama, Crater Lake, Oregon. *Contribution to Mineralogy and Petrology*, **98**, 224–256.

Beier, C., Turner, S. P., Sinton, J. M. & Gill, J. B. 2010. Influence of subducted components on back-arc melting dynamics in the Manus Basin. *Geochemistry, Geophysics, Geosystems: G3*, **11**.

Berlo, K., Turner, S., Blundy, J. & Hawkesworth, C. 2004. The extent of U-series disequilibria produced during partial melting of the lower crust with implications for the formation of the Mount St. Helens dacites. *Contributions to Mineralogy and Petrology*, **148**, 122–130.

Blake, S. & Rogers, N. 2005. Magma differentiation rates from (^{226}Ra/^{230}Th) and the size and power output of magma chambers. *Earth and Planetary Science Letters*, **236**, 654–669.

Blundy, J. & Wood, B. 2003. Mineral-melt partition of uranium, thorium and their daughters. *In*: Bourdon, B., Henderson, G. M., Lundstrom, C. C. & Turner, S. P . (eds) *Uranium-Series Geochemistry*. Geochemical Society – Mineralogical Society of America, Washington, DC, **52**, 59–123.

Bourdon, B. & Elliott, T. 2010. Melt production in the mantle: constraints from U-series. *In*: Dosseto, A., Turner, S. & Van Orman, J. A. (eds) *Timescales of Magmatic Processes: from Core to Atmosphere*. Wiley-Blackwell, Oxford.

Bourdon, B., Wörner, G. & Zindler, A. 2000. U-series evidence for crustal involvement and magma residence times in the petrogenesis of Parinacota volcano, Chile. *Contributions to Mineralogy and Petrology*, **139**, 458–469.

Bourdon, B., Henderson, G. M., Lundstrom, C. C. & Turner, S. P. 2003*a*. *Uranium-series Geochemistry*. Geochemical Society–Mineralogical Society of America, Washington, DC.

Bourdon, B., Turner, S. & Dosseto, A. 2003*b*. Dehydration and partial melting in subduction zones: constraints from U-series disequilibria. *Journal of Geophysical Research B: Solid Earth*, **108**, ECV 3-1–3-19.

Cooper, K. M. & Reid, M. R. 2003. Re-examination of crystal ages in recent Mount St. Helens lavas: implications for magma reservoir processes. *Earth and Planetary Science Letters*, **213**, 149–167.

Cooper, K. M., Reid, M. R., Murrell, M. T. & Clague, D. A. 2001. Crystal and magma residence at Kilauea Volcano, Hawaii: ^{230}Th–^{226}Ra dating of the 1955 east rift eruption. *Earth and Planetary Science Letters*, **184**, 703–718.

DeMets, C. R., Gordon, G., Argus, D. F. & Stein, S. 1990. Current plate motions. *Geophysical Journal International*, **101**, 425–478.

DePaolo, D. J. 1981. Trace element and isotopic effects of combined wall rock assimilation and fractional crystallization. *Earth and Planetary Science Letters*, **53**, 189–202.

Dosseto, A. & Turner, S. 2010. Magma cooling and differentiation – U-series isotopes. *In*: Dosseto, A., Turner, S. & Van Orman, J. A. (eds) *Timescales of Magmatic Processes: From Core to Atmosphere*. Wiley-Blackwell, Oxford, 160–180.

Dosseto, A., Bourdon, B., Joron, J.-L. & Dupré, B. 2003. U–Th–Pa–Ra study of the Kamchatka arc: new constraints on the genesis of arc basalts. *Geochimica et Cosmochimica Acta*, **67**, 2857–2877.

DOSSETO, A., TURNER, S. & DOUGLAS, G. B. 2006. Uranium-series isotopes in colloids and suspended sediments: timescale for sediment production and transport in the Murray–Darling River system. *Earth and Planetary Science Letters*, **246**, 418–431.

DOSSETO, A., TUNER, S. P., SANDIFORD, M. & DAVIDSON, J. 2008. Uranium-series isotope and thermal constraints on the rate and depth of silicic magma genesis. *In*: ANNEN, C. & ZELLMER, G. (eds) *Dynamics of Crustal Magma Transfer, Storage and Differentiation*. Geological Society, London, Special Publications, **304**, 169–181.

DOSSETO, A., TURNER, S. & VAN ORMAN, J. A. (eds) 2010. *Timescales of Magmatic Processes: From Core to Atmosphere*. Wiley-Blackwell, Oxford.

DUFRANE, S. A., TURNER, S., DOSSETO, A. & VAN SOEST, M. 2009. Reappraisal of fluid and sediment contributions to Lesser Antilles magmas. *Chemical Geology*, **265**, 272–278.

ELLIOTT, T., PLANK, T., ZINDLER, A., WHITE, W. & BOURDON, B. 1997. Element transport from subducted slab to juvenile crust at the Mariana arc. *Journal of Geophysical Research*, **102**, 14 991–15 019.

ENGLAND, R. N. & DAVIES, H. L. 1973. Mineralogy of ultramafic cumulates and tectonites from eastern Papua. *Earth and Planetary Science Letters*, **17**, 416–425.

HAWKESWORTH, C. J., TURNER, S. P., MCDERMOTT, F., PEATE, D. W. & VAN CALSTEREN, P. 1997. U–Th isotopes in arc magmas: implications for element transfer from the subducted crust. *Science*, **276**, 551–555.

HUANG, F., GAO, L. & LUNDSTROM, C. C. 2008. The effect of assimilation, fractional crystallization, and ageing on U-series disequilibria in subduction zone lavas. *Geochimica et Cosmochimica Acta*, **72**, 4136–4145.

HUANG, F., LUNDSTROM, C. C., SIGURDSSON, H. & ZHANG, Z. 2011. U-series disequilibria in Kick'em Jenny submarine volcano lavas: a new view of time-scales of magmatism in convergent margins. *Geochimica et Cosmochimica Acta*, **75**, 195–212.

KERSTING, A. B. & ARCULUS, R. J. 1995. Pb isotope composition of Klyuchevskoy volcano, Kamchatka and North Pacific sediments: implications for magma genesis and crustal recycling in the Kamchatka arc. *Earth and Planetary Science Letters*, **136**, 133–148.

LANDWEHR, D., BLUNDY, J., CHAMORRO-PEREZ, E. M., HILL, E. & WOOD, B. 2001. U-series disequilibria generated by partial melting of spinel lherzolite. *Earth and Planetary Science Letters*, **188**, 329–348.

LUNDSTROM, C. C., SHAW, H. F., RYERSON, F. J., PHINNEY, D. L., GILL, J. B. & WILLIAMS, Q. 1994. Compositional controls on the partitioning of U, Th, Ba, Pb, Sr and Zr between clinopyroxene and haplobasaltic melts: implications for uranium series disequilibria in basalts. *Earth and Planetary Science Letters* **28**, 407–423.

MCKENZIE, D. 1985. ^{230}Th–^{238}U disequilibrium and the melting processes beneath ridge axes. *Earth and Planetary Science Letters.*, **72**, 149–157.

MELEKESTSEV, I., KHRENOV, A. P. & KOZHEMYAKA, N. N. 1991. Tectonic position and general description of volcanoes of Northern group and Sredinny range. *Active volcanoes of Kamchatka*. Academy of Sciences, Moscow, n°1, partie II (avant-propos), 79–97.

MIYASHIRO, A. 1974. Volcanic rock series in island arcs and active continental margins. *Journal of Science*, **274**, 321–355.

PEATE, D. W. 2005. U series disequilibria: insights into mantle melting and the timescales of magma differentiation. *Reviews of Geophysics*, **43**.

PETFORD, N., KOENDERS, M. A. & TURNER, S. 2008. Channelized melt flow in downwelling mantle: implications for ^{226}Ra–^{210}Pb disequilibria in arc magmas. *Journal of Geophysical Research*, **113**.

PINEAU, F., SEMET, M. P., GRASSINEAU, N., OKRUGIN, V. M. & JAVOY, M. 1999. The genesis of the stable isotope (O,H) record in arc magmas: the Kamtchatka case. *Chemical Geology*, **153**, 93–124.

PRICE, R. C. *ET AL.* 2007. U–Th–Ra fractionation during crustal-level andesite formation at Ruapehu volcano, New Zealand. *Chemical Geology*, **244**, 437–451.

REAGAN, M. K., MORRIS, J. D., HERSTROM, E. A. & MURRELL, M. T. 1994. Uranium series and beryllium isotope evidence for an extended history of subduction modification of the mantle below Nicaragua. *Geochimica et Cosmochimica Acta*, **58**, 4199–4212.

REAGAN, M. K., TEPLEY III, F. J., GILL, J. B., WORTEL, M. & GARRISON, J. 2006. Timescales of degassing and crystallization implied by ^{210}Po–^{210}Pb–^{226}Ra disequilibria for andesitic lavas erupted from Arenal volcano. *Journal of Volcanology and Geothermal Research*, **157**, 135–146.

ROGERS, N. W., EVANS, P. J., BLAKE, S., SCOTT, S. C. & HAWKESWORTH, C. J. 2004. Rates and timescales of fractional crystallization from ^{238}U–^{230}Th–^{226}Ra disequilibria in trachyte lavas from Longonot volcano, Kenya. *Journal of Petrology*, **45**, 1747–1776.

RUBIN, K. H. & MACDOUGALL, J. D. 1988. ^{226}Ra excesses in mid-ocean-ridge basalts and mantle melting. *Nature*, **335**, 158–161.

RUBIN, K. H., VAN DER ZANDER, I., SMITH, M. C. & BERGMANIS, E. C. 2005. Minimum speed limit for ocean ridge magmatism from ^{210}Pb–^{226}Ra–^{230}Th disequilibria. *Nature*, **437**, 534–538.

SIGMARSSON, O. 1996. Short magma chamber residence time at an Icelandic volcano inferred from U-series disequilibria. *Nature*, **382**, 440–442.

SIGMARSSON, O., CONDOMINES, M. & BACHÉLERY, P. 2005. Magma residence time beneath the Piton de la Fournaise Volcano, Reunion Island, from U-series disequilibria. *Earth and Planetary Science Letters*, **234**, 223–234.

SIMS, K. W. W., DEPAOLO, D. J., MURELL, M. T., BALDRIDGE, W. S., GOLDSTEIN, S. J. & CLAGUE, D. A. 1995. Mechanisms of magma generation beneath Hawaii and mid-ocean ridges: uranium/thorium and samarium/neodymium isotopic evidence. *Science*, **267**, 508–512.

SIMS, K. W. W., DEPAOLO, D. J., MURELL, M. T., BALDRIDGE, W. S., GOLDSTEIN, S., CLAGUE, D. & JULL, M. 1999. Porosity of the melting zone and variations in the solid mantle upwelling rate beneath Hawaii: inferences from ^{238}U–^{230}Th–^{226}Ra and ^{235}U–^{231}Pa disequilibria. *Geochimica et Cosmochimica Acta*, **63**, 4119–4138.

SIMS, K. W. W. *ET AL.* 2002. Chemical and isotopic constraints on the generation and transport of magma beneath the East Pacific Rise. *Geochimica et Cosmochimica Acta*, **66**, 3481–3504.

SIMS, K. W. W. *ET AL.* 2008*a*. An inter-laboratory assessment of the thorium isotopic composition of synthetic and rock reference materials. *Geostandards and Geoanalytical Research*, **32**, 65–91.

SIMS, K. W. W. *ET AL.* 2008*b*. $^{238}U-^{230}Th-^{226}Ra-^{210}Pb-^{210}Po$, $^{232}Th-^{228}Ra$, and $^{235}U-^{231}Pa$ constraints on the ages and petrogenesis of Vailulu'u and Malumalu Lavas, Samoa. *Geochemistry Geophysics Geosystems*, **9**.

SPIEGELMAN, M. 2000. UserCalc: a web-based uranium series calculator for magma migration problems. *Geochemistry, Geophysics, Geosystems*, **1**.

SPIEGELMAN, M. & ELLIOTT, T. 1993. Consequences of melt transport for uranium series disequilibrium in young lavas. *Geochemistry, Geophysics, Geosystems*, **118**, 1–20.

STRACKE, A., BOURDON, B. & MCKENZIE, D. 2006. Melt extraction in the Earth's mantle: constraints from U–Th–Pa–Ra studies in oceanic basalts. *Earth and Planetary Science Letters*, **244**, 97–112.

STRAUB, S. M., GOMEZ-TUENA, A., STUART, F. M., ZELLMER, G. F., ESPINASA-PERENA, R., CAI, Y. & IIZUKA, Y. 2011. Formation of hybrid arc andesites beneath thick continental crust. *Earth and Planetary Science Letters*, **303**, 337–347.

TEGNER, C., WILSON, J. R. & ROBINS, B. 2005. Crustal assimilation in basalt and jotunite: constraints from layered intrusions. *Lithos*, **83**, 299–316.

TEPLEY, F. J., III, LUNDSTROM, C. C., GILL, J. B. & WILLIAMS, R. W. 2006. U–Th–Ra disequilibria and the time scale of fluid transfer and andesite differentiation at Arenal volcano, Costa Rica (1968–2003). *Journal of Volcanology and Geothermal Research*, **157**, 147–165.

THOMAS, R. B., HIRSCHMANN, M. M., CHENG, H., REAGAN, M. K. & EDWARDS, R. L. 2002. ($^{231}Pa/^{235}U$)–($^{230}Th/^{238}U$) of young mafic volcanic rocks from Nicaragua and Costa Rica and the influence of flux melting on U-series systematics of arc lavas. *Geochimica et Cosmochimica Acta*, **66**, 4287–4309.

TOUBOUL, M., BOURDON, B., VILLEMANT, B., BOUDON, G. & JORON, J.-L. 2007. $^{238}U-^{230}Th-^{226}Ra$ disequilibria in andesitic lavas of the last magmatic eruption of Guadeloupe Soufriere, french Antilles: processes and timescales of magma differentiation. *Chemical Geology*, **246**, 181–206.

TURNER, S. & HAWKESWORTH, C. 1997. Constraints on flux rates and mantle dynamics beneath island arcs from Tonga-Kermadec lava geochemistry. *Nature*, **389**, 568–573.

TURNER, S., MCDERMOTT, F., HAWKESWORTH, C. & KEPEZHINSKAS, P. 1998. A U-series study of lavas from Kamchatka and the Aleutians: constraints on source composition and melting processes. *Contributions to Mineralogy and Petrology*, **133**, 217–234.

TURNER, S., PEATE, D. W., HAWKESWORTH, C. J., EGGINS, S. M. & CRAWFORD, A. J. 1999. Two mantle domains and the time scales of fluid transfer beneath the Vanuatu arc. *Geology*, **27**, 963–966.

TURNER, S., BOURDON, B., HAWKESWORTH, C. & EVANS, P. 2000. $^{226}Ra-^{230}Th$ evidence for multiple dehydration events, rapid melt ascent and time scales of differentiation beneath the Tonga–Kermadec island arc. *Earth and Planetary Science Letters*, **179**, 581–593.

TURNER, S., EVANS, P. & HAWKESWORTH, C. 2001. Ultrafast source-to-surface movement of melt at island arcs from $^{226}Ra-^{230}Th$ systematics. *Science*, **292**, 1363–1366.

TURNER, S., BOURDON, B. & GILL, J. 2003. Insights into magma genesis at convergent margins from U-series isotopes. *In*: BOURDON, B., HENDERSON, G. M., LUNDSTROM, C. C. & TURNER, S. P. (eds) *Uranium-Series Geochemistry*. Geochemical Society–Mineralogical Society of America, Washington, DC, **52**, 255–315.

TURNER, S., REGELOUS, M., HAWKESWORTH, C. & ROSTAMI, K. 2005. Partial melting processes above subducting plates: constraints from $^{231}Pa-^{235}U$ disequilibria. *Geochimica et Cosmochimica Acta*, **70**, 480–503.

TURNER, S., SIMS, K. W. W., REAGAN, M. & COOK, C. 2007. A $^{210}Pb-^{226}Ra-^{230}Th-^{238}U$ study of Klyuchevskoy and Bezymianny volcanoes, Kamchatka. *Geochimica et Cosmochimica Acta*, **71**, 4771–4785.

WILLIAMS, R. W. & GILL, J. B. 1989. Effects of partial melting on the uranium decay series. *Geochimica et Cosmochimica Acta* **53**, 1607–1619.

ZOU, H. & ZINDLER, A. 2000. Theoretical studies of $^{238}U-^{230}Th-^{226}Ra$ and $^{235}U-^{231}Pa$ disequilibria in young lavas produced by mantle melting. *Geochimica et Cosmochimica Acta*, **64**, 1809–1817.

High-magnesium andesites: the example of the Papuan Volcanic Arc

IAN E. M. SMITH

Geology Programme, School of Environment, University of Auckland, PB 92019 Auckland Mail Centre 1142, Auckland, New Zealand (e-mail: ie.smith@auckland.ac.nz)

Abstract: The late Cenozoic arc-type volcanic arc in southeastern Papua New Guinea developed in an environment of complex tectonic processes including obduction, subduction, rifting and sea floor spreading. The volcanic arc extends from the Papuan Peninsula south-eastward through the D'Entrecasteaux Islands into the Louisiade Archipelago. Lithologies are predominantly basaltic andesite and andesite, but include basalt, dacite and rhyolite. The rocks have typical arc-type geochemical features but include a group ranging from basalt to dacite which, although comparable in most other aspects of their compositions, are higher in MgO, Cr and Ni. These high-Mg rocks are less porphyritic and have simple olivine- or clinopyroxene- dominated phenocryst assemblages compared with the associated low-Mg rocks. The low-Mg rocks are plagioclase-phyric and contain augite and hypersthene with or without olivine, hornblende and biotite phenocrysts. Boninites are spatially associated with, but genetically unrelated to the arc-type rocks in Papua. The high-Mg rocks represent magmas derived by partial melting of subduction-modified mantle which rose rapidly from their source. In contrast, the low-Mg lavas represent magmas which were modified by shallow processes. The unusual abundance of high-Mg lavas in southeastern Papua is related to extensional tectonics which allowed deep sourced magmas to rise without significant modification.

Supplementary material: Major and trace element analyses of lavas from the Papuan volcanic arc are available at http://www.geolsoc.org.uk/SUP18644

Magmas erupted at convergent plate boundaries constitute the most complex of the major igneous associations. This reflects the range of tectonic possibilities and the variety of components which variably contribute to magma generation during active plate convergence and consequent subduction of a lithospheric slab. Although there is not yet universal agreement about the petrogenesis of subduction-related magmas, there is a general consensus that they are the result of multi-component systems to which the down-going slab, the overlying mantle wedge and, in some cases, crust and ocean-floor sediments variably contribute. Their ultimate origin is in the mantle, the result of melting caused by migration of fluids and/or melts from the subducting plate (e.g. Hawkesworth *et al.* 1979; Arculus & Powell 1986; Grove & Kinzler 1986; McCulloch & Gamble 1991; Hawkesworth *et al.* 1993; Brenan *et al.* 1995; Elliot *et al.* 1997; Kessel *et al.* 2005). These major petrogenetic components tend to have a unique blend in any particular geological setting. However, it is now recognized that the interaction of mantle-derived magmas with the overlying lower and/or middle crust is a significant, possibly the dominant, factor in the evolution of continental subduction-related magmas (e.g. Price *et al.* 2005; Annen *et al.* 2006). Thus, the nature of the sub-arc crust and, importantly, the tectonic environment can influence the chemistry of the magmas erupted at the surface.

One of the major obstacles to resolving the question of the origin of subduction-related magmas has been the difficulty of identifying the characteristics of primary magmas in a magmatic association in which the rocks are commonly profoundly modified by fractionation, crustal interaction and magma mixing processes associated with recharge in open systems. However, if the ultimate origin of subduction-related magmas lies within the mantle wedge overlying the subducting lithospheric slab, then a first-order criterion for recognizing primary magmas will be Mg/Fe ratios appropriate to equilibrium with mantle olivine defined as magnesium numbers (Mg# $= 100 \times$ mol.Mg (mol.MgO + mol. FeO)) greater than *c.* 68 corresponding to equilibrium with residual olivine with magnesium number greater than *c.* 88. That such criteria are rarely met in suites of arc-type volcanic rocks testifies to the efficiency of fractionation, mixing and assimilation processes in the supra subduction setting in filtering out primitive chemical characteristics (e.g. Smith *et al.* 1997).

Typically the most primitive rock compositions in volcanic arc associations represent magmas which have undergone significant evolution in terms of Mg/Fe ratio and which show depletion of

From: GÓMEZ-TUENA, A., STRAUB, S. M. & ZELLMER, G. F. (eds) 2014. *Orogenic Andesites and Crustal Growth.* Geological Society, London, Special Publications, **385**, 117–135.
First published online July 31, 2013, http://dx.doi.org/10.1144/SP385.7

mantle compatible elements (e.g. Cr, Ni). These 'parental' magmas can be modelled as derivative of high magnesium primary magmas. Examples of primitive high-Mg lavas are known from volcanic arc systems (e.g. Perfit *et al.* 1980; Tatsumi 1983; Falloon & Green 1986; Myers *et al.* 1986; Nye & Reid 1986; Crawford *et al.* 1987; Saunders *et al.* 1987; Smith & Mitchell 1989; Eggins 1993; Monzier *et al.* 1993; Yogodzinski *et al.* 1995; Heyworth *et al.* 2007; Straub *et al.* 2011) but are rare and often appear to require unusual tectonic conditions for their appearance at the Earth's surface. Such rocks may provide a key to resolving the debate about the origin of arc type magmas. High-Mg basalt magma may in fact be the primitive progenitor of arc-type volcanic associations and, as suggested by Gust & Perfit (1987), evolve into the more common high-Al basalt by removal of pyroxene and olivine at pressures less than 10 kbar.

This paper presents the petrology of a suite of arc-type volcanic rocks in which high-Mg lavas are unusually common. These volcanics form a part of the late Cenozoic volcanic province of southeastern Papua New Guinea, a complex region in which contrasting and overlapping volcanic associations represent a magmatic response to rapidly changing tectonic patterns. The high-Mg lavas are interpreted as representing primitive magmas and their occurrence as due to an unusual tectonic situation rather than to unique processes of magma generation.

The Papuan arc

Tectonic setting

Papua New Guinea is a complex region of the circum-Pacific rim where late Cenozoic tectonism involves several minor plates caught up in a major zone of interaction between the Pacific and Australian plates (Johnson & Molnar 1972; Curtis 1973; Baldwin *et al.* 2012) (Fig. 1). Arc-type volcanism has occurred in widely separated areas of Papua New Guinea during the late Cenozoic, principally in the Bismarck volcanic arc, the Highlands province and the Papuan province (Johnson 1979). The Bismarck volcanic arc is clearly associated with active subduction resulting from interaction between the Solomon Sea and south Bismarck micro-plates (Johnson 1979). Although not obviously linked to present-day subduction, the Highlands province straddles a suture between the Australian plate and the Bismarck micro-plate that has been interpreted as a Cenozoic collision zone (Hamilton 1979). The Papuan Peninsula extends southeastward from the Highlands province, following the boundary between the Indo-Australian plate and Solomon Sea micro-plate, and is bounded on either side by lower Tertiary oceanic crust. The onshore geological record of this area is complex and fragmentary, and the sequence and nature of events are open to multiple interpretations (Smith 1982*a*, 2013; Davies *et al.* 1984; Smith & Milsom 1984). An important component of this record is a widespread episode of late Miocene to Recent arc-type volcanic activity which produced a variety of high-K calc-alkaline rocks (the Papuan arc) as well as volcanic and plutonic shoshonite suites (Smith 1972, 1982*a*).

The tectonic evolution of southeastern Papua New Guinea involves subduction (possibly with a reversal in polarity), obduction and rifting, the complexities of which are not totally resolved (Davies *et al.* 1984; Smith & Milsom 1984; Smith 2013). This has resulted in the juxtaposition of an unusual variety of late Mesozoic and Cenozoic rock associations, which include metamorphic core complexes (Davies & Warren 1988; Hill *et al.* 1995), ultramafic, mafic and felsic plutonic igneous rocks (Davies & Smith 1971; Smith & Davies 1976) and tholeiitic to peralkaline volcanic rocks (Smith *et al.* 1977; Smith 1982*a*). The essential features of the geological basement of the Papuan Peninsula are a core of moderate to high-grade metamorphic rocks overlain by an obducted sheet of ultramafic and associated mid-ocean ridge type basalts; these rocks are Late Mesozoic and lower Tertiary in age (Davies & Smith 1971). The archipelagos lying to the east and SE of the Papuan Peninsula show, to varying degrees, the fragments of this general sequence (Zirakparvar *et al.* 2012).

Tectonic activity in southeastern Papua began during late Mesozoic or early Tertiary times in response to collision between the Australian and Pacific plates. A simplified interpretation of the major tectonic events inferred from the geology of the Papuan Peninsula and offshore archipelagos includes, (1) formation of the Coral Sea Basin and accumulation of a thick volcano-sedimentary prism (D'Entrecasteaux Complex) during the Eocene (Davies 1973; Smith & Milsom 1984), (2) burial metamorphism of the sedimentary prism as a consequence of the emplacement (obduction) of the Papuan Ultramafic Belt during Eocene–Oligocene time (Davies 1980) and (3) late Cenozoic extension, uplift and unroofing of the D'Entrecasteaux Complex (Hill *et al.* 1992, 1995), calc-alkaline and shoshonitic volcanism (Smith 1982*a*) and, in Quaternary times, the eruption of peralkaline rhyolites accompanying opening of the Woodlark Basin (Luyendyk *et al.* 1973; Smith *et al.* 1977; Smith & Milsom 1984).

Active normal faulting in the Trobriand Platform to the north of the D'Entrecasteaux Islands and in Goodenough and Milne Bay graben to the south

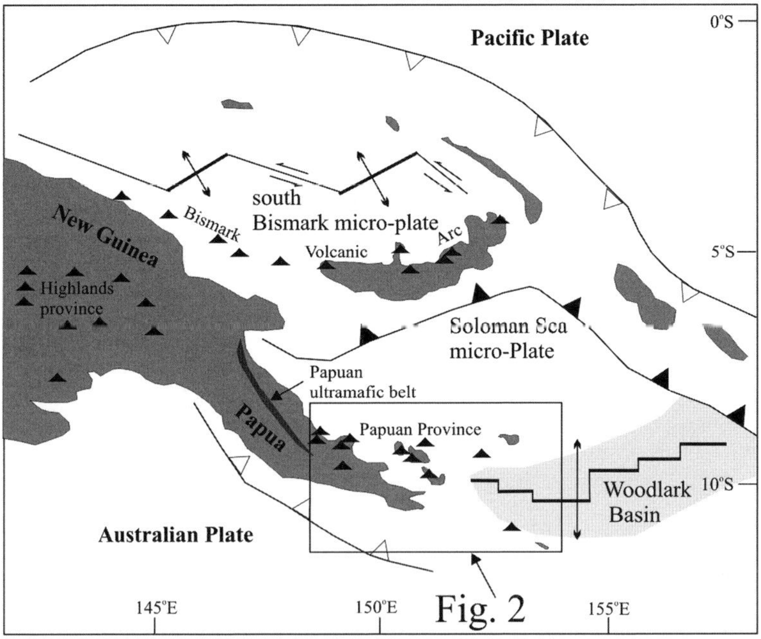

Fig. 1. Map of the Papua New Guinea region showing the principal features mentioned in the text. Subduction directions are indicated by triangles (solid, active; open, extinct).

shows that the crust in southeastern Papua is undergoing widespread horizontal extension (Ripper 1982). The east–west or ESE–WNW orientation of these faults indicates approximately north–south-directed extension which is at right angles to the trend of the volcanic arc. An east–west-trending sea floor spreading system lies immediately to the east of the D'Entrecasteaux Islands (Fig. 2). Studies of sea floor magnetic anomalies in the Woodlark Basin suggest that this spreading system has been active since at least 3.5 Ma and is propagating westward (Weissel *et al.* 1982). At the point where this spreading system intersects continental crust, rifting is occurring (Benes *et al.* 1994) and is accompanied by recent peralkaline volcanism (Smith 1976) and hot spring activity (Binns *et al.* 1987).

The western Calvados Islands lie to the south, and Egum Atoll to the north of the Woodlark basin spreading zone and are subaerial remnants of volcanoes representing the known eastern end of the Papuan arc (Fig. 2). The available dates show that volcanism in the Calvados islands preceded opening of the Woodlark Basin and that the Egum Atoll volcano developed on the northern side of the propagating rift at about 2 Ma. Westward, the data show that the volcanoes of the D'Entrecasteaux and Amphlett Islands were erupted in a tectonic environment dominated by crustal extension and uplift rates of the order of 8 mm a^{-1} (Hill *et al.* 1995). The localization of some recent volcanoes along faults in western Fergusson Island and on Goodenough Island suggests that there was some structural control of magma conduits around the margins of the rising metamorphic core complexes.

Although the oldest volcanoes in the eastern part of the Papuan arc may have erupted in a subduction-related tectonic setting, it is clear that most of the arc-type rocks presently exposed were erupted in an environment of crustal extension. In recent times, arc-type volcanic activity in the western D'Entrecasteaux Islands has occurred at the same time that rift-related peralkaline volcanoes erupted in the east.

Despite the fact that there is no suggestion of active subduction in the present-day pattern of seismicity beneath southeastern Papua, there is a continuing moderate level of arc-type volcanic activity in the western part of the Papuan arc (Smith 1982*b*). There is no doubt that after the late Mesozoic, the area underwent major convergence which has resulted in crustal thickening. This convergent event is logically correlated with emplacement of the Papuan ophiolites (Hamilton 1979; Davies & Jacques 1984; Smith & Milsom 1984) and may have developed into a subduction zone after ophiolite emplacement, as in northern New Zealand (Malpas *et al.* 1992). Initially subduction may

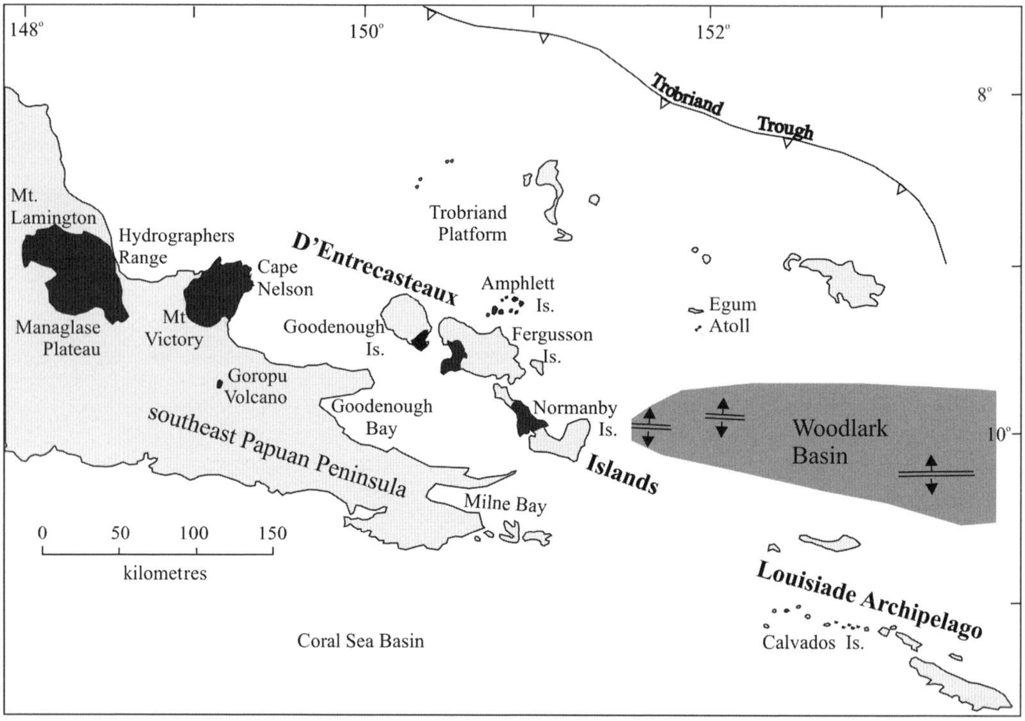

Fig. 2. Map of southeastern Papua showing the localities mentioned in the text. Areas of major outcrop of the volcanoes of the Papuan Arc are shown in black. Arrows indicate spreading directions in the Woodlark Basin.

have been toward the north, as evidenced by the occurrence of mid-Miocene shoshonitic plutonic and volcanic rocks in southeastern Papua (Smith 1972, 1982*a*, 2013) However, interpretation of the Trobriand Trough, north of the Trobriand Platform as a relic trench suggests that southward-oriented subduction could have occurred in late Tertiary times and was terminated by the westward migration of sea floor spreading in the Woodlark Basin. In this model the subducting oceanic lithosphere dipped southward beneath the D'Entrecasteaux Islands and Papuan mainland (Davies *et al.* 1984; Smith 2013).

The volcanic arc

The Papuan arc comprises late Cenozoic (11 Ma to Recent; Smith & Compston 1982) high-K arc-type volcanic rocks which form a belt of volcanic centres on the NE Papuan mainland, in the D'Entrecasteaux Islands and in the archipelago to the SE. There is a general progression of ages from the oldest centres (11 Ma) in the SE of the arc to the recently active volcanoes (Mt Victory, 1880's, Goropu Volcano, 1943–4, Mt Lamington, 1951; Smith 1982*b*) in the NW (Smith & Compston 1982). The volcanic association comprises two fundamental suites; those with compositions that are typical of arc-type rocks and those with distinctively high abundances of Mg, Cr and Ni (Smith & Mitchell 1989). The volcanic rocks are exposed in six areas representing distinct eruptive centres or clusters of eruptive centres (Fig. 2). These areas are the western Calvados Islands, Egum Atoll, Normanby Island, Amphlett Islands, western Fergusson Island and Goodenough Island. Together with the Cape Nelson volcanic complex (Jakes & Smith 1970) and the Mount Lamington–Hydrographers Range–Managlase Plateau area (Smith & Davies 1976) which lie to the west on the Papuan Peninsula, these volcanic centres define the Papuan volcanic arc.

The Papuan arc is a curvilinear feature which extends approximately 300 km SE from Mount Lamington on the north coast of the Papuan Peninsula. The eastern centres of the arc diverge from this trend because of the effects of sea floor spreading in the western Woodlark Basin. Volcanic activity within the arc has migrated westward through the late Cenozoic (Smith & Compston 1982) with Quaternary activity centred in the western D'Entrecasteaux Islands and the Papuan Peninsula.

The most recent eruptions are those of Mount Victory in the 1880s (Smith 1982*b*), nearby Goropu Volcano in 1943–44 (Baker 1946) and Mount Lamington in 1951 (Taylor 1958). Although the volcanic centres differ with respect to age and/or the relative abundance of rock types, there is no systematic spatial pattern to variations in the distribution of rock types (Smith 1982*a*; Smith & Milsom 1984).

The Calvados Islands, part of the Louisiade Archipelago, lie about 200 km to the SE of the Papuan mainland. Most of the islands are pre-Tertiary metamorphic and ultramafic rocks (Smith 1973; Zirakparvar *et al.* 2012), but at the western end of the group, volcanogenic conglomerate, agglomerate and tuff, together with subordinate lava flows, crop out on several islands within a radius of about 8 km. These rocks represent the easternmost and oldest volcanic centre in the arc (11.4 Ma; Smith & Compston 1982). The maximum aggregate thickness of the exposed volcanic rocks in the Calvados Islands is approximately 200 m. Rock types include aphyric to slightly porphyritic pyroxene and hornblende andesite and pyroxene dacite.

Egum Atoll lies about 175 km to the NW of the Calvados Islands. It is a typical atoll made up of reefs and low-lying coral islands which rise from ocean depths of 2000 m to enclose a large lagoon containing small (0.5 km^2 or less) volcanic islets. The islets are composed of jointed, generally porphyritic, hornblende andesites which are the emergent tip of a much larger andesitic volcano. Radiometric dating of one lava flow gave a late Pliocene age (2.85 Ma; Smith & Compston 1982).

Normanby Island is the easternmost of the three main islands in the D'Entrecasteaux Group and is formed mainly of pre-Tertiary metamorphic and ultramafic rocks. A variety of volcanic rocks with an aggregate thickness of 200 m overlie the pre-Tertiary basement in the central and western parts of the island. Microfossils from interbedded limestone indicate a late Miocene age (Davies 1973), and limited radiometric dating has yielded a Pliocene age (3 Ma; Smith & Compston 1982) for the volcanic rocks. The dominant volcanic lithologies are pyroxene-bearing basalt and andesite which are associated with subordinate hornblende andesite and dacite. The oldest of the volcanic rocks are typically porphyritic, whereas the younger sequences consist of aphyric to sparsely porphyritic basalt to andesite lavas overlain by porphyritic alkali-rich dacite.

The Amphlett Islands lie to the north of Fergusson Island in the D'Entrecasteaux Islands and are entirely volcanic. The islands are the remnants of an andesitic stratovolcano marked by a large positive Bouguer gravity anomaly (Davies 1973) centred on the group. The exposed volcanic sequence is early Pliocene in age (3.6–4.0 Ma; Smith & Compston 1982). Rock types, exposed as massive to vesicular lava flows and intercalated agglomerate beds, include basaltic andesite, andesite and dacite. The lavas are typically porphyritic, but relatively aphyric varieties also occur on most of the islands of the group.

Arc-type volcanic rocks occur extensively on western Fergusson Island and on Goodenough Island. On Fergusson Island, volcanic rocks occur mainly in the southwestern and west–central portions of the island, but isolated outcrops also occur near the centre of the island and along the north coast. The largest contiguous area of volcanic rocks is on at the southwestern end of the island where they attain their maximum exposed thickness of about 600 m. The volcanic rocks rest on an undulating, pre-Tertiary metamorphic basement of moderate relief. No volcanic land forms have been recognized although Davies & Ives (1965) have suggested that some of the peaks in the area may be volcanic plugs or necks. Radiometric ages for these volcanic rocks span a range from 6.27 to 0.4 Ma (Smith & Compston 1982).

At the western end of Fergusson Island is an alluvium-covered lowland and several small steep-sided cumulodomes and volcanic plateaux. Here, thin basaltic andesite and andesite lava flows are intercalated with thick dacite and rhyolite ash flow deposits. These volcanic rocks are relatively flat lying with an aggregate thickness exceeding 300 m and are the youngest eruptive rocks on the island. Geothermal features are prevalent and are among the most active in southeastern Papua New Guinea. The presence of youthful volcanic land forms together with geothermal activity suggests that these volcanic rocks are late Pleistocene to Holocene in age.

Volcanic rocks crop out along the margins of the metamorphic core of Goodenough Island. Their most extensive development is in the SE where they attain an aggregate exposed thickness of about 600 m. Elsewhere the volcanic rocks occur as small cones or flows adjacent to the major faults, or form isolated hills on the extensive alluvial apron surrounding the central metamorphic complex. These volcanic rocks may be as old as Pleistocene, but many of the small eruptive centres on Goodenough Island are undoubtedly very recent (Smith 1982*b*).

Petrography

The arc-type volcanic association in southeastern Papua comprises a diverse suite of rocks which includes basalt (10%), basaltic andesite (25%), andesite (45%), dacite (15%) and rhyolite (5%). The rhyolites are closely associated with andesite and dacite in western Fergusson Island and have been described in detail by Smith & Johnson (1981). The basalts, basaltic andesites and some andesites are typically porphyritic vesicular rocks containing sparse to abundant phenocrysts of plagioclase (An_{80-45}), olivine, pale-green clinopyroxene and orthopyroxene, and in some lavas, microphenocrysts of spinel and/or iron-titanium oxide minerals. The phenocrysts are set in a groundmass of labradorite, clinopyroxene, iron-titanium oxide minerals, less commonly orthopyroxene and, in a few samples, olivine. Hornblende-bearing lavas ranging from andesite to dacite are also an important component of the association. The hornblende-bearing lavas are highly porphyritic and contain phenocrysts of plagioclase (An_{60-30}), brown hornblende, biotite and iron-titanium oxide minerals in a groundmass dominated by plagioclase, but also contain clinopyroxene, orthopyroxene and rare olivine or quartz.

Phenocryst assemblages are generally restricted to three or four phases, but a few andesites and dacites contain complex phenocryst assemblages including hornblende, biotite, two pyroxenes and olivine, in addition to plagioclase. Many of the andesites are characterized by the presence of complexly zoned and commonly corroded feldspar phenocrysts and, less commonly, clusters (autoliths) of one or more of the phenocryst phases. Hornblendite inclusions are an additional feature of some lavas. The absence of ultramafic inclusions in the lavas from the offshore part of the arc is in contrast with abundant inclusions from Mount Lamington reported by Arculus *et al.* (1983).

An integral component of the Papuan arc is a suite of rocks with relatively high MgO contents and high Mg# (for definitions see the geochemistry section below). These high-Mg lavas are petrographically distinct from the remainder of the suite, even though they are spatially and temporally inseparable. Compared with other lavas, they are more sparsely porphyritic, contain a higher proportion of ferromagnesian minerals and show notably little petrographic variation. They are typically aphyric to sparsely porphyritic and contain less than 10% phenocrysts and/or microphenocrysts. The phenocrysts which do occur are limited to four phases:- olivine (Fo_{93-71}), clinopyroxene (Di_{91-72}), orthopyroxene (En_{91-81}) and plagioclase (An_{88-45}), with the latter being much less abundant than the other phases. The following phenocryst assemblages are observed:

(1) olivine > clinopyroxene > plagioclase (the dominant assemblage);
(2) clinopyroxene > olivine;
(3) orthopyroxene > clinopyroxene > olivine (an uncommon assemblage);
(4) olivine or clinopyroxene alone (a rare assemblage).

Microphenocrysts of chrome spinel and/or iron-titanium oxide minerals occur in many of the high-Mg lavas. The phenocrysts are dispersed in a hypocrystalline to pilotaxitic trachytoid groundmass of labradorite, clinopyroxene and calcium-poor pyroxene with both inclined and straight extinction (pigeonite), iron-titanium oxide minerals, and in some samples olivine. Fresh, dark brown glass is also a common constituent.

Clusters (glomerocrysts and autoliths) of two or more of the microphenocryst phases olivine, clinopyroxene, plagioclase and, rarely, chromite are common in many of the high-Mg lavas and may account for 15% of the phenocryst total. In some of the larger autoliths, the microphenocrysts are intergrown with, and surrounded by, very fine networks of acicular calcic plagioclase.

Olivine occurs as euhedral and rare skeletal phenocrysts, which may contain inclusions of clinopyroxene. In a few samples olivine is rimmed by, and in the case of smaller grains, completely replaced by, iddingsite, indicating that oxidizing conditions existed in some of the magmas; iddingsite is the only oxidized phase present in the high-Mg lavas.

Compositionally zoned euhedral nickel-bearing olivine (Fo_{92-78}) occurs in nearly all of the high-Mg lavas. In zoned crystals the cores tend to have higher Mg/(Mg + Fe) ratios and higher Ni contents than the rims, but olivine phenocrysts with reverse chemical zoning patterns are always present. The phenocryst phases are more magnesian than associated groundmass grains (Fo_{80-70}). The calcium content of olivine phenocrysts increases slightly from core to rim, generally varying from *c.* 0.15 to *c.* 0.30 wt%, although a few values up *c.* 0.60 wt% were measured. Olivine composition appears to be unrelated to the SiO_2 content of the rock.

Olivine generally appears to have crystallized in equilibrium with magma of approximately the composition of its host rock. As a test of this, olivine compositions (in terms of Fo content) are plotted against the Mg-number of their host rock in Figure 3. Also plotted is the line representing the locus of compositions in equilibrium with liquids assuming the crystal/liquid Fe/Mg partitioning experimentally determined by Roeder & Emslie (1970). Equilibrium liquidus olivine phenocrysts should plot on the line, and olivine crystallizing below the liquidus should plot below the line, reflecting rising Fe/Mg ratios in the residual liquid as crystallization proceeds. In general, mineral textures and the composition of coexisting minerals suggest equilibrium conditions prevailed.

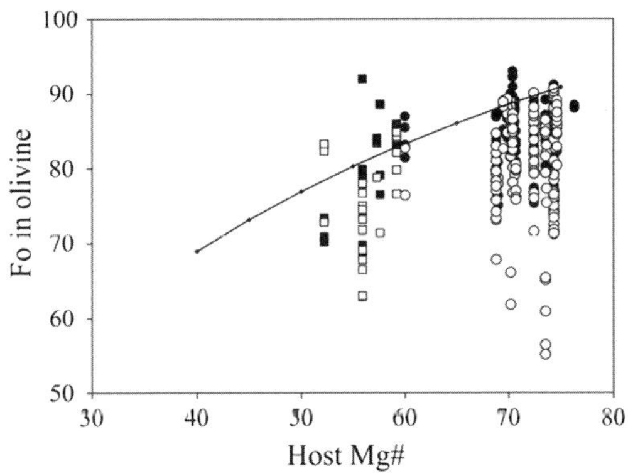

Fig. 3. Plot of olivine compositions *vs* Mg-number of host rock. The line is the locus of compositions in equilibrium following the experimentally determined partition coefficients of Roeder & Emslie (1970). High-Mg samples are represented by circles, low-Mg samples are represented by squares, filled symbols are crystal cores, open symbols crystal rims.

Clinopyroxene phenocrysts and microphenocrysts are typically pale green euhedral to subhedralcrystals. Endiopside and diopside are the most common clinopyroxene phenocrysts, whereas groundmass grains show a wider compositional range extending to Ca-poor pyroxene. Most phenocrysts contain between 0.5 and 1.0 wt% Cr_2O_3, and many contain trace amounts of NiO. Additional characteristics of clinopyroxene are moderate to high Al_2O_3 (4–7 wt%), low TiO_2 (<2 wt%) and relatively constant CaO (19.7–22.6 wt%) abundances. The Mg/(Mg + Fe) ratio of clinopyroxene is slightly lower than that of the associated olivine but calculations of Fe–Mg partitioning between clinopyroxene phenocrysts and their host magmas indicate that the clinopyroxene was in equilibrium with the magma. The Mg–Fe compositional variation between clinopyroxene phenocrysts and groundmass is greater than that of olivine, which is consistent with later crystallization.

Whole rock geochemistry

Representative major and trace element analyses of lavas from the Papuan arc are presented in Table 1. Geochemical data for the arc are also available in Smith (1982*a*), Smith & Compston (1982), Smith & Mitchell (1989), and Hegner & Smith (1992). Sample numbers in Table 1 refer to material archived in the School of Environment, University of Auckland. The complete dataset comprising 72 major and trace element analyses is provided in the Supplementary material. The dataset from the Papuan arc represents all of the eruptive centres from the part of the arc which lies off shore of the Papuan Peninsula; it has two significant limitations. Firstly, outcrop is extremely limited in the eastern (older) part of the arc where volcanoes are represented only by small island remnants. In the western part of the arc there is a greater volume of volcanic rocks, but exposures are limited by dense vegetation. Secondly, there are no sample suites representing detailed collections from continuous stratigraphic sections of individual volcanoes. For these reasons the dataset is used to address general problems of arc petrogenesis and detailed petrologic modelling has not been attempted.

Sample preparation and analytical methods

External surfaces of fresh rock samples were removed and the samples then split into chips and ground to a fine powder in a tungsten carbide ring grinder (Note that high field strength element content of the tungsten carbide grinder used is <0.01 ppm and contamination during crushing is not significant). H_2O^+, H_2O^- and CO_2 were determined by gravimetry, FeO measured by ammonium metavanadate titration. Reported major element abundances were measured by X-ray fluorescence (Siemens SR3000 spectrometer) at the University of Auckland using standard techniques on glass fusion discs prepared with SPECTRACHEM 12–22 flux. Trace elements (apart from V) were analysed by LA-ICP-MS at the Research School of Earth Sciences, Australian National University, using Excimer LPX120 laser and Agilent 7500 series mass spectrometer. For this work, the same fused glass discs as for XRF were used. Detection limits are < 1 ppb and analytical errors are < 1%.

Mineral compositions were determined on a JEOL JXA-5A microprobe (Otago University) using WDS and EDS techniques and on a JEOL 733A Superprobe (Stanford University) using WDS techniques. Replicate analyses provided cross checks for the different instruments and methods to ensure that the mineral compositions are an internally consistent dataset.

Geochemical features of the arc

All of the volcanic centres in the Papuan arc have erupted andesite accompanied in most areas by basaltic andesite with or without subordinate basalt and dacite. These rocks form a suite characterized by high total alkali contents and comparatively high K_2O/Na_2O ratios. In terms of Gill's (1981) classification they form a medium- to high-K calc-alkaline association (Fig. 4). The ferromagnesian elements (Ti, Fe, Mn, Mg, V, Cr and

Table 1. *Chemical analyses of representative samples from the Papuan Arc*

High-Mg suite

	33645	33611	33647	33628	33630	33631	33669	33670	33636	33638	33639
Rock type	Basalt				Basaltic andesite			Andesite			
(wt%)											
SiO_2	47.51	48.80	49.39	52.39	54.64	55.07	57.45	57.67	58.98	59.55	59.98
TiO_2	1.64	1.68	1.43	1.15	0.94	1.30	1.00	0.69	0.95	0.75	1.00
Al_2O_3	15.00	13.70	15.53	14.67	15.20	15.27	14.52	14.65	15.74	14.58	14.47
Fe_2O_3	2.39	4.35	3.03	4.51	4.50	2.69	2.78	2.27	4.24	2.65	2.18
FeO	6.41	5.30	5.11	3.04	2.19	4.05	2.62	4.29	1.03	2.93	3.19
MnO	0.15	0.15	0.13	0.12	0.12	0.11	0.09	0.12	0.07	0.08	0.08
MgO	11.77	9.65	9.76	7.94	6.90	6.79	5.85	7.16	3.94	6.64	5.45
CaO	9.37	8.05	9.34	7.66	7.34	6.20	7.59	7.11	4.94	4.92	4.82
Na_2O	3.40	2.50	3.16	3.56	3.92	4.01	3.35	3.14	4.08	4.20	3.98
K_2O	0.52	2.25	0.98	2.66	2.33	2.57	2.59	1.68	3.28	2.05	2.99
P_2O_5	0.37	0.84	0.76	0.52	0.40	0.45	0.44	0.20	0.39	0.20	0.32
H_2O^+	0.48	1.28	0.42	0.56	0.50	0.41	0.93	0.41	1.13	0.47	0.50
H_2O^-	0.46	1.04	0.38	0.46	0.38	0.33	0.25	0.33	0.57	0.85	0.53
CO_2	0.23	0.07	0.03	0.05	0.08	0.11	0.05	0.05	0.04	0.00	0.06
Total	99.70	99.66	99.45	99.29	99.44	99.36	99.51	99.77	99.38	99.87	99.55
(ppm)											
Cs	0.9	1.0	0.2	0.7	1.1	0.8	0.8	1.7	2.8	1.0	3.4
Ba	272.4	995.6	411.7	882.4	812.3	755.2	911.8	473.6	1086.7	569.8	814.3
Rb	8.6	35.7	6.7	37.7	25.9	30.5	25.4	27.5	48.1	27.2	41.4
Sr	597.8	700.9	663.4	1049.2	907.4	826.2	1231.7	642.9	1042.7	650.3	887.4
Pb	1.0	6.8	4.1	17.2	13.4	11.5	12.4	11.7	26.4	10.8	25.7
Th	1.7	7.7	3.2	15.6	8.9	6.7	8.6	5.6	19.0	5.8	14.2
U	0.4	1.4	0.4	2.9	1.9	1.5	1.9	1.2	3.7	1.2	3.3
Zr	162.5	248.8	133.3	185.6	167.7	178.4	192.0	107.2	210.5	132.4	177.3
Nb	3.9	16.0	5.1	7.1	7.3	14.1	5.4	3.6	7.9	4.7	8.3
Hf	4.2	7.5	3.7	5.4	5.0	5.2	6.0	3.8	6.4	4.1	5.4
Ta	0.4	1.1	0.4	0.5	0.6	1.1	0.6	0.5	0.6	0.4	0.9
Y	30.2	28.2	22.9	19.2	18.7	17.5	18.3	43.1	18.2	22.3	18.0
La	37.3	65.6	42.9	64.9	57.6	54.4	59.6	54.0	68.5	51.8	58.5
Ce	35.8	82.0	47.6	82.6	70.1	62.1	72.9	37.6	86.4	44.5	66.2
Pr	5.2	11.4	6.5	10.1	8.8	7.8	9.1	6.8	10.4	6.4	8.1
Nd	24.1	48.9	29.2	42.0	37.1	32.3	39.1	29.7	42.6	27.5	32.7
Sm	5.7	9.4	6.7	7.6	7.3	6.2	7.6	6.6	8.0	5.5	6.3
Eu	1.8	2.4	1.9	2.0	1.9	1.8	2.1	1.9	2.0	1.5	1.6
Gd	5.9	8.0	5.9	6.1	5.9	5.4	6.2	7.2	6.2	5.2	5.1
Tb	0.9	1.1	0.8	0.8	0.8	0.7	0.8	1.0	0.8	0.7	0.7
Dy	5.8	6.5	5.1	4.5	4.2	4.0	4.3	6.5	4.2	4.1	3.7
Ho	1.1	1.2	1.0	0.8	0.7	0.7	0.7	1.4	0.8	0.8	0.7
Er	3.3	3.1	2.6	2.1	2.0	1.9	2.0	4.1	2.0	2.3	1.9
Tm	0.4	0.4	0.3	0.3	0.3	0.3	0.3	0.6	0.3	0.3	0.2
Yb	2.9	2.8	2.3	1.8	1.8	1.6	1.7	3.6	1.8	2.1	1.5
Lu	0.4	0.4	0.3	0.3	0.3	0.2	0.2	0.6	0.3	0.3	0.2
Sc	16.0	18.3	15.3	13.2	12.5	10.0	8.1	12.0	8.1	9.0	8.1
V	156.0	213.0	193.0	159.0	126.0	128.0	126.0	171.0	90.0	91.0	100.0
Cr	460.0	336.9	388.2	382.3	303.4	218.4	144.1	318.6	157.1	329.5	235.7
Ni	247.6	203.2	194.7	213.4	144.0	144.9	130.2	90.6	63.1	205.3	116.0
Cu	15.0	17.2	15.0	21.2	32.2	25.5	34.2	28.2	12.1	13.3	9.9
Zn	24.8	42.3	26.1	39.0	40.8	44.9	67.1	31.3	30.6	30.8	29.0
Ga	15.0	19.8	16.1	18.8	19.0	19.4	20.6	17.9	22.5	18.4	19.4

Ni) and Ca and P show a positive correlation with each other and a clearly defined inverse relationship with the abundances of Si, K and incompatible trace elements. The abundances of large ion lithophile elements (LILE) are high but fall within the expected range for high-K andesites. Normalized trace element abundances (Fig. 5) are characterized by the features considered to be distinctive of subduction-related magmas (e.g. Pearce 1982; McCulloch & Gamble 1991) such as LILE enrichment coupled with strong depletions in Nb relative to Ba and Sr.

Table 1. *Continued*

Low-Mg suite

	33646	33649	33654	33657	33658	33660	33664	33633	33671	33625	33626
Rock type	Basalt		Basaltic andesite					Andesite		Dacite	
(wt%)											
SiO_2	47.91	50.47	53.70	54.22	55.22	55.45	56.30	57.79	60.29	65.82	66.10
TiO_2	2.00	1.64	1.67	1.02	1.46	1.18	1.16	1.09	1.15	0.61	0.78
Al_2O_3	17.13	19.21	17.15	17.62	17.66	17.84	17.65	16.49	16.54	15.97	16.81
Fe_2O_3	2.90	4.33	2.34	1.79	3.45	2.49	2.36	3.46	2.54	2.27	1.66
FeO	6.28	2.54	5.40	5.87	3.83	3.84	3.93	2.57	2.55	1.21	0.85
MnO	0.15	0.11	0.14	0.14	0.13	0.11	0.11	0.09	0.12	0.07	0.02
MgO	7.69	4.59	3.99	4.51	3.08	3.93	3.80	3.71	2.64	0.94	1.01
CaO	10.03	8.91	8.30	9.02	6.92	7.71	6.82	5.71	4.03	1.98	2.41
Na_2O	3.89	3.76	3.97	3.09	4.36	4.00	4.26	3.92	5.19	5.58	4.79
K_2O	0.48	1.58	1.53	1.32	2.29	1.67	1.78	2.82	3.14	4.18	4.17
P_2O_5	0.37	0.51	0.34	0.30	0.56	0.30	0.32	0.41	0.46	0.18	0.28
H_2O^+	0.44	1.27	0.66	0.44	0.31	0.91	0.71	0.94	0.60	0.34	0.44
H_2O^-	0.29	0.78	0.34	0.17	0.17	0.33	0.32	0.65	0.23	0.34	0.35
CO_2	0.23	0.12	0.26	0.00	0.10	0.08	0.16	0.01	0.01	0.08	0.02
Total	99.79	99.82	99.79	99.51	99.54	99.84	99.68	99.66	99.49	99.57	99.69
(ppm)											
Cs	0.1	0.8	1.7	0.4	1.0	1.1	0.9	1.9	2.5	1.6	1.7
Ba	163.4	531.1	504.1	461.3	786.5	464.8	491.8	827.4	845.9	492.9	958.9
Rb	3.7	21.1	22.1	11.0	23.3	19.7	22.9	42.6	56.9	60.7	69.4
Sr	466.5	863.6	485.3	907.3	981.0	597.2	543.1	777.0	498.8	188.6	470.6
Pb	1.3	5.3	7.8	6.9	10.8	7.1	9.0	13.9	17.8	16.4	17.7
Th	1.0	4.2	5.4	4.0	7.6	5.1	5.4	11.2	12.8	16.7	17.1
U	0.2	1.0	1.5	0.7	1.8	1.0	3.2	2.1	2.9	3.5	3.3
Zr	170.5	156.5	182.0	118.1	194.4	160.8	167.1	220.5	326.0	412.1	320.2
Nb	3.1	7.3	5.5	2.9	6.6	5.3	5.5	7.8	11.4	20.6	11.9
Hf	4.5	4.3	5.4	3.8	5.3	4.7	4.9	6.6	9.4	11.6	9.3
Ta	0.3	0.6	0.5	0.3	0.5	0.5	0.4	0.6	1.1	1.8	1.1
Y	30.3	23.7	31.4	19.0	26.3	22.3	24.8	25.5	35.6	34.4	20.4
La	32.0	45.3	43.2	40.8	54.4	40.9	44.2	58.6	69.7	65.8	61.2
Ce	30.4	51.3	47.9	39.9	68.9	40.5	45.2	69.8	91.4	85.0	74.4
Pr	4.7	6.9	6.7	5.4	9.1	5.2	5.8	9.2	11.7	9.7	8.6
Nd	22.6	30.5	31.0	24.2	39.3	23.2	25.2	38.5	47.9	36.8	32.3
Sm	6.0	6.8	7.3	5.2	8.2	5.2	5.3	7.6	9.8	7.1	5.6
Eu	1.9	2.1	2.1	1.5	2.3	1.6	1.6	1.9	2.5	1.1	1.4
Gd	6.2	6.0	7.2	4.7	7.1	4.9	5.1	6.4	8.5	6.1	4.7
Tb	1.0	0.9	1.0	0.7	0.9	0.7	0.8	0.9	1.2	1.0	0.7
Dy	6.3	5.1	6.6	4.1	5.7	4.5	5.0	5.3	7.6	6.7	4.2
Ho	1.3	1.0	1.3	0.8	1.1	0.9	1.0	1.0	1.5	1.3	0.8
Er	3.6	2.6	3.7	2.1	2.9	2.5	2.7	2.8	4.1	4.1	2.3
Tm	0.5	0.4	0.5	0.3	0.4	0.4	0.4	0.4	0.6	0.6	0.3
Yb	3.3	2.5	3.4	2.0	2.7	2.3	2.6	2.4	4.1	4.4	2.3
Lu	0.5	0.4	0.5	0.3	0.4	0.3	0.4	0.4	0.6	0.7	0.4
Sc	18.7	11.8	15.2	16.9	11.1	12.7	9.2	10.4	6.1	4.4	4.4
V	206.0	176.0	180.0	215.0	194.0	153.0	139.0	123.0	84.0	38.0	69.0
Cr	147.4	80.6	27.9	22.3	7.0	40.1	79.8	111.6	59.7	14.5	15.5
Ni	77.9	48.3	28.7	12.3	5.1	19.9	22.9	54.2	28.9	6.1	11.2
Cu	13.2	15.9	21.6	41.3	12.2	13.4	14.1	25.5	5.9	35.7	38.0
Zn	26.8	28.5	33.3	34.3	33.1	36.7	42.8	47.1	33.2	52.3	46.1
Ga	16.8	19.0	19.7	19.1	20.7	18.8	19.0	21.2	22.8	22.7	20.0

In terms of $MgO-SiO_2$ covariance of the basaltic and intermediate lavas (Fig. 4), a high-Mg group can be clearly separated from the remainder of the suite by an empirically-derived line with the formula $MgO = (30 - (0.43 \times SiO_2))$ (Smith & Mitchell 1989; Smith *et al.* 1997). Thus defined, high-Mg lavas comprise approximately half of the samples analysed in this study. Hereafter we refer to high-Mg and low-Mg groups while recognizing that the compositions designated as low-Mg are in fact typical of those found in many arc-type volcanic associations.

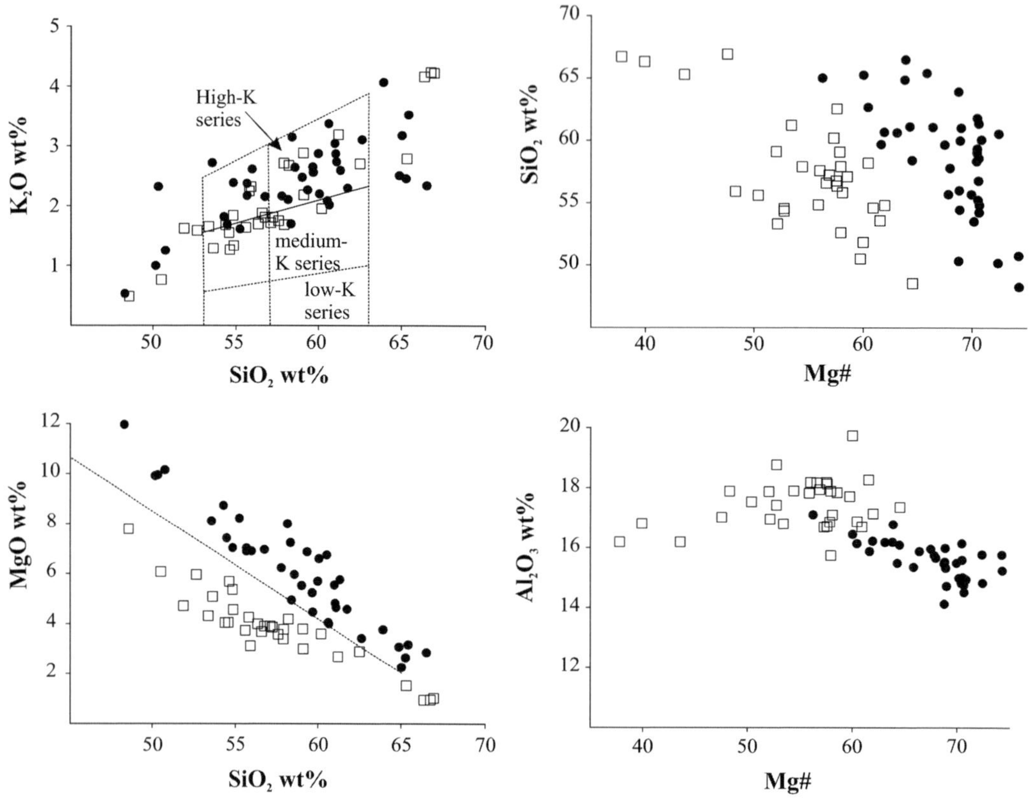

Fig. 4. Major element variation in samples from the Papuan arc. Solid dots denote high-Mg compositions, open squares are low-Mg compositions. Field boundaries on the K_2O v. SiO_2 plot are from Gill (1981), the line separating high- from low-Mg samples ($MgO = [30 - (0.43 \times SiO_2)]$) is from Smith & Mitchell (1989).

The high-Mg lavas vary in chemical composition from basalt to dacite, but the majority are basaltic andesite and andesite. Characteristic of these lavas are anomalously high, but variable abundances of Cr (158 to 592 ppm) and Ni (80–343 ppm), and Mg# (mol 100 MgO/[MgO + FeO], Fe_2O_3/FeO = 0.20) in the range of 64–76 (Fig. 6). The compositional trends of the high-Mg lavas converge at high silica values, on the trends shown by the remainder of the suite (Fig. 4). It is significant that apart from higher abundances of Mg, Cr and Ni, and lower abundances of Ti, Al and Na, the high-Mg lavas have virtually the same trace element abundances as those forming the remainder of the suite (Fig. 5).

The general pattern of rare earth element (REE) abundances in samples from the Papuan arc is one of moderate to strong enrichment of light REE relative to chondritic abundances, whereas the heavy REE are comparatively enriched but unfractionated (Fig. 7). There are no significant Eu anomalies, as observed in Kermadec and New Zealand andesites by Smith *et al.* (1997) and interpreted as the result of plagioclase fractionation. Figure 8 plots ratios of REE that have been used by Davidson *et al.* (2007, 2013) to define the shape of normalized patterns. These plots demonstrate small but distinct differences between the high- and low-Mg suites. High-Mg lavas have slightly higher La/Yb, and Dy/Yb (steeper normalized patterns) than low-Mg lavas although the general shape as defined by Dy/Dy* ratios is comparable for both suites.

Isotopic data from the Papuan arc have been presented by Smith & Compston (1982) and Hegner & Smith (1992); these data are illustrated in Figure 9. $^{87}Sr/^{86}Sr$ ratios range from 0.7035 to 0.7043 indicating a relatively primitive source. $^{143}Nd/^{144}Nd$ and $^{87}Sr/^{86}Sr$ plot along a mantle array; they are more evolved than Pacific MORB, but overlap SE Indian Ridge MORB. High-Mg and low-Mg suites have overlapping ranges. These data are significant because they indicate that (1) there is a mantle component to the isotopic

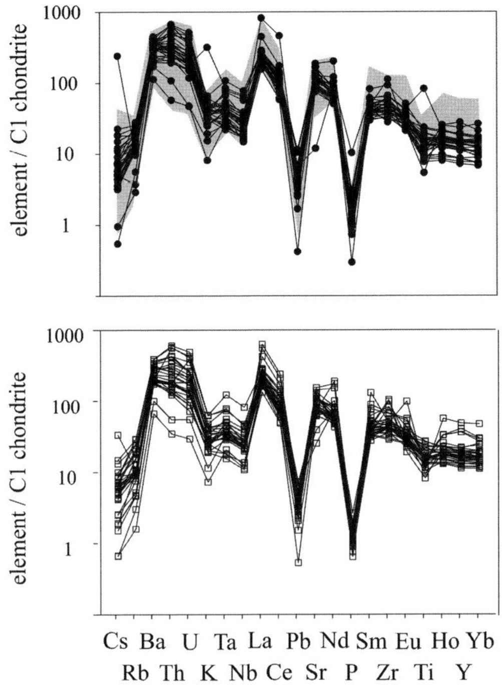

Fig. 5. Multi element plot of representative samples from the high-Mg (upper panel solid dots) and low-Mg (lower panel open squares) groups of the Papuan arc. Elemental abundances are normalized to the chondritic values of Sun & McDonough (1989). The shaded field in the upper diagram is the area of low-Mg samples from the lower diagram.

component of Nd and Sr, (2) that very little if any contamination of the magmas occurred as they rose through the thick sequence of isotopically evolved metamorphic rocks of the core complexes, and (3) the high-Mg and low-Mg suites are isotopically the same.

Papuan arc lavas have high ratios of $^{206}Pb/^{204}Pb$ (18.30 to 18.91), $^{207}Pb/^{204}Pb$ (15.51 to 15.60) and $^{208}Pb/^{204}Pb$ (38.09 to 38.70), which plot well outside the range of MORB and define trends which also plot well away from the trend for mantle-derived Pacific Ocean Basin basalts. The lead isotope ratios are more radiogenic than typical MORB lead, and considering that the fields for ocean floor sediment and for the Papuan arc overlap indicate significant contamination of a mantle source by ocean floor sediment.

Although limited, the isotope data are compatible with the hypothesis that both high-Mg and low-Mg suites share a common origin and that they are the result of partial melting of a mantle source which has been modified by interaction with fluids from subducting oceanic lithosphere.

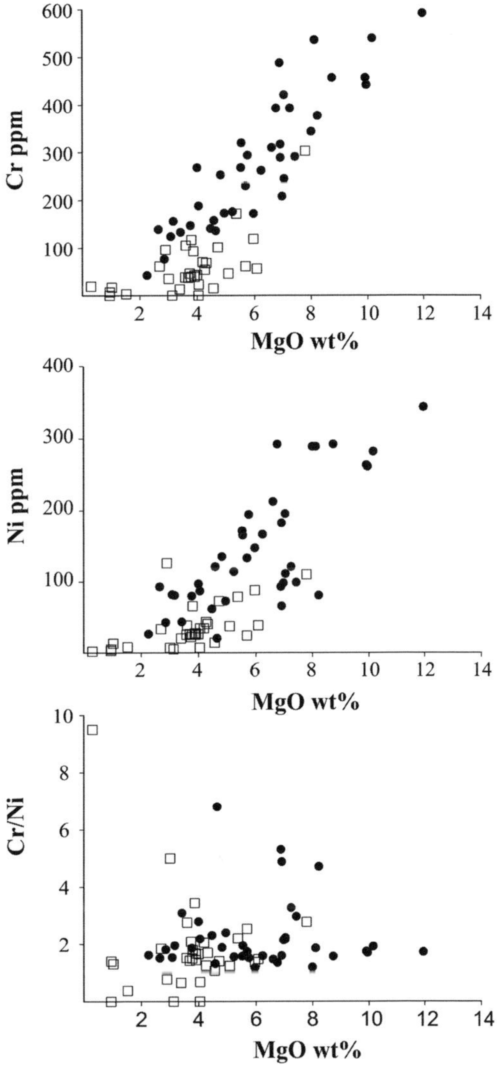

Fig. 6. Cr and Ni abundances in volcanic rocks from the Papuan arc. High-Mg samples shown as solid dots and low-Mg samples as open squares.

Discussion

The Papuan arc is unusual in several aspects; it is not associated with an active Wadati–Benioff zone, it is closely associated with extreme uplift rates, it shows clear migration of eruptive centres through time, it overlaps peralkaline volcanism which does not normally occur in subduction settings, and it contains a significant proportion of high-Mg rock compositions. These unusual aspects are related and together provide an important link in our understanding of magma generation processes in subduction settings.

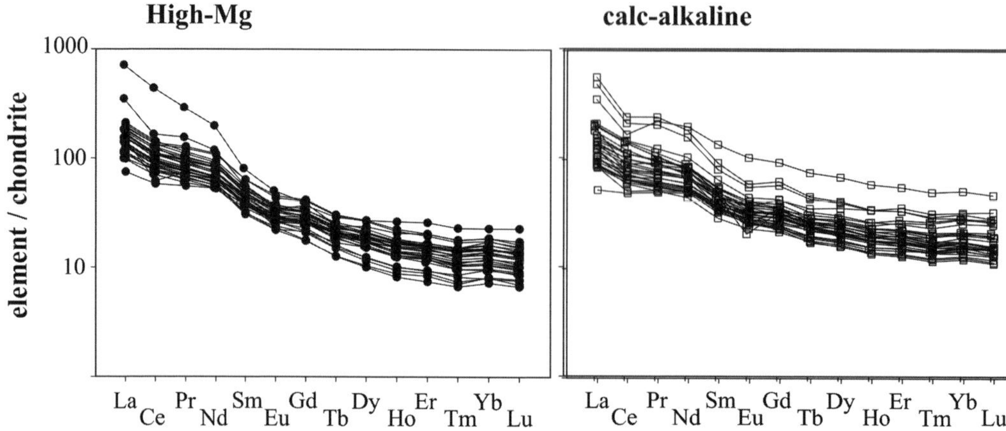

Fig. 7. Chondrite normalized rare earth element abundances in samples from the Papuan arc. High-Mg samples shown as solid dots and low-Mg samples as open squares. Normalizing values are from Sun & McDonough (1989).

The essential precursor to arc-type magmatism is generally agreed to be interaction between subducting oceanic lithosphere and overlying peridotitic mantle wedge to produce a chemically modified fertile source capable of yielding magmas with arc-type chemical characteristics. If subduction-modified mantle is not immediately involved in a melting event it may remain as a fallow source until a later tectonic event initiates melting. In the Papuan case, creation of the source may have been linked to mid-Tertiary subduction of the Solomon Sea Plate southward from the Trobriand Trough, although the orientation and the timing of this event is poorly constrained (Hamilton 1979; Davies *et al.* 1984; Smith & Milsom 1984). In this model, tapping the source to produce arc-type magmas occurs in response to the change in tectonic setting which presaged sea floor spreading to the east. This delayed magmatism model resolves the apparent conflict in the juxtaposition of late Cenozoic magmatic associations recognized by Smith *et al.* (1977) and explains the occurrence of young, typically arc-type volcanic rocks in an area not undergoing active plate convergence. Similar models have been proposed for the arc-type volcanoes in northern Vanuatu (Barsdell *et al.* 1982) and Calmus *et al.* (2003) have suggested a delay between ridge subduction and the eruption of high-Mg andesites in Baja California. An alternative but not exclusive hypothesis is that the subduction of young hot Soloman Sea crust was aseismic.

The significance of high-Mg andesites

If it is assumed that arc-type magmas originate in the mantle overlying the subducting slab, then primary magmas will have equilibrated with peridotitic residue and, therefore, will have high Mg-numbers (about 70) and high abundances of mantle compatible elements such as Ni (>200 ppm) and Cr (>400 ppm), as well as high MgO contents (cf. Tatsumi & Eggins 1995). Lavas that fulfill these criteria are relatively rare in most volcanic arc associations, although they do have widespread occurrence (Smith & Mitchell 1989). A simple explanation for this observation is that primary arc magmas are usually modified by processes within the crust. Only in unusual tectonic settings do they pass through this crustal filter unmodified as they ascend from their source.

High-Mg lavas are unusually abundant in the Papuan arc. They occur in all centres and are not separable in time and space from the low-Mg lavas. It is suggested that their unusual abundance is linked to the atypical tectonic setting of volcanism in southeastern Papua and that crustal extension allowed relatively rapid transit of magmas through the crust, so that modification of primitive magmas by crustal processes was minimal. Do the high-Mg lavas in the Papuan arc therefore represent primitive arc magmas?

Arculus *et al.* (1983) have argued convincingly that high MgO, Cr and Ni in some of the lavas forming Mount Lamington on the mainland west of the D'Entrecasteaux Islands are the result of contamination from the subjacent Papuan Ultramafic Belt. They cite the presence of ultramafic mineral clusters in the rocks and present numerical models in support of this hypothesis. Mafic mineral clusters also occur in some of the high-Mg lavas in the D'Entrecasteaux Islands, but petrographic and geochemical data indicate that in this case the mineral clusters are comagmatic in origin (autoliths).

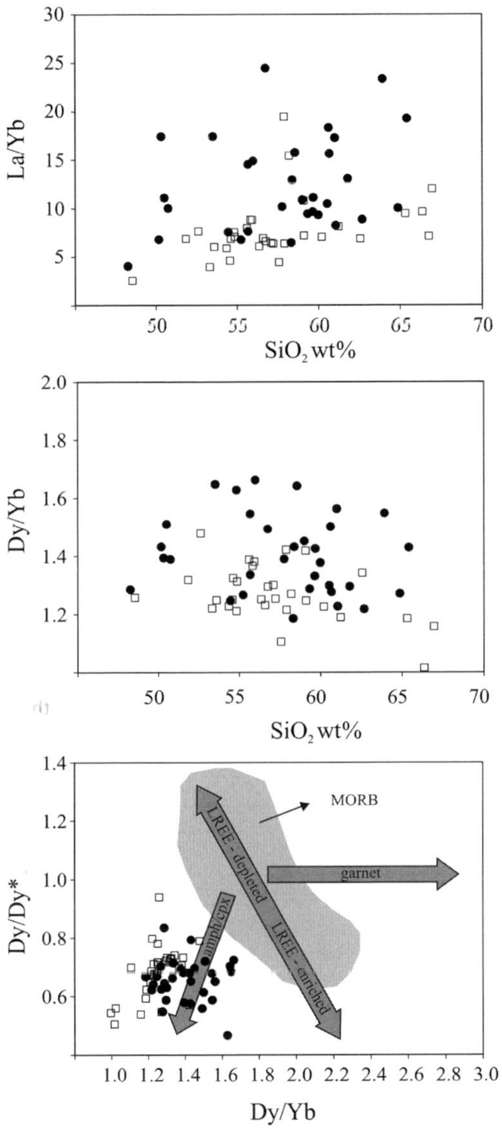

Fig. 8. REE ratio diagrams comparing high-Mg and low-Mg samples of the Papuan Arc to. High-Mg samples shown as solid dots and low-Mg samples as open squares. The two upper diagrams are plots suggested by Davidson *et al.* (2007) to distinguish garnet and amphibole fractionation trajectories. The lower diagram follows Davidson *et al.* (2013). See text for discussion.

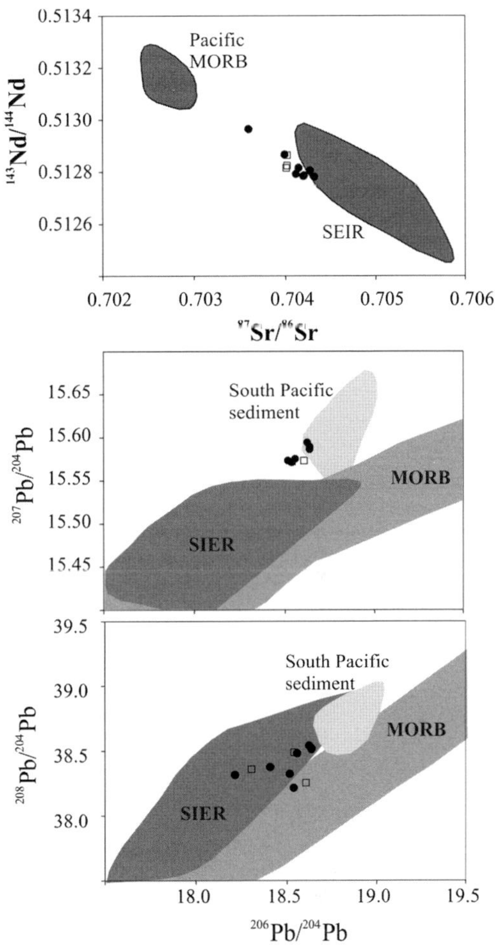

Fig. 9. Isotopic abundances in samples from the Papuan Arc. Data from Hegner & Smith (1992). High-Mg samples shown as solid dots and low-Mg samples as open squares. Comparative data fields are shown for Pacific MORB and the southeast Indian Ridge (Michard *et al.* 1986; Hart 1988; Dosso *et al.* 1988; Mahoney *et al.* 1989).

Olivine and pyroxene contained in the mineral clusters are optically and compositionally identical to the phenocrysts with which they coexist, and no remnant metamorphic textures were found, which would indicate that olivine or pyroxene were derived from underlying ultramafic tectonites. Normal and reverse chemical zoning within these crystals in terms of Mg/Mg + Fe, albeit minor, are characteristic features of the larger crystals of olivine and pyroxene, whether they occur in clusters or as phenocrysts in the high-Mg lavas. These data are interpreted to indicate that the olivine and pyroxene are magmatic in origin.

The compositional range and extent of chemical zoning of olivine and pyroxene in the high-Mg lavas are similar to those of the same minerals in the ultramafic cumulates from the Papuan Ultramafic Belt (England & Davies 1973). Contamination thus remains a possibility, but ultramafic cumulates comprise less than 10% of the Papuan Ultramafic

Belt (England & Davies 1973) and it is unlikely that their spatial distribution is as extensive as that of the high-Mg lavas in the Papuan arc. Further, the Al_2O_3 content of clinopyroxene contained in the high-Mg lavas (0.8–7.1 wt%) is significantly higher than that of clinopyroxene in the ultramafic cumulates (0.8–2.6 wt%). Finally, the overlapping compositions for most chemical components of the lavas and the separation of the compositional trends into two discrete groups on a $MgO-SiO_2$ plot would not be expected if the high magnesium contents in the lavas were the result of a random process such as contamination. It follows that that contamination cannot be a general explanation for the occurrence of high-Mg lavas in southeastern Papua.

The accumulation of olivine and pyroxene by crystal settling is a viable means of increasing the MgO, Cr and Ni content of a magma. The occurrence of mafic autoliths in some of the high-Mg lavas from the east Papuan volcanic province indicates this as an explanation for high-Mg lavas. However, on the basis of petrographic evidence and Fe–Mg partitioning, the olivine ($Fo_{\sim 87}$) and clinopyroxene phenocrysts contained in the high-Mg lavas were in equilibrium with a melt composition equivalent to the erupted whole rock composition. This observation implies that minimal crystal accumulation (or fractionation) occurred. Further, Mg-numbers in the range of 66 to 75 are similar to those of a melt in equilibrium with the upper mantle.

An important consideration in regard to a general petrogenetic model is that the petrographic data and qualitative geochemical arguments do permit the derivation of the low-Mg lavas from high-Mg parent magmas by crystal fractionation. As noted earlier, the only significant difference between the high-Mg and the low-Mg lavas is the higher abundance of MgO, Cr, and Ni in the former. Further, these elements are concentrated in olivine and pyroxene phenocrysts, and fractionation of these phases would substantially lower the MgO, Cr and Ni content of the resulting magma.

If the mafic phases, melt and upper mantle are in equilibrium, then a two-component mixing hypothesis comparable to the restite model for the origin of granites (Chappell & White 1974) may be applicable. Two-component mixing is not contamination by pre-existing material, but the mixing of a partial melt with components of the crystalline residuum with which it is in equilibrium. The mixing process offers a way by which chemical components which are normally not associated can become correlated. If the high MgO, Cr and Ni content of the high-Mg lavas is related to inclusion of residual source material, then the incompatible element content of the melt fraction is independent of the Ni and Cr content of the whole rock (melt plus some crystalline residuum). The major argument against two-component mixing is that the olivine and pyroxene crystals appear to be equilibrium crystallization products based on (1) the existence of chemical zoning and (2) the occurrence of chromite, spinel and rare clinopyroxene inclusions in olivine, and olivine and rare spinel inclusions in clinopyroxene.

Whole-rock and mineral chemical compositions of the Papuan high-Mg lavas indicate that olivine and clinopyroxene phenocrysts have crystallized from a melt with approximately the composition of their hosts, and that both could have equilibrated with a mantle source. The primitive nature of the lavas is also reflected in the common occurrence of chromite inclusions in olivine, the occurrence of chromite microphenocrysts in some samples, and the near-MORB values for FeO*/MgO, Al_2O_3/TiO_2 and CaO/TiO_2 ratios. However, comparatively high K_2O/TiO_2 ratios and incompatible element concentrations indicate the involvement of a more evolved source. From this discussion it is suggested that the high-Mg lavas in the Papuan arc represent relatively unmodified primitive magmas and are close to primary arc-type magmas in their compositional characteristics. These primary magmas are hydrous partial melts of subduction-modified mantle.

The naming question

Mg-rich rocks in a number of circum-Pacific arcs have been identified and referred to by a variety of different names: magnesian andesites (Kay 1980; Tatsumi & Ishizaka 1982; Johnson *et al.* 1983); sanukitioids (Tatsumi & Ishizaka 1981); high-magnesium andesite (Meen & Eggler 1987); and bajharites (Saunders *et al.* 1987). Smith & Mitchell (1989) suggested the use of the term high-Mg to recognize the existence of rocks with relatively high MgO contents among arc-type igneous associations in the same way that high-K has been used in the literature to identify suites of rocks with that particular geochemical characteristic.

Magnesium-rich volcanic rocks similar to the late Cenozoic high-Mg lavas in southeastern Papua occur in many other parts of the circum-Pacific region (Smith & Mitchell 1989). An important feature of all of these areas is their intimate association with rocks which are typical of arc environments generally. Boninites are also relatively Si-rich rocks which are high in MgO that occur in a number of circum-Pacific regions (Kuroda *et al.* 1978; Sun & Nesbitt 1978; Jenner 1981; Bloomer & Hawkins 1987). It is important to recognize that the mineralogy and chemistry of the high-Mg rocks described and referred to in

Table 2. *Mean composition of rock types from the Papuan Arc together with boninites from southeastern Papua*

	Papuan Arc		
	Low-Mg suite ($n = 40$)	High-Mg suite ($n = 43$)	Boninite ($n = 21$)
(wt%)			
SiO_2	58.55	57.81	57.61
TiO_2	1.17	1.00	0.25
Al_2O_3	17.47	15.47	8.38
FeO^{tot}	5.99	6.04	9.59
MnO	0.11	0.10	0.20
MgO	3.40	6.51	17.83
CaO	6.45	6.47	4.94
Na_2O	4.26	3.70	0.83
K_2O	2.24	2.27	0.33
P_2O_5	0.36	0.36	0.04
Mg-number	54.4	69.40	79.60
(ppm)			
Ba	795	873	48
Rb	51	51	6
Sr	692	821	100
Th	6	8	0.6
Zr	228	194	0.2
Nb	6	5	2
Y	29	21	5
La	44	49	3
Ce	70	74	7
Sc	14	14	30
V	137	132	156
Cr	49	299	1775
Ni	30	155	378

Total Fe calculated as FeO.
n = number of analyses used in the mean calculation. Boninite data from Jenner (1981).

this paper are quite distinct from boninites. For example, in the comparison of Papuan data (Table 2) the only chemical similarity between arc-type high-Mg lavas and boninites is their intermediate silica content. High-Mg lavas and boninites are unrelated rock types, and it is confusing to refer to arc-type magnesium-rich rocks as having 'boninitic' affinities.

Petrogenetic observations from the Papuan arc lavas

Using petrographic and geochemical criteria, discrete high- and low- Mg groups of lavas in the Papuan arc have been identified. These groups are not separable in time or space and, therefore, represent different parts of a single magmatic system.

The high-Mg group lavas are relatively unfractionated and show compositions which range from basalt to dacite. The high-Mg lavas are enriched in Mg, Cr and Ni, but have Cr/Ni ratios and incompatible element abundances typical of arc-type volcanic rocks, suggesting their derivation by a subduction-like process. The FeO^*/MgO and Al_2O_3/TiO_2 ratios, the abundance of transition elements, and the high incompatible element contents of the high-Mg lavas require that they be generated from a previously undepleted mantle source. K_2O/TiO_2 ratios indicate chemical modification of the mantle source by a potassium-bearing hydrous fluid which was probably derived from dehydrating, subducted lithosphere.

Arguably, the basaltic end of this compositional range represents primary arc-type magma and the higher-Si compositions are the products of fractionation processes in which the fractionating phase assemblage is low-Si. However, if fractionation took place at high pressures the amphibole stability field would be intersected at high temperatures (Foden & Green 1992). There is a general increase in La/Yb ratio correlated with SiO_2 (Fig. 8) content that supports this concept and the geological history of southeastern Papua which involves obduction and crustal thickening is compatible with the notion of a deeply-rooted magmatic system.

The dominance of olivine and clinopyroxene and corresponding paucity of orthopyroxene and plagioclase in the high-Mg magmas is consistent with their derivation from undepleted mantle peridotite modified by hydrous fluids evolving from dehydrating lithosphere (Wood & Turner 2009). Hydrous solutions containing K^+, Na^+ and Ca^{2+} enhance the stability of olivine and clinopyroxene relative to orthopyroxene and plagioclase, respectively, by lowering the activity of SiO_2 (Ewart 1979; Mysen 1982).

Variations in CaO content of olivine suggest that olivine crystallized under relatively constant temperature and pressure at moderate to deep levels within the lithosphere (Simkin & Smith 1970), interpretations that are supported by the Al_2O_3 content of clinopyroxene in the high-Mg rocks. Since the Al_2O_3 content of the clinopyroxene phenocrysts from the high-Mg lavas suggests that they formed at moderate lithospheric depths and that the parental magmas or magma chambers were either generated or existed at relatively similar depths throughout southeastern Papua. Further, the similarity in clinopyroxene compositions in all of the high-Mg rocks suggests the existence of a relatively uniform source region for magma generation throughout the area, since the bulk composition of the melt is also a controlling factor in the composition of clinopyroxene (Campbell & Borley 1974). Although there is very little variation in the chemical composition of individual phenocrysts within any of the phenocryst species which characterize the high-Mg lavas, the abundances of certain trace elements, most notably the rare earth elements, may vary markedly between the lavas, even when erupted from the same vent or cluster of vents.

The low-Mg lavas of the Papuan arc are more typical of arc-type magmas. A consistent explanation is that they are the result of fractionation from high-Mg parents at lower pressures where olivine and pyroxene are the predominating fractionating phases. However, it is likely that their compositions are also influenced by open system recharge, mixing and entrainment processes that have become widely recognized as important in the petrogenesis of andesitic rocks (e.g. Price *et al.* 2012). Because of the limitations of the dataset, discussed earlier, there is no attempt to model these processes in detail. However, I suggest that the range of compositions observed in the Papuan arc represent a combination of deep-seated and shallow fractionation processes stemming from primitive high-Mg basaltic magma. The unusually high proportion of high-Mg compositions in Papua is directly a consequence of the extensional tectonic regime which allows magma to rise rapidly, thus tapping the deeper regions of the magmatic system.

Conclusion

An unusual proportion of high-Mg lavas in the Papuan arc represent liquids tapped from a deep-seated magmatic system in an extensional tectonic regime. What is unusual in the Papuan arc is a tectonic environment which has allowed rapid ascent of magmas, rather than unusual petrogenetic conditions. If primary magmas in subduction systems are high-Mg basalts comparable to those in the Papuan arc, then in general these are profoundly modified above their source in typical arc settings.

The presence of a high-pressure fractionation series (the high-Mg andesites and dacites) in the Papuan arc is also a reflection of the unusual tectonic environment. However, it provides evidence that the crust also plays a role in arc settings in defining the pressure (depth) at which fractionation processes can take place, given that the crust/mantle interface is a major density barrier (at least beneath immature crust) where magmas can pond. Where the crust is thick, deep-seated fractionation involving amphibole may occur, whereas beneath thin crust fractionation is dominated by olivine and orthopyroxene.

The Papuan arc is characterized by medium to high-K compositions throughout the composition range. The high-Mg primary magmas share this characteristic together with high LILE contents. This chemical characteristic is not a product of fractionation or crustal interaction and is, therefore, a primary characteristic derived from the mantle wedge.

From this model of magma genesis in the Papuan arc, it is suggested that arc-type magmas in general are the products of independent processes involving subduction-related metasomatism of the mantle wedge, partial melting of this source and interaction with the overlying lithosphere in which crustal thickness, tectonic setting and crustal composition are independent variables which further complicate the processes of magma genesis. The relationship between subduction, crustal extension and eruption of magmas with arc-type chemical characteristics in southeastern Papua has been explained by Johnson *et al.* (1978) and Smith (1982*a*) as a delayed melting process in which magmas were generated by partial melting of a source which was metasomatized during an earlier convergent event.

The manuscript has benefited from the careful reviews of G. Yogodzinski, and T. Feeley and the editorial handling by S. M. Straub. J. Wilmshurst (University of Auckland) and C. Allen (Australian National University) assisted with the analytical work.

References

Annen, C., Blundy, J. D. & Sparks, R. S. J. 2006. The genesis of intermediate and silicic magmas in deep crustal hot zones. *Journal of Petrology*, **47**, 505–539, http://dx.doi.org/10.1093/petrology/egi084.

Arculus, R. J. & Powell, R. 1986. Source component mixing in regions of arc magma generation. *Journal of Geophysical Research*, **91**, 5913–5926, http://dx.doi.org/10.1029/JB091iB06p05913.

Arculus, R., Johnson, R. W., Chappell, B. W., McKee, C. O. & Sakai, H. 1983. Ophiolite-contaminated andesites, trachybasalts and cognate inclusions of Mount Lamington Papua New Guinea: anhydrite-amphibole bearing lavas and the 1951 cumulodome. *Journal of Volcanology and Geothermal Research*, **4**, 215–247.

Baker, G. 1946. Preliminary note on volcanic eruptions in the Goropu Mountains southeastern Papua during the period December 1943 to August 1944. *Journal of Geology*, **54**, 19–31.

Baldwin, S. L., Fitzgerald, P. G. & Webb, L. E. 2012. Tectonics of New Guinea. *Annual Reviews in Earth and Planetary Sciences*, **40**, 495–520.

Barsdell, M., Smith, I. E. M. & Sporli, K. B. 1982. The origin of reversed geochemical zoning in the northern New Hebrides volcanic arc. *Contributions to Mineralogy and Petrology*, **81**, 148–155.

Benes, V., Scott, S. D. & Binns, R. A. 1994. Tectonics of rift propagation into a continental margin: Western Woodlark Basin, Papua New Guinea. *Journal of Geophysical Research*, **99**, 4439–4455.

BINNS, R. A., SCOTT, S. D. & PACLARK-PARTICIPANTS 1987. Western Woodlark Basin: potential analogue setting for volcanogenic massive sulphide deposits. *Proceedings of Pacific Rim Congress* **87 II**, 531–535, Australian Institute of Mining and Metallurgy, Parkville, Victoria.

BLOOMER, S. H. & HAWKINS, J. W. 1987. Petrology and geochemistry of boninite series volcanic rocks from the Mariana Trench. *Contributions to Mineralogy and Petrology*, **97**, 361–377.

BRENAN, J. M., SHAW, H. F., RYERSON, F. J. & PHINNEY, D. L. 1995. Mineral-aqueous fluid partitioning of trace elements at 900 °C and 2.0 Gpa: Constraints on the trace element chemistry of mantle and deep crustal fluids. *Geochimica et Cosmochimica Acta*, **59**, 3331–3350.

CALMUS, T., AGUILLON-ROBLESB, A. *ET AL.* 2003. Spatial and temporal evolution of basalts and magnesian andesites ('bajaites') from Baja California, Mexico: the role of slab melts. *Lithos*, **66**, 77–105.

CAMPBELL, I. H. & BORLEY, G. D. 1974. The geochemistry of the pyroxenes from the Lower Layered Series of the Jimberlana Intrusion, Western Australia. *Contributions to Mineralogy and Petrology*, **47**, 281–297.

CHAPPELL, B. W. & WHITE, A. J. R. 1974. Two contrasting granite types. *Pacific Geology*, **8**, 173–174.

CRAWFORD, A. J., FALLOON, T. J. & EGGINS, S. 1987. The origin of island arc high-alumina basalts. *Contributions to Mineralogy and Petrology*, **97**, 417–430.

CURTIS, J. W. 1973. Plate tectonics and the Papua–New Guinea–Solomon Islands region. *Journal of the Geological Society of Australia*, **20**, 21–36.

DAVIDSON, J., TURNER, S., HANDLEY, H., MACPHERSON, C. & DOSSETO, A. 2007. An amphibole 'sponge' in arc crust. *Geology*, **35**, 787–790.

DAVIDSON, J., TURNER, S. & PLANK, T. 2013. Dy/Dy*: variations arising from mantle sources and petrogenetic processes. *Journal of Petrology*, **54**, 525–537.

DAVIES, H. L. 1973. *Fergusson Island, Papua New Guinea – 1:250 000 Geological Series*. Bureau of Mineral Resources Australia Explanatory notes, **SC/56-5**.

DAVIES, H. L. 1980. Folded thrust fault and associated metamorphics in the Suckling–Dayman Massif, Papua New Guinea. *American Journal of Science*, **280-A**, 171–191.

DAVIES, H. L. & IVES, D. J. 1965. *The geology of Fergusson and Goodenough Islands, Papua New Guinea*. Bureau of Mineral Resources Australia Report, **82**.

DAVIES, H. L. & JACQUES, A. L. 1984. Emplacement of ophiolite in Papua New Guinea. *In*: GASS, I. G., LIPPARD, S. J. & SHELTON, A. W. (eds) *Ophiolites and Oceanic Lithosphere*. Geological Society, London, Special Publications, **13**, 341–350.

DAVIES, H. L. & SMITH, I. E. M. 1971. Geology of eastern Papua. Bulletin of the Geological Society of America, **82**, 3299–3312.

DAVIES, H. L. & WARREN, R. G. 1988. Origin of eclogite-bearing domed layered metamorphic complexes ('core complexes') in the D'Entrecasteaux Islands, Papua New Guinea. *Tectonics*, **7**, 1–21.

DAVIES, H. L., SYMONDS, P. A. & RIPPER, I. A. 1984. Structure and evolution of the southern Solomon Sea region. *Bureau of Mineral Resources Journal of Geology and Geophysics*, **9**, 49–68.

DOSSO, L., BOUGAULT, H., BEUZERT, P., CALVEZ, J-Y. & JORON, J-L. 1988. The geochemical structure of the southeast Indian Ridge. *Earth and Planetary Science Letters*, **88**, 47–59.

EGGINS, S. M. 1993. Origin and differentiation of picritic arc magmas, Ambae (Aoba), Vanuatu. *Contributions to Mineralogy and Petrology*, **114**, 79–100.

ELLIOTT, T., PLANK, T., ZINDLER, A., WHITE, W. & BOURDON, B. 1997. *Journal of Geophysical Research*, **102**, 14991–15019.

ENGLAND, R. N. & DAVIES, H. L. 1973. Mineralogy of ultramafic cumulates and tectonites from eastern Papua. *Earth and Planetary Science Letters*, **17**, 416–425.

EWART, A. 1979. A review of the mineralogy and chemistry of Tertiary–Recent dacitic, latitic, rhyolitic and related salic volcanic rocks. *In*: BAKER, F. (ed.) *Trondjemites, Dacites and Related Rocks*. Elsevier, Amsterdam, 13–121.

FALLOON, T. J. & GREEN, D. H. 1986. Anhydrous partial melting of MORB pyrolite and other peridotite compositions at 10 Kb: implications for the origin of primitive MORB glasses. *Mineralogy and Petrology*, **37**, 181–219.

FODEN, J. D. & GREEN, T. H. 1992. Possible role of amphibole in the origin of andesite: some experimental and natural evidence. *Contributions to Mineralogy and Petrology*, **109**, 479–493.

GILL, J. B. 1981. *Orogenic Andesite and Plate Tectonics*. Springer Verlag, New York.

GROVE, T. L. & KINZLER, R. J. 1986. Petrogenesis of andesites. *Annual Reviews of Earth and Planetary Science*, **14**, 417–454.

GUST, D. A. & PERFIT, M. R. 1987. Phase relations of a high-Mg basalt from the Aleutian island arc: implications for primary island arc basalts and high-Al basalts. *Contributions to Mineralogy and Petrology*, **97**, 7–18.

HAMILTON, W. 1979. *Tectonics of the Indonesian region*. US Geological Survey Professional paper, **1078**.

HART, S. R. 1988. Heterogeneous mantle domains signatures, genesis and mixing chronologies. *Earth and Planetary Science Letters*, **90**, 273–296.

HAWKESWORTH, C. J., NORRY, M. J., RODDICK, J. C., BAKER, P. E., FRANCIS, P. W. & THORPE, R. S. 1979. $^{143}Nd/^{144}Nd$, $^{87}Sr/^{86}Sr$, and incompatible element variations in calc-alkaline andesites and plateau basalts from South America. *Earth and Planetary Science Letters*, **42**, 45–57.

HAWKESWORTH, C. J., GALLAGHER, K., HERGT, J. M. & MCDERMOTT, F. 1993. Mantle and slab contributions in arc magmas. *Annual Reviews of Earth and Planetary Science*, **21**, 175–204.

HEGNER, E. & SMITH, I. E. M. 1992. Isotopic compositions of late Cenozoic volcanics from southeast Papua New Guinea: evidence for multi-component sources in arc and rift environments. *Chemical Geology*, **97**, 233–249.

HEYWORTH, Z., TURNER, S. *ET AL.* 2007. ^{238}U-^{230}Th-^{226}Ra-^{210}Pb constraints on the genesis of high-Mg andesites at White Island, New Zealand. *Chemical Geology*, **243**, 105–121.

HILL, E. J., BALDWIN, S. L. & LISTER, G. S. 1992. Unroofing of active metamorphic core complexes in the D'Entrecasteaux Islands, Papua New Guinea. *Geology*, **20**, 907–910.

HILL, E. J., BALDWIN, S. L. & LISTER, G. S. 1995. Magmatism as an essential driving force for formation of active metamorphic complexes in eastern Papua New Guinea. *Journal of Geophysical Research*, **100**, 10441–10451.

JAKES, P. & SMITH, I. E. M. 1970. High potassium calc–alkaline rocks from Cape Nelson, Eastern Papua. *Contributions to Mineralogy and Petrology*, **28**, 259–271.

JENNER, G. A. 1981. Geochemistry of high-Mg andesites from Cape Vogel, Papua New Guinea. *Chemical Geology*, **33**, 301–332.

JOHNSON, R. W. 1979. Geotectonics and volcanism in Papua New Guinea: a review of the late Cenozoic. *Bureau of Mineral Resources Journal of Geology and Geophysics*, **4**, 181–207.

JOHNSON, T. & MOLNAR, P. 1972. Focal mechanisms and plate tectonics of the southwest Pacific. *Journal of Geophysical Research*, **77**, 5000–5032.

JOHNSON, R. W., MACNAB, R. P., ARCULUS, R. J., RYBURN, R. J., COOKE, R. J. S. & CHAPPELL, B. W. 1983. Bamus Volcano, Papua New guinea: dormant neighbour of Uluwan, and magnesian andesite locality. *Geologische Rundschau*, **72**, 207–237.

JOHNSON, R. W., MACKENZIE, D. E. & SMITH, I. E. M. 1978. Delayed partial melting of subduction modified mantle in Papua New Guinea. *Tectonophysics*, **46**, 197–216.

KAY, R. W. 1980. Aleutian magnesian andesites: melts from subducted Pacific Ocean crust. *Journal of Volcanology and Geothermal Research*, **4**, 117–132.

KESSEL, R., SCHMIDT, M. W., UKMER, P. & PETTKE, T. 2005. Trace element signature of subduction-zone fluids, melts and supercritical liquids at 120–180 km depth. *Nature*, **437**, 724–727.

KURODA, M., SHIRAKI, K. & URANO, H. 1978. Boninite as a possible calc-alkaline magma. *Bulletin of Volcanology*, **41**, 563–575.

LUYENDYK, B. P., MACDONALD, K. C. & BRYAN, W. B. 1973. Rifting history of the Woodlark Basin in the southwest Pacific. *Geological Society of America Bulletin*, **84**, 1125–1134.

MAHONEY, J. J., NATLAND, J. H., WHITE, W. M., POREDA, R., BLOOMER, S. H., FISHER, R. L. & BAXTER, A. N. 1989. Isotopic and geochemical provinces of the western Indian Ocean spreading centres. *Journal of Geophysical Research*, **94**, 4033–4052.

MALPAS, J., SPORLI, K. B., BLACK, P. M. & SMITH, I. E. M. 1992. Northland Ophiolite, New Zealand, and implications for plate-tectonic evolution of the southwest Pacific. *Geology*, **20**, 149–152.

MCCULLOCH, M. T. & GAMBLE, J. A. 1991. Geochemical and geodynamical constraints on subduction zone magmatism. *Earth and Planetary Science Letters*, **102**, 358–374.

MEEN, J. K. & EGGLER, D. H. 1987. Petrology and geochemistry of the Cretaceous Independence volcanic suite, Absaroka Mountains, Montana: clues to the composition of the Archean sub-Montana mantle. *Geological Society of America Bulletin*, **98**, 238–247.

MICHARD, A., MONTIGNY, R. & SCHLICH, R. 1986. Geochemistry of the mantle beneath the Rodriguez Trople Junction and the southeast Indian Ridge. *Earth and Planetary Science Letters*, **78**, 104–114.

MONZIER, M., DANYUSHEVSHY, L. V., CRAWFORD, A. J., BELTON, H. & COTTON, J. 1993. High-Mg andesites from the southern termination of the New Hebrides Island arc (S.W. Pacific). *Journal of Volcanology and Geothermal Research*, **57**, 193–217.

MYERS, J. D., MARSH, B. D. & SINHA, A. K. 1986. Geochemical and strontium isotopic characteristics of parental Aleutian Arc magmas: evidence from the basaltic lavas of Atka. *Contributions to Mineralogy and Petrology*, **94**, 1–11.

MYSEN, B. O. 1982. The role of mantle anatexis. *In*: THORPE, R. S. (ed.) *Andesites, Orogenic Andesites and Related Rocks*. John Wiley & Sons, Chichester, 489–522.

NYE, C. J. & REID, M. R. 1986. Geochemistry of primary and least fractionated lavas from Okmok Volcano, central Aueutians: implications for arc magma genesis. *Journal of Geophysical Research*, **91**, 10271–10287.

PEARCE, J. A. 1982. Trace element characteristics of lavas from destructive plate boundaries. *In*: THORPE, R. S. (ed.) *Andesites, Orogenic Andesites and Related Rocks*. John Wiley & Sons, Chichester, 252–248.

PERFIT, M. R., GUST, D. A., BENCE, A. E., ARCULUS, R. J. & TAYLOR, S. R. 1980. Chemical characteristics of island arc basalts; implications for mantle sources. *Chemical Geology*, **30**, 227–256.

PRICE, R. C., GAMBLE, J. A., SMITH, I. E. M., STEWART, R. B., EGGINS, S. & WRIGHT, I. C. 2005. An integrated model for the temporal evolution of andesites and rhyolites and crustal development in New Zealand's North Island. *Journal of Volcanology and Geothermal Research*, **140**, 1–24.

PRICE, R. C., GAMBLE, J. A., SMITH, I. E. M., MASS, R., WAIGHT, T., STEWART, R. B. & WOODHEAD, J. 2012. The anatomy of an andesite volcano: a time-stratigraphic study of andesite petrogenesis and crustal evolution at Ruapehu Volcano, New Zealand. *Journal of Petrology*, **53**, 2139–2189.

RIPPER, I. D. 1982. Seismicity of the Indo-Australian/Solomon Sea plates boundary in the southeast Papua region. *Tectonophysics*, **87**, 355–369.

ROEDER, P. L. & EMSLIE, R. F. 1970. Olivine-liquid equilibrium. *Contributions to Mineralogy and Petrology*, **29**, 275–298.

SAUNDERS, A. D., ROGERS, G., MARRINER, G. F., TERRELL, D. J. & VERMA, S. P. 1987. Geochemistry of Cenozoic to Recent volcanic rocks, Baja California, Mexico: implications for the petrogenesis of post-subduction magmas. *Journal of Volcanology and Geothermal Research*, **32**, 223–246.

SIMKIN, T. & SMITH, J. V. 1970. Minor-element distribution in olivine. *Journal of Geology*, **78**, 304–325.

SMITH, I. E. M. 1972. High–potassium intrusives from southeastern Papua. *Contributions to Mineralogy and Petrology*, **34**, 167–176.

SMITH, I. E. M. 1973. Geology of the Calvados Islands. *Bureau of Mineral Resources Australia Bulletin*, **139**, 59–66.

SMITH, I. E. M. 1976. Peralkaline rhyolites from the D'Entrecasteaux Islands, Papua New Guinea. *In*:

Johnson, R. W. (ed.) *Volcanism in Australasia*. Elsevier, New York, 275–286.

Smith, I. E. M. 1982*a*. Volcanic evolution in eastern Papua. *Tectonophysics*, **87**, 333–365.

Smith, I. E. M. 1982*b*. Young volcanoes in eastern Papua. *Geological Survey of Papua New Guinea Memoir*, **10**, 257–265.

Smith, I. E. M. 2013. The chemical characterization and tectonic significance of ophiolite terrains in southeastern Papua New Guinea. *Tectonics*, **32**, 159–170, http://dx.doi.org/10.1002/tect.20023

Smith, I. E. M. & Compston, W. 1982. Strontium isotopes in Cenozoic volcanic rocks from southeastern Papua New Guinea. *Lithos*, **15**, 199–206.

Smith, I. E. M. & Davies, H. L. 1976. The geology of the southeast Papuan mainland. *Bureau of Mineral Resources Australia Bulletin*, **165**, 1–85.

Smith, I. E. M. & Johnson, R. W. 1981. Contrasting rhyolite suites in the late Cenozoic of Papua New Guinea. *Journal of Geophysical Research*, **86**, 10257–10272.

Smith, I. E. M. & Milsom, J. S. 1984. Late Cenozoic volcanism and extension in eastern Papua. *In*: Kokelaar, B. P. & Howells, M. F. (eds) *Processes in Modern and Ancient Marginal Basins*. Geological Society, London, Special Publications, **16**, 163–171.

Smith, I. E. M. & Mitchell, P. 1989. High–magnesium lavas in the late Cenozoic volcanic arc associations of southeastern Papua New Guinea. *Nues Jahrbuch fur Mineralogie*, **H.11**, 481–497.

Smith, I. E. M., Chappell, B. W., Ward, G. K. & Freeman, R. S. 1977 Peralkaline rhyolites associated with andesitic arcs of the southwest Pacific. *Earth and Planetary Science Letters*, **7**, 230–236.

Smith, I. E. M., Worthington, T. J., Price, R. C. & Gamble, J. A. 1997. Primitive magmas in arc-type volcanic associations: examples from the southwest Pacific. *The Canadian Mineralogist*, **35**, 257–273.

Straub, S. M., Gomez-Tuena, A., Stuart, F. M., Zellmer, G. F., Espinasa-Perena, R., Cai, Y. & Lizuka, Y. 2011. Formation of hybrid arc andesites beneath thick continental crust. *Earth and Planetary Science Letters*, **303**, 337–347.

Sun, S. S. & McDonough, W. F. 1989. Chemical and isotopic systematic of oceanic basalts: implications for mantle composition and processes. *In*: Saunders, A. D. & Norry, M. J. (eds) *Magmatism in the Ocean Basins*. Geological Society, London, Special Publications, **42**, 313–345.

Sun, S. S. & Nesbitt, R. W. 1978. Geochemical regularities and genetic significance of ophiolitic basalts. *Geology*, **6**, 689–693.

Tatsumi, Y. 1983. High-magnesian andesites in the Setouchi belt, southwest Japan and their possible relation to the evolutionary history of the inter-arc basin. *In*: Hilde, T. W. C. & Uyeda, S. (eds) *Geodynamics of the Western Pacific – Indonesian Region*. American Geophysical Union, Washington, D.C., 331–341.

Tatsumi, Y. & Eggins, S. 1995. *Subduction Zone Magmatism*. Blackwell Science Inc., Cambridge, MA.

Tatsumi, Y. & Ishizaka, K. 1981. Existence of andesitic primary magma: an example from southwest japan. *Earth and Planetary Science Letters*, **53**, 124–130.

Tatsumi, Y. & Ishizaka, K. 1982. Magnesian andesite and basalt from Shodo-Shima Island, southwest Japan, and their bearing on the genesis of calc-alkaline andesites. *Lithos*, **15**, 161–172.

Taylor, G. A. M. 1958. *The 1951 eruption of Mount Lamington, Papua*. Bureau of Mineral Resources Australia, Canberra, Bulletin **38**.

Weissel, J. K., Taylor, B. & Karner, G. D. 1982. The opening of the Woodlark Basin, subduction of the Woodlark spreading system and evolution of northern Melanesia since mid-Pliocene time. *Tectonophysics*, **87**, 253–277.

Wood, B. J. & Turner, S. P. 2009. Origin of primitive high-Mg andesite: Constraints from natural examples and experiments. *Earth and Planetary Science Letters*, **263**, 59–66.

Yogodzinski, G. M., Kay, R. M., Volynets, O. N., Koloskov, A. V. & Kay, S. M. 1995. Magnesian andesite in the western Aleutian Komandorsky region: implications for slab melting and processes in the mantle wedge. *Geological Society of America Bulletin*, **107**, 505–519.

Zirakparvar, N. A., Baldwin, S. L. & Vervoort, J. D. 2012. The origin and geochemical evolution of the Woodlark Rift of Papua New Guinea. *Gondwana Research*, http://dx.doi.org/10.1016/j.gr.2012.06.013

Influence of source materials and fractionating assemblage on magmatism along the Aegean Arc, and implications for crustal growth

MARLINA A. ELBURG[1,2]*, INGRID SMET[1] & ELIEN DE PELSMAEKER[1]

[1]*Department of Geology and Soil Science, Ghent University, Krijgslaan 281 S8, 9000 Ghent, Belgium*

[2]*Geology Division, SAEES, University of KwaZulu-Natal, Private Bag x54001, 4000 Durban, South Africa*

**Corresponding author (e-mail: elburg@ukzn.ac.za)*

Abstract: Volcanic products from Methana, Santorini and Nisyros show mineralogical and geochemical characteristics that can be linked to their differentiation at different crustal levels, and varying sources of subducted sediments. The products from Methana, in the western part of the arc, where the overriding plate is thicker, are affected by amphibole fractionation and crustal contamination. Santorini volcano, in the central part of the arc, is located on an extensively thinned section of the overriding plate; it shows high Y-contents, a-typical for average continental crust. Crustal contamination is minor, similar to Nisyros volcano. Products of the latter, easternmost volcanic centre have been variably affected by amphibole fractionation. Its sub-arc magma source is different from the central and western part of the arc, reflecting sedimentary input from the Nile.

Amphibole fractionation, caused by crystallization at greater depths, seems necessary to lower Y-contents and yield volcanic products that resemble average continental crust in this respect. In detail, however, none of the magmatic products of the Aegean Arc displays MREE–HREE patterns or levels of compatible elements that are an exact match for average continental crust. Garnet stability and mixing processes appear crucial to obtain the REE and compatible element characteristics of continental crust.

Few volcanoes have been studied to the same extent as Santorini, located in the central part of the Aegean Arc. It is one of the sixteen ‘Decade Volcanoes’, named after the International Decade of Natural Disaster Reduction (1990–2000), which have been targeted for research because of their potential to cause significant environmental destruction. This is thought to have been reflected in the demise of the Minoan civilization in the wake of Santorini’s last big eruption around 1625 BC. The research performed in the last century has been summarized in an excellent monograph (Druitt *et al.* 1999). The fact that so much work has been spent on the volcano may give the, as we will show mistaken, impression that Santorini could be regarded as a very ‘typical’ volcano, and a good model for continental arc magmatism. In this contribution we compare the mineralogy and geochemistry of Santorini with that of the two other nominally active volcanoes in the Aegean Arc: Methana in the west, and Nisyros in the east. These data indicate that Santorini is actually aberrant in its geochemical behaviour, which might be a reflection of its location on thinned continental crust with an active fault system, inhibiting the crystallization of amphibole. None of our data from the Aegean Arc are exact geochemical matches for modern estimates of the average continental crust. The comparison suggests that garnet fractionation and mixing processes are necessary to produce the chemical signature of the average continental crust.

Geological setting

Magmatism in the Aegean Arc (Fig. 1a) is a result of the northward subduction of the African plate under the Aegean microplate. The process has been ongoing in the Aegean region since at least the Jurassic (Van Hinsbergen *et al.* 2005), with repeated back-stepping of the trench, and accretion of continental slivers onto the overriding plate. The present-day South Aegean Volcanic Arc is, therefore, built upon a basement of African derivation. This basement underwent significant stretching and thinning from the Eocene onwards (Jolivet 2001), which was most pronounced in the north-central part of the present Aegean Sea. Therefore, the crustal thickness is appreciably less underneath Santorini (<25 km), in the middle of the arc, than underneath the western side of the arc (approximately 30 km), where Methana peninsula is located (Papazachos *et al.* 1995). The eastern side, to which Nisyros belongs, has an intermediate crustal thickness. The

From: Gómez-Tuena, A., Straub, S. M. & Zellmer, G. F. (eds) 2014. *Orogenic Andesites and Crustal Growth.* Geological Society, London, Special Publications, **385**, 137–160.
First published online May 30, 2013, http://dx.doi.org/10.1144/SP385.1

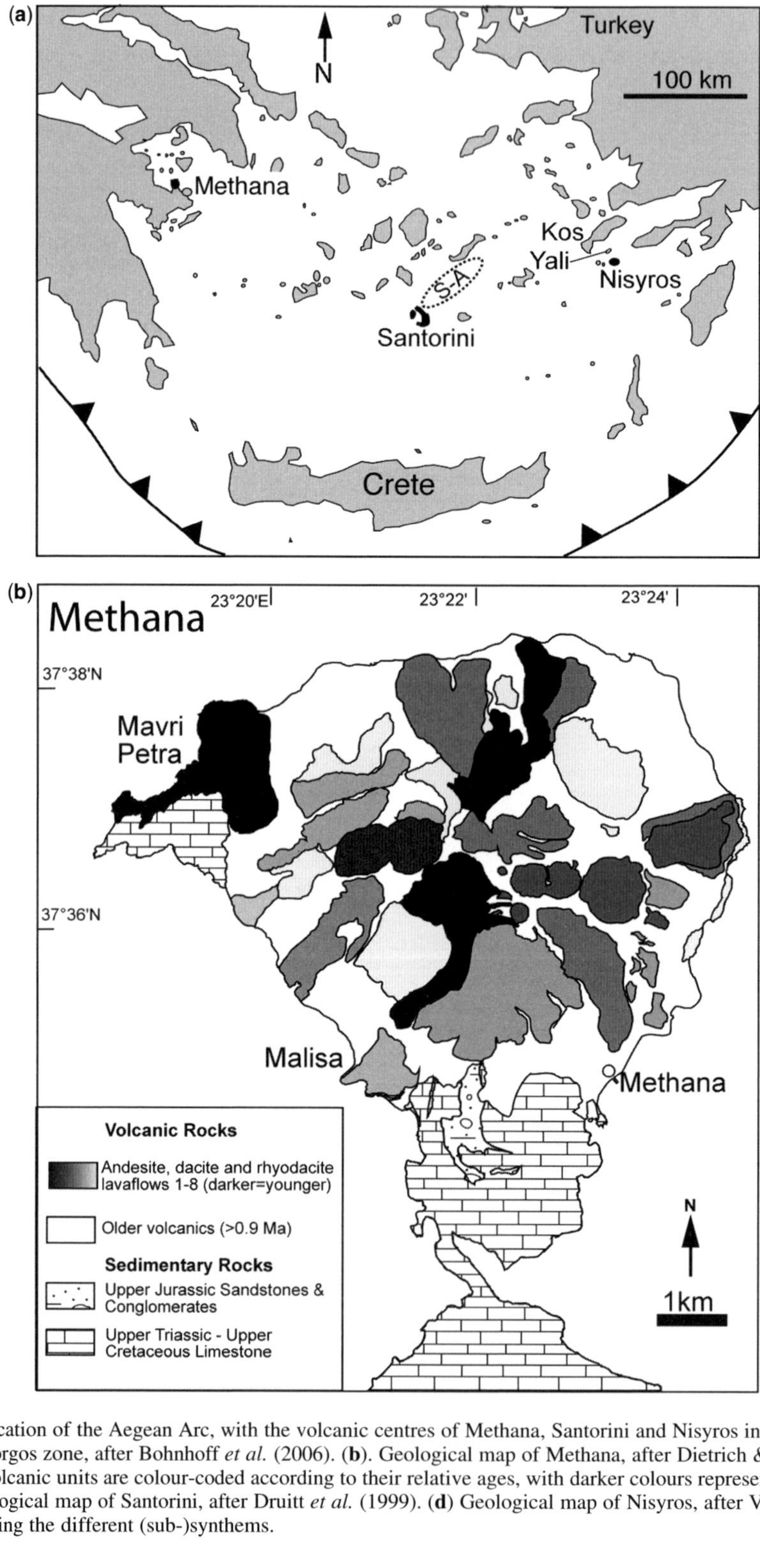

Fig. 1. (**a**) Location of the Aegean Arc, with the volcanic centres of Methana, Santorini and Nisyros indicated. SA, Santorini-Amorgos zone, after Bohnhoff *et al.* (2006). (**b**). Geological map of Methana, after Dietrich & Gaitanakis (1995). The volcanic units are colour-coded according to their relative ages, with darker colours representing younger units; (**c**) Geological map of Santorini, after Druitt *et al.* (1999). (**d**) Geological map of Nisyros, after Volentik *et al.* (2005*a*), showing the different (sub-)synthems.

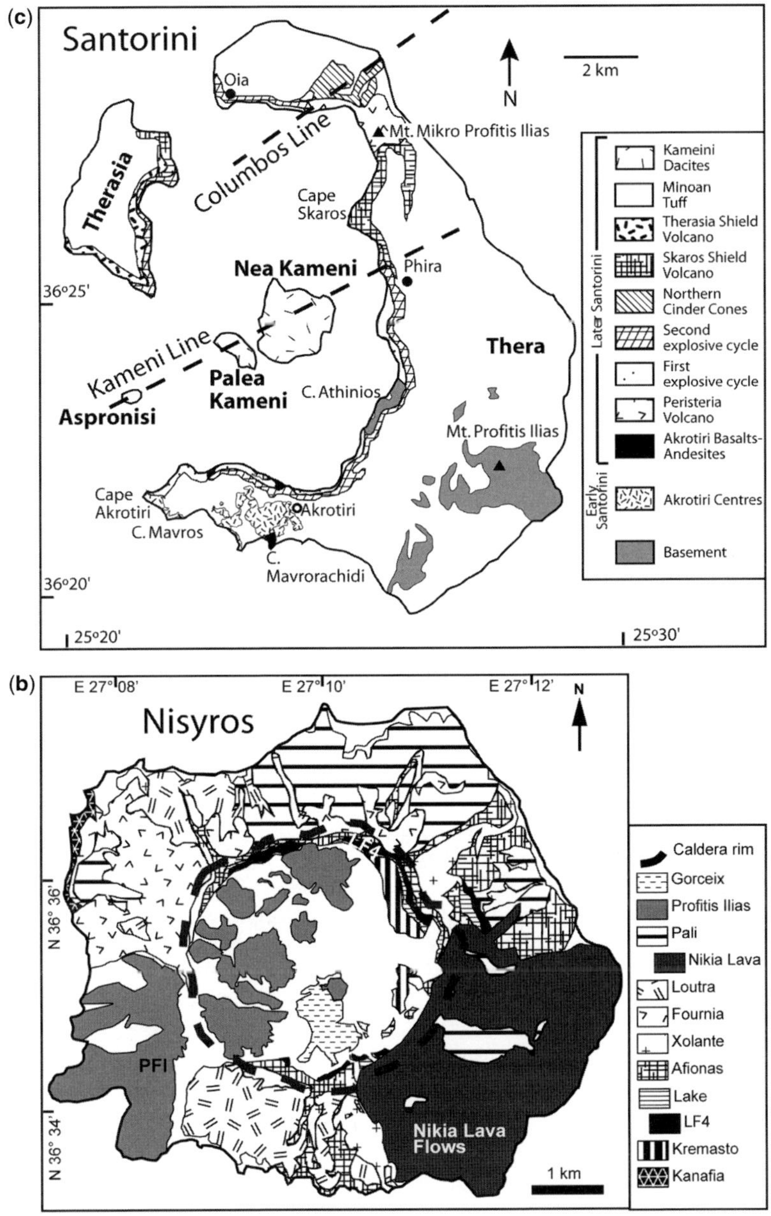

Fig. 1. *Continued*

tectonic regime responsible for extension of the Aegean crust may also influence the exact timing and location of volcanism in the present-day active centres (Pe-Piper & Piper 2002).

Methana

Of the three volcanic centres studied, Methana records the longest extrusive activity: the first volcanic deposits are likely older than 900 ka (Dietrich & Gaitanakis 1995), whereas the most recent volcanic eruption took place around 2200 yr BP. Methana's basement consists of Mesozoic sedimentary rocks, mainly limestones with subordinate siliciclastic rocks, which crop out in the southernmost and northwestern part of the peninsula (Fig. 1b). The remaining central part is covered with numerous small-scale volcanic domes and flows of andesitic to rhyodacitic composition and varying age, characterized by the presence of numerous more mafic, igneous enclaves. Small volumes of basaltic-andesites are only found in two locations, namely the southern Malisa volcano (probably >0.5 Ma) and the pyroclastic deposits and lavas linked to the historic Mavri Petra complex in the north.

Santorini

The exposed basement of Santorini volcano consists of limestones and metapelites, the latter of Paleocene–Eocene age (Tataris 1956). Granitoids of approximately 10 Ma have been found in boreholes (Skarpelis & Kyriakopoulos 1992). Volcanism in Santorini started around 600 ka (Druitt *et al.* 1999) on Akrotiri peninsula (Fig. 1c), with the deposition of intermediate to felsic hyaloclastite deposits and tuffs, plus small-scale domes of rhyodacites. This was followed by the establishment of the mafic to intermediate Peristeria volcano in the north, and small basaltic centres on Akrotiri peninsula. Volcanic activity continued in two large-scale eruptive cycles containing both lava flows and pyroclastic deposits, geochemically ranging from basalts to rhyodacites (Druitt *et al.* 1999). These two explosive cycles ended with the felsic, so-called Minoan eruption around 1625 BC (Friedrich *et al.* 2006). More recent activity has seen the gradual infilling of the caldera with the emergence of the dacitic Kameini Islands, on which the most recent volcanic activity dates from 1950 (Pyle & Elliott 2006). The location of the individual eruptive centres is strongly influenced by the presence of the NW–SE Santorini–Amorgos fault (Druitt *et al.* 1999; Bohnhoff *et al.* 2006; Dimitriadis *et al.* 2009).

In this paper we will be referring to the approximately 600 000-year-old intermediate to felsic deposits from the Akrotiri peninsula, that predate both Peristeria volcano and the related basaltic to basaltic-andesitic Akrotiri cinder cones, as 'Early Santorini'. The 'younger' series includes everything else (also the Akrotiri cinder cones) and will be denoted as 'Later Santorini'.

Nisyros

The island of Nisyros is the most recent of the three volcanic centres discussed here, as the edifice was only established after the massive eruption that deposited the rhyolitic Kos Plateau Tuff at 161 ka (Smith *et al.* 1996). The latest activity, producing felsic plugs within the caldera, may have been synchronous with the eruption of the Yali centre (Volentik *et al.* 2005*b*) around 24 ka (Wagner *et al.* 1976), but could be as young as 10 ka (Zellmer & Turner 2007). The island consists of volcanic deposits only, ranging from the earliest hyaloclastite deposits and pillow lavas through a section of lava flows and associated pyroclastic deposits (Fig. 1d). These range from basaltic andesites to rhyolites, with more felsic pyroclastic deposits and domes towards the top of the sequence (Volentik *et al.* 2005*b*).

Petrography

Methana

Despite their range in silica contents, most of Methana's volcanic deposits show a remarkable similarity in the identity of their mineral phases. Plagioclase and amphibole are dominant in the porphyritic host rocks and more crystalline, equigranular enclaves, both as large phenocrysts and as groundmass crystals. Clinopyroxene is the third most common mineral; orthopyroxene and biotite are not uncommon and Fe–Ti-oxides are a ubiquitous minor component; apatite is an accessory mineral. Euhedral zircon grains are sometimes found as inclusions within plagioclase, biotite or amphibole phenocrysts of (basaltic-)andesitic to dacitic volcanics. Olivine (sometimes with Cr-rich spinel inclusions) can usually be found in the more mafic lithologies, whereas quartz is more commonly present in the felsic volcanics. Some intermediate samples contain both quartz and olivine, with disequilibrium indicated by the quartz being mantled by fine-grained pyroxene, or olivine overgrown by amphibole. Amphibole can be replaced and/or overgrown by clinopyroxene, which sometimes also overgrows orthopyroxene. Both amphibole and biotite can show thick rims of opacite (Fig. 2a). Plagioclase crystals often show (dis)continuous or oscillatory zoning and sieve textures.

Santorini

Santorini's lavas and pyroclastic deposits are dominated by anhydrous mineral assemblages. Olivine, clinopyroxene and plagioclase characterize the basalts to basaltic andesites, followed by orthopyroxene in the intermediate to felsic rocks (Fig. 2b). High-forsterite olivine crystals in basalts can contain inclusions of Cr-rich spinel. Ti-bearing magnetite is ubiquitous as a groundmass phase, but only appears as larger crystals in andesitic compositions. Apatite is typically found as small acicular or prismatic crystals enclosed within plagioclase, but larger crystals are not observed. In all rock types, plagioclase phenocrysts show a variety of (sometimes very complex) inclusion and zoning patterns, with common sieve-textured zones, as is typical for arc volcanics. Overgrowths of clinopyroxene

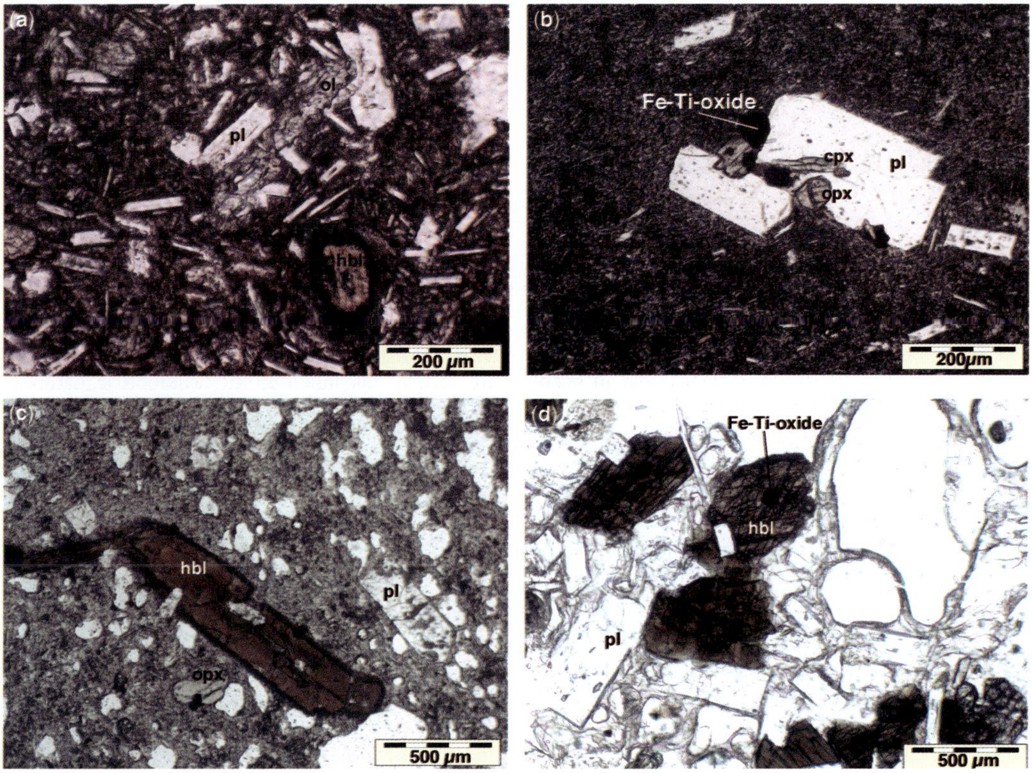

Fig. 2. Photomicrographs of the volcanic products from the Aegean Arc. (**a**) Sample IM38 from Methana, containing plagioclase (pl), olivine (ol) plus amphibole (hbl), with the latter showing an opacite rim, and a fine-grained overgrowth of pyroxene; (**b**) Dacite sample SE205 from Palea Kameini (Later Santorini) with plagioclase, orthopyroxene (opx), clinopyroxene (cpx) and Fe–Ti-oxide. Hydroxyl-bearing minerals are absent in all of the Later Santorini samples, except for rare enclaves; (**c**) Sample SE234a from the Akrotiri Lavas (Early Santorini) with plagioclase, amphibole and orthopyroxene (and irregular versicles). Amphibole has an opacite rim, indicating destabilization due to degassing, but indications for magma mixing are not obvious; (**d**) Vesicular andesitic enclave IN204G from Upper Pumice (Pali sub-synthem) of Nisyros. Amphibole-bearing samples on this island appear to display earlier crystallization of Fe–Ti-oxide than amphibole-free samples.

on orthopyroxene, or vice versa, are occasionally observed (Nicholls 1971). Hydroxyl-bearing minerals such as amphibole are exceedingly scarce on Santorini, apart from the Rhyodacites and Tuffs of Akrotiri (all unit names following Druitt *et al.* 1999) of Early Santorini, in which they constitute the most abundant mafic mineral (Fig. 2c). In younger pyroclastic deposits (Later Santorini series), amphibole has only been observed within mafic enclaves (Cottrell *et al.* 1999); they are not found in any of the lava flows. Quartz has been observed in some of the older deposits on the Akrotiri peninsula, but is interpreted to be xenocrystal, and associated with the disaggregation of granitoid xenoliths. Very minor amounts of biotite, and exceedingly rare accessory zircon are found in some of the most felsic samples of the Rhyodacites and Tuffs of Akrotiri.

Nisyros

The basalts to basaltic-andesites of Nisyros contain plagioclase (often with sieve textures and intricate zoning patterns), olivine and clinopyroxene phenocrysts in a fine-grained matrix. The most magnesium-rich samples typically display high modal olivine and clinopyroxene contents. Both amphibole-bearing and amphibole-free intermediate to felsic volcanic deposits are found on the island, but a clear-cut temporal division between them, as seen on Santorini, is not observed. If there is a temporal trend, it is opposite from that on Santorini, with amphibole becoming more common in the younger deposits. The amphibole-free deposits contain plagioclase, clinopyroxene, with orthopyroxene appearing in the more felsic rocks, while minor amounts of olivine are still found in intermediate samples. The crystallization of amphibole appears to be associated with the early fractionation of Fe–Ti-oxides (Fig. 2d) and also apatite; plagioclase, orthopyroxene and clinopyroxene are typically additional phenocryst phases. Zircon occurs sporadically within the most felsic samples; quartz or biotite have not been observed.

Analytical techniques

Whole rock (WR) major and trace element compositions were determined by Inductively Coupled Plasma–Optical Emission Spectroscopy (ICP–OES) at Ghent University, using the equipment housed at the Department of Analytical Chemistry. Approximately 4 g of each powder was dried at 105 °C and Loss on Ignition (LOI) determined by igniting the samples at 1000 °C overnight. Sample and flux were mixed in a ratio of 1:4 using 65/35 lithiummeta-/-tetraborate flux (Breitländer), and were fused in high purity graphite crucibles at 1050 °C for 20 min. The resulting glass was dissolved in 2% HNO_3 for analysis with a Spectro Arcos ICP–OES. Calibration lines were produced by analysis of a full-procedure blank and geological reference materials, dissolved following the same procedure (BHVO-2, AGV-2, QLO-1 and GSP-2 from the US Geological Survey, JSy-1 and JB-2 from the Japanese Geological Survey, and NIM-L from Mintek, South Africa). Accuracy of the analysis was monitored by the analysis of USGS standard BCR-2. The accuracy for major elements, Ba, Sr, Zr and V is better than 2% relative; Sc, Co, Y and Zn are better than 5%; Ce, Ni and Cr around 10% for concentrations over 30 ppm.

Trace element analyses were performed by ICP–Mass Spectrometry (MS), using a Perkin Elmer Elan 5000 or ThermoFinnigan X-series, after dissolution in HF/HNO_3 in capped beakers at approximately 110 °C, followed by conversion of fluorides to nitrates by dry-down with aqua regia and HNO_3. Calibration standards consisted of three different dissolutions of the BHVO-2 and two of the AGV-2 USGS reference material, plus a full procedure blank. Corrections were applied for drift, using both internal standards (B, In and Tl) and interpolation between 'dummy' standards. Interferences of Light Rare Earth Elements (LREE)-oxides on Heavy Rare Earth Elements (HREE) were corrected for, using the measured ThO^+/Th^+ ratio and run-specific oxide production ratios for the elements in question. For all elements used in this study, accuracy and precision was equal to or better than 5% relative, apart from Li and Ta (precision 6 and 8%, respectively). For zircon-bearing samples, Zr values are those from ICP–OES, and Hf values are not used due to incomplete digestion of zircon during low pressure dissolution. Ratios of M(iddle)REE/HREE such as Ho/Lu, crucial for this study, show a standard deviation of better than 3% relative and an accuracy that is better than 1% relative, as judged from our database of BCR-2 analyses ($n = 14$) obtained during the analytical sessions for this study.

Geochemistry

Our own geochemical data set for Methana (Smet in prep.), Santorini and Nisyros was combined with published analyses for the latter two volcanic centres, as taken from the GeoRoc database (http://georoc.mpch-mainz.gwdg.de/georoc/) in 2011, plus the data from Huijsmans (1985), Druitt *et al.* (1999) and Kirchenbaur *et al.* (2012) for Santorini and Vanderkluysen *et al.* (2005) for Nisyros. The data from Buettner *et al.* (2005) were not used, since some of their trace element analyses (e.g. for Rb) are almost an order of magnitude different from other workers' results. The composition of the total Continental Crust (CC) (Rudnick & Gao 2003) is also shown in the diagrams, as is the field for the Kos Plateau Tuff (Bachmann *et al.* 2007; Zouzias & St. Seymour 2008), which serves as an example of an extreme felsic end-member, in this case of the Nisyros system. Average compositions were calculated for Methana, Later Santorini and Nisyros, to facilitate comparison between the islands, and with average continental crust. Representative analyses are given in Table 1.

In terms of major elements (Fig. 3), Methana shows the most restricted range, with SiO_2 concentrations ranging from 50.5 to 68 wt%. This contrasts with the wider range seen on the other islands (Santorini: typically 50–75.5 wt%; Nisyros: 52.5–73.5). This extension to more felsic compositions is unlikely to result from more comprehensive sampling only, but is also reflected by the scarcity of true pyroclastic deposits on Methana. Methana forms one end-member with respect to major element trends, and Later Santorini the other. Early Santorini, however, shows greater similarities to Methana. Some of the intermediate deposits on Nisyros follow the Later Santorini trend, while the more felsic ones largely overlap with Early Santorini. This is exemplified by the diagrams of TiO_2, Na_2O, P_2O_5, FeO^*/MgO (lower for Methana) and to a lesser extent Al_2O_3 (lower for Later Santorini) v. SiO_2. The Later Santorini trend shows a clear inflection in the diagrams for TiO_2 and P_2O_5, whereas Methana shows a monotonous decrease for TiO_2 and fairly constant levels of P_2O_5. Methana shows very limited iron enrichment with increasing SiO_2 content, and, thereby, follows the trend for 'calc-alkaline' rocks from Miyashiro (1974), whereas Later Santorini is closer to a tholeiitic trend. Another very clear distinction is seen in the trend for Na_2O, where this oxide shows a slightly positive correlation with increasing silica contents in Methana, but increases strongly in Later Santorini. Potassium on the other hand does not show remarkable distinctions between the different islands. Some of the samples from Early Santorini have somewhat lower K_2O contents, but this can also be due to interaction with seawater, as hyaloclastite deposits are not uncommon.

The distinctions and similarities seen in the major element diagrams do not show a clear correspondence to the trace element trends (Fig. 4). Nisyros stands out with respect to its high Ba, Sr and Nb (and Ta) contents; as also noted by Mitropoulos *et al.* (1987), Later Santorini shows by far the greatest enrichment in Y (and MREE to HREE), while Rb and Zr are also on the high side. Both the Early and Later Santorini samples have higher maximum Th contents (and U, not shown) than Methana or Nisyros. The behaviour of Sr is particularly notable. Later Santorini has the lowest concentration (typically <300 ppm) at all stages of differentiation. The more mafic Methana and Early Santorini samples contain 300–400 ppm and, like Later Santorini, these concentrations decrease with increasing SiO_2 content. However, there is a small group of Methana samples at approximately 65% SiO_2 that contain about 500 ppm Sr; these also have high Ni (and Cr, not shown) contents. This somewhat erratic behaviour of Sr is also seen on Nisyros, where a group of samples at 54–57% SiO_2 show Sr levels that can be as high as 1150 ppm (but without the enrichment in Ni or Cr as seen on Methana), whereas others contain 450–650 ppm at this differentiation stage. Although the high-Sr samples from Methana all come from a single deposit, this is not the case for those from Nisyros, which can be found in both the older (LF4, following Vanderkluysen *et al.* 2005) and younger (PFI) deposits on the island. The high Sr contents are not accompanied by elevated concentrations of other elements such as Ca or Ba. Although the Methana data show approximately linear trends for major elements, a more curved trend is seen in the highly compatible elements Ni and Cr.

Average compositions for the three islands cluster around the SiO_2-content of the total Continental Crust around 60 wt% SiO_2. The averages also illustrate that they sometimes represent a composition which is not present as an actual sample – this is most pronounced for strongly curved arrays, such as the SiO_2–TiO_2 diagram for later Santorini, where the average lies at lower TiO_2 contents than the intermediate samples forming the array.

Primitive mantle-normalized trace element diagrams (Fig. 5) show the hallmarks of subduction-related magmatism for all three volcanic centres, with negative anomalies for Nb, Ta and Ti, and positive anomalies for Pb and K, and for the mafic samples (Fig. 5a) also for Sr. With increasing differentiation (Fig. 5b, c), the concentrations of the most incompatible elements increase; only the Later Santorini series also shows an increase in the less incompatible elements such as the HREE. This may also be the reason for a more pronounced negative Eu-anomaly in the felsic samples of this series compared to the other volcanic centres. All series show an increasingly 'spoon-shaped' REE pattern with increasing SiO_2 contents, with decreasing Ho/Lu and Dy/Yb (Davidson *et al.* 2007) ratios (Figs 6 & 7).

The three volcanic centres show clear distinctions in their radiogenic isotope signatures (Fig. 8). Methana shows the highest $^{87}Sr/^{86}Sr$ (0.7051–0.7095) and lowest $^{143}Nd/^{144}Nd$ (0.51264–0.51226) ratios, thereby displaying the most 'crustal' characteristics (Fig. 8a). The range in Sr and Nd isotopic characteristics for Santorini and Nisyros is similar to each other, but the majority of the Later Santorini samples is shifted towards higher Sr and/or Nd isotopic ratios compared to the majority of the Nisyros samples, thereby shifting the Santorini array towards the right of the Nisyros array. Early Santorini lies closer to the Nisyros array. The majority of the Methana samples lies on the extension of the Santorini trend, but the high-Sr Methana samples form the extension of the Nisyros trend. Also plotted are our data for sedimentary basement lithologies and xenoliths included in the Methana eruptives, which fall broadly into the same field as the volcanic Methana samples. Granitoid xenoliths included in the Akrotiri basalts (Later Santorini), which were analysed to see whether they could represent crustal contaminants, fall within the range of Later Santorini, suggesting that these may be cumulates, while the metapelitic basement of this island displays significantly higher Sr and lower Nd isotopic characteristics.

Later Santorini displays a trend where the minimum $^{87}Sr/^{86}Sr$ ratio increases with SiO_2 content (Fig. 8b), whereas this is more poorly developed for the other two centres. This is an indication that crustal contamination occurred in Santorini, but was minor, as also observed by Mitropoulos & Tarney (1992). For the other datasets, there appears to be a better correlation between Sr concentration and Sr isotopic compositions (Fig. 8c), with the high-Sr samples of a given centre having the lower Sr isotopic ratios.

Table 1. *Representative whole rock analyses. Oxides in wt%, trace elements in ppm*

Series	Early Santorini	Early Santorini	Early Santorini	Later Santorini	Later Santorini	Later Santorini	Methana	Methana	Methana	Nisyros	Nisyros	Nisyros
Unit #	Akrotiri Rhyodacite enclave	C. Mavros	Akrotiri Rhyodacite	Skaros	Cape Riva	Minoan	Delta-5	Malisa	Delta-2	LF4	LF5	Nikia lava
Sample	SE351	SE341	SE352	SE232	SE229	SE218	IM296	IM23	DPM-42	IN215	IN222	IN242H
SiO_2	53.10	59.24	71.28	51.17	61.13	70.88	62.75	61.20	68.13	55.74	59.86	70.79
TiO_2	1.03	0.85	0.41	0.82	0.81	0.40	0.52	0.60	0.41	0.70	1.13	0.34
Al_2O_3	18.25	17.38	15.00	17.66	16.33	14.87	16.78	16.61	16.11	15.33	16.67	14.97
Fe_2O_3*	9.75	6.89	2.47	9.13	6.88	3.08	5.03	5.79	3.83	6.86	7.26	2.71
MnO	0.18	0.14	0.09	0.16	0.15	0.08	0.11	0.11	0.10	0.12	0.13	0.07
MgO	4.12	3.29	0.93	6.86	2.72	0.61	2.75	3.47	1.69	7.35	2.49	0.79
CaO	8.99	6.91	2.83	11.03	5.66	2.18	5.99	6.61	4.69	9.92	5.63	2.48
Na_2O	3.63	3.55	4.16	2.64	4.41	4.66	3.40	2.98	3.27	2.89	4.39	4.64
K_2O	0.73	1.60	2.73	0.48	1.78	3.17	2.04	1.78	2.52	0.97	2.18	3.13
P_2O_5	0.20	0.15	0.09	0.09	0.16	0.08	0.12	0.12	0.05	0.13	0.26	0.09
LOI	1.01	1.46	3.79	−0.10	1.21	3.63	1.03	1.15	1.57	1.27	-0.14	0.66
Total	99.77	100.06	99.23	99.85	99.05	100.95	99.49	99.27	100.81	98.86	98.94	98.73
Li	6.1	4.8	3.6	6.6	15.5	16.2	23.4	20.0	23.9	7.4	18.9	22.7
Be	0.9	1.3	2.1	0.5	1.2	1.9	1.4	1.2	1.5	0.9	1.5	1.3
Sc	31.4	21.5	6.5	32.9	22.0	8.5	13.5	19.3	9.7	24.9	19.0	4.3
V	248	191	45	231	125	25	102	127	72	166	211	37
Cr	13	8	3	62	22	2	41	60	16	233	8	4
Co	20	18	4	34	14	4	13	15	8	29	14	4
Ni	5	7	2	57	12	3	15	13	8	76	4	3
Cu	23	23	5	NA	NA	NA	15	10	8	59	38	5
Zn	80	64	44	NA	NA	NA	53	56	54	51	76	47

Ga	19	17	15	NA	NA	NA	17	17	16	NA	NA	16
Rb	11	40	72	10	52	96	65	49	75	18	55	88
Sr	320	262	196	233	177	96	254	342	252	570	389	267
Y	33	29	22	19	38	41	19	21	16	14	28	16
Zr	144	137	194	65	188	292	128	143	134	87	181	230
Nb	4.5	7.1	9.9	3.5	7.7	10.5	8.0	8.0	9.5	5.6	15.3	16.6
Mo	0.7	2.9	3.3	0.5	1.3	2.1	1.4	1.3	2.0	0.9	2.0	3.1
Cs	0.2	0.5	0.9	0.3	1.8	3.0	3.1	2.1	3.6	0.6	1.7	3.2
Ba	194	469	599	115	317	516	398	385	522	230	508	787
La	16.4	20.0	28.2	7.1	20.2	29.0	22.2	22.2	26.9	10.0	28.0	32.7
Ce	34.4	37.2	50.3	15.7	42.6	56.0	40.6	43.4	48.0	21.0	53.9	56.3
Pr	4.8	4.2	5.5	2.2	5.2	6.4	4.4	4.9	5.0	2.7	6.0	5.6
Nd	21.1	16.7	19.5	9.3	21.9	24.1	16.4	18.7	17.7	11.2	23.2	18.6
Sm	4.9	3.7	3.6	2.5	5.2	5.2	3.2	3.8	3.2	2.7	4.7	3.0
Eu	1.5	1.1	0.9	0.9	1.3	1.0	0.9	1.0	0.8	0.8	1.2	0.7
Gd	5.3	4.1	3.3	3.3	5.9	5.7	3.1	3.6	3.0	2.9	5.1	2.7
Tb	0.86	0.66	0.53	0.56	1.03	0.99	0.49	0.57	0.45	0.44	0.74	0.41
Dy	5.54	4.38	3.37	3.50	6.58	6.33	3.08	3.52	2.75	2.61	4.76	2.51
Ho	1.17	0.96	0.72	0.73	1.42	1.39	0.64	0.73	0.55	0.56	0.99	0.53
Er	3.48	2.92	2.27	2.29	4.56	4.61	1.92	2.14	1.68	1.62	3.00	1.64
Tm	0.52	0.44	0.36	0.34	0.70	0.76	0.29	0.32	0.26	0.23	0.47	0.26
Yb	3.31	2.99	2.59	2.23	4.44	4.92	2.03	2.16	1.80	1.47	2.83	1.85
Lu	0.50	0.46	0.41	0.33	0.69	0.79	0.31	0.32	0.28	0.22	0.43	0.30
Hf	3.37	3.17	4.02	1.55	5.09	7.76	2.29	2.25	NA	2.30	4.47	3.40
Ta	0.23	0.52	0.92	0.21	0.57	0.83	0.60	0.46	0.72	0.46	0.90	1.15
Pb	4.6	6.3	14.7	3.4	10.7	16.1	11.2	9.3	13.3	4.3	8.7	13.1
Th	2.16	9.14	17.74	1.55	9.53	18.47	8.90	7.61	10.86	1.83	6.67	10.42
U	0.62	2.86	6.37	0.45	2.82	5.45	2.12	1.49	2.53	0.60	1.90	2.96

LOI, Loss on ignition; na, not analysed. $Fe_2O_3^*$ = all iron reported as Fe_2C_3.

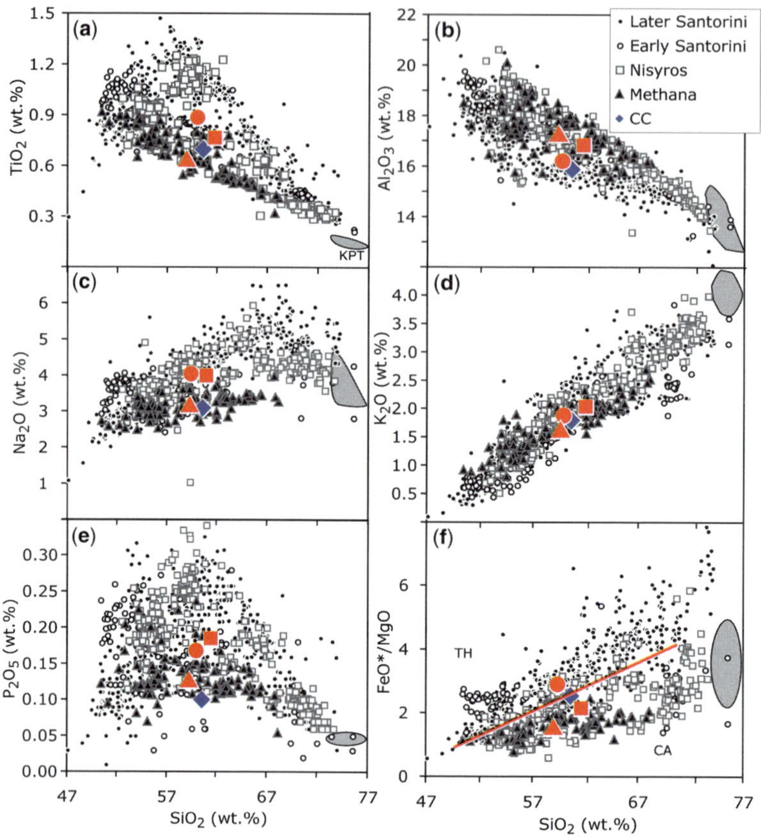

Fig. 3. Major element variation diagrams, with SiO_2 as fractionation index, for Methana, Santorini (Early and Later) and Nisyros. Data sources given in text. Field for Kos Plateau Tuff (KPT) from Bachmann *et al.* (2007) and Zouzias & St. Seymour (2008) and average continental crust (CC) from Rudnick & Gao (2003). The large circle, square and triangle are the average compositions of Later Santorini, Nisyros and Methana, respectively. The large group of mafic samples from Early Santorini are mafic igneous enclaves within the rhyodacites (Mortazavi & Sparks 2004); these may not all represent liquid compositions. (**a**) TiO_2; (**b**) Al_2O_3; (**c**) Na_2O; (**d**) K_2O; (**e**) P_2O_5; (**f**) FeO*/MgO, with FeO* = all iron expressed as FeO. Line dividing the trends for calc-alkaline (CA) and tholeiitic (TH) series from Miyashiro (1974).

The three centres are also distinct in terms of Pb isotopes (Fig. 8d, e), with the clearest grouping seen in the $^{208}Pb/^{204}Pb$ v. $^{206}Pb/^{204}Pb$ diagram. Nisyros and Santorini lie on the same trend, but Nisyros displays lower $^{206}Pb/^{204}Pb$ ratios than Santorini; Early Santorini falls towards the high end of the range defined by Later Santorini. Apart from the samples of the >0.9 Ma undifferentiated volcanics (Dietrich & Gaitanakis 1995), the Methana samples have a similar range of $^{206}Pb/^{204}Pb$ ratios as Santorini, but their $^{208}Pb/^{204}Pb$ ratio is offset to higher values. The Santorini granitoid xenoliths fall towards the high side of the range for Santorini as a whole, and the Santorini basement falls within the field for the Methana samples. The Methana xenoliths and basement samples are widely scattered in Pb isotope space.

Discussion

Source or process

Since subduction-related magmas seldom come to the surface without undergoing some modification in the form of crystal fractionation/accumulation and/or magma mixing, with or without concomitant crustal contamination, it is important to assess to which extent the signature we analyse reflects the (mantle) sources, and to which extent this original signature has been modified; and if so, by which processes.

Sub-arc source. Contributions from slab-derived fluids v. (partial melts of) subducted continental material are often assessed using the correlation of ratios of fluid mobile/fluid immobile elements and radiogenic isotope ratios (Turner *et al.* 1997).

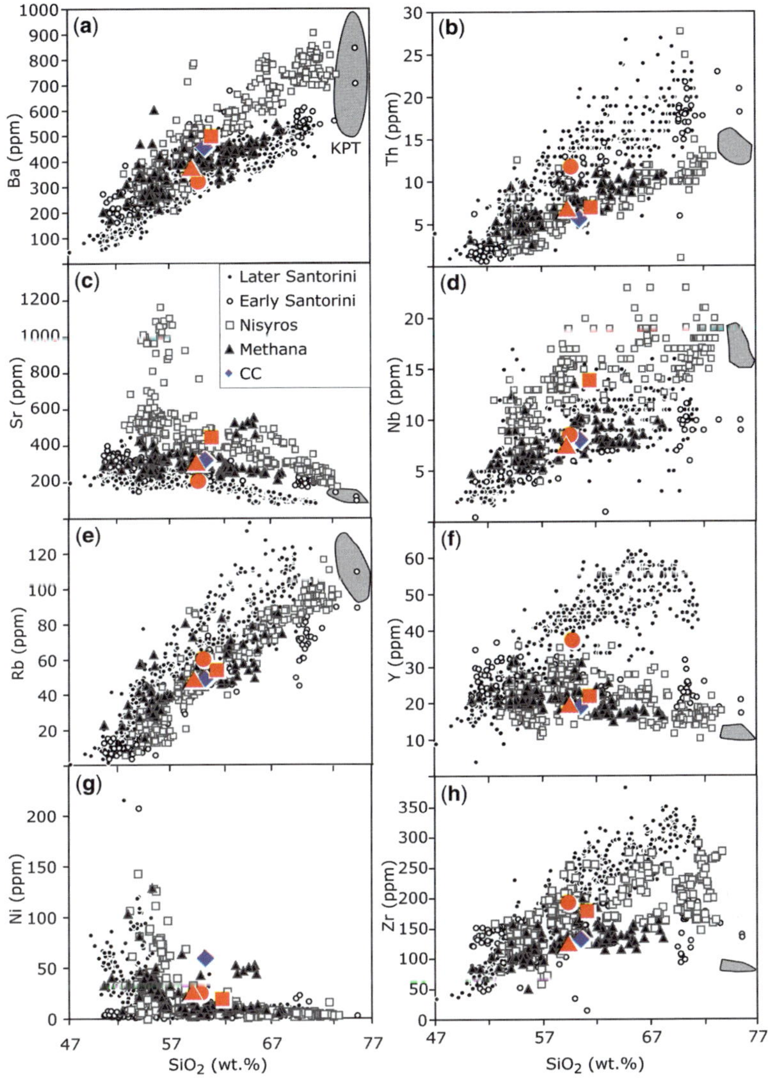

Fig. 4. Trace elements diagrams, with SiO_2 as a differentiation index; legend as in Figure 3. (**a**) Ba; (**b**) Th; (**c**) Sr; (**d**) Nb; (**e**) Rb; (**f**) Y; (**g**) Ni; (**h**) Zr.

Figure 9 shows the results for the most mafic samples of the suites (MgO > 5% and Mg# > 55, which meant that none of the Early Santorini samples qualified). Fluids from the subducted slab are thought to have high ratios of fluid-mobile over fluid-immobile elements (e.g. Ba/Th, Sr/Nd, Pb/Ce, Rb/Nb) and low $^{87}Sr/^{86}Sr$ ratios, whereas subducted sediments impart high $^{87}Sr/^{86}Sr$ and low $^{143}Nd/^{144}Nd$ ratios combined with low mobile/immobile element ratios, but high Th/Nd ratios; this is indicated by the arrows in Figure 9. These classic correlations are quite poor for the samples studied here. Methana has intermediate Ba/Th contents, but the highest $^{87}Sr/^{86}Sr$ ratios. Nisyros shows the highest Ba/Th and Sr/Nd ratios, combined with the lowest $^{87}Sr/^{86}Sr$ ratios, but this could also result from a change in the source of the subducted material, as discussed below, rather than the medium (fluid v. melt) by which the material is transferred to the mantle wedge. The lack of correlation between Pb/Ce or Rb/Nb and $^{87}Sr/^{86}Sr$ supports this idea. No correlation is observed between Pb/Ce and $^{206}Pb/^{204}Pb$ ratios or Th/Nd v. $^{143}Nd/^{144}Nd$ ratios.

Therefore, the dataset discussed here does not allow us to distinguish between 'fluid' and 'melt'

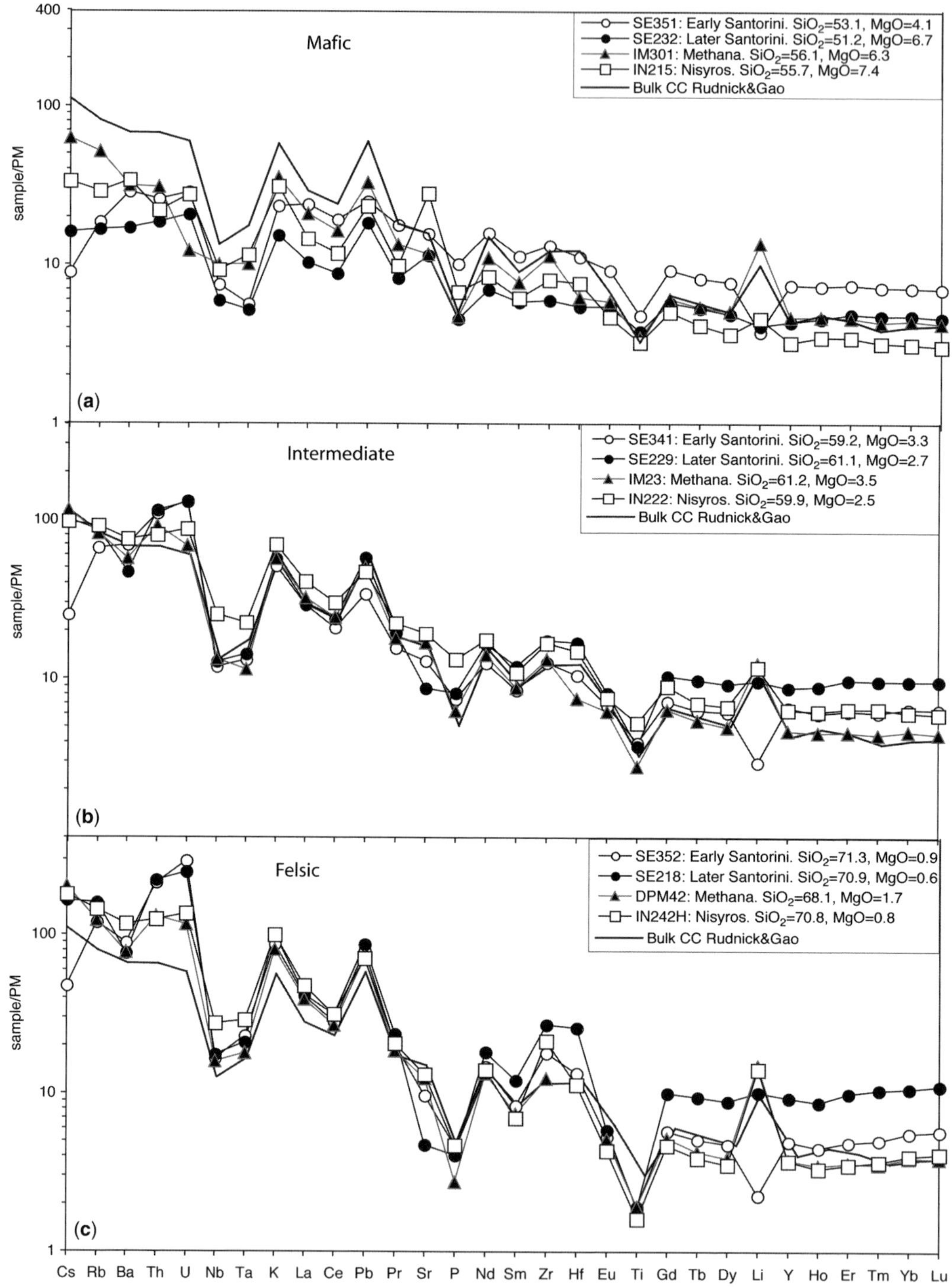

Fig. 5. Primitive-mantle normalized diagrams for selected (**a**) mafic, (**b**) intermediate and (**c**) felsic samples from the various volcanic centres, compared to values for average Continental Crust from Rudnick & Gao (2003). Normalizing values from Palme & O'Neill (2003).

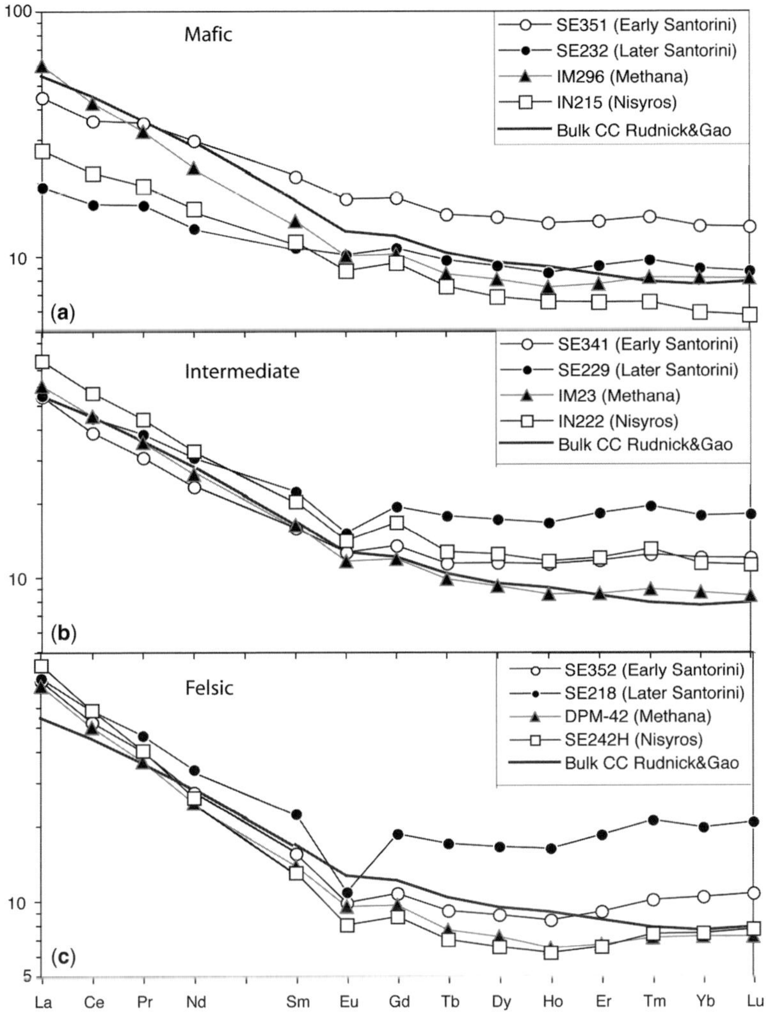

Fig. 6. Chondrite-normalized rare earth element diagrams for selected (**a**) mafic, (**b**) intermediate and (**c**) felsic samples from the various volcanic centres, compared to average Continental Crust from Rudnick & Gao (2003). Normalizing values from Taylor & McLennan (1985).

components in the source of the magmas, but mainly shows the changing source of the slab component. High concentrations of Ba, Sr and Nb for Nisyros and its low $^{206}Pb/^{204}Pb$ ratios may reflect that the sediment being subducted on the eastern side of the arc is influenced by input from the Nile. This river drains the Ethiopian flood basalts, which are characterized by relatively unradiogenic Pb isotopic compositions, and high concentrations of Ba, Sr and Nb (Pik *et al.* 1999; George & Rogers 2002). Data for these Afar basalts are shown in Figure 8, as the Pb isotopic composition of the Nile sediments is not known. However, the relatively low Sr (Talbot *et al.* 2000) and high Nd (Scrivner *et al.* 2004) isotopic ratios for the Nile sediments suggest a significant influence of the basaltic province on its isotopic signature. The fact that this along-arc change in Pb isotope signature seems to be related to a change in the composition of the subducted sediment is at odds with models for magmatism on Methana and Nisyros that only see a role for melting of the lithospheric mantle, as proposed by Mitropoulos & Tarney (1992).

Crystal fractionation. The major element trends are a reflection of the modifications taking place during differentiation. These are obviously different for the three volcanic centres. The trend for Later

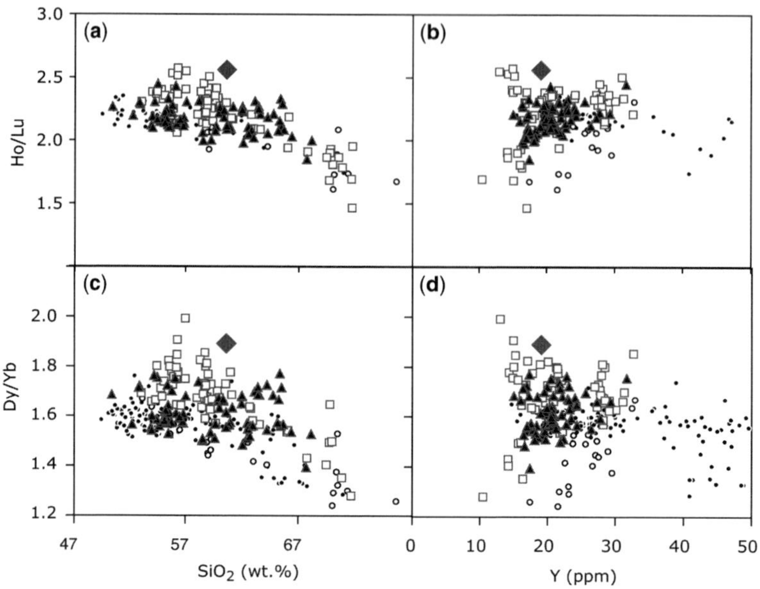

Fig. 7. Ho/Lu and Dy/Yb ratio v. SiO_2 (**a**, **c**) and Y (**b**, **d**) for the different volcanic centres, compared to average continental crust. For Ho/Lu, only our own analyses are shown here, to avoid uncertainty about precision and accuracy for some of the older published analyses. Legend as in Figure 3.

Santorini shows the classical hallmarks of crystal fractionation, with inflections in the Harker diagrams for TiO_2 and P_2O_5. Although such inflections can also be explained by mixing between an intermediate magma with both a more mafic and a more felsic one, the maximum concentrations for the two oxides occur at different SiO_2 contents (approximately 59 wt% SiO_2 for TiO_2 and 61 wt% for P_2O_5), which would not be expected if the inflection was caused by having two mixing arrays. Such inflections are not seen for Methana and Early Santorini; some deposits on Nisyros show increasing TiO_2 and P_2O_5 concentrations, while others show similar straight trends as seen for Methana and Early Santorini. From the samples' petrography, we can be sure that magma mixing has taken place on Methana. However, intermediate samples from Early Santorini (Cape Mavros) are not characterized by the disequilibrium assemblages that are seen on Methana. Therefore, the low TiO_2 and P_2O_5 concentrations for Methana and Early Santorini are unlikely to only reflect mixing between a basaltic and a rhyolitic end-member. Fe–Ti-oxide is observed as a phenocryst phase at SiO_2 contents below 59%, so the depletion in TiO_2 and P_2O_5 may be caused by the earlier onset of Ti-bearing magnetite and apatite fractionation. This can be caused by higher concentrations of water in the magma, as this depresses the liquidus temperatures of the silicates more than that of non-silicate minerals (Sisson & Grove 1993; Parat *et al.* 2008). Fe–Ti-oxide fractionation during more advanced stages of differentiation also explains the more 'tholeiitic' trend in the FeO*/MgO diagram seen in Later Santorini compared to the other groups (Elburg 2010), although the fractionation of amphibole may also help to deplete the residual liquids in iron (Alonso-Perez *et al.* 2009).

A similar grouping of samples is seen in the Harker variation diagram for Na_2O, where Methana and Early Santorini show hardly any enrichment in sodium with increasing SiO_2, whereas Nisyros display higher values, and Later Santorini higher still. Later Santorini is furthermore distinguished by the lowest Al_2O_3 contents; and higher Y (plus the MREE and HREE), Zr and Rb. Since the ranges for these elements overlap at low SiO_2 contents for the different groups, it is most likely that fractionation is also responsible for the behaviour of these elements. A depletion in aluminium reflects a greater role for fractionation of plagioclase, and this is in agreement for the more negative Eu-anomaly (lower Eu/Eu*) for the Later Santorini samples. The trend for Y does not mimic that of P, so fractionation of apatite is not the dominant control here. The depletion is likely to result from amphibole fractionation, which must have been less important in the Later Santorini samples than in the other centres, as also suggested by Mortazavi & Sparks (2004). However, the

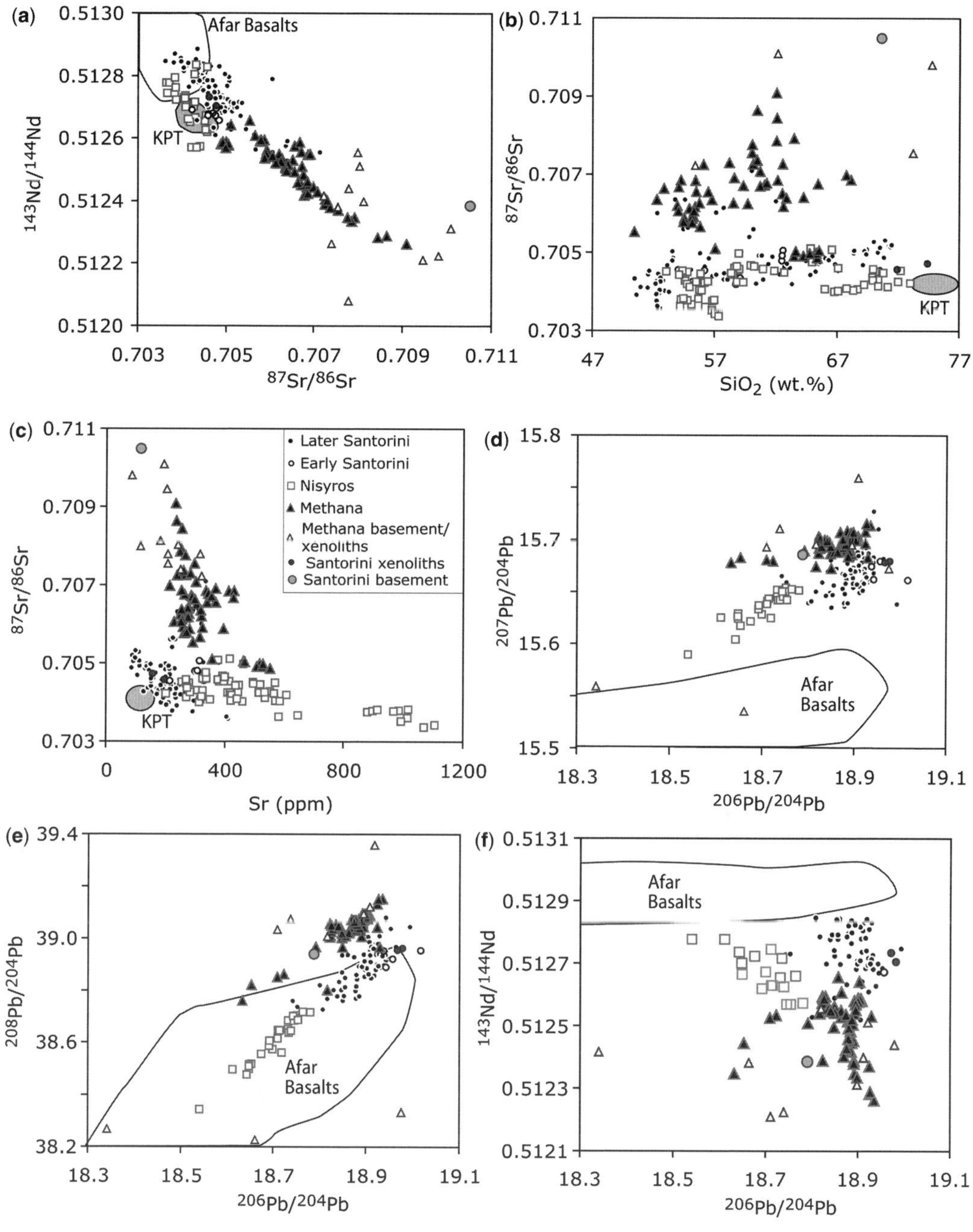

Fig. 8. Radiogenic isotope diagrams for Methana, Santorini and Nisyros. Field for Afar basalts from Pik *et al.* (1999), as a proxy for the composition of Nile sediment. (**a**) $^{143}Nd/^{144}Nd$ v. $^{87}Sr/^{86}Sr$; (**b**) $^{87}Sr/^{86}Sr$ v. SiO_2 (wt%); (**c**) $^{87}Sr/^{86}Sr$ v. Sr concentration (ppm); (**d**): $^{207}Pb/^{204}Pb$ v. $^{206}Pb/^{204}Pb$. The Methana samples with lower Pb isotopic ratios belong to the oldest volcanic deposits; (**e**) $^{208}Pb/^{204}Pb$ v. $^{206}Pb/^{204}Pb$; (**f**) $^{143}Nd/^{144}Nd$ v. $^{206}Pb/^{204}Pb$.

MREE–HREE pattern for these samples is, nevertheless, somewhat spoon-shaped, suggesting that amphibole fractionation did occur and influenced the shape of the Later Santorini pattern, but that it was significantly more important for the other centres. Rubidium and Zr concentrations may also

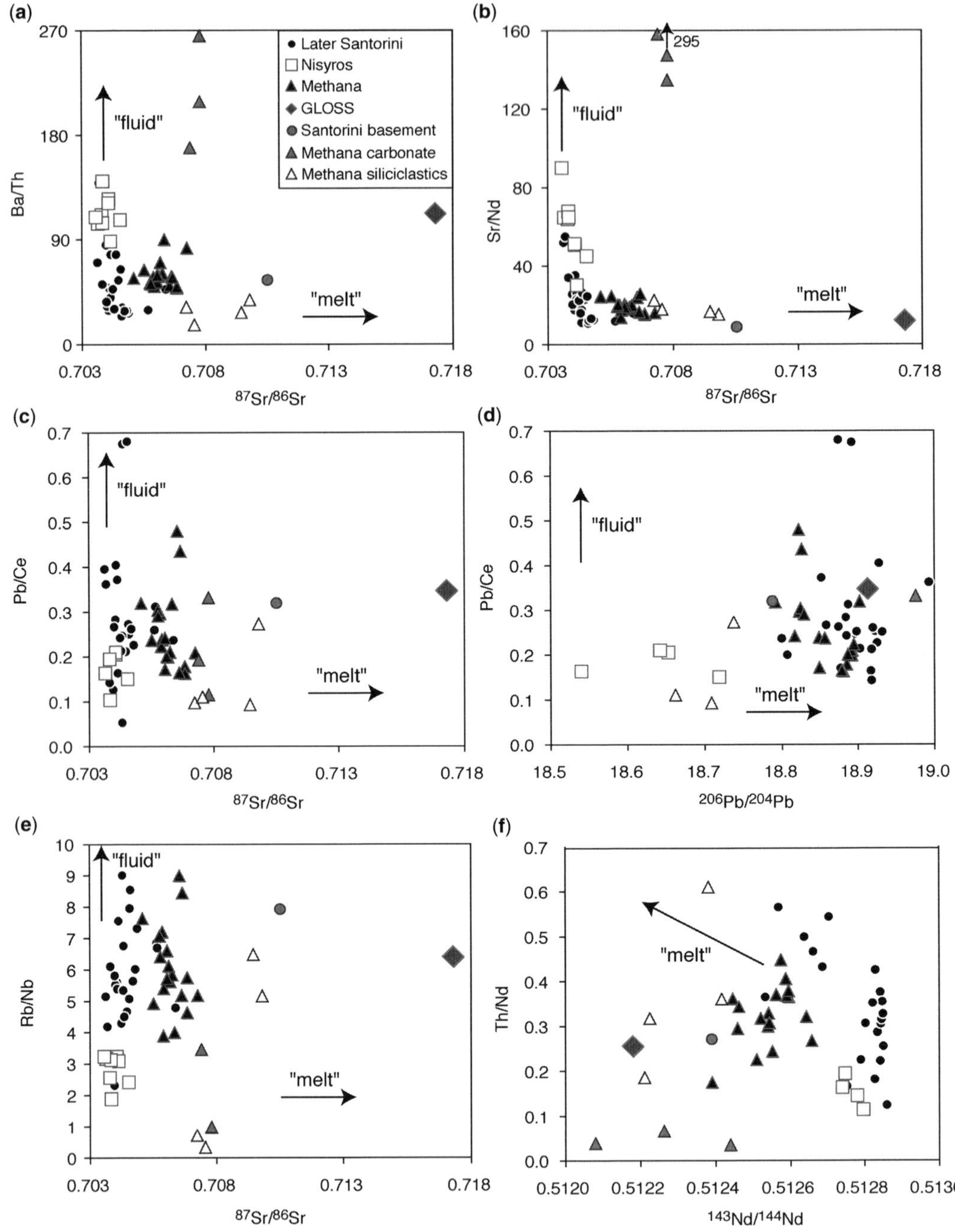

Fig. 9. (**a**) Ba/Th, (**b**) Sr/Nd and (**c**) Pb/Ce v. $^{87}Sr/^{86}Sr$ ratio; (**d**) Pb/Ce v. $^{206}Pb/^{204}Pb$ ratio; (**e**) Rb/Nb v. $^{87}Sr/^{86}Sr$ ratio; and (**f**) Th/Nd v. $^{143}Nd/^{144}Nd$ ratio for the most mafic samples (MgO > 5%, Mg# > 55) of the volcanic suites, analysed crustal end-members and Global Subducting Sediment (GLOSS; Plank & Langmuir 1998). Legend as in Figure 3. Note that the imposed restrictions have excluded any samples from the Early Santorini suite. The arrows indicate the general composition of the 'slab fluid' and 'sediment melt' component as these are generally identified in these type of diagrams. The fact that the position of the suites relative to the 'fluid' and 'melt' trends varies suggests that this may not be the most appropriate interpretation of the dataset. It is more likely that the identity of the slab component changes along arc.

reflect this process, but to a lesser extent, as these elements are less compatible in amphibole than Y is (GERM Partition Database: http://earthref.org/KDD/).

Fractionation depth. Previous authors have tried to estimate the depth at which crystal fractionation occurred in Nisyros and Santorini, and previous published microprobe data can also be used to provide pressure estimates with the geothermobarometers of Ridolfi *et al.* (2010) for amphibole, Nimis & Ulmer (1998) and Nimis (1999) for clinopyroxene, and Putirka *et al.* (2003) for clinopyroxene-liquid.

Mortazavi & Sparks (2004) estimate the pressure of differentiation for the Early Santorini deposits on Akrotiri peninsula as 1.8–2.5 kbar (180–250 MPa), based on maximum water contents of the mafic enclaves. The geothermobarometer of Ridolfi *et al.* (2010) for the published amphibole compositions and our own unpublished data yields a range of pressures and temperatures for the mafic enclaves from 980 °C and 480 MPa to 930 °C and 280 MPa, and for the rhyodacitic host rocks of 960 °C–350 MPa down to 820 °C–120 MPa for hornblende rims. A clinopyroxene core (own unpublished data) from a rhyodacite yielded 210 MPa and 980 °C with the Putirka *et al.* (2003) clinopyroxene-liquid geothermobarometer.

Experimental data (Cottrell *et al.* 1999) indicate storage at shallow depths (<50 MPa) for the main batch of magma of the Minoan eruption of the Later Santorini series, but a prior stage in its evolution at >200 MPa. The presence of groundmass hornblende in cauliform andesitic scoria clasts indicate water pressures greater than 75 MPa. However, applying the amphibole geothermo-barometer of Ridolfi *et al.* (2010) to the amphibole compositions published by Cottrell *et al.* (1999) yields pressures of 300–400 MPa. The lower value of 300 MPa comes from the microlites in these clasts. Considering that the microlites must have been formed during chilling of the andesite magma against the cooler rhyodacitic host magma, this pressure estimate seems at odds with the 50 MPa deduced from experimental data.

Gertisser *et al.* (2009) provide pressure estimates of 4.3 kbar for the Lower Pumice 2 of the Later Santorini series, based on the assumption of quartz saturation. Applying the amphibole geothermobarometer on their published analyses yields lower pressures of 2.5 kbar.

Clinopyroxene analyses from mafic samples of the Later Santorini series (Huijsmans 1985) yield quite variable pressures of 1.5–4 kbar and 1140–1160 °C according to the clinopyroxene-liquid geothermobarometer. The clinopyroxene-only geobarometer indicates significantly lower pressures, including negative ones, reflecting the rather large uncertainty (2.6 kbar) on these calculations.

Wyers & Barton (1989) give pressure estimates for Nisyros of 8 kbar for felsic rocks and 3.5–4 kbar for mafic lavas, based on the then available olivine-liquid and plagioclase-liquid geothermobarometers, with the caveat that the error on the pressures is about 2 kbar. Applying the Ridolfi *et al.* (2010) amphibole barometer on their two published amphibole analyses (from dacite and rhyodacite) yields pressures of 2.8–4.4 kbar. Using the same geobarometer, Bachmann *et al.* (2012) report pressures of 2.5–1.5 kbar for the amphiboles from Nisyros Upper and Lower Pumice, but suspect those pressures to be an underestimate. Magmas erupted just after caldera collapse were found to have fractionated at shallower depths, as a result of the decompression that took place.

Amphibole compositions for Methana were published by Mitropoulos & Tarney (1992), and yield pressures of 445–120 MPa and temperatures of 990–814 °C. This range is similar to what is observed for the Early Santorini deposits.

In summary, polybaric fractionation seems to have been important in Nisyros as well as Later Santorini, but the dominant magma storage location for the felsic Later Santorini magmas appears to have been at shallow depth, whereas the felsic magmas from Nisyros spent most time at deeper levels. Early Santorini and Methana are similar with the earlier stages of amphibole fractionation having taken place at around 4 kbar. The differences in depth of fractionation are likely to have influenced amphibole stability, and thereby the differentiation trends for the different volcanic centres.

Crustal contamination. The correlation between Sr isotopic signature and Sr concentrations (Fig. 8c) is likely to be a combined source, differentiation and contamination effect, with samples that have high Sr concentrations being less sensitive to the influence of crustal contamination on this isotopic system. This interpretation agrees with the offset of the Nisyros, high-Sr Methana and Early Santorini samples towards lower $^{87}Sr/^{86}Sr$ ratios for their $^{143}Nd/^{144}Nd$ ratios compared to those of the Later Santorini and Methana samples.

Nisyros is the only centre that shows a reasonable negative correlation between Nd and $^{206}Pb/^{204}Pb$ isotopic signature (Fig 8f), with the high $^{143}Nd/^{144}Nd$–low $^{206}Pb/^{204}Pb$ samples generally being the most mafic (cf. Fig. 9f). This implies that the trend may be due to crustal contamination. The fact that no trend is visible for Santorini or Methana in Figure 8f can be explained by a similar $^{206}Pb/^{204}Pb$ isotopic signature for the uncontaminated magma, and the contaminant. There is a rough anti-correlation between $^{208}Pb/^{204}Pb$ (as well as $^{207}Pb/^{204}Pb$) and $^{143}Nd/^{144}Nd$ for Santorini and Methana combined, suggesting a higher $^{208,7}Pb/^{204}Pb$ ratio for the upper crustal contaminant than for the primary magma.

We performed a simple Assimilation-Fractional Crystallization one-stage model (DePaolo 1981) for the Methana, Santorini and Nisyros data. These models are only a very rough first-order indication of what could have happened, considering that we do not distinguish individual suites on the islands, that we do not take polybaric fractionation into account and that the solutions are generally not unique. It is, therefore, most instructive to see what scenarios are not permitted within the imposed constraints. End-members of this modelling process are given in Table 2. We tried to limit the isotopic variation in the end-members as much as possible, testing the assumption that all variation was a result of crustal contamination only. The primitive magma end-member for Nisyros has a lower Pb and Nd isotopic signature than for the two other islands. The initial concentrations of Sr, Nd and Pb vary per suite, with Later Santorini having the lowest concentrations. The local Santorini basement or the Methana xenoliths turned out to be unsuitable contaminants, as their Pb and Sr isotopic compositions were too low. We therefore used a component based on the data for the Ios basement (Buettner *et al.* 2005). The rate of assimilation over crystallization (R) was kept low for Later Santorini (0.3) to mimic upper crustal crystallization conditions, and this explains the general trend of the data in Sr–Nd and Pb–Nd isotope diagrams (Fig. 10a, b). For Nisyros, the rate of assimilation over fractionation was set higher, and the considerably higher Sr contents of the primitive magma causes the modelled trend to coincide with the island's Sr–Nd isotope array. The most interesting result is that the shape of the $^{143}Nd/^{144}Nd$ v. $^{87}Sr/^{86}Sr$ array for the Methana data could not be modelled starting from the primitive magma composition used for the other two suites. It is possible to obtain the least contaminated end-member in this way, but a different second stage is needed, using a crustal end-member with a lower Sr/Nd elemental ratio (Fig. 10a). The failure of the single stage model could also mean that the primary magma for Methana is close in isotopic composition to the least evolved sample analysed, and that the high Sr and low Nd isotopic composition reflects source contamination, that is crustal material added to the mantle wedge. The group of Methana samples with high Sr concentrations shows lower Sr isotopic ratios, which could indicate the presence of a second type of magma, and shows more resemblance to the Nisyros model in terms of Sr and Nd isotopes.

The amount of contamination necessary to explain the analysed samples with the lowest Nd and highest Sr isotope ratios is 4% and 6% for

Table 2. *Parameters for Assimilation-Fractional Crystallization models after DePaolo (1981)*

	Later Santorini	Nisyros	Methana	Methana stage 2
$^{87}Sr/^{86}Sr$ magma	0.7035	0.7035	0.7035	0.7055
Sr ppm magma	300	1000	800	400
D Sr	1.5	1.5	1.3	1.3
$^{87}Sr/^{86}Sr$ crust	0.739	0.739	0.739	0.739
Sr ppm crust	180	180	180	120
$^{143}Nd/^{144}Nd$ magma	0.51290	0.51285	0.51290	0.51266
Nd ppm	4	9	10	12
D Nd	0.1	0.1	0.1	0.1
$^{143}Nd/^{144}Nd$ crust	0.51205	0.51205	0.51205	0.51205
Nd ppm crust	18	18	18	60
$^{206}Pb/^{204}Pb$ magma	18.75	18.35	18.75	18.75
$^{207}Pb/^{204}Pb$ magma	15.62	15.54	15.62	15.62
$^{208}Pb/^{204}Pb$ magma	38.7	38.2	38.7	38.7
Pb ppm magma	2	3	5	8
D Pb	0.06	0.06	0.06	0.06
$^{206}Pb/^{204}Pb$ crust	19.1	19.1	19.1	19.1
$^{207}Pb/^{204}Pb$ crust	15.73	15.73	15.73	15.73
$^{208}Pb/^{204}Pb$ crust	39.2	39.2	39.2	39.2
Pb ppm crust	22	22	22	22
R	0.30	0.70	0.70	0.75
F most evolved	0.80	0.93	0.91	0.90
Maximum % contaminant	0.06	0.04	0.06	0.08

D, bulk distribution coefficient; R, ratio of assimilation/fractionation; F, fraction of liquid remaining.

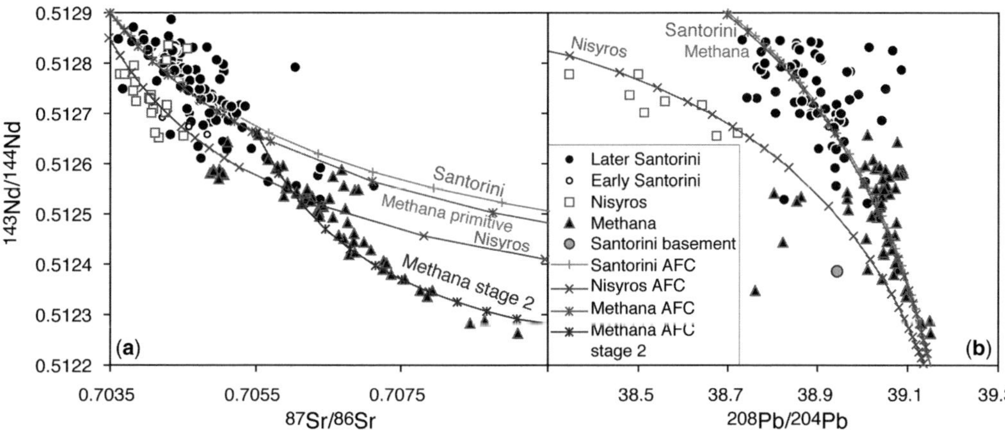

Fig. 10. Result of Assimilation–Fractional Crystallization models for (**a**) $^{143}Nd/^{144}Nd$ v. $^{87}Sr/^{86}Sr$ and (**b**) $^{208}Pb/^{204}Pb$. End-members are given in Table 2. Note that the second stage for Methana only results in a separate curve in panel (a), and that the curves for Methana and Santorini virtually coincide in panel (b).

Nisyros and Later Santorini, respectively (Table 2). If the primitive magma for Methana had Sr and Nd isotopic characteristics similar to the other two islands, 14% of contaminant needs to be added. If the primary magma has Sr and Nd isotopic characteristics similar to the most primitive sample analysed, 8% contamination is indicated by the model. These amounts are, of course, very dependent on the exact Sr and Nd isotopic composition of the crustal end-member, for which we do not have any clear indication.

Differentiation and eruption mechanisms

Although Methana shows more evidence for crustal contamination in terms of its isotopic signature, the major and trace element differentiation trends observed are not significantly different from Early Santorini and the Nisyros samples that have undergone amphibole fractionation without extensive crustal contamination. Modelling the trace element and isotopic trends for Methana assuming that Assimilation combined with Fractional Crystallization (AFC) has been the differentiation mechanism is, however, too simplistic: field and petrographical evidence shows that magma mixing and mingling have also been operational. It is remarkable that amphibole-rich intermediate volcanic deposits, with mafic enclaves resulting from magma mingling, are generally lava domes, or flows with limited lateral extent, such as the whole of Methana, Early Santorini, and the Nikia and Profitis Ilias units of Nisyros. The difference between lava domes, and lava flows (for which mingling seems less obvious), was already observed for Nisyros and Later Santorini by Zellmer & Turner (2007). We propose that the crystallization of amphibole may be the key to this phenomenon. When magmas stall at sufficient depth (>100–200 MPa, Carmichael 2002), amphibole will start to crystallize, but this phase is not stable at shallow depths (Carmichael 2002) – this explains the contrasting differentiation trends for Methana and Later Santorini, while Nisyros seems to have two crustal differentiation levels (Wyers & Barton 1989). With the onset of amphibole crystallization, there is a steep increase in the crystallinity of the magma (Barclay & Carmichael 2004). The increasing crystallinity will lead to the magma's 'uneruptibility', and it will remain stalled at depth – unless it is remobilized by input of a less evolved, hotter magma. Thereby, the input of a more mafic magma, and the concomitant chance of magma mixing and magma mingling is a prerequisite for the eruption of these magmas, as also proposed by Tamura & Tatsumi (2002) for the Izu-Bonin arc. In the case of shallower magmatic systems, where amphibole crystallization does not play a role, input of a more mafic magma may be the eruption trigger (Martin *et al.* 2008), but it is not a necessity.

This stalling of magmas at depth, creating a crystal mush, of which either the interstitial liquid may be extracted to give rise to more silicic volcanic products (Bachmann & Bergantz 2004; Bachmann *et al.* 2007), or which erupt wholesale (Bachmann *et al.* 2002), is increasingly recognized as being an important process in igneous petrogenesis (Charlier *et al.* 2005; Deering *et al.* 2011). In the case of Methana, the resulting volcanic deposits are rather small in scale, suggesting that the size of the magma chambers is quite limited. This agrees well with the important role for crustal contamination, as inferred from the Sr and Nd isotopic signature: small magma chambers have a relatively large area where interaction with host rocks can take place. This contrasts with the limited amount of crustal contamination seen in Nisyros, or in the related magmatic system that gave rise to the Kos Plateau Tuff (Bachmann *et al.* 2007); the reduced crustal thickness underneath Nisyros–Kos compared to Methana is likely to have played a role too, and the thermal structure or the composition of the host rocks may also have had an influence. This process of crystallization and remobilization seems to have been the main eruption mechanism for Methana, and of lesser importance for Nisyros and Santorini. This is also recorded in the observed phenocryst assemblage on the islands: many of the Methana deposits contain biotite crystals, while these are exceedingly rare on the other islands, even though potassium contents are similar. This suggests that late-crystallizing biotite was formed on Methana, but that fractionation did not reach this stage on the other two islands, even though the volcanic deposits there extend to more felsic compositions.

On a global scale, the Aegean arc is classified as having a high proportion of lava domes, indicating high viscosity of the magmas, and this lava dome percentage shows a good anti-correlation with convergence rate (Zellmer 2008). The relatively slow convergence rate between the African and Aegean plates is likely related to the presence of crustal fragments on the downgoing plate (Van Hinsbergen *et al.* 2005), and the partial subduction of crustal fragments can also be the source of the elevated $^{87}Sr/^{86}Sr$ ratios of the Methana magmas.

Arc magmatism and crustal growth

When comparing the volcanic products from Methana, Santorini and Nisyros with the composition of the continental crust, it is clear that Later Santorini shows the least geochemical similarity. Santorini is an end-member volcano, where magmatic differentiation dominantly took place in a shallow magma chamber, suppressing the importance of amphibole fractionation and the necessity for magma mingling and mixing. Similar trends can be seen in water-poor magmatic systems, such as mid-ocean ridges, as these also do not fractionate amphibole. This amphibole-free fractionation leads to high Y and HREE concentrations, which are clearly different from what is seen for the average continental crust, as is the high FeO*/MgO ('tholeiitic') trend. Comparison of the Santorini data with published analyses of volcanic rocks from other convergent margins, such as the Sunda-Banda, Mariana or Honshu arcs, shows that the strong enrichment in Y observed in Santorini is rather unusual. Apart from its location on the most thinned area of continental crust, the presence of large-scale faults may also influence this reduced depth of fractionation. Similar high-Y samples are found in the Taupo Volcanic Zone, New Zealand, where they also have been interpreted as a result from shallow fractionation processes (Deering *et al.* 2008).

Amphibole-dominated differentiation can only occur in water-bearing magmas at sufficient depth, and this is most easily achieved when the crust is already thickened – either because it is continental crust already, or because it is thickened oceanic crust. If amphibole fractionation is a necessity to keep HREE levels low, and the magmatic products similar to continental crust, then continental crust formation is a prolonged process, taking place in long-lived volcanic arcs, whereby previous magmatic additions have thickened the crust sufficiently to promote deep, amphibole-dominated fractionation.

However, if we look at the MREE–HREE pattern of the amphibole-dominated Aegean magmas and compare that to the patterns for average continental crust from Rudnick & Gao (2003), there is a subtle difference. The Aegean magmas have lower MREE/HREE ratios than the continental crustal estimates. At similar SiO_2 or Y concentrations (Fig. 7), the continental crust estimate has a higher Ho/Lu and Dy/Yb ratio than even the amphibole-dominated Aegean volcanics. If this difference is considered to be significant, then the continental crust has not been formed by magmatic processes involving amphibole as the controlling fractionating phase or residue, but rather with garnet as the mineral determining the MREE–HREE characteristics (Larocque & Canil 2010), since this mineral has higher distribution coefficients for HREE than for MREE, whereas this is the other way around for amphibole.

Garnet signatures within modern subduction-related volcanics are relatively rare, being restricted to adakites (Martin *et al.* 2005) and high-magnesian andesites (Kelemen 1995), and this means that the average continental crust has not been formed by subduction processes that are exactly the same as those happening in present-day arcs. The fractionation products of modern arcs do not have the Ho/Lu ratios that characterize the average continental crust, so the process of fractionation, followed by foundering of mafic cumulates (Plank 2005), as has been proposed to happen in present-day arcs, cannot have generated the continental crust, unless garnet was involved at some stage.

In present-day subduction-related magmas, the 'adakitic' signature can be generated by melting of the subducted oceanic crust with garnet as a residual phase during magma generation (Defant &

Drummond 1990) or during magma differentiation, by fractional crystallization of a garnet-bearing assemblage (Macpherson *et al.* 2006), which is stabilized by relatively low water contents (Zellmer *et al.* 2012). Melting of modern subducted oceanic crust is likely to be more extensive when the subducted slab is still young and hot (Peacock 2003). The decrease of mantle potential temperature over Earth's history (Van Hunen *et al.* 2008) implies that melting of subducting slabs would have been more pronounced in the past (Martin 1986; Taylor & McLennan 1995). This is in agreement with the geochemical signature of Archean trondhjemites and tonalites (Rapp *et al.* 1991), which are likely to have formed by melting of amphibolite or eclogite. Since there is broad agreement that the continental crust has been formed over most of Earth's history, with about half of it having formed prior to the Proterozoic (Collerson & Kamber 1999), it is likely that the average composition of the continental crust will still bear the imprint of this hotter geotherm.

The relatively high Ni (and Cr) contents in the continental crust are not matched by most of our Aegean data either, apart from the small group of high-Sr Methana samples. Their detailed petrogenesis will be discussed in more detail elsewhere, but there is evidence that entrainment of magnesian mineral phases has played a role in their enrichment in compatible elements. Some high-Mg andesites, which are viewed as being modern analogues for continental crust formation (Kelemen *et al.* 2007), have also been interpreted to have formed by mixing of andesite with ultramafic xenocrysts (Streck *et al.* 2007). This is an alternative explanation for the high MgO, Ni and Cr contents, which are often interpreted as reflecting interaction of a slab melt with the mantle, of these high-Mg andesites. Calc-alkaline (low FeO*/MgO) magmas, of which the Methana samples are an excellent example and which are more similar to the continental crust than tholeiites, show ubiquitous evidence of magma mixing and mingling processes, although seldom leading to the enhanced Cr and Ni contents at intermediate SiO_2 contents seen in continental crust. Perhaps the idea that the average composition of the continental crust can be explained by a single (albeit multi-stage) process should be abandoned. Interpreting the geochemical signature of the continental crust as a mixture of felsic and mafic components may be more fruitful. In this case the garnet signature comes from the felsic component, and reflects melting of a mafic precursor at depth, either in the subducted slab (Martin 1986; Defant & Drummond 1990; Foley *et al.* 2002), or in overthickened mafic crust (Smithies 2000; Nagel *et al.* 2012). The high levels of incompatible elements, such as Ni, Cr and MgO, reflect a mafic component within the continental crust. This view is supported by the fact that most Archean terranes are bimodal in character (Barker & Arth 1976). One of the arguments in favour of the slab-melting scenario for generating continental crust, rather than melting of thickened crust, is the high level of compatible elements in the high-Mg andesites generated during slab melting (Kelemen 1995), which matches the composition of the continental crust; concentrations of these compatible elements are lower when melting happens at the base of overthickened crust. If we consider the average crust as a mechanical mixture of felsic and mafic components, then the model in which melting happens at the base of the crust may be viewed more favourably. However, the incompatible trace element characteristics of the earliest continental are best explained by remelting of a tectonically overthickened crust, with the geochemical characteristics of arc-type basalt (Nagel *et al.* 2012), so, also in this scenario, subduction-related magmatism appears to be instrumental in forming the continental crust. As the felsic part of the crust would be formed by remelting of a broadly basaltic precursor (or fractionation from a basaltic melt), the more mafic residues still need to be removed from the crust via foundering or delamination (Houseman *et al.* 1981; Kay & Kay 1993; Plank 2005) in order to achieve the average andesitic composition of the continental crust. The higher level of Ni, Cr and Mg in the continental crust compared to the averages of modern arc rocks reflects the higher contents of these elements in the largely komatiitic composition of the mafic crustal end-member in the Archaean, contrasting with the broadly basaltic flux seen in present-day arcs.

Conclusions

The datasets for Methana, Santorini and Nisyros form a good example of the influence of along-arc changes in input as well as the influence of crustal thickness on the mineralogy and, thereby, geochemistry of eruptive products. Amphibole is shown to be an important fractionating phase, imparting a characteristic spoon-shaped MREE-HREE patterns and low Y contents, and necessitating the input of mafic magma for remobilisation, promoting magma mixing and mingling to operate. Despite the wide range in magmatic products, reflecting varying amounts and depths of fractionation, magma mixing and crustal contamination, none of the whole rock compositions approaches that of the average continental crust in terms of Ho/Lu ratios or the level of compatible elements. This implies that garnet, rather than amphibole, was a residual or fractionating phase during early continental crust formation. The high level of compatible elements in average continental crust are most easily explained by viewing the crust as a mixture of mafic and felsic components.

G. Zellmer and V. Troll are thanked for forcing Elburg to think and write about this subject to a greater extent than would have been done otherwise. This research was supported by FWO grants G.A078.11 and 1.5.044.10N to Elburg, and a Ghent University BOF grant 01D37710 to Smet. The constructive comments by reviewers C. Deering and O. Bachmann were greatly appreciated.

References

ALONSO-PEREZ, R., MÜNTENER, O. & ULMER, P. 2009. Igneous garnet and amphibole fractionation in the roots of island arcs: experimental constraints on andesitic liquids. *Contributions to Mineralogy and Petrology*, **157**, 541–558.

BACHMANN, O. & BERGANTZ, G. W. 2004. On the origin of crystal-poor rhyolites: extracted from batholithic crystal mushes. *Journal of Petrology*, **45**, 1565–1582.

BACHMANN, O., DUNGAN, M. A. & LIPMAN, P. W. 2002. The Fish Canyon magma body, San Juan volcanic field, Colorado: rejuvenation and eruption of an upper-crustal batholith. *Journal of Petrology*, **43**, 1469–1503.

BACHMANN, O., CHARLIER, B. L. A. & LOWENSTERN, J. B. 2007. Zircon crystallization and recycling in the magma chamber of the rhyolitic Kos Plateau Tuff (Aegean arc). *Geology*, **35**, 73–76.

BACHMANN, O., DEERING, C. D., RUPRECHT, J. S., HUBER, C., SKOPELITIS, A. & SCHNYDER, C. 2012. Evolution of silicic magmas in the Kos-Nisyros volcanic center, Greece: a petrological cycle associated with caldera collapse. *Contributions to Mineralogy and Petrology*, **163**, 151–166.

BARCLAY, J. & CARMICHAEL, I. S. E. 2004. A hornblende basalt from western Mexico: water-saturated phase relations constrain a pressure-temperature window of eruptibility. *Journal of Petrology*, **45**, 485–506.

BARKER, F. & ARTH, J. G. 1976. Generation of trondhjemites-tonalitic liquids and Archean bimodal trondhjemite-basalt suites. *Geology*, **4**, 596–600.

BOHNHOFF, M., RISCHE, M., MEIER, T., BECKER, D., STAVRAKAKIS, G. & HARJES, H. P. 2006. Microseismic activity in the Hellenic Volcanic Arc, Greece, with emphasis on the seismotectonic setting of the Santorini-Amorgos zone. *Tectonophysics*, **423**, 17–33.

BUETTNER, A., KLEINHANNS, I. C., RUFER, D., HUNZIKER, J. C. & VILLA, I. M. 2005. Magma generation at the easternmost section of the Hellenic arc: Hf, Nd, Pb and Sr isotope geochemistry of Nisyros and Yali volcanoes (Greece). *Lithos*, **83**, 29–46.

CARMICHAEL, I. S. E. 2002. The andesite aqueduct: perspectives on the evolution of intermediate magmatism in west-central (105–99°W) Mexico. *Contributions to Mineralogy and Petrology*, **143**, 641–663.

CHARLIER, B. L. A., WILSON, C. J. N., LOWENSTERN, J. B., BLAKE, S., VAN CALSTEREN, P. W. & DAVIDSON, J. P. 2005. Magma generation at a large, hyperactive silicic volcano (Taupo, New Zealand) revealed by U–Th and U–Pb systematics in zircons. *Journal of Petrology*, **46**, 3–32.

COLLERSON, K. D. & KAMBER, B. S. 1999. Evolution of the continents and the atmosphere inferred from Th–U–Nb systematics of the depleted mantle. *Science*, **283**, 1519–1522.

COTTRELL, E., GARDNER, J. E. & RUTHERFORD, M. J. 1999. Petrologic and experimental evidence for the movement and heating of the pre-eruptive Minoan rhyodacite (Santorini, Greece). *Contributions to Mineralogy and Petrology*, **135**, 315–331.

DAVIDSON, J., TURNER, S., HANDLEY, H., MACPHERSON, C. & DOSSETO, A. 2007. Amphibole 'sponge' in arc crust? *Geology*, **35**, 787–790.

DEERING, C. D., COLE, J. W. & VOGEL, T. A. 2008. A rhyolite compositional continuum governed by lower crustal source conditions in the Taupo Volcanic Zone, New Zealand. *Journal of Petrology*, **49**, 2245–2276.

DEERING, C. D., BACHMANN, O., DUFEK, J. & GRAVLEY, D. M. 2011. Rift-related transition from andesite to rhyolite volcanism in the Taupo Volcanic Zone (New Zealand) controlled by crystal-melt dynamics in mush zones with variable mineral assemblages. *Journal of Petrology*, **52**, 2243–2263.

DEFANT, M. J. & DRUMMOND, M. S. 1990. Derivation of some modern arc magmas by melting of young subducted lithosphere. *Nature*, **347**, 662–665.

DEPAOLO, D. 1981. Trace element and isotopic effects of combined wallrock assimilation and fractional crystallisation. *Earth and Planetary Science Letters*, **53**, 189–202.

DIETRICH, V. & GAITANAKIS, P. 1995. *Geological Map of Methana Peninsula (Greece)*. ETH, Zürich, Switzerland.

DIMITRIADIS, I., KARAGIANNI, E. *ET AL.* 2009. Seismicity and active tectonics at Coloumbo Reef (Aegean Sea, Greece): monitoring an active volcano at Santorini Volcanic Center using a temporary seismic network. *Tectonophysics*, **465**, 136–149.

DRUITT, T. H., EDWARDS, L. *ET AL.* 1999. *Santorini Volcano*. The Geological Society, London, Memoir 19.

ELBURG, M. A. 2010. Sources and processes in arc magmatism: the crucial role of water. *Geologica Belgica*, **13**, 119–134.

FOLEY, S., TIEPOLO, M. & VANNUCCI, R. 2002. Growth of early continental crust controlled by melting of amphibolite in subduction zones. *Nature*, **417**, 837–840.

FRIEDRICH, W. L., KROMER, B., FRIEDRICH, M., HEINEMEIER, J., PFEIFFER, T. & TALAMO, T. 2006. Santorini eruption radiocarbon dated to 1627–1600 B.C. *Science*, **312**, 548.

GEORGE, R. M. & ROGERS, N. W. 2002. Plume dynamics beneath the African plate inferred from the geochemistry of the Tertiary basalts of southern Ethiopia. *Contributions to Mineralogy and Petrology*, **144**, 286–304.

GERTISSER, R., PREECE, K. & KELLER, J. 2009. The Plinian Lower Pumice 2 eruption, Santorini, Greece: magma evolution and volatile behaviour. *Journal of Volcanology and Geothermal Research*, **186**, 387–406.

Houseman, G. A., McKenzie, D. P. & Molnar, P. 1981. Convective instability of a thickened boundary layer and its relevance for the thermal evolution of continental convergent belts. *Geophysical Research Letters*, **86**, 6135–6155.

Huijsmans, J. P. P. 1985. *Calc-alkaline lavas from the volcanic complex of Santorini, Aegean Sea, Greece*. PhD thesis, Utrecht University.

Jolivet, L. 2001. A comparison of geodetic and finite strain pattern in the Aegean, geodynamic implications. *Earth and Planetary Science Letters*, **187**, 95–104.

Kay, R. W. & Kay, S. M. 1993. Delamination and delamination magmatism. *Tectonophysics*, **219**, 177–189.

Kelemen, P. B. 1995. Genesis of high Mg# andesites and the continental crust. *Contributions to Mineralogy and Petrology*, **120**, 1–19.

Kelemen, P. B., Hanghøj, K. & Greene, H. R. 2007. One view of the geochemistry of subduction-related magmatic arcs, with an emphasis on primitive andesite and lower crust. *In*: Holland, H. D. & Turekian, K. K. (eds) *Treatise on Geochemistry: The Crust*. Elsevier, Oxford, 1–70.

Kirchenbaur, M., Münker, C., Schuth, S., Garbe-Schonberg, D. & Marchev, P. 2012. Tectonomagmatic of Eastern Mediterranean K-rich lavas. *Journal of Petrology*, **53**, 27–65.

Larocque, J. & Canil, D. 2010. The role of amphibole in the evolution of arc magmas and crust: the case from the Jurassic Bonanza arc section, Vancouver Island, Canada. *Contributions to Mineralogy and Petrology*, **159**, 475–492.

Macpherson, C. G., Dreher, S. T. & Thirlwall, M. F. 2006. Adakites without slab melting: high pressure differentiation of island arc magma, Mindanao, the Philippines. *Earth and Planetary Science Letters*, **243**, 581–593.

Martin, H. 1986. Effect of steeper Archean geothermal gradient on geochemistry of subduction-zone magmas. *Geology*, **14**, 753–756.

Martin, H., Smithies, R. H., Rapp, R., Moyen, J.-F. & Champion, D. 2005. An overview of adakite, tonalite-trondhjemite-granodiorite (TTG) and sanukitoid: relationships and some implications for crustal evolution. *Lithos*, **79**, 1–24.

Martin, V. M., Morgan, D. J., Jerram, D. A., Caddick, M. J., Prior, D. J. & Davidson, J. P. 2008. Bang! Month-scale eruption triggering at Santorini volcano. *Science*, **321**, 1178–1178.

Mitropoulos, P. & Tarney, J. 1992. Significance of mineral compositions variations in the Aegean Island arc. *Journal of Volcanology and Geothermal Research*, **51**, 283–303.

Mitropoulos, P., Tarney, J., Saunders, A. D. & Marsh, N. G. 1987. Petrogenesis of Cenozoic volcanic rocks from the Aegean Island Arc. *Journal of Volcanology and Geothermal Research*, **32**, 177–193.

Miyashiro, A. 1974. Volcanic rock series in island arcs and active continental margins. *American Journal of Science*, **274**, 321–355.

Mortazavi, M. & Sparks, R. S. J. 2004. Origin of rhyolite and rhyodacite lavas and associated mafic inclusions of Cape Akrotiri, Santorini: the role of wet basalt in generating calc-alkaline silicic magmas. *Contributions to Mineralogy and Petrology*, **146**, 397–413.

Nagel, T. J., Hoffmann, J. E. & Münker, C. 2012. Generation of Eoarchaean tonalite-trondhjemite-granodiorite series from thickened mafic arc crust. *Geology*, **40**, 375–378.

Nicholls, I. A. 1971. Petrology of Santorini Volcano, Cyclades, Greece. *Journal of Petrology*, **12**, 67–119.

Nimis, P. 1999. Clinopyroxene geobarometry of magmatic rocks. Part 2. Structural geobarometers for basic to acid, tholeiitic and mildly alkaline systems. *Contributions to Mineralogy and Petrology*, **135**, 62–74.

Nimis, P. & Ulmer, P. 1998. Clinopyroxene geobarometry of magmatic rocks Part 1: an expanded structural geobarometer for anhydrous and hydrous, basic and ultrabasic systems. *Contributions to Mineralogy and Petrology*, **133**, 122–135.

Palme, H. & O'Neill, H. S. C. 2003. Cosmochemical estimates of mantle composition. *In*: Carlson, R. W., Holland, H. D. & Turekian, K. K. (eds) *Treatise on Geochemistry, vol.2, The Mantle and Core*. Elsevier, New York.

Papazachos, B. C., Hatzidimitriou, P. M., Panagiotopoulos, D. G. & Tsokas, G. N. 1995. Tomography of the crust and upper mantle in southeast Europe. *Journal of Geophysical Research*, **100**, 12 405–12 422.

Parat, F., Holtz, F. & Feig, S. 2008. Pre-eruptive conditions of the Huerto Andesite (Fish Canyon system, San Juan volcanic field, Colorado): influence of volatiles (COHS) on phase equilibria and mineral composition. *Journal of Petrology*, **49**, 911–935.

Peacock, S. M. 2003. Thermal structure and metamorphic evolution of subducting slabs. *In*: Eiler, J. M. (ed.) *Inside the Subduction Factory*. AGU, Washington, 7–22.

Pe-Piper, G. & Piper, D. J. W. 2002. *The Igneous Rocks of Greece. The Anatomy of an Orogen*. Schweizerbart science publishers, Stuttgart.

Pik, R., Deniel, C., Coulon, C., Yirgu, G. & Marty, B. 1999. Isotopic and trace element signatures of Ethiopian flood basalts: evidence for plume–lithosphere interactions. *Geochimica et Cosmochimica Acta*, **63**, 2263–2279.

Plank, T. 2005. Constraints from thorium/lanthanum on sediment recycling at subduction zones and the evolution of the continents. *Journal of Petrology*, **46**, 921–944.

Plank, T. & Langmuir, C. H. 1998. The chemical composition of subducting sediment and its consequences for the crust and mantle. *Chemical Geology*, **145**, 325–394.

Putirka, K., Ryerson, F. J. & Mikaelian, H. 2003. New igneous thermobarometers for mafic and evolved lava compositions, based on clinopyroxene + liquid equilibria. *American Mineralogist*, **88**, 1542–1554.

Pyle, D. M. & Elliott, J. R. 2006. Quantitative morphology, recent evolution, and future activity of the Kameni Islands volcano, Santorini, Greece. *Geosphere*, **2**, 253–268.

Rapp, R. P., Watson, E. B. & Miller, C. F. 1991. Partial melting of amphibolite/eclogite and the origin of Archean trondhjemites and tonalites. *Precambrian Research*, **51**, 1–25.

Ridolfi, F., Renzulli, A. & Puerini, M. 2010. Stability and chemical equilibrium of amphibole in calc-alkaline magmas: an overview, new thermobarometric formulations and application to subduction-related volcanoes. *Contributions to Mineralogy and Petrology*, **160**, 45–66.

RUDNICK, R. L. & GAO, S. 2003. Composition of the continental crust. *In*: HOLLAND, H. D. & TUREKIAN, K. K. (eds) *Treatise on Geochemistry: The Crust*. Elsevier, Oxford, 1–64.

SCRIVNER, A. E., VANCE, D. & ROHLING, E. J. 2004. New neodymium isotope data quantify Nile involvement in Mediterranean anoxic episodes. *Geology*, **32**, 565–568.

SISSON, T. W. & GROVE, T. L. 1993. Experimental Investigations of the role of H_2O in calc-alkaline differentiation and subduction zone magmatism. *Contributions to Mineralogy and Petrology*, **113**, 143–166.

SKARPELIS, N. & KYRIAKOPOULOS, K. 1992. Occurrence and $^{40}Ar/^{39}Ar$ dating of a granite in Thera (Santorini, Greece). *Geologische Rundschau*, **81**, 729–735.

SMET, I. In prep. *Spatial and temporal petrological-geochemical variations in the West Aegean arc: influence of local geodynamic parameters on magma genesis*. PhD thesis, Ghent University.

SMITH, P. E., YORK, D., CHEN, Y. & EVENSEN, N. M. 1996. Single crystal Ar-40–Ar-39 dating of a late Quaternary paroxysm on Kos, Greece: concordance of terrestrial and marine ages. *Geophysical Research Letters*, **23**, 3047–3050.

SMITHIES, R. H. 2000. The Archaean tonalite–trondhjemite–granodiorite (TTG) series is not an analogue of Cenozoic adakite. *Earth and Planetary Science Letters*, **182**, 115–125.

STRECK, M. J., LEEMAN, W. P. & CHESLEY, J. 2007. High-magnesian andesite from Mount Shasta: a product of magma mixing and contamination, not a primitive mantle melt. *Geology*, **35**, 351–354.

TALBOT, M. R., WILLIAMS, M. A. J. & ADAMSON, D. A. 2000. Strontium isotope evidence for late Pleistocene reestablishment of an integrated Nile drainage network. *Geology*, **28**, 343–346.

TAMURA, Y. & TATSUMI, Y. 2002. Remelting of an andesitic crust as possible origin for rhyolitic magma in oceanic arcs: an example from the Izu–Bonin arc. *Journal of Petrology*, **43**, 1029–1047.

TATARIS, A. A. 1956. The Eocene in the semimetamorphosed basement of Thera Island. *Bulletin of the Geological Society of Greece*, **III**, 232–238.

TAYLOR, S. R. & MCLENNAN, S. M. 1985, *The Continental Crust: Its Composition and Evolution*. Blackwell Scientific, Oxford.

TAYLOR, S. R. & MCLENNAN, S. M. 1995. The geochemical evolution of the continental crust. *Reviews of Geophysics*, **33**, 241–265.

TURNER, S., HAWKESWORTH, C. ET AL. 1997. $^{238}U–^{230}Th$ disequilibria, magma petrogenesis, and flux rates beneath the depleted Tonga–Kermadec island arc. *Geochimica et Cosmochimica Acta*, **61**, 4855–4884.

VAN HINSBERGEN, D. J. J., HAFKENSCHEID, E., SPAKMAN, W., MEULENKAMP, J. E. & WORTEL, R. 2005. Nappe stacking resulting from subduction of oceanic and continental lithosphere below Greece. *Geology*, **33**, 325–328.

VAN HUNEN, J., VAN KEKEN, P., HYNES, A. & DAVIES, G. F. 2008. Tectonics of early Earth: some geodynamic considerations. *In*: CONDIE, K. C. & PEASE, V. (eds) *When did Plate Tectonics Begin on Planet Earth*? Geological Society of America, Boulder, 157–171.

VANDERKLUYSEN, L., VOLENTIK, A. C. M., PRINCIPE, C., HUNZIKER, J. C. & HERNANDEZ, J. 2005. Nisyros' volcanic evolution: the growth of a stratovolcano. *In*: HUNZIKER, J. C. & MARINI, L. (eds) *The Geology, Geochemistry and Evolution of Nisyros Volcano (Greece). Implications for the Volcanic Hazards*. Section des Sciences de la Terre, Université de Lausanne, Lausanne, 100–106.

VOLENTIK, A. C. M., PRINCIPE, C., VANDERKLUYSEN, L. & HUNZIKER, J. C. 2005*a*. *Geological Map of Nisyros Volcano (Greece)*. Université de Lausanne, Lausanne.

VOLENTIK, A. C. M., PRINCIPE, C., VANDERKLUYSEN, L. & HUNZIKER, J. C. 2005*b*. Explanatory notes on the 'Geological Map of Nisyros Volcano (Greece)'. *In*: HUNZIKER, J. C. & MARINII, L. (eds) *The Geology, Geochemistry and Evolution of Nisyros Volcano (Greece). Implications for the Volcanic Hazards*. Université de Lausanne, Lausanne, 7–25.

WAGNER, G. J., STORZER, D. & KELLER, J. 1976. Spaltspurendatierungen quartärer Gesteinslgläser aus dem Mittelmeerraum. *Neues Jahrbuch für Mineralogie, Monatshefte*, **2**, 84–94.

WYERS, G. P. & BARTON, M. 1989. Polybaric evolution of calc-alkaline magmas from Nisyros, Southeastern Hellenic Arc, Greece. *Journal of Petrology*, **30**, 1–37.

ZELLMER, G. F. 2008. Some first-order observations on magma transfer from mantle wedge to upper crust at volcanic arcs. *In*: ANNEN, C. & ZELLMER, G. (eds) *Dynamics of Crustal Magma Transfer, Storage and Differentiation*. Geological Society, London, Special Publications, **304**, 15–31.

ZELLMER, G. F. & TURNER, S. P. 2007. Arc dacite genesis pathways: evidence from mafic enclaves and their hosts in Aegean lavas. *Lithos*, **95**, 346–362.

ZELLMER, G. F., IIZUKA, Y., MIYOSHI, M., TAMURA, Y. & TATSUMI, Y. 2012. Lower crustal H_2O controls the formation of adakitic melts. *Geology*, **40**, 487–490.

ZOUZIAS, D. & ST. SEYMOUR, K. 2008. Consanguineous geochemistry of the Kos Plateau and Tilos D and E Pumices, Aegean Volcanic Arc, Hellas. *Neues Jahrbuch für Mineralogie, Abhandlungen*, **184**, 231–241.

Crystal uptake into aphyric arc melts: insights from two-pyroxene pseudo-decompression paths, plagioclase hygrometry, and measurement of hydrogen in olivines from mafic volcanics of SW Japan

GEORG F. ZELLMER[1,2]*, NAOYA SAKAMOTO[3], YOSHIYUKI IIZUKA[2], MASAYA MIYOSHI[4], YOSHIHIKO TAMURA[5], HUI-HO HSIEH[2] & HISAYOSHI YURIMOTO[3]

[1]*Institute of Natural Resources, Massey University, Palmerston North 4442, New Zealand*

[2]*Institute of Earth Sciences, Academia Sinica, 128 Academia Road Sec. 2, Nankang, Taipei 11529, Taiwan, ROC*

[3]*Isotope Imaging Laboratory, Hokkaido University (HokuDai), Sapporo 060-0810, Japan*

[4]*Department of Education and Regional Studies, University of Fukui, 3-9-1 Bunkyo, Fukui City, Fukui 910-8507, Japan*

[5]*IFREE, JAMSTEC, 2-15 Natsushima, Yokosuka, Kanagawa 237-0061, Japan*

**Corresponding author (e-mail: g.f.zellmer@massey.ac.nz)*

Abstract: Minerals of mafic rocks from the SW Japan arc have been studied to deduce $P-T-X_{H2O}$ conditions and their variations in mafic arc magmas. Two-pyroxene thermobarometry of magmas from several volcanoes yields constant temperatures and variable pressures. MELTS fractional crystallization modelling is employed to show that such 'pseudo-decompression paths' (PDPs) are artefacts that derive from uptake of pyroxene antecrysts formed at a range of crustal levels by isobaric cooling of previously intruded mafic melts. It is shown that PDPs can be used to constrain oxygen fugacities and initial water contents of the intruded magmas. These constraints, and plagioclase hygrometry, indicate that initial melt H_2O contents change systematically along the SW Japan arc. Direct determination of hydrogen in olivine by secondary ion mass spectrometry yields consistently low olivine H_2O contents of 11 ± 4 ppm (1σ), with little, if any, along-arc variations. MELTS modelling indicates that crystallization of calcic plagioclase and olivine dominantly occurs during upper crustal differentiation of mafic melts. The combined data indicate that aphyric melts are released from the mantle wedge, taking up most if not all crystals from previously intruded plutonic rocks during rapid magma ascent to the surface.

Until the recent recognition that some primary arc magmas released from the mantle wedge may be of quite evolved, up to dacitic, compositions (Straub *et al.* 2011), investigating mafic arc magmas was the only feasible option to obtain information about source processes and conditions (such as melting and the origin of magmatic H_2O) at destructive plate margins. Compared to the magmas erupted at constructive plate margins, mafic arc magmas are enriched in H_2O and fluid mobile large ion lithophile elements, features commonly attributed to contributions from the dehydrating subducting slab to the mantle wedge below volcanic arcs (e.g. Gill 1981; Class *et al.* 2000; Zellmer *et al.* 2000). H_2O-bearing melts are characterized by significantly lower solidus and liquidus temperatures compared to their dry counterparts (Pertermann & Hirschmann 2003; Grove *et al.* 2006; Green *et al.* 2010), and H_2O also affects the relative stability of different mineral phases during crystallization, destabilizing plagioclase (Housh & Luhr 1991) and, at pressures of the lower crust, expanding the stability of amphibole at the expense of garnet (Müntener *et al.* 2001; Alonso-Perez *et al.* 2009). Investigating potential variations in H_2O, which may also be linked to eruptive explosivity (Cashman 2004), is, therefore, paramount in understanding differences both in the fundamental processes of melt generation and differentiation and in the eruptive style of arc volcanoes through space and time.

Another distinctive characteristic of hydrous arc magmas is their variable and complex petrographic

From: Gómez-Tuena, A., Straub, S. M. & Zellmer, G. F. (eds) 2014. *Orogenic Andesites and Crustal Growth.* Geological Society, London, Special Publications, **385**, 161–184.
First published online June 11, 2013, http://dx.doi.org/10.1144/SP385.3

features compared to dry magmas. Arc magmas may be aphyric, monomineralic, or multi-mineralic (Tatsumi & Ishizaka 1981). Their crystal cargo typically shows an abundance of disequilibrium textures, with individual crystals containing multiple resorption zones or zones of partial resorption (particularly at intermediate compositions) and distinct changes in their mineral chemistry (e.g. Zellmer *et al.* 1999; Charlier & Zellmer 2000; Zellmer & Clavero 2006; Zellmer & Turner 2007; Jerram & Martin 2008; Ramos & Tepley 2008; Streck 2008). Frequently, some mineral phases are in chemical disequilibrium with their host melt (resulting in reaction rims) and/or with other minerals within the same magma (e.g. co-occurrence of quartz and olivine, Ohta *et al.* 1992). These disequilibrium textures are commonly attributed to open-system processes, such as mixing between magmas (e.g. Koyaguchi 1986; Snyder 1997; Clynne 1999) or between magma and crystal mush (Pyle *et al.* 1988; Nakamura 1995), or uptake of crustal xenocrysts into the melt (e.g. Ohta *et al.* 1992; Stewart *et al.* 1996; Dungan 2005; Streck *et al.* 2007). Recent work has also identified the potential for some crystals to have in fact crystallized from a previous batch of magma of similar composition, generated within the same overall magmatic system, and to have been taken up into its present host melt just prior to eruption. Such 'antecrysts' (a term coined by Wes Hidreth in 2003) may appear to be in chemical equilibrium with their host melt, but may be in isotopic disequilibrium (Davidson *et al.* 2007; Jerram & Martin 2008). Clearly, complex crystal zoning patterns and disequilibria need to be carefully evaluated in all mafic arc magmas, particularly if the chemistry of minerals and melts is to be used to put constraints on melt H_2O contents and crystallization pressures and temperatures through mineral hygrometry (Putirka 2008; Lange *et al.* 2009) and thermobarometry (Putirka 2008), respectively.

In this contribution, the crystal cargo of mafic samples from the volcanically active arc front volcanoes of SW Japan is investigated in conjunction with thermodynamic crystallization modelling (Ghiorso & Sack 1995; Smith & Asimow 2005) to elucidate the origin of the crystals and to characterize along-arc changes in the H_2O content of the melts and the effect of such changes on magmatic evolution.

Geological background

In SW Japan, characterized by a normal crustal thickness of *c.* 35 km (Oda & Ushio 2007; Ueno *et al.* 2008), the subducting Philippine Sea slab consists of old (37–115 Ma) West Philippine Basin lithosphere subducting steeply under Kyushu, and younger (15–26 Ma), more buoyant Shikoku Basin lithosphere subducting shallowly beneath western Honshu (cf. Miyoshi *et al.* 2010). In western Honshu, the slab terminates with its leading edge at a depth of about 200 km, just behind the active volcanic arc (Kimura *et al.* 2005; Abdelwahed & Zhao 2007). Based on the occurrence of deep long-period tremors, the young slab is thought to dehydrate beneath Shikoku in the forearc (Obara 2002). Figure 1 represents a simplified map of the region, indicating slab surface isodepth contours and the location of the volcanic centres, from which the mafic samples of the present study originate. Sample details are provided in Table 1. Morphologically, the volcanic arc changes northeastwards from the large, stratocone and caldera volcanoes in southern and central Kyushu to typically (with the exception of Daisen) much smaller monogenetic edifices in northern Kyushu and western Honshu. Some mafic magmas from SW Japan have recently been investigated for pyroxene thermometry (Lindsley 1983) and plagioclase hygrometry (Lange *et al.* 2009), indicating approximately constant lower crustal temperatures of *c.* 1080 °C, but decreasing H_2O contents from Aso volcano in central Kyushu to Daisen volcano in western Honshu (Zellmer *et al.* 2012*b*). These observations were used to explain the along-arc geochemical changes in intermediate eruptive products towards adakitic compositions, which dominate in western Honshu (Morris 1995; Zellmer *et al.* 2012*b*). The present study provides further details on the major oxide bulk rock and mineral chemistry of the mafic samples, extends the previously published data southwards to Kirishima and eastwards to Kannabe, and discusses thermobarometric and hygrometric constraints in conjunction with alphaMELTS thermodynamic modelling. We also provide the first ion microprobe data on H_2O content in volcanic olivines from Japan, in an attempt to use these as a proxy for the H_2O content of the mafic melts from which they crystallized.

Analytical techniques

XRF major oxide analyses of whole rocks

Powder samples of whole rocks were prepared using a jaw crusher and an agate ball mill at the Institute of Earth Sciences, Academia Sinica. Rock powders were fused to make glass beads and analysed for major oxides by X-ray fluorescence using a Rigaku RIX 2000 spectrometer at the Department of Geosciences, National Taiwan University. The analytical procedures were the same as those described by Lee *et al.* (1997) and Wang

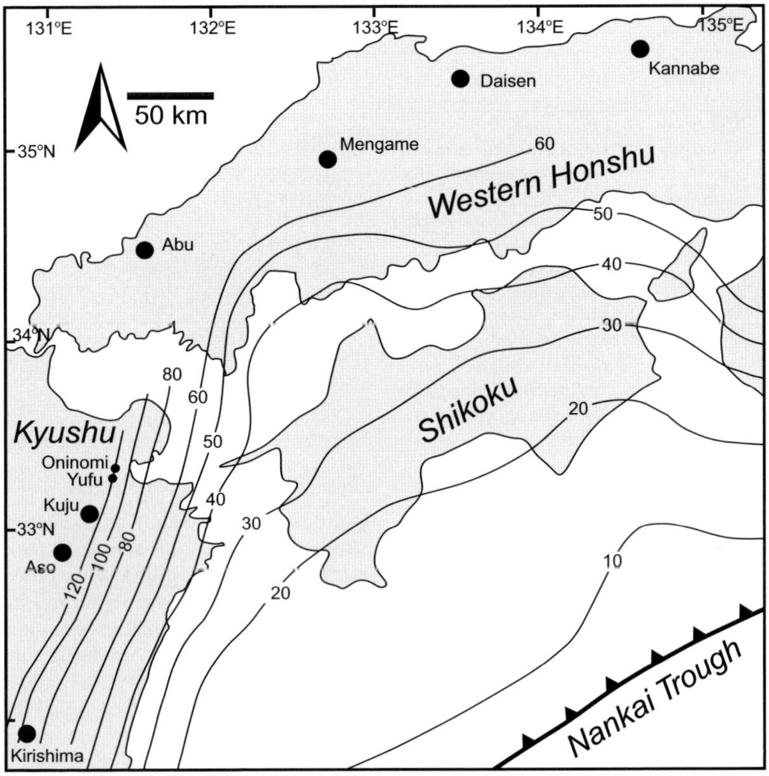

Fig. 1. Regional location map of SW Japan volcanoes referred to in the text. The axis of the Nankai Trough is indicated by a barbed line. Contour lines represent isodepth (in km) contours of the upper surface of the subducted Philippine Sea plate, estimated from the results by Baba *et al.* (2002), Hirose *et al.* (2008) and Nakajima & Hasegawa (2007), accurate to better than ± 5 km. In Western Honshu, subduction beyond ~60 km depth occurs aseismically (Abdelwahed & Zhao 2007).

et al. (2004), yielding analytical uncertainties of better than $\pm 5\%$ (2σ) for all major elements.

Analysis of mineral and groundmass compositions by electron probe micro-analysis (EPMA)

Chemical compositions of phenocrysts, glasses and groundmass were obtained with a JEOL electron probe (W-EPMA: JXA-8900R) and a JEOL field emission electron probe (FE-EPMA: JXA-8500F), equipped with four and five wavelength dispersive spectrometers, respectively, at the Institute of Earth Sciences, Academia Sinica. Secondary- and back-scattered electron images were used to guide the analysis on target positions of minerals. Primary beam conditions of quantitative analyses were as follows. The W-EPMA was operated with an acceleration voltage of 15 kV. Minerals and glass were analysed using a beam current of 12 nA and a beam diameter of 2 μm, while groundmass bulk composition was determined by averaging between *c.* 100 and 300 analyses per sample, obtained with a beam current of 20 nA and a beam diameter of 20 μm. The FE-EPMA was operated with an acceleration voltage of 12 kV. Minerals and glass were analysed using beam currents of 6 and 3 nA, respectively, and a beam diameter of 2 μm. On both instruments, the measured X-ray intensities were corrected by the ZAF method using the standard calibration of chemical-known standard minerals. Counting times for each element and both upper and lower baselines were 10 s and 5 s, respectively. Standards run as unknowns yielded relative standard deviations of less than 1% for Si, Na and K, and less than 0.5% for other elements. Detection limits were less than 500 ppm for all elements.

Analysis of H_2O in olivines by secondary ion mass spectrometry (SIMS)

Following identification of suitable olivine grains by electron microscopy, samples were prepared for ion microbeam analysis using a Cameca

Table 1. *Sample details*

Sample	Volcano	Location	Latitude (N)	Longitude (E)	Sample type
09KIR01	Kirishima	Ohachidake	31°52′45.2″	130°53′48.6″	Lava flow
09KIR02	Kirishima	Karakunidake	31°57′13.7″	130°52′15.8″	Lava flow
09ASO01	Aso	Kami-Komezuka cone	32°53′53.3″	131°02′47.5″	Scoria block
09KUJ01A	Kuju	Hiijidake	33°06′41.4″	131°16′39.0″	Lava block
09YUF01	Yufu	*River bed on north slope*	33°17′31″	131°23′55″	Mafic enclave
09ONI02	Oninomi	*Next to motorway*	33°18′59″	131°24′34″	Lava flow
09ABU01	Abu	Mt Fusumayama	34°26′46.9″	131°35′58.3″	Lava block
09ABU02	Abu	Mt Fusumayama	34°27′4.8″	131°36′5.3″	Scoria pieces
09ABU03	Abu	Mt Kasayama	34°26′50.9″	131°23′52.4″	Lava flow
09MEN01	Mengame	*Next to road*	34°56′14.8″	132°43′6.3″	Lava flow
09DAI01	Daisen	*same as sample 56 from Tamura* et al. *(2000)*	35°20′38.9″	133°25′37.3″	Lava flow
09KAN04	Kannabe	Kannabe scoria cone	35°30′12.5″	134°40′25.3″	Scoria block

IMS-6f ion probe at the Isotope Imaging Laboratory of the University of Hokkaido. Polished thick sections of volcanic samples were cut into *c.* 7 mm × 7 mm × 2 mm blocks in order to mount all samples in the same sample holder for measurement during the same analysis session without requiring sample exchange. A 50 nm thick gold coat was applied to all samples. Gold-coated standard olivines were mounted in the same sample holder. The small size of each sample restricted the number of olivine grains that could be analysed to three to six crystals per sample, ranging in size from 100–200 μm to 400–1000 μm, depending on the sample. An exception was one symplectic olivine grain from Aso with >2 mm diameter.

To determine the hydrogen ($^{1}H^{-}$) contents within the cores of selected olivine grains, six samples and three standards (SC1, SC3 and SC4 from San Carlos, cf. Kurosawa *et al.* 1992) were mounted in a 9-hole holder with 6 mm × 6 mm windows, and pumped down in the airlock at a pressure of 10^{-8} Torr, remaining at vacuum for >7 days before analyses in order to remove terrestrial adsorbed water. During the analysis, a liquid nitrogen cold trap placed near the samples was used to reduce the background for secondary ions, by maintaining pressure at less than 4×10^{-9} Torr in the sample chamber.

A $^{133}Cs^{+}$ primary ion beam of 14.5 keV was focused to a beam spot of *c.* 20 μm diameter on the surface of the sample with a beam current of *c.* 40 nA. In order to avoid crater edge effects and any remaining adsorbed water on the surface, a relatively large primary beam and a 150 μm contrast aperture were used, and secondary ions generated from the centre of the primary irradiation area of *c.* 10 μm diameter were collected using a field aperture. The intense primary beam used is effective in improving the detection limit of sample hydrogen during SIMS analysis (Yurimoto *et al.* 1989). We carefully selected analysis positions using stigmatic imaging during pre-sputtering with 20 μm × 20 μm beam rastering for 30–60 s, to avoid water contamination stemming from grain boundaries, cracks and hydrous micro-inclusions. We occasionally observed a sudden increase of the hydrogen secondary ion signal during measurement. This was typically due to water contamination at depth during the excavation of the sputtered crater. In these cases, we immediately evaluated the reason for the sudden increase in secondary hydrogen intensity by stigmatic imaging and rejected the disturbed data. Thus, *c.* 20% of a total of 76 analyses were discarded, ranging from less than 8% in samples with few microinclusions to almost 40% in samples with many inclusions. A normal incident electron gun was used to compensate for positive charging of the sputtered region due to the primary beam. Secondary ions of $^{1}H^{-}$ and $^{30}Si^{-}$ were collected sequentially in an electron multiplier for 2 s and 1 s, respectively. The sequence was performed for 20 cycles and averaged to calculate the $^{1}H/^{30}Si$ ratio of the ablated material.

H_2O contents were corrected against three olivine standards (SC1, SC3 and SC4 from San Carlos) of known compositions previously determined by SIMS (Kurosawa *et al.* 1992). Standards SC1, SC3 and SC4 contain, respectively, 50 wt ppm H_2O, 44 wt ppm H_2O, and 6 wt ppm H_2O. The typical ion intensities from standard SC4 were *c.* 100 cps and *c.* 10^6 cps for ^{1}H and ^{30}Si, respectively. H_2O contents of olivine samples were determined using daily calibration curves obtained from the standard olivine grains, which were linear and went through the origin. Standard reproducibility over the course of a day ranged between *c.* ±10% (1σ) for SC1 and *c.* ±25% (1σ) for SC4.

Results

Major oxide chemistry of bulk rocks and groundmass

Whole rock data determined by XRF and groundmass data analysed by EPMA is provided in Table 2 for the fresh, unaltered mafic SW Japan samples. With the exception of the andesite sample from Kuju volcano, all studied rocks are basalts or basaltic andesites. They are of low-K, calc-alkalic to calcic compositions (Peacock 1931; Peccerillo & Taylor 1976), and fall squarely into the calc-alkaline field of Kuno (1968). These samples therefore are in many ways typical for mafic eruption products at arc front volcanoes.

Relative to the bulk rocks, groundmass compositions are typically enriched in SiO_2 and the alkalies, but depleted in MgO, MnO, FeO and TiO_2, while Al_2O_3 and CaO are either mildly enriched or depleted. This indicates that mafic mineral phases dominate the crystal cargo (see below). Note that sample 09YUF01, a mafic enclave from Yufu volcano, and sample 09ABU02 were, respectively, too coarsly crystalline and too vesicular to obtain groundmass analyses.

Petrography and mineral chemistry

The petrography of mafic to intermediate volcanic samples from the SW Japan arc has been previously described (Furuyama 1976; Koyaguchi 1986; Matsuura 1986; Ohta *et al.* 1990, 1992; Imura 1994; Kamata 1997; Tamura *et al.* 2000; Imura & Kobayashi 2001; Harada 2003; Miyoshi *et al.* 2005, 2012). A detailed account of crystal zoning textures within individual samples is beyond the scope of the present along-arc study. It suffices to say that most volcanoes display the complex zoning and mineral-melt disequilibria that are characteristic for arc magmas, and that samples range from monomineralic to multi-mineralic with typically 10 to >30% crystals set in a fine-grained groundmass. Some notable features include patchy zoning and corrosion of plagioclase crystals in most samples, the presence of symplectic olivine with abundant exsolution of magnetite at Aso, the co-occurrence of olivine and quartz in basalts from Oninomi, and olivines as a single 'phenocryst' phase in some Daisen basalts, both in and out of Fe/Mg equilibrium with their host melt. Table 3 provides a summary of the mineral chemistry of major phases from representative mafic rock samples studied here, with the number of analyses (*n*) indicated for each mineral phase, and Figure 2 shows pyroxene compositions in samples that contain both clinopyroxene and orthopyroxene.

H_2O in olivine

H_2O contents of olivine cores are given in Table 4 together with their forsterite contents

Table 2. *Bulk rock and groundmass major oxide chemistry*

Sample	wr/gm	SiO_2	TiO_2	Al_2O_3	FeO^T	MgO	MnO	CaO	Na_2O	K_2O	P_2O_5	*n*
09KIR01	wr	53.66	0.68	20.81	6.61	4.13	0.16	10.87	2.05	0.87	0.11	
	gm	59.24	0.79	16.83	7.02	2.89	0.10	7.19	3.60	1.53	n. d.	116
09KIR02	wr	51.67	0.84	17.78	8.62	6.94	0.18	11.48	1.66	0.71	0.11	
	gm	54.69	0.79	18.00	7.88	4.54	0.14	9.08	2.83	1.19	n. d.	287
09ASO01	wr	51.65	0.92	19.36	8.66	5.27	0.17	10.35	2.24	1.17	0.21	
	gm	55.82	0.55	17.47	6.72	4.10	0.14	8.20	3.26	1.52	n. d.	164
09KUJ01A	wr	60.96	0.68	18.90	5.29	2.32	0.17	6.26	3.03	2.10	0.29	
	gm	67.14	0.41	15.79	2.45	0.46	0.06	3.72	3.61	3.29	n. d.	137
09YUF01	wr	54.81	0.95	18.93	6.90	4.77	0.14	9.56	2.76	1.00	0.17	
09ONI02	wr	54.41	1.94	16.89	8.10	5.38	0.16	8.13	2.99	1.51	0.49	
	gm	59.22	1.22	18.95	3.90	1.86	0.06	6.93	4.40	1.91	n. d.	170
09ABU01	wr	49.44	1.59	17.04	8.28	8.74	0.16	10.60	2.54	1.14	0.47	
	gm	52.36	1.28	17.66	6.12	5.34	0.09	11.38	3.55	1.25	n. d.	113
09ABU02	wr	46.40	1.77	19.46	9.92	10.48	0.19	9.30	1.39	0.68	0.41	
09ABU03	wr	56.95	1.63	15.77	7.27	5.65	0.13	7.75	3.03	1.42	0.39	
	gm	58.41	0.76	17.62	4.12	3.05	0.06	8.12	4.06	1.54	n. d.	208
09MEN01	wr	51.86	1.02	17.51	7.03	9.37	0.15	8.63	2.83	1.30	0.29	
	gm	53.23	0.74	18.27	5.36	5.63	0.10	9.64	3.72	1.39	n. d.	216
09DAI01	wr	51.45	1.24	17.38	7.61	8.39	0.15	9.24	3.03	1.19	0.30	
	gm	54.48	1.14	18.72	4.41	4.14	0.08	9.43	4.29	1.68	n. d.	152
09KAN04	wr	48.52	1.86	18.11	10.17	7.48	0.20	9.22	3.22	0.76	0.45	
	gm	49.42	1.56	18.08	9.18	5.66	0.15	9.74	3.87	0.79	n. d.	161

n, number of averaged EPMA groundmass analyses to constrain groundmass composition.

Table 3. *Summary of mineral chemistry for selected samples from SW Japan*

Sample	Crystal cargo	Compositional range	*n*	Groundmass crystals (typically <20 μm*)	Compositional range	*n*
09KIR01	ol	Fo_{64-76}	111	plag	An_{46-90}	159
	cpx	$En_{41-49}Fs_{10-21}Wo_{36-46}$	56	ol, opx, cpx, oxides	n.d.	
	opx	$En_{63-74}Fs_{23-34}Wo_{2-6}$	202			
	plag	An_{43-92}	431			
09ASO01	symplectic ol/mt	Fo_{50-99} (Fo_{59-68} & Fo_{93-99} peaks)	72	plag	An_{50-73}	59
	cpx	$En_{38-50}Fs_{11-21}Wo_{35-40}$	157		(single grain at An_{89})	
	opx	$En_{63-70}Fs_{26-33}Wo_4$	38	cpx, minor ol, oxides	n.d.	
	plag	An_{50-93}	141			
09KUJ01	cpx	$En_{38-46}Fs_{14-20}Wo_{37-44}$	41	plag	An_{40-62}	38
	opx	$En_{53-70}Fs_{26-43}Wo_{2-7}$	145	cpx, opx, oxides	n.d.	
	plag	An_{83-90} & An_{50-68} (bimodal)	165			
09YUF01	ol	Fo_{72-84}	35	plag	An_{34-85}	118
(enclave)	cpx	$En_{40-48}Fs_{8-16}Wo_{40-48}$	93	opx	$En_{65-73}Fs_{24-33}Wo_{2-4}$	3
		(single grain at $En_{37}Fs_{19}Wo_{44}$)		ol, cpx, oxides	n.d.	
	plag	An_{40-89}	74			
	minor amph	n.d.				
09ONI02	plag	An_{38-85}	454	fsp	$An_{20-70}Ab_{29-67}Or_{1-13}$	118
	ol	Fo_{70-84}	249	cpx	$En_{38-49}Fs_{12-22}Wo_{31-46}$	60
	minor qz			opx	$En_{66-67}Fs_{28-32}Wo_{2-5}$	3
				minor ol, qz, oxides	n.d.	
09ABU01 &	ol	Fo_{50-88}	357	plag	An_{35-77}	58
09ABU02	cpx	$En_{45-50}Fs_{11-16}Wo_{45-50}$	29	ol, cpx, oxides	n.d.	
(combined)	plag	An_{51-82} (single grain at An_{37})	113	minor ksp	n.d.	
09ABU03	opx	$En_{61-81}Fs_{15-36}Wo_{3-5}$	49	cpx	$En_{46-50}Fs_{11-16}Wo_{36-41}$	
				opx, oxides	n.d.	
09MEN01	ol	Fo_{62-91}, cores typically $Fo_{>80}$	134	fsp	$An_{5-71}Ab_{27-68}Or_{1-46}$	38
	minor fsp	$An_{14-51}Ab_{47-56}Or_{2-25}$	52	ol, cpx, phl	n.d.	
09DAI01	ol	Fo_{65-89}, cores typically $Fo_{>80}$	179	fsp	$An_{6-66}Ab_{33-67}Or_{1-44}$	103
				ol, cpx, oxides	n.d.	
09KAN04	ol	Fo_{72-85}	168	plag	An_{51-68}	93
	plag	$An_{29-82}Ab_{17-70}Or_{0-3}$	117	ol		

*In shortest dimension.

Abbreviations: ol, olivine; cpx, clinopyroxene; opx, orthopyroxene; plag, plagioclase; amph, amphibole; qz, quartz; fsp, feldspar; ksp, orthoclase; phl, phlogopite; n.d., not determined.

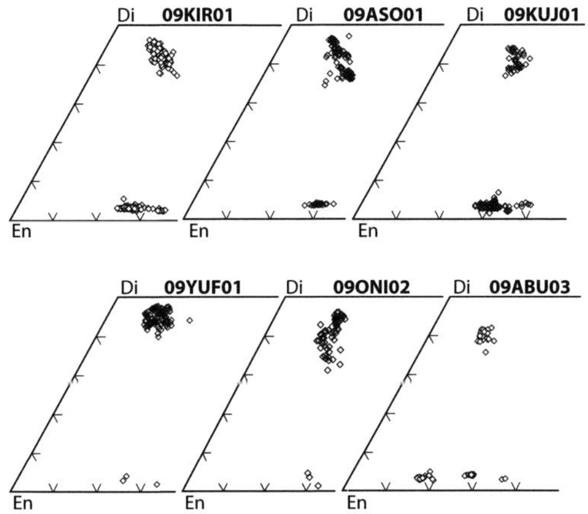

Fig. 2. Compositions of pyroxenes from selected samples that contain both clinopyroxene and orthopyroxene.

($0.80 < X_{Fo} < 0.90$). With the exception of sample 09ASO01, where only one large olivine grain was available, we analysed three to six olivine crystals per sample, some repeatedly, from a total of six samples (one from each volcano: Aso, Yufu, Oninomi, Abu, Mengame, and Daisen). In total, we obtained over 60 analyses, or eight to twelve from each sample. Figure 3 shows average H_2O contents of those crystals that were analysed between two and six times, with uncertainties representing $1\sigma_m$ standard errors, and H_2O contents of those crystals that were analysed only once, with 25% standard errors based on standard reproducibility. In the Aso sample, with only one large olivine crystal, we probed two spots within the core and one closer to the rim, avoiding zones of symplectite. H_2O contents are ubiquitously low, varying between *c.* 5 and *c.* 20 ppm, with standard errors typically of the order of ± 5 ppm or less. Evidently, crystals from individual rock samples are indistinguishable in their H_2O contents. Further, olivines from different volcanoes are identical in their H_2O content at the 95% confidence limit, perhaps with the exception of Abu (>10 ppm H_2O) and Mengame (mostly <10 ppm H_2O). We note that during analysis of Abu samples, hydrous microinclusions were often encountered, and although analyses with spikes in the hydrogen signal were discarded, it cannot be precluded that the analysis of Abu olivines (and possibly some other samples) suffered from contamination by very small microinclusions that were not obvious as irregularities in the hydrogen signal. Given the uncertainties, the volcanic olivines from SW Japan are best considered as a single group, yielding ubiquitously low H_2O contents of *c.* 11 ± 4 ppm (1σ).

Table 4. *H_2O concentrations in volcanic olivines*

Sample	Crystal #	X_{Fo}	Analysis #	H_2O (ppm)	Mean	Std error
09ASO01	A1	0.81	1	5	7.7	2.1
			2	12		
			3	6		
	A2	0.81	1	14	8	2
			2	6		
			3	7		
			4	5		
	A3	0.88	1	13	12	3.4
			2	7		
			3	21		
			4	7		
09YUF01	A	0.82	1	12	13.1	1.4
			2	16		
			3	12		
	B	0.82	1	8	9.7	2.9
			2	6		
			3	18		
			4	6		
	C	0.82	1	19*		
09ONI02	A	0.82	1	13	10.3	1.4
			2	8		
			3	12		
			4	8		
	B	0.81	1	18	10.5	2.5
			2	7		
			3	10		
			4	7		
	C	0.81	1	8	12	1.8
			2	14		
			3	10		
			4	16		
09ABU02	A	0.87	1	14	16.7	1.3
			2	16		
			3	13		
	B	0.86	1	13*		
	C	0.87	1	24	19.3	1.8
			2	16		
			3	21		
			4	17		
09MEN01	A	0.89	1	6	7	0.7
			2	8		
	B	0.87	1	17*		
	C	0.86	1	6*		
	D	0.89	1	3	6.1	0.8
			2	7		
			3	8		
			4	6		
			5	7		
			6	6		
	E	n. d.	1	6*		
	F	0.86	1	5*		
09DAI01	A	0.86	1	7	9.9	1.3
			2	13		
			3	11		
			4	8		
	B	0.78	1	15*		
	C	0.86	1	11	10.5	0.9
			2	10		
	D	0.88	1	12	11	1
			2	10		
	E	0.85	1	12*		

*Single analysis with ±25% (1σ) uncertainty based on daily standard reproducibility.
n.d., not determined.

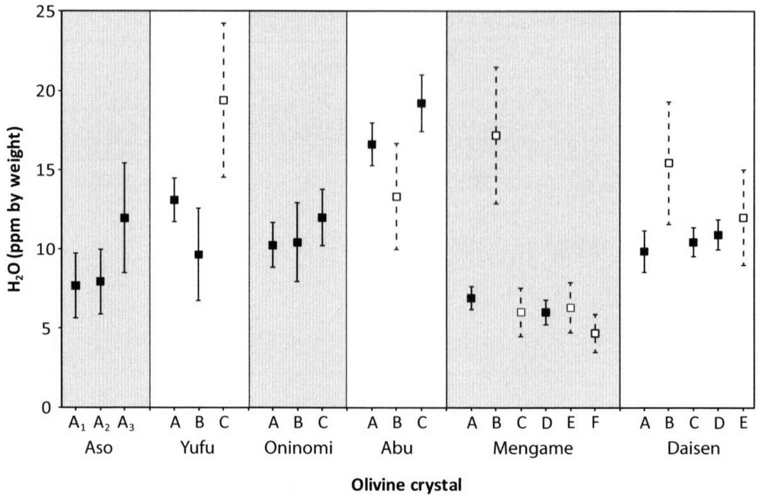

Fig. 3. Averages (solid squares) with σ_m standard errors of H_2O contents in repeatedly ($2 \leq n \leq 6$) analysed olivine crystal cores and one crystal rim (Aso A_3) from six along-arc mafic volcanic rocks from SW Japan. Open squares are crystal cores analysed once only, with $\pm 25\%$ error bars (dashed lines) based on standard reproducibility. Background shading is used to help visualize different eruption locations.

Discussion

In the following, each of the most abundant major mineral phases (pyroxenes, plagioclase and olivine) in mafic SW Japan arc volcanics is considered in turn, providing insights into the conditions of magmatic evolution and the origin of the crystal cargo. For pyroxenes, it is shown that thermobarometric results are defective. Pyroxene pseudo-decompression paths (PDPs) are introduced, and alphaMELTS thermodynamic modelling is used to show how they may elucidate oxygen fugacities and along-arc variations in water content. PDPs also provide evidence against closed-system polybaric magma evolution, and show that crystal uptake into aphyric melts is an important process. Conventional magma mixing models cannot explain the observed range in pyroxene chemistry in these mafic magmas. For plagioclase, hygrometry is used to assess along-arc variations in H_2O content, extending previous work (Zellmer *et al.* 2012*b*) southwards to Kirishima and northeastwards to Kannabe. Evidence is provided for plagioclase crystal uptake into the melt. Again, magma mixing processes appear insufficient to explain the range in plagioclase crystal chemistry. For olivine, the measured H_2O contents are discussed, and again, evidence for crystal uptake into initially aphyric melts is provided. Finally, we discuss the conditions of mafic magma ascent and storage at volcanic arcs on the basis of the evidence from SW Japan, concluding that rapid melt transfer from source to surface with minimal crustal storage times is indicated for mafic arc melts.

Pyroxenes, their implications for conditions of magmatic evolution, and the origin of the pyroxene crystal cargo: introducing pseudo-decompression paths

Approach. As there is ample evidence for crustal assimilation in mafic rocks from SW Japan (Ohta *et al.* 1992; Hunter 1998), contamination of the melt phase of the studied samples was considered likely. Therefore, the two-pyroxene thermobarometer of Putirka (2008) was chosen in an attempt to constrain pressures and temperatures of crystallization, because in this approach P–T constraints are independent of melt composition. Hundreds of EPMA analyses of minerals from the mafic samples studied yielded both orthopyroxenes and clinopyroxenes with significant compositional variations within samples from Kirishima, Aso, Kuju, Yufu, Oninomi, and Abu. Orthopyroxene was not found in samples from Mengame and Daisen, and no pyroxenes were found in the sample from Kannabe (cf. Table 3).

The approach to P–T determination is illustrated in the following using the example of Aso basalt 09ASO01. Each orthopyroxene was combined with all clinopyroxenes that could be considered to be in Fe–Mg exchange equilibrium with that orthopyroxene on chemical grounds [$0.95 \leq K_D(Fe-Mg)^{cpx-opx} = (X_{Fe}^{cpx}/X_{Mg}^{cpx})/(X_{Fe}^{opx}/X_{Mg}^{opx}) \leq 1.23$, cf. Putirka 2008]. Temperatures were computed using equation (36) of Putirka (2008), appropriate for the SW Japan clinopyroxene compositions. Pressures were derived using the T-independent barometer (equation (38) of Putirka 2008). We excluded pyroxene pairs where the pressure estimate of the T-independent barometer was different from the T-dependent pressure estimate (equation (39) of Putirka 2008) at the 1σ significance level (i.e. >4.6 kbar difference). Notably, a large number of clinopyroxenes are typically in equilibrium with each orthopyroxene, and these crystal pairs form individual $P-T$ trends with constant slopes of about 35 bars K^{-1} (Fig. 4a). Average P and T were computed for each of these individual $P-T$ trends. The cross in Figure 4 indicates the maximum standard deviation in $P-T$ for all of these $P-T$ trends (± 0.7 kbar and ± 28.5 K, strongly correlated). Note that at the 2σ level, these uncertainties are comparable to the accuracy of the thermometer (± 56 K), but are significantly smaller than the quoted accuracy of the barometer (± 3.7 kbar), cf. Putirka (2008).

Using the same approach for all samples for which two pyroxenes were available, we obtained only few $P-T$ estimates for Yufu and Oninomi basalts, but a fairly large number of $P-T$ estimates for the mafic rocks from Kirishima, Aso and Abu (Fig. 4b). The latter samples appear to display isothermal decompression paths, assuming that the relative pressure variations recorded by different pyroxene pairs within each sample are real. In the following, we will attempt to evaluate the reliability of these $P-T$-paths and interpret their meaning in terms of magma ascent and crystallization dynamics.

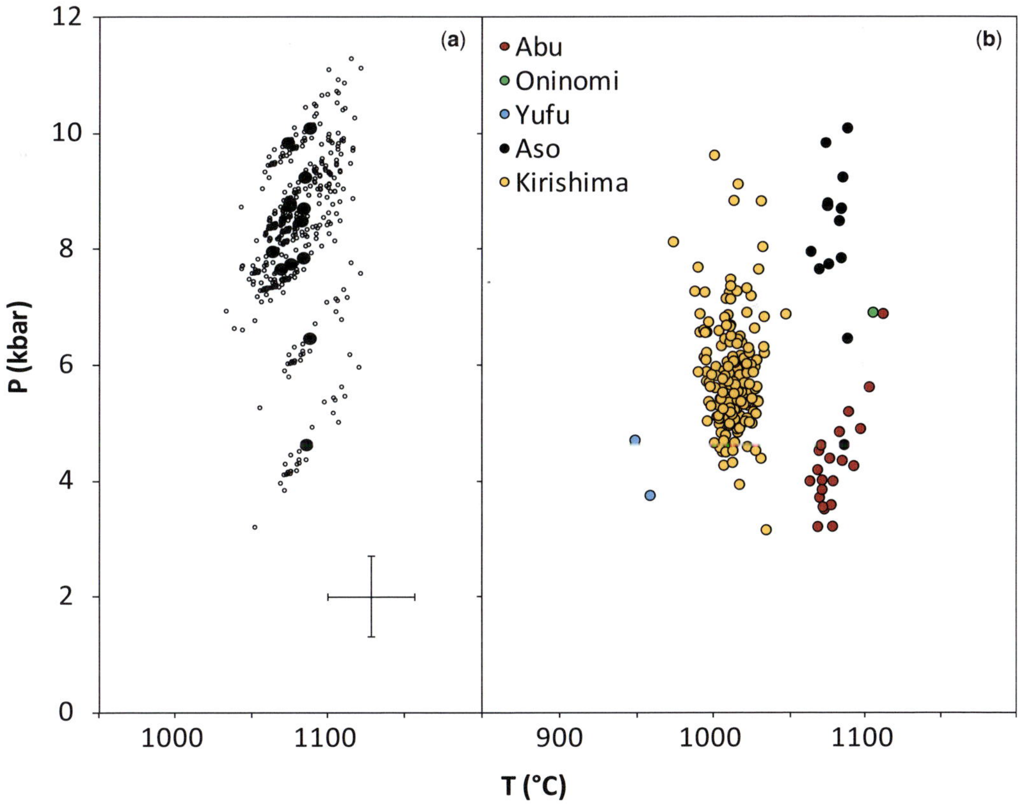

Fig. 4. Pressure–temperature conditions estimated from orthopyroxene–clinopyroxene pairs in Fe–Mg equilibrium, as derived from the Putirka (2008) thermobarometer. (**a**) In sample 09ASO01, each orthopyroxene is in equilibrium with a number of clinopyroxenes, yielding several trends of positive slope (small circles). Large symbols are the $P-T$ averages from each of these trends. The cross represents one standard deviation for each of these $P-T$ estimates. (**b**) $P-T$ estimates for all studied SW Japan samples that contain orthopyroxenes and clinopyroxenes. Kirishima, Aso and Abu yield large pressure variations at roughly constant temperatures. These trends are referred to as pseudo-decompression paths (PDPs, see text for discussion).

AlphaMELTS crystal fractionation modelling and PDPs. A recent release of the MELTS (Ghiorso & Sack 1995) thermodynamic algorithm, alphaMELTS Version 1.2 (29 February 2012), built on the original Adiabat_1ph front end of Smith & Asimow (2005), was used to assess pyroxene crystallization pressures and temperatures by forward modelling of crystal fractionation. Given the apparent roughly isothermal decompression paths, alphaMELTS was first run in MELTS isothermal and isentropic modes, to model isothermal and adiabatic decompression, respectively, from 10 kbar to the surface, in order to test if rapidly rising melts would crystallize during their ascent to the surface. Bulk rock compositions with 0.3 to 8 wt% initial H_2O content were used as starting compositions, with crystallization occurring during decompression and associated degassing of H_2O at 1010 °C (Kirishima) or 1080 °C (Aso and Abu). Depending on H_2O content and bulk composition, subliquidus conditions at 10 kbar yielded garnet ± clinopyroxene ± feldspar in the initial melt. However, during decompression, minor amounts of plagioclase and/or spinel-structured oxides (Ti–magnetite) ± olivine are the only phases crystallizing above 2 kbar in all model runs. Evidently, clinopyroxene and orthopyroxene are not stable at the pressures and temperatures suggested by the thermobarometer.

It therefore appears that the suggested isothermal decompression paths are some sort of an artefact, resulting from pairing of clinopyroxenes and orthopyroxenes that formed under different conditions (but happen to be in Fe–Mg exchange equilibrium). To explore this further, alphaMELTS was run in MELTS isobaric cooling mode (in 10 K steps) from superliquidus conditions (1300 °C) down to 900 °C at pressures of 1 kbar and 2–10 kbar (in 2 kbar intervals). This assumes a two-step model: first, crystallization of various magma batches occurs at different crustal levels; second, phenocrysts are then scavenged by later ascending aphyric melts that pass to the surface. Typically, clinopyroxenes formed at all pressures investigated, while in hydrous compositions orthopyroxenes were restricted to mid- to upper crustal levels. Further, when orthopyroxenes formed, they typically did not crystallize at the same temperatures as clinopyroxenes. To model uptake of previously formed pyroxenes from a range of crustal pressures into the melt, all pyroxene crystals formed in this way were paired with each other, filtered for Fe–Mg exchange disequilibrium, and evaluated for P–T conditions using the thermobarometer in the same manner as described for the rock samples in the previous section. Results are shown in Figure 5a for the 09ASO01 bulk composition and a range of initial water contents and oxygen fugacities. The thermobarometer yields P–T paths similar to those observed for our samples under a variety of conditions. In the following, these P–T paths will be referred to as PDPs because they give the illusion of rapid melt ascent at constant temperature, whereas in reality they may have resulted from a very different process. Evidently, PDPs do not represent pressure and temperature conditions under which the pyroxene crystals formed. However, alphaMELTS modelling may be valuable in the generation of different PDPs by forward modelling in order to assess the conditions under which polybaric uptake of antecrysts formed by isobaric cooling occurred.

As evident from Figure 5a, initial H_2O contents appear to affect the pseudo-temperatures of PDPs, with increasing H_2O generally resulting in decreasing pseudo-temperatures. Further, at fairly low H_2O contents (2 wt%) the total pseudo-pressure range displayed by the PDPs appears to increase through an expansion of maximum pseudo-pressures to upper mantle values (despite crystallization modelled to occur only at crustal levels). Oxygen fugacities also have a strong effect on pseudo-pressure ranges and some effect on pseudo-temperatures: Figure 5a shows PDPs at oxygen fugacities buffered at the nickel–nickel oxide (NNO) buffer and one and two log-units above (NNO + 1, NNO + 2). Notably, orthopyroxenes only form above the quartz–fayalite–magnetite (QFM ≅ NNO-0.7) buffer. At the NNO buffer, PDPs typically display large pseudo-pressure ranges of 5 kbar or more. Increasing oxygen fugacities result in decreasing pseudo-pressure ranges for crystallization of hydrous magmas, and in overall decreasing pseudo-pressures. Note that at high oxygen fugacities, the number of orthopyroxenes in equilibrium with clinopyroxenes typically decreases significantly, and in hydrous conditions PDPs essentially contract to a single pseudo-P–T value (given the uncertainties) at NNO+2 and above (Fig. 5a).

Polybaric uptake of previously formed crystals v. polybaric melt evolution. The range of isobaric crystallization pressures required to produce large pseudo-pressure ranges is assessed on Figure 5b–d using polybaric crystallization of sample 09ASO01 at 5 wt% initial H_2O and the NNO buffer as an example. Under these conditions, orthopyroxene is only stable at pressures of <4 kbar. We consider three scenarios: (i) uptake of orthopyroxene antecrysts crystallized at 1 kbar, combined with clinopyroxene crystallized at 1 kbar and at one mid- to lower crustal level (4, 6, 8 or 10 kbar, cf. Fig. 5b); (ii) uptake of orthopyroxene antecrysts crystallized at 2 kbar, combined with clinopyroxene crystallized at 2 kbar and at one mid- to lower crustal level (4, 6, 8 or 10 kbar, cf.

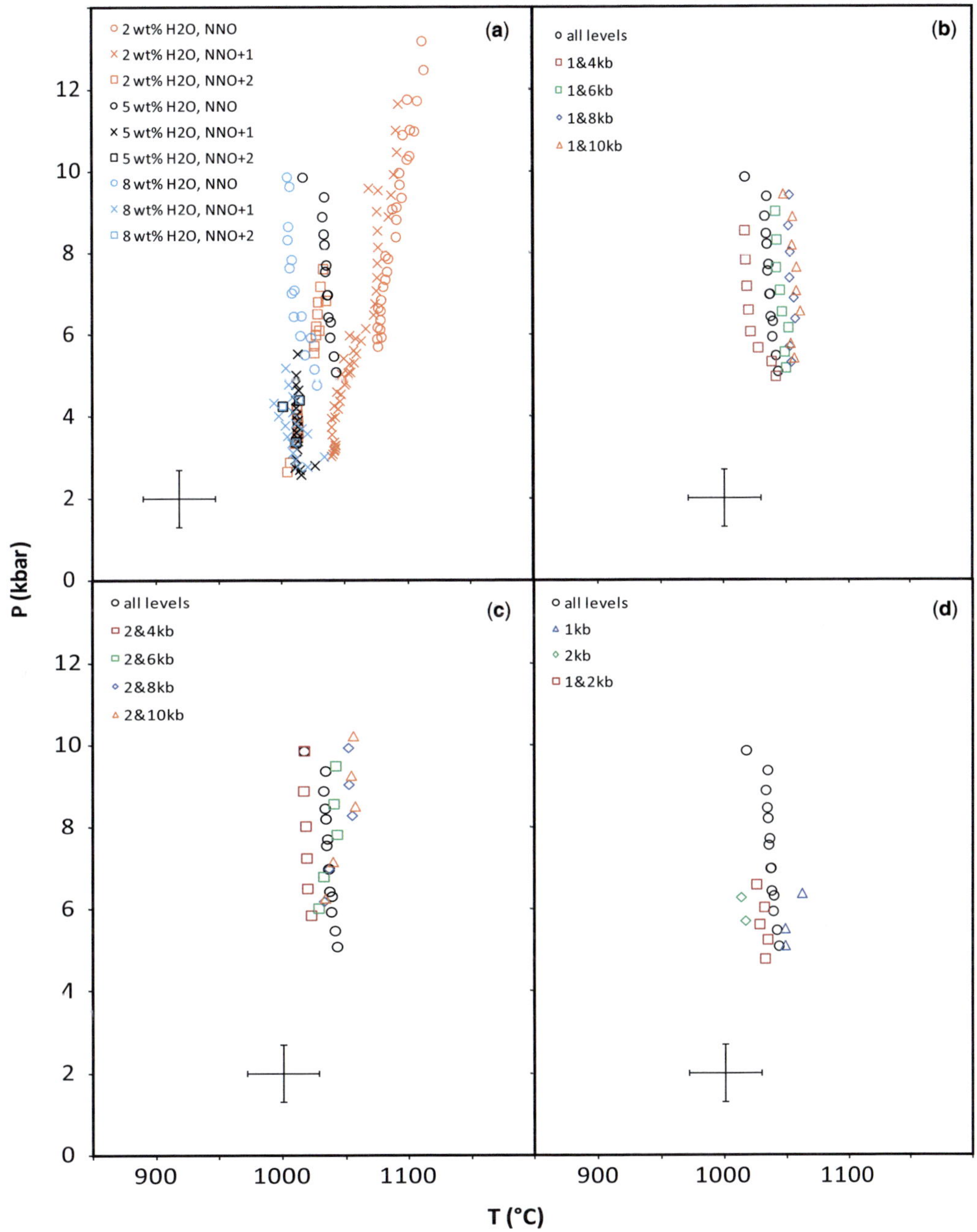

Fig. 5. Modelled two-pyroxene pseudo-decompression paths (PDPs) generated by combining alphaMELTS results with two-pyroxene thermobarometry, using the 09ASO01 bulk rock as starting composition. (**a**) Combining all equilibrium pyroxene pairs formed by isobaric cooling at 10 to 1 kbars, at different initial H_2O contents and oxygen fugacity buffers. (**b–d**) Using the 5 wt% H_2O, NNO model as example, the PDP formed by pairing pyroxenes across all crustal levels is compared to PDPs formed by selective pairing of pyroxenes from (b) 1 kbar and one mid- to lower crustal level, (c) 2 kbar and one mid- to lower crustal level and (c) only upper crustal levels. See text for discussion.

Fig. 5c); (iii) uptake of pyroxene antecrysts from shallow crustal levels only (1 kbar only, 2 kbar only, or 1 and 2 kbar, cf. Fig. 5d). It is found that uptake of crystals formed at a single mid- to lower crustal level and a single shallow crustal level suffices to produce pseudo-pressure ranges comparable to those produced by crystal uptake at all levels (Fig. 5b, c). However, uptake of crystals formed only at shallow crustal levels will not yield significant pseudo-pressure variations (Fig. 5d).

The possibility that crystals are derived from only two crustal levels (upper and middle to lower) brings to mind concepts of polybaric melt evolution, which invoke one or more deep crustal magma reservoirs feeding into one or more upper crustal, subvolcanic chambers (e.g. Turner *et al.* 2000; Putirka & Condit 2003; Jerram & Martin 2008; Hamada *et al.* 2011; Dahren *et al.* 2012). In such scenarios, most crystals are not considered to represent assimilated antecrysts from previously cooled magmas, but instead are formed as part of the same magma, which first fractionates at a deep level and then ascends into a shallow chamber where it continues to crystallize prior to eruption. Using alphaMELTS, we can test if such alternative scenarios may also be successful in reproducing the observed two-pyroxene PDPs. Again, sample 09ASO01 with 5 wt% H_2O at the NNO buffer may serve as an example. We consider isobaric crystal fractionation at 10 kbar from superliquidus temperatures to some subliquidus temperature (between 1100 and 950 °C), followed by magma ascent (with the melt carrying much of its previously fractionated crystal load) to 1 kbar and further isobaric fractionation at this shallower level due to degassing and cooling down to 900 °C. In the deep reservoir, the alphaMELTS model suggests crystallization of clinopyroxene below 1200 °C and clinopyroxene + garnet from 1100 down to 950 °C, and concomitant melt evolution through slight enrichment in SiO_2 and Al_2O_3 and strong depletion in MgO, and CaO. At the point of release from the deep crust, the melt carries between *c.* 6 and 8.5 wt% H_2O. When decompressed to shallow crustal levels, degassing reduces the H_2O content to less than 4 wt% and plagioclase and spinel-structured oxides (Ti-magnetite) form. Orthopyroxenes only crystallize when the melt has cooled by an additional *c.* 100 K or more, and most of these low temperature orthopyroxenes are not in Fe/Mg-equilibrium with the clinopyroxenes formed in the deep reservoir. A few equilibrium pairs yield a restricted pseudo-pressure range of <3 kbar. This simple experiment suggests that a single magma, fractionating during intermittent magma ascent and associated melt evolution due to cooling- and degassing-induced crystallization, cannot produce the observed two-pyroxene PDPs. In order for two-pyroxene PDPs with large pseudo-pressure ranges to be generated, uptake of antecrysts formed by cooling of mafic melts at several crustal levels appears to be required.

Polybaric uptake of previously formed crystals v. magma mixing. Disequilibrium between crystals and melt has previously been taken as evidence for open system processes operating in arc magmas, with magma mixing being the process invoked conventionally. There is ample evidence for recharge magma mixing in the rock record, and mixing has also been called upon as an eruption trigger (e.g. Pallister *et al.* 1992; Murphy *et al.* 1998). In the case of pyroxene, it may be argued that apparent equilibrium pairs are derived from mixing of variably differentiated magmas originating from crustal magma reservoirs under different $P-T$ conditions. In order to assess the viability of such mixing in producing the observed PDPs, alphaMELTS modelling may again be employed with particular attention to the melt compositions in equilibrium with the crystallized pyroxenes. On Figure 6, the melt compositions in equilibrium with the pyroxene pairs that produce the PDP of sample 09ASO01 (with 5 wt% H_2O at the NNO buffer) are plotted as an example, and compared to the groundmass composition of this sample. It is evident that the modelled pyroxenes form from melts with significantly lower Mg# than this groundmass. This indicates that there needs to be an additional, more mafic mixing component, as is also required by the presence of magnesian mafic minerals within the whole rock. The Mg# of potential additional mixing components to generate the groundmass composition of 09ASO01 partially overlaps with mafic whole rock samples from SW Japan. However, the modelled average melts with low Mg# and low silica that crystallize clinopyroxene require high Mg# intermediates as mixing components (cf. Fig. 6). In order for two-pyroxene PDPs with large pseudo-pressure ranges to be generated by mixing, the existence of such high Mg# intermediates beneath the currently active SW Japan arc has to be invoked. Such melts might be similar to the silica-rich wedge melts or their derivatives seen elsewhere (e.g. Straub *et al.* 2011), and our data cannot preclude that magma mixing alone is capable of producing the observed trends. However, it is questionable whether such enigmatic high Mg# intermediates can be called upon as upper crustal (1–2 kbar, as orthopyroxene is present) mixing components in SW Japan, given that such melts are not observed in the eruptive record. We therefore suggest that uptake of antecrysts formed by cooling of mafic melts at several crustal levels is a more reasonable explanation for the observed two-pyroxene PDPs.

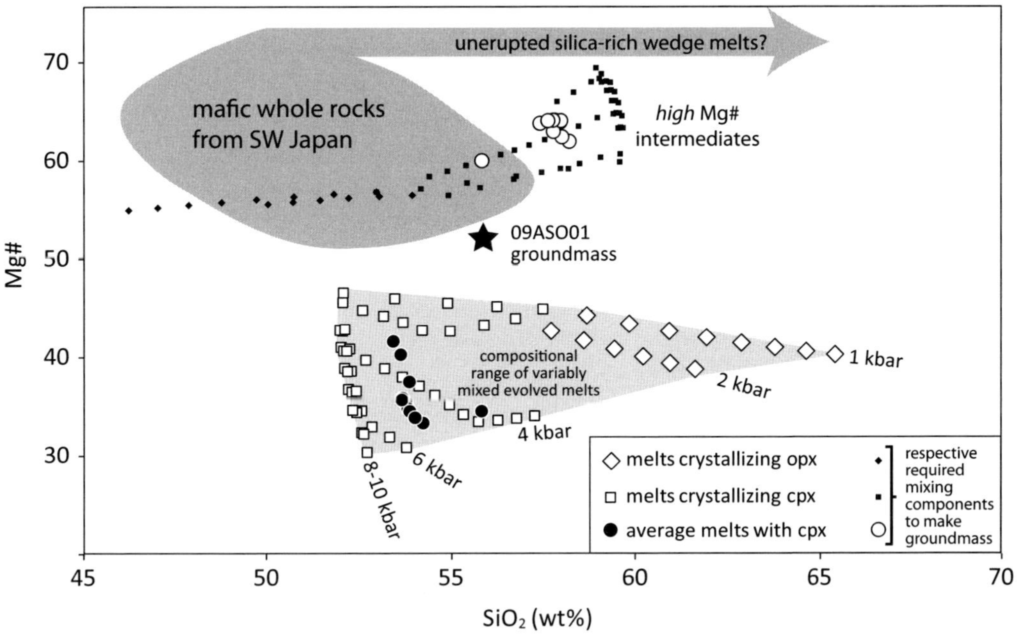

Fig. 6. AlphaMELTS fractionation models of melt compositions that crystallize clinopyroxene (between 1 and 10 kbar, open squares) and orthopyroxene (in evolved melts at 1 and 2 kbar only, open diamonds) that produce the PDP of sample 09ASO01. The models use sample 09ASO01 with 5 wt% H_2O as starting composition, fractionating during cooling down to 900 °C at the NNO buffer. Average melts with clinopyroxene that form the modelled PDP in Figure 5a are indicated. Smaller solid symbols are calculated mixing compositions to obtain the 09ASO01 groundmass, indicated by the star. Such mixing components overlap partially with the field of mafic whole rocks from SW Japan. However, high Mg# intermediates would also be required, suggesting that if mixing is the prominent mechanism, unerupted high Mg# intermediates similar in composition to silica-rich wedge melts or their derivatives (cf. Straub *et al.* 2011) must be invoked.

Estimating oxygen fugacities and H_2O contents from two-pyroxene PDPs in SW Japan. Two-pyroxene PDPs of Kirishima, Aso, and Abu are now considered separately, and are reproduced by alphaMELTS models that inform the conditions of crystallization. Given the considerable pseudo-pressure ranges of Kirishima, Aso, and Abu, it is concluded that a large proportion of crystals in these volcanoes represents antecrysts, and that the original melt compositions are more closely resembled by the groundmass composition of the samples investigated. Figure 7 compares observed two-pyroxene PDPs at Kirishima, Aso and Abu with PDP models of polybaric uptake of pyroxene antecrysts crystallized by isobaric cooling from melts with compositions of the groundmasses of samples 09KIR01, 09ASO01 and 09ABU03. This assumes that groundmass compositions so closely resemble original melt compositions that they may be used as starting compositions in thermodynamic crystallization models. Note that melt compositions may have slightly changed due to assimilation by partial resorption of antecrysts and/or xenocrysts taken up into the melt during its pre-eruptive history. Thus, while it may not be possible to infer quantitatively accurate magmatic conditions from these models, they are nevertheless useful in comparing along-arc variations in SW Japan qualitatively.

In considering oxygen fugacities, it is noted that samples from Kirishima and Aso yield two-pyroxene PDPs with pseudo-pressure ranges of 6 kbar, and the sample from Abu yields a PDP with a pseudo-pressure range of about 4 kbar. Thermodynamic crystallization modelling (alpha-MELTS) shows that the existence of orthopyroxene requires an oxygen fugacity above the QFM buffer, and that pseudo-pressure ranges of this magnitude are typically not produced at oxygen fugacities more than two log-units above the NNO buffer in hydrous compositions (Fig. 5a). It may therefore be concluded that the magmas in SW Japan are typically buffered at NNO to NNO + 2. Therefore, while the two-pyroxene thermobarometer may not be useful for determining actual P–T conditions in hydrous arc magmas, it may have some use in the

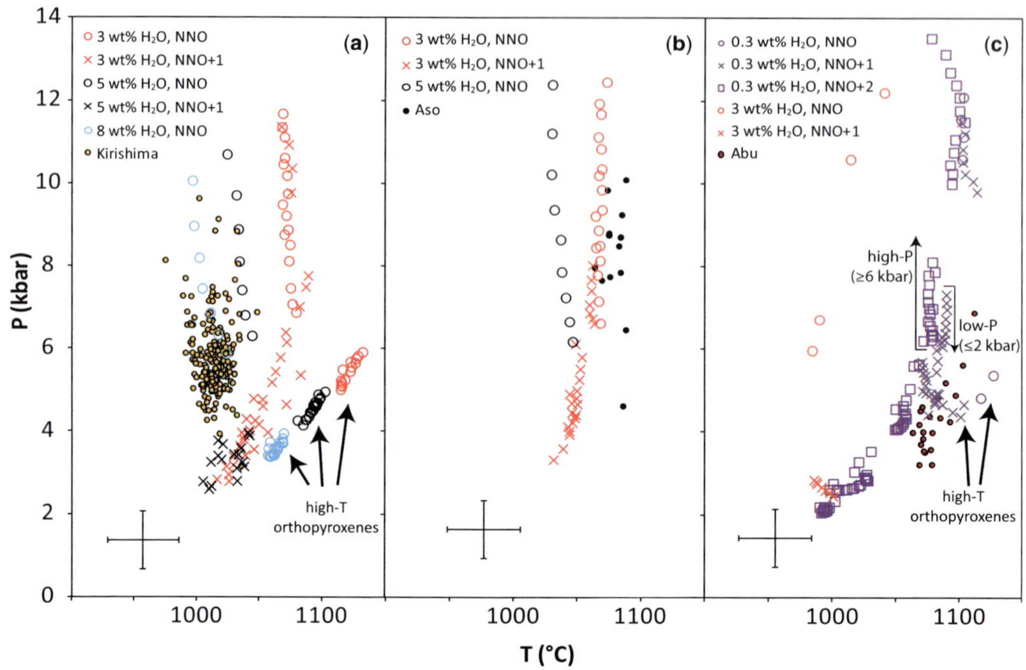

Fig. 7. Modelled two-pyroxene pseudo-decompression paths (PDPs), generated by combining alphaMELTS results with two-proxene thermobarometry, using groundmass compositions of samples (**a**) 09KIR01, (**b**) 09ASO01 and (**c**) 09ABU03. Oxygen fugacity buffers and H_2O contents were chosen to approach the observed PDPs from Kirishima, Aso and Abu (reproduced from Fig. 4b). High-temperature orthopyroxenes form in the less primitive samples, and in hydrous compositions are clearly offset from the PDPs formed by low-temperature orthopyroxenes at the NNO buffer. In models with low water contents required to reproduce Abu compositions, the low-*P* arrow refers to the alphaMELTS pressures of orthopyroxenes formed at NNO + 1, while the high-*P* arrow refers to the NNO + 2 model. (See text for discussion.)

determination of oxygen fugacities. It is noted that the indicated fO_2 conditions significantly above the QFM buffer have been considered typical for primitive arc magmas (e.g. Eggins 1993; Grove *et al.* 2005; Mallmann & O'Neil 2007; Kelley & Cottrell 2009), but are in contrast to recent notions of lower fO_2 conditions around the QFM buffer in the source region of primitive arc magmas (Mallmann & O'Neil 2009) followed by oxygenation of these magmas during their passage through the crust (cf. Lee *et al.* 2012). The two-pyroxene data from Kirishima may be used to assess potential changes in magmatic oxidation states in SW Japan. The groundmass composition of the sample is andesitic, that is, quite evolved, and at the NNO buffer this results in the formation of some high-T orthopyroxenes, which yield tight clusters in *P*–*T* space when paired with clinopyroxenes formed at lower temperatures (Fig. 7a). Clearly, such high-*T* orthopyroxenes are not observed in our dataset, suggesting that the high-T crystallization of the melts progressed at lower fO_2, preventing early orthopyroxene formation, and that magmas became more oxygenated at a later stage, upon cooling.

The two-pyroxene PDPs may also be used to inform the initial H_2O content of the magmas. At Kirishima, low pseudo-temperatures require relatively high initial H_2O contents of at least 5–8 wt% (Fig. 7a). More moderate initial H_2O contents of $\leq$3 wt% are suggested for the Aso PDP (Fig. 7b). In contrast, very low initial H_2O contents of *c.* 0.3 wt% are suggested by the high pseudo-temperatures of the Abu PDP (Fig. 7c). At such low H_2O contents, orthopyroxenes crystallize at all modelled pressures at oxygen fugacities above the QFM buffer. However, the Abu data preclude assimilation of high pressure orthopyroxenes. This may indicate that elevated oxygen fugacities are restricted to upper crustal levels (and lower temperature fractionation, see above), so that fractionation in the mid- to lower crust occurs around the QFM buffer and orthopyroxene formation is therefore inhibited. Finally, it should be noted that the relatively large range of initial H_2O contents suggested by the two-pyroxene PDPs becomes less pronounced in the actual magmas at upper crustal conditions, where initially high-H_2O melts degas, while initially low-H_2O melts become enriched through fractionation of anhydrous minerals.

Plagioclase hygrometry and the origin of the plagioclase crystal cargo

Plagioclase hygrometry (Lange *et al.* 2009) was recently employed on SW Japan arc volcanoes by Zellmer *et al.* (2012*b*), who showed that the maximum plagioclase anorthite contents monotonically decreased from central Kyushu towards western Honshu. On the basis of approximately constant maximum pyroxene temperatures (Lindsley 1983), it was argued that the decrease in maximum anorthite contents of plagioclase reflected decreasing H_2O contents of primary arc magmas from *c.* 6 wt% to less than 2 wt% along the arc. Because the Moho is thought to represent a plagioclase saturation boundary (Müntener & Ulmer 2006), Zellmer *et al.* (2012*b*) assumed plagioclase crystallization at lower crustal pressures of 10 kbar and temperatures of 1080 ± 30 °C. When the same approach is adopted here, the X_{An} and H_2O variations can be extended southwards to Kirishima and eastwards to Kannabe, where H_2O may be increasing due to contributions from deep dehydration of the underlying old Pacific slab that subducts in this area (Fig. 8).

However, the notion that the most calcic plagioclase crystals form at the crust–mantle boundary is challenged by the alphaMELTS thermodynamic crystallization models run as part of the present study. At high pressures, hydrous arc melts have clinopyroxene on the liquidus (Hamada & Fujii 2008), resulting in rapidly decreasing Ca/Na ratios of evolving magmas prior to plagioclase saturation. The alphaMELTS results suggest that the most calcic plagioclases are formed during crystallization of mafic magmas in the upper crust (1–2 kbar), when H_2O solubility in the melts is typically less than 5 wt% and plagioclase enters the crystallizing phase assemblage earlier. Further, the temperature of plagioclase crystallization is a function of H_2O content. At shallow crustal pressures, hydrous melts with about 4 wt% H_2O crystallize plagioclase below about 1050 ± 30 °C, while at low H_2O contents (0 wt% $< H_2O \leq 1$ wt%) plagioclase becomes stable at *c.* 100 K higher temperatures. If these constraints are taken into account, maximum H_2O ranges between 4.6 wt% and 0.2 wt% during plagioclase fractionation in the upper crust (Fig. 8), and becomes more consistent with the very low H_2O contents in western Honshu suggested by the Abu two-pyroxene PDP (Fig. 7c). It should be noted that garnet crystallizes at lower crustal pressures (10 kbar) in almost all alphaMELTS runs at even very low amounts of water (0.3 wt%). Further, the temperature of garnet stability decreases with increasing H_2O contents. The notion of Zellmer *et al.* (2012*b*) that garnet becomes an important lower crustal phase that may control melt differentiation to adakitic intermediates in Western Honshu, where H_2O contents of mafic melts are lower than in Kyushu, thus remains valid.

Our crystallization models also suggest that the compositions of many of the samples studied here have too low Ca/Na ratios to have generated the highest anorthite contents of *c.* 0.90 (± 0.03) observed in Kyushu. The highest anorthite contents yielded by crystallization models of our samples are typically 0.80 (± 0.03), with the exception of the 09KIR02 whole rock, which at initial water contents of 4–5 wt% yields highly calcic plagioclase ($X_{An} \geq 0.90$) during low pressure crystal fractionation. This indicates that primary melts from the mantle wedge beneath Kyushu can have higher Ca/Na ratios than most of the rocks we have sampled.

Considering the range of plagioclase compositions at individual volcanoes, it is evident that plagioclase crystals display a highly diverse crystal chemistry within each of the studied rock samples, with ΔX_{An} typically exceeding 0.4 units (Fig. 8a). In contrast, alphaMELTS modelling suggests that closed-system isobaric cooling from superliquidus temperatures to 900 °C, when the melt has already evolved to much more felsic compositions than observed, leads to precipitation of plagioclase with a lower compositional range. Further, as already pointed out, the rock samples studied have too low Ca/Na ratios to explain the most calcic plagioclase crystals observed within them by closed-system processes. Both of these observations point to open-system behaviour. This notion is further supported by the comparable compositional range of large plagioclase crystals and plagioclase microlites in the groundmass hosting those crystals. If the large crystals had formed from their host melt, their anorthite content would be consistently higher than that of the microlites. This is clearly not observed. Quite in contrast, for example, the Mengame basalt yields low X_{An} plagioclase crystals hosted within a groundmass with generally higher X_{An} plagioclase microlites (Fig. 8a). Finally, clear compositional gaps within the plagioclase crystal load, most evident at Kuju volcano, are equally inconsistent with simple closed system crystallization processes.

The above observations may again be considered in the context of conventional magma mixing models v. polybaric uptake of previously crystallized plagioclase antecrysts and xenocrysts. Magma mixing and associated microlite transfer

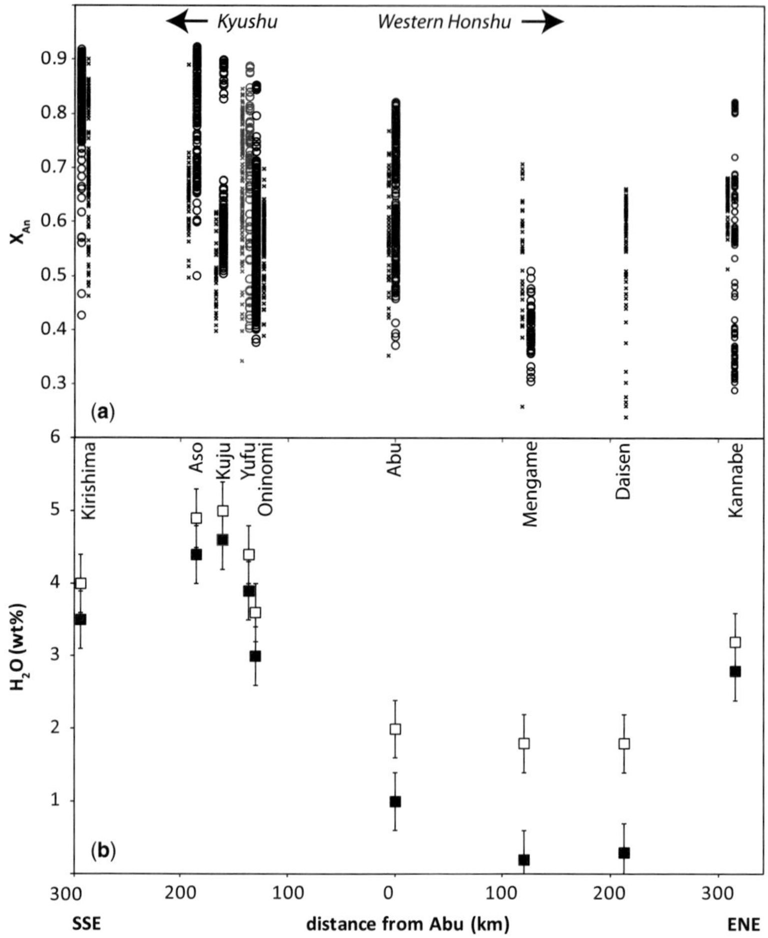

Fig. 8. (**a**) Along-arc variation of anorthite contents of large crystals (circles) and groundmass microlites (crosses), horizontally separated for clarity. For details, see Zellmer *et al.* (2012*b*). (**b**) Along-arc variation of H_2O in the lower crust, derived by plagioclase hygrometry, assuming plagioclase crystallization in the lower crust (10 kbar) at 1080 $\pm$ 30 °C (open squares, inconsistent with thermodynamic crystallization modelling), or in the upper crust (1 kbar) at 1050 $\pm$ 30 °C or 1150 $\pm$ 30 °C, for 3–5 wt% and $\leq$1 wt% H_2O, respectively (solid squares, consistent with thermodynamic crystallization modelling). See text for discussion.

from mafic enclaves into more evolved host lavas has previously been invoked to explain the large compositional range in microlites in andesitic melts (Humphreys *et al.* 2009). However, alpha-MELTS modelling indicates that the most sodic plagioclase crystals found in SW Japan are formed from evolved melts with SiO_2 well in excess of 65 wt%, irrespective of H_2O content. If such compositions constituted component melts in mixing processes that yield mafic magmas with SiO_2 contents of less than 55 wt%, ultramafic mixing components would also be required. Such ultramafic compositions are not present in recent SW Japan volcanics. The combined evidence therefore suggests that many if not all plagioclase crystals are antecrysts or xenocrysts taken up into mafic, aphyric or sparsely phyric, and likely rapidly ascending melts.

H_2O in olivine and the origin of the olivine crystal cargo

Due to ample evidence for crustal assimilation in many of the mafic rock samples, including the co-occurrence of olivine and quartz (Ohta *et al.* 1992), olivines from SW Japan were originally targeted for analysis as they were thought to have crystallized from the mafic component, possibly prior to significant crustal assimilation of the magmas.

Initially, olivines were separated and screened for melt inclusions large enough for analysis of H_2O by Fourier transform infrared (FTIR) spectroscopy. However, given the small crystal size (typically <100 μm), the low number of melt inclusions, and the small size of the few available inclusions (typically <25 μm), it became clear that this approach was analytically too challenging. It was therefore decided to investigate potential variations in the H_2O contents of the olivines themselves, focusing on the olivine cores that would have crystallized earliest and would have been least prone to re-equilibration with the host melt due to rapid hydrogen diffusion in olivine (Mackwell & Kohlstedt 1990; Gaetani *et al.* 2012). There is some controversy about the degree of potential re-equilibration of hydrogen between olivine and the host melt in natural basaltic rocks. Experimental evidence points to diffusional equilibration within a few days at magmatic temperatures (Portnyagin *et al.* 2008; Gaetani *et al.* 2012), and gradients in hydrogen concentrations within olivine xenocrysts indicate similar timescales (Peslier & Luhr 2006). However, complete equilibration of olivine crystals and their host melt may not always be achieved (Kurosawa *et al.* 1997), particularly if crystal residence times are short, that is if melt ascent is fast (≥ 0.2 m s^{-1}, cf. Peslier & Luhr 2006; Gaetani *et al.* 2012) and cooling upon eruption is rapid. Hydrogen partitioning into the olivine crystals may be treated in the same way as the inclusion of other incompatible trace elements into the olivine crystal structure, which depends on trace element concentration in the melt and an olivine/liquid partition coefficient. If the parental magmas displayed significant variations in H_2O, as suggested by the two-pyroxene PDPs and plagioclase hygrometry (this study, and Zellmer *et al.* 2012*b*), this might then be reflected in concomitant variations in the H_2O contents of olivines grown from these melts.

To our knowledge, these are the first published data of H_2O in volcanic olivines from Japan. The hydrogen contents in these olivines are low and comparable to previously published data of H_2O in volcanic olivines (Peslier 2010). This implies that (i) analytical uncertainties are too large to distinguish potential differences in H_2O contents of olivines erupted from different volcanic centres (Fig. 3), and (ii) contamination by hydrous microinclusions may significantly affect the observed signal, suggesting that our data may still be overestimates of actual H_2O concentration within the olivine crystal structure. As a consequence, given the present analytical capabilities, the H_2O content in volcanic olivines cannot be successfully used to constrain potential variations in H_2O of mafic melts along the SW Japan arc. However, we note that Mengame olivines from Western Honshu, which yielded the most consistently reproducible low H_2O contents of *c.* 6 ppm, would imply a magmatic H_2O content of *c.* 0.4 wt% using reasonable partition coefficients (Hauri *et al.* 2006), which compares well with the low H_2O estimate from plagioclase hygrometry for this volcano.

Turning to the origin of the olivine crystals, alphaMELTS modelling shows that significant olivine fractionation is restricted to mid- to upper crustal levels in wet magmas, and only to the upper crust (<4 kbar) for melts with <1 wt% H_2O. Mafic lavas from Mengame and Daisen carry significant amounts of olivine (with some samples exceeding 10% crystal cargo), and given the low suggested water contents in these centres (Fig. 8b), upper crustal melt evolution is indicated if these olivines formed from the melts they are hosted in, as suggested previously for Daisen (Tamura *et al.* 2000). However, even at 1 or 2 kbar, alphaMELTS model runs for Mengame and Daisen bulk compositions at 1 wt% H_2O and the NNO buffer suggest only 5–10% olivine crystallization before feldspar and clinopyroxene become stable, and much greater amounts of feldspar and clinopyroxene (up to *c.* 3 times more) than olivine by the time olivine is exhausted (at $<15\%$ olivine crystallization). The crystal assemblages in the studied basalts from Western Honshu therefore could not have formed by crystallization from their host melts.

Our model results indicate that most of these olivines, and potentially the entire crystal load, got taken up into the host melt prior to eruption. Olivine crystal rims from Mengame and Daisen (cf. Fig. 9) are slightly enriched in Al_2O_3 (up to *c.* 1 wt%, which is more than can be hosted within the olivine crystal structure), and TEM data that will be published elsewhere show nm-sized veinlets of an Al-bearing amorphous phase penetrating into the rims. We interpret these veinlets to have formed during alteration by interaction with hydrothermal, metasomatic fluids at subsolidus temperatures, altering previously emplaced olivine-bearing intrusions, from which the olivine crystals hosted by these basalts originated. This is another piece of evidence for the uptake of olivine crystals into an initially aphyric or sparsely phyric mafic melt.

Can mafic magmas be erupted from long-lived mid- to upper crustal magma reservoirs?

Figure 10 compares two endmember scenarios for the origin of the crystal cargo in arc magmas. Conventional models of magma mixing and ensuing disequilibrium growth of crystals in subvolcanic magma chambers (Fig. 10a, b) are likely applicable in evolved systems where large magma chambers can develop through long-term thermal incubation

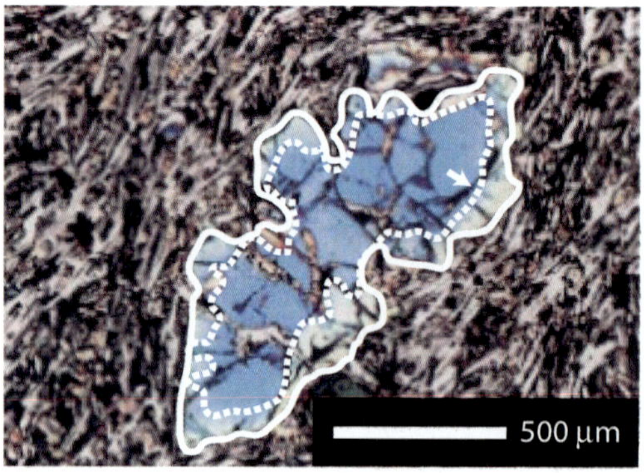

Fig. 9. Typical olivine crystal from a Daisen basalt seen in cross-polarized light, set in a groundmass of plagioclase, clinopyroxene and iron–titanium oxide (reproduced from Tamura *et al.* 2000). Solid white line outlines crystal–groundmass boundary, stippled white line indicates transition between core to rim.The higher birefringence at the rim indicates a higher iron content, as noted earlier (Tamura *et al.* 2000). However, note that this Fe-rich rim zone is finely structured in places (indicated by arrow), of variable width, and laterally discontinuous.

by repeated injections of magma into the upper crust, which may ultimately lead to large caldera-forming eruptions. In contrast, the combined evidence from the mafic samples of SW Japan points to uptake of antecrystic or xenocrystic pyroxene, plagioclase and olivine into initially aphyric, rapidly ascending melt, likely just prior to eruption (Fig. 10c, d). In fact, it appears that the majority of crystals carried in SW Japan arc magmas have not grown from their host melts, but instead originate from small, previously intruded and solidified melt batches. Uptake of previously formed crystals by aphyric or sparsely phyric melts may be a common phenomenon in many mafic arc magmas. This process can explain the large variety of petrographic features found in mafic arc volcanic rocks (e.g. Tatsumi & Ishizaka 1982). Uptake of antecrysts and/or xenocrysts has been invoked previously to account for U-series age dichotomies in mineral separates from mafic arc lavas (Zellmer *et al.* 2005), the high MgO content of some arc andesites (e.g. Streck *et al.* 2007; Shellnutt & Zellmer 2010), and significant chemical disequilibrium between the crystal load and its host melt even in simpler, oceanic arc settings, in which mafic magmas dominate (Smith *et al.* 2010).

Uptake of most if not all crystals implies that the host melt is transferred from its source to the surface without cooling during prolonged storage in mid- to shallow crustal magma reservoirs. In contrast to less hydrous magmas erupted at constructive plate margins, where much of the crystal cargo precipitates from the melt upon cooling during shallow crustal melt ascent (Zellmer *et al.* 2011, 2012*a*), hydrous arc melts will not initially crystallize during decompression: they may leave their source close to the anhydrous mantle solidus (England & Katz 2010) and, hence, at super-liquidus temperatures, or reach superliquidus temperature during ascent (Annen *et al.* 2006) and potentially undergo considerable degassing before crystallization will commence (Blundy *et al.* 2010). The viscosity η_m of such hydrous melts will be very low, irrespective of their silica contents. Buoyant ascent of such melts through a dyke of width w will occur with a velocity of $v = \Delta\rho g w^2/12\eta_m$, where $\Delta\rho$ is the density difference between the melt and the wall rocks and g is the acceleration due to gravity (e.g. Clemens 1998). Reasonable values ($\Delta\rho =$ 100 kg m^2, $w = 1$ m, $\eta_m = 10^2$ Pa s) yield ascent rates of the order of 1 m s^{-1}, suggesting hydrous arc melts may traverse the crust in less than one day if unimpeded. Intriguingly, some of the PDPs yielded by two-pyroxene thermobarometry, although entirely an artefact of antecryst uptake into ascending melts, may in fact mirror the adiabatic decompression of initially aphyric hydrous mafic melts immediately prior to their eruption.

The results from this study indicate that the ultimate fate of batches of mafic melts ascending beneath active arc volcanoes may be two-fold: they may erupt following rapid ascent from source to surface (e.g. Blot 1972; Fedorov & Markhinin 1983; Blatter & Carmichael 1998), picking up antecrysts and/or xenocrysts along the way; or they may stall at depth, either fractionating by cooling and degassing to form mafic crystal mushes or plutonic bodies, or triggering remobilization of more evolved protoliths (e.g. Bacon 1986; Clynne 1999; Zellmer & Turner 2007). Scenarios of mafic melts' ascent and storage in mid- to upper crustal subvolcanic reservoirs, which are subsequently tapped at the onset of an eruption, appear to be precluded when considering the mineral chemistry of the mafic SW Japan arc samples investigated here.

These results are in line with several recent observations from other volcanic arcs:

(1) Arc melts can directly ascend from the mantle wedge through a thick overriding crust to the surface without being chemically modified by assimilation of partial melts of old crustal materials (Straub *et al.* 2008, 2011, 2013).

(2) Significant chemical (major oxide and trace element) and isotopic variations in arc magmas can originate from partial melting of wedge mantle that has been variably metasomatized by slab contributions (Straub

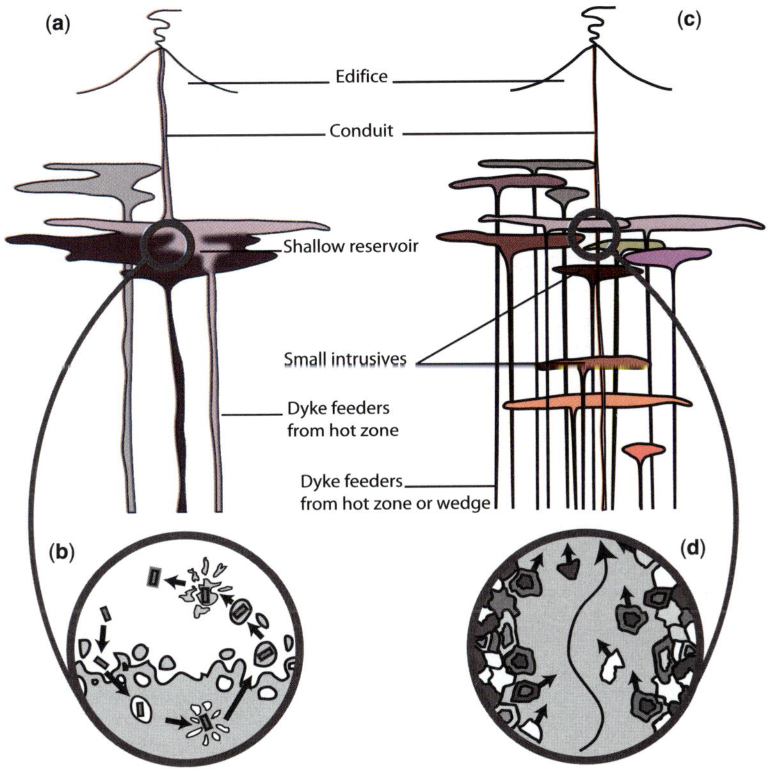

Fig. 10. Comparison of petrogenetic models of magma mixing and antecryst uptake to generate complex zoning and disequilibrium textures in arc magmas. (**a**) Magmas accumulate and mix in the upper crust through repeated injection of sills from the deep crustal hot zone. Adapted from Annen *et al.* (2006). (**b**) Crystal growth records the compositional variations of the mixing and crystallizing magmas. Adapted from Troll & Schmincke (2002). (**c**) Alternative model of rapid ascent of small batches of initially aphyric melts with a wide compositional range from the deep crustal hot zone or directly from the wedge. Most melts stall and solidify rapidly, forming intrusives. (**d**) During melt ascent, crystals are taken up into the melt from previously formed intrusions. As a result, upper crustal intrusions may contain a variety of complexly zoned crystals that originally formed at deeper levels. Erupting magmas sample crystals formed at a large range of crustal pressures and temperatures, and from melts with a large compositional range. Note that different colours in (a) and (c) indicate a range of magma compositions and temperatures, rather than being indicative of melt fraction or source as in Annen *et al.* (2006).

et al. 2008, 2011, 2013; Straub & Zellmer 2012).

(3) In extensional settings, sparsely phyric mafic arc magmas are frequently erupted, implying rapid magma ascent from source to surface without storage and cooling in crustal magma reservoirs (Zellmer *et al.* 2013).

(4) Frequent ascent of small melt batches appears to dominate over less frequent ascent and storage of larger volumes of magma (Zellmer & Annen 2008).

When small melt batches stall in the crust, they lose heat and solidify rapidly (Zellmer 2008, 2009). This may be one of the reasons why to date it has not been possible to unambiguously detect mafic magma reservoirs by geophysical means.

Crystals in highly porphyritic intermediate arc magmas are known to be derived through remobilization of igneous protoliths formed by multiple preceding intrusive episodes (e.g. Murphy *et al.* 2000; Zellmer *et al.* 2003*a*, 2003*b*; Claiborne *et al.* 2010). The data presented here extends the notion of remobilization of previously formed crystals to the mafic regime, and suggest that reliable insights into the source of mafic arc melts can rarely be gained from whole rocks, but require careful removal of the crystal cargo from its host groundmass or glass prior to major oxide, trace element, or isotopic analysis. Crystals may still be used as recorders of magmatic processes, but in mafic arc magmas these may typically be processes experienced by previous intrusions prior to the passage of the melt batch that carried the crystal cargo to the surface at the onset of eruption (with a likely exception being some outermost overgrowth rims).

Conclusions

The following are the principal conclusions of this study.

(1) In hydrous mafic arc magmas, orthopyroxene and clinopyroxene do not precipitate at the same time. In less evolved samples, orthopyroxene typically forms at lower temperatures than clinopyroxene, while in more evolved samples it may also form at higher temperatures than clinopyroxene. In each of these scenarios, two-pyroxene thermobarometric estimates are defective, despite an abundance of pyroxene pairs that appear to be in Fe/Mg-equilibrium. In *P–T* space, such equilibrium pyroxene pairs may yield iso-thermal pseudo-decompression paths (PDPs), which can be used to estimate H_2O contents and oxygen fugacities during crystallization. Arc magmas begin to fractionate around the QFM buffer, but oxygen fugacities increase as cooling proceeds in the upper crust, typically yielding conditions between the NNO buffer and up to two log-units above. Detailed analysis of PDPs indicates polybaric uptake of pyroxene antecrysts formed by isobaric cooling of previously intruded melts at various crustal pressures.

(2) To form highly calcic plagioclase ($X_{An} \geq 0.9$), fractionating melts need to be hydrous. However, at the high pressures of the lower crust, clinopyroxene is on the liquidus, resulting in depletion of Ca in the melt and the crystallization of more sodic plagioclase at lower temperatures. Highly calcic plagioclase is therefore a product of upper crustal differentiation of mafic melts with high Ca/Na ratios at <5 wt% H_2O. Plagioclase hygrometry can be used to quantify melt H_2O contents if magmatic temperatures can be determined by thermometry or thermodynamic modelling. Further, the observed range of plagioclase compositions exceeds that which may be generated by crystallization from most of the SW Japan samples investigated. This and the similar compositional range of large crystals and groundmass microlites indicate uptake of plagioclase antecrysts.

(3) Volcanic olivines in SW Japan have low water contents of *c.* 5 to 20 ppm, not distinguishable between volcanic centres given the uncertainties associated with the presently available analytical techniques. Partitioning of hydrogen into volcanic olivines is too low to provide constraints on potential variations in the H_2O contents of the melts from which these crystals formed. However, the high proportion of olivine compared to other crystal phases in some volcanic rocks from Western Honshu, and preliminary evidence for metasomatic alteration of olivine rims, suggests that most or all of these olivines are antecrysts from previously intruded mafic crystal mushes or plutons.

(4) Uptake of antecrysts and xenocrysts into aphyric or sparsely phyric mafic melts appears to be very common at volcanic arcs, and implies that mafic magmas are not being sourced from large, long-lived mid- to upper magma reservoirs. Instead, they are commonly sourced from the mantle or the deep crust, from where they rapidly ascent to the surface at the onset of eruption. Crystals may be used as recorders of magmatic processes experienced by previous intrusions prior to the passage of the melt batch that carried the crystal cargo to the surface.

GFZ thanks Shyh-Lung Hwang, Shoichi Ito, Jun-Ichi Kimura, George Ma, Paul Morris and Tzen-Fu Yui for discussions, and Liz Cottrell and Ben Andrews for support through a Global Volcanism fellowship at the Smithsonian Institution. Luiza-Maria Cioflec, Madoka Shimono and Hiroshi Shukuno assisted during fieldwork, and Ya-Ting Hsu and Rong-Yi Yan helped with sample processing. This contribution was improved through reviews by Morihisa Hamada and an anonymous reviewer, and constructive editorial comments by Susanne Straub. We acknowledge funding by the National Science Council of Taiwan (Grants 98-2918-I-001-014 and 100-2116-M-001-007 to GFZ, and 99-2116-M-001-013 to YI) while the analytical part of this study was conducted.

References

Abdelwahed, M. F. & Zhao, D. 2007. Deep structure of the Japan subduction zone. *Physics of the Earth and Planetary Interiors*, **162**, 32–52.

Alonso-Perez, R., Müntener, O. & Ulmer, P. 2009. Igneous garnet and amphibole fractionation in the roots of island arcs: experimental constraints on andesitic liquids. *Contributions to Mineralogy and Petrology*, **157**, 541–558.

Annen, C., Blundy, J. D. & Sparks, R. S. J. 2006. The genesis of intermediate and silicic magmas in deep crustal hot zones. *Journal of Petrology*, **47**, 505–539.

Baba, T., Tanioka, Y., Cummins, P. R. & Uhira, K. 2002. The slip distribution of the 1946 Nankai earthquake estimated from tsunami inversion using a new plate model. *Physics of the Earth and Planetary Interiors*, **132**, 59–73.

Bacon, C. R. 1986. Magmatic inclusions in silicic and intermediate volcanic rocks. *Journal of Geophysical Research*, **91**, 6091–6112.

BLATTER, D. L. & CARMICHAEL, I. S. E. 1998. Hornblende peridotite xenoliths from central Mexico reveal the highly oxidized nature of subarc upper mantle. *Geology*, **26**, 1035–1038.

BLOT, C. 1972. Volcanisme et séismes du manteau supérieur dans l'Archipel des Nouvelles-Hébrides. *Bulletin of Volcanology*, **36**, 446–461.

BLUNDY, J., CASHMAN, K. V., RUST, A. & WITHAM, F. 2010. A case for CO_2-rich arc magmas. *Earth and Planetary Science Letters*, **290**, 289–301.

CASHMAN, K. V. 2004. Volatile controls on magma ascent and eruption. *In*: SPARKS, R. S. J. & HAWKESWORTH, C. J. (eds) *The State of the Planet: Frontiers and Challenges in Geophysics*. American Geophysical Union, Washington, DC, 109–124.

CHARLIER, B. L. A. & ZELLMER, G. F. 2000. Some remarks on U–Th mineral ages from igneous rocks with prolonged crystallisation histories. *Earth and Planetary Science Letters*, **183**, 457–469.

CLAIBORNE, L. L., MILLER, C. F., FLANAGAN, D. M., CLYNNE, M. A. & WOODEN, J. L. 2010. Zircon reveals protracted magma storage and recycling beneath Mount St. Helens. *Geology*, **38**, 1011–1014.

CLASS, C., MILLER, D. M., GOLDSTEIN, S. L. & LANGMUIR, C. H. 2000. Distinguishing melt and fluid subduction components in Umnak Volcanics, Aleutian Arc. *Geochemistry Geophysics Geosystems*, **1**, 1999GC000010.

CLEMENS, J. D. 1998. Observations on the origins and ascent mechanisms of granitic magmas. *Journal of the Geological Society, London*, **155**, 843–851.

CLYNNE, M. A. 1999. A complex magma mixing origin for rocks erupted in 1915, Lassen Peak, California. *Journal of Petrology*, **40**, 105–132.

DAHREN, B., TROLL, V. R., ANDERSSON, U. B., CHADWICK, J. P., GARDNER, M. F., JAXYBULATOV, K. & KOULAKOV, I. 2012. Magma plumbing beneath Anak Krakatau volcano, Indonesia: evidence for multiple magma storage regions. *Contributions to Mineralogy and Petrology*, **163**, 631–651.

DAVIDSON, J. P., MORGAN, D. J. & CHARLIER, B. L. A. 2007. Frontiers in textural and microgeochemical analysis: isotopic microsampling of magmatic rocks. *Elements*, **3**, 253–259.

DUNGAN, M. A. 2005. Partial melting at the earth's surface: implications for assimilation rates and mechanisms in subvolcanic intrusions. *Journal of Volcanology and Geothermal Research*, **140**, 193–203.

EGGINS, S. M. 1993. Origin and differentiation of picritic arc magmas, Ambae (Aoba), Vanuatu. *Contributions to Mineralogy and Petrology*, **114**, 79–100.

ENGLAND, P. C. & KATZ, R. F. 2010. Melting above the anhydrous solidus controls the location of volcanic arcs. *Nature*, **467**, 700–703.

FEDOROV, S. A. & MARKHININ, Y. K. (eds) 1983. *The Great Tolbachik Fissure Eruption*. Cambridge University Press, Cambridge.

FURUYAMA, K. 1976. Petrology of the Kannabe Volcano Group. *Journal of the Geological Society of Japan*, **82**, 327–336.

GAETANI, G. A., O'LEARY, J. A., SHIMIZU, N., BUCHOLZ, C. E. & NEWVILLE, M. 2012. Rapid reequilibration of H_2O and oxygen fugacity in olivine-hosted melt inclusions. *Geology*, **40**, 915–918.

GHIORSO, M. S. & SACK, R. O. 1995. Chemical mass transfer in magmatic processes IV. A revised and internally consistent thermodynamic model for the interpolation and extrapolation of liquid-solid equilibria in magmatic systems at elevated temperatures and pressures. *Contributions to Mineralogy and Petrology*, **119**, 197–212.

GILL, J. B. 1981. *Orogenic Andesites and Plate Tectonics*. Springer, Heidelberg.

GREEN, D. H., HIBBERSON, W. O., KOVÁCS, I. & ROSENTHAL, A. 2010. Water and its influence on the lithosphere–asthenosphere boundary. *Nature*, **467**, 448–451.

GROVE, T. L., BAKER, M. B., PRICE, R. C., PARMAN, S. W., ELKINS-TANTON, L. T., CHATTERJEE, N. & MÜNTENER, O. 2005. Magnesian andesite and dacite lavas from Mt. Shasta, northern California: products of fractional crystallization of H_2O-rich mantle melts. *Contributions to Mineralogy and Petrology*, **148**, 542–565.

GROVE, T. L., CHATTERJEE, N., PARMAN, S. W. & MÉDARD, E. 2006. The influence of H_2O on mantle wedge melting. *Earth and Planetary Science Letters*, **249**, 74–89.

HAMADA, M. & FUJII, T. 2008. Experimental constraints on the effects of pressure and H_2O on the fractional crystallization of high-Mg island arc basalt. *Contributions to Mineralogy and Petrology*, **155**, 767–790.

HAMADA, M., KAWAMOTO, T., TAKAHASHI, E. & FUJII, T. 2011. Polybaric degassing of island arc low-K tholeiitic basalt magma recorded by OH concentrations in Ca-rich plagioclase. *Earth and Planetary Science Letters*, **308**, 259–266.

HARADA, H. 2003. *Geology and petrology of Quaternary volcanic rocks at Mengameyama area, Hiroshima Prefecture*. Graduation thesis, Shimane University.

HAURI, E. H., GAETANI, G. A. & GREEN, D. H. 2006. Partitioning of water during melting of the Earth's upper mantle at H_2O-undersaturated conditions. *Earth and Planetary Science Letters*, **248**, 715–734.

HIROSE, F., NAKAJIMA, J. & HASEGAWA, A. 2008. Three-dimensional velocity structure and configuration of the Philippine Sea slab in southwestern Japan estimated by double-difference tomography. *Journal of Geophysical Research*, **113**, B09315.

HOUSH, T. B. & LUHR, J. F. 1991. Plagioclase-melt equilibria in hydrous systems. *American Mineralogist*, **76**, 477–492.

HUMPHREYS, M. C. S., CHRISTOPHER, T. & HARDS, V. 2009. Microlite transfer by disaggregation of mafic inclusions following magma mixing at Soufrière Hills volcano, Montserrat. *Contributions to Mineralogy and Petrology*, **157**, 609–624.

HUNTER, A. G. 1998. Intracrustal controls on the coexistence of tholeiitic and calc-alkaline magma series at Aso volcano, SW Japan. *Journal of Petrology*, **39**, 1255–1284.

IMURA, R. 1994. Geology of Kirishima Volcano. *Bulletin of the Earthquake Research Institute*, University of Tokyo, Tokyo, **69**, 189–209.

IMURA, R. & KOBAYASHI, T. 2001. *Geological map of Kirishima volcano (1:50 000)*. Geological Map of Volcanoes 11. Geological Survey of Japan.

Jerram, D. A. & Martin, V. M. 2008. Understanding crystal populations and their significance through the magma plumbing system. *In*: Annen, C. & Zellmer, G. F. (eds) *Dynamics of Crustal Magma Transfer, Storage and Differentiation*. Geological Society, London, Special Publications, **304**, 133–148.

Kamata, H. 1997. *Geology of the Miyanohara District. With Geological Sheet Map at 1:50 000*. Geological Survey of Japan, Tokyo.

Kelley, K. A. & Cottrell, E. 2009. Water and the oxidation state of subduction zone magmas. *Science*, **325**, 605–607.

Kimura, J.-I., Stern, R. J. & Yoshida, T. 2005. Reinitiation of subduction and magmatic responses in SW Japan during Neogene time. *Geological Society of America Bulletin*, **117**, 969–986.

Koyaguchi, T. 1986. Textural and compositional evidence for magma mixing and its mechanism, Abu volcano group, Southwestern Japan. *Contributions to Mineralogy and Petrology*, **93**, 33–45.

Kuno, H. 1968. Differentiation of basalt magmas. *In*: Hess, H. H. & Poldervaart, A. (eds) *Basalts: The Poldervaart Treatise on Rocks of Basaltic Composition*. Wiley Interscience, New York, 623–688.

Kurosawa, M., Yurimoto, H., Matsumoto, K. & Sueno, S. 1992. Hydrogen analysis of mantle olivine by secondary ion mass spectrometry. *In*: Syono, Y. & Manghnani, M. H. (eds) *High-Pressure Research: Application to Earth and Planetary Sciences*. Terra Scientific Publishing, Tokyo, 283–287.

Kurosawa, M., Yurimoto, H. & Sueno, S. 1997. Patterns in the hydrogen and trace element compositions of mantle olivines. *Physics and Chemistry of Minerals*, **24**, 385–395.

Lange, R. A., Frey, H. M. & Hector, J. 2009. A thermodynamic model for the plagioclase-liquid hygrometer/thermometer. *American Mineralogist*, **94**, 494–506.

Lee, C. Y., Tsai, J. H., Ho, H. H., Yang, T. F., Chung, S. L. & Chen, C.-H. 1997. Quantitative analysis in rock samples by an X-ray fluorescence spectrometer (I) major elements. *Annual Meeting of the Geological Society of China*. Geological Society of China, Taipei, 418–420.

Lee, C.-T. A., Luffi, P. et al. 2012. Copper systematics in arc magmas and implications for crust–mantle differentiation. *Science*, **336**, 64–68.

Lindsley, D. H. 1983. Pyroxene thermometry. *American Mineralogist*, **68**, 477–493.

Mackwell, S. J. & Kohlstedt, D. L. 1990. Diffusion of hydrogen in olivine: implications for water in the mantle. *Journal of Geophysical Research*, **95**, 5079–5088.

Mallmann, G. & O'Neil, H. S. C. 2007. The effect of oxygen fugacity on the partitioning of Re between crystals and silicate melt during mantle melting. *Geochimica et Cosmochimica Acta*, **71**, 2837–2857.

Mallmann, G. & O'Neil, H. S. C. 2009. The crystal/melt partitioning of V during mantle melting as a function of oxygen fucacity compared with some other elements (Al, P, Ca, Sc, Ti, Cr, Fe, Ga, Y, Zr and Nb). *Journal of Petrology*, **50**, 1765–1794.

Matsuura, H. 1986. K–Ar ages of an alkali basalt from Mt. Mengame, Hiroshima Prefecture. *Journal of the Geological Society of Japan*, **92**, 235–237.

Miyoshi, M., Hasenaka, T. & Sano, T. 2005. Genetic relationships of the compositionally diverse magmas from Aso post-caldera volcanism. *Bulletin of the Volcanological Society of Japan*, **50**, 269–283.

Miyoshi, M., Shimono, M., Hasenaka, T., Sano, T., Mori, Y. & Fukuoka, T. 2010. Boron systematics of Hisatsu and Kirishima basaltic rocks from southern Kyushu, Japan. *Geochemical Journal*, **44**, 359–396.

Miyoshi, M., Sumino, H. et al. 2012. K–Ar ages determined for post-caldera volcanic products from Aso volcano, central Kyushu, Japan. *Journal of Volcanology and Geothermal Research*, **229–230**, 64–73.

Morris, P. A. 1995. Slab melting as an explanation of Quaternary volcanism and aseismicity in southwest Japan. *Geology*, **23**, 395–398.

Müntener, O. & Ulmer, P. 2006. Experimentally derived high-pressure cumulates from hydrous arc magmas and consequences for the seismic velocity structure of lower arc crust. *Geophysical Research Letters*, **33**, 1–5.

Müntener, O., Kelemen, P. B. & Grove, T. L. 2001. The role of H_2O during crystallization of primitive arc magmas under uppermost mantle conditions and genesis of igneous pyroxenites: an experimental study. *Contributions to Mineralogy and Petrology*, **141**, 643–658.

Murphy, M. D., Sparks, R. S. J. et al. 1998. The role of magma mixing in triggering the current eruption at the Soufriere Hills volcano, Montserrat, West Indies. *Geophysical Research Letters*, **25**, 3433–3436.

Murphy, M. D., Sparks, R. S. J., Barclay, J., Carroll, M. R. & Brewer, T. S. 2000. Remobilization of andesite magma by intrusion of mafic magma at the Soufriere Hills volcano, Montserrat, West Indies. *Journal of Petrology*, **41**, 21–42.

Nakajima, J. & Hasegawa, A. 2007. Subduction of the Philippine Sea plate beneath southwestern Japan: slab geometry and its relationship to arc magmatism. *Journal of Geophysical Research*, **112**, 1–18.

Nakamura, M. 1995. Continuous mixing of crystal mush and replenished magma in the ongoing Unzen eruption. *Geology*, **23**, 807–810.

Obara, K. 2002. Nonvolcanic deep tremor associated with subduction in southwest Japan. *Science*, **269**, 1679–1681.

Oda, H. & Ushio, T. 2007. Topography of the Moho and Conrad discontinuities in the Kyushu district, Southwest Japan. *Journal of Seismology*, **11**, 221–233.

Ohta, T., Hasenaka, T. & Fujimaki, H. 1990. Geology and petrography of Yufu–Tsurumi volcano group, Oita Prefecture. *Journal of Mineralogy, Petrology and Economic Geology*, **85**, 113–129.

Ohta, T., Hasenaka, T., Ban, M. & Sasaki, M. 1992. Characteristic geology and petrology of non-arc type volcanism at Oninomi Monogenetic Volcano, Yufu–Tsurumi Graben. *Bulletin of the Volcanological Society of Japan*, **37**, 119–131.

Pallister, J. S., Hoblitt, R. P. & Reyes, A. G. 1992. A basalt trigger for the 1991 eruptions of Pinatubo volcano? *Nature*, **356**, 426–428.

Peacock, M. A. 1931. Classification of igneous rock series. *Journal of Geology*, **39**, 54–67.

PECCERILLO, A. & TAYLOR, S. R. 1976. Geochemistry of Eocene calc-alkaline volcanic rocks from the Kastamonu area, northern Turkey. *Contributions to Mineralogy and Petrology*, **58**, 63–81.

PERTERMANN, M. & HIRSCHMANN, M. M. 2003. Partial melting experiments on a MORB-like pyroxenite between 2 and 3 GPa: constraints on the presence of pyroxenite in basalt source regions from solidus location and melting rate. *Journal of Geophysical Research*, **108**, 2125.

PESLIER, A. H. 2010. A review of water contents of nominally anhydrous natural minerals in the mantles of Earth, Mars and the Moon. *Journal of Volcanology and Geothermal Research*, **197**, 239–258.

PESLIER, A. H. & LUHR, J. F. 2006. Hydrogen loss from olivines in mantle xenoliths from Simcoe (USA) and Mexico: Mafic alkalic magma ascent rates and water budget of the sub-continental lithosphere. *Earth and Planetary Science Letters*, **242**, 302–319.

PORTNYAGIN, M., ALMEEV, R., MATVEEV, S. & HOLTZ, F. 2008. Experimental evidence for rapid water exchange between melt inclusions in olivine and host magma. *Earth and Planetary Science Letters*, **272**, 541–552.

PUTIRKA, K. D. 2008. Thermometers and barometers for volcanic systems. *In*: PUTIRKA, K. D. & TEPLEY, F. (eds) *Minerals, Inclusions and Volcanic Processes*. American Mineralogical Society, Washington, DC Reviews in Mineralogy and Geochemistry, **69**, 61–120.

PUTIRKA, K. & CONDIT, C. D. 2003. Cross section of a magma conduit system at the margin of the Colorado Plateau. *Geology*, **31**, 701–704.

PYLE, D. M., IVANOVICH, M. & SPARKS, R. S. J. 1988. Magma cumulate mixing identified by U–Th disequilibrium dating. *Nature*, **331**, 157–159.

RAMOS, F. C. & TEPLEY, F. J. 2008. Inter- and intracrystalline isotopic disequilibria: techniques and applications. *In*: PUTIRKA, K. & TEPLEY, F. J. (eds) American Mineralogical Society, Washington, DC Reviews in Mineralogy and Geochemistry, **69**, 403–444.

SHELLNUTT, J. G. & ZELLMER, G. F. 2010. High-Mg andesite genesis by upper crustal differentiation. *Journal of the Geological Society, London*, **167**, 1081–1088.

SMITH, P. M. & ASIMOW, P. D. 2005. Adiabat_1ph: a new public front-end to the MELTS, pMELTS, and pHMELTS models. *Geochemistry Geophysics Geosystems*, **6**, Q02004.

SMITH, I. E. M., STEWART, R. B., PRICE, R. C. & WORTHINGTON, T. J. 2010. Are arc-type rocks the products of magma crystallisation? Observations from a simple oceanic arc volcano: Raoul Island, Kermadec Arc, SW Pacific. *Journal of Volcanology and Geothermal Research*, **190**, 219–234.

SNYDER, D. 1997. The mixing and mingling of magmas. *Endeavour*, **27**, 19–22.

STEWART, R. B., PRICE, R. C. & SMITH, I. E. M. 1996. Evolution of high-K arc magma, Egmont volcano, Taranaki, New Zealand: evidence from mineral chemistry. *Journal of Volcanology and Geothermal Research*, **74**, 275–295.

STRAUB, S. M. & ZELLMER, G. F. 2012. Volcanic arcs as archives of plate tectonic change. *Gondwana Research*, **21**, 495–516.

STRAUB, S. M., LAGATTA, A. B., MARTIN-DEL POZZO, A. L. & LANGMUIR, C. H. 2008. Evidence from high-Ni olivines for a hybridized peridotite/pyroxenite source for orogenic andesites from the central Mexican Volcanic Belt. *Geochemistry Geophysics Geosystems*, **9**, 1–33.

STRAUB, S. M., GOMEZ-TUENA, A., STUART, F. M., ZELLMER, G. F., CAI, Y. & ESPINASA-PERENA, R. 2011. Formation of hybrid arc andesites beneath thick continental crust. *Earth and Planetary Science Letters*, **303**, 337–347.

STRAUB, S. M., GOMEZ-TUENA, A. *ET AL*. 2013. The processes of melt differentiation in arc volcanic rock: insights from OIB-type arc magmas in the central Mexican Volcanic Belt. *Journal of Petrology*, **54**, 665–701.

STRECK, M. J. 2008. Mineral textures and zoning as evidence for open system processes. *In*: PUTIRKA, K. & TEPLEY, F. (eds) *Minerals, Inclusions and Volcanic Processes*. American Mineralogical Society, Washington, DC Reviews in Mineralogy and Geochemistry, **69**, 595–622.

STRECK, M. J., LEEMAN, W. P. & CHESLEY, J. 2007. High-magnesian andesite from Mount Shasta: a product of magma mixing and contamination, not a primitive mantle melt. *Geology*, **35**, 351–354.

TAMURA, Y., YUHARA, M. & ISHII, T. 2000. Primary arc basalts from Daisen Volcano, Japan: equilibrium crystal fractionation versus disequilibrium fractionation during supercooling. *Journal of Petrology*, **41**, 431–448.

TATSUMI, Y. & ISHIZAKA, K. 1981. Existence of andesitic primary magma: an example from southwest Japan. *Earth and Planetary Science Letters*, **53**, 124–130.

TATSUMI, Y. & ISHIZAKA, K. 1982. Origin of high-magnesian andesites in the Setouchi volcanic belt, southwest Japan, I. Petrographical and chemical characteristics. *Earth and Planetary Science Letters*, **60**, 293–304.

TROLL, V. R. & SCHMINCKE, H.-U. 2002. Magma mixing and crustal recycling recorded in ternary feldspar from compositionally zoned peralkaline Ignimbrite 'A', Gran Canaria, Canary Islands. *Journal of Petrology*, **43**, 243–270.

TURNER, S. P., GEORGE, R. M. M., EVANS, P. J., HAWKESWORTH, C. J. & ZELLMER, G. F. 2000. Time-scales of magma formation, ascent and storage beneath subduction-zone volcanoes. *Philosophical Transactions of the Royal Society of London A*, **358**, 1443–1464.

UENO, T., SHIBUTANI, T. & ITO, K. 2008. Configuration of the continental Moho and Philippine Sea slab in southwest Japan derived from receiver function analysis: relation to subcrustal earthquakes. *Bulletin of the Seismological Society of America*, **98**, 2416–2427.

WANG, K. L., CHUNG, S. L., O'REILLY, S. Y., SUN, S. S., SHINJO, R. & CHEN, C. H. 2004. Geochemical constraints for the genesis of post-collisional magmatism and the geodynamic evolution of the northern Taiwan region. *Journal of Petrology*, **45**, 975–1011.

YURIMOTO, H., KUROSAWA, M. & SUENO, S. 1989. Hydrogen analysis in quartz crystals and quartz glasses by secondary ion mass spectrometry. *Geochimica et Cosmochimica Acta*, **53**, 751–755.

ZELLMER, G. F. 2008. Some first order observations on magma transfer from mantle wedge to upper crust at volcanic arcs. *In*: ANNEN, C. & ZELLMER, G. F. (eds) *Dynamics of Crustal Magma Transfer, Storage and Differentiation*. Special Publications, Geological Society, London, 15–31.

ZELLMER, G. F. 2009. Petrogenesis of Sr-rich adakitic rocks at volcanic arcs: insights from global variations of eruptive style with plate convergence rates and surface heat flux. *Journal of the Geological Society*, **166**, 725–734.

ZELLMER, G. F. & ANNEN, C. 2008. An introduction to magma dynamics. *In*: ANNEN, C. & ZELLMER, G. F. (eds) *Dynamics of Crustal Magma Transfer, Storage and Differentiation*. Geological Society, London, Special Publications, **304**, 1–13.

ZELLMER, G. F. & CLAVERO, J. 2006. Using trace element correlation patterns to decipher a sanidine crystal growth chronology: an example from Taapaca volcano, Central Andes. *Journal of Volcanology and Geothermal Research*, **156**, 291–301.

ZELLMER, G. F. & TURNER, S. P. 2007. Arc dacite genesis pathways: evidence from mafic enclaves and their hosts in Aegean lavas. *Lithos*, **95**, 346–362.

ZELLMER, G. F., BLAKE, S., VANCE, D., HAWKESWORTH, C. & TURNER, S. 1999. Plagioclase residence times at two island arc volcanoes (Kameni islands, Santorini, and Soufriere, St. Vincent) determined by Sr diffusion systematics. *Contributions to Mineralogy and Petrology*, **136**, 345–357.

ZELLMER, G. F., TURNER, S. P. & HAWKESWORTH, C. J. 2000. Timescales of destructive plate margin magmatism: new insights from Santorini, Aegean volcanic arc. *Earth and Planetary Science Letters*, **174**, 265–282.

ZELLMER, G. F., HAWKESWORTH, C. J., SPARKS, R. S. J., THOMAS, L. E., HARFORD, C., BREWER, T. S. & LOUGHLIN, S. 2003*a*. Geochemical evolution of the Soufrière Hills volcano, Montserrat, Lesser Antilles Volcanic Arc. *Journal of Petrology*, **44**, 1349–1374.

ZELLMER, G. F., SPARKS, R. S. J., HAWKESWORTH, C. J. & WIEDENBECK, M. 2003*b*. Magma emplacement and remobilization timescales beneath Montserrat: insights from Sr and Ba zonation in plagioclase phenocrysts. *Journal of Petrology*, **44**, 1413–1431.

ZELLMER, G. F., ANNEN, C., CHARLIER, B. L. A., GEORGE, R. M. M., TURNER, S. P. & HAWKESWORTH, C. J. 2005. Magma evolution and ascent at volcanic arcs: constraining petrogenetic processes through rates and chronologies. *Journal of Volcanology and Geothermal Research*, **140**, 171–191.

ZELLMER, G. F., RUBIN, K. H., DULSKI, P., IIZUKA, Y., GOLDSTEIN, S. L. & PERFIT, M. R. 2011. Crystal growth during dike injection of MOR basaltic melts: evidence from preservation of local Sr disequilibria in plagioclase. *Contributions to Mineralogy and Petrology*, **161**, 153–173.

ZELLMER, G. F., DULSKI, P., IIZUKA, Y. & PERFIT, M. 2012*a*. Rates and processes of crystallization in on-axis and off-axis MOR basaltic melts. *Lithos*, **154**, 1–15.

ZELLMER, G. F., IIZUKA, Y., MIYOSHI, M., TAMURA, Y. & TATSUMI, Y. 2012*b*. Lower crustal H_2O controls on the formation of adakitic melts. *Geology*, **40**, 487–490.

ZELLMER, G. F., FREYMUTH, H., CEMBRANO, J. M., CLAVERO, J. E., VELOSO, E. A. E. & SIELFELD, G. G. 2013. Altered mineral uptake into fresh arc magmas: insights from U–Th isotopes of samples from Andean volcanoes under differential crustal stress regimes. *In*: GÓMEZ-TUENA, A., STRAUB, S. M. & ZELLMER, G. F. (eds) *Orogenic Andesites and Crustal Growth*. Geological Society, London, Special Publications, **385**. First published online July 18, 2013, http://dx.doi.org/10.1144/SP385.9

Altered mineral uptake into fresh arc magmas: insights from U–Th isotopes of samples from Andean volcanoes under differential crustal stress regimes

GEORG F. ZELLMER[1,2]*, HEYE FREYMUTH[3], JOSÉ M. CEMBRANO[4], JORGE E. CLAVERO[4], EUGENIO A. E. VELOSO[5] & GERD G. SIELFELD[6]

[1]*Institute of Natural Resources, Massey University, Palmerston North 4442, New Zealand*

[2]*Institute of Earth Sciences, Academia Sinica, 128 Academia Road Sec. 2, Nankang, Taipei 11529, Taiwan, ROC*

[3]*Department of Earth Sciences, University of Bristol, Wills Memorial Building, Queens Road, Bristol BS8 1RJ, UK*

[4]*Departamento de Ingeniería Estructural y Geotécnica, Pontificia Universidad Católica de Chile, Santiago, Vicuña Mackenna 4860, Macul, Santiago, Chile*

[5]*Departamento de Ciencias Geológicas, Facultad de Ingeniería y Ciencias Geológicas, Universidad Católica del Norte, Antofagasta, Chile*

[6]*Magma Energy Corp, Carlos Antunes 1940, Providencia, Santiago, Chile*

**Corresponding author (e-mail: g.f.zellmer@massey.ac.nz)*

Abstract: Uranium–thorium (U–Th) isotope compositions of whole rocks, groundmasses and minerals from mafic to intermediate Andean arc magmas were determined to assess the influence of crustal stress on rates of pre-eruptive crystallization and the significance of crystal uptake. Volcanoes investigated include Lascar in the central Andes, situated in a compressional regime, and Callaqui and Lonquimay in the southern Andes, situated in a transtensional regime. In the southern Andes, Calbuco volcano, which experienced variations in the local stress field due to volcano unloading triggered by large debris avalanches, was also targeted. In U–Th equiline space, whole rock and groundmass coincide in each studied sample, and are close to secular equilibrium or in slight U-excess. No isochronal relationships are displayed by the mineral separates, although many are out of secular equilibrium. Further, ($^{234}U/^{238}U$) activity ratios of some mineral separates from the southern Andes are in disequilibrium and vary significantly between different phases of individual rock samples. The combined data indicate that many crystals are foreign to the melts they are carried in, and that some mineral phases have experienced incipient weathering before their incorporation into the melt. Using Calbuco as an example, we speculate that volcano loading may affect secondary alteration processes at depth.

The principles that govern magma transport from the Earth's mantle through the crustal lid by fracture propagation have been known for decades (e.g. Shaw 1980; Spera 1980, and references therein). Propagation is controlled by the orientation and magnitude of the principal stresses within the crust, as they are modified by the internal pressure of magma. Igneous dykes and sills are injected along planes perpendicular to the axis of least principal stress, σ_3 (cf. Fig. 1). In turn, magma transport pathways and times govern the location and extent of thermal anomalies within the crust, and the chemical evolution of the melts. Furthermore, it has been shown that steady state creep below the upper seismogenic zone leads to steady state permeability and continuous fluid flow (Cox 2002), so that magma ascent by dyking events leading up to volcanic eruptions will be controlled by the thickness of the seismogenic zone, and will be largely independent of overall crustal thickness.

While interactions between magmatism and tectonism are quite well understood from a qualitative viewpoint, quantitative data obtained from the natural laboratory are essentially absent. The last three decades have seen a multitude of studies that have constrained crystal and magma residence

From: GÓMEZ-TUENA, A., STRAUB, S. M. & ZELLMER, G. F. (eds) 2014. *Orogenic Andesites and Crustal Growth*. Geological Society, London, Special Publications, **385**, 185–208.
First published online July 18, 2013, http://dx.doi.org/10.1144/SP385.9

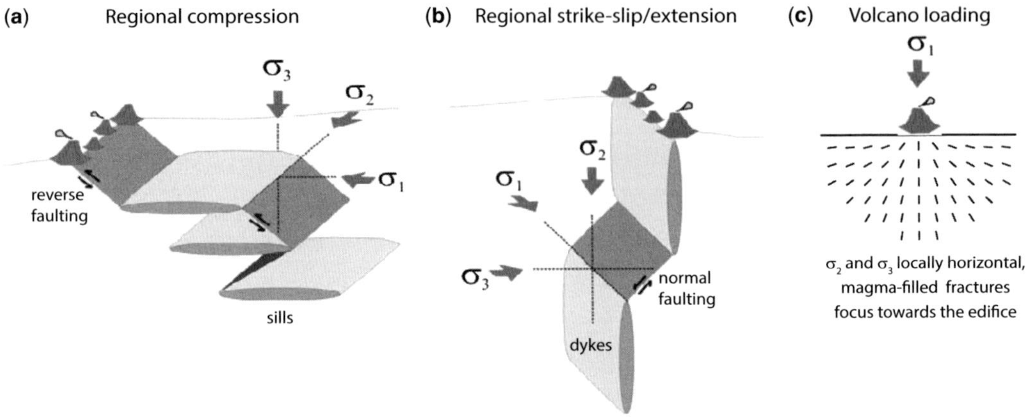

Fig. 1. Stress-field control on magma ascent, adapted from Cembrano & Lara (2009) and Muller *et al.* (2001). (**a**) In compressive settings, horizontal sills will allow longer residence times. (**b**) In strike-slip/extensional settings, vertical dykes will enhance rapid transport of magma. (**c**) Local effect of melt focusing by edifice loading. See text for discussion.

times, using both radiometric and geospeedometric methods (e.g. Allègre & Condomines 1976; Pyle *et al.* 1988; Volpe & Hammond 1991; Schaefer *et al.* 1993; Nakamura 1995; Heath *et al.* 1998; Zellmer *et al.* 1999, 2000, 2003; Turner *et al.* 2000, 2003; Cooper & Reid 2003; Costa *et al.* 2003; Morgan *et al.* 2004; Costa & Dungan 2005; Donnelly & Cooper 2006; Morgan & Blake 2006; Zellmer & Clavero 2006; Turner & Costa 2007). Yet, these have been isolated attempts to understand subvolcanic processes and conditions, such as melt generation, transport and storage, and magmatic pressures, temperatures, compositions and volatile contents at individual volcanoes, without considering tectonic controls. The potential of relating residence times to the crustal stress field in hazardous continental arc volcanoes is one motivation for the present study.

One typical feature of arc volcanics is their complex petrography compared to magmas erupted in other tectonic settings. Crystals often show disequilibrium textures, such as resorption zones associated with distinct chemical variations (e.g. Zellmer *et al.* 1999; Charlier & Zellmer 2000; Zellmer & Clavero 2006; Zellmer & Turner 2007; Jerram & Martin 2008; Ramos & Tepley 2008; Streck 2008). Frequently, minerals are in chemical disequilibrium with their host melt, as evident from reaction rims. This is typically attributed to open-system processes, such as mixing between magmas (e.g. Snyder 1997; Clynne 1999) or between magma and crystal mush (Pyle *et al.* 1988; Nakamura 1995), or crystal uptake into the melt during crustal magma transfer or storage (e.g. Stewart *et al.* 1996; Dungan 2005; Davidson *et al.* 2007; Streck *et al.* 2007; Jerram & Martin 2008; Zellmer *et al.* 2013). How stresses in the overriding crust may influence such processes is unknown. Magmas that cannot ascend easily due to an unfavourable tectonic condition might be expected to be more strongly affected by antecryst or xenocryst uptake, which may then be reflected by an older average age of the crystal cargo. Again, relating crystal residence times to the crustal stress field may therefore provide valuable insights into the complex petrogenetic processes operating at destructive plate margins.

Approach

Selection of appropriate samples

The stress fields at convergent margins are highly variable, resulting in volcanic arcs with strain characteristics ranging from compressional to extensional (Jarrard 1986). However, closer inspection shows that even within individual arcs, differences in crustal stress fields may be considerable. In the Central Andes of northern Chile, for example, the crust is under strong compression, while in the southern Andes, oblique convergence results in both transpressional and transtensional settings for individual volcanic edifices in the arc. In the present study, we investigate samples erupted under compression from Lascar volcano in the Central Andes, and samples erupted under transtension from Callaqui and Lonquimay volcano in the southern Andes (Table 1 and Fig. 2a). The hypothesis to be tested is that magmas ascending through

Table 1. *Sample details*

Volcano name	Sample name	Southern latitude	Western longitude	Comment
Lascar	LA1	23°19′27.76″	67°64′41.57″	Western edge of historical lava flow east of Tumbre
	LA2	23°19′39.19″	67°46′14.94″	Historical lava flow sampled above blocks of welded ignimbrite
	LA4	23°23′47.39″	67°44′48.67″	Young lava flow towards SE of the volcano
Callaqui	08CA01	38°00′03.01″	71°29′18.48″	70 ka columnar basalts
	08CA02	38°00′04.39″	71°28′56.18″	Holocene lava flow, estero Ñireco, by the road
	08CA03	37°54′36.66″	71°23′44.38″	Lava flow from NW-trending crater alignment of Quillayoahue
Lonquimay	08LO01	38°20′53.65″	71°31′18.74″	Lava flow from Navidad crater (1988–1989 AD)
	08LO02	38°17′46.52″	71°29′24.82″	Lava block from distal end of northeastern flow field
	08LO03	38°21′03.44″	71°31′11.70″	Lava flow from eruptive centre belonging to NE-Chain (historical eruption, 1887?)
	08LO04	38°21′37.22″	71°31′44.35″	Lava flow from eruptive centre belonging to NE-Chain (historical eruption, 1887?)
Calbuco	CAJ-10A	41°19′03.56″	72°41′31.98″	Block from block-and-ash flow deposit dated at 9615 ± 60 years BP by ^{14}C
	CAJ-24	41°19′03.31″	72°40′53.73″	Block from immediately underneath a surge bed dated at 6915 ± 80 years BP by ^{14}C
	CAJ-18	41°14′39.49″	72°33′21.71″	Block from block-and-ash flow deposit dated at 445 years BP by ^{14}C
	CAJ-23	41°14′09.45″	72°34′03.98″	Prismatically jointed block from 1961 AD hot lahar

crust under compression tend to pond in crustal magma reservoirs and thus will yield longer residence times and greater degrees of differentiation and, potentially, assimilation than magmas ascending through crust in transtensional settings (cf. Fig. 1a, b).

Further, the regional stress field in volcanic arcs may be significantly modulated through local effects, such as volcano loading and pressurization of upper crustal magma reservoirs, as has been shown by analogue experiments and theoretical work (e.g. van Wyk de Vries & Merle 1998; Muller *et al.* 2001; Karlstrom *et al.* 2009). In the case of volcano loading, σ_1 may locally be vertical, resulting in focusing of propagating magma-filled fractures towards the growing edifice (e.g. van Wyk de Vries & Merle 1998; Muller *et al.* 2001, cf. Fig. 1c). However, edifice loading will also increase the vertical stress above background lithostatic stress, thereby impeding the vertical propagation of fissures of buoyant magma and thus decreasing eruption potential (Pinel & Jaupart 2000). As a result, it may be expected that crustal magma residence times will increase with increasing edifice load. Correspondingly, large debris avalanche events that result in volcano unloading may lead to reduced magma residence times during subsequent eruptive activity. Whether volcano loading is, indeed, a significant factor in the amount of pressure build-up required to trigger an eruption from a subvolcanic reservoir, thus potentially affecting magma residence times, has yet to be rigorously tested in nature, although several studies have previously suggested that loading may play a role on the basis of distinct changes in eruption volumes or rates in response to small ambient stress variations (McNutt & Beavan 1987; Bacon & Lanphere 2006). This study investigates samples erupted from Calbuco volcano in the southern Andes (Fig. 2a) both prior to and following two large Holocene debris avalanche events that destroyed most of the volcanic edifice (cf. Table 1).

Uranium–thorium isochron dating

To determine residence times and the origin of the crystal cargo at the selected volcanoes, U–Th mineral isochron dating was employed (e.g. Allègre & Condomines 1976; Pyle *et al.* 1988; Condomines 1997; Heath *et al.* 1998; Turner *et al.* 2000, 2003; Zellmer *et al.* 2000, 2008; Jicha *et al.* 2007; Turner & Costa 2007). Mineral separates of erupted rock samples have previously been successfully used to obtain absolute ages of crystallization,

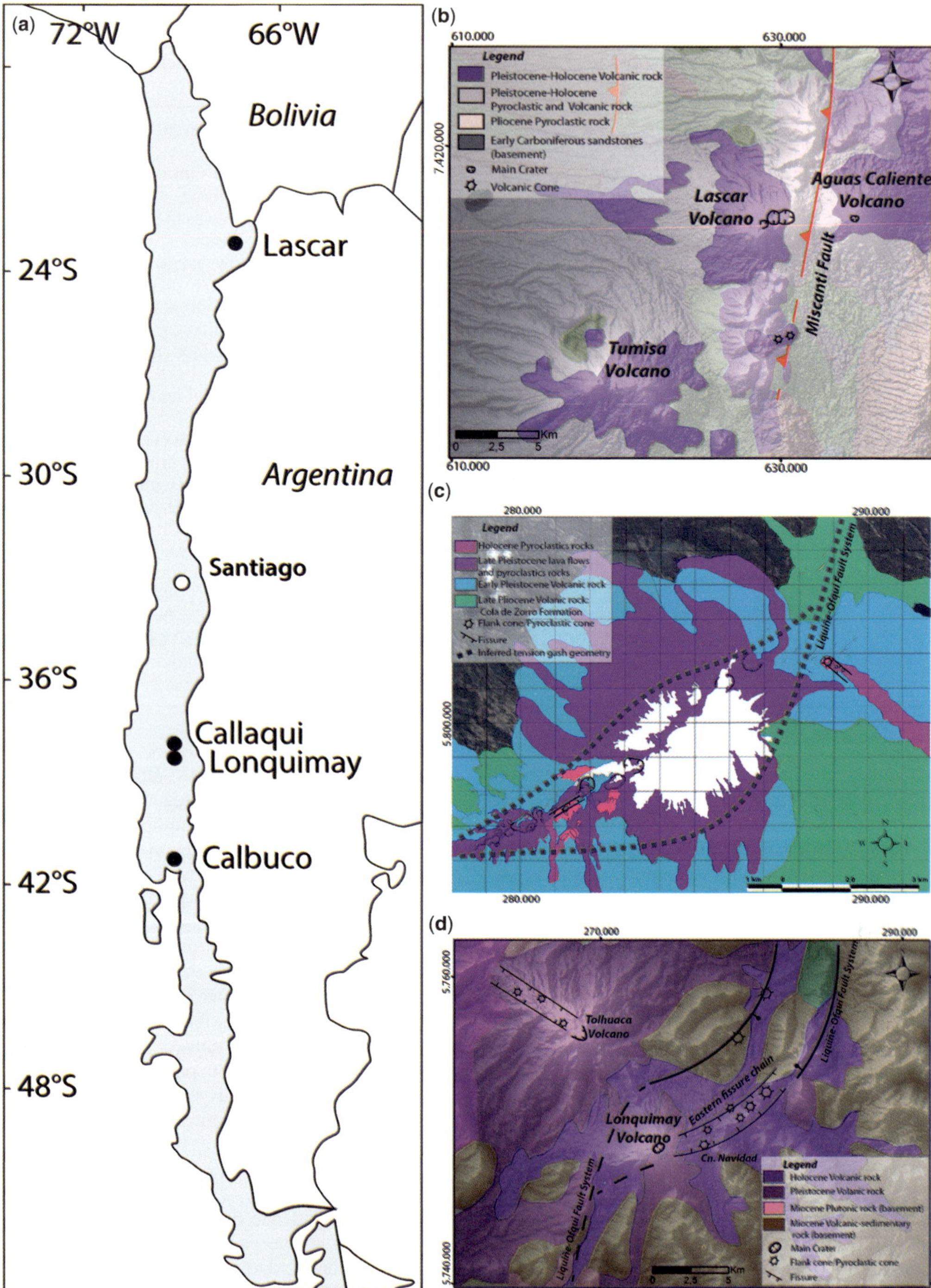

Fig. 2. (**a**) Location map of the volcanoes studied. (**b**) Simplified regional geological map of Lascar Volcano and adjacent edifices, with Universal Transverse Mercator (UTM) coordinates. (**c**) Simplified geological map of Callaqui Volcano, with UTM coordinates. (**d**) Simplified regional geological map of Lonquimay and Tolhuaca volcanoes, with UTM coordinates.

which can then be compared to independently constrained eruption ages (e.g. ^{14}C ages, or historic eruption ages) to derive pre-eruptive magma residence times within the crust. We analysed several phases (minerals plus whole rock and/or groundmass) from each of three samples from Lascar, three samples from Callaqui, two samples from Lonquimay and four samples from Calbuco (two pre-avalanche and two post-avalanche samples), in order to elucidate potential differences in absolute residence times, and how these might be affected by the crustal stress field.

Geological background of the investigated volcanoes

Láscar Volcano, Northern Volcanic Zone

Lascar Volcano in Northern Chile (23°22′S 67°44′W; 5590 m) is currently the most active volcano of the Central Volcanic Zone of the Andes. It is a small-volume system located at the border of the Chilean Altiplano, adjacent to the Salar de Atacama (30 km to the east) and has been active for more than 43 ka (Gardeweg *et al.* 1998). Lascar is a twin strato-volcano (Fig. 2b) consisting of two adjacent cones (Lascar I and II), which are arranged in an ENE orientation. These two cones emitted numerous more-or-less interlaced lava flows on the western slope. The flows in the eastern flank of the volcano are short and are completely covered with thick pyroclastic deposits. Six craters occupy the summit area (Deruelle *et al.* 1995).

The structural geology of the area is still poorly known. However, Post-Pliocene reverse faults have been mapped close to the volcano (Ramírez & Gardeweg 1982) and a prominent north–south-trending lineament appears to crosscut part of the Quaternary volcanic deposits (Fig. 2b). This lineament, which extends for tens of kilometres north and south of Lascar volcano, has been called the 'Miscanti fault'. Until recently, little was known regarding its precise nature and kinematics. However, recent work by González *et al.* (2009) suggests that the Miscanti lineament corresponds to the north–south-trending hinge zone of an asymmetrical fault-propagated fold. Volcano-tectonic relationships a few kilometres to the south of Láscar, at the latitude of Lejía Volcano, indicate that Pleistocene volcanism in this part of the Central Volcanic Zone of the Andes has been coeval with intra-arc compression (e.g. Cembrano *et al.* 2008; González *et al.* 2009).

The Láscar basement consists of Devonian to Early Carboniferous sandstones (Lila Formation), Permian volcanic rocks and granites (Cas Formation) and volcanic Permo-Triassic rocks (Peine Strata and Cerro Negro Strata) (Ramírez & Gardeweg 1982). Basement rocks occur as fault-bound inliers surrounded by Tertiary rocks.

Callaqui and Lonquimay volcanoes, Southern Volcanic Zone

Callaqui Volcano represents a NE–SW-trending Late Pleistocene to Holocene fissure system, the younger emissions of which were fed from three NE-trending parallel alignments of pyroclastic cones. It is located to the west of the main trace of the Liquiñe-Ofqui Fault System (Fig. 2c). Basalts to basaltic andesites predominate, although dacites were recognized as dykes in an older unit (*c.* 120 ka), which was deeply eroded during the last glacial maxima. Holocene magmas extruded as pyroclastics or lavas are olivine–basalts to augite–basalts. At least five historic eruptions have been recorded for this volcano. The last one was a small event that took place in 1980. Currently, Callaqui Volcano has strong solfataric activity on the south flank near the summit. Based on morphostructural features and both K–Ar and ^{14}C dates, six chronostratigraphic units, with ages ranging from 171 ka to Holocene, have been mapped by Naranjo *et al.* (1999). Small volume pyroclastic flow deposits have been dated between 9950 and 320 years BP.

The clear tension-gash-like structure of this NE-elongated volcano (Fig. 2c), along with independent structural and seismic evidence showing long-term and short-term NE-trending direction of the maximum principal stress associated with the dextral transpressional Liquiñe-Ofqui fault zone (e.g. Cembrano *et al.* 1996; Lavenu & Cembrano 1999; Melnick *et al.* 2006; Cembrano & Lara 2009), makes this an excellent case study for potentially rapid ascent of primitive magmas under transtension.

Around 50 km south of Callaqui, Lonquimay Volcano, or Lonquimay Volcanic Complex (LVC), consists of the Lonquimay stratovolcano itself and a NE-trending chain of small vents, which occur to the east along a fissure called the 'Eastern fissure chain' (EFC, Moreno & Gardeweg 1989, cf. Fig. 2d). Lonquimay volcano was built mainly during the last postglacial period, and its last activity corresponds to the formation of the 'Navidad' pyroclastic cone in 1988–1989, assigned to the EFC and described by Moreno & Gardeweg (1989), Gardeweg *et al.* (1990) and Naranjo *et al.* (1991). Products ejected by the LVC consist of very sparsely phyric basaltic to andesitic lava and pyroclastic flows of calc-alkaline affinity. Explosive activity was documented in 1853, 1887, 1889–1890, 1933, 1940 and 1988–1990 (25 December 1988 to January 1990) and was dominated by Strombolian or Vulcanian eruptions (Suárez & Emparan 1997). Lonquimay Volcano itself consists of a small truncated cone-shaped stratovolcano, slightly elongated in NE–SW orientation and with steep slopes (30–40°). Using morphostructural criteria, Thiele *et al.* (1987) identified five stratigraphic units for the main stratovolcano. These units include glaciated andesitic–basaltic lava flows and pyroclastic deposits covered in part by post-glacial andesitic and pyroclastic flows, and by mafic to intermediate lava flows erupted from flank vents and from numerous craters, cinder cones and domes of the EFC.

The LVC has developed along an ENE-striking dilational jog located at a right step between two overlapping NNE-striking segments of the Liquiñe-Ofqui fault system. Crustal seismicity recorded during the 1988–1989 eruptive event (e.g. Barrientos & Acevedo-Aránguiz 1992) is consistent with a bulk dextral strike–slip setting, which characterizes the tectonics of the volcanic arc between 38°S and 46°S (e.g. Cembrano & Lara 2009).

The basement in this part of the Andes consists of the thick Eocene to Lower Miocene sequence of lavas, volcaniclastic deposits, and fluvial and lacustrine sediments of the Cura-Mallín Formation, conformably overlain by the Middle to Upper Miocene continental volcaniclastic sequence composed of andesitic lavas and minor epiclastic and sedimentary rocks of the Trapa-Trapa Formation, which locally displays weak to moderate propylitic alteration (Niemeyer & Muñoz 1983; Muñoz & Niemeyer 1984). These strata are unconformably overlain by younger Quaternary volcanics and sedimentary rocks (Campanario and Cola de Zorro formations). Recently, Jordan *et al.* (2001) reported $^{40}Ar/^{39}Ar$ ages of 22.8 ± 0.7 Ma and 24.6 ± 1.8 Ma for lavas from the Cura-Mallín Formation. Andesites from the Trapa-Trapa Formation yielded K–Ar ages of 11.8 and 18.6 Ma (Muñoz & Niemeyer 1984) and an $^{40}Ar/^{39}Ar$ age of 16.2 ± 0.2 Ma (Jordan *et al.* 2001).

Calbuco Volcano, Southern Volcanic Zone

Calbuco is an active and hazardous volcano located in the southern Andes of Chile (Stern 2004), and has had ten explosive eruptions in the last century, last erupting in 1961. Its evolution is characterized mainly by the extrusion of silicic andesite lavas and domes and their associated pyroclastic flows (mainly block-and-ash flows and lateral blasts), as well as cold and hot lahars (López-Escobar *et al.* 1992; Hickey-Vargas *et al.* 1995; Castruccio *et al.* 2010). A simplified geological map of Calbuco is provided in Figure 3. A distinctive hummocky terrain is well developed on its lower northern flank, associated with two sector collapses that affected the volcano in postglacial times. Radiocarbon dating of some pyroclastic deposits and palaeosoil horizons indicate that the two sector collapses occurred at *c.* 6500 and *c.* 1200 years BP (Clavero *et al.* 2008). The extremely young age of one of the deposits, distributed towards increasingly populated areas, poses important problems in terms of volcanic hazards related to possible future sector collapses.

The older debris avalanche deposit is distributed mainly on the northwestern flank, covers a minimum area of 55 km^2 and has a maximum extension of 16 km from the current vent. The deposits reveal incorporation of flank volcanoclastics during collapse, some of which have preserved internal structures. The younger debris avalanche deposit is distributed mainly towards the northern and northeastern flanks of the volcano, covering a minimum area of 25 km^2 and having a maximum extension of 14 km from the current vent (Clavero *et al.* 2008).

The Calbuco basement is composed of Early Pleistocene volcaniclastic material and porphyritic basalts, known as the 'Huenuhuenu' strata, Miocene plutonic rocks of the extensive Northern Patagonian Batholith (gabbros to granodiorites that have been grouped into the 'Ralun' superunit), and older (Upper Palaeozoic?) metasedimentary rocks (López-Escobar *et al.* 1992). The plutonic rocks have been dated at 16–10 Ma (K–Ar in biotite, Munizaga *et al.* 1988).

Analytical techniques

Whole-rock major oxide geochemistry

Powder samples of whole rocks were prepared using a jaw crusher and an agate ball mill at the Institute of Earth Sciences, Academia Sinica. Rock powders were fused to make glass beads and analysed for major oxides by X-ray fluorescence using a Rigaku RIX 2000 spectrometer at the Department of Geosciences, National Taiwan University. The analytical procedures were the same as those described by Lee *et al.* (1997) and Wang *et al.* (2004), yielding analytical uncertainties better than $\pm 5\%$ (2σ) for all major elements. Loss on ignition was determined by routine procedures.

Whole-rock trace element geochemistry

A large suite of trace elements, including the rare earth elements (REEs), was analysed at National Taiwan University on dissolutions of the samples by inductively coupled plasma mass spectrometry (ICPMS), using an Agilent 7500cx spectrometer. Sample powders were dissolved in a HF/HNO_3 (1:1) mixture in capped Savilex beakers at

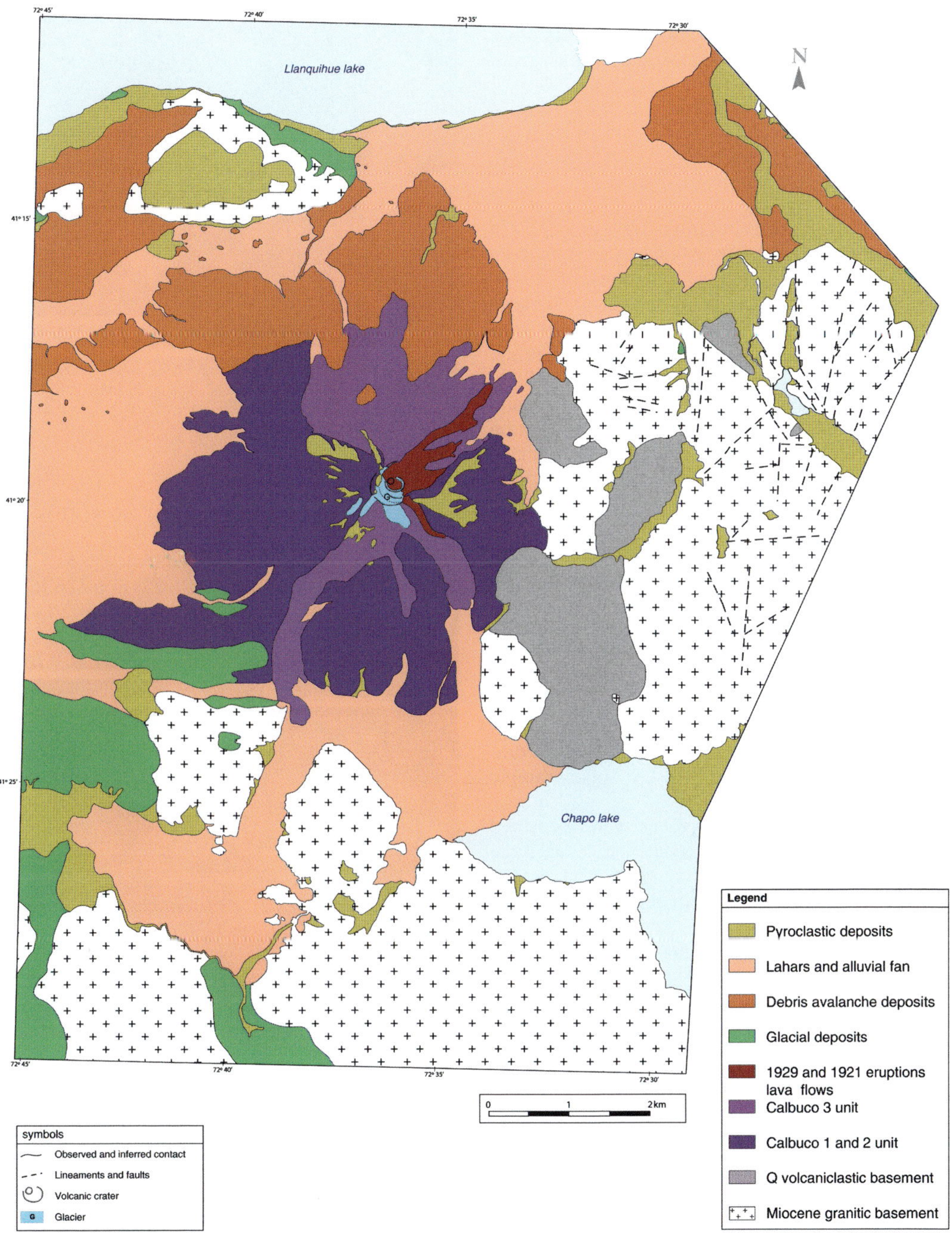

Fig. 3. Geological map of Calbuco Volcano and surroundings.

c. 100 °C, followed by evaporation to dryness, refluxing in 7N HNO_3, followed by evaporation to dryness and final dissolution of the sample cake in 2% HNO_3. The solution was diluted with 2% HNO_3 and an internal standard solution of 10 ppb Rh and Bi to a sample/solution weight ratio of 1:4000. The internal standard was used for monitoring signal shift during ICPMS measurements. The analytical precision for trace elements is better than 5% (2σ) (Yang *et al.* 2005). Also run as an unknown was a solution of the BCR-2 standard, which agreed well with literature data.

Analysis of U and Th isotopes in whole rocks, groundmasses and mineral separates

Several kilograms of each sample were processed at Chengxin Geological Services Co. Ltd, Langfang City, PRC, to obtain gram quantities of highly pure groundmass and mineral separates. Samples were crushed using mechanical crushers and grinders, and the crushate was washed prior to sieving. This was followed by a combination of Wifley table, magnetic, electromagnetic, and heavy liquid (tribromomethane and diiodomethane) separation procedures, and hand-picking under the microscope. Separates were cleaned in alcohol and dried at 80 °C. The extremely high purity of minerals was confirmed using binocular microscopy at the Institute of Earth Sciences, Academia Sinica.

Analysis of whole rocks, groundmasses and mineral separates for uranium and thorium isotopes was undertaken at the University of Bristol. Samples were spiked with a ^{229}Th–^{236}U spike and decomposed in $HF/HClO_4$ followed by a repeated treatment with $HClO_4$ as described by Yokoyama *et al.* (1999). They were subsequently dissolved in 6N HCl, which was repeated until the sample was completely dissolved in HCl to ensure spike–sample equilibration.

Uranium and thorium isotopes were measured largely following the procedure of Hoffmann *et al.* (2007) with the following exceptions: U and Th isotopes were measured in separate sequences and U was measured with the RPQ grounded, while Th was measured with its own cup configuration and without using a dispersion lens. The in-house standard ‘Thosi’ with ^{230}Th/^{232}Th similar to silicate rocks was used as a bracketing standard for Th. The BCR-2 standard was routinely dissolved together with the samples. Harwell Uraninite was measured in U sequences and gave results of (^{234}U/^{238}U) = 1.0004 ± 0.0017 (2σ, n = 7), and IRMM036 was measured in Th sequences with ^{230}Th/^{232}Th = $3.057 \times 10^{-6} \pm 1.0028 \times 10^{-8}$ (2σ, n = 7). Typical procedural blanks were 10 pg U and 15 pg Th.

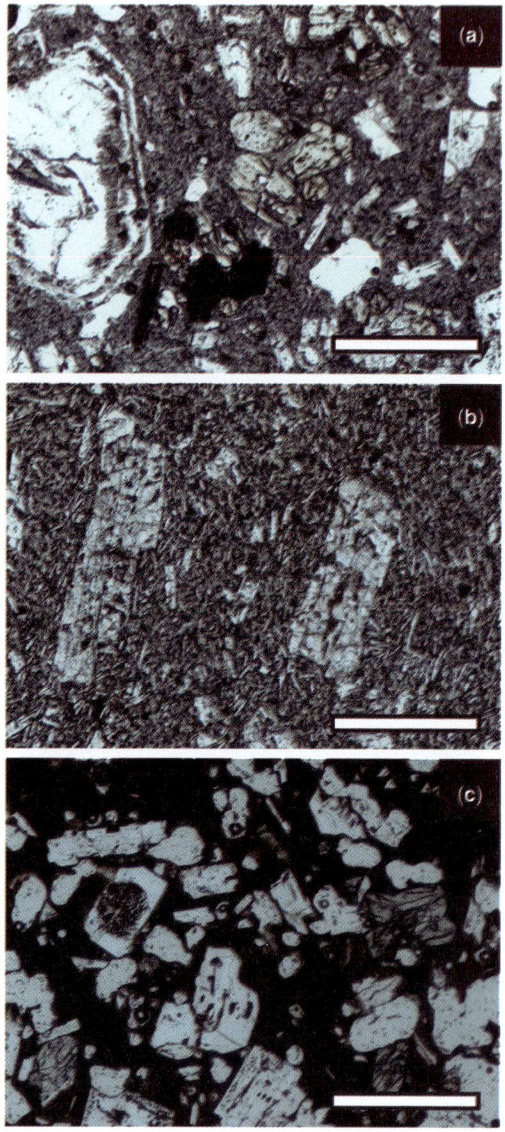

Fig. 4. Plain-polarized micrographs of selected samples, with 500 μm scale bars: (**a**) Sample LA1 from Lascar, showing sieve-textured plagioclase, rounded pyroxense and magnetite set in a fine-grained groundmass. (**b**) Sample 08CA01 from Callaqui, with a coarser-grained groundmass hosting crystals of plagioclase with abundant microfractures. (**c**) Sample CAJ10A from Calbuco, with an abundant anhedral crystal cargo set in a dark, glassy groundmass. Lonquimay sample petrography has not been studied as rocks are so sparsely phyric that thin sections would mostly show groundmass only.

Results

Petrography

The petrography of the samples studied here is typical for mafic to intermediate composition samples from arc volcanoes. Variations in crystal phase proportions, sizes, and petrographic textures are as large between individual samples from a

Table 2. *Whole-rock major oxide compositions (in wt%)*

Volcano name	Lascar			Callaqui			Lonquimay				Calbuco			
Sample name	LA1	LA2	LA4	08CA01	08CA02	08CA03	08LO01	08LO02	08LO03	08LO04	CAJ-10A	CAJ-24	CAJ-18	CAJ-23
SiO_2	57.89	63.14	65.82	55.88	54.28	54.74	58.07	56.97	57.04	60.35	54.92	55.35	60.55	54.92
TiO_2	0.89	0.60	0.48	1.14	1.29	1.22	1.39	1.61	1.32	0.89	0.98	0.88	0.63	0.87
Al_2O_3	17.85	16.98	16.63	18.63	18.75	17.44	16.36	16.34	16.76	16.68	19.11	20.32	18.84	19.59
$Fe_2O_3^T$	6.94	4.94	3.99	8.10	8.88	8.89	10.61	10.55	10.08	9.16	9.26	8.07	6.35	8.21
MnO	0.11	0.09	0.07	0.16	0.15	0.15	0.27	0.25	0.23	0.25	0.16	0.15	0.14	0.14
MgO	3.61	1.92	1.31	2.34	3.12	4.11	2.03	2.41	2.18	1.20	3.68	2.88	2.33	3.60
CaO	7.25	4.55	3.69	6.84	7.96	7.25	5.41	5.79	5.80	4.69	7.84	8.29	6.45	8.25
Na_2O	3.68	3.65	3.84	4.44	4.01	3.82	5.16	5.06	5.00	5.59	3.25	3.48	4.03	3.52
K_2O	1.59	2.72	3.07	1.71	1.52	1.75	1.05	0.95	0.95	1.18	0.59	0.61	0.86	0.59
P_2O_5	0.33	0.22	0.21	0.31	0.27	0.29	0.49	0.36	0.47	0.35	0.18	0.19	0.17	0.16
LOI	0.68	0.96	0.83	1.08	0.64	0.51	−0.18	0.11	0.09	−0.08	0.30	0.63	0.83	0.58
Total	100.83	99.76	99.94	100.63	100.87	100.16	100.66	100.40	99.93	100.24	100.28	100.84	101.18	100.42

LOI, loss on ignition.

Table 3. *Whole-rock trace element compositions (in ppm)*

Volcano name	Lascar			Callaqui			Lonquimay				Calbuco				Standard	
Sample name	LA1	LA2	LA4	08CA01	08CA02	08CA03	08LO01	08LO02	08LO03	08LO04	CAJ-10A	CAJ-18	CAJ-23	CAJ-24	BCR2	BCR2
P	1493	1052	985	1374	1211	1286	2046	1656	2057	1595	900	869	863	889	1631	1462
Sc	15.1	10.9	8.47	19.1	22.0	21.2	22.5	30.1	26.2	25.6	18.8	12.6	19.3	16.8	32.9	28.0
Ti	4963	3315	2724	6308	7197	6928	7330	9044	7353	4803	5539	3487	5286	4900	13310	12400
V	148	87.6	60.4	153	218	192	44.1	81.0	42.3	5.56	141	89	185	113	411	385
Cr	22.1	6.95	3.8	1.87	12.9	34.7	7.02	6.79	2.84	4.35	5.7	3.4	8.3	5.6	13.1	12.3
Mn	865	670	552	1139	1121	1110	1868	1936	1750	1805	1222	992	1114	1066	1514	1431
Co	18.8	10.8	6.73	14.8	18.7	22.3	13.4	16.1	10.8	5.89	18.9	12.5	21.2	15.7	36.8	35.4
Ni	22.1	5.89	3.07	3.30	9.65	20.4	3.87	4.02	2.54	2.92	4.8	4.0	9.4	4.5	12.1	11.5
Cu	38.3	15.1	13.1	46.5	65.4	63.2	34.1	47.3	19.9	19.1	21.1	20.1	35.5	27.8	21.2	19.8
Zn	75.7	64.6	60.9	81.1	82	81.2	110.2	123.6	110	113.1	67.3	70.4	84.3	60.8	128.1	119.9
Ga	21.0	19.9	19.6	22.3	21.9	20.3	21.7	22.3	21.8	23.2	18.5	19.3	20	19	22.4	21.5
As	4.36	12.2	13.7	5.26	5.83	6.79	8.19	8.04	8.64	11.1	4.06	6.11	5.28	4.12	1.09	1.05
Rb	52.1	123	141	46.8	43.8	54.1	22.4	20.8	20.7	26.7	16.2	24.6	16.9	16.7	48.4	47.4
Sr	735	402	372	572	598	551	440	467	485	436	342	348	368	358	340	335
Y	19.4	21.9	19.8	32.2	27.7	28.9	39.4	35.7	36.7	41.1	23	19.6	19.9	22.4	36.6	36.4
Zr	152	167	151	180	162	190	131	119	119	148	92	111	85	91	189	187
Nb	7.98	11.1	11.8	4.92	4.33	5.15	3.04	2.90	2.81	3.30	3.34	3.48	2.82	3.34	12.06	11.80
Mo	1.38	1.85	2.06	1.56	1.36	1.57	1.20	1.37	1.18	1.48	1.09	1.22	0.98	1.15	297	295
Ag	0.060	0.064	0.059	0.066	0.063	0.071	0.054	0.051	0.050	0.057	0.043	0.048	0.042	0.043	0.067	0.069
Cd	0.059	0.062	0.060	0.064	0.061	0.066	0.056	0.056	0.057	0.061	0.051	0.051	0.050	0.050	0.130	0.128

Sb	0.496	1.08	1.21	0.407	0.436	0.504	0.602	0.576	0.567	0.709	0.290	0.518	0.392	0.294	0.355	0.352
Cs	2.95	8.45	9.64	3.19	3.06	3.71	2.21	2.03	2.08	2.62	1.13	1.65	1.14	1.14	1.14	1.18
Ba	400	508	560	435	384	419	380	345	343	418	163	222	162	160	678	701
La	22.8	27.6	29.9	20.3	17.6	20.4	15.6	14.1	14.9	16.5	7.1	8.9	6.9	7.2	24.7	23.8
Ce	48.25	56.4	61.1	46.8	41.3	46.9	36.9	32.9	35.1	40	17.6	21.1	16.8	17.7	52.0	50.8
Pr	6.06	6.78	7.19	6.42	5.67	6.36	5.42	4.87	5.18	5.85	2.61	2.91	2.44	2.62	6.87	6.62
Nd	23.9	24.9	25.9	27.4	24.4	26.8	25.5	22.7	24.2	26.7	12.2	12.7	11.2	12.2	28.6	27.7
Sm	4.73	4.86	4.96	6.44	5.73	6.18	6.73	5.98	6.39	6.96	3.4	3.21	3.09	3.34	6.69	6.42
Eu	1.34	1.21	1.20	1.72	1.55	1.51	2.09	1.96	2.03	2.19	1.10	1.03	1.02	1.10	2.13	2.09
Gd	4.61	4.95	4.96	6.24	5.52	5.94	6.84	6.03	6.43	7.01	3.49	3.19	3.13	3.44	6.94	6.82
Tb	0.593	0.651	0.624	0.908	0.794	0.838	1.07	0.960	1.01	1.10	0.585	0.504	0.513	0.573	1.043	1.013
Dy	3.32	3.67	3.37	5.48	4.72	4.94	6.79	6.11	6.33	7.05	3.8	3.16	3.29	3.73	6.41	6.2
Ho	0.679	0.741	0.659	1.17	1.00	1.04	1.46	1.32	1.36	1.52	0.83	0.69	0.72	0.82	1.36	1.32
Er	1.86	2.05	1.80	3.24	2.75	2.86	4.01	3.67	3.71	4.22	2.33	1.94	2.00	2.29	3.69	3.62
Tm	0.269	0.302	0.256	0.480	0.400	0.424	0.594	0.550	0.551	0.637	0.359	0.306	0.306	0.35	0.538	0.53
Yb	1.72	1.93	1.65	3.07	2.58	2.70	3.75	3.48	3.47	4.06	2.32	2.01	1.96	2.26	3.42	3.32
Lu	0.262	0.291	0.243	0.470	0.388	0.411	0.581	0.535	0.534	0.625	0.358	0.315	0.305	0.346	0.518	0.504
Hf	3.65	4.23	3.96	4.66	4.23	4.94	3.49	3.26	3.25	4.02	2.413	2.784	2.22	2.373	4.794	4.649
Ta	0.614	1.20	1.35	0.376	0.330	0.394	0.216	0.212	0.206	0.234	0.242	0.266	0.209	0.275	0.886	0.861
W	0.812	2.14	2.42	0.634	0.617	0.701	0.617	0.700	0.596	0.686	0.48	0.637	0.555	0.679	0.553	0.54
Pb	12.0	18.0	22.2	17.1	20.0	18.5	17.6	17.8	14.6	17.7	8.6	10.8	5.8	4.9	10.1	10.1
Th	5.54	12.3	14.1	6.78	6.57	8.32	3.07	2.76	2.75	3.26	1.171	1.56	1.095	1.206	5.985	6.062
U	1.58	4.39	5.00	1.77	1.75	2.16	0.973	0.858	0.870	1.03	0.35	0.47	0.33	0.35	1.68	1.70

single volcanic edifice as they are between different volcanoes, with the exception of samples from Lonquimay, which contain so few large crystals that no attempt was made to study their petrography under the microscope (cf. Moreno & Gardeweg 1989). Due to the similar crystal assemblages and abundant disequilibrium textures that are common to all studied samples, only general findings are summarized here. A few examples are provided in Figure 4. Samples are typically porphyritic with a phenocryst assemblage dominated by plagioclase, pyroxenes and oxides set in a glassy to variably fine-grained groundmass. Minor olivine and hornblende, when present, are generally corroded, minor quartz is rounded and embayed. Large crystals are typically subhedral to anhedral, and occasionally fragments of broken crystals are observed (cf. Fig. 4a). Plagioclase crystals frequently display complex zoning and sieve-textures (e.g. Figs 4a, c) and, in some samples, contain abundant microfractures (Fig. 4b). Oxides are typically abundant in the groundmass, but some larger (>100 μm) crystals also occur, in exceptional cases exceeding 500 μm in size. In some samples, glomeroporphyritic aggregates or xenolithic clots of variable size and dominated by plagioclase and pyroxene have been observed (not shown, but see e.g. Hickey-Vargas *et al.* 1995). Overall, our findings are consistent with the petrographic observations published in previous petrogenetic studies of Lascar, Callaqui and Calbuco (Stone & Ingerson 1934; López-Escobar *et al.* 1992, 1995; Matthews *et al.* 1994; Rea 2009), to which the reader is referred for further details, including the observed large range in mineral chemistry that is equally characteristic for subduction zone magmas in general.

Whole-rock major and trace element chemistry

Whole-rock major element oxide compositions are reported in Table 2, and trace element compositions are reported in Table 3. Following the classification schemes of Peacock (1931), Peccerillo & Taylor (1976) and Arculus (2003), we find that samples from Lascar are medium to high-K, low-Fe, calc-alkalic andesites and dacites; samples from Callaqui are medium to high-K, medium to high-Fe, calc-alkalic basaltic andesites; samples from Lonquimay are medium-K, high-Fe, calc-alkalic andesites; and samples from Calbuco are low to medium-K, medium-Fe, calcic basaltic andesites and andesites. On a plot of total alkalis v. silica (Fig. 5), each volcano forms a distinct field with different initial alkali contents, at least for the limited number of samples investigated here.

In terms of trace elements, all samples are enriched in large ion lithophile elements and in light REEs relative to heavy REEs, and show negative anomalies in Ta, Nb, and Ti (Fig. 6). Such patterns are observed in many arc lavas and are taken to be indicative of subduction zone magmatism. Thus, although some variations in major and trace element compositions of the studied samples are evident in detail, all can be considered typical for magmas generated at an active continental margin.

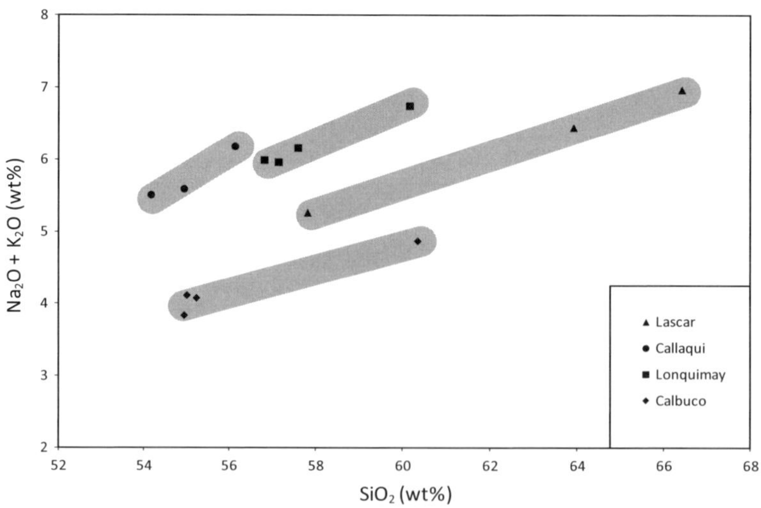

Fig. 5. Total alkali v. silica diagram of the whole-rock samples studied here. Note that the fields are not meant to imply any specific petrogenetic process that might relate the samples from each volcano.

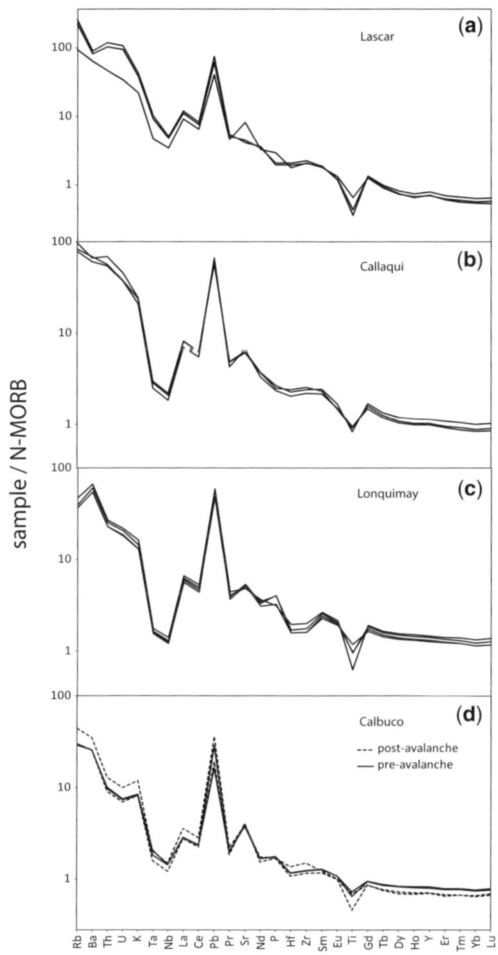

Fig. 6. Multi-element spidergram of the studied whole-rock samples, normalized to N-MORB (Sun & McDonough 1989).

Uranium and thorium isotopes in whole rocks, groundmasses, and mineral separates

Uranium–thorium isotope systematics are provided in Table 4, and samples from Lascar, Callaqui, Lonquimay and Calbuco are plotted on U–Th equiline diagrams in Figures 7–10. Only two of the sparsely phyric Lonquimay samples had sufficient crystals to attempt internal isochron work. Without exception, whole rock and groundmass show almost identical $^{238}U-^{230}Th$ disequilibria in each of the samples investigated. Whole rocks yield small ^{238}U excesses of 10–15% in Lonquimay and Calbuco, and are close to secular equilibrium in Callaqui, but reach up to 36% ^{238}U excess in Lascar. Although most mineral separates display variable amounts of disequilibria, with a total range of $0.71 < (^{238}U/^{230}Th) < 1.46$, there are no systematic variations. ^{238}U excesses dominate, but none of the samples displays linear relationships that provide meaningful ages (internal isochrons). The two pre-avalanche samples from Calbuco form linear arrays with slopes >1 on the U–Th equiline diagram (Fig. 10).

The activity ratios of $(^{234}U/^{238}U)$ for whole rocks and groundmasses are all close to equilibrium, but, with the exception of Lascar Volcano, some of the mineral separates display significant ^{234}U excesses ($15‰ \leq \delta^{234}U \leq 142‰$) that are variable for different minerals within individual rock samples (Table 4). This is the case for sample 08CA01 from Callaqui, which displays the largest $^{238}U-^{230}Th$ disequilibria in the mineral separates studied from this volcano (Fig. 8); olivine and magnetite from the Lonquimay sample 08LO03, which plot in ^{238}U excess; plagioclase from Lonquimay sample 08LO04, which plots in $^{238}U-^{230}Th$ equilibrium (Fig. 9); and most plagioclase and magnetite separates of all but the oldest sample from Calbuco.

Finally, $(^{238}U/^{232}Th)$ activity ratios of most separates are in the range that is typical for convergent margin magmas (e.g. Condomines & Sigmarsson 1993). Notable exceptions are the two pre-avalanche samples from Calbuco. Sample CAJ10A yields a magnetite $(^{238}U/^{232}Th)$ activity ratio of 7.30, while sample CAJ24 yields a plagioclase $(^{238}U/^{232}Th)$ activity ratio of 2.83. Both of these separates are in ^{230}Th excess relative to ^{238}U and form the end points of the linear arrays displayed by the separates of these rocks.

Discussion

To date, very little work has been published on the U-series systematics of the Central and Southern Andean arc magmas. An early along-arc study by Sigmarsson *et al.* (1990) focused on a few mafic to intermediate whole-rock samples from the Southern Volcanic Zone and identified small to moderate ^{238}U excesses that were attributed to fluids originating from the dehydration of recently subducted sediments. In contrast, some intermediate melts in the Central Andes and the southernmost Southern and Austral Andes have been shown to display significant Th excesses, interpreted to be due to residual garnet in the lower overriding crust or in the subducting slab (Sigmarsson *et al.* 1998; Wörner *et al.* 2004*a*, *b*; Jicha *et al.* 2007). To our knowledge, only two Andean volcanoes have been studied in detail in terms of U–Th systematics, namely Parinacota in the Central Andes (Bourdon *et al.* 2000; Hora *et al.* 2009) and Puyehue-Cordón Caulle in the Southern Volcanic Zone (Jicha *et al.*

Table 4. *U–Th composition of whole rocks and separates*

Volcano	Sample	Phase	U (ng g^{-1})	U $\pm 2\sigma$	Th (ng g^{-1})	Th $\pm 2\sigma$	($^{234}U/^{238}U$)	$\pm 2\sigma$ (‰)	($^{238}U/^{232}Th$)	$\pm 2\sigma$ (‰)	($^{230}Th/^{232}Th$)	$\pm 2\sigma$ (‰)
Lascar	LA1	wr	1614	2.4	5418	13.5	1.001	0.8	0.904	1.5	0.804	1.9
		gm	1860	4.8	6267	26.2	1.002	1.7	0.900	3.7	0.803	2.3
		px	241	0.5	800	7.6	1.003	1.1	0.915	9.3	0.849	8.5
	LA2	wr	4505	9.4	12194	36.8	1.000	1.2	1.121	2.2	0.822	2.8
		gm	5247	9.4	14198	58.6	1.001	1.3	1.121	3.7	0.822	1.9
		mt	1234	3.0	4070	88.3	1.006	1.1	0.920	21.6	0.809	19.9
	LA4	wr	5189	26.2	14188	71.0	1.002	1.5	1.110	5.1	0.822	4.1
		plag	446	1.8	1156	3.9	1.003	3.2	1.171	4.0	0.884	1.8
Callaqui	08CA01	wr	1776	4.5	6600	23.0	1.000	2.0	0.816	3.2	0.799	4.2
		gm	2013	5.2	7530	18.1	1.004	2.6	0.811	3.2	0.799	2.6
		ol	35	0.5	76	0.7	1.025	14.7	1.386	17.4	1.099	6.7
		plag	49	0.1	143	1.3	1.046	1.1	1.036	8.8	1.147	8.8
		mt	75	0.1	152	1.3	1.133	1.3	1.493	8.2	1.313	7.2
	08CA02	wr	1725	3.9	6367	19.7	1.000	1.2	0.822	2.0	0.789	2.7
		gm	2022	0.9	7507	23.4	1.003	1.5	0.817	2.8	0.790	2.1
		ol	47	0.1	162	2.1	1.004	2.0	0.879	13.1	0.847	7.4
		px	117	0.4	435	4.6	1.002	3.0	0.818	10.8	0.801	10.2
		plag	21	0.1	78	0.4	1.030	4.6	0.814	6.3	0.794	4.2
	08CA03	wr	2132	4.0	8012	22.6	1.001	1.0	0.808	1.8	0.789	2.4
		gm	2763	2.0	10408	34.7	1.003	1.5	0.806	3.1	0.790	2.1
		px	100	0.7	368	2.9	1.003	7.0	0.822	10.0	0.800	5.1
		plag	29	0.1	97	0.3	1.009	1.6	0.901	2.0	0.889	1.9
		mt	411	1.1	1433	10.2	1.019	1.1	0.870	7.2	0.786	5.9

Lonquimay	08LO03	wr	882	1.8	2710	6.9	1.001	1.1	0.988	1.7	0.856	2.3
		gm	946	1.8	2906	8.8	1.004	1.3	0.988	2.4	0.856	2.9
		ol	73	0.6	119	0.6	1.021	4.2	1.860	9.6	1.713	2.7
		mt	104	0.2	147	1.4	1.052	1.8	2.145	9.3	1.470	9.1
	08LO04	wr	1051	1.9	3226	9.7	1.001	0.5	0.989	2.1	0.859	2.1
		gm	1099	2.5	3367	12.3	1.003	1.3	0.991	3.0	0.864	2.0
		ol	175	1.1	394	1.6	1.002	5.8	1.347	6.6	1.188	2.7
		plag	10	0.02	28	0.3	1.032	0.7	1.082	9.2	1.092	9.6
Calbuco	CAJ10A	wr	355	0.7	1170	3.4	1.001	1.4	0.921	2.3	0.827	3.3
		gm	400	0.9	1332	3.9	1.005	1.4	0.911	2.4	0.819	1.5
		px1	115	1.4	221	5.9	1.000	12.4	1.580	29.1	1.516	23.1
		px2	89	0.1	134	0.6	1.005	1.0	2.020	3.9	2.017	1.5
		mt	296	0.8	123	0.4	0.995	1.2	7.298	2.3	8.419	2.3
	CAJ24	wr	367	0.7	1217	3.6	1.001	0.9	0.915	2.0	0.825	3.4
		gm	460	0.9	1495	4.9	1.005	1.2	0.933	2.7	0.847	1.8
		plag	37	0.1	39	0.2	1.144	1.6	2.830	4.2	4.007	4.5
		mt1	109	0.2	240	2.0	1.073	1.3	1.373	8.1	1.633	7.1
		mt2	107	0.3	221	1.0	1.073	2.5	1.474	4.2	1.720	2.9
	CAJ18	wr	475	1.8	1563	6.2	1.002	3.3	0.922	4.2	0.831	2.4
		px	57	0.2	105	1.0	1.000	3.0	1.628	9.5	1.516	5.5
		plag	20	0.1	47	0.9	1.016	1.4	1.294	18.6	1.140	19.9
		mt	121	0.4	187	4.7	1.007	3.9	1.963	25.4	1.344	22.3
	CAJ23	wr	330	0.6	1072	4.1	1.001	0.7	0.935	3.2	0.835	3.8
		gm	455	1.6	1436	17.5	1.004	3.5	0.961	12.3	0.857	9.0
		px	44	0.5	130	1.8	1.008	11.3	1.013	17.6	0.892	12.8
		mt1	58	0.1	99	1.8	1.084	1.3	1.793	17.8	1.580	17.1
		mt2	59	0.3	97	1.7	1.085	9.1	1.839	17.8	1.611	18.3

Abbreviations: gm, groundmass; mt, magnetite; ol, olivine; plag, plagioclase; px, pyroxene; wr, whole rock.

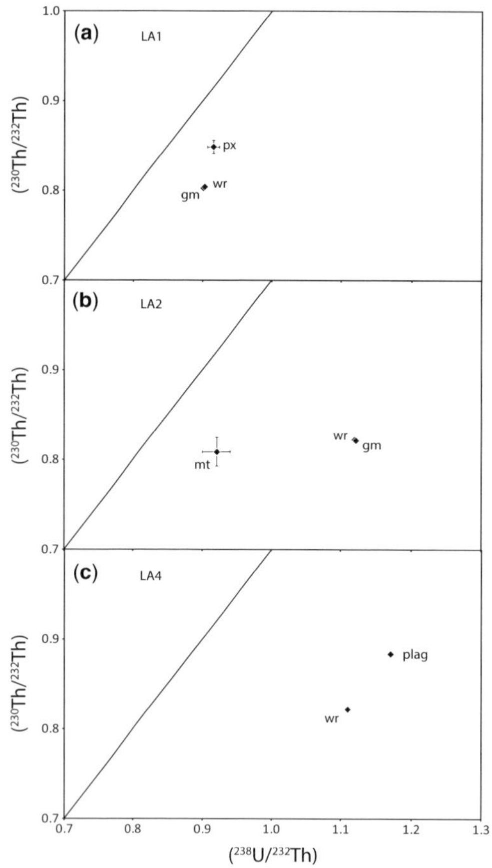

Fig. 7. U–Th equiline diagrams for samples from Lascar. wr, whole rock; gm, groundmass; px, pyroxene; mt, magnetite; plag, plagioclase. Analytical uncertainties (2σ) are smaller than symbol size, unless indicated otherwise.

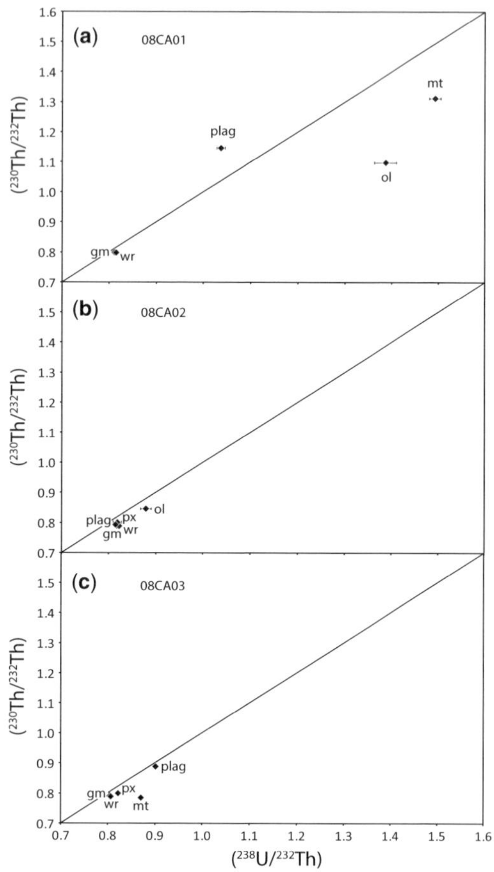

Fig. 8. U–Th equiline diagrams for samples from Callaqui. wr, whole rock: gm, groundmass; mt, magnetite; plag, plagioclase; ol, olivine; px, pyroxene. Analytical uncertainties (2σ) are smaller than symbol size, unless indicated otherwise.

2007), with whole-rock samples yielding variable ^{238}U and ^{230}Th excesses in both cases. While the number of whole-rock samples studied here is small, the data yield slight to moderate ^{238}U excesses in all samples, including some of the recent eruptives from Lascar in the Central Andes of northern Chile.

The present study is the first comprehensive U–Th isotope investigation of mineral separates that compares arc volcanoes erupting under different regimes of crustal stress. Over the last couple of decades, U–Th mineral isochron studies have yielded some insights into the rates and processes of crystallization at a number of arcs (for an extensive review, see Condomines *et al.* 2003; for some more recent studies see Reagan *et al.* 2003; Turner *et al.* 2003; Jicha *et al.* 2005, 2007, 2009; Zellmer *et al.* 2005), with pre-eruptive residence times ranging from 0 to >100 ka. A review of crystal residence times inferred by U–Th–Ra mineral isochrons by Zellmer *et al.* (2005) showed that many of these ages had large associated uncertainties of typically more than ±10 ka (2σ), which may in part be attributed to significant scatter of the mineral data in U–Th equiline space. It was also shown that age dichotomies existed between U–Th and Ra–Th ages, pointing either to prolonged crystal growth until the time or eruption, or mixing of crystal populations, for example, of young crystals formed just prior to eruption with older cumulate xenocrysts.

The effects of magma mixing on ^{238}U–^{230}Th disequilibria and U–Th mineral isochrons have been modelled in some detail by Hughes & Hawkesworth (1999). Repeated magma recharge into a crystallizing open system reservoir is capable of

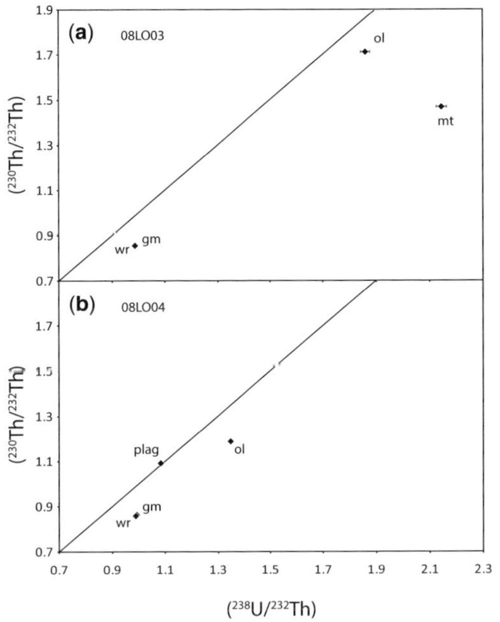

Fig. 9. U–Th equiline diagrams for samples from Lonquimay. wr, whole rock; gm, groundmass; mt, magnetite; ol, olivine; plag, plagioclase. Analytical uncertainties (2σ) are smaller than symbol size, unless indicated otherwise.

producing significant scatter in separated mineral phases. Hughes & Hawkesworth (1999) pointed out that one indication of such recharge processes may be relatively low groundmass (^{230}Th/^{232}Th) activity ratios compared to those of the other phase separates, as a result of unaged, sparsely phyric melts with high (^{238}U/^{230}Th) recharging aging magma reservoirs containing older mineral phases with U–Th isotopes closer to the equiline. However, such systematic groundmass offsets are not observed in the present dataset. Although this does not necessarily preclude magma mixing by open system recharge in the genesis of some of the samples studied here, it is evident that other processes are necessary to account for the unsystematic scatter of the U–Th mineral data, particularly at Callaqui and Lonquimay (Figs 8 & 9), and for the linear but non-isochronal relationships in the Calbuco samples (Fig. 10).

On the U–Th equiline diagrams, for each sample we studied, whole rock and groundmass typically form one end of the data array and are compositionally almost identical to each other. Note that in most cases this does not represent a mass balance problem; at Callaqui and particularly at Lonquimay, where samples are very sparsely phyric, the groundmass dominates the U–Th budget of the rocks, being enriched in U and Th by *c.* 1–2 orders of magnitude relative to the mineral separates (cf. Table 4);

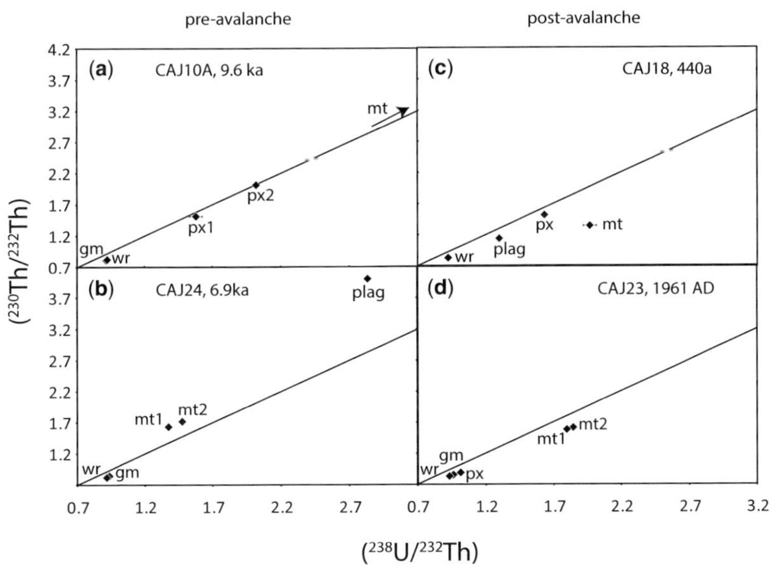

Fig. 10. U–Th equiline diagrams for samples from Calbuco, with eruption ages indicated for each sample. wr, whole rock; gm, groundmass; px, pyroxene; mt, magnetite; plag, plagioclase. Numbers 1 and 2 mark redissolutions and reanalyses of the same mineral separate batch. Analytical uncertainties (2σ) are smaller than symbol size, unless indicated otherwise.

and at Lascar, clean separates were obtained from only one mineral phase per sample, plotting variably at lower, similar, or higher U–Th ratios than the whole rock (Fig. 7). Only at Calbuco, where the groundmass is typically only 5- to 10-fold enriched in U and Th relative to the minerals, may the presence of a low U–Th phase (e.g. trace amounts of unseparated monazite or apatite) have to be invoked (Fig. 10).

The significant deviations of ($^{234}U/^{238}U$) activity ratios of some of the mineral separates from secular equilibrium and the variations in these deviations even within individual rock samples may provide some insights into the source of the observed scatter on the equiline diagrams. In the study of igneous rocks, samples that display ($^{234}U/^{238}U$) disequilibria of more than a few per mille are frequently discarded, as such offsets are taken as evidence for secondary alteration (e.g. Yokoyama *et al.* 2003), because many weathered continental rocks and associated waters display significant $^{234}U/^{238}U$ disequilibria (^{234}U excesses and deficits) acquired through alpha recoil and leaching processes (e.g. Ivanovich & Harmon 1992; Bourdon *et al.* 2003). Several previous studies have documented the effectiveness of fluid–rock interaction in changing uranium concentrations in igneous rocks and, thereby, fractionating uranium from thorium (e.g. Buntebarth 1976; Gosnold 1987; Villemant & Fléhoc 1989; Guthrie 1991; Villemant *et al.* 1996).

However, all whole rocks and groundmasses studied here are in $^{234}U/^{238}U$ equilibrium. The ($^{234}U/^{238}U$) disequilibria of some mineral separates from fresh samples with groundmass ($^{234}U/^{238}U$) activity ratios in secular equilibrium may indicate that at Callaqui, Lonquimay and Calbuco altered minerals were taken up into and preserved within ascending Andean arc magmas prior to their eruption (cf. Fig. 11). Uptake of antecrysts (a term coined by Wes Hidreth in 2003) and/or xenocrysts into magmas prior to their eruption has been suggested previously (e.g. Streck *et al.* 2007; Jerram & Martin 2008; Smith *et al.* 2010). Zellmer *et al.* (2013) argues for crystal uptake into initially aphyric arc melts to be one of the dominant processes by which many mafic arc magmas acquire their crystal cargo in SW Japan.

In the subduction zone setting, one might expect the process of crystal uptake into ascending melts to dominate over crystallization of these melts in general: the volatile rich character of arc magmas can result in superheating upon ascent, because the adiabatic dT/dP that such melts follow during rapid decompression is smaller than the Clapeyron slope of their liquidus (e.g. Annen *et al.* 2006; Blundy *et al.* 2010). If indeed the majority of crystals in such melts are of antecrystic or xenocrystic

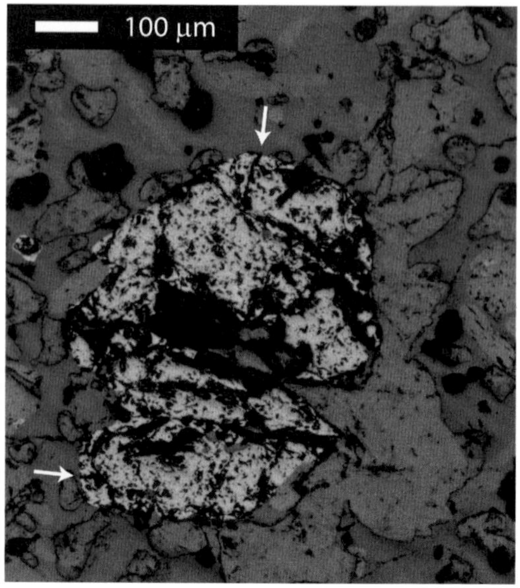

Fig. 11. Reflected light photograph of a fractured magnetite crystal from sample CAJ24, with clear alteration textures visible along some of the fractures. Arrows point to grain edges where fractures clearly do not propagate into the fresh groundmass glass, indicating that alteration occurred prior to uptake of the crystal into the melt.

origin, as the results of Zellmer *et al.* (2013) suggest, the implication would be that magma ascent from mantle wedge to eruption site is very rapid, with ascent velocities of the order of 1 m s^{-1}, resulting in a timescale of less than 1 day for erupted melts to traverse the overriding crust (cf. Zellmer *et al.* 2013). Such rapid magma ascent is consistent with evidence from some geophysical studies that indicate links between deep seismicity and volcanic eruptions (Blot 1972; Fedorov *et al.* 1983), and from preservation of mantle xenoliths in some arc andesites (Blatter & Carmichael 1998).

A remarkable implication of the mineral separate data presented here is that the alteration of the crystal cargo must have occurred recently. We note that large ($^{234}U/^{238}U$) disequilibria are found in some mineral separates from all southern Andean volcanoes, where continuous hydrothermal alteration below active volcanoes may be fostered by the abundance of meteoric water in the temperate and rainy climate of this region. In contrast, all samples from Lascar Volcano are in ^{234}U–^{238}U secular equilibrium. Lascar is situated on the edge of the arid Atacama plateau, known as one of the driest deserts on Earth, and hydrothermal circulation of meteoric water will thus be strongly inhibited. We do not have enough data from Lascar to argue for crystal uptake into rapidly ascending melts over crystal fractionation in long-lived open system (cf. Figs 7a, c, and Hughes & Hawkesworth 1999) subvolcanic magma reservoirs. However, the similarly complex petrographic features of central and southern Andean samples (Fig. 4) may indicate that crystal uptake in Lascar is as common as at the other volcanoes studied here.

The introduction of a variety of (incipiently weathered) mineral grains from upper crustal plutonic rocks into fresh magmas prior to their eruption is consistent with the scatter observed on some of the U–Th equiline diagrams. With ($^{234}U/^{238}U$) > 1 in most separates, a case may be made for uranium assimilation by precipitation of uranium from hydrothermal fluids that are variably enriched in ^{234}U. This process may also explain the high U–Th ratios of some of the mineral separates studied here. Concomitant introduction of ^{230}Th-excesses may involve solution-to-solid recoil gain within some minerals of the plutons (e.g. Rosholt 1982; Osmond & Ivanovich 1992), prior to crystal uptake into the magma. Such a recoil implantation process is most effective if the surface area to volume ratio is high (Romer & Rocholl 2004), for example, when minerals contain abundant microfractures (e.g. Figs 4b & 11). One striking aspect of our dataset from Calbuco (Fig. 10) is the strong linearity on the equiline diagram displayed by the phases of most samples, with slopes >1 for pre-avalanche samples and slopes <1 for post-avalanche deposits ($R^2 > 0.999$ for samples CAJ10A, CAJ24 and CAJ23). These likely represent mixing trends between the host magma (represented by the groundmass) and high U–Th alteration endmembers that variably contribute to the U–Th geochemistry of each phase within the crystal cargo. The transition from slopes >1 to slopes <1 may then be due to changes in the composition of the alteration endmember from ($^{230}Th/^{238}U$) >1 to ($^{230}Th/^{238}U$) <1, given that whole-rock and groundmass U–Th compositions are approximately constant and plot close to the equiline in all samples. Prior to the Calbuco debris avalanche events, the 9.6 ka eruption may have taken up crystals that experienced alteration through a ^{230}Th predominant recoil gain process from a U-rich fluid in $^{238}U/^{234}U$ equilibrium, while the 6.9 ka shows evidence for uptake of crystals that were also enriched in ^{234}U, possibly due to the presence of younger fluids or renewed hydrothermal activity. Following volcano unloading, crystals taken up into the melt remain altered in terms of ($^{234}U/^{238}U$) activity ratios (although compared to the sixteenth century eruption, the 1961 AD event appears to carry some fresh pyroxene crystals, that may have grown from new, recharging melts). In contrast to the pre-avalanche samples, there is no evidence for recoil gain of ^{230}Th by the crystal cargo in either of the post-avalanche deposits. During incipient weathering, enrichment of ^{230}Th by solution-to-solid recoil gain may be facilitated when microfractures within the crystals are narrow enough to ensure efficient daughter implantation into grain boundaries (Osmond & Ivanovich 1992). It is possible that a pressure drop resulting from volcano unloading by debris avalanche events leads to widening of pre-existing microfractures in subvolcanic lithologies, and promotes weathering by low-temperature mineral precipitation rather than recoil gain processes. Additional samples from other volcanic edifices that have been affected by sector collapses will have to be investigated to test this hypothesis.

Concluding remarks

The data presented here provide several new insights into the petrogenesis of arc magmas. Irrespective of the crustal stress field, uptake of variably altered antecrystic and/or xenocrystic material seems to be common in volcanic arcs, and it is possible that many if not all crystals in some mafic to intermediate arc melts are derived from previous intrusions by remobilization. Crystal uptake may be less prominent in transtensional settings, where some initially aphyric melts can traverse the crust largely uninhibited and, therefore, remain comparatively crystal poor (as, for example, at Lonquimay).

The igneous protoliths that represent the source of crystal cargo taken up into arc magmas appear to be variably affected by incipient weathering processes in the subsurface, resulting in changes of their primary U-series isotope systematics through hydrothermal alteration by precipitation or recoil gain processes along crystal fractures. ^{234}U–^{238}U and ^{230}Th–^{238}U disequilibria suggest that the processes affecting these isotope systems occurred within the last *c.* 1 Ma and 350 ka, respectively. Thus, ancient crustal materials would have to have experienced recent alteration to account for the data presented here. However, with the exception of volcanic edifices situated in very arid regions, hydrothermal circulation involving meteoric water may, in fact, be operating continuously in areas of active volcanism. Whether crustal stresses, such as volcano loading, affect subvolcanic weathering processes systematically, as suggested by our data from Calbuco, remains to be tested at other sites.

Crystal uptake as a common process in orogenic andesites may explain the large uncertainties in many published internal U–Th isochrons. From this perspective, future U–Th mineral isochron studies need to provide sound evidence to show unequivocally that all minerals have grown from their host melt, for example through detailed petrographic and mineral chemical and/or isotopic (Sr, Pb, U) analyses. The tightness of internal U–Th isochrons yielding precise crystallization ages (e.g. Bourdon *et al.* 2000; Condomines *et al.* 2003; Jicha *et al.* 2007) may on their own not be sufficient in ensuring the reliability of the obtained ages. Independent evidence is particularly important in cases where ages are yielded by what are effectively two-point isochrons, where different mineral phases cluster tightly due to insufficient U–Th fractionation, or because selected minerals (often plagioclase) may be associated with analytical uncertainties that are large enough to question the usefulness of the phase in providing additional age constraints. Extreme caution is warranted when U-series isotope data are employed to provide age constraints on preconceived ideas of magmatic evolution. Procuring abundant constraints on the petrogenetic processes operating beneath active arc volcanoes, independent from U-series constraints, is paramount.

Georg F. Zellmer thanks T. Elliott, G. Ma, M. Nakamura, K. Rubin, G. Shellnutt and G. Wörner for useful discussions, and Liz Cottrell and Ben Andrews for support through a Global Volcanism fellowship at the Smithsonian Institution. L.-M. Cioflec assisted during fieldwork, and Y.-T. Hsu and R.-Y. Yan helped with sample processing. We are grateful to editor A. Gómez-Tuena, and to B. Jicha and an anonymous reviewer, whose constructive and thoughtful comments allowed us to improve this paper. GFZ acknowledges funding by the National Science Council of Taiwan (Grants 100-2116-M-001-007 and 101-2116-M-001-009) for fieldwork and while the analytical part of this study was conducted. HF was funded by a University of Bristol postgraduate research scholarship. We also acknowledge funding through FONDECYT (Grants 1060187 to JMC, and 1090387 to JEC). JMC thanks FONDAP-CEGA 15090013 (Centro de Excelencia en Geotermia de los Andes).

References

Allègre, C. J. & Condomines, M. 1976. Fine chronology of volcanic processes using $^{238}U-^{230}Th$ systematics. *Earth and Planetary Science Letters*, **28**, 395–406.

Annen, C., Blundy, J. D. & Sparks, R. S. J. 2006. The genesis of intermediate and silicic magmas in deep crustal hot zones. *Journal of Petrology*, **47**, 505–539.

Arculus, R. J. 2003. Use and abuse of the terms calcalkaline and calcalkalic. *Journal of Petrology*, **44**, 929–935.

Bacon, C. R. & Lanphere, M. A. 2006. Eruptive history and geochronology of Mount Mazama and the Crater Lake region, Oregon. *Geological Society of America Bulletin*, **118**, 1331–1359.

Barrientos, S. E. & Acevedo-Aránguiz, P. S. 1992. Seismological aspects of the 1988–1989 Lonquimay (Chile) volcanic eruption. *Journal of Volcanology and Geothermal Research*, **53**, 73–87.

Blatter, D. L. & Carmichael, I. S. E. 1998. Hornblende peridotite xenoliths from central Mexico reveal the highly oxidized nature of subarc upper mantle. *Geology*, **26**, 1035–1038.

Blot, C. 1972. Volcanisme et séismes du manteau supérieur dans l'Archipel des Nouvelles-Hébrides. *Bulletin of Volcanology*, **36**, 446–461.

Blundy, J., Cashman, K. V., Rust, A. & Witham, F. 2010. A case for CO_2-rich arc magmas. *Earth and Planetary Science Letters*, **290**, 289–301.

Bourdon, B., Wörner, G. & Zindler, A. 2000. U-series evidence for crustal involvement and magma residence times in the petrogenesis of Parinacota volcano, Chile. *Contributions to Mineralogy and Petrology*, **139**, 458–469.

Bourdon, B., Henderson, G. M., Lundstrom, C. C. & Turner, S. P. 2003. *Uranium-Series Geochemistry*. Reviews in Mineralogy and Geochemistry, **52**. The Mineralogical Society of America, Washington, DC.

Buntebarth, G. 1976. Distribution of uranium in intrusive bodies due to combined migration and diffusion. *Earth and Planetary Science Letters*, **32**, 84–90.

Castruccio, A., Clavero, J. & Rivera, A. 2010. Comparative study of lahars generated by the 1961 and 1971 eruptions of Calbuco and Villarrica volcanoes, Southern Andes of Chile. *Journal of Volcanology and Geothermal Research*, **190**, 297–311.

Cembrano, J. & Lara, L. 2009. The link between volcanism and tectonics in the southern volcanic zone of the Chilean Andes: a review. *Tectonophysics*, **471**, 96–113.

Cembrano, J., Herve, F. & Lavenu, A. 1996. The Liquine Ofqui fault zone: a long-lived intra-arc fault system in southern Chile. *Tectonophysics*, **259**, 55–66.

Cembrano, J., González, G. et al. 2008. The interplay between crustal tectonics and volcanism in the Central and Southern Volcanic zones of the Chilean Andes. Paper presented at the International Symposium on Andean Geodynamics, Nice, 2–4 September 2008.

Charlier, B. L. A. & Zellmer, G. F. 2000. Some remarks on U–Th mineral ages from igneous rocks with prolonged crystallisation histories. *Earth and Planetary Science Letters*, **183**, 457–469.

Clavero, J., Godoy, E., Arancibia, G., Rojas, C. & Moreno, H. 2008. Multiple Holocene sector collapses at Calbuco volcano, Southern Andes. Paper presented at the IAVCEI General Assembly, Reykjavik, Iceland, 18–25 August 2008.

Clynne, M. A. 1999. A complex magma mixing origin for rocks erupted in 1915, Lassen Peak, California. *Journal of Petrology*, **40**, 105–132.

Condomines, M. 1997. Dating recent volcanic rocks through $^{230}Th-^{238}U$ disequilibrium in accessory minerals: example of the Puy de Dome (French Massif Central). *Geology*, **25**, 375–378.

Condomines, M. & Sigmarsson, O. 1993. Why are so many arc magmas close to $^{238}U-^{230}Th$ radioactive equilibrium? *Geochimica et Cosmochimica Acta*, **57**, 4491–4497.

Condomines, M., Gauthier, P.-J. & Sigmarsson, O. 2003. Timescales of magma chamber processes and dating of young volcanic rocks. *In*: Bourdon, B., Henderson, G. M., Lundstrom, C. C. & Turner, S. P. (eds) *Uranium-Series Geochemistry*. Reviews in

Mineralogy and Geochemistry, **52**. The Mineralogical Society of America, Washington, DC, 125–174.

Cooper, K. M. & Reid, M. R. 2003. Re-examination of crystal ages in recent Mount St. Helens lavas: implications for magma reservoir processes. *Earth and Planetary Science Letters*, **213**, 149–167.

Costa, F. & Dungan, M. 2005. Short time scales of magmatic assimilation from diffusion modeling of multiple elements in olivine. *Geology*, **33**, 837–840.

Costa, F., Chakraborty, S. & Dohmen, R. 2003. Diffusion coupling between trace and major elements and a model for calculation of magma residence times using plagioclase. *Geochimica et Cosmochimica Acta*, **67**, 2189–2200.

Cox, S. F. 2002. Fluid flow in mid- to deep crustal shear systems: experimental constraints, observations on exhumed high fluid flux shear systems, and implications for seismogenic processes. *Earth Planets Space*, **54**, 1121–1125.

Davidson, J. P., Morgan, D. J. & Charlier, B. L. A. 2007. Frontiers in textural and microgeochemical analysis: isotopic microsampling of magmatic rocks. *Elements*, **3**, 253–259.

Deruelle, B., Medina, E. T., Figueroa, O. A., Maragano, M. C. & Viramonte, J. G. 1995. The recent eruption of Lascar volcano (Atacama-Chile, April 1993): petrological and volcanological relationships. *Comptes Rendus – Academie des Sciences, Serie II: Sciences de la Terre et des Planetes*, **321**, 377–384.

Donnelly, C. & Cooper, K. M. 2006. Comparison of U–Th–Ra Disequilibria in multiple crystal populations in lava from the current eruption of Mt. St. Helens. *EOS Transactions of the AGU*, **87**, Fall Meeting Suppl., Abstract V54B-03.

Dungan, M. A. 2005. Partial melting at the earth's surface: implications for assimilation rates and mechanisms in subvolcanic intrusions. *Journal of Volcanology and Geothermal Research*, **140**, 193–203.

Fedorov, S. A., Gorel'chik, V. I., Stepanov, V. V. & Garbuzova, V. T. 1983. The development of the Great Tolbachik Fissure Eruption in 1975 from seismological data. *In*: Fedotov, S. A. & Markhinin, Y. K. (eds) *The Great Tolbachik Fissure Eruption*. Cambridge University Press, Cambridge, 189–203.

Gardeweg, M., Moreno, H. & Naranjo, J. A. 1990. *Comportamiento del volcán Lonquimay en su fase eruptiva 1988–1990 y su influencia en el medio ambiente*. Servicio Nacional de Geología y Minería, Santiago.

Gardeweg, M. C., Sparks, R. S. J. & Matthews, S. J. 1998. Evolution of Lascar volcano, northern Chile. *Journal of the Geological Society, London*, **155**, 89–104.

González, G., Cembrano, J., Aron, F., Veloso, E. E. & Shyu, J. B. H. 2009. Coeval compressional deformation and volcanism in the central Andes, case studies from northern Chile (23°S–24°S). *Tectonophysics*, **28**, 1–18.

Gosnold, W. D. 1987. Redistribution of U and Th in shallow plutonic environments. *Geophysical Research Letters*, **14**, 291–294.

Guthrie, V. 1991. Determination of recent ^{238}U, ^{234}U and ^{230}Th mobility in granitic rocks: application of a natural analogue to the high-level waste repository environment. *Applied Geochemistry*, **6**, 63–74.

Heath, E., Turner, S., Macdonald, R., Hawkesworth, C. J. & van Calsteren, P. 1998. Long magma residence times at an island arc volcano (Soufriere, St. Vincent) in the Lesser Antilles: evidence from ^{238}U–^{230}Th isochron dating. *Earth and Planetary Science Letters*, **160**, 49–63.

Hickey-Vargas, R., Abdollah, M. J., Parada, M. A., López, L. & Frey, F. 1995. Crustal xenoliths from Calbuco Volcano, Andean Southern Volcanic Zone: implications for crystal composition and magma-crust interaction. *Contributions to Mineralogy and Petrology*, **119**, 331–344.

Hoffmann, D. L., Prytulak, J., Richards, D. A., Elliott, T., Coath, C. D., Smart, P. L. & Scholz, D. 2007. Procedures for accurate U and Th isotope measurements by high precision MC-ICPMS. *International Journal of Mass Spectrometry*, **264**, 97–109.

Hora, J. M., Singer, B. S., Wörner, G., Beard, B. L., Jicha, B. R. & Johnson, C. M. 2009. Shallow and deep crustal control on differentiation of calc-alkaline and tholeiitic magma. *Earth and Planetary Science Letters*, **285**, 75–86.

Hughes, R. D. & Hawkesworth, C. J. 1999. The effects of magma replenishment processes on ^{238}U–^{230}Th disequilibrium. *Geochimica et Cosmochimica Acta*, **63**, 4101–4110.

Ivanovich, M. & Harmon, R. S. 1992. *Uranium Series Disequilibrium: Applications to Earth, Marine, and Environmental Sciences*. Oxford Science Publications, Oxford.

Jarrard, R. 1986. Relations among subduction parameters. *Reviews of Geophysics*, **24**, 217–284.

Jerram, D. A. & Martin, V. M. 2008. Understanding crystal populations and their significance through the magma plumbing system. *In*: Annen, C. & Zellmer, G. F. (eds) *Dynamics of Crustal Magma Transfer, Storage and Differentiation*. Geological Society, London, Special Publications, **304**, 133–148.

Jicha, B. R., Singer, B. S., Beard, B. L. & Johnson, C. M. 2005. Contrasting timescales of crystallization and magma storage beneath the Aleutian Island arc. *Earth and Planetary Science Letters*, **236**, 195–210.

Jicha, B. R., Singer, B. S., Beard, B. L., Johnson, C. M., Moreno-Roa, H. & Naranjo, J. A. 2007. Rapid magma ascent and generation of ^{230}Th excesses in the lower crust at Puyehue–Cordón Caulle, Southern Volcanic Zone, Chile. *Earth and Planetary Science Letters*, **255**, 229–242.

Jicha, B. R., Johnson, C. M., Hildreth, W., Beard, B. L., Hart, G. L., Shirey, S. B. & Singer, B. S. 2009. Discriminating assimilants and decoupling deep- vs. shallow-level crystal records at Mount Adams using ^{238}U–^{230}Th disequilibria and Os isotopes. *Earth and Planetary Science Letters*, **277**, 38–49.

Jordan, T. E., Burns, W. M., Veiga, R., Pángaro, F., Copeland, P., Kelley, S. & Mpodozis, C. 2001. Extension and basin formation in the southern Andes caused by increased convergence rate: a mid-Cenozoic trigger for the Andes. *Tectonics*, **20**, 308–324.

Karlstrom, L., Dufek, J. & Manga, M. 2009. Organization of volcanic plumbing through magmatic lensing by magma chambers and volcanic loads. *Journal of Geophysical Research*, **114**, B10204.

LAVENU, A. & CEMBRANO, J. 1999. Compressional and transpressional-stress pattern for Pliocene and Quaternary brittle deformation in fore arc and intra-arc zones (Andes of Central and Southern Chile). *Journal of Structural Geology*, **21**, 1669–1691.

LEE, C. Y., TSAI, J. H., HO, H. H., YANG, T. F., CHUNG, S. L. & CHEN, C.-H. 1997. Quantitative analysis in rock samples by an X-ray fluorescence spectrometer (I) major elements. *Annual Meeting of the Geological Society of China*. Geological Society of China, Taipei, 418–420.

LÓPEZ-ESCOBAR, L., PARADA, M. A., MORENO, H., FREY, F. & HICKEY-VARGAS, R. 1992. A contribution to the petrogenesis of Osorno and Calbuco volcanoes, Southern Andes (41°00′–41°30′S): comparative study. *Revista Geologica de Chile*, **19**, 211–226.

LÓPEZ-ESCOBAR, L., PARADA, M. A., HICKEY-VARGAS, R., FREY, F., KEMPTON, P. D. & MORENO, H. 1995. Calbuco volcano and minor eruptive centers distributed along the Liquiñe-Ofqui fault zone, Chile (41°–42°S): contrasting origin of andesitic and basaltic magma in the Southern Volcanic Zone of the Andes. *Contributions to Mineralogy and Petrology*, **119**, 345–361.

MATTHEWS, S. J., JONES, A. P. & GARDEWEG, M. C. 1994. Lascar volcano, northern Chile; evidence for steady-state disequilibrium. *Journal of Petrology*, **35**, 401–432.

MCNUTT, S. R. & BEAVAN, R. J. 1987. Eruptions of Pavlof volcano and their possible modulation by ocean load and tectonic stresses. *Journal of Geophysical Research*, **92**, 11 509–11 523.

MELNICK, D., FOLGUERA, A. & RAMOS, V. A. 2006. Structural control on arc volcanism: the Caviahue-Copahue complex, Central to Patagonian Andes transition (38°S). *Journal of South American Earth Sciences*, **22**, 66–88.

MORENO, H. & GARDEWEG, M. 1989. La erupción reciente en el Complejo Volcánico Lonquimay (Diciembre 1988), Andes del Sur. *Revista Geológica de Chile*, **16**, 93–117.

MORGAN, D. J. & BLAKE, S. 2006. Magmatic residence times of zoned phenocrysts: introduction and application of the binary element diffusion modelling (BEDM) technique. *Contributions to Mineralogy and Petrology*, **151**, 58–70.

MORGAN, D. J., BLAKE, S., ROGERS, N. W., DE VIVO, B., ROLANDI, G., MACDONALD, R. & HAWKESWORTH, C. J. 2004. Timescales of crystal residence and magma chamber volume from modelling of diffusion profiles in phenocrysts: Vesuvius 1944. *Earth and Planetary Science Letters*, **222**, 933–946.

MULLER, J. R., ITO, G. & MARTEL, S. J. 2001. Effects of volcano loading on dike propagation in elastic half-space. *Journal of Geophysical Research*, **106**, 11 101–11 113.

MUNIZAGA, F., HERVE, F., DRAKE, R., PANKHURST, R. J., BROOK, M. & SNELLING, N. 1988. Geochronology of the Lake Region of south-central Chile (39°–42°S): preliminary results. *Journal of South American Earth Sciences*, **1**, 309–316.

MUÑOZ, J. & NIEMEYER, H. 1984. *Hoja Laguna del Maule, Regiones del Maule y Biobío*. Carta Geológica de Chile No. 64, Scale 1:250 000, Servicio Nacional de Geología y Minería, Santiago, Chile.

NAKAMURA, M. 1995. Continuous mixing of crystal mush and replenished magma in the ongoing Unzen eruption. *Geology*, **23**, 807–810.

NARANJO, J. A., MORENO, H. & GARDEWEG, M. 1991. Erupción de 1988–1990 del volcán Lonquimay, Andes del Sur (38°20′S), Chile. *In*: *VI Congreso Geológico Chileno*, Viña del Mar. Servicio Nacional de Geología y Minería, Santiago, Chile, 445–448 (resúmenes en extenso).

NARANJO, J. A., HALLER, M. J., OSTERA, H. A., PESCE, A. H. & SRUOGA, P. 1999. Geología y peligros del Complejo Volcánico Planchón-Peteroa, Andes del Sur (35°15′S), Región del Maule, Chile-Provincia de Mendoza, Argentina. *Servicio Nacional de Geología y Minería, Boletín*, **52**, 1–55.

NIEMEYER, H. & MUÑOZ, J. 1983. *Hoja Laguna de la Laja, Region del Bío Bío. Carta Geología de Chile No. 57, Scale 1:250 000*. Servicio Nacional de Geología y Minería, Santiago, Chile.

OSMOND, J. K. & IVANOVICH, M. 1992. Uranium-series mobilization and surface hydrology. *In*: IVANOVICH, M. & HARMON, R. S. (eds) *Uranium-Series Disequilibrium: Applications to Earth, Marine, and Environmental Sciences*. Clarendon Press, Oxford, 259–289.

PEACOCK, M. A. 1931. Classification of igneous rock series. *Journal of Geology*, **39**, 54–67.

PECCERILLO, A. & TAYLOR, S. R. 1976. Geochemistry of Eocene calc-alkaline volcanic rocks from the Kastamonu area, northern Turkey. *Contributions to Mineralogy and Petrology*, **58**, 63–81.

PINEL, C. & JAUPART, C. 2000. The effect of edifice load on magma ascent beneath a volcano. *Philosophical Transactions of the Royal Society of London A*, **358**, 1515–1532.

PYLE, D. M., IVANOVICH, M. & SPARKS, R. S. J. 1988. Magma cumulate mixing identified by U–Th disequilibrium dating. *Nature*, **331**, 157–159.

RAMÍREZ, C. & GARDEWEG, M. 1982. *Carta Geológica de Chile, Hoja Toconao*, mapa escala 1:250 000. Servicio Nacional de Geología y Minería, Santiago, Chile.

RAMOS, F. C. & TEPLEY, F. J. 2008. Inter- and intracrystalline isotopic disequilibria: techniques and applications. *In*: PUTIRKA, K. & TEPLEY, F. J. (eds) *Mineral Inclusions, Volcanic Processes*. Reviews in Mineralogy and Geochemistry, **69**. The Mineralogical Society of America, Washington, DC, 403–444.

REA, J. C. 2009. *The petrology and geochemistry of Volcan Callaqui, Chile*. Honours thesis, Wesleyan University, Connecticut.

REAGAN, M. K., SIMS, K. W. W. ET AL. 2003. Timescales of differentiation from mafic parents to rhyolite in North American continental arcs. *Journal of Petrology*, **44**, 1703–1726.

ROMER, R. L. & ROCHOLL, A. 2004. Activity disequilibrium of ^{230}Th, ^{234}U, and ^{238}U in old stilbite: effects of young U mobility and α-recoil. *Geochimica et Cosmochimica Acta*, **68**, 4705–4719.

ROSHOLT, J. N. 1982. Mobilization and weathering. *In*: IVANOVICH, M. & HARMON, R. S. (eds) *Uranium Series Disequilibrium: Applications to Environmental Problems*. Clarendon Press, Oxford, 167–180.

SCHAEFER, S. J., STURCHIO, N. C., MURRELL, M. T. & WILLIAMS, S. N. 1993. Internal ^{238}U-series systematics of pumice from the November 13, 1985, eruption of Nevado del Ruiz, Colombia. *Geochimica et Cosmochimica Acta*, **57**, 1215–1219.

SHAW, H. R. 1980. Fracture mechanisms of magma transport from the mantle to the surface. *In*: HARGRAVES, R. B. (ed.) *Physics of Magmatic Processes*. Princeton University Press, Princeton, NJ, 201–246.

SIGMARSSON, O., CONDOMINES, M., MORRIS, J. D. & HARMON, R. S. 1990. Uranium and ^{10}Be enrichments by fluids in Andean arc magmas. *Nature*, **346**, 163–165.

SIGMARSSON, O., MARTIN, H. & KNOWLES, J. 1998. Melting of a subducting oceanic crust from U–Th disequilibria in austral Andean lavas. *Nature*, **394**, 566–569.

SMITH, I. E. M., STEWART, R. B., PRICE, R. C. & WORTHINGTON, T. J. 2010. Are arc-type rocks the products of magma crystallisation? Observations from a simple oceanic arc volcano: Raoul Island, Kermadec Arc, SW Pacific. *Journal of Volcanology and Geothermal Research*, **190**, 219–234.

SNYDER, D. 1997. The mixing and mingling of magmas. *Endeavour*, **27**, 19–22.

SPERA, F. J. 1980. Aspects of magma transport. *In*: HARGRAVES, R. B. (ed.) *Physics of Magmatic Processes*. Princeton University Press, Princeton, NJ, 265–323.

STERN, C. R. 2004. Active Andean volcanism: its geologic and tectonic setting. *Revista Geologica de Chile*, **31**, 161–206.

STEWART, R. B., PRICE, R. C. & SMITH, I. E. M. 1996. Evolution of high-K arc magma, Egmont volcano, Taranaki, New Zealand: evidence from mineral chemistry. *Journal of Volcanology and Geothermal Research*, **74**, 275–295.

STONE, J. B. & INGERSON, E. 1934. Some volcanoes of southern Chile. *American Journal of Science*, **28**, 269–287.

STRECK, M. J. 2008. Mineral textures and zoning as evidence for open system processes. *In*: PUTIRKA, K. & TEPLEY, F. J. (eds) *Mineral Inclusions, Volcanic Processes*. Reviews in Mineralogy and Geochemistry, **69**. The Mineralogical Society of America, Washington, DC, 595–622.

STRECK, M. J., LEEMAN, W. P. & CHESLEY, J. 2007. High-magnesian andesite from Mount Shasta: a product of magma mixing and contamination, not a primitive mantle melt. *Geology*, **35**, 351–354.

SUÁREZ, M. & EMPARAN, C. 1997. *Hoja Curacautín, Regiones de la Araucanía y del Biobío*, Carta Geológica de Chile, 71, 1:250 000. Servicio Nacional de Geología y Minería, Santiago, Chile.

SUN, S. S. & MCDONOUGH, W. F. 1989. Chemical and isotopic systematics of oceanic basalts: implications for mantle composition and processes. *In*: SAUNDERS, A. D. & NORRY, M. J. (eds) *Magmatism in Ocean Basins*. Geological Society, London, Special Publications, **42**, 313–345.

THIELE, R., LAHSEN, A., MORENO, H., VARELA, J., VERGARA, M. & MUNIZAGA, F. 1987. *Estudio Geológico Regional a Escala 1:100 000 de la Hoya superior y curso medio del río Biobío, ENDESA-Departamento de Geología y Geofísica (Inédito)*. Universidad de Chile, Departamento de Geología, Santiago.

TURNER, S. P. & COSTA, F. 2007. Measuring timescales of magmatic evolution. *Elements*, **3**, 267–272.

TURNER, S. P., GEORGE, R. M. M., EVANS, P. J., HAWKESWORTH, C. J. & ZELLMER, G. F. 2000. Time-scales of magma formation, ascent and storage beneath subduction-zone volcanoes. *Philosophical Transactions of the Royal Society of London A*, **358**, 1443–1464.

TURNER, S. P., GEORGE, R. M. M., JERRAM, D. A., CARPENTER, N. & HAWKESWORTH, C. J. 2003. Case studies of plagioclase growth and residence times in island arc lavas from Tonga and the Lesser Antilles, and a model to reconcile discordant age information. *Earth and Planetary Science Letters*, **214**, 279–294.

VAN WYK DE VRIES, B. & MERLE, O. 1998. Extension induced by volcanic loading in regional strike-slip zones. *Geology*, **26**, 983–986.

VILLEMANT, B. & FLÉHOC, C. 1989. U–Th fractionation by fluids in K-rich magma genesis: the Vico volcano, Central Italy. *Earth and Planetary Science Letters*, **91**, 312–326.

VILLEMANT, B., BOUDON, G. & KOMOROWSKI, J. C. 1996. U-series disequilibrium in arc magmas induced by water-magma interaction. *Earth and Planetary Science Letters*, **140**, 259–267.

VOLPE, A. M. & HAMMOND, P. E. 1991. ^{238}U–^{230}Th–^{226}Ra disequilibria in young Mount St. Helens rocks: time constraint for magma formation and crystallisation. *Earth and Planetary Science Letters*, **107**, 475–486.

WANG, K. L., CHUNG, S. L., O'REILLY, S. Y., SUN, S. S., SHINJO, R. & CHEN, C. H. 2004. Geochemical constraints for the genesis of post-collisional magmatism and the geodynamic evolution of the northern Taiwan region. *Journal of Petrology*, **45**, 975–1011.

WÖRNER, G., MERCIER, R., KILIAN, R. & MAMANI, M. 2004*a*. Geochemistry, radiogenic and U-series isotopes in Andean magmas: crustal assimilation and adakite-like magmatism in the Central and Southern Andes. Paper presented at the IAVCEI General Assembly, Volcanism and its impact on society, Pucon, Chile, 14–19 November 2004.

WÖRNER, G., WEGNER, W., KIEBALA, A., SINGER, B. S., HEUMANN, A., KRONZ, A. & HORA, J. 2004*b*. Evolution of Taapaca volcano, N. Chile, evidence from major and trace element, Sr-, Nd-, Pb-, and U-series isotopes, age dating and chemical zoning in sanidine megacrysts. Paper presented at the IAVCEI General Assembly, Volcanism and its impact on society, Pucon, Chile, 14–19 November 2004.

YANG, J.-H., CHUNG, S.-L., WILDE, S. A., WU, F.-Y., CHU, M.-F., LO, C.-H. & FAN, H.-R. 2005. Petrogenesis of post-orogenic syenites in the Sulu Orogenic Belt, East China: geochronological, geochemical and Nd-Sr isotopic evidence. *Chemical Geology*, **214**, 99–125.

YOKOYAMA, T., MAKISHIMA, A. & NAKAMURA, E. 1999. Evaluation of the coprecipitation of incompatible trace elements with fluoride during silicate rock dissolution by acid digestion. *Chemical Geology*, **157**, 175–187.

YOKOYAMA, T., KOBAYASHI, K., KURITANI, T. & NAKAMURA, E. 2003. Mantle metasomatism and rapid ascent of slab components beneath island arcs:

evidence from $^{238}U-^{230}Th-^{226}Ra$ disequilibria of Miyakejima volcano, Izu arc, Japan. *Journal of Geophysical Research*, **108**, 2329.

ZELLMER, G. F. & CLAVERO, J. 2006. Using trace element correlation patterns to decipher a sanidine crystal growth chronology: an example from Taapaca volcano, Central Andes. *Journal of Volcanology and Geothermal Research*, **156**, 291–301.

ZELLMER, G. F. & TURNER, S. P. 2007. Arc dacite genesis pathways: evidence from mafic enclaves and their hosts in Aegean lavas. *Lithos*, **95**, 346–362.

ZELLMER, G. F., BLAKE, S., VANCE, D., HAWKESWORTH, C. & TURNER, S. 1999. Plagioclase residence times at two island arc volcanoes (Kameni islands, Santorini, and Soufriere, St. Vincent) determined by Sr diffusion systematics. *Contributions to Mineralogy and Petrology*, **136**, 345–357.

ZELLMER, G. F., TURNER, S. P. & HAWKESWORTH, C. J. 2000. Timescales of destructive plate margin magmatism: new insights from Santorini, Aegean volcanic arc. *Earth and Planetary Science Letters*, **174**, 265–282.

ZELLMER, G. F., SPARKS, R. S. J., HAWKESWORTH, C. J. & WIEDENBECK, M. 2003. Magma emplacement and remobilization timescales beneath Montserrat: insights from Sr and Ba zonation in plagioclase phenocrysts. *Journal of Petrology*, **44**, 1413–1431.

ZELLMER, G. F., ANNEN, C., CHARLIER, B. L. A., GEORGE, R. M. M., TURNER, S. P. & HAWKESWORTH, C. J. 2005. Magma evolution and ascent at volcanic arcs: constraining petrogenetic processes through rates and chronologies. *Journal of Volcanology and Geothermal Research*, **140**, 171–191.

ZELLMER, G. F., RUBIN, K. H., GRÖNVOLD, K. & JURADO-CHICHAY, Z. 2008. On the recent bimodal magmatic processes and their rates in the Torfajökull-Veidivötn area, Iceland. *Earth and Planetary Science Letters*, **269**, 387–397.

ZELLMER, G. F., SAKAMOTO, N., IIZUKA, Y., MIYOSHI, M., TAMURA, Y., HSIEH, H.-H. & YURIMOTO, H. 2013. Crystal uptake into aphyric arc melts: insights from two-pyroxene pseudo-decompression paths, plagioclase hygrometry, and measurement of hydrogen in olivines from mafic volcanics of SW Japan. *In*: GÓMEZ-TUENA, A., STRAUB, S. M. & ZELLMER, G. F. (eds) *Orogenic Andesites and Crustal Growth*. Geological Society, London, Special Publications, **385**. First published online June 11, 2013, http://dx.doi.org/10.1144/SP385.3

Lying in wait: deep and shallow evolution of dacite beneath Volcán de Santa María, Guatemala

B. S. SINGER[1]*, B. R. JICHA[1], J. H. FOURNELLE[1], B. L. BEARD[1], C. M. JOHNSON[1], K. E. SMITH[1], S. E. GREENE[1], N. T. KITA[1], J. W. VALLEY[1], M. J. SPICUZZA[1] & N. W. ROGERS[2]

[1]*Department of Geoscience, University of Wisconsin-Madison, Madison, USA*

[2]*Department of Earth and Environmental Science, The Open University, Milton Keynes, UK*

**Corresponding author (e-mail: bsinger@geology.wisc.edu)*

Abstract. The Plinian eruption in October 1902 of 8.5 km^3 of dacitic pumice and minor basaltic andesite scoria and ash at Volcán de Santa María, Guatemala violently interrupted a 25 kyr period of repose that had followed ~75 kyr of cone-growth via extrusion of 8 km^3 of basaltic andesite lava. Two-oxide and pyroxene thermometry reveal an oxidized (Ni-NiO + 2 log units) and thermally-zoned magma body in which basaltic andesite with 54 wt% SiO_2 at 1020 °C and dacite with 65 wt% SiO_2 at 870 °C coexisted. Plagioclase in dacite pumice and basaltic andesite scoria shows remarkably similar zoning characterized by repeated excursions toward high anorthite and increases in Mg, Fe, and Sr associated with resorption surfaces along which dacitic to rhyolitic melt inclusions are trapped. The melt inclusions increase slightly in K_2O as SiO_2 increases from 69 to 77 wt%, whereas H_2O contents between 5.2 and 1.4 wt% drop with increasing K_2O. These observations suggest that crystallization of the plagioclase, and evolution of a high-silica rhyolitic residual melt, occurred mainly in the conduit as the compositionally-zoned magma body decompressed and degassed from >180 MPa, or >5 km depth, toward the surface. The similarity of plagioclase composition, zoning, and melt inclusion compositions in pumice and scoria suggests that crystals which grew initially in the cooler dacite, were exchanged between dacitic and basaltic andesite magma as the two magmas mingled and partially mixed *en route* to the surface. Since 1922 > 1 km^3 of dacitic magma similar to the 1902 pumice has erupted effusively to form the Santiaguito dome complex in the 1902 eruption crater.

Trace element and Sr–Nd–Pb–O and U–Th isotope data indicate that cone-forming basaltic andesite lavas record processes operating in the deep crust in which wallrock heating sufficient to induce partial melting and assimilation involved several pulses of recharging mantle-derived basalt over at least 50 kyr. A fundamental shift in process coincides with the termination of cone-building at 25 ka: the 1902 dacite reflects >40% fractional crystallization of plagioclase + amphibole + clinopyroxene + magnetite from ~20 km^3 of basaltic andesite magma left-over following cone-building that cooled slowly without assimilating additional crust. Small contrasts in Sr–Nd–Pb ratios, a modest contrast in $\delta^{18}O$(WR), and a large difference in the ($^{238}U/^{230}Th$) activity ratio between the 1902 scoria and dacite indicate that these two magmas are not consanguineous, rather this basaltic andesite is likely a recent arrival in the system. A glass–whole rock–magnetite–amphibole ^{238}U–^{230}Th isochron of 9.5 ± 2.5 ka for a 1972 Santiaguito dacite lava suggests that deeper, occluded portions of the silicic magma body, not erupted in 1902, incubated in the crust for at least 10 kyr prior to the 1902 eruption. Basaltic andesite inclusions in the Santiaguito dacite lava domes are interpreted to be modified remnants of the cone-forming magma parental to the 1902 dacite.

Supplementary material: Electron probe analyses of glass standards, and SIMS data from standards and melt inclusions for the hydrogen measurements are available at http://www.geolsoc.org.uk/SUP18606

The origin of andesitic to dacitic magma in subduction zones remains a major focus of igneous petrology owing to the central role it plays in crustal evolution and the hazards posed by explosive, often deadly, eruptions. Many eruptions of dacite also contain lesser amounts of more mafic components, basalt to basaltic andesite in composition, that beg the question of whether the dacite is genetically related to the mafic magma, or if the ascent of mafic magma into the dacite may have promoted eruption (e.g. Sparks *et al.* 1977; Pallister *et al.* 1992). Whereas geochemical observations (e.g. Hildreth & Moorbath 1988) and theoretical models (e.g. Dufek & Bergantz 2005; Annen *et al.* 2006) provide a conceptual framework for understanding how intermediate magmas result from the

From: Gómez-Tuena, A., Straub, S. M. & Zellmer, G. F. (eds) 2014. *Orogenic Andesites and Crustal Growth*. Geological Society, London, Special Publications, **385**, 209–234.
First published online June 11, 2013, http://dx.doi.org/10.1144/SP385.2

interaction of hydrous mantle-derived basalt with lower crustal rocks that partially melt and thus blend with the basalt, many details of the processes involved, including their timescales, remain obscure. Recent advances in $^{40}Ar/^{39}Ar$ geochronology have made it possible to date precisely latest Pleistocene and Holocene volcanic rocks in subduction zones (e.g. Jicha & Singer 2006; Hora *et al.* 2007; Jicha *et al.* 2012). In turn, this makes several tens of thousands of years of pre-historic eruptions amenable to U-series isotope disequilibrium study, thereby linking volcano growth episodes, large explosive eruptions, or collapse events with specific magmatic processes (e.g. Jicha *et al.* 2005, 2007, 2009; Hora *et al.* 2009). Moreover, the resulting estimates for the timing of mixing or recharge events, and duration of crystallization/pluton formation or partial melting episodes provide unique insights on the dynamics of intermediate magma formation (e.g. Hawkesworth *et al.* 2000; Bacon & Lowenstern 2005; Jicha *et al.* 2005; Bacon *et al.* 2007). In this light, $^{40}Ar/^{39}Ar$ dating (Escobar-Wolf *et al.* 2010; Singer *et al.* 2011) and U-series isotope (Jicha *et al.* 2010) studies of pre-historic basaltic andesite lavas that comprise the Santa María composite volcano, Guatemala provide a backdrop for understanding more fully the origin of voluminous dacitic magma that has erupted historically from the same magmatic system.

The Santa María–Santiaguito volcanic complex has a relatively simple history that includes a ~75 kyr period of basaltic andesite cone-growth, followed by repose for ~25 kyr and explosive eruption of dacitic pumice and minor basaltic andesitic scoria and ash in 1902 (Rose 1987*a*; Escobar-Wolf *et al.* 2010; Singer *et al.* 2011). By some estimates the 24 October 1902 eruption of Santa María was the third most deadly of the twentieth century, with more than 8750 casualties (Witham 2005). Subsequent eruptions that began in 1922, and which continue to the present day, comprise dacitic lava and ash that occasionally contain quenched mafic inclusions (Rose 1987*a*, *b*; Harris *et al.* 2003; Escobar-Wolf *et al.* 2008). Whereas Singer *et al.* (2011) used whole rock major and trace element geochemical data, as well as Sr, Pb, Nd, and U–Th isotope compositions to investigate the evolution of basaltic andesite lavas that formed the Santa María cone, this paper focuses on the origin of the historically-erupted dacitic tephra and lava and its relationship to cone-forming mafic magmas that preceded the dacite. In addition to new whole rock major and trace element and Sr–Nd–Pb–O isotope compositions from these historically erupted materials, we present the composition of phenocryst minerals including zoning in plagioclase obtained via electron probe micro-analysis (EPMA), and H_2O concentrations of plagioclase-hosted melt inclusions in dacite pumice determined using secondary ion mass spectrometry (SIMS). These datasets are used collectively to quantify processes that led to the catastrophic eruption of 1902.

Geological and petrologic background

Subduction of the Cocos plate at 7–8 cm/yr beneath the Caribbean plate has generated the Central American Volcanic Arc (CAVA) between Guatemala and Costa Rica. Volcán de Santa María, near the western end of the CAVA, is one of 39 centers that define the arc front which is 165–190 km from the Middle America Trench (Fig. 1). Crustal thickness beneath the arc varies from >40 km at its northwestern (Guatemala) and southeastern (Costa Rica) ends, to ~30 km beneath its centre in Nicaragua (Carr 1984; Carr *et al.* 2003, 2007). Volcán de Santa María sits atop the Chortis block that includes mainly poorly exposed Palaeozoic metamorphic rocks, but also Cretaceous ophiolites which crop out along the Motagua fault north of the volcano (Fig. 1; Beccaluva *et al.* 1995). The surface of the Chortis block is widely capped with several hundred metres of undivided volcanic rocks of late Cenozoic age that include deeply eroded remnants of Pleistocene andesite volcanoes which crop out beneath the Santa María cone (Fig. 2).

The majority of geochemical data within the CAVA are from lavas erupted in the central and southern parts of the arc. Beneath Guatemala the crust is thicker and more lithologically diverse which allows for a potentially more complex evolution of magma (e.g. Rose 1987*a*; Halsor & Rose 1988; Vogel *et al.* 2006; Walker *et al.* 2007). It has also been proposed by Heydolph *et al.* (2012) that the lithospheric mantle beneath Guatemala is exceptionally heterogeneous, leading to a distinctively 'crustal' isotopic composition of mafic lavas in this part of the CAVA. The Santa María–Santiaguito complex is strongly bimodal both chemically and temporally, having erupted 8 km^3 of basaltic andesite lava (51–56% SiO_2) between 103 and 25 ka, but with dacite (63–67% SiO_2) dominating the >9 km^3 of historically erupted material (Rose 1987*a*; Escobar-Wolf 2010; Singer *et al.* 2011; Fig. 2). The 1902 eruption was itself bimodal with >98% of the airfall deposit comprising dacite pumice and ash overlain by a minor layer of basaltic andesite ash and scoria (Rose 1972; Williams & Self 1982). Lithic fragments of granitoid and metamorphic rocks also comprise several percent by mass of the airfall deposit (Rose 1972). Despite the thick crust beneath Volcán de Santa María, Jicha *et al.* (2010) used Sr and U–Th isotope data to show that the flux of slab fluid into the mantle below Guatemala is equal to, or greater than, in

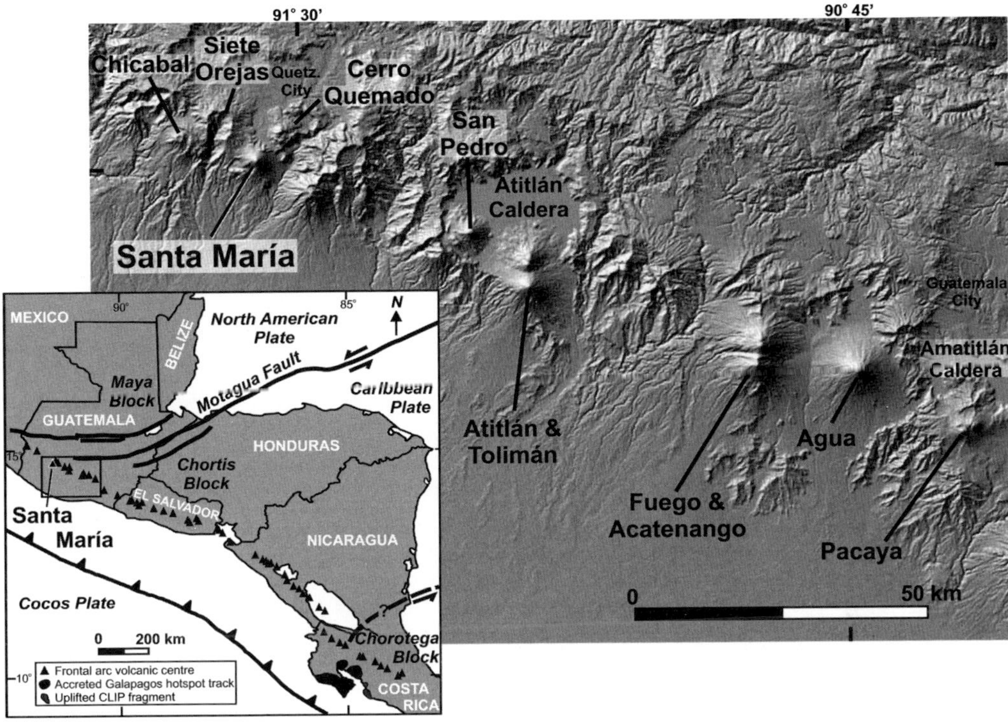

Fig. 1. Digital elevation map of western Guatemala adapted from Escobar-Wolf *et al.* (2010). Quetz., Quetzaltenango. Inset: tectonic setting of Santa María volcano modified from Carr *et al.* (2007*b*) and Geldmacher *et al.* (2008). CLIP, Caribbean Large Igneous Province.

other parts of CAVA where crust is relatively thin (e.g. Nicaragua), a finding that conflicts with other models (Patino *et al.* 2000; Carr *et al.* 2003; Heydolph *et al.* 2012). Moreover, Jicha *et al.* (2010) inferred that a portion of the crust beneath Santa María comprises ancient mid-ocean ridge basalt (MORB) and is not limited to metamorphic and granitic rocks. Singer *et al.* (2011) present major and trace element compositions and Sr, Nd and Pb isotope ratios from 27 basaltic andesite lava flows that comprise a sequence of four cone-building phases, each erupted from a common central summit vent or nearby flank vents, that have been dated using the $^{40}Ar/^{39}Ar$ method at between 103–72, 72, 60–46 and 35–25 ka. Thus the entire period of mafic cone-building spanned about 75 kyr. Energy-constrained, recharge, assimilation-fractional crystallization modelling (EC-RAFC; Spera & Bohrson 2004; Bohrson & Spera 2007) reveals that during the initial phases of cone growth between about 103 and 35 ka, repeated recharge of a deep magma reservoir system with basalt was likely an important process that modulated the trace element and isotopic composition of erupted lavas. Moreover, the same EC-RAFC modelling suggests that the episodic addition of recharge magma to the mass of magma evolving within the deep crustal storage zone augmented latent heat being released by crystallization, thereby amplifying the role of partial melting and assimilation during the final phase of cone growth 35–25 ka (Singer *et al.* 2011).

Samples and analytical methods

During field campaigns totalling 20 days in 2005, 2007 and 2009, samples were collected from the 1902 airfall deposit and the Santiaguito dome complex. Dacite pumice blocks, scoria lapilli, and angular lithic fragments of granitoid were collected in stratigraphic sequence from three road cuts through the 1902 airfall deposit, including site SM-07-09 that lies 4 km NW of the 1902 crater (14°46.560′N, 91°34.480′W, 2631 masl) where the deposit is 95 cm thick, site SM-09-16 that is 8 km SW of the crater (14°42.905′N, 91°37.321′W, 1190 masl) where the deposit is 160 cm thick, and at site SM-05-1902 about 3 km from the crater (14°46.239′N, 91°35.014′W, 2570 masl) where only dacite pumice was collected (Fig. 2). In addition, samples were collected from seven of the

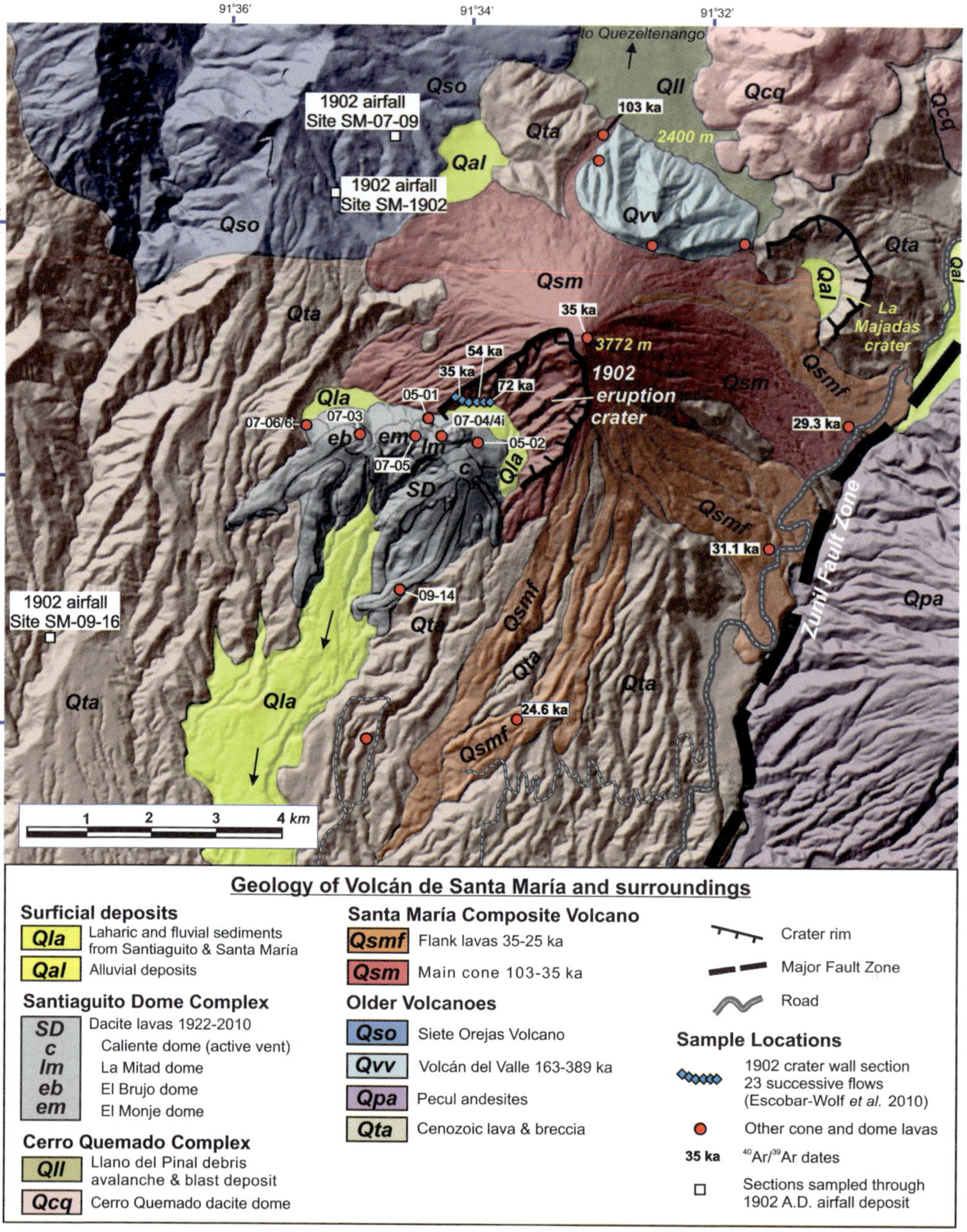

Fig. 2. Geological map of the Santa María-Santiaguito volcanic system, modified from Escobar-Wolf *et al.* (2010).

more than 30 dacite lava flows that comprise the Santiaguito dome complex, including three lavas erupted between 1939 and 1942 on the La Mitad dome, one from a 1950 eruption of the El Monje dome, two lavas from the El Brujo dome (1972 and 1973 flows), and two samples from the Caliente dome, including a 1972 flow and one from a flow resulting from an eruption in 2002 at the currently active vent (Fig. 2). Portions of the 1939 dacite flow on La Mitad contained dark coloured, fine-grained inclusions of basaltic andesite, 1–10 cm across and with crenulated margins that appear to have quenched glassy rims (Rose 1987*a*), as did the 1972 flow from El Brujo. We collected several inclusions from these two localities for analysis.

Whereas Santa María cone lavas are highly porphyritic, two-pyroxene olivine-bearing basaltic andesites, dacite pumice in 1902 airfall deposit contains up to 20% phenocrysts of clinopyroxene, orthopyroxene, amphibole, plagioclase, iron titanium oxides, and trace zircon. The basaltic andesite scoria is petrographically similar to the cone-forming lavas with the exception that some scoria lapilli contain minor amphibole phenocrysts. Dacites from Santiaguito dome complex are slightly less SiO_2-rich than the 1902 pumice and contain a higher modal percent of the same phenocryst minerals found in the pumice (Rose 1972, 1987*a*).

Review of Santa María geochemistry

The major and trace element compositions and Sr, Nd and Pb isotope ratios from 27 basaltic andesite lava flows that comprise the Santa María composite cone were presented by Singer *et al.* (2011). In order to emphasize the profound compositional and temporal bimodality of the entire Santa María–Santiaguito complex, Singer *et al.* (2011, figs 5 & 6) compared these 27 mafic cone-forming lavas to the historically-erupted dacite tephra and lavas in variation diagrams. The major and trace element compositions from whole rock samples of the historically-erupted materials described in the previous section, including pumice, scoria and lithic fragments collected from the 1902 airfall deposit, and from each of the major phases of the Santiaguito dome complex, plus quenched mafic inclusions in two of the dacite lava domes, are presented in this paper. From a subset of these whole rock samples we also present new Sr, Nd and Pb isotope ratios and $\delta^{18}O$(WR) values. Jicha *et al.* (2010) present whole rock $^{238}U-^{230}Th$ isotope compositions from Santa María cone-forming lavas, the 1902 dacite pumice and basaltic andesite scoria, and several Santiaguito dacite dome lavas. Here we report new $^{238}U-^{230}Th$ isotope compositions from minerals and glass separated from the 1902 dacite pumice and from a dacitic lava that erupted in 1972 to form part of the El Monje dome in the Santiaguito dome complex.

Whole rock major and trace element analyses

Twenty-nine whole rock samples (including three glass separates from 1902 pumice) of 250–500 g each were crushed (tungsten-carbide piston crusher), powdered (aluminum oxide ceramic shatterbox and puck), and analysed for major (X-ray fluorescence) and trace elements (inductively-coupled plasma mass spectrometry, ICP-MS) at the Open University, UK following methods of Rogers *et al.* (2006). Details on procedures including precision and accuracy are provided in Singer *et al.* (2011).

Pb, Sr and Nd isotopes

A representative subset of the whole rock samples were analysed for $^{87}Sr/^{86}Sr$ (18 samples), $^{208}Pb/^{204}Pb$, $^{207}Pb/^{204}Pb$, $^{206}Pb/^{204}Pb$ (14 samples) and $^{143}Nd/^{144}Nd$ (17 samples) at the UW-Madison Radiogenic Isotope Laboratory via thermal ionization mass spectrometry (TIMS, Micromass Sector 54 instrument). Analytical procedures follow those described in Singer *et al.* (2011). For Pb isotope analyses, instrumental mass bias was corrected empirically based on analyses of NIST SRM-981 ($n = 10$) and SRM-982 ($n = 11$) run under similar conditions as samples. The pooled average mass fractionation correction based on the measured $^{207}Pb/^{206}Pb$ analysis of SRM-981 and the measured $^{208}Pb/^{206}Pb$ ratio of SRM-982 was 0.00148 ± 0.00045 per amu. Strontium and neodymium isotope measurements were exponentially corrected for mass fractionation using $^{86}Sr/^{88}Sr = 0.1194$ and $^{146}Nd/^{144}Nd = 0.7219$, respectively. Repeat analysis of NIST SRM-981 yielded an average $^{87}Sr/^{86}Sr$ ratio of 0.710267 ± 0.000018 (2-SD; $n = 30$). Duplicate samples of the USGS rock standard BCR-1 were processed through the entire analytical procedure and the measured $^{87}Sr/^{86}Sr$ ratios were 0.705021 and 0.705028. The measured $^{143}Nd/^{144}Nd$ for two in-house Nd standards (Ames I and Ames II) are 0.512149 ± 0.000014 ($n = 15$; 2-SD) and 0.511974 ± 0.000016 ($n = 15$; 2-SD), respectively. The measured $^{143}Nd/^{144}Nd$ for La Jolla Nd was 0.511844 ± 0.000015 ($n = 6$; 2-SD). Duplicate samples of the USGS rock standard BCR-1 measured $^{143}Nd/^{144}Nd$ ratios were 0.512920 ± 0.000010 and 0.512908 ± 0.000008 (errors are 2-SE in-run statistics).

U and Th isotopes

In addition to two whole rock samples of dioritic lithic fragments in the 1902 airfall deposit, minerals and glass separated from 1902 dacite pumice sample SM-07-09B and dacite lava sample SG-05-01 from the Mitad dome, were analysed for U–Th isotope composition. Solution-based U and Th isotope measurements were obtained using a Micromass *Isoprobe* multi-collector ICP-MS at UW-Madison's Radiogenic Isotope Laboratory using procedures of Jicha *et al.* (2009). Rock standards including AThO, BCR-1, and AGV-1 were analysed along with variable concentration IRMM-035 and IRMM-036 solutions to monitor accuracy, reproducibility and external precision (details in Singer *et al.* 2011). The mean $^{232}Th/^{230}Th$ values for IRMM 035 ($87\,799 \pm 1006$; 2-SD; $n = 15$) and IRMM 036 ($326\,300 \pm 2126$; 2-SD; $n = 8$) are indistinguishable from consensus values (Sims *et al.* 2008).

O isotopes

Whole rock powders were analysed at the University of Wisconsin-Madison by laser fluorination (Valley *et al.* 1995) using an air-lock sample chamber (Spicuzza *et al.* 1998) with bromine pentafluoride as the reagent and a 25W CO_2 laser. Isotope ratios were measured using a Finnigan MAT 251 dual inlet mass spectrometer. Whole rock $\delta^{18}O$ values were standardized using standard UWG-2 garnet (accepted $\delta^{18}O = 5.80‰$ VSMOW; Valley *et al.* 1995). Eight standards run on each day of analysis (three days total) yield a reproducibility of $< \pm 0.16‰$ (2 std. dev.). The average difference in $\delta^{18}O$ value between duplicate analyses of whole rock powders was $<0.08‰$.

Electron probe microanalysis of phenocryst minerals

Epoxy mounts were prepared with separates of plagioclase, iron–titanium oxides, pyroxenes and amphibole phenocrysts from 1902 dacite pumice sample SM-07-09B and scoria sample SM-7-09E. Phenocrysts and their melt inclusions were imaged in back-scattered electron (BSE) mode on the Hitachi S-3400N Variable Pressure Scanning Electron Microscope (VP-SEM) at the University of Wisconsin-Madison. Energy Dispersive Spectrometry (EDS) located grains of specific minerals (e.g. orthopyroxene v. clinopyroxene or ilmenite v. titanomagnetite) which then were analysed by Wavelength Dispersive Spectrometry (WDS) on the Cameca SX51 5-spectrometer electron microprobe at UW-Madison. Major element compositions of minerals were obtained using a 1 μm diameter beam with an accelerating potential of 15 keV. Natural and synthetic standards of olivine, pyroxene, plagioclase, amphibole and iron–titanium oxides were used for calibration. Automation and data reductions used Probe for Windows software with the matrix correction of Armstrong (1988). In anticipation of element migration under a focused electron beam (Morgan & London 1996), analyses of plagioclase-hosted melt inclusions employed time-integrated intensities collected at 2 s intervals for 20 s, allowing Na, K and Si concentrations in the glass to be extrapolated to a time-zero value. To obtain a large number of anorthite content profiles of plagioclase phenocrysts, we used greyscale values from the BSE images as a proxy for An content generally following the procedures outlined by Ginibre *et al.* (2002) and Triebold *et al.* (2006). The An profiles were calibrated via WDS analyses of 10 points along each profile.

Secondary Ion Mass Spectrometry (SIMS) analysis of trace elements in plagioclase

Polished gold-coated epoxy mounts of plagioclase crystals were analysed for K, Mg, Ti, Fe, Sr and Ba using the Cameca ims-1280 ion microprobe at the WiscSIMS Laboratory, UW-Madison. The analysis conditions are after Kita *et al.* (2004) using high mass resolution mode (mass resolving power of 5000) without energy filtering. The primary beam of O^- ions with 1 nA intensity was focused to a spot of *c.* 10 μm in diameter at the sample surface. For each element, secondary ion intensity was normalized to the intensity of the $^{30}Si^+$ ions. A natural plagioclase standard with known trace element concentrations (Lab1, Kita *et al.* 2004) was analysed multiple times in the same analytical session, from which relative sensitivity factors of each element were determined. Reproducibility of ratios between respective ions and $^{30}Si^+$ of repeated analyses of plagioclase standard was better than 3% (1σ).

SIMS analysis of plagioclase-hosted melt inclusions

In polished, gold-coated epoxy mounts 2 mm thick, melt inclusions ranging from 30–100 μm across in plagioclase from 1902 dacite pumice sample SM-07-09B were analysed for hydrogen using the Cameca IMS-1280 ion microprobe at the WiscSIMS Laboratory, UW-Madison. A primary beam of Cs^+ (20 keV impact energy and 0.6 nA beam current) was focused to a spot of ~8 μm at the sample surface and a normal incidence electron gun was used for charge compensation. General instrument parameters are similar to that of oxygen isotope analysis (Kita *et al.* 2004). The negative secondary ions of H^- and $^{30}Si^-$ were measured using the axial electron multiplier in single collection mode. In order to precisely aim at these melt inclusions, we obtained secondary H^- ion images of the sample surface by rastering the primary beam over a 50 × 50 μm areas for three minutes and identified the location of melt inclusions from a hot spot of high H^- count rates. Subsequently, we used a stationary primary beam centred on the melt inclusion and detected secondary H^- and $^{30}Si^-$

ions that were emitted from only the central 2 μm squares of the sputtered area by using a field aperture with opening of 400 × 400 μm (magnification of transfer optics was ×200 from sample to the field aperture). Mass resolving power was set to 3000. The secondary H^- ion intensities were normalized to that of the $^{30}Si^-$ ion and the $H^-/^{30}Si^-$ ratios were obtained after correction for hydrogen background as measured several times on the host plagioclase crystals between measurement of the melt inclusions.

The seven rhyolite glasses used as standards for the hydrogen measurements were produced experimentally from a sample of a natural obsidian flow, and contain between 0.1 and 6.1 wt% H_2O (Tenner *et al.* 2010). Electron probe major element compositions, and SIMS data from these glasses can be found in the Supplementary material. The glass standards were analysed multiple times during the analytical session to generate a calibration curve for $H^-/^{30}Si^-$ ratios (Supplementary material). Following SIMS analysis, the SiO_2 content of each analysed inclusion was measured by electron probe and the calibration curve was used to estimate the wt% H_2O with a precision of about ±3% (1σ; Supplementary material).

Results

Major and trace elements and Sr–Nd–Pb–O isotopes

The whole rock major and trace element concentrations and Sr, Nd, and Pb isotope ratios are summarized in Table 1. The O isotope results from the whole rock samples of this study, as well as samples of the cone-forming basaltic andesite lavas measured for elemental concentrations and Sr–Nd–Pb isotope ratios by Singer *et al.* (2011), are summarized in Table 2. Representative plots of K_2O, MgO, Th, Sr, $^{87}Sr/^{86}Sr$, $^{143}Nd/^{144}Nd$, $^{206}Pb/^{204}Pb$ and $\delta^{18}O$(WR) v. SiO_2 serve to illustrate the major geochemical and isotopic features of the 1902 eruptive products and to contrast these with the preceding Santa María cone-forming lavas and the subsequently erupted Santiaguito dome lavas (Fig. 3). In the 1902 airfall deposit, dacite pumice contains 64.6 to 67.2 wt% SiO_2, whereas the basaltic andesite scoria has 53.9 to 55.9 wt% SiO_2, thus revealing the strongly bimodal nature of the 1902 magma body distinguished by a >10% gap in SiO_2. Rhyodacitic glass separated from three dacite pumice blocks contains 67.8 to 70.4 wt% SiO_2 and is richer in K_2O and Th (and other incompatible elements) and poorer in MgO and Sr on average (and other compatible elements) than the bulk dacite pumice (Fig. 3). The 1902 scoria is similar in major and trace element composition to the more SiO_2-rich of the basaltic andesite lava flows that erupted during the latest cone-forming phase of Santa María volcano between 35 and 25 ka. Lavas comprising the Santiaguito dome complex contain less SiO_2, 62.4 to 64.7 wt%, compared to the 1902 dacite pumice and the quenched inclusions we have measured from two of these flows are remarkably similar to the 1902 scoria with respect to major and trace element composition (Fig. 3). The six fine- to medium-grained diorite to granodiorite lithic fragments we chose to analyse from the 1902 deposit range from 48 to 60% SiO_2 and, with the exception of MgO, are not collinear with the basaltic andesitic cone-lavas of Santa María volcano, instead being scattered with no simple relationship to the volcanic rocks. These results are consistent with, and reinforce, the observations of Rose (1987*a*) regarding the bimodality of the volcano, and the similarity among mafic and silicic magmas over time. However, owing to vast improvements in analytical precision during the last three decades, and a larger number of measurements than obtained by Rose *et al.* (1977), the Sr, Nd and Pb isotope results we have obtained reveal several previously unappreciated features of this system.

The $^{87}Sr/^{86}Sr$ ratios of 0.703848 and 0.703880 obtained from two blocks of 1902 dacite pumice are similar to that of 0.703837 in the glass separated from a third block; however, these ratios are significantly lower than ratios that range from 0.703897 to 0.703950 obtained from four scoria lapilli from the same eruption (Fig. 3). The Santiaguito dacite lavas (and quenched mafic inclusions) erupted in 1939 from the Mitad dome, 1972 from El Brujo, 1972 and 2002 from Caliente, have $^{87}Sr/^{86}Sr$ ratios similar to those of the 1902 dacite pumice; however, lavas from the eruptions in 1950 of El Monje, and 1973 of El Brujo are significantly higher at 0.703905 and 0.703977, respectively (Fig. 3). Notably, the entire range of $^{87}Sr/^{86}Sr$ ratios found in the rhyodacite glass, dacite pumice and dacite lava flows erupted in 1902 and afterward is comparable to that of the basaltic andesite lavas that erupted 35–25 ka during the late phase of Santa María cone growth (Fig. 3). With one exception, the diorite–granodiorite lithic fragments span this same range. For the most part $^{143}Nd/^{144}Nd$ ratios mirror those of $^{87}Sr/^{86}Sr$. The 1902 rhyodacite glass has a higher $^{143}Nd/^{144}Nd$ ratio than the dacite pumice, whereas pumice has a similar ratio to three of the four scoria samples. The 1902 scoria and quenched basaltic andesite inclusion from the 1939 La Mitad dome lava span a range of $^{143}Nd/^{144}Nd$ ratios comparable to those of the basaltic andesitic lavas that erupted 35–25 ka during the late phase of Santa María cone growth.

Table 1. *Major and trace element, and Sr, Nd and Pb isotope data from whole rock and glass samples in 1902 airfall deposit and Santiaguito dome lavas*

	1902 Plinian eruption								
Sample	SM-07-09A	SM-07-09B	SM-1902	SM-09-16A	SM-09-16B	SM-07-09G	SM-07-09Bg	SM-09-16Ag	SM-09-16Bg
unit name	1902 airfall	1902 airfall	1902 airfall	1902 airfall	1902 airfall	1902 airfall	1902 airfall	1902 airfall	1902 airfall
type	pumice block	pumice block	pumice block	pumice block	pumice block	pumice block	pumice glass	pumice glass	pumice glass
SiO_2 (wt%)	66.19	64.69	66.32	67.23	66.89	64.58	70.43	68.96	67.82
TiO_2	0.32	0.38	0.33	0.30	0.30	0.38	0.27	0.24	0.20
Al_2O_3	16.30	17.11	16.70	16.56	16.46	16.95	14.68	16.16	18.06
Fe_2O_3	3.63	4.40	3.83	3.34	3.34	4.06	2.27	2.10	1.64
MnO	0.14	0.16	0.15	0.14	0.14	0.17	0.12	0.10	0.09
MgO	1.17	1.61	1.24	1.09	1.04	1.49	0.63	0.62	0.49
CaO	3.58	4.48	3.86	3.52	3.49	4.24	2.27	3.04	3.98
Na_2O	4.90	4.93	5.02	4.94	5.00	4.87	4.88	5.08	5.33
K_2O	1.86	1.67	1.80	1.84	1.86	1.70	2.30	2.07	1.80
P_2O_5	0.18	0.19	0.19	0.18	0.18	0.18	0.14	0.15	0.13
LOI	1.29	0.70	0.98	0.97	0.78	0.82	1.34	1.06	0.73
Total	99.56	100.33	100.41	100.09	99.48	99.46	99.33	99.59	100.27
Li (ppm)	24.9	22.8	24.0	24.6	25.1	22.6	29.9	27.4	24.7
Sc	4.26	6.68	4.34	3.53	3.44	6.43	3.39	3.19	2.46
Ti	1973	2363	2025	1766	1775	2317	1627	1473	1150
V	31.9	51.3	35.2	25.8	25.8	53.5	15.1	12.7	9.7
Cr	0.8	2.0	0.6	0.5	0.7	2.7	0.4	6.4	5.4
Mn	1171	1224	1158	1153	1183	1440	937	901	738
Co	5.5	7.3	5.3	4.6	4.4	24.9	2.9	2.8	2.2
Ni	0.6	1.6	0.4	0.2	0.4	2.2	0.7	0.7	0.4
Cu	8.8	16.8	8.1	10.2	11.2	51.1	14.0	14.8	8.6
Zn	90	99	90	89	91	91	79	73	57
Ga	17.7	18.8	17.8	18.0	18.2	18.1	16.4	17.2	18.4
Rb	37.1	32.1	34.9	36.7	37.3	33.1	46.6	41.8	35.5
Sr	409	460	434	421	427	451	257	358	495
Y	17.1	17.0	16.8	16.5	16.5	16.7	18.8	17.5	14.4
Zr	148	135	144	147	147	136	188	167	141
Nb	4.76	4.50	4.56	4.70	4.85	4.37	5.92	5.27	4.48
Mo	0.60	1.53	0.57	0.59	0.61	0.65	0.77	0.68	0.57
Sn	1.09	1.08	0.85	0.97	1.01	0.98	1.45	1.14	0.96
Sb	0.479	0.489	0.452	0.488	0.483	0.211	0.589	0.535	0.484
Cs	1.87	1.69	1.74	1.83	1.83	1.64	2.33	2.10	1.78
Ba	855	828	827	861	873	800	1028	953	860
La	15.22	14.09	14.71	15.12	15.13	14.19	17.27	16.12	14.37
Ce	31.22	29.23	30.26	30.80	30.92	30.12	35.51	33.06	28.79
Pr	4.11	3.93	4.00	4.02	4.06	3.86	4.59	4.28	3.67
Nd	15.87	15.34	15.56	15.60	15.55	15.19	17.46	16.32	13.94
Sm	3.19	3.15	3.09	3.07	3.07	3.11	3.41	3.24	2.69
Eu	0.89	0.91	0.89	0.87	0.87	0.92	0.81	0.85	0.85
Gd	2.82	2.79	2.79	2.73	2.74	2.79	3.00	2.82	2.36
Tb	0.43	0.44	0.44	0.43	0.42	0.44	0.47	0.44	0.37
Dy	2.57	2.60	2.55	2.50	2.48	2.57	2.78	2.58	2.16
Ho	0.55	0.56	0.54	0.54	0.53	0.55	0.60	0.56	0.46
Er	1.66	1.64	1.62	1.60	1.60	1.63	1.83	1.69	1.40
Yb	1.81	1.76	1.75	1.78	1.76	1.74	2.07	1.90	1.57
Lu	0.30	0.28	0.28	0.29	0.29	0.29	0.34	0.32	0.26
Hf	3.59	3.45	3.49	3.55	3.58	3.29	4.56	4.00	3.38
Ta	0.29	0.28	0.28	0.29	0.29	0.26	0.35	0.31	0.39
Pb	9.12	9.60	8.73	9.15	9.22	8.33	12.74	10.16	9.06
Th	2.43	2.24	2.33	2.42	2.43	2.18	3.08	2.74	2.31
U	1.00	0.94	0.96	1.00	1.01	0.89	1.26	1.17	0.97
$^{206}Pb/^{204}Pb$		18.717			18.702				18.700
$^{207}Pb/^{204}Pb$		15.592			15.593				15.590
$^{208}Pb/^{204}Pb$		38.432			38.450				38.438
$^{87}Sr/^{86}Sr$		0.7038799			0.7038477				0.7038372
$\pm 2\sigma$		0.000011			0.000020				0.000020
$^{143}Nd/^{144}Nd$		0.5129115			0.5129195				0.5129372
$\pm 2\sigma$		0.000009			0.000010				0.000008

The lithic fragments have higher $^{143}Nd/^{144}Nd$ ratios than any of the volcanic rocks (Fig. 3). The $^{206}Pb/^{204}Pb$ ratios of 18.70–18.72 for the 1902 dacite pumice and rhyodacite glass separated from it are slightly higher on average than those of the two scoria samples, 18.68–18.70, and as seen for the other isotopes, all the historically-erupted materials fall within the range spanned by the latest cone-forming basaltic andesite lavas erupted 35–25 ka on Volcán de Santa María (Fig. 3). The lithic fragments have $^{206}Pb/^{204}Pb$ ratios that fall at the higher end of the range defined by the volcanic rocks.

The $\delta^{18}O$(WR) values of 6.25 to 6.42‰ for most of the cone-forming basaltic andesite lavas correlate positively with SiO_2, whereas the youngest of these lavas from the summit of the cone that is $^{40}Ar/^{39}Ar$-dated at ~35 ka (Escobar-Wolf *et al.*

1902 Plinian eruption										
SM-07-09E 1902 airfall scoria	SM-09-07 1902 airfall scoria	SM-09-16D 1902 airfall scoria	SM-09-16F 1902 airfall scoria	SM-09-16H 1902 airfall scoria	SM-07-09F 1902 airfall lithic frag	SM-07-09H 1902 airfall lithic frag	SM-09-16C1 1902 airfall lithic frag	SM-09-16C2 1902 airfall lithic frag	SM-09-16C3 1902 airfall lithic frag	SM-09-16C4 1902 airfall lithic frag
53.52	54.44	54.40	55.90	53.93	54.52	55.73	53.38	52.96	48.04	60.02
0.92	0.87	0.90	0.81	0.89	1.18	1.26	0.97	0.87	0.79	0.71
19.25	18.66	18.63	18.26	18.44	17.59	16.73	18.63	19.61	16.37	17.60
8.54	8.35	8.46	7.67	8.55	8.86	9.46	8.67	8.56	10.94	6.03
0.14	0.15	0.14	0.13	0.14	0.15	0.19	0.16	0.16	0.18	0.11
4.27	4.04	3.99	3.73	4.47	3.94	3.63	4.10	4.39	9.05	2.40
8.32	7.83	7.84	7.07	7.94	6.81	6.67	7.01	7.55	8.21	5.17
3.55	3.60	3.55	3.74	3.63	4.66	4.13	4.42	4.66	2.50	4.90
1.01	1.11	1.15	1.31	1.10	1.84	1.71	1.40	1.36	2.02	2.08
0.18	0.20	0.20	0.20	0.20	0.40	0.51	0.29	0.27	0.17	0.22
0.41	0.70	0.68	0.62	0.67	0.31	0.17	0.34	0.17	1.29	0.35
100.10	99.94	99.95	99.44	99.97	100.27	100.18	99.36	100.55	99.55	99.58
10.6	11.8	12.2	14.8	11.5	14.0	13.5	19.2	8.4	25.7	16.7
26.69	23.74	23.75	21.00	23.71	23.63	29.67	24.89	26.36	26.70	16.12
5988	5113	5444	5003	5270	7139	8367	7041	5159	4579	4221
250.8	197.6	210.9	176.6	206.4	227.7	243.2	224.9	199.2	219.7	114.8
36.1	31.6	42.7	31.1	48.2	46.7	48.8	39.9	46.6	129.9	6.4
1107	1178	1107	1091	1168	1165	1589	1359	1295	1458	912
28.9	28.5	28.8	24.5	30.3	25.4	26.9	28.7	26.1	59.5	16.2
22.6	15.5	21.0	14.8	20.4	27.5	27.8	17.9	19.6	163.0	6.2
78.7	88.2	74.9	59.8	67.2	93.8	232.3	156.0	38.4	101.8	34.3
96	95	92	92	97	117	147	100	95	98	74
20.8	19.5	19.6	19.7	19.5	20.9	20.3	20.8	22.4	17.2	20.3
19.1	20.9	22.0	27.5	19.9	37.8	48.6	27.4	29.3	50.0	41.0
545	521	513	516	528	424	381	486	525	356	413
19.4	19.8	18.9	18.6	18.6	26.4	38.1	23.6	25.0	21.3	19.9
105	104	108	116	110	5	9	5	8	21	4
3.25	3.12	3.13	3.80	3.12	7.13	8.98	6.51	4.38	4.11	4.74
0.98	0.38	0.38	0.41	0.37	1.10	1.90	0.24	0.37	0.10	1.26
1.12	1.05	0.70	1.15	1.04	1.38	1.83	0.82	1.66	0.97	1.09
0.245	0.236	0.212	0.252	0.207	0.325	0.133	0.427	0.034	0.114	0.389
0.88	1.01	0.96	1.26	0.87	1.31	1.15	1.49	0.78	2.18	1.98
524	511	520	586	510	728	688	666	512	465	837
9.24	9.73	9.78	11.15	9.58	13.71	19.99	13.76	13.93	8.01	12.95
20.65	21.03	21.15	23.82	20.70	28.65	45.36	29.28	30.79	20.43	26.46
2.98	3.05	3.08	3.38	3.00	4.85	6.51	4.11	4.36	3.08	3.63
13.03	13.37	13.45	14.30	13.28	21.49	27.93	17.64	18.61	13.09	15.35
3.26	3.29	3.32	3.35	3.25	5.15	6.51	4.12	4.30	3.18	3.61
1.01	1.01	1.02	1.01	1.00	1.23	1.48	1.16	1.21	0.90	1.08
3.28	3.35	3.38	3.34	3.31	5.00	6.48	4.18	4.22	3.25	3.54
0.53	0.54	0.55	0.53	0.52	0.78	1.03	0.66	0.68	0.54	0.56
3.18	3.23	3.17	3.09	3.10	4.68	6.13	3.89	4.00	3.28	3.34
0.67	0.67	0.66	0.65	0.64	0.99	1.29	0.82	0.86	0.72	0.70
1.89	1.93	1.87	1.83	1.82	2.77	3.64	2.33	2.47	2.12	1.99
1.75	1.83	1.74	1.75	1.68	2.49	3.31	2.15	2.33	2.28	1.84
0.26	0.28	0.27	0.27	0.27	0.36	0.49	0.33	0.36	0.36	0.28
2.71	2.58	2.72	2.89	2.67	0.31	0.32	0.28	0.50	1.04	0.21
0.19	0.18	0.18	0.22	0.18	0.40	0.59	0.37	0.33	0.27	0.25
6.12	5.36	5.43	6.30	5.31	7.91	8.41	5.59	5.47	2.63	6.64
1.19	1.34	1.57	1.89	1.24	1.72	2.39	0.79	1.37	1.00	0.55
0.47	0.54	0.62	0.75	0.51	0.42	0.86	0.31	0.40	0.48	0.24
18.708		18.685			18.751	18.720	18.713			
15.605		15.588			15.594	15.587	15.587			
38.494		38.425			38.502	38.463	38.462			
0.7039014		0.7039406	0.7039497	0.7038965	0.7039923	0.7038872	0.7037453			0.7038652
0.000010		0.000020	0.000020	0.000020	0.000014	0.000010	0.000023			0.000023
0.5129212		0.5128925	0.5129082	0.5129150	0.5129415	0.5129410	0.5129679			0.5129680
0.000008		0.000008	0.000008	0.000008	0.000010	0.000009	0.000008			0.00009

2010; Singer *et al.* 2011) has a much higher value of 6.95‰ (Fig. 3). In the 1902 deposit the basaltic andesite scoria has a $\delta^{18}O$ of 6.54‰ and the dacite pumice is 6.95‰; the latter is identical to that of the youngest cone lava. The Santiaguito dacite dome lavas have $\delta^{18}O$ between 6.55 and 6.82‰, whereas the quenched basaltic andesite inclusions in two of these lavas are 6.50 and 6.60‰ (Fig. 3), similar to the 1902 scoria. The $\delta^{18}O$ values are inversely correlated with $^{87}Sr/^{86}Sr$ such that the early cone lavas have lower $\delta^{18}O$ and higher $^{87}Sr/^{86}Sr$ than the youngest cone lavas and the dacites that erupted long after cone growth ceased (Fig. 4).

U–Th isotopes

Whole rock ($^{238}U/^{232}Th$) and ($^{230}Th/^{232}Th$) activity ratios of nine historically erupted dacite

Table 1. *Major and trace element, and Sr, Nd and Pb isotope data from whole rock and glass samples in 1902 airfall deposit and Santiaguito dome lavas (Continued)*

	Santiaguito Dome lavas							Mafic inclusions	
Sample unit name type	SM-07-04 Mitad 1939 lava	SG-05-01 Brujo 1947 lava	SM-07-05 Monje 1950 lava	SM-07-06 Brujo 1972 lava	SG-05-02 Cal 1972 lava	SM-07-03 Brujo 1975 lava	SM-09-14 2002 flow lava	SM-07-04i Mitad 1939 inclusion	SM-07-06i Brujo 1972 inclusion
SiO_2 (wt%)	63.01	64.44	64.16	63.75	64.65	63.39	62.38	55.28	54.90
TiO_2	0.52	0.39	0.44	0.49	0.51	0.46	0.55	0.89	0.88
Al_2O_3	17.59	17.23	17.32	17.41	16.80	17.40	17.79	18.90	18.69
Fe_2O_3	5.24	4.31	4.69	4.98	4.92	4.79	5.33	8.25	8.19
MnO	0.14	0.14	0.14	0.15	0.14	0.14	0.14	0.15	0.15
MgO	1.88	1.50	1.67	1.83	1.68	1.71	2.12	3.23	3.23
CaO	5.23	4.46	4.74	4.93	4.47	4.83	5.48	7.83	7.72
Na_2O	4.85	4.86	4.82	4.82	4.97	4.77	4.71	4.08	3.91
K_2O	1.60	1.71	1.68	1.67	1.82	1.67	1.58	1.20	1.09
P_2O_5	0.24	0.21	0.22	0.22	0.24	0.22	0.23	0.32	0.23
LOI	0.07	0.23	0.13	0.07	0.06	0.10	0.01	0.12	0.94
Total	100.35	99.49	100.03	100.32	100.27	99.47	100.31	100.25	99.93
Li (ppm)	20.7	22.4	21.3	21.2	22.0	21.3	20.4	31.7	24.9
Sc	10.36	6.32	7.98	8.97	8.68	8.37	11.12	21.58	22.83
Ti	3215	2392	2731	2942	3191	2824	3178	5579	5404
V	84.6	48.0	61.3	70.6	76.6	65.6	89.2	200.1	204.5
Cr	1.3	1.6	1.1	1.3	2.3	1.6	4.3	0.6	1.3
Mn	1151	1142	1146	1167	1153	1137	1183	1156	1181
Co	9.4	7.2	8.1	9.2	8.8	8.6	11.8	18.2	19.1
Ni	1.2	3.7	1.0	1.2	1.9	1.1	2.2	1.8	3.7
Cu	10.6	11.0	8.9	13.1	14.5	9.2	34.8	21.1	41.6
Zn	95	90	91	92	93	91	92	101	100
Ga	19.3	18.3	18.6	18.9	19.6	18.6	19.7	20.6	20.2
Rb	21.0	33.7	33.2	32.5	35.5	32.6	31.1	14.8	19.4
Sr	481	465	466	476	436	476	491	509	545
Y	16.8	17.7	18.2	18.5	19.8	18.1	18.5	17.5	19.0
Zr	143	149	150	151	167	149	143	113	109
Nb	6.78	4.52	4.49	4.42	5.16	4.39	4.25	3.53	3.36
Mo	1.44	1.71	0.55	0.53	6.70	0.53	0.50	5.37	5.77
Sn	1.04	1.08	0.91	1.03	1.21	1.02	1.07	1.31	1.11
Sb	0.430	0.441	0.418	0.263	0.505	0.434	0.386	0.325	0.290
Cs	1.47	1.66	1.61	1.56	1.81	1.58	1.46	1.02	1.03
Ba	784	802	778	766	905	775	736	558	543
La	10.84	14.66	14.36	14.17	15.21	14.25	13.66	8.51	9.87
Ce	23.68	30.32	29.90	29.68	33.00	29.81	28.46	15.74	21.79
Pr	3.44	4.02	4.01	3.95	4.38	3.99	3.84	2.95	3.09
Nd	14.20	15.97	15.92	15.83	17.60	15.91	15.58	13.01	13.39
Sm	3.10	3.30	3.31	3.36	3.72	3.34	3.35	3.23	3.22
Eu	0.90	0.96	0.96	0.97	1.01	0.97	0.99	0.95	0.97
Gd	2.89	3.00	3.04	3.11	3.38	3.05	3.18	3.21	3.23
Tb	0.45	0.47	0.49	0.49	0.53	0.49	0.49	0.52	0.52
Dy	2.76	2.73	2.82	2.89	3.16	2.83	2.91	3.11	3.13
Ho	0.59	0.58	0.60	0.61	0.68	0.60	0.62	0.65	0.67
Er	1.71	1.72	1.77	1.80	1.98	1.77	1.82	1.85	1.88
Yb	1.78	1.82	1.84	1.83	2.03	1.82	1.88	1.75	1.77
Lu	0.28	0.30	0.30	0.29	0.32	0.30	0.29	0.26	0.27
Hf	3.58	3.59	3.59	3.58	4.17	3.56	3.40	2.91	2.80
Ta	0.28	0.27	0.27	0.26	0.32	0.26	0.33	0.20	0.20
Pb	8.94	8.39	8.11	8.11	10.13	8.59	7.86	7.43	6.59
Th	2.07	2.24	2.24	2.19	2.50	2.20	2.04	1.01	1.35
U	0.80	0.92	0.91	0.89	1.06	0.89	0.82	0.48	0.55
$^{206}Pb/^{204}Pb$	18.685		18.692	18.704	18.684	18.724		18.690	
$^{207}Pb/^{204}Pb$	15.572		15.580	15.595	15.589	15.615		15.574	
$^{208}Pb/^{204}Pb$	38.386		38.407	38.464	38.433	38.526		38.391	
$^{87}Sr/^{86}Sr$	0.7038585		0.7039050	0.7038864	0.7038754	0.7039773	0.7038624	0.7038992	
$\pm 2\sigma$	0.000017		0.000008	0.000010	0.000008	0.000010	0.000020	0.000010	
$^{143}Nd/^{144}Nd$	0.5129132			0.5128839	0.5129188	0.5129197	0.5129113	0.5129041	
$\pm 2\sigma$	0.000009			0.000010	0.000011	0.000009	0.000008	0.000009	

pumice, scoria, dacite dome lavas and their inclusions obtained by Jicha *et al.* (2010) are summarized in Table 3 to provide context for mineral data discussed below. These activity ratios, plus those obtained from 13 basaltic andesite lavas erupted on Santa María volcano between 72 and 25 ka (each corrected for time since eruption using the $^{40}Ar/^{39}Ar$ ages; Singer *et al.* 2011), are all in U-excess by between 3 and 26% (Fig. 5). The older cone lavas are on average further from the equiline, lower in the ($^{230}Th/^{232}Th$) activity ratio, and thus exhibit a greater degree of U-excess than most of the younger cone lavas and the historically-erupted materials. The 1902 dacite pumice and five Santiaguito dacite dome lavas have remarkably similar ($^{238}U/^{232}Th$) and ($^{230}Th/^{232}Th$) activity ratios of about 1.26 and 1.16, respectively, whereas the 1902 scoria and basaltic andesite inclusions from two of the dome lavas have slightly lower ($^{230}Th/^{232}Th$) activity ratios

Table 2. *$\delta^{18}O_{SMOW}$ values of whole rock samples from Volcán de Santa María lava flows and tephra and Santiaguito dome lavas*

Sample	Material	wt% SiO_2	$\delta^{18}O_{SMOW}$
SM-05-01	Early cone-lava	52.1	6.25
SM-05-02	Early cone-lava	51.6	6.29
SM-05-04	Early cone-lava	51.4	6.31
SM-05-05	Early cone-lava	52.9	6.35
SM-05-06	Early cone-lava	52.9	6.37
SM-05-07	Early cone-lava	52.7	6.40
SM-05-09	Early cone-lava	53.4	6.42
SM-05-13	Late cone-lava	52.6	6.38
SM-05-15	Late cone-lava	54.1	6.38
SM-05-20	Late cone-lava	54.1	6.44
SM-05-22	Late cone-lava	54.1	6.43
SM-07-01	Late cone-lava	54.7	6.95
SM-07-09E	1902 scoria	53.5	6.54
SM-09-09B	1902 pumice	64.7	6.95
SM-07-04	1939 dome lava	63.0	6.82
SM-07-04i	1939 dome inclusion	55.3	6.50
SM-07-06	1972 dome lava	63.8	6.60
SM-06-06i	1972 dome inclusion	54.9	6.60
SG-05-02	1972 dome lava	64.7	6.68
SM-07-03	1975 dome lava	63.4	6.55

Complete chemical analyses of early and late cone-lavas are in Singer *et al.* (2011). Complete data for historically erupted materials are in Table 1.

of 1.14. Notably the two magmas that plot nearest to the equiline with the lowest U-excesses include those represented by the 1902 scoria, and one of the youngest basaltic andesite lavas of the Santa María cone (sample SM-07-01 of Singer *et al.* 2011) that erupted from a vent near the summit of the volcano at ~35 ka (Fig. 5).

The ($^{238}U/^{232}Th$) and ($^{230}Th/^{232}Th$) activity ratios of glass, magnetite, orthopyroxene, and amphibole from the 1902 dacite pumice and from the 1972 Caliente dacite dome lava are reported in Table 3. The activity ratios of glass, magnetite and orthopyroxene in the 1902 dacite pumice yield an isochron of 1.9 ± 1.2 ka (Fig. 6a), whereas glass, whole rock, magnetite and amphibole from the 1972 Caliente dome lava define an isochron of 9.5 ± 2.5 ka (Fig. 6b). Note that the whole rock samples of the 1902 dacite pumice and mafic scoria plot below the 1.9 ka isochron in such a way as to be consistent with mixing of the mafic and silicic melts as discussed below. We interpret these internal isochrons as estimates of the time since crystallization began in the two dacite magmas (e.g. Condomines *et al.* 2003; Jicha *et al.* 2005).

When acquiring data from basaltic andesite lavas of the Santa María cone, we discovered that $^{87}Sr/^{86}Sr$ ratios are inversely correlated with the activity ratio ($^{238}U/^{230}Th$) which is a measure of U-excess in the magmas prior to eruption (Jicha *et al.* 2010; Singer *et al.* 2011). Moreover, we observe that over the last ~75 kyr $^{87}Sr/^{86}Sr$ ratios have declined, and $\delta^{18}O$ values increased as the magmas evolved in composition such that they appear to be closer to a state of secular equilibrium in the U–Th isotope system. With the exception of the dacite lava from the 1973 El Brujo flow, the historically-erupted 1902 dacite pumice (and glass) and other Santiaguito dacite lavas erupted between 1939 and 1972 plot in the group with the smallest U-excesses, lowest $^{87}Sr/^{86}Sr$ ratios, and highest $\delta^{18}O$ values (Fig. 6). Basaltic andesite magmas that also fall in this group are represented by the 1902 scoria and one of the youngest cone-forming lavas on Santa María that erupted between 35 and 25 ka.

Oxide and pyroxene composition, oxygen fugacity and thermometry

Titanomagnetite and rare ilmenite in the 1902 dacite pumice and basaltic andesite scoria typically occur as small (<30 μm) inclusions within phenocrysts of amphibole and plagioclase. Like Fe–Ti-oxides from some other well-studied volcanoes (Santorini, Mt St Helens, Mt Pelée), ilmenite and titanomagnetite in the 1902 ejecta are relatively homogenous and show none of the zoning observed at the Soufrière Hills Volcano of Montserrat or Mt Unzen of Japan (Nakada & Motomura 1999; Venezky & Rutherford 1999; Pichavant *et al.* 2002; Devine *et al.* 2003). Titanomagnetite–ilmenite pairs in plagioclase or amphibole host crystals that have log(Mg/Mn) satisfying the equilibrium criteria of Bacon & Hirschmann (1988) were used for estimating temperature and fO_2 (Table 4). Hematite contents in ilmenite above 30 wt% led us to use the algorithm of Ghiorso & Sack (1991) to estimate equilibrium temperature and fO_2. Nine oxide pairs from the dacite pumice yield tightly clustered temperatures between 860 and 885 °C and log fO_2 values between −10.2 and −10.6, whereas four pairs in the basaltic andesite scoria reveal a wider range of temperatures between 925 and 1040 °C with log fO_2 values of −8.0 and −9.5. Together, these T–fO_2 estimates form an array comparable to the Ni–Ni–O+2 buffer curve and are thus similar to dacite from Shiveluch and Pinatubo volcanoes, but significantly more oxidized than dacite from Mt St Helens (Hattori 1993; Evans & Scaillet 1997; Blundy *et al.* 2006) and the Valley of 10 000 Smokes (Hildreth 1983) (Fig. 7).

Pyroxene phenocrysts in 1902 dacite pumice include augite (En_{34-44} Fs_{13-20} Wo_{37-48}) and hypersthene (En_{34-44} Fs_{27-37} $Wo_{0.2-2.3}$), whereas

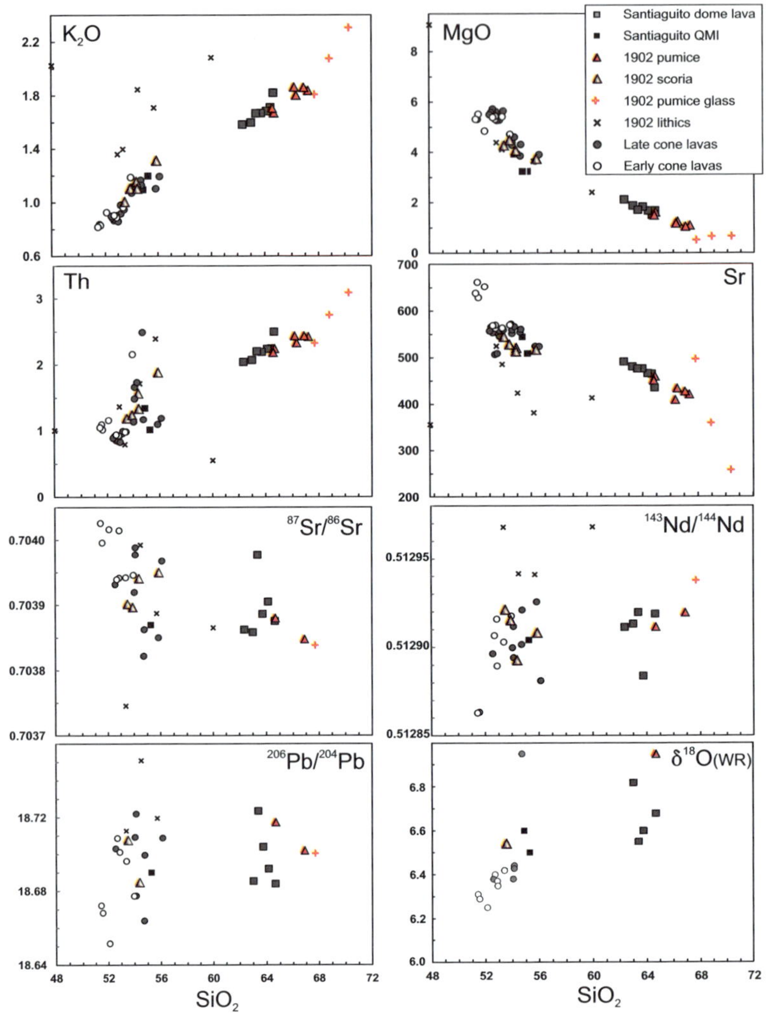

Fig. 3. Variation of selected major elements K_2O and MgO (in wt%), trace elements (in ppm) and Sr, Nd and Pb isotope ratios, and $\delta^{18}O$(WR) values with wt% SiO_2. Diorite and granodiorite fragments from the 1902 airfall deposit are denoted as lithic fragments; quenched mafic inclusions found in Santiaguito dome dacite lavas are denoted as QMI. The major and trace element and radiogenic isotope ratios for the early and late mafic cone lavas (51–55% SiO_2) are from Singer *et al.* (2011).

the basaltic andesite scoria contains augite (En_{38-46} Fs_{10-19} Wo_{39-46}) and bronzite–hypersthene (En_{59-80} Fs_{20-39} $Wo_{0.25-4.0}$); zoning is subtle, <2 mol%, in both minerals (representative analyses in Table 5). As a check on the T–fO_2 estimates from the oxides, temperatures of clinopyroxene and orthopyroxene pairs were estimated using two calibrations of experimental data–one that includes clinopyroxenes with Mg# >0.75 and one that excludes them (Putirka 2008, equations 36 and 37, respectively). Compositions from pyroxene pairs that did not have K_D (Fe–Mg) values within three standard deviations of the equilibrium value of 1.09 ± 0.014 were not used to calculate temperatures, and if the two calibrations were within the ±30 °C uncertainty of one another, the average temperature was taken as representative (Table 5). The two-pyroxene temperatures are consistent with those estimated from the two-oxide pairs, suggesting that pyroxene crystallized between 750 and 870 °C in the dacite and between 890 and 1000 °C in the basaltic andesite (Fig. 7). The range of two-pyroxene temperatures to values lower than those recorded by the oxides likely reflects Fe–Mg diffusion and partial re-equilibration during magma ascent that did not affect the iron–titanium oxide

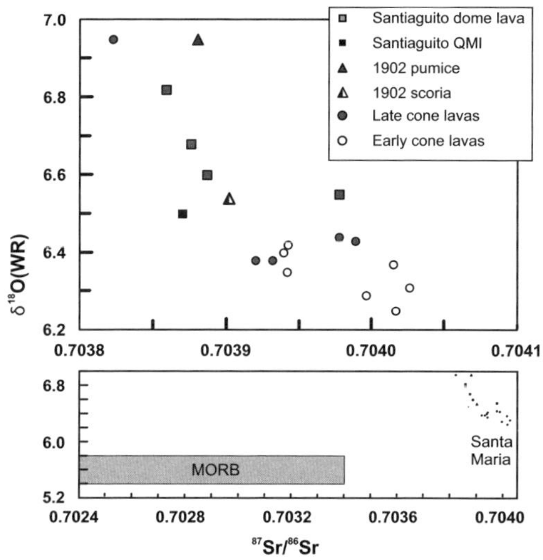

Fig. 4. Plot of $\delta^{18}O(WR)$ v. $^{87}Sr/^{86}Sr$ for the Santa María cone lavas, 1902 dacite pumice and basaltic andesite scoria and Santiaguito dacite dome lavas and their quenched mafic inclusions (QMI). The lower panel shows these data relative to the range in mid-ocean ridge basalt (MORB; from data in Eiler 2001; Hofmann & Hart 1978; Cohen & O'Nions 1982).

pairs because they were encased within larger host phenocrysts. Using the two-oxide data, we take the temperature of the dacite, 880 °C, and 1020 °C of the scoria (using the three highest temperature pairs in Table 4) as our best estimates of pre-eruptive storage conditions for the compositionally zoned magma.

Plagioclase zoning and melt inclusions in 1902 dacite and scoria

The 450 plagioclase crystals imaged and analysed by EMPA display several patterns of zoning, but by far the most abundant are andesine (An_{44-36}) that include one or more irregularly sculpted resorption surfaces characterized by an abrupt rimward increase of 1020 mol% anorthite. About 80% of crystals in the dacite pumice and 70% in the basaltic andesite scoria contain more than one resorption surface, several contain as many as five of these 'saw-tooth' zones which are common in mixed or mingled calc-alkaline magmas (e.g. Singer *et al.* 1995; Humphreys *et al.* 2006). Trace element profiles show that about half of the anorthite 'spikes' are accompanied by significant changes in FeO, TiO_2, K_2O, MgO, MnO, Ba, and Sr (Fig. 8). Mantles of these phenocrysts – the outermost 10–30% of the radial distance – are not disturbed by resorption surfaces, and commonly contain low-frequency, low-amplitude oscillatory zoning, or are homogenous. About 15% of phenocrysts in the scoria are very calcic, weakly zoned (An_{93-82}), and free of resorption surfaces.

Table 3. *U–Th isotope results for Santiaguito dacite lavas and 1902 tephra*

Sample	Material	($^{230}Th/^{232}Th$)	2σ	($^{238}U/^{232}Th$)	2σ	U ppm	Th ppm
Santiaguito dome lavas (post-1902)							
SM-07-03	lava wr	1.167	0.007	1.262	0.008	0.914	2.196
SM-07-04	lava wr	1.166	0.007	1.258	0.008	0.871	2.101
SM-07-04i	inclusion wr	1.141	0.007	1.247	0.007	0.560	1.362
SM-07-06	lava wr	1.162	0.007	1.264	0.008	0.905	2.172
SM-07-06i	inclusion wr	1.138	0.007	1.256	0.008	0.551	1.330
SG-05-02	lava wr	1.172	0.007	1.258	0.008	0.882	2.127
SG-05-01	lava wr	1.174	0.007	1.279	0.008	0.954	2.261
	amphibole	1.200	0.007	1.659	0.010	0.374	0.206
	magnetite	1.172	0.007	1.375	0.008	0.561	0.256
	gmass	1.165	0.005	1.276	0.005	1.133	2.693
1902 eruption products							
SM-07-09B	wr pumice	1.155	0.007	1.277	0.008	1.054	2.503
	glass	1.182	0.005	1.276	0.005	1.304	3.102
	opx	1.204	0.013	2.541	0.015	0.178	0.213
	magnetite	1.185	0.007	1.421	0.009	0.182	0.389
SM-07-09E	wr scoria	1.143	0.007	1.193	0.007	0.559	1.420
SM-07-09F	wr lithic fragment	0.886	0.005	0.890	0.005	0.447	1.523
SM-07-09H	wr lithic fragment	0.888	0.005	0.880	0.005	0.700	2.412

Abbreviations: wr, whole rock; opx, orthopyroxene; gmass, groundmass.
The whole rock data from lavas and bulk tephra were originally reported in Jicha *et al.* (2010). They are listed here to provide context for the isotope compositions of minerals and lithic fragments.

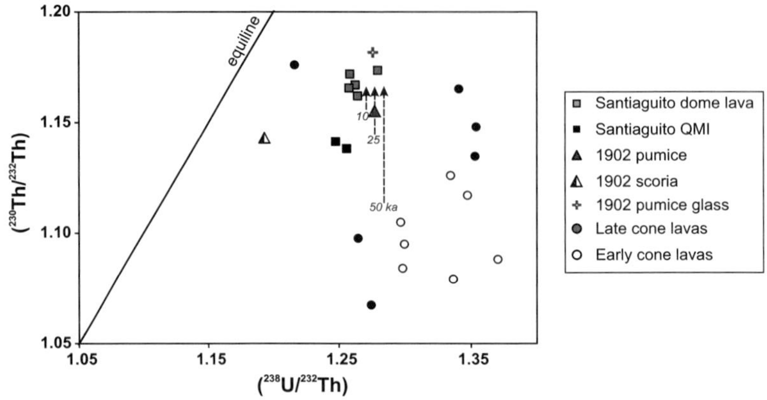

Fig. 5. Plot of ($^{238}U/^{232}Th$) and ($^{230}Th/^{232}Th$) activity ratios for basaltic andesite cone forming lavas of Santa María volcano, 1902 dacite pumice and basaltic andesite scoria, and Santiaguito dacite dome lavas and their quenched mafic inclusions (QMI). The dashed arrows indicate 10, 25 and 50 kyr of closed-system ingrowth of ^{230}Th via radioactive decay.

Approximately one-third of plagioclase crystals in the 1902 dacite pumice and one-fifth of those in the scoria contain optically clear melt inclusions. The anorthite variations and oscillatory zoning in the crystals that host these melt inclusions are similar, regardless of whether they are found in the dacite or the scoria. The melt inclusions are typically ovoid, up to 50 μm across, are commonly located along resorption surfaces, but can also be found in the cores and, less commonly, the mantles of crystals (Fig. 8). The inclusions range from dacite to high silica rhyolite with between 69 and 79 wt%, SiO_2 (anhydrous) and 1.5–3.5% K_2O (Fig. 9a). Interestingly, the melt inclusions in the dacite pumice overlap the composition of the rhyodacitic glass separated from the bulk pumice, whereas those in plagioclase from the scoria extend to significantly higher SiO_2 and K_2O contents (Fig. 9a).

The H_2O contents of 20 plagioclase-hosted melt inclusions in the dacite pumice determined by SIMS are between 1.4 and 6.9 wt% (Table 6) and in general H_2O declines as K_2O increases from about 2 to 3 wt% (Fig. 9b). Given that typical intermediate composition arc magmas contain very low amounts of dissolved CO_2 (e.g. below 400 ppm; Bacon *et al.* 1992; Wallace & Gerlach 1994; Saito *et al.* 2001; Roman *et al.* 2006), we assume that the volatile budget in the plagioclase-hosted melt inclusions in the 1902 dacite pumice is 100% H_2O. Accordingly, the pressure at which the melt inclusions were trapped was calculated using the program VolatileCalc (Newman & Lowenstern 2002) indicating minimum values for crystallization of the plagioclase between 170 and 15 MPa, corresponding to depths of about 5–1 km, or greater (Fig. 9c).

Discussion

Deep origin of dacite during volcanic repose

The decline in $^{87}Sr/^{86}Sr$, rise in $^{143}Nd/^{144}Nd$, and drop in the degree of U-excess during *c.* 75 kyr of mafic cone-building volcanism (Figs 3 & 5), led Jicha *et al.* (2010) and Singer *et al.* (2011) to infer that during this period several batches of mantle-derived basalt cooled, crystallized, partially melted lower crustal rocks that include a MORB-like component, and that blending of these partial melts with the basalt created hybrid basaltic andesite magmas. In contrast to Heydolph *et al.* (2012), who propose assimilation of mantle pyroxenite, Singer *et al.* (2011) interpret the 'old' $^{207}Pb/^{204}Pb$ ratios of the cone-forming mafic lavas to reflect fluid derived from sediment subducted into the mantle wedge. A large fluid flux beneath Santa María helps to explain the exceptionally high ^{238}U-excesses (Fig. 5) and the strongly oxidized state of these magmas (Fig. 7). The Sr, Nd, Pb, O and U–Th isotope ratios of the historically-erupted dacite pumice and lava and basaltic andesite scoria and quenched mafic inclusions are similar to the youngest cone-forming basaltic andesite lavas that, according to Jicha *et al.* (2010) and Singer *et al.* (2011), reflect a greater extent of blending basalt with lower crustal partial melts than do earlier erupted cone lavas owing to the sustained and repeated flux of basalt that progressively heated the lower crust. Development of a MASH-like zone at the base of the crust beneath Santa María was long ago proposed by Rose (1987*a*) on the basis of major element and sparse trace element data that suggested replenishment of a basaltic

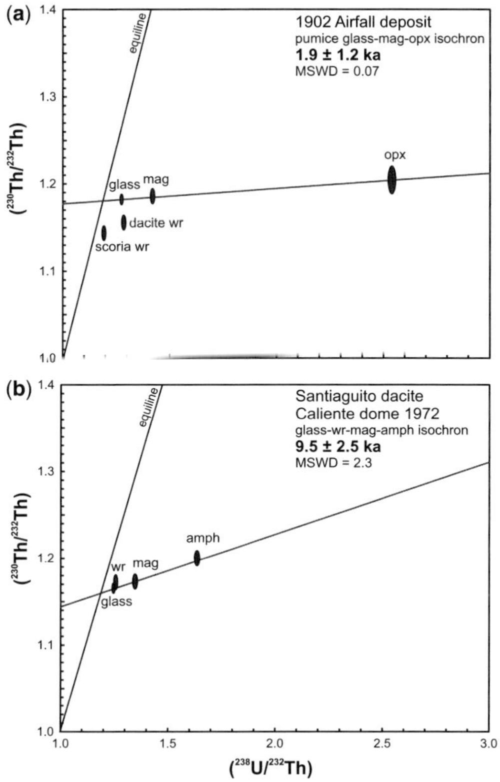

Fig. 6. Plots of ($^{238}U/^{232}Th$) and ($^{230}Th/^{232}Th$) activity ratios for: (**a**) phases from the 1902 airfall deposit including whole rock (wr) dacite pumice and glass, magnetite (mag) and orthopyroxene (opx) separated from the pumice, and whole rock scoria. The glass, magnetite and orthopyroxene yield an internal isochron of 1.9 ± 1.2 ka; (**b**) phases from the 1972 AD Caliente dacite flow, including whole rock, glass, magnetite and amphibole. These phases define an internal isochron of 9.5 ± 2.5 ka.

reservoir system in which magmas fractionally crystallized to a greater extent during the last phase of cone-growth. The similarity between the major, trace element and isotopic composition of the 1902 scoria and the last-erupted cone lavas dated between 35 and 25 ka (e.g. Figs 3 & 10) strongly argues that the historically-erupted basaltic andesite underwent similar processing in the deep crust. Rose (1987*a*) hypothesized that as the cone grew larger, the crustal column increased in thickness, thereby retarding the ability of the hybrid basaltic andesite magma to ascend to the surface.

The $^{40}Ar/^{39}Ar$ chronology of cone-building, in parallel with new trace element and isotopic data from both the composite cone (Escobar-Wolf *et al.* 2010; Singer *et al.* 2011) and now the historically-erupted dacites, enable us to further evaluate the Rose (1987*a*) hypothesis and document a major shift in processes that took place below the volcano. The $^{40}Ar/^{39}Ar$ geochronology indicates that the final, rapid pulse of cone-growth, fuelled by eruptions of hybrid basaltic andesite magma, ceased by 25 ka. We propose that at 25 ka the basaltic andesite magma began to pond deep within the crust, and having either exhausted the supply of readily fusible phases in the crustal wall rocks, or by becoming insulated from these crustal rocks as rinds or envelopes of partly solidified crystal-rich mush formed along the periphery of the MASH zone, or both, it cooled slowly such that fractional crystallization occurred without significant assimilation.

Modelling of four trace elements: K, Rb, Dy and Yb, using the EC-RAFC algorithm of Bohrson & Spera (2007), helps quantify key aspects of how basaltic andesite fractionally crystallized to form the 1902 dacite and illustrates that this behaviour contrasts sharply with that of the earlier period of open-system crystallization, melting and assimilation that affected cone-forming lavas. We chose one of the youngest cone-forming basaltic andesite lavas, sample SM-07-01 from the 3772 m summit of Santa María volcano (Singer *et al.* 2011) as a parental composition with 9681 ppm K, 22.3 ppm Rb, 3.62 ppm Dy and 2.11 ppm Yb. The liquidus temperature of this parent magma was set at 1020 °C based on two-oxide thermometry of the remarkably similar scoria in the 1902 airfall deposit (Fig. 7). The equilibration temperature of the modelled daughter magma was set at 880 °C, based also on two-oxide thermometry of the 1902 dacite pumice. A solidus temperature for hydrous hornblende dacite at lower crustal pressure was estimated at 725 °C based on experiments of Conrad *et al.* (1988). Heat capacities of the magma and wallrock are identical to those used by Singer *et al.* (2011) and recommended by Bohrson & Spera (2007) and the initial temperature of the wallrock was assumed to be 775 °C. Given these boundary conditions, the 140 °C drop in temperature from 1020 to 880 °C results in 43 wt% of the basaltic andesite parent magma crystallizing to form cumulate solids. The bulk distribution coefficients that best predict the concentrations of K, Rb, Dy and Yb in the residual melt are 0.09, 0.005, 1.75 and 1.35, respectively, consistent with fractionation of a plagioclase + hornblende + pyroxene + magnetite cumulate (Fig. 10; Davidson *et al.* 2007).

The O isotope data are consistent with the modelling presented by Singer *et al.* (2011) wherein the cone-forming basaltic andesite lavas reflect evolution of a deep crustal reservoir through both recharge with basalt and increasing extents of assimilation of a MORB-like component that, relative to the fluid-modified mantle wedge below

Table 4. *Magnetite and ilmenite compositions and T–fO_2 estimates from 1902 tephra calculated using Ghiorso & Sack (1991)*

	Pairs in dacite pumice block SM-07-09G																		Pairs in basaltic andesite scoria SM-07-09E							
	mag 1	ilm 1	mag 2	ilm 2	mag 3	ilm 3	mag 4	ilm 4	mag 5	ilm 5	mag 6	ilm 6	mag 7	ilm 7	mag 8	ilm 8	mag 9	ilm 9	mag 1	ilm 1	mag 2	ilm 2	mag 3	ilm 3	mag 4	ilm 4
SiO_2	0.08	0.05	0.06	0.04	0.07	0.02	0.07	0.03	0.09	0.05	0.07	0.08	0.06	0.05	0.07	0.07	0.07	0.03	0.10	0.10	0.06	0.02	0.10	0.08	0.06	0.07
TiO_2	5.68	35.84	5.79	36.50	5.59	35.94	5.80	36.51	5.75	36.42	5.52	35.82	5.65	36.49	5.63	35.60	5.63	35.13	8.07	35.34	6.75	36.14	8.09	35.85	5.92	34.69
Al_2O_3	2.41	0.31	2.31	0.33	2.11	0.30	2.29	0.35	2.28	0.33	2.43	0.35	2.45	0.32	2.26	0.34	2.26	0.04	1.68	0.12	2.62	0.42	1.70	0.37	1.96	0.12
FeO	82.34	54.17	84.17	54.42	84.40	54.10	82.24	54.69	82.02	55.41	82.32	55.51	82.63	54.30	81.32	54.44	81.32	55.12	79.15	53.04	82.48	56.16	79.20	54.49	84.00	54.80
MnO	0.89	1.47	0.93	1.34	0.92	1.41	0.94	1.30	1.03	1.27	1.00	1.02	0.85	1.49	0.90	1.16	0.90	1.19	0.90	1.03	0.79	0.60	0.90	1.02	1.04	0.61
MgO	1.63	2.28	1.51	2.20	1.49	2.27	1.57	2.22	1.54	2.17	1.36	1.76	1.62	2.21	1.50	2.17	1.50	2.20	1.73	3.48	1.83	2.53	1.84	1.74	1.57	1.96
CaO	0.01	0.02	0.02	0.02	0.00	0.02	0.03	0.05	0.01	0.00	0.03	0.03	0.01	0.03	0.01	0.03	0.01	0.05	0.09	0.21	0.00	0.02	0.11	0.02	0.02	0.08
Cr_2O_3	0.02	0.01	0.00	0.00	0.00	0.02	0.03	0.01	0.02	0.02	0.01	0.01	0.00	0.00	0.03	0.00	0.03	0.00	0.03	0.01	0.06	0.01	0.03	0.01	0.03	0.04
ZnO	0.18	0.12	0.15	0.06	0.15	0.07	0.15	0.04	0.18	0.09	0.19	0.11	0.16	0.14	0.20	0.05	0.15	0.08	0.17	0.07	0.06	0.05	0.19	0.12	0.13	0.06
V_2O_3	0.10	0.02	0.14	0.00	0.15	0.00	0.09	0.02	0.02	0.03	0.25	0.78	0.18	0.22	0.28	0.32	0.20	0.01	0.16	0.06	0.26	0.00	0.18	0.78	0.00	0.00
Total	93.35	94.28	95.08	94.91	94.89	94.15	93.21	95.21	92.94	95.79	93.18	95.47	93.61	95.25	92.20	94.18	92.07	93.85	92.07	93.47	94.91	95.94	92.34	94.48	94.73	92.43
XIlm		0.662		0.674		0.665		0.670		0.667		0.668		0.673		0.674		0.677		0.633		0.658		0.674		0.670
XUsp	0.158		0.158		0.152		0.161		0.159		0.158		0.158		0.156		0.156		0.232		0.190		0.233		0.163	
T (°C)	877		869		864		876		876		864		871		878		876		1009		929		1009		1039	
log fO_2	−10.35		−10.54		−10.60		−10.41		−10.39		−10.59		−10.49		−10.35		−10.41		−8.47		−9.51		−8.39		−7.89	

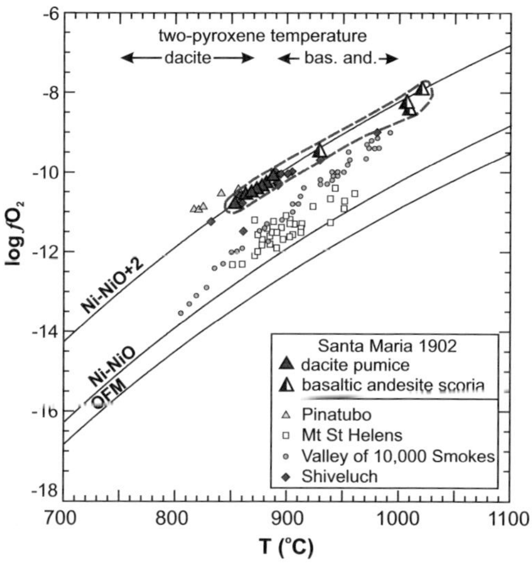

Fig. 7. Calculated equilibration temperature and oxygen fugacity (fO_2) for Fe–Ti oxide pairs. Each pair of grains is located within a single host phenocryst in the 1902 dacite pumice or scoria. The range of temperature estimates for co-existing pyroxenes in the 1902 dacite and scoria are indicated (representative data from Table 5). For comparison, Fe–Ti results from Pinatubo (Hattori 1993), Mt St Helens, Shiveluch (Blundy *et al.* 2006) and the Valley of Ten Thousand Smokes (Hildreth 1983) are illustrated.

Santa María, is characterized by lower $^{87}Sr/^{86}Sr$ ratios and higher $\delta^{18}O$ values (Fig. 4).

The O isotope data are also consistent with a closed-system crystallization origin of the dacites from basaltic andesite magma left-over from the cone-building period. It is well known that extensive high temperature crystallization of calc-alkaline magmas has little effect, perhaps increasing by a few tenths permil, on $\delta^{18}O$ of derivative magmas (Matsuhisa 1979; Singer *et al.* 1992). Specifically, crystal fractionation of parent magmas similar to the cone-forming lavas at Santa María with between 50 and 54% SiO_2 and $\delta^{18}O$ values of between 6.50 and 6.95‰ could produce the historically erupted dacites. Conspicuously, the youngest cone-forming basaltic andesite lava that we have measured (sample SM-07-01 that was used as a parent magma in the fractional crystallization model) has a $\delta^{18}O$ value identical to the 1902 dacite and thus represents a potential parent magma for this dacite (Fig. 3). If assimilation has also contributed to the origin of the dacites, the assimilant must have a $\delta^{18}O$(WR) value similar to that of the late cone-forming lavas.

The $^{238}U/^{230}Th$ isochron of 1.9 ± 1.2 ka defined by glass, magnetite and orthopyroxene in the 1902 dacite pumice (Fig. 6a) implies that crystals carried to the surface in this magma began to form very shortly before the eruption. That the whole rock sample of 1902 dacite pumice falls below the isochron may reflect mixing between the basaltic andesite, which plots well below the isochron, and the bulk dacite, that lies between the basaltic andesite and the isochron (Fig. 6a). Mixing of the basaltic andesite with the dacite is recorded in detail in the zoning of plagioclase phenocrysts in both of these magmas (Fig. 9) that will be discussed below. More surprising is the 9.5 ± 2.5 ka $^{238}U/^{230}Th$ isochron defined by glass, magnetite, amphibole and bulk material from the 1972 lava flow on the north flank of the Caliente dacite dome (Figs 2 & 6b). From this isochron, we infer that portions of the dacite magma which now comprise the Santiaguito dome complex that grew via effusive and mildly explosive activity in the decades following the explosive onset of silicic volcanism at Santa María in 1902, began to crystallize several thousand years prior to the 1902 eruption. In this scenario, deeper, occluded reaches of the dacite magma body, perhaps including crystal-rich zones near the cooler, more viscous, periphery of the silicic reservoir, had begun to solidify no later than 10 kyr prior to the Plinian eruption of 1902, but were unable to erupt until decades after the eruption in 1902 of the more silica- and H_2O-rich melt layer that capped the system.

The volume of dacite erupted explosively in 1902 is estimated at 8.5 km^3 dense rock equivalent (Williams & Self 1982). If our model is correct, the minimum volume of basaltic andesite parent magma required to create the 1902 dacite is about 20 km^3. Notably, this is well within the 40–67 km^3 of magma estimated to have fuelled the preceding growth of the 8 km^3 cone of basaltic andesite that comprises Santa María volcano, based on EC-RAFC modelling that accounts for recharging of the system with mantle-derived magma, assimilation of partially melted crust, and fractional crystallization (Fig. 10c; Singer *et al.* 2011). Using the thermal model of Hawkesworth *et al.* (2000) and a cooling time of 25 kyr suggests that a 20 km^3 mass of basaltic andesite which drops 140 °C and crystallizes 43% would generate a power output of about 25 MW, which is perhaps an order of magnitude lower than power generated during the preceding period of open-system behaviour during the later period of cone growth. The correspondingly modest and protracted release of heat during 25 kyr of cooling in the deep crust may also explain why evidence that partial melting and assimilation were important in creating a large volume of dacite is absent. If the 9.5 ± 2.5ka $^{238}U/^{230}Th$ isochron from the 1972 Caliente flow is

Table 5. *Representative pairs of pyroxene core compositions with T estimates using equations 36 and 37 from Putirka (2008)*

	1902 dacite pumice SM-07-09G											1902 basaltic andesite scoria SM-07-09E						
Clinopyroxenes																		
SiO_2	51.76	52.82	53.13	52.69	52.81	51.14	52.56	53.25	53.31	52.99	52.38	51.43	50.05	49.13	53.45	50.64	49.13	52.44
TiO_2	0.32	0.38	0.15	0.17	0.22	0.27	0.27	0.12	0.11	0.16	0.26	0.62	0.80	0.84	0.17	0.73	0.84	0.54
Al_2O_3	1.51	2.04	0.87	1.23	1.17	3.75	1.77	0.93	0.94	1.05	1.36	3.54	4.14	4.92	0.95	3.75	4.92	3.63
FeO	10.91	10.51	10.11	11.18	11.82	10.78	11.01	11.70	10.55	11.14	11.71	8.79	8.95	9.51	9.62	8.27	9.51	8.80
MnO	0.37	0.28	0.46	0.43	0.32	0.32	0.31	0.44	0.32	0.55	0.32	0.21	0.24	0.26	0.45	0.21	0.26	0.25
MgO	11.93	14.01	13.08	12.12	12.21	11.40	12.01	12.30	12.55	11.90	12.13	14.93	14.20	14.21	13.57	14.38	14.21	15.55
CaO	21.68	20.12	21.38	21.85	21.52	21.53	22.04	21.29	22.20	21.83	21.39	19.27	20.10	19.21	21.83	20.91	19.21	18.97
Na_2O	0.32	0.31	0.36	0.28	0.31	0.33	0.39	0.23	0.23	0.39	0.30	0.49	0.45	0.44	0.31	0.35	0.44	0.48
K_2O	0.03	0.01	0.01	0.00	0.00	0.18	0.00	0.01	0.01	0.00	0.01	0.00	0.00	0.00	0.01	0.01	0.00	0.00
Cr_2O_3	0.02	0.01	0.00	0.02	0.01	0.05	0.02	0.04	0.01	0.01	0.02	0.01	0.03	0.06	0.02	0.01	0.06	0.06
Total	98.86	100.49	99.54	99.96	100.40	99.76	100.38	100.31	100.24	100.02	99.88	99.29	98.96	98.58	100.38	99.26	98.58	100.71
Orthopyroxenes																		
SiO_2	53.82	53.77	54.04	53.93	53.57	53.87	53.93	53.58	53.73	53.27	53.52	53.36	53.86	53.68	52.94	53.55	53.90	53.78
TiO_2	0.13	0.10	0.12	0.11	0.19	0.11	0.11	0.03	0.12	0.12	0.13	0.36	0.22	0.11	0.11	0.24	0.16	0.19
Al_2O_3	1.01	1.01	0.96	0.92	0.89	1.00	0.96	1.03	1.09	0.96	1.02	2.38	1.58	1.16	0.56	2.07	0.64	0.70
FeO	18.78	18.38	18.38	18.45	21.84	18.02	18.53	18.58	18.47	18.26	18.80	15.95	16.48	18.24	23.76	16.31	19.53	19.62
MnO	2.64	1.92	2.50	2.18	0.59	1.86	1.92	2.50	1.91	1.91	1.84	0.52	0.59	2.68	0.87	0.40	1.02	0.92
MgO	23.16	23.51	23.06	23.41	21.88	23.45	23.51	23.17	23.47	23.43	23.39	24.96	25.12	23.31	20.42	25.31	23.19	23.08
CaO	0.43	0.70	0.44	0.52	0.65	0.69	0.69	0.44	0.66	0.70	0.61	1.90	1.70	0.37	0.96	1.45	0.96	1.13
Na_2O	0.02	0.01	0.00	0.00	0.00	0.03	0.03	0.01	0.00	0.02	0.02	0.04	0.04	0.03	0.02	0.03	0.01	0.00
K_2O	0.01	0.02	0.00	0.00	0.00	0.01	0.00	0.01	0.00	0.00	0.00	0.00	0.00	0.00	0.00	0.00	0.00	0.01
Cr_2O_3	0.02	0.02	0.00	0.01	0.01	0.00	0.01	0.01	0.01	0.00	0.00	0.03	0.02	0.00	0.00	0.01	0.00	0.01
Total	100.02	99.44	99.51	99.53	99.63	99.04	99.70	99.34	99.48	98.67	99.33	99.49	99.62	99.58	99.64	99.36	99.42	99.44
K_D (Cpx/Opx)	1.13	0.96	0.97	1.17	0.97	1.23	1.16	1.19	1.07	1.20	1.20	0.92	0.96	0.86	0.89	0.82	0.80	0.67
T(°C) eq. 36	737	855	761	759	896	788	767	779	766	761	805	997	972	860	961	951	958	964
T(°C) eq. 37	724	883	761	739	850	775	738	779	739	730	786	1002	966	927	940	965	999	1022
T (°C) ave.	731	869	761	749	873	781	752	779	753	746	795	1000	969	893	951	958	978	993

Oxides in wt%.

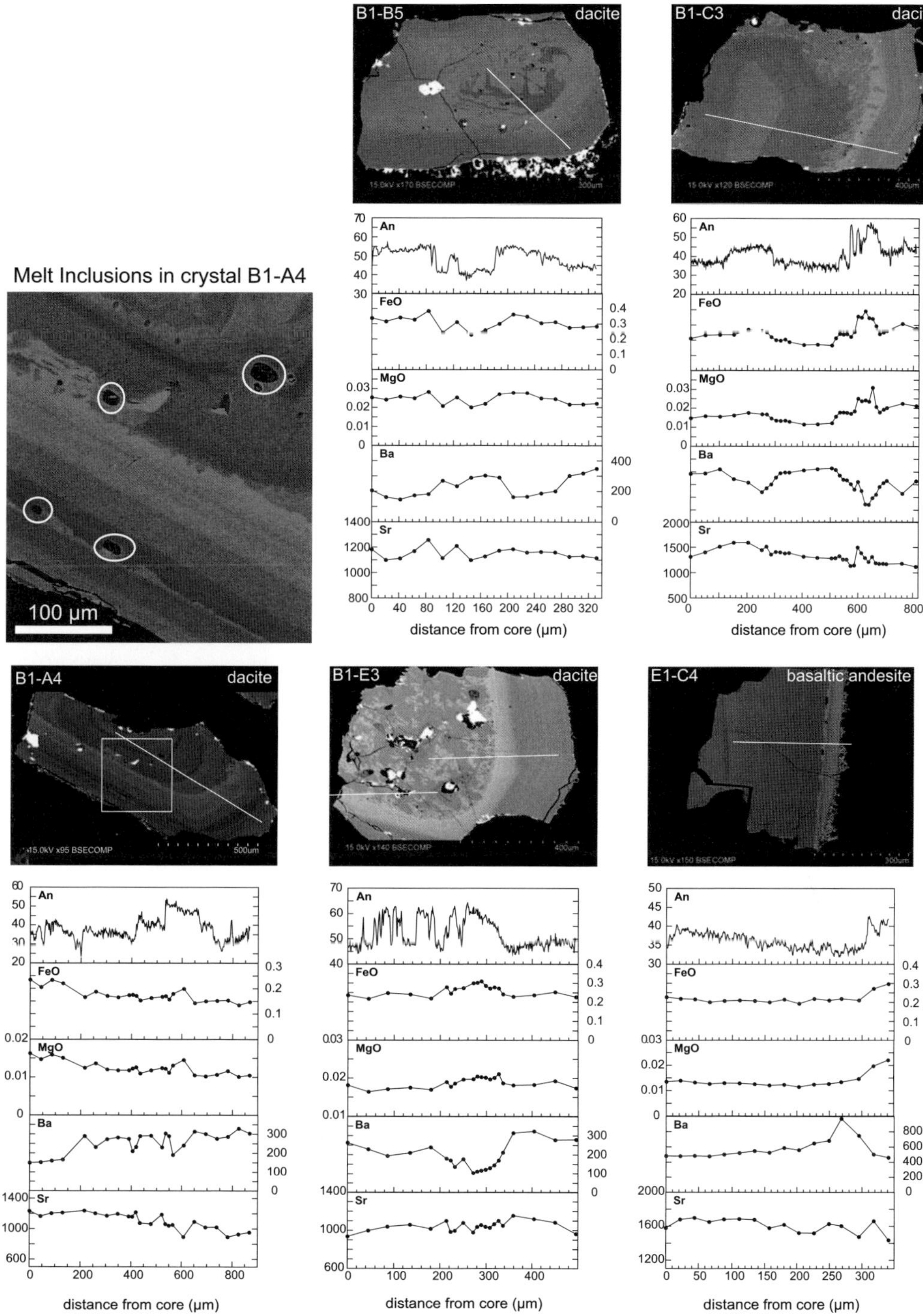

Fig. 8. Representative illustrations of zoning and melt inclusions in plagioclase revealed by backscattered electron images. Core-to-rim profiles of measured anorthite content determined by wavelength dispersive electron probe microanalysis, and FeO, MgO, Ba and Sr contents determined by secondary ion mass spectrometry. The box in the lower left panel with a backscattered electron image is enlarged above to show four melt inclusions typical of those measured for H_2O contents. See text for discussion.

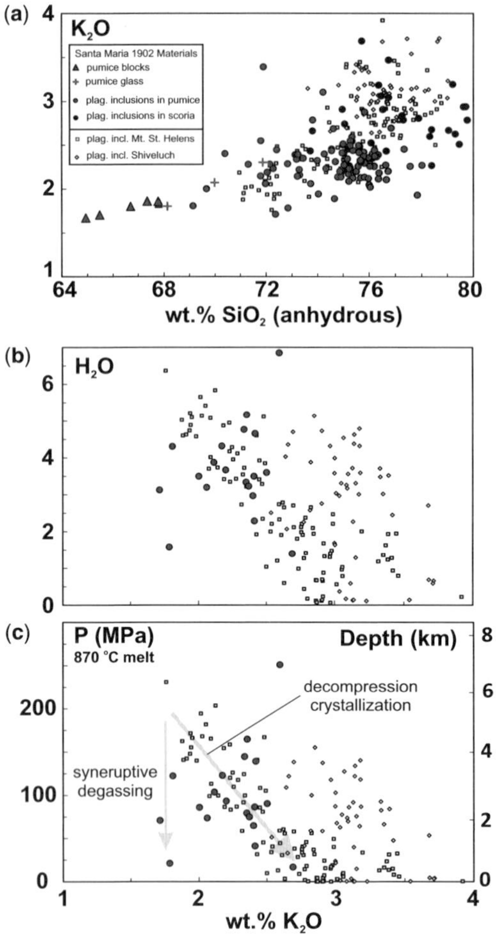

Fig. 9. Compositions of plagioclase-hosted melt inclusions. (**a**) K_2O v. SiO_2 plot shows that inclusions in 1902 dacite pumice overlap the composition of bulk glass separated from pumice blocks and that together inclusions in the pumice and scoria span a range of composition similar to that of phenocryst-hosted melt inclusions from Mt St Helens and Shiveluch dacite and andesite (data from Blundy *et al.* 2006). Inclusion compositions obtained by electron probe microanalysis, dacite pumice and glass by X-ray fluorescence. (**b**) H_2O (wt%) v. K_2O plot for 20 melt inclusions in 1902 dacite pumice. The vast majority of inclusions show a decline in H_2O with increasing SiO_2 comparable to the inclusions in Mt St Helens dacite (data from Blundy *et al.* 2006). (**c**) Interpretation of the variation in melt inclusion compositions (see text for details).

representative of when a large portion of the dacite began to crystallize, the 25 MW estimate of power output is a maximum. Alternatively, it seems likely that portions of the dacite magma may have crystallized at various times during the past 25 kyr, depending on where–within the plexus of mushy-to-molten environments of the deep crustal MASH zone–they began to cool.

An additional clue that may bear on the duration of the closed-system fractional crystallization process is the U–Th isotope composition of the quenched basaltic andesite inclusions found in the 1972 El Brujo dacite dome lava and the 1939 La Mitad dacite dome lava (Samples SM-07-04i and -06i; Tables 1 and 3). The two inclusions have nearly identical ($^{230}Th/^{232}Th$) and ($^{238}U/^{232}Th$) activity ratios of 1.14 and 1.25, respectively, despite being erupted 33 years apart and from nearly opposite ends, about 2 km apart, of the Santiaguito dome complex (Fig. 2). Their ($^{238}U/^{232}Th$) activity ratio is slightly lower than that of the 1902 dacite pumice and Santiaguito dacite dome lavas, consistent with fractionation of amphibole + clinopyoxene in which $D_{Th} > D_U$ and generation of a slight ^{238}U excess is to be expected (Brenan *et al.* 1995). Moreover, the ($^{230}Th/^{232}Th$) ratio is significantly lower than the dacites by an amount equivalent to 25 kyr of closed-system ingrowth of ^{230}Th (Fig. 5). These two inclusions are remarkably similar, not only in U–Th isotope composition, but also in major and most trace element concentrations, to the basaltic andesite lava, SM-07-01, used as a parent magma in the fractional crystallization model illustrated in Figure 10.

New fuel to the fire: the 1902 basaltic andesite and shallow processes

The Sr, Nd and Pb isotope ratios of the 1902 basaltic andesite scoria are, on average, slightly different than the average ratio of the 1902 dacite pumice or glass (Fig. 3). However, the scoria is lower in $\delta^{18}O$ than the dacite. Moreover, the scoria has a U–Th isotope composition that is slightly lower in ($^{230}Th/^{232}Th$) than the 1902 dacite pumice, but significantly lower in ($^{238}U/^{232}Th$) than either the 1902 pumice or the basaltic andesite inclusions in the 1939 and 1972 Santiaguito dacite dome lavas (Figs 5 & 6a). Collectively, these observations indicate that the basaltic andesite scoria produced mainly at the end of the 1902 Plinian eruption is not directly related to the much larger mass of dacite erupted at the onset. Instead, we propose that this basaltic andesite represents an influx of new magma into the MASH zone that, shortly prior to the 1902 eruption, spread out beneath the dacite magma, but was prohibited from mixing across the interface by the contrasts in both density and viscosity of the two magmas. The basaltic andesite supplied heat and perhaps volatiles via diffusion across the interface to the dacite, but, as two-oxide thermometry shows, thermal equilibration, and thus extensive

Table 6. *Representative plagioclase-hosted melt inclusion compositions*

1902 Dacite pumice sample SM-07-09G																			
Crystal#	**B1-A4-1**	**B1-A4-2**	**B1-B3-1**	**B1-B3-2**	**B1-B5-2**	**B1-B5-3**	**B1-B5-4**	**B1-B5-5**	**B1-B5-7**	**B1-B7-5**	**B1-C3-1**	**B1-C3-2**	**B1-C3-3**	**B1-C3-4**	**B1-C3-5**	**B1-E3-1**	**B1-E3-2**	**B1-E3-3**	**B1-E3-4**
Position	mantle	rim	rim	rim	core	core	core	core	mantle	mantle	rim	rim	rim	rim	rim	core	core	core	core
SiO_2	70.95	70.82	71.05	69.90	67.15	66.41	63.37	66.56	69.98	70.74	68.23	71.86	64.54	69.58	69.97	72.23	73.53	72.72	71.17
Al_2O_3	13.74	13.45	13.84	13.50	16.61	14.31	15.19	15.28	14.23	13.64	14.93	15.68	17.14	13.07	13.29	14.26	14.00	13.91	13.50
TiO_2	0.13	0.19	0.24	0.29	0.16	0.27	0.32	0.06	0.25	0.23	0.14	0.28	0.11	0.07	0.16	0.28	0.35	0.23	0.20
MgO	0.37	0.40	0.41	0.37	0.72	0.65	0.75	0.77	0.58	0.44	0.41	0.33	0.36	0.49	0.44	0.44	0.41	0.41	0.46
FeO	1.14	1.03	1.46	1.39	2.25	1.83	2.22	2.18	1.82	1.44	1.16	1.24	1.25	1.47	1.45	1.46	1.30	1.58	1.50
CaO	1.44	1.32	1.64	1.60	3.00	2.47	3.06	2.64	2.02	1.53	2.34	2.87	4.03	1.61	1.46	1.62	1.80	1.73	1.70
Na_2O	2.95	2.77	3.50	2.33	4.31	4.01	1.81	4.41	4.08	3.28	3.51	4.55	4.05	2.90	3.09	3.57	4.08	4.07	3.79
K_2O	2.11	2.20	2.50	2.36	2.00	2.06	1.71	2.40	2.35	2.60	2.34	1.78	1.81	2.42	2.37	2.17	2.41	2.69	2.41
MnO	0.08	0.12	0.11	0.26	0.19	0.25	0.06	0.26	0.07	0.00	0.18	0.05	0.07	0.19	0.09	0.03	0.15	0.12	0.12
Total	92.92	92.31	94.76	91.99	96.39	92.28	94.48	94.56	95.39	93.89	93.23	98.64	93.36	91.80	92.32	96.05	98.04	97.46	94.85
H_2O 1	3.22	3.67	3.60	5.17	3.50	3.20	3.13	2.97	3.23	6.85	4.77	1.73	4.31	4.66	3.23	4.45	2.29	1.48	3.82
H_2O 2	4.53		3.60						3.45			1.44				4.19		1.32	3.18
P (MPa)	104	93	90	165	86	74	65	75	80	251	145	21	122	139	73	123	41	17	86

1902 Basaltic andesite scoria sample SM-07-09E																			
Crystal#	**E1-A10-1**	**E1-B10-1**	**E1-C4-1**	**E1-C4-2**	**E1-D9-1**	**E1-E10-1**	**E2-A3-1**	**E2-B9-1**	**E3-C3-1**	**E2-C3-2**	**E2-C9-2**	**E3-A1-1**	**E3-A1-2**	**E3-B1-1**	**E3-B8-4**	**E3-C1-1**	**E3-D8-1**	**E3-E4-1**	**E3-E6-1**
Position	rim	mantle	rim	rim	rim	core	core	mantle	rim	rim	rim	rim	rim	rim	mantle	mantle	rim	mantle	mantle
SiO_2	75.52	74.71	73.13	72.62	71.48	72.49	74.00	72.80	71.61	73.88	73.65	71.40	70.27	72.77	74.99	75.35	74.44	73.85	74.70
Al_2O_3	13.13	12.27	11.13	11.76	11.38	11.79	12.40	12.38	13.41	12.34	11.92	10.77	11.41	11.89	12.20	11.03	10.54	11.01	13.45
TiO_2	0.19	0.13	0.28	0.26	0.29	0.28	0.25	0.25	0.26	0.22	0.36	0.34	0.41	0.12	0.10	0.17	0.11	0.26	0.22
MgO	0.28	0.25	0.23	0.27	0.87	0.29	0.14	0.32	0.04	0.04	0.80	0.68	1.00	0.17	0.08	0.11	0.31	0.40	0.36
FeO	1.13	0.92	1.29	1.17	3.69	1.27	0.51	0.90	0.29	0.26	1.96	3.87	4.78	0.35	0.28	0.26	1.67	1.27	1.60
CaO	0.88	0.97	0.70	0.71	0.74	0.93	1.42	0.99	1.36	1.09	0.69	0.64	0.53	0.97	0.83	0.65	0.69	0.63	0.82
Na_2O	4.57	3.87	2.14	2.15	2.04	1.99	3.18	1.46	2.12	1.11	3.87	1.87	2.41	2.09	2.22	1.58	1.97	1.86	3.82
K_2O	3.06	2.81	2.94	2.63	2.91	2.67	2.60	2.51	2.27	4.47	3.19	3.47	2.91	2.94	2.33	3.14	2.93	2.72	3.69
MnO	0.00	0.00	0.00	0.00	0.00	0.00	0.00	0.00	0.00	0.00	0.00	0.00	0.00	0.00	0.00	0.00	0.00	0.00	0.00
Total	98.77	95.93	91.84	91.57	93.40	91.71	94.49	91.62	91.35	93.42	96.44	93.04	93.73	91.30	93.03	92.29	92.66	91.99	98.66

Oxides in wt%. H_2O determined by SIMS. Minimum pressure calculated using Newman & Lowenstern (2002). Where two independent H_2O analyses (H_2O 1 & 2) are reported from 6 of the 19 inclusions in the dacite pumice, the average minimum pressure is calculated.

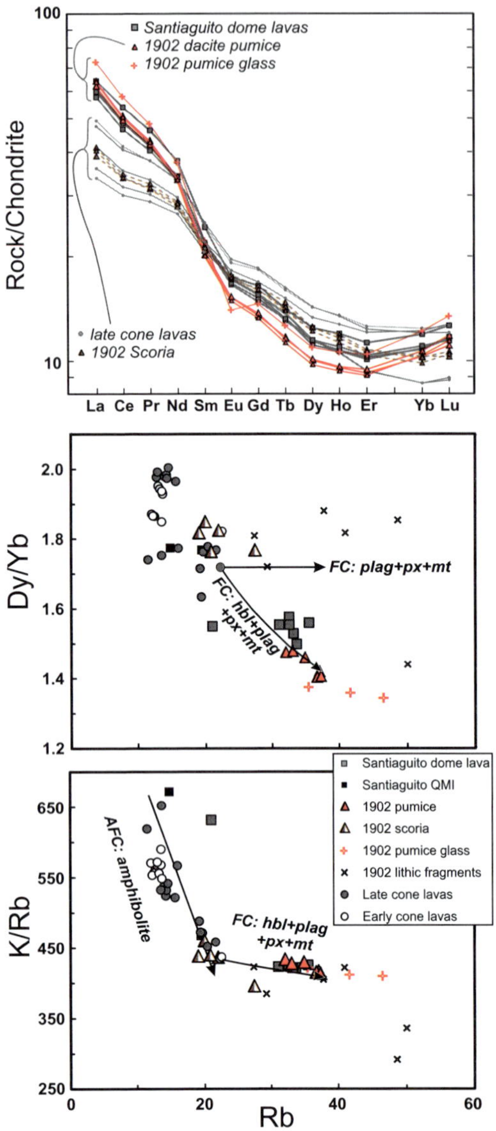

Fig. 10. Trace element variation in Santa María dacites. In upper panel chondrite-normalized rare earth element (REE) contents indicate that dacites are depleted in middle to heavy REE relative to cone-forming and 1902 basaltic andesite compositions (normalizing values from Sun & McDonough 1989). The process diagrams of Dy/Yb and K/Rb v. Rb in the middle and lower panels illustrate fractional crystallization models discussed in the text.

chemical mixing of the two magmas, did not take place (Sparks & Marshall 1986). However, the resulting increase in buoyancy, even if small, may have rapidly promoted convection within the dacite and exsolution of H_2O across the upper part of the dacite magma cap (Snyder 2000). In turn, exsolution and vesiculation could have initiated the ascent of magma that led to the Plinian eruption.

Plagioclase zoning and the compositions of melt inclusions provide additional clues about processes that attended the latest stages of evolution of the two magmas that erupted in 1902. The abrupt, large amplitude reversals in plagioclase zoning that occur rimward of resorption surfaces and are commonly accompanied by increases in FeO, MgO, and Sr (Fig. 8) in phenocrysts contained in both the 1902 dacite pumice and scoria, argue that these two magmas mingled physically and that some limited chemical mixing had also begun to take place (e.g. Singer *et al.* 1995; Humphreys *et al.* 2006). The remarkable similarity in zoning–including multiple reversals of composition–within the majority of phenocrysts in the two magmas, together with the overlapping composition of melt inclusions trapped in these phenocrysts (Fig. 9a), imply that many plagioclase crystals were exchanged back-and-forth between rhyolitic melt and basaltic andesite magma.

The co-variation of K_2O and SiO_2 in the plagioclase-hosted melt inclusions (Fig. 9a) is consistent with fractional crystallization of the 1902 dacite magma to produce rhyodacite to high-silica rhyolite melt compositions. Moreover, the drop in H_2O with increasing K_2O recorded by the vast majority of these melt inclusions is remarkably similar to the behaviour recorded by plagioclase and amphibole-hosted melt inclusions in the dacite pumice and lava erupted during May to October, 1980 at Mt St Helens (Fig. 9b; Blundy & Cashman 2005). The drop in H_2O with increasing K_2O is also paralleled by andesite pumice erupted between 2001 and 2004 at Shiveluch volcano, Kamchatka (Blundy *et al.* 2006). Blundy & Cashman (2005) and Blundy *et al.* (2006) have proposed that this co-variation reflects decompression-driven crystallization of the melt under water-saturated conditions at both Mt St Helens and Shiveluch volcanoes (Fig. 9c). The co-variation of H_2O and K_2O in the 1902 Santa María dacite strongly suggests that rapid decompression-driven crystallization of plagioclase from pressures of at least 180 MPa or >5 km depth occurred *en route* to the surface such that few inclusions experienced syneruptive degassing and most record evolution of residual high silica rhyolite melt (Fig. 9c). If correct, this conclusion suggests that the model of Blundy & Cashman (2005) may apply widely to dacitic magma that erupts explosively in subduction zones around the world.

The exchange of plagioclase crystals between the thermally and chemically contrasting melts led to the repeated reversals in zoning during mingling and mixing of the two magmas that occurred mainly during magma ascent. It is remarkable that the interaction of these two contrasting magmas *en route* to the surface – a complication that did not affect the Mt St Helens dacite (Blundy & Cashman 2005; Blundy *et al.* 2006) – led to complex zoning in plagioclase, but did not obliterate the signature of H_2O-saturated decompression of the silicic melts. Perhaps, more importantly, the plagioclase crystals that dominate the phenocryst modes of the 1902 dacite and scoria record mainly late-stage, shallow-level, fractional crystallization that occurred after the large body of volatile-rich dacite had accumulated during an earlier, and likely far more protracted, period of fractional crystallization deep within the crust. Thus, andesitic to dacitic magma erupted explosively at frontal arc composite volcanoes including Mt St Helens (Berlo *et al.* 2007), Shiveluch (Blundy *et al.* 2006; Humphreys *et al.* 2006), and Santa María appears to share a common two-stage crystallization history. At Santa María volcano, a 'snap-shot' recording of the early stage of fractional crystallization – corresponding to slow cooling in the deep crust – is reflected in the 9.5 ± 2.5 ka $^{238}U-^{230}Th$ isochron of the 1972 dome lava and the U–Th isotope composition of quenched mafic inclusions in the dome lavas (Figs 5 & 6b), whereas the later and more rapid decompression-driven crystallization in the conduit during magma ascent to the surface is consistent with the 1.9 ± 1.2 ka isochron from the 1902 dacite pumice (Fig. 6a).

Conclusions

Although in terms of eruptive history and magma compositions, Volcán de Santa María is relatively simple, and strongly bimodal, with 8 km^3 of basaltic andesite erupted between about 100 and 25 ka, and >9 km^3 of dacite erupted historically, generation of the highly explosive dacite was complex and took place over tens of thousands of years, mainly at depths greater than 5 km during a period of volcanic repose. During *c.* 75 kyr of cone-building, trace element and Sr–Nd–Pb–O and U–Th isotope data indicate that lower crustal heating sufficient to induce partial melting and assimilation involved several pulses of recharging mantle-derived basalt. A fundamental shift in process coincides with the termination of cone-building at 25 ka: the 1902 dacite reflects $>40\%$ fractional crystallization of plagioclase + amphibole + pyroxene + magnetite from ~ 20 km^3 of basaltic andesite magma left-over from cone-building that cooled slowly without assimilating significant amounts of additional crust.

Small contrasts in Sr–Nd–Pb isotope ratios, a modest contrast in $\delta^{18}O$, and a significant difference in the $(^{238}U/^{230}Th)$ activity ratio between the 1902 scoria and dacite indicate that these two magmas are not strictly consanguineous, rather this basaltic andesite is likely a recent arrival in the system. A glass–whole rock–magnetite–amphibole $^{238}U-^{230}Th$ isochron of 9.5 ± 2.5 ka for a 1972 Santiaguito dacite lava suggests that deeper, occluded portions of the silicic magma body not erupted in 1902 incubated in the crust for at least 10 kyr, and perhaps longer, prior to the 1902 eruption.

Following fractional crystallization deep in the crust, the final phase of creating the 1902 dacite involved rapid decompression and crystallization of the magma *en route* to the surface, as has been proposed at other intermediate composition subduction-zone volcanoes. Importantly, the phenocrysts that grew during this late stage of fractional crystallization had little to do with the formation of the dacite itself. Instead, the composition of the dacite reflects mainly slow, hydrous crystallization of amphibole over tens of thousands of years from basaltic andesite ponded deep within the crust. Arrival of an isotopically distinct mass of basaltic andesite into this deep crustal MASH zone may have supplied heat that promoted buoyancy, convection and vesiculation within dacite that, in turn, initiated ascent and the Plinian eruption of 1902.

We thank Bill Rose, Rüdiger Escobar-Wolf and the geologists and staff of INSIVUMEH, Guatemala for guidance in the field, and John Hora for laboratory assistance. Reviews by Bill Rose and especially Simon Turner, as well as the editorial comments of Georg Zellmer, helped us clarify several issues and are much appreciated. This study was supported by NSF Grant EAR-0738007 (Singer and Jicha). The WiscSIMS ion microprobe lab is partly supported by NSF EAR (grants 0319230, 0744079, 1053466).

References

ANNEN, C., BLUNDY, J. D. & SPARKS, S. J. 2006. The genesis of intermediate and silicic magmas in deep crustal hot zones. *Journal of Petrology*, **47**, 505–539.

ARMSTRONG, J. 1988. Quantitative analysis of silicate and oxide materials: comparison of Monte Carlo, ZAF, and $\phi(\rho z)$ procedures. *In*: NEWBURY, D. E. (ed.) *Microbeam Analyses*. San Francisco Press, San Francisco, 239–246.

BACON, C. & HIRSCHMANN, M. 1988. Mg/Mn partitioning as a test for equilibrium between coexisting Fe–Ti oxides. *American Mineralogist*, **73**, 57–61.

BACON, C. R. & LOWENSTERN, J. B. 2005. Late Pleistocene granodiorite source for recycled zircon and phenocrysts in rhyodacite lava at Crater Lake, Oregon. *Earth and Planetary Science Letters*, **233**, 277–293.

BACON, C., NEWMAN, S. & STOLPER, E. 1992. Water, CO_2, Cl, and F in melt inclusions in phenocrysts from three

Holocene explosive eruptions, Crater Lake, Oregon. *American Mineralogist*, **77**, 1021–1030.

Bacon, C. R., Sissson, T. W. & Mazdab, F. K. 2007. Young cumulate complex beneath Veniaminof caldera, Aleutian arc, dated by zircon in erupted plutonic blocks. *Geology*, **35**, 491–494.

Beccaluva, L., Bellia, S. *et al.* 1995. The northwestern border of the Caribbean plate in Guatemala: new geological and petrological data on the Motagua ophiolitic belt. *Ophioliti*, **20**, 1–15.

Berlo, K., Blundy, J., Turner, S. & Hawkesworth, C. 2007. Textural and chemical variation in plagioclase phenocrysts from the 1980 eruptions of Mount St. Helens, USA. *Contributions to Mineralogy and Petrology*, **154**, 291–308.

Blundy, J. D. & Cashman, K. 2005. Rapid decompression-driven crystallization recorded by melt inclusions from Mount St. Helens volcano. *Geology*, **33**, 793–796.

Blundy, J. D., Cashman, K. & Humphreys, M. 2006. Magma heating by decompression-Driven crystallization beneath andesite volcanoes. *Nature*, **443**, 76–80.

Bohrson, W. A. & Spera, F. J. 2007. Energy-Constrained Recharge, Assimilation, and Fractional Crystallization (EC-RA_xFC): a Visual Basic computer code for calculating trace element and isotope variations of open-system magmatic systems. *Geochemistry, Geophysics, Geosystems*, **8**, 2007GC001781.

Brenan, J. M., Shaw, H. F., Ryerson, F. J. & Phinney, D. L. 1995. Experimental-determination of trace-element partitioning between pargasite and a synthetic hydrous andesitic melt. *Earth and Planetary Science Letters*, **135**, 1–11, doi: http://dx.doi.org/10.1016/0012-821X(95)00139-4.

Carr, M. J. 1984. Symmetrical and segmented variation of physical and geochemical characteristics of the Central American volcanic front. *Journal of Volcanology and Geothermal Research*, **20**, 231–252.

Carr, M. J., Feigenson, M. D., Patino, L. C. & Walker, J. A. 2003. Volcanism and geochemistry in Central America: progress and problems. *In*: Eiler, J. (ed.) *Inside the Subduction Factory*. American Geophysical Union, Washington, DC, 153–174.

Carr, M. J., Patino, L. C. & Feigenson, M. D. 2007. Petrology and geochemistry of Lavas. *In*: Bundschuh, J. & Alvarado, G. E. (eds) *Central America: Geology, Resources, and Hazards*. Taylor & Francis, London, 565–590.

Cohen, R. S. & O'Nions, R. K. 1982. The lead, neodymium and strontium isotopic structure of ocean ridge basalts. *Journal of Petrology*, **23**, 299–324.

Condomines, M., Gauthier, P.-J. & Sigmarsson, O. 2003. Timescales of magma chamber processes and dating of young volcanic rocks. *In*: Bourdon, B., Henderson, G. M., Lundstrom, C. C. & Turner, S. P. (eds) *Uranium Series Geochemistry,* Reviews in Mineralogy and Geochemistry. Mineralogical Society of America, Washington, DC, 125–174.

Conrad, W. K., Nicholls, I. A. & Wall, V. J. 1988. Water-saturated and -undersaturated melting of metaluminous and peraluminous crustal compositions at 10 kb: evidence for the origin of silicic magmas in the Taupo Volcanic Zone, New Zealand, and other occurrences. *Journal of Petrology*, **29**, 765–803.

Davidson, J., Turner, S., Handley, H., Macpherson, C. & Dosseto, A. 2007. An amphibole 'sponge' in arc crust. *Geology*, **35**, 787–790, doi: http://dx.doi.org/10.1130/G23637A.1

Devine, J., Rutherford, M., Norton, G. & Young, S. 2003. Magma storage region processes inferred from geochemistry of Fe-Ti oxides in andesitic magma, Soufrière Hills Volcano, Montserrat, W.I. *Journal of Petrology*, **44**, 1375–1400.

Dufek, J. & Bergantz, W. 2005. Lower crustal magma genesis and preservation: a stochastic framework for the evaluation of basalt-crust interaction. *Journal of Petrology*, **46**, 2167–2195.

Eiler, J. M. 2001. Oxygen isotope variations of basaltic lavas and upper mantle rocks. *In*: Valley, J. W. & Cole, D. R. (eds) *Stable Isotope Geochemistry*. Mineralogical Society of America, Washington, DC, Reviews in Mineralogy series, **43**, 319–364.

Escobar-Wolf, R., Matias-Gomez, O. & Rose, W. 2008. Updated geologic map of the Santiaguito dome complex, Guatemala. Abstract presented at the IAVCEI 2008 General Assembly in Reykjavik, Iceland.

Escobar-Wolf, R. P., Diehl, J. F., Singer, B. S. & Rose, W. I. 2010. $^{40}Ar/^{39}Ar$ and paleomagnetic constraints on the evolution of Volcan Santa María, Guatemala. *Geological Society of America Bulletin*, **122**, 757–771.

Evans, B. & Scaillet, B. 1997. The redox state of Pinatubo dacite and the ilmenite–hematite solvus. *American Mineralogist*, **82**, 625–629.

Geldmacher, J., Hoernle, K., Van Den Bogaard, P., Hauff, F. & Klügel, A. 2008. Age and geochemistry of the Central American forearc basement (DSDP Leg 67 and 84): insights into Mesozoic arc volcanism and seamount accretion on the fringe of the Caribbean LIP. *Journal of Petrology*, **49**, 1781–1815.

Ginibre, C., Kronz, A. & Wörner, G. 2002. High-resolution quantitative imaging of plagioclase composition using accumulated backscattered electron images: new constraints on oscillatory zoning. *Contributions to Mineralogy and Petrology*, **142**, 436–448.

Ghiorso, M. & Sack, R. 1991. Fe–Ti oxide geothermometry: thermodynamic formulation and the estimation of intensive variables in silicic magmas. *Contributions to Mineralogy and Petrology*, **119**, 224–238.

Halsor, S. P. & Rose, W. I. 1988. Common characteristics of paired volcanoes in northern Central America. *Journal of Geophysical Research*, **93**, 4467–4476.

Harris, A. J. L., Rose, W. I. & Flynn, L. P. 2003. Temporal trends in lava dome extrusion at Santiaguito 1922–2000. *Bulletin of Volcanology*, **65**, 77–89.

Hattori, K. 1993. High-sulfur magma, a product of fluid discharge from underlying mafic magma: evidence from Mount Pinatubo, Philippines. *Geology*, **21**, 1083–1086.

Hawkesworth, C. J., Blake, S. *et al.* 2000. Time scales of crystal fractionation in magma chambers – integrating physical, isotopic and geochemical perspectives. *Journal of Petrology*, **41**, 991–1006.

Heydolph, K., Hoernle, K., Hauff, F., van den Bogaard, P., Portnyagin, M., Bindeman, I. & Garbe-Schonberg, D. 2012. Along and across arc geochemical variations in NW Central America:

evidence for involvement of lithospheric pyroxenite. *Geochimica et Cosmochimica Acta*, **84**, 459–491.

HILDRETH, W. 1983. The compositionally zoned eruption of 1912 in the Valley of Ten Thousand Smokes, Katmai National Park, Alaska. *Journal of Volcanology and Geothermal Research*, **18**, 1–56.

HILDRETH, W. & MOORBATH, S. 1988. Crustal contributions to arc magmatism in the Andes of central Chile. *Contributions to Mineralogy and Petrology*, **98**, 455–489.

HOFMANN, A. W. & HART, S. M. 1978. An assessment of local and regional isotopic equilibrium in the mantle. *Earth and Planetary Science Letters*, **38**, 44–62.

HORA, J. M., SINGER, B. S. & WORNER, G. 2007. Eruptive flux through thick crust of the Andean central volcanic zone: $^{40}Ar/^{39}Ar$ constraints from Volcán Parinacota, Chile. *Geological Society of America Bulletin*, **119**, 343–362.

HORA, J. M., SINGER, B. S., WORNER, G., BEARD, B. L., JICHA, B. R. & JOHNSON, C. L. 2009. Shallow and deep control on differentiation of calc-alkaline and tholeiitic magma. *Earth and Planetary Science Letters*, **285**, 75–86.

HUMPHREYS, M., BLUNDY, J. & SPARKS, R. 2006. Magma evolution and open-system processes at Shiveluch Volcano: insights from phenocryst zoning. *Journal of Petrology*, **47**, 2303–2334.

JICHA, B. R. & SINGER, B. S. 2006. Volcanic history and magmatic evolution of Seguam Island, Aleutian Island arc, Alaska. *Geological Society of America Bulletin*, **118**, 805–822.

JICHA, B. R., SINGER, B. S., BEARD, B. L. & JOHNSON, C. L. 2005. Contrasting timescales of crystallization and magma storage beneath the Aleutian Island arc. *Earth and Planetary Science Letters*, **236**, 195–210.

JICHA, B. R., SINGER, B. S., BEARD, B. L., JOHNSON, C. L., MORENO-ROA, H. & NARANJO, J. A. 2007. Rapid magma ascent and generation of ^{230}Th excesses in the lower crust at Puyehue-Cordón Caulle, Southern Volcanic Zone, Chile. *Earth and Planetary Science Letters*, **255**, 249–242.

JICHA, B. R., JOHNSON, C. M., HILDRETH, W., BEARD, B. L., HART, G. L., SHIREY, S. B. & SINGER, B. S. 2009. Discriminating assimilants and decoupling deep- vs. shallow-level crystal records at Mount Adams using $^{238}U-^{230}Th$ disequilibria and Os isotopes. *Earth and Planetary Science Letters*, **277**, 38–49.

JICHA, B. R., SMITH, K. E., SINGER, B. S., BEARD, B. L. & JOHNSON, C. M. 2010. Crustal assimilation no match for slab-fluids beneath Volcán de Santa María, Guatemala. *Geology*, **38**, 859–862.

JICHA, B. R., COOMBS, M. L., CALVERT, A. T. & SINGER, B. S. 2012. Geology and $^{40}Ar/^{39}Ar$ geochronology of the medium- to high-K Tanaga volcanic cluster, western Aleutians. *Geological Society of America Bulletin*, **124**, 842–856.

KITA, N., IKEDA, Y., TOGASHI, S., LIU, Y., MORISHITA, Y. & WEISBERG, M. 2004. Origin of ureilites inferred from a SIMS oxygen isotopic and trace element study of clasts in the Dar al Gani 319 polymict ureilite. *Geochimica et Cosmochimica Acta*, **68**, 4213–4235.

MATSUHISA, Y. 1979. Oxygen isotope compositions of volcanic rocks from the east Japan island arcs and their bearing on petrogenesis. *Journal of Volcanology and Geothermal Research*, **5**, 271–296.

MORGAN, G. & LONDON, D. 1996. Optimizing the electron microprobe analysis of hydrous alkali aluminosilicate glasses. *American Mineralogist*, **81**, 1176–1185.

NAKADA, S. & MOTOMURA, Y. 1999. Petrology of the 1991–1995 eruption at Unzen: effusion pulsation and groundmass crystallization. *Journal of Volcanology and Geothermal Research*, **89**, 173–196.

NEWMAN, S. & LOWENSTERN, J. 2002. VolatileCalc: a silicate melt–H_2O–CO_2 solution model written in Visual Basic for excel. *Computers and Geosciences*, **28**, 597–604.

PALLISTER, J., HOBLITT, R. & REYES, A. 1992. A basalt trigger for the 1991 eruptions of Pinatubo volcano? *Nature*, **356**, 426–428.

PATINO, L. C., CARR, M. J. & FEIGENSON, M. D. 2000. Local and regional variations in Central American arc lavas controlled by variations in subducted sediment input. *Contributions to Mineralogy and Petrology*, **138**, 265–283.

PICHAVANT, M., MARTEL, C., BOURDIER, J. & SCAILLET, B. 2002. Physical conditions, structure, and dynamics or a zoned magma chamber: Mount Pelée (Martinique, Lesser Antilles Arc). *Journal of Geophysical Research*, **107**, B5, doi: http://dx.doi.org/10.1029/2001JB000315

PUTIRKA, K. 2008. Thermometers and barometers for volcanic systems. *In*: PUTIRKA, K. & TEPLEY, F. (eds) *Minerals, Inclusions and Volcanic Processes*. Mineralogical Society of America, Washington, DC, Reviews in Mineralogy and Geochemistry, **69**, 61–120.

ROGERS, N., THOMAS, L., MACDONALD, R., HAWKESWORTH, C. & MOKADEM, F. 2006. $^{238}U-^{230}Th$ disequilibrium in recent basalts and dynamic melting beneath the Kenya rift. *Chemical Geology*, **234**, 148–168.

ROMAN, D., CASHMAN, K., GARDNER, C., WALLACE, P. & DONOVAN, J. 2006. Storage and interaction of compositionally heterogeneous magmas from the 1986 eruption of Augustine Volcano, Alaska. *Bulletin of Volcanology*, **68**, 240–254.

ROSE, W. I. 1972. Santiaguito Volcanic Dome, Guatemala. *Geological Society of America Bulletin*, **83**, 1413–1434.

ROSE, W. I. 1987*a*. Santa María, Guatemala: a bimodal soda-rich calc-alkalic Stratovolcano. *Journal of Volcanology and Geothermal Research*, **33**, 109–129.

ROSE, W. I. 1987*b*. Volcanic activity at Santiaguito volcano, 1976-1984. *In*: FINK, J. H. (ed.) *The Emplacement of Silicic Domes and Lava Flows*, Geological Society of America Special Papers, **212**, 17–27.

ROSE, W. I., GRANT, N. K., HAHN, G. A., LANGE, I. M., POWELL, J. L., EASTER, J. & DEGRAFF, J. M. 1977. The evolution of Santa María Volcano, Guatemala. *Geology*, **85**, 63–87.

SAITO, G., KAZAHAYA, K., SHINOHARA, H., STIMAC, J. & KAWANABE, Y. 2001. Variation of volatile concentration in a magma system of Satsuma-Iwojima volcano deduced from melt inclusions analyses. *Journal of Volcanology and Geothermal Research*, **108**, 11–31.

SIMS, K. W. W., GILL, J. B. ET AL. 2008. An inter-laboratory assessment of the thorium isotopic composition of synthetic and rock reference materials. *Geostandards and Geoanalytical Research*, **32**, 65–91.

Singer, B. S., O'Neil, J. R. & Brophy, J. G. 1992. Oxygen isotope constraints on the petrogenesis of Aleutian arc magmas. *Geology*, **20**, 367–370.

Singer, B., Dungan, M. & Layne, G. 1995. Textures and Sr, Ba, Mg, Fe, K, and Ti compositional profiles in volcanic plagioclase: clues to dynamics of calc-alkaline magma chambers. *American Mineralogist*, **80**, 776–798.

Singer, B. S., Smith, K. E., Jicha, B. R., Beard, B. L., Johnson, C. M. & Rogers, N. W. 2011. Tracking open-system differentiation during growth of Santa María volcano, Guatemala. *Journal of Petrology*, **52**, 2335–2363.

Snyder, D. 2000. Thermal effects of the intrusion of basaltic magma into a more silicic magma chamber and implications for eruption triggering. *Earth and Planetary Science Letters*, **175**, 257–273.

Sparks, R. S. J. & Marshall, L. A. 1986. Thermal and mechanical constraints on mixing between mafic and silicic magmas. *Journal of Volcanology and Geothermal Research*, **29**, 99–124.

Sparks, R. S. J., Sigurdsson, H. & Wilson, L. 1977. Magma mixing: a mechanism for triggering acid explosive eruptions. *Nature*, **267**, 315–318.

Spera, F. J. & Bohrson, W. A. 2004. Open-system magma chamber evolution: an energy-constrained geochemical model incorporating the effects of concurrent eruption, recharge, variable assimilation and fractional crystallization (EC-RA_xFC). *Journal of Petrology*, **45**, 2459–2480.

Spicuzza, M. J., Valley, J. W. & McConnell, V. S. 1998. Oxygen isotope analysis of whole rock via laser fluorination: an air-lock approach. *Geological Society of America Abstracts with Program*, **30**, 80.

Sun, S. S. & McDonough, W. F. 1989. Chemical and isotopic systematics of oceanic basalts: implications for mantle composition and processes. *In*: Sauders, A. D. & Norry, M. J. (eds) *Magmatism in the Ocean Basins*. Geological Society, London, Special Publications, **42**, 313–345.

Tenner, T. J., Hirschmann, M. M., Withers, A. C. & Hervig, R. L. 2010. Hydrogen partitioning between nominally anhydrous upper mantle minerals and melt between 3 and 5 GPa and applications to hydrous peridotite melting. *Chemical Geology*, **262**, 42–56.

Triebold, S., Wörner, G. & Kronz, A. 2006. Anorthite- and celsian-calibrated backscattered electron profiles, trace elements, and growth textures in feldspars from the Teide-Pico Viejo volcano complex (Tenerife). *Journal of Volcanology and Geothermal Research*, **154**, 117–130.

Valley, J. W., Kitchen, N., Kohn, M. J., Niendorf, C. R. & Spicuzza, M. J. 1995. UWG-2, a garnet standard for oxygen isotope ratios: strategies for high precision and accuracy with laser heating. *Geochimica et Cosmochimica Acta*, **59**, 5223–5231.

Venezky, D. & Rutherford, M. 1999. Petrology and Fe–Ti oxide reequilibration of the 1991 Mount Unzen mixed magma. *Journal of Volcanology and Geothermal Research*, **89**, 213–230.

Vogel, T. A., Patino, L. C., Eaton, J. K., Valley, J. V., Rose, W. I., Alvarado, G. E. & Viray, E. L. 2006. Origin of silicic magmas along the Central American volcanic front: genetic relationship to mafic melts. *Journal of Volcanology and Geothermal Research*, **156**, 217–228.

Walker, J. A., Carr, M. J., Patino, L. C., Johnson, C. M., Feigenson, M. D. & Ward, R. L. 1995. Abrupt change in magma generation processes across the Central American arc in southeastern Guatemala: flux dominated melting near the base of the wedge to decompression melting near the top of the wedge. *Contributions to Mineralogy and Petrology*, **120**, 378–390.

Walker, J. A., Mickelson, J. E. *et al.* 2007. U-series disequilibria in Guatemalan lavas, crustal contamination and implications for magma genesis along the Central American subduction zone. *Journal of Geophysical Research*, **112**, 2006JB004589.

Wallace, P. & Gerlach, T. 1994. Magmatic vapor source for sulfur dioxide released during volcanic eruptions: evidence from Mount Pinatubo. *Science*, **265**, 497–499.

Williams, S. & Self, S. 1982. The October 1902 Plinian eruption of Santa María Volcano, Guatemala. *Journal of Volcanology and Geothermal Research*, **16**, 33–56.

Witham, C. S. 2005. Volcanic disasters and incidents: a new database. *Journal of Volcanology and Geothermal Research*, **148**, 191–233.

Volcanoes of the Diamante cross-chain: evidence for a mid-crustal felsic magma body beneath the Southern Izu–Bonin–Mariana arc

ROBERT J. STERN[1]*, YOSHI TAMURA[2], OSAMU ISHIZUKA[3], HIROSHI SHUKANO[2], SHERMAN H. BLOOMER[4], ROBERT W. EMBLEY[5], MATTHEW LEYBOURNE[6], HIROSHI KAWABATA[2], AKIKO NUNOKAWA[2], ALEXANDER R. L. NICHOLS[2], EDWARD KOHUT[7] & IGNACIO PUJANA[1]

[1]*Geosciences Department, University of Texas at Dallas, Box 830688, Richardson, TX 75083-0688, USA*

[2]*Institute for Research on Earth Evolution (IFREE), Japan Agency for Marine-Earth Science and Technology (JAMSTEC), Yokosuka 237-0061, Japan*

[3]*Institute of Geoscience, Geological Survey of Japan/AIST, Tsukuba 305-8567, Japan*

[4]*Department of Geosciences, Oregon State University, 128 Kidder Hall, Corvallis, OR 97331, USA*

[5]*Pacific Marine Environmental Laboratory/NOAA, Newport, OR 97365-5258, USA*

[6]*ALS Geochemistry, 2103 Dollarton Hwy, North Vancouver, BC Canada V7H 0A7*

[7]*Department of Geological Sciences, University of Delaware, Newark, DE 19716, USA*

**Corresponding author (e-mail: rjstern@utdallas.edu)*

Abstract: Three submarine Diamante cross-chain volcanoes in the southern Mariana arc mark a magma-healed zone of along-arc (north–south) extension that allows either mafic mantle-derived basalts or felsic magmas from the middle of thickened arc crust to erupt. The largest volcano is East Diamante, with a well-developed (5 × 10 km) caldera that formed via violent felsic submarine eruptions beginning nearly 0.5 Ma. One or more of these eruptions also formed a giant submarine dune field extending 30 km to the NW of the volcano. Felsic igneous activity continues at least as recently as *c.* 20 000 years ago, with emplacement of resurgent dacite domes, some hot enough to power the only black smoker hydrothermal system known in the Mariana arc. In contrast, felsic eruptions do not occur on the two volcanoes to the west, implying that the mid-crustal felsic zone does not underlie the thinner crust of the Mariana Trough back-arc basin. Diamante cross-chain lavas define a medium K suite; mafic lava phenocryst assemblages show arc-like associations of anorthite-rich plagioclase with Fe-rich olivine. Magmatic temperatures for a basaltic andesite and three dacites are *c.* 1100 °C and *c.* 800 °C, respectively, typical for cool, wet, subduction-related felsic magmas. Felsic magmas formed under low-P crustal conditions. The Diamante cross-chain is the southernmost of at least seven and perhaps eight Mariana arc volcanoes in a *c.* 115 km long arc segment characterized by felsic eruptions. This is the 'Anatahan Felsic Province', which may have formed above a mid-crustal tonalite body that formed by fractionation or was re-melted when heated by *c.* 1200 °C mafic, mantle-derived magmas. Across- and along-arc variations suggest that felsic eruptions and dome emplacement occurred when midcrustal tonalite was remobilized by intrusions of mafic magma, while north–south extension facilitated the development of conduits to the surface.

Supplementary material: Detailed Hyperdolphin ROV dive tracks, Cook 7 dredge locations, $^{40}Ar/^{39}Ar$ analytical data, analytical methods, major and selected trace element analyses of whole rock samples, and compositional data for minerals are available at http://www.geolsoc.org.uk/SUP18611

The Izu–Bonin–Mariana arc system (IBM) extends over 2800 km south from near Tokyo, Japan, to beyond Guam, USA (Fig. 1a), and is an excellent example of an intra-oceanic arc system (IOAS; Stern 2010). IOASs are built on oceanic crust and contrast in several ways with arcs built on continental crust, such as Japan or the Andes. Because IOAS crust is thinner, denser and more refractory than that beneath Andean- or Japan-type margins, study of IOAS melts and fluids allows more confident

From: Gómez-Tuena, A., Straub, S. M. & Zellmer, G. F. (eds) 2014. *Orogenic Andesites and Crustal Growth.* Geological Society, London, Special Publications, **385**, 235–255.
First published online June 27, 2013, http://dx.doi.org/10.1144/SP385.6

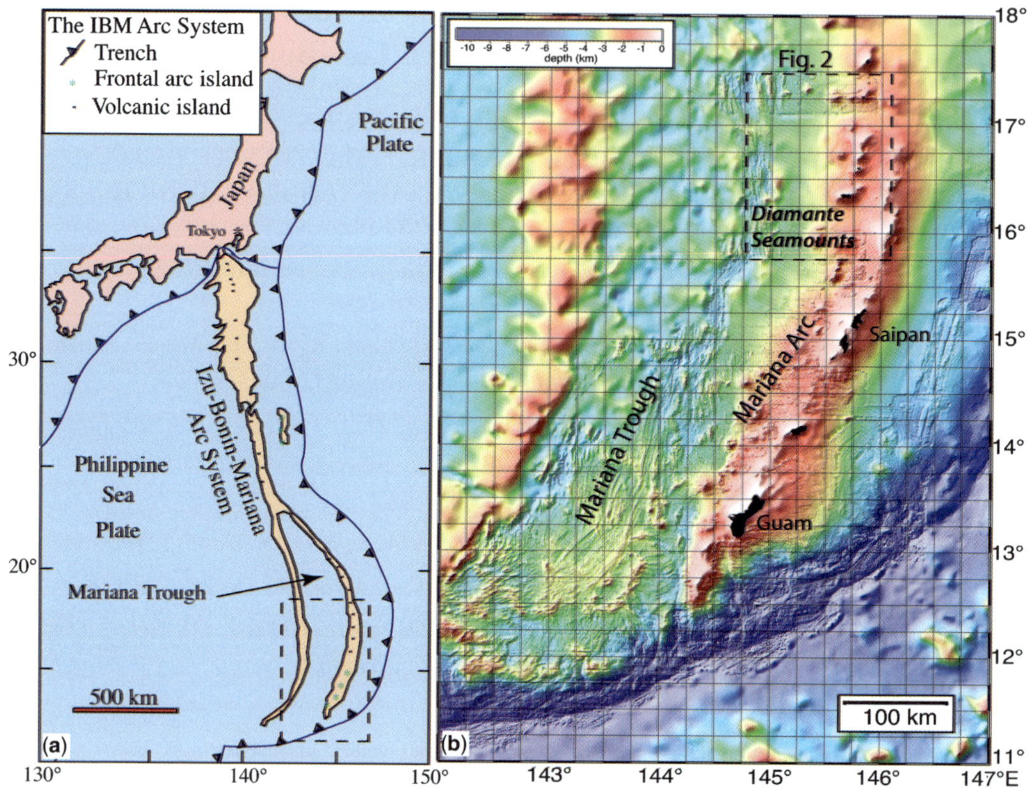

Fig. 1. Locality maps. (**a**) The Izu–Bonin–Mariana arc system in the western Pacific, dashed box shows the location of (b). (**b**) The southern Mariana arc system, showing the area of Figure 2 in dashed box.

assessment of mantle-to-crust fluxes and processes than is possible for continental arcs. We are particularly interested in understanding felsic IOAS rocks better because the latter are microcosms of continental crust formation that must be generated by fractionation or melting of mafic arc crust. The IBM arc system is an excellent natural laboratory for studying these rocks, because felsic rocks are common, even though basalt and basaltic andesite (<57 wt% SiO_2) are clearly the predominant eruptive products (Tamura & Tatsumi 2002; Straub 2008). In this contribution, we advance our understanding of IOAS felsic magmagenesis by describing a new occurrence of these rocks in the southern Mariana arc (Fig. 1b) and explore several interesting aspects of these rocks.

The Mariana arc system shows important along-strike variations in tectonic and magmatic behaviour, including the fact that arc volcanism south of Anatahan (Fig. 2) is entirely submarine (Stern *et al.* 2003). A distinctive east–west-trending set of three volcanoes was first investigated during the 1979 Mariana expedition by Dixon & Stern (1983), who named the seamounts in consonance with other nearby Mariana arc seamounts, Esmeralda and Ruby. Over the past decade, southern Mariana submarine volcanoes have become foci of increasingly detailed studies. During the Cook 7 expedition in 2001 aboard R/V *Melville,* this region was bathymetrically mapped and surveyed with HAWAII MR-1 towed sonar and extensively dredge-sampled, including several dredges around the study area, the Diamante cross-chain. In 2003 and 2004, a NOAA team investigated submarine hydrothermal activity in the Mariana arc, including seafloor studies using the Canadian ROV ROPOS. The many discoveries during the NOAA cruises included an extensive, vigorous hydrothermal field in the caldera of East Diamante (Baker *et al.* 2008). Japanese and US scientists, using the JAMSTEC ROV *Hyper-Dolphin* aboard the R/V *Natsushima* (NT09–08, June 2009), began studying the Diamante cross-chain during eight dives (HPD1011–1018), returning to the study area in 2010 during NT10–12.

The study area is of particular petrological interest because it is part of a 115 km-long Mariana arc segment characterized by felsic volcanism (Fig. 2; Stern & Hargrove 2003). Seven edifices within

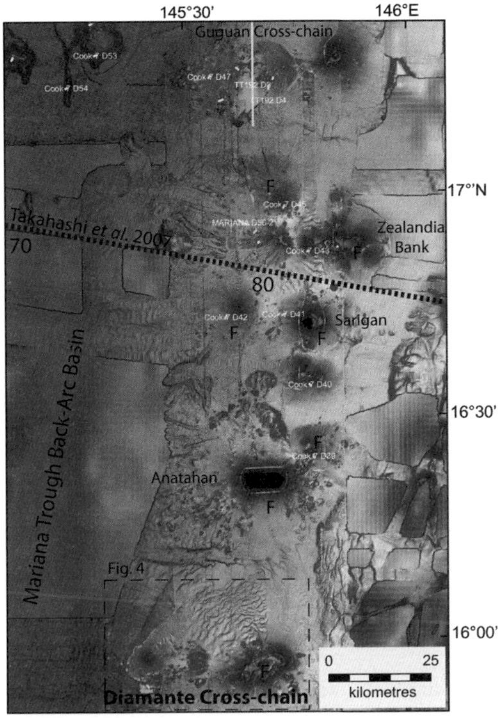

Fig. 2. Anatahan felsic province of the Mariana arc showing the location of the Diamante cross-chain. 'F' indicates volcano with felsic (>60 wt% SiO_2) lavas. Short dashed line 'Takahashi *et al.* 2007' is deep seismic profile of Takahashi *et al.* (2007), summarized in Figure 3. '70' and '80' are OBS locations. Dashed box shows location of detailed study area in Figure 4.

this region, including the 2003 Anatahan eruption (Wade *et al.* 2005), have erupted felsic lavas. Volcano size – proxy for age and, hence, the presence of evolved magmatic storage systems – does not seem important in controlling the concentration of felsic eruptions within this segment. Instead, felsic magmas may be tapped from a 3–5 km thick 'tonalite layer' in the mid-crust, with $V_p = 6.1–6.5$ km s^{-1}, beneath the volcanic front edifices (Fig. 3; Takahashi *et al.* 2007). This layer tapers away to the west, where rear-arc cross-chain volcanoes erupting only basalts are found. Crustal structure beneath the volcanic front edifices has also been studied by Calvert *et al.* (2008). These studies show that the middle crust is thickest beneath volcanic front edifices.

This report focuses on the results from Cook 7 dredges, ROPOS diving during TT-167, and NT09–08 and NT10–12 *Hyper-Dolphin* ROV dives, with an emphasis on characterizing the distribution of mafic and felsic igneous rocks and presenting a broad petrological outline based on new major element and limited trace element geochemical data. Sixteen dredge or ROV samplings over a 10 × 35 km area provide information on Diamante cross-chain volcanic rocks. Sampling locations are shown on Figure 4 and individual ROV dive tracks with sampling localities are provided in the Supplementary material. A subsequent report will focus on more complete trace element and isotopic data for these lavas and investigate their petrogenesis in more detail.

Submarine volcanoes of the Diamante cross-chain

Three principal volcanoes 10–15 km apart make up the Diamante cross-chain: East Diamante, Central Diamante and West Diamante (Fig. 4). East Diamante marks the Mariana arc volcanic front and lies about 145 km above the subducted slab (Syracuse & Abers 2006). A small volcano sampled by Cook 7 D34 (SE Diamante) lies *c.* 135 km above the slab. Central Diamante lies *c.* 180 km and West Diamante lies *c.* 250 km above the subducted slab. Locations for samplings are shown in Figure 5 and detailed charts for individual dives can be found in the Supplementary material. Details of Cook 7 dredges can also be found in the Supplementary material. These results allow us to sketch the geology of these three volcanoes, at least in broad strokes (Fig. 4).

West and Central Diamante are relatively simple basaltic volcanoes. Our dive results on these volcanoes (HPD1016 and HPD1018, respectively) and the parasitic cone on the eastern slope of West Diamante (HPD1017) are consistent with previous results from dredging during the MARA and Cook 7 expeditions indicating that these are mostly basaltic edifices. Central Diamante is the site of a 2003 earthquake swarm, which Heeszel *et al.* (2008) concluded was due to faulting along an east–west-striking, steeply dipping normal fault plane, with a dominantly N–S-oriented tension axis. Normal faulting is also expressed morphologically at Central Diamante, where east–west-trending normal faults with scarps facing the summit region indicate that this volcano formed over a zone of strong north–south extension. The east–west orientation of the Diamante cross-chain and the similar elongation of East Diamante volcano are also consistent with formation of the cross-chain in a region dominated by north–south extension.

East Diamante is complex, both volcanologically and petrologically. It is also the only Mariana arc volcano known to have a vigorous 'smoker' hydrothermal system (Baker *et al.* 2008), so understanding its geological and petrological evolution is of special interest. Figure 5 depicts

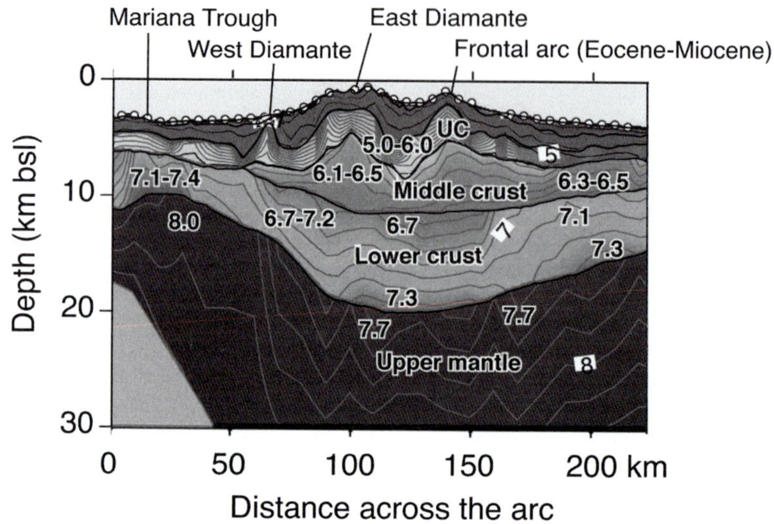

Fig. 3. Crustal structure beneath southern Mariana arc from seismic profile of Takahashi *et al.* (2007). Location of profile is shown on Figure 2; this has been modified to show approximate relative positions of East Diamante and West Diamante volcanoes. Note presence of thick low-V_p layer beneath the volcanic front; this probably also underlies East Diamante.

the summit region of East Diamante volcano, which has the form of a complex caldera, *c.* 10 × 5 km, elongated ENE–WSW and breached on its northern and southwestern sides. The caldera floor is irregular, with several resurgent domes in the middle of the caldera; some of these host hydrothermal activity. The NE caldera wall is the simplest, with a steep inner wall, gentler outer slope,

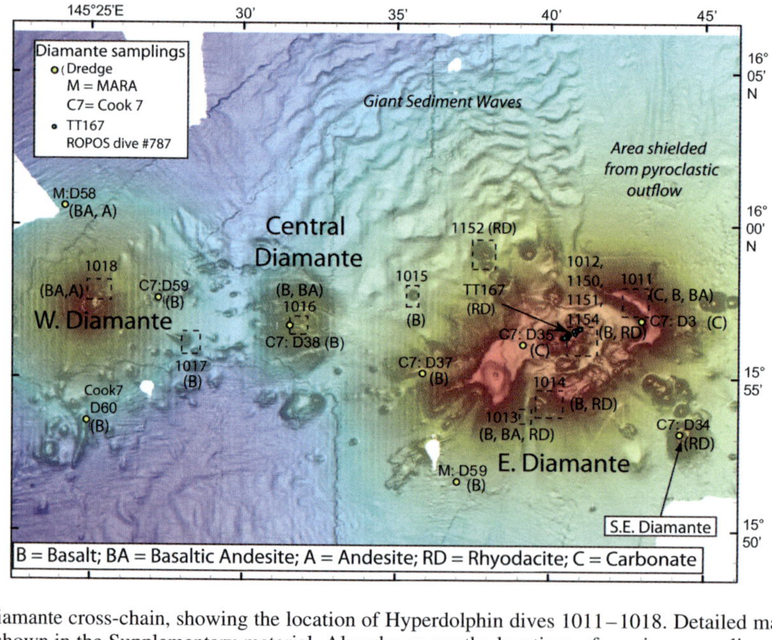

Fig. 4. The Diamante cross-chain, showing the location of Hyperdolphin dives 1011–1018. Detailed maps for each dive are shown in the Supplementary material. Also shown are the locations of previous samplings: MARA dredges 58 and 59 (Dixon & Stern 1983); Cook 7 dredges 34, 37, 38, 59 and 60; and TT167 ROPOS dive 787. Compositions of samples are also indicated.

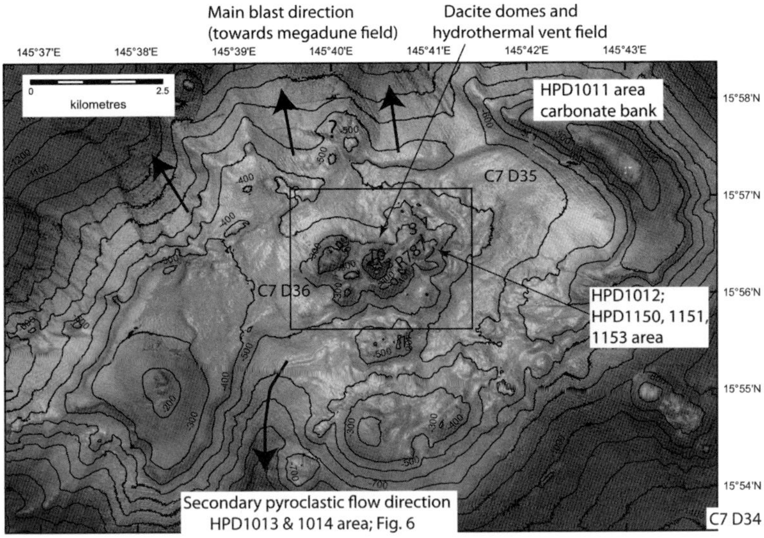

Fig. 5. Bathymetric map of East Diamante volcano, showing general location of 4 Hyperdolphin dives (detailed locations shown in the Supplementary material), Cook 7 dredges D34–D36 (red lines except for D34 which is slightly off the map), and major geological features. Approximate path of TT167 R787 and locations of samples 1, 3 and 10 are also shown.

and a boomerang-shaped outline. This is the only sector of the volcano that is so regular, and we thought before sampling it during Cook 7 D35 and HPD1011 that it would expose pre-caldera collapse volcanic stratigraphy. Results of one dredge and two ROV transects demonstrate that the *c.* 400 m thick section exposed in this scarp is composed of biogenic carbonate sediments and shelly detritus (with a few lavas, apparently interbedded, near the top). Biostratigraphic studies of these carbonates indicate that these are shallow-water deposits of Plio-Pleistocene age. Similar material was recovered from the region west of the resurgent domes in the caldera Cook 7 D36 (Fig. 5; Supplementary material). Because horizontal bedding is exposed in near-vertical exposures – probably associated with faulting accompanying caldera formation – this carbonate sequence probably predates caldera formation. The entire NE sector of the volcano as far south as the morphological change near 15°56′N (from high, steep and smooth to low, gentle and rough) is composed of carbonate rocks. The thickness of these carbonates and paucity of volcanic material indicate that the volcano experienced a significant period of quiescence and subsidence marked by the growth of the carbonate platform. Magmatic resurgence of the volcano culminating in caldera collapse may be indicated by the mafic lavas recovered near the top of the NE caldera wall during HPD1011. We wonder if the unsampled broad, smooth, high west of the caldera might also be dominated by pre-caldera carbonate rocks, but a dredge just to the west of Figure 5 (Cook 7 D37; Supplementary material) recovered pumice, basalts and Mn crust and no carbonate sediments.

Caldera formation marked an important episode of magmatic resurgence of East Diamante. Dive results from HPD1013 in the SW breach provide our best perspective on felsic eruptions from East Diamante caldera. This dive traversed a downcut section of several submarine pyroclastic flows (Fig. 5). Figure 6 is a simplified section based on observations during HPD1013. The *c.* 150 m thick section of felsic pyroclastics and lapilli- and ash-sized is characterized by two *c.* 30 m thick units of coarse pumice breccia separated by finer ash, suggesting at least two violent felsic eruptions. Other than two mafic samples (R01, R02) collected as float near the start of the dive, all samples collected were felsic and quite pumiceous, with 20–25 vol% vesicles. The felsic samples contain phenocrysts of quartz and/or pyroxene near the base of the section and become aphyric upsection. R02 and R04–06 contain 5–10 vol% quartz phenocrysts, R05, R06, and R08 contain 5 vol% pyroxene phenocrysts.

The sequence of felsic pyroclastics observed during HPD1013 probably was deposited as a result of chaotic outflow from the caldera due to voluminous felsic eruptions, perhaps leading to caldera formation. These pyroclastic deposits are

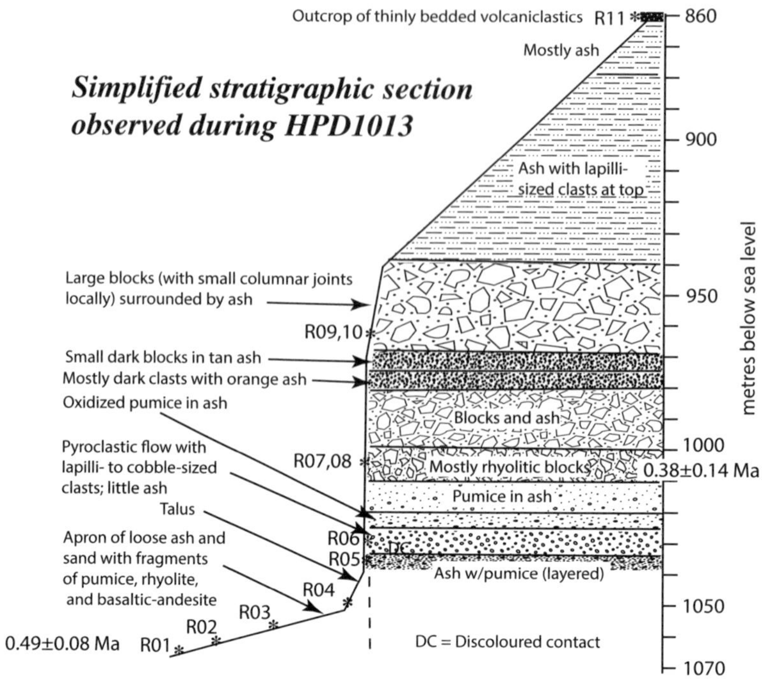

Fig. 6. Simplified stratigraphic column of felsic pyroclastic beds observed on breached southern wall of East Diamante volcano, as traversed during HPD1013. Profile shows approximate slopes, location of rock samplings, (R01–R11), lithostratigraphy and relative thickness of beds.

similar to the youngest units exposed at West Rota volcano (Stern *et al.* 2008). It is unlikely that these erupted from existing dacitic domes, because coarse pyroclastics should have buried these domes if the eruption produced 100 m thick deposits at the site of HPD1013. These are probably resurgent domes, emplaced after caldera formation.

Even more striking evidence of violent felsic eruptions that may be linked to formation of the East Diamante caldera formation is seen north of the northern breached caldera wall, where a well-preserved field of giant sediment waves extends at least 20 km north and is as wide as 30 km. These sediment waves have unusually long wavelengths (1–2 km) and amplitudes (*c.* 100 m tall). The waves are crudely concentric around the northern part of East Diamante and the field broadens with distance from the volcano, indicating that these probably resulted from unusually strong currents emanating from the region now occupied by the caldera. The impression that these sediment waves may be a far-field manifestation of a violent submarine eruption ('Neptunian'; Allen & McPhie 2009) – perhaps some kind of a submarine pyroclastic flow – is strengthened by the observation that such waves are missing on the seafloor NE of the volcano, which was shielded by the high-standing carbonate ridge on the caldera's eastern margin. The presence of the waves immediately downslope from where the northern flank of the caldera is missing is striking circumstantial evidence that the field of giant sediment waves is linked to one or more violent eruptions that accompanied caldera formation. A similar interpretation of the giant sediment waves on the flanks of major arc volcanoes was made by Draut & Clift (2006), although they emphasize bottom currents whereas we emphasize Neptunian eruptions (Allen & McPhie 2009). During one of these eruptions, a 100 m-thick pyroclastic flow (studied during HPD1013) was directed south from the caldera. During or after the last caldera-forming event, rhyodacitic domes were intruded into the caldera and minor dacitic lava flows issued from them. These felsic magmatic systems provided sufficient heat to allow the development of a vigorous hydrothermal system, which is still active.

^{40}Ar/^{39}Ar geochronology

Because they are built on Mariana Trough crust, the Diamante volcanoes must have grown after the back-arc basin opened beginning *c.* 5 Ma (Stern *et al.* 2003). Table 1 and Figure 7 summarize three

Table 1. *Results of stepwise-heating analyses of volcanic rocks from the studied area*

Analysis no.	Station & sample no.	Total age ($\pm 1\sigma$)				Plateau age ($\pm 1\sigma$)				
		Integrated age (Ma)	Inverse isochron age (Ma)	$^{40}Ar/^{36}Ar$ intercept	MSWD	Weighted average (Ma)	Inverse isochron age (Ma)	$^{40}Ar/^{36}Ar$ intercept	MSWD	Fraction of ^{39}Ar (%)
U11037	HPD1012R7	0.040 ± 0.008	0.021 ± 0.004	301.1 ± 2.1	0.86	0.020 ± 0.004	0.013 ± 0.009	312 ± 23	0.36	71.4
U11302	HPD1013R1	0.47 ± 0.09	0.61 ± 0.15	291 ± 7	1.32	0.47 ± 0.08	0.61 ± 0.15	291 ± 7	1.32	100.0
U11357	HPD1013R8	0.36 ± 0.16	0.37 ± 0.12	283 ± 39	0.60	0.37 ± 0.14	0.37 ± 0.12	283 ± 39	0.60	100.0

MSWD, mean square of weighted deviates ($\sqrt{SUMS/(n-2)}$) in York (1969).
Integrated ages were calculated using sum of the total gas released.
$l_b = 4.962 \times 10^{-10}\ a^{-1}$, $l_e = 0.581 \times 10^{-10}\ a^{-1}$, $^{40}K/K = 0.01167\%$ (Steiger & Jäger 1977).

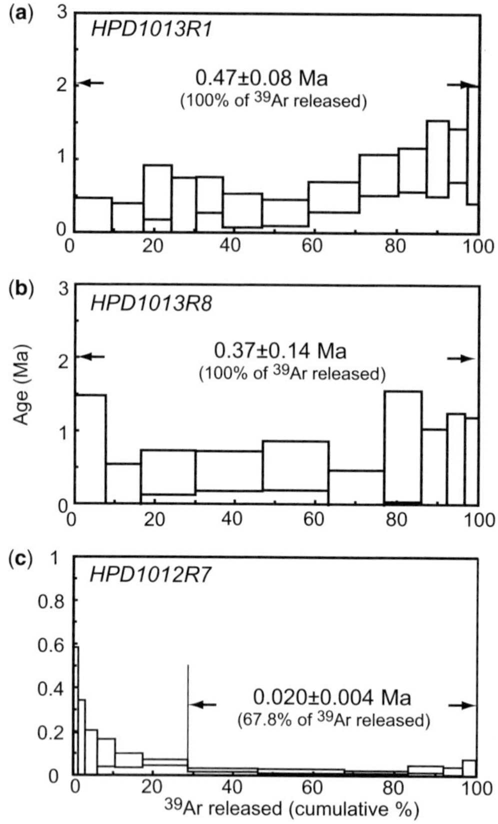

Fig. 7. $^{40}Ar/^{39}Ar$ age spectrum for (**a**) East Diamante basaltic andesite sample HPD1013 R01; (**b**) East Diamante felsic pyroclast sample HPD1013 R08; (**c**) Dacite sample HPD1012-R07 from a felsic dome in East Diamante caldera.

new $^{40}Ar/^{39}Ar$ whole-rock ages for East Diamante samples (techniques and results can be found in the Supplementary material). The oldest one is for a basaltic andesite recovered as float from the base of the section sampled during HPD1013 (R01; Fig. 6), which yields a plateau age of 0.47 ± 0.08 Ma (Fig. 7a). A felsic sample (HPD1013R8) from *c.* 70 m higher (Fig. 6) is from the first coarse pyroclastic layer in this section. This yields a plateau age of 0.37 ± 0.14 Ma (Fig. 7b). Finally, we have one $^{40}Ar/^{39}Ar$ age for a felsic whole-rock sample (HPD1012 R7; Fig. 5) from a dacite (68.6 wt% SiO_2; 1.6 wt% K_2O) collected from a small knoll inside East Diamante caldera. This yields an age of $20\,000 \pm 4000$ years (Fig. 7a; Supplementary material).

We interpret these three ages to indicate the following sequence of events. Construction of a stratovolcano began before *c.* 0.5 Ma and culminated in violent felsic volcanism beginning *c.* 0.38 Ma. The present East Diamante caldera began to form about this time or later. Felsic igneous activity including emplacement of resurgent dacite domes followed caldera formation, continuing up to *c.* 20 ka. Such a youthful age is consistent with other evidence that East Diamante is a dormant volcano, including continued vigorous hydrothermal activity associated with felsic domes inside the caldera.

Diamante cross-chain petrology and chemical composition

Basalts were recovered from all three volcanoes. Basalts from East Diamante are porphyritic, with phenocrysts of 20–30 vol% plagioclase, 0.2–0.7 vol% olivine, and 0.5–5 vol% clinopyroxene (Fig. 8d). More differentiated basalts contain orthopyroxene and magnetite instead of olivine. Basalts from Central Diamante and West Diamante contain more olivine (1–4 vol%) and less plagioclase (10–20 vol%) than do East Diamante basalts. Clinopyroxene is rare in Central Diamante lavas, but 7–10 vol% clinopyroxene is typical in West Diamante basalts. East Diamante andesites contain 4–6 vol% plagioclase and <1 vol% clinopyroxene $\pm$ orthopyroxene (Fig. 8a). Dacites and rhyolites are found only from East Diamante and comprise two general types. One is porphyritic, containing plagioclase (17–30 vol%), quartz (4–10 vol%), *c.* 1 vol% clinopyroxene, 1–2 vol% orthopyroxene and 0.2–0.3 vol% magnetite (Fig. 8b, c). The other type is aphyric, with <3 vol% phenocrysts and is two-pyroxene plagioclase dacite.

Eighty-one volcanic rocks from the Diamante cross-chain were analysed for major and limited trace element abundances. Data in the compilation include five analyses from Dixon & Stern (1983), 15 analyses of Cook 7 (D34, D37, D38 D59, and D60) samples, three analyses of samples collected during TT167 ROPOS dive 787 (Fig. 5), and 58 analyses of samples collected during NT09–08 HPD1011–1018 and NT10–12 HPD1150–1154 dives. Only the data of Dixon & Stern (1983) have been previously published; the other 76 analyses are listed in the Supplementary material. These 81 analyses are distributed between the three major volcanic centres: 51 samples from East Diamante (including 4 parasitic cones on its flanks), eight samples from Central Diamante, and 21 samples from West Diamante. For ease of comparison, major element data discussed below have been normalized to 100wt% anhydrous, but sums of the original analyses are also reported in the Supplementary material.

Fig. 8. Representative photomicrographs (crossed-polarized light) of East Diamante igneous rocks, 100 μm-scale bar (black) in lower left applies to all images. P, plagioclase; Qz, quartz; O, orthopyroxene; C, clinopyroxene; Ol, olivine; white V, void. Note that plagioclase phenocrysts in felsic samples (**b**, **c**) show oscillatory zoning whereas those in basalt from NNW parasitic cone sampled by HPD1015 (**d**) show cloudy interiors, evidence of magma mixing.

These lavas range from primitive basalts (*c.* 10 wt% MgO and maximum Mg # (=100 Mg/ Mg + Fe) = 68) to rhyolite (maximum SiO_2 = 75.4 wt%). The three volcanoes show different extents of differentiation, with West Diamante erupting the most primitive lavas (mean SiO_2 = 52.2 ± 2.2 wt%; mean Mg# = 55 ± 8.3) through mostly mafic but fractionated Central Diamante (mean SiO_2 = 54.8 ± 7.1 wt%; mean Mg# = 39.9 ± 5.5) to East Diamante (mean SiO_2 = 63.1 ± 9.0 wt%; mean Mg# = 40 ± 10.2). The lavas are strongly bimodal in silica Fig. 9), with mafic lavas containing 50–58 wt% SiO_2 and felsic samples containing 66–76 wt% SiO_2. There is a marked silica gap between 58 and 66 wt% SiO_2, which is best seen in the data for East Diamante (Fig. 9a). This bimodal variation is similar to lavas from Sumisu caldera in the Izu arc (Tamura *et al.* 2005). All 21 West Diamante samples are mafic, with 49–57 wt% SiO_2 (Fig. 9c). All but one of the analysed lavas from Central Diamante are mafic (50–53 wt% SiO_2; Fig. 9b).

Insights into the nature of Diamante cross-chain lavas can be gained from examining plots of FeO*/MgO v. SiO_2 (Fig. 10a) and Na_2O v. MgO (Fig. 10b). The FeO*/MgO v. SiO_2 diagram shows that East Diamante lavas have similar FeO*/MgO of 2–4 in spite of very different SiO_2 contents. Most mafic Diamante lavas plot in or near the tholeiitic field, whereas most felsic lavas plot in the calc-alkalic field. Arculus (2003) has criticized the tholeiitic/calc-alkalic diagram and argued that it makes more petrological sense to distinguish between high-Fe, medium-Fe and low-Fe suites (Fig. 10a). East Diamante mafic lavas mostly plot in the high-Fe field, well within the tholeiitic field whereas East Diamante felsic lavas are all calc-alkalic, plotting near the medium-Fe–low-Fe divide. Arculus (2003, p. 932) emphasized that his 'boundaries have been drawn as far as possible to ensure that various suites currently asserted to be dominantly formed by fractional crystallization remain within a given field'. Because East Diamante mafic and felsic lavas plot in different fields,

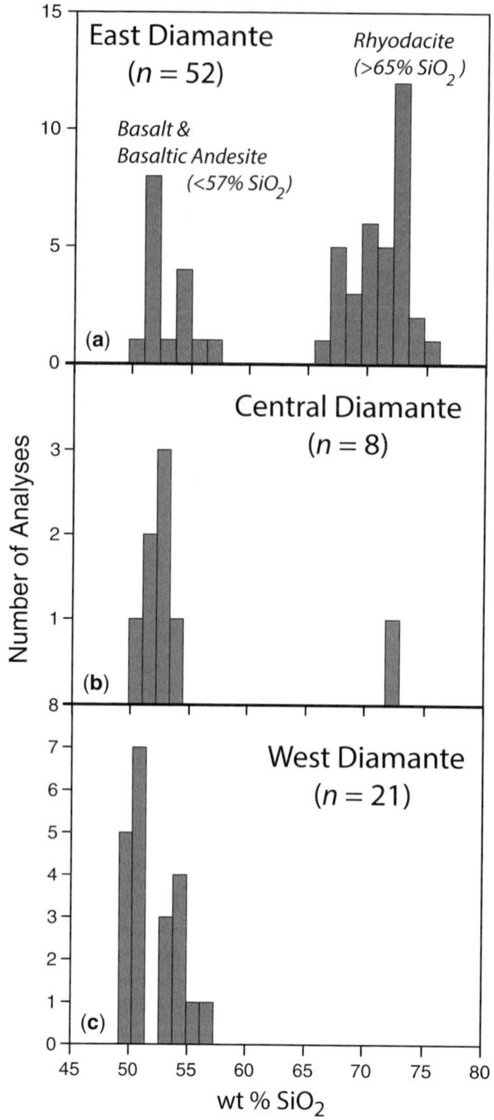

Fig. 9. Silica histograms for Diamante cross-chain lavas (recalculated 100 wt% anhydrous). (**a**) East Diamante volcano; (**b**) Central Diamante volcano; (**c**) West Diamante volcano. *n*, number of analyses.

they appear to be unrelated by simple fractional crystallization. In contrast, Central Diamante lavas cluster tightly around the high-Fe/medium-Fe boundary, with one felsic lava plotting similarly but at higher silica content. West Diamante lavas are mostly low-Fe suites but some also plot in the medium-Fe field.

The Na_2O v. MgO relationships of Diamante cross-chain lavas are shown in Figure 10b. All data define Na_6 (Na_2O content at 6 wt% MgO; Plank & Langmuir 1988) *c.* 2.46 wt%, and moderately correlate ($r^2 = 0.83$). This is similar to the average Na_6 value of 2.4 ± 0.42 wt% that Plank & Langmuir (1988) calculated for three Mariana arc volcanoes (Maug, Pagan and Sarigan). Treating each of the three Diamante cross-chain volcanoes separately yields East Diamante Na_6 ~2.5 wt% ($r^2 = 0.64$), Central Diamante Na_6 ~1.62 wt% ($r^2 = 0.94$) and West Diamante Na_6 ~2.5 wt% ($r^2 = 0.82$). Samples from East Diamante have a range in Na_2O content at the same (3–5 wt%) MgO contents. This may reflect different primary basalt magmas resulting from different degrees of mantle melting as discussed by Tamura *et al.* (2005). Plank & Langmuir (1988) related Na_6 to the height of the melting column (*c.* depth to subduction zone minus crustal thickness); if this were the important control for Diamante cross-chain lavas, Na_6 should be highest (shortest melting column) beneath East Diamante and lowest (longest melting column beneath West Diamante. Clearly there are other important controls on the extent of melting manifested by Na_2O–MgO relationships for Diamante cross-chain lavas.

The only primitive lavas (i.e. those with Mg# >65, generally >8 wt% MgO) are from Cook7 D60 from West Diamante, although HPD1015 parasitic on East Diamante has Mg# = 62 and erupts lavas with >8 wt% MgO. Fe_6 (FeO* content at 6 wt% MgO) relationships for Diamante cross-chain lavas suggest cross-arc variations in the depth of mantle melting. Fe_6 varies from *c.* 13 wt% beneath Central Diamante to *c.* 8.5 wt% beneath West Diamante (Fig. 11). The correlation coefficient for West Diamante lavas is low but a small variation in FeO* over the interval of 3–10 wt% MgO allows Fe_6 to be estimated with confidence at 8.5 ± 1 wt%. East Diamante lavas show two trends: one suite comprising older lavas from the NE caldera wall whereas samples from the parasitic cone on its WNW flank show lower Fe_6. Other East Diamante mafic lavas define a high-Fe suite, with Fe_6 ~15 wt%. These estimates of Fe_6 encompass the average of 10.6 ± 1.65 wt% for three Mariana arc volcanoes (Maug, Pagan and Sarigan; Plank & Langmuir 1988). Higher Fe_6 beneath the volcanic front is interpreted as deeper mean melting depth due to destabilization of olivine (Kelley *et al.* 2010).

Major element abundances for Diamante cross-chain lavas are plotted against silica in Figure 12. Concentrations of TiO_2, MgO, FeO* and CaO decrease with increasing silica, but the large silica gap breaks the continuity of these trends. The behaviour of TiO_2 (Fig. 12a) and FeO* (Fig. 12d) suggests that a phase rich in these elements, probably magnetite, is residual or crystallized early

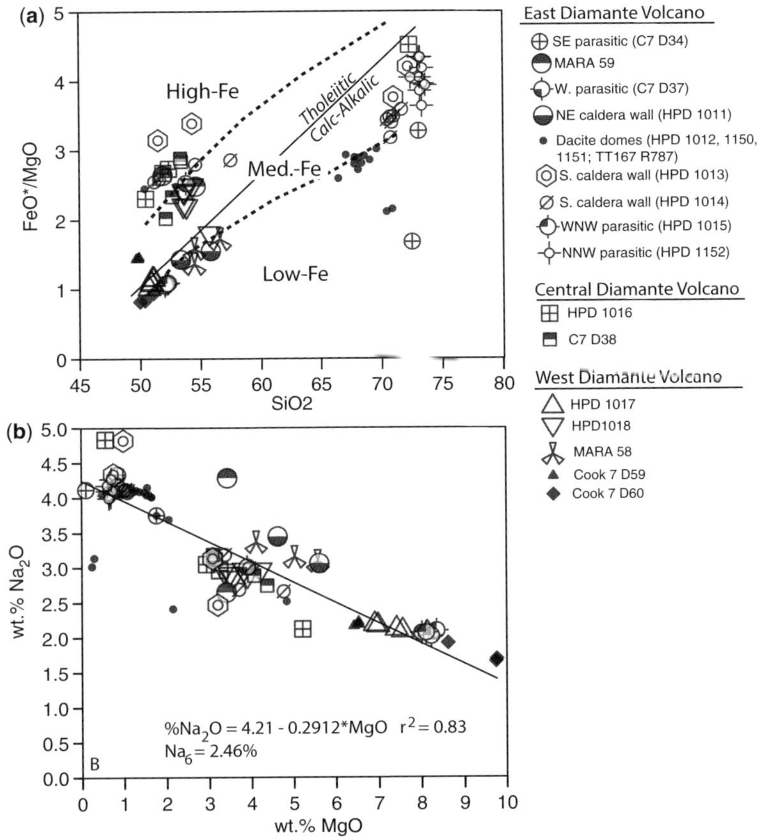

Fig. 10. Geochemical variations observed in Diamante cross-chain igneous rocks. (**a**) wt% FeO*/MgO v. wt% SiO_2 plot of Diamante volcanic rocks. Straight diagonal line separates tholeiitic and calc-alkalic fields of Miyashiro (1974), curved dashed lines separate high-Fe, medium-Fe, and low-Fe fields of Arculus (2003). (**b**) Plot of Diamante volcanic rock wt% Na_2O v. wt% MgO.

during Diamante magmatic evolution. Most Diamante mafic lavas are high-Al basalt, although the Cook 7 D59 and HPD1015 primitive lavas are not (Fig. 12b). The decrease in Al_2O_3 with increasing silica indicates an important role for plagioclase as a fractionating or residual phase. Rapid decrease of MgO with increasing silica (Fig. 12c) indicates an important role for fractionating and/or residual olivine and pyroxene. Monotonically decreasing CaO with silica (Fig. 12e) further indicates control by clinopyroxene and/or calcic plagioclase. Na_2O increases rapidly with increasing silica in mafic rocks, indicating that it is a strongly incompatible element over this silica range; however, as shown in Figure 9b, East Diamante magmas had varying contents of Na_2O, reflecting that of their primary magmas. The K_2O–SiO_2 diagram (Fig. 12g) shows that Diamante cross-chain lavas mostly plot in the medium-K field, although some felsic samples fall in the low-K field. These are typical Mariana arc lavas for the most part, which are characteristically a medium-K suite (Stern *et al.* 2003).

Figure 12 also shows fields for hydrous melting experiments on IBM basalts, intended to capture the composition of liquids generated by melting of mafic IBM arc crust. The dashed fields show the results of Nakajima & Arima (1998), over a P range of 1–1.5 GPa and 2–5 wt% H_2O, whereas the star shows the ‘inferred felsic melt’ of Tatsumi & Suzuki (2009), determined at 0.3 GPa in the presence of 0.49–2.8 wt% H_2O. These experimental results are similar to the composition of East Diamante felsic lavas except that Na_2O and K_2O of the experimental starting materials are lower. This may be accounted for by the fact that the basaltic starting materials (from the Izu arc) used in these experiments contain less Na_2O and K_2O than do Mariana arc lavas (Stern *et al.* 2003).

Figure 13 summarizes the phenocryst compositions in Diamante cross-chain mafic lavas as

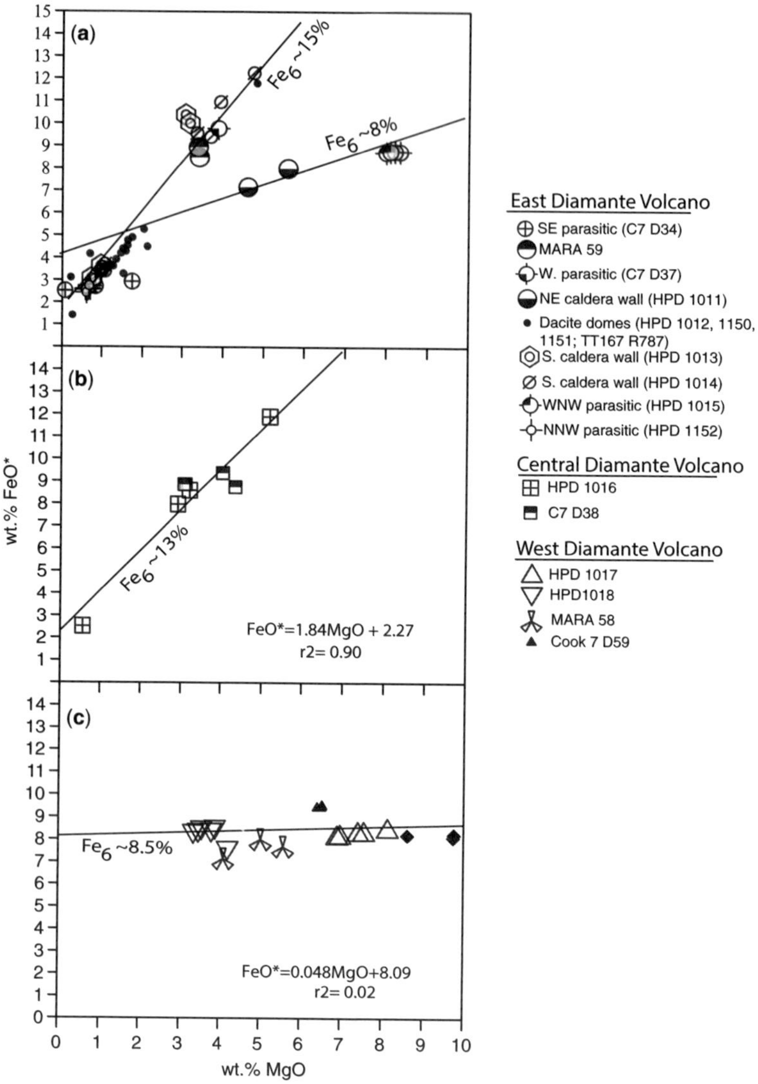

Fig. 11. Plots of wt% FeO* v. wt% MgO for whole-rock samples from (**a**) East Diamante (except HPD1015 samples from parasitic cone and HPD1011older edifice samples); (**b**) Central Diamante; and (**c**) West Diamante.

determined from electron microprobe analyses (Supplementary material). Samples show modest variations in Fo content of olivine and An content of plagioclase except for 1015R01 (from a satellite cone on the WNW flank of East Diamante). This basalt has a primitive major element composition (*c.* 8 wt% MgO) but shows a very wide range of olivine, clinopyroxene and plagioclase compositions (Fig. 13g–i), suggesting mixing of primitive and fractionated magmas. Further evidence of mixing is manifested by the presence of many resorbed plagioclase phenocrysts in these lavas. Central Diamante samples show slightly more magnesian olivines (~Fo75) in equilibrium with slightly less calcic plagioclase (~An85), but the range of plagioclase compositions (Fig. 13l, o) suggests magma mixing. Lavas from West Diamante volcano show a range of olivine compositions (~Fo78–90) in association with calcic plagioclase compositions (~An80–90). From these results, there is evidence of mixing of mafic and very evolved magmas beneath Central Diamante and the western flank of East Diamante volcano.

Electron microprobe results for coexisting olivine and plagioclase in the Diamante cross-chain mafic lavas are summarized in Figure 14.

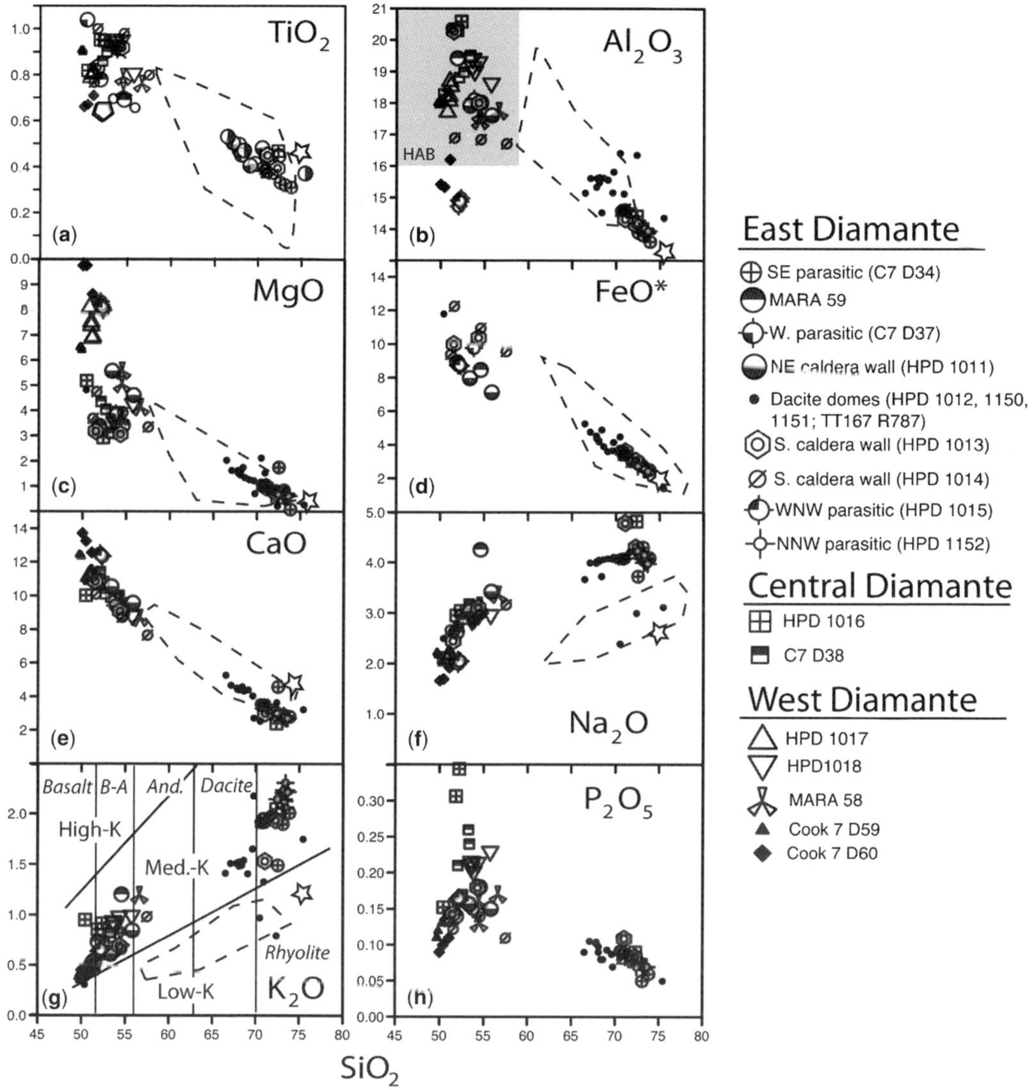

Fig. 12. Harker variation diagrams for Diamante cross-chain volcanic rocks (all oxides in wt%). (**a**) TiO_2 v. SiO_2. (**b**) Al_2O_3 v. SiO_2. Grey field labelled 'HAB' shows field of high-alumina basalt (Crawford *et al.* 1987). (**c**) MgO v. SiO_2. (**d**) Total iron as FeO v. SiO_2. (**e**) CaO v. SiO_2. (**f**) Na_2O v. SiO_2. (**g**) K_2O v. SiO_2. East Diamante lavas mostly plot within the medium-K (calc-alkaline) field, although some samples plot in the low-K field. (**h**) P_2O_5 v. SiO_2. Dashed fields in (a)–(g) encompass experimental melts of tholeiitic metabasalt, 1–1.5 GPa, 2–5 wt% H_2O (Nakajima & Arima 1998). Star in (a)–(g) indicates composition of 'inferred felsic melt' generated by hydrous melting of Izu arc tholeiite at 0.3 GPa (*c.* 10 km depth; Tatsumi & Suzuki 2009).

Most Diamante cross-chain basalts contain coexisting Fe-rich olivine (~Fo70) and Ca-rich plagioclase (~An90), characteristic of arc basalts (Stern 2010). Even the volcanoes behind the volcanic front – Central and West Diamante volcanoes – show Fo–An relations that are arc-like, in contrast to the situation for the Guguan cross-chain to the north. Interestingly, only HPD1015 primitive 'mixed' basalt plots in the field defined by OIB and MORB.

Figure 15 summarizes equilibrium temperatures for four East Diamante igneous rocks that contain two pyroxenes, which permit magmatic temperatures to be estimated using the two-pyroxene thermometer of Lindsley & Andersen (1983). These samples are fractionated, with one basaltic andesite

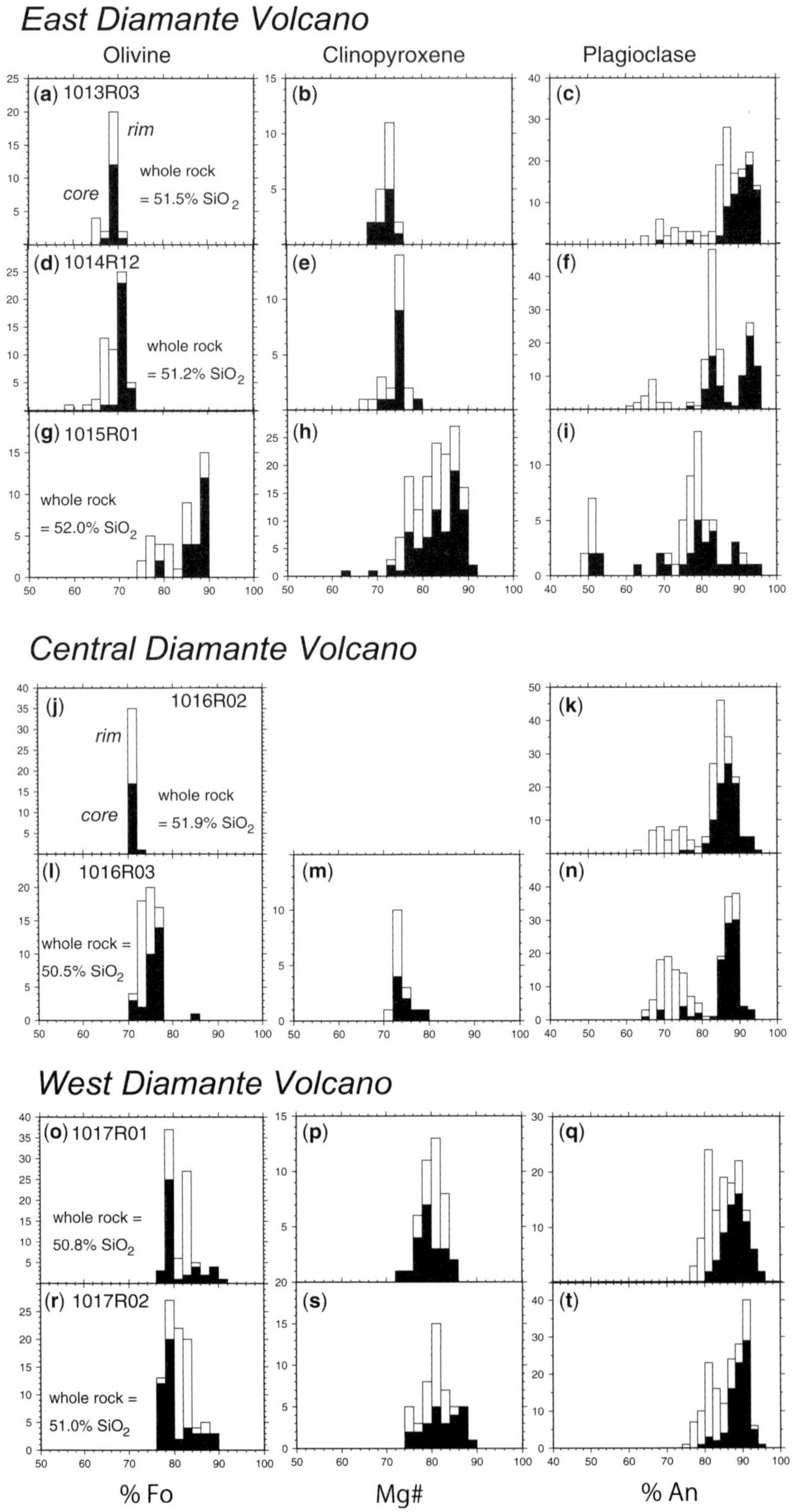
East Diamante Volcano
Olivine
Clinopyroxene
Plagioclase
(a) 1013R03
rim
core
whole rock
= 51.5% SiO2
(b)
(c)
(d) 1014R12
whole rock
= 51.2% SiO2
(e)
(f)
(g) 1015R01
whole rock
= 52.0% SiO2
(h)
(i)
Central Diamante Volcano
(j)
1016R02
rim
core
whole rock
= 51.9% SiO2
(k)
(l) 1016R03
whole rock =
50.5% SiO2
(m)
(n)
West Diamante Volcano
(o) 1017R01
whole rock =
50.8% SiO2
(p)
(q)
(r) 1017R02
whole rock =
51.0% SiO2
(s)
(t)
% Fo
Mg#
% An

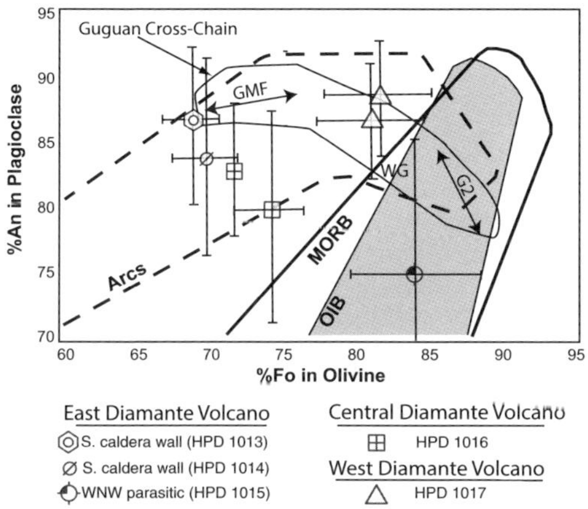

Fig. 14. Plagioclase mol% An content v. olivine mol% Fo for lavas from Diamante cross-chain volcanoes ($\pm 1\sigma$), modified after Stern *et al.* (2006). Note strong cross-arc variation for Guguan samples, from low Fo/high An for samples from Guguan volcanic front lavas (GMF) to OIB- and MORB-like assemblages (high Fo/low An compositions) for Guguan 2 (G2), farthest from the trench; W Guguan (WG) which lies between the two volcanoes has intermediate compositions. Such strong cross-arc variations in Fo–An compositional relations are not seen for Diamante cross-chain samples which (except HPD1015) plot in the arc field.

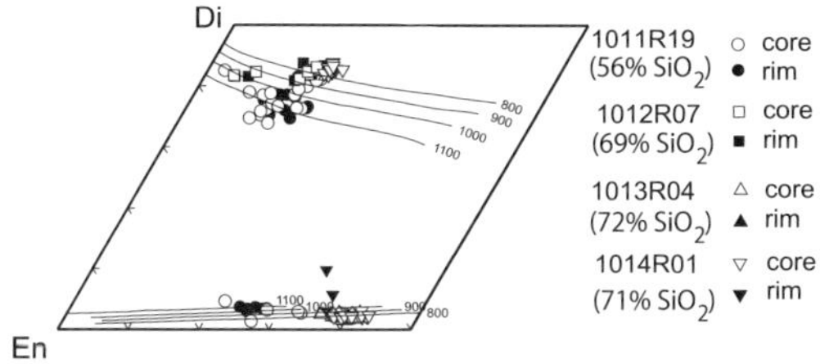

Fig. 15. Coexisting pyroxene compositions for East Diamante lavas, one basaltic andesite and three rhyodacites, plot is left side of pyroxene quadrilateral; Di, diopside; En, entatite. Tick marks are at 10 mol%. Whole-rock SiO_2 abundances (in wt%) are given in parentheses for each sample. Magmatic equilibration temperatures for both cores and rims are estimated using the two-pyroxene thermometer of Lindsley & Andersen (1983). Both core and rim compositions of individual phenocrysts are plotted in each sample.

and three dacites. This is a graphical thermometer; temperature of pyroxenes can be inferred from plotting compositions on a Ca–Mg–Fe diagram (Fig. 15). Both core and rim compositions of individual phenocrysts are plotted for each sample (compositional data can be found in the Supplementary material). Basaltic andesite sample HPD1011-R19 gives higher temperatures of *c.* 1000–1100 °C, whereas temperatures calculated for the three felsic rocks indicate relatively low temperatures of *c.* 800 °C. Augite in sample 1012-R07 scatter because some may have been derived from mafic inclusions, but many others plot close to T ~ 800 °C.

Data for a limited number of trace elements are summarized in Figure 16. Figure 16a shows Rb variations, which scatter to somewhat higher concentrations with silica. Ba increases more consistently with silica (Fig. 16b). Sr decreases with SiO_2, indicating that plagioclase exerts an important control. Zr also scatters to higher values with silica (Fig. 16d). Y defines two groups for the felsic samples, a low-Y (15–30 ppm) group defined by East Diamante samples from most of the resurgent domes, HPD1014 samples from the southern caldera wall, and the NNW parasitic cone, and a high-Y group defined by the SE parasitic cone, HPD samples from the southern caldera wall, and the felsic sample from Central Diamante volcano (Fig. 16e). Rb/Ba decreases with increasing silica (Fig. 16f), which is difficult to explain by crystal fractionation because both elements are highly incompatible. Sr/Y decreases strongly with silica (Fig. 16g), further indicating that Diamante magmas evolved by low-P (garnet-free) fractionation or anatexis. Zr/Y is generally lower in mafic samples than felsic samples (Fig. 16h). Among the felsic samples, samples from the resurgent domes, HPD 1014 samples from the S. caldera wall, and the NNW parasitic cones have higher Zr/Y than samples from the SE parasitic edifice and HPD 1013 samples from the southern caldera wall.

Discussion

Felsic volcanism in intra-oceanic arcs is increasingly recognized, but how and why these magmas form is controversial. The Diamante cross-chain provides some useful constraints for answering this question. One important feature of this felsic magmatic system seems to be that it has persisted for a significant length of time; in the case of East Diamante, a magmatic history encompassing at

Fig. 13. Summary of electron microprobe results for phenocrysts in seven Diamante cross-chain lavas: mol% Fo in olivine (left column), atomic Mg# in clinopyroxene (centre column) and mol% An in plagioclase (right column). Vertical axes are numbers of analyses. Whole-rock silica contents are also given in left panels (in wt%). Core and rim compositions are given as dark and light bars.

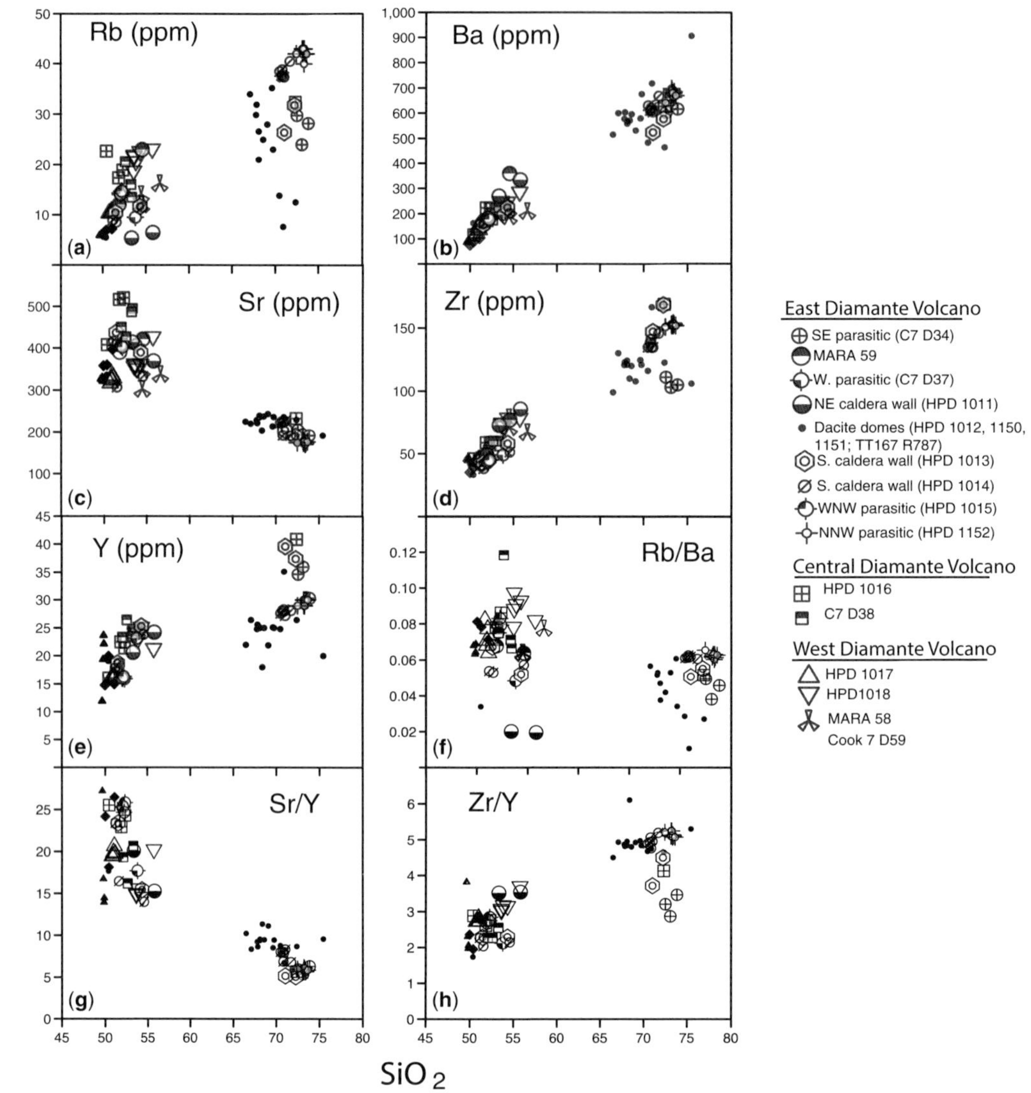

Fig. 16. Selected trace element concentrations (in ppm) and ratios in Diamante volcanic rocks as a function of wt% SiO_2: (**a**) Rb v. SiO_2; (**b**) Ba v. SiO_2; (**c**) Sr v. SiO_2; (**d**) Zr v. SiO_2; (**e**) Y v. SiO_2; (**f**) Rb/Ba v. SiO_2; (**g**) Sr/Y v. SiO_2; (**h**) Zr v. SiO_2. See text for further discussion.

least 0.5 Ma is indicated, although the volcano was dormant long enough to allow a thick carbonate bank to form on its flanks. Violent eruption of volatile-rich felsic magmas beginning at *c.* 0.38 Ma shaped the modern caldera and affected the surrounding seafloor. We suspect that the giant dune fields on the north flank of East Diamante formed during one or more of these events. Quieter emplacement of degassed felsic magma followed, including emplacement of resurgent dacite domes followed caldera formation, continuing up to *c.* 20 ka. But age of the volcano cannot be the only important control on whether or not felsic magmas are present, there are other Mariana volcanoes, such as Pagan and Asuncion, which are larger and thus are likely to be of similar age but which do not erupt felsic lavas.

We do not definitively resolve how East Diamante felsic magmas formed here because we have limited trace element and no isotopic data; those results and analysis will be presented elsewhere. There are several pertinent observations from our data that are noteworthy. First, these are not slab melts (adakites), because Sr/Y of felsic

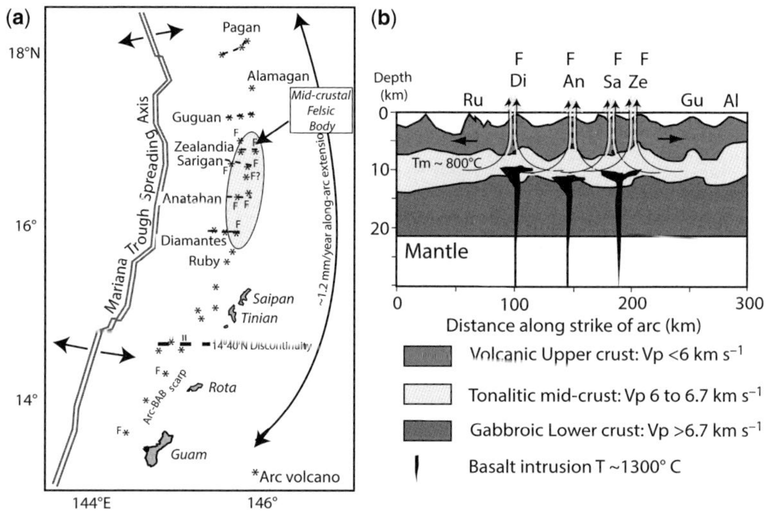

Fig. 17. Anatahan Felsic Province location and how these melts might be generated. (**a**) Simplified map showing the distribution of arc volcanoes (*) and back-arc basin spreading axis (double solid line 'Mariana Trough Spreading Axis'). Location of volcanoes with significant felsic volcanism 'F' in the south-central Mariana arc, defining the Anatahan Felsic Province. Along-arc extension rate of 1.2 mm a^{-1} occurs between Guam and Agrigan (not shown, just north of Pagan; Kato *et al.* 2003). The distribution of felsic igneous rocks in the AFP suggests that this region is underlain by a felsic body in the middle crust that is able to escape to the surface because of cross-arc extensional faults (dashed lines) at the Diamantes, Anatahan, Sarigan and Zealandia. Cross-arc extensional zones associated with the Guguan cross-chain to the north and 14°40′N discontinuity to the south do not involve mid-crustal felsic magmas.
(**b**) Cross-section (based on Calvert *et al.* 2008, fig. 12a). Mantle-derived mafic magmas *c.* 1300 °C intrude and pond near or within felsic middle crust, which melts *c.* 800 °C. Mafic injection remobilizes felsic middle crust, which escapes to the surface as a result of along-arc extension.

lavas is low (<10; Fig. 16g). Adakites have >56 wt% SiO_2 and Sr/Y >30 (e.g. Castillo 2006). The low Sr/Y characteristic of East Diamante felsics indicates these are part of an 'andesite–dacite rhyolite' (ADR) suite formed by low-P processes dominated by feldspar, including magmatic fractionation and crustal anatexis.

East Diamante felsic lavas are good examples of cool, wet, oxidized felsic melts. Felsic magmas can be generated either by fractionation of dry mafic magmas generated by decompression mantle melting to yield hot felsic melts, or above subduction zones dominated by flux melting of the mantle, producing cooler, more oxidized and wetter felsic magmas (Bachmann & Bergantz 2009). Christiansen (2005) identified two different types of large volume (>1000 km^3) felsic eruptions of Cenozoic age in the western USA. 'Yellowstone-type' ignimbrites are sparsely phyric and characterized by anhydrous mineral assemblages. These magmas formed by fractional crystallization of more mafic precursors under low oxygen fugacity conditions. Such lavas show crystallization temperatures of 830 to 1050 °C and are characteristic of hotspot magmatic provinces, such as the Snake River Plain. In contrast, 'Fish Canyon-type' ignimbrites are phenocryst-rich and evolved under high water and oxygen fugacity conditions characteristic of continental magmatic arcs. These felsic magmas have pre-eruption temperatures of 730–820 °C and probably formed by crustal melting. Bachmann & Bergantz (2009) summarized evidence that eruption of these magmas reflect interstitial melt trapped within large, upper crustal mush zones, and that interstitial melt extraction is most efficient when mush zones contain 50–60 vol% crystals that produce a semi-rigid framework that hinder chamber-wide convection currents but the permeability is still high enough that interstitial melt can flow between the crystals.

Diamante magmas clearly formed in the crust, at pressures where plagioclase controlled magmatic evolution, via magmatic differentiation or by partial melting of juvenile arc crust. We do not have a full spectrum of geochemical and isotopic data needed to show how these melts formed but several lines of evidence indicates formation mostly by crustal melting, not fractionation. First, East Diamante felsic igneous rocks are part of a bimodal population, with a marked silica gap between 58 and 66 wt% SiO_2. Similar to the conclusions reached by Shukuno *et al.* (2006) for the bimodal igneous

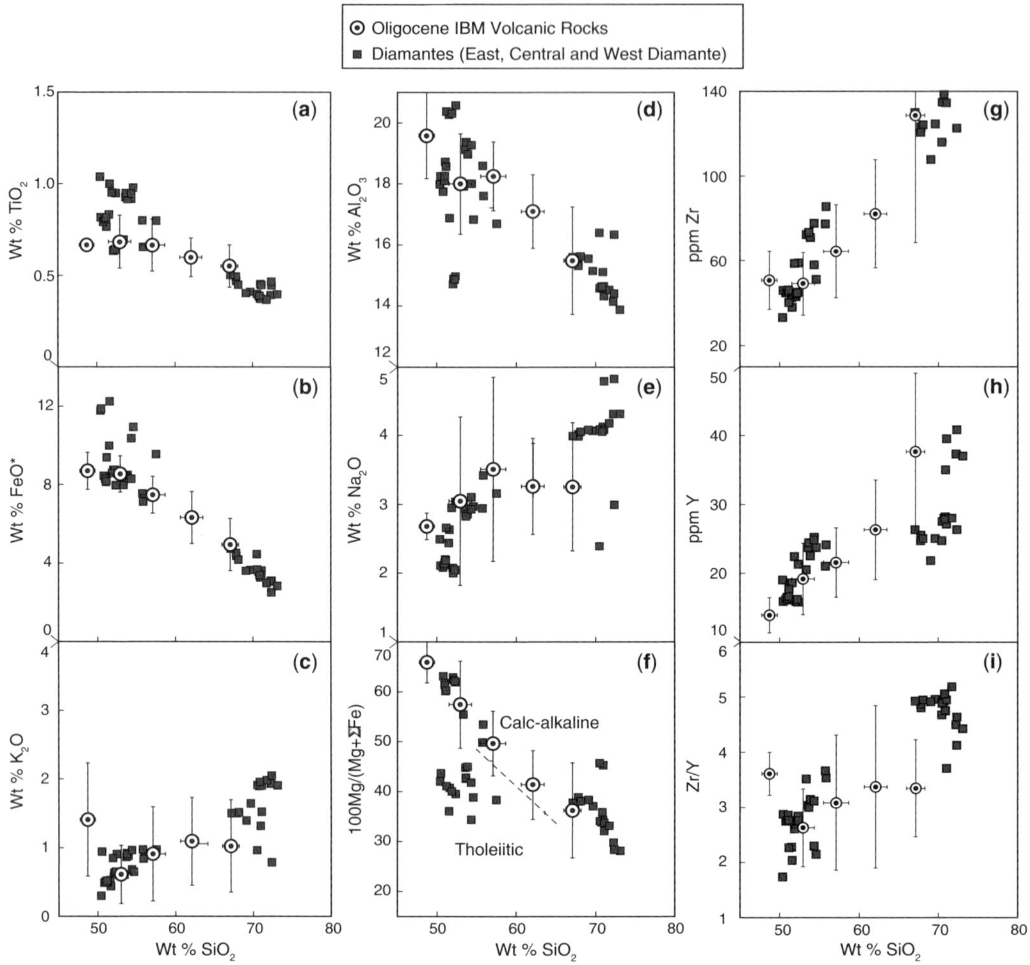

Fig. 18. Comparison of lava compositions from Diamante cross-chain volcanoes with IBM Oligocene rocks and collision zone plutons (Tamura *et al.* 2010). Diamante lavas are compositionally similar to Oligocene volcanic rocks of the Izu arc. Tamura *et al.* (2010) proposed that IBM arc middle crust involved in the collision with Honshu was partially melted during the collision and then intruded into the overlying upper crust of the Honshu and IBM arcs. The similarity of Oligocene middle crust to Diamante and other felsic AFP lavas suggests that remobilization of this crust may also occur under extension.

rocks of Sumisu volcano in the Izu arc, it is difficult to understand how fractionation of mafic parental magma would cause such a silica gap. It is correspondingly easier to understand how felsic melts formed as part of a bimodal suite by crustal melting at 800 °C as a result of heating by emplacement of *c.* 1200 °C mafic magmas into the crust.

The widespread occurrence of felsic lavas in the region including and north of the Diamante seamounts indicates that felsic magmas do not form in magma chambers associated with individual volcanoes and that instead there is a regional source for these magmas. East Diamante is the southernmost of at least seven volcanoes that have recently erupted significant volumes of felsic magmas. This region encompasses *c.* 110 km along the Mariana arc and was identified as the Anatahan Felsic Province (AFP) by Stern & Hargrove (2003; Fig. 17). East Diamante is the southernmost representative of AFP volcanism. To the north in the AFP, Anatahan erupted dacites with up to 66 wt% SiO_2, most recently in 2003 (Wade *et al.* 2005). Dacite with up to 66 wt% SiO_2 was also recovered during Cook 7 from a submarine volcano *c.* 15 km NE of Anatahan. Sarigan erupted lavas with up to 61 wt% SiO_2 (Woodhead 1989). Dacite with up to 66 wt% SiO_2 was recovered during Cook 7 from a submarine volcano west of Sarigan. NT0908 explored Zealandia crater and recovered dacites with up to 72 wt% SiO_2. Dacite with up to 61 to 66 wt% SiO_2 was also recovered from a seamount west of Zealandia Bank during Cook 7. In addition to these seven AFP sites where felsic lavas are known, a seamount south of Sarigan seems to have exploded felsic pyroclastic material from 300 m below sea level up to 12 000 m in the atmosphere in May 2010 (Green *et al.* 2013). Such an explosive eruption was likely to have been a felsic eruption, but there are no analyses of this material to confirm or refute this suggestion.

Volcano size – proxy for age and, hence, evolved magmatic storage systems – does not control the distribution of felsic lavas in the AFP. This is also seen for small felsic plugs or cones on the flanks of East Diamante, including those sampled by Cook 7 D34 and HPD1152. We suggest that AFP felsic magmas are derived from the 3–5 km thick 'tonalite layer' imaged in the mid-crust, with $V_p =$ 6.1–6.5 km s^{-1}, beneath volcanic front edifices (Fig. 3; Takahashi *et al.* 2007). Crustal structure beneath Mariana volcanic front edifices has also been studied by Calvert *et al.* (2008). These studies show that the middle crust is thickest beneath the volcanic front. Bimodal volcanism at the volcanic front (East Diamante) suggests that interaction of hot (*c.* 1200 °C), wet (3 wt% H_2O), mafic magmas with tonalitic middle crust helps generate East Diamante felsic magmas. Evidence of this interaction may be present in the disequilibrium textures in basaltic lavas sampled during HPD1015. The mid-crustal tonalitic layer tapers away to the west, where basaltic rear-arc cross-chain volcanoes are found, and this may be why felsic lavas are rare in Central Diamante and unknown in West Diamante.

Structural controls may also be important for AFP volcanism. Christiansen (2005) argued that viscous low-T felsic magmas could erupt only when the roof over the magma body collapsed. Thus eruption and emplacement of AFP felsic magmas may also have been favored by strong along-arc extension affecting the Marianas. GPS measurements indicate that extension between Agrigan and Guam is occurring at nearly 12 mm a^{-1} (Kato *et al.* 2003). There is insufficient spatial resolution in the GPS data to reveal whether extension is uniform or localized, but several lines of evidence indicate that this strain is localized in discrete zones. These include east–west orientations of individual volcanoes, such as East Diamante and Anatahan, alignments of volcanic chains, such as Guguan and Diamante cross-chains, and crustal earthquake swarms near 14°30′N, beneath Central Diamante, and west of Sarigan (Heeszel *et al.* 2008). These approximately east–west-oriented zones of extension allow felsic magmas, if present, to erupt. This is the case beneath East Diamante but the felsic magma body does not seem to extend far enough west to be tapped by West Diamante volcano.

The idea that AFP felsic volcanism is the result of re-melting Mariana middle crust is supported when the composition of Diamante cross-chain lavas are compared with the similar composition of Oligocene igneous rocks studied by Tamura *et al.* (2010; Fig. 18). They proposed that IBM arc middle crust involved in the collision with Honshu was partially melted during the collision and then intruded into the overlying upper crust of the Honshu and IBM arcs. The similarity of Oligocene middle crust to Diamante and other felsic AFP lavas suggests that remobilization of this crust may also occur under extension.

The regional distribution of felsic volcanism in the AFP and its association with along-arc extension are the most convincing observations to us that this felsic volcanism is due to re-melting of tonalitic middle arc crust, not due to fractionation of mafic melts. Future studies to test these ideas are in progress.

Conclusions

Lavas erupted from three submarine volcanoes of the Diamante volcanic cross-chain in the Mariana arc provide useful perspectives on intra-oceanic arc felsic magmagenesis and eruption. New data from five dredges and 13 ROV dives during four research expeditions indicate that cross-chain magmas formed by interaction of subduction-related basaltic flux from the mantle focused along an east–west extension zone that traversed thick arc crust in the east and refractory back-arc basin crust in the west. East Diamante volcano is by far the largest of the three volcanoes and has a well-developed (5 × 10 km), structurally controlled (elongate east–west) caldera. East Diamante mafic volcanism began at least as early as 0.47 ± 0.08 Ma ($^{40}Ar/^{39}Ar$ age) and the volcano had quiescent intervals that were long enough to construct a *c.* 400 m thick sequence of Plio-Pleistocene carbonate sediments. Timing of caldera formation is approximated by $^{40}Ar/^{39}Ar$ age of 0.37 ± 0.14 Ma for a felsic tuff fragment in a thick pyroclastic sequence. Caldera formation was followed by emplacement of resurgent dacite domes, one of which is young enough ($^{40}Ar/^{39}Ar$ age of 20 000 ± 4000 years) to provide sufficient heat to power a vigorous hydrothermal system. In contrast to the large and complex East Diamante volcano, West Diamante is a simple basaltic edifice, whereas the smallest edifice, Central Diamante has erupted minor felsic magmas and is affected by east–west normal faulting.

Eighty-one analyses of rock major element and selected trace element compositions reveal that Diamante cross-chain lavas define a medium K suite. These include abundant basalts and dacites, with a silica gap between 58 and 66 wt% SiO_2, suggesting that felsic magmas did not form by fractional crystallization of mantle-derived basalt. Phenocryst assemblages in mafic rocks show arc-like associations of anorthite-rich plagioclase with Fe-rich olivines. Magmatic temperatures calculated for coexisting ortho- and clinopyroxene indicate *c.* 1100 °C for a basaltic andesite and *c.* 800 °C for three dacites, typical for cool, wet, subduction-related felsic magmas. Low Sr/Y ratios indicate that felsic magma formed under low-P crustal conditions. Diamante cross-chain volcanoes mark a zone of along-arc (north) extension which focuses mantle melts into E–W magma-healed rifts, and leads to variable interaction with overlying crust. West Diamante Volcano lies on thin, refractory crust of the Mariana Trough back-arc basin so mantle-derived basalts erupt with little crustal interaction. In contrast, East Diamante is built on thickened arc crust with tonalitic middle crust that can be remobilized by basaltic intrusion. Once re-melted by mafic intrusions, this mid-crustal felsic reservoir could erupt because localization of north–south extension along the Diamante volcanic cross-chain facilitated the development of conduits to the surface.

Finally, the presence of at least seven, or perhaps eight volcanoes that erupt felsic magmas in a *c.* 115 km long arc segment called the 'Anatahan Felsic Province' indicate that a mid-crustal tonalite body exists beneath this part of the Mariana arc. Felsic eruptions can be expected anywhere in this region following intrusion of mafic magma at *c.* 1200 °C into thickened arc crust. The midcrustal tonalite layer can then be melted or remobilized at *c.* 800 °C and erupt.

NSF grant SGER-0909398 supported US involvement in NT0908 research cruise. NSF grant SGER-1026150 supported US involvement in NT1012 research cruise. Cook 7 expedition and sample investigation was supported by NSF grant OCE 0001827. We appreciate thoughtful reviews by Amy Draut and Karsten Haase. This is UTD Geosciences contribution #1243.

References

Allen, S. R. & McPhie, J. 2009. Products of Neptunian eruptions. *Geology*, **37**, 639–642.

Arculus, R. J. 2003. Use and abuse of the terms calcalkaline and calcalkalic. *Journal of Petrology*, **44**, 929–935.

Bachmann, O. & Bergantz, G. W. 2009. Rhyolites and their source mushes across tectonic settings. *Journal of Petrology*, **49**, 2277–2285.

Baker, E. T., Embley, R. W. *et al.* 2008. Hydrothermal activity and volcano distribution along the Mariana arc. *Journal of Geophysical Research, Solid Earth*, **113**, doi: http://dx.doi.org/10.1029/2007JB005423

Calvert, A. J., Klemperer, S. L., Takahashi, N. & Kerr, B. C. 2008. Three-dimensional crustal structure of the Mariana island arc from seismic tomography. *Journal of Geophysical Research*, **113**, B01406, doi: http://dx.doi.org/10.1029/2007JB004939

Castillo, P. R. 2006. An overview of adakite petrogenesis. *Chinese Science Bulletin*, **51**, 257–268.

Christiansen, E. H. 2005. Contrasting processes in silicic magma chambers: evidence from very large volume ignimbrites. *Geological Magazine*, **142**, 669–681.

Crawford, A. J., Falloon, T. J. & Eggins, S. 1987. Origin of island arc high-alumina basalts. *Contributions to Mineralogy and Petrology*, **97**, 417–430.

Dixon, T. H. & Stern, R. J. 1983. Petrology and geochemistry of submarine volcanoes in the Southern Mariana Arc. *Geological Society of America Bulletin*, **94**, 1159–1172.

Draut, A. E. & Clift, P. D. 2006. Sedimentary processes in modern and ancient oceanic arc settings: evidence from the Jurassic Talkeetna Formation of Alaska and the Mariana and Tonga arcs, Western Pacific. *Journal of Sedimentary Research*, **76**, 493–514.

Green, D. N., Evers, J. G., Fee, D., Matoza, R. S., Snellen, M., Smets, P. & Simons, D. 2013. Hydroacoustic, infrasonic and seismic monitoring of the submarine eruptive activity and sub-aerial plume generation at South Sarigan, May 2010. *Journal of Volcanology and Geothermal Research*, **257**, 31–43.

Heeszel, D. S., Wiens, D. A., Shore, P. J., Shiobara, H. & Sugioka, H., 2008. Earthquake evidence for along-arc extension in the Mariana Islands. *Geochemistry, Geophysics, Geosystems*, **9**, doi: http://dx.doi.org/10.1029/2008GC002186

Kato, T., Beavan, J., Matsushima, T., Kotake, Y., Camacho, J. T. & Nakao, S. 2003. Geodetic evidence of back-arc spreading in the Mariana Trough. *Geophysical Research Letters*, **30**, doi: http://dx.doi.org/10.1029/2002GL016757

Kelley, K. A., Plank, T., Newman, S., Stolper, E. M., Gove, T. L., Parman, S. & Hauri, E. H. 2010. Mantle melting as a function of water content beneath the Mariana Arc. *Journal of Petrology*, **51**, 1711–1738.

Lindsley, D. H. & Andersen, D. J. 1983. A two-pyroxene thermometer. Proceedings of 13th Lunar and Planetary Science Conference. *Journal of Geophysical Research*, **88**(Suppl.), 887–906.

Miyashiro, A. 1974. Volcanic rock series in island arcs and active continental margins. *American Journal of Science*, **274**, 321–355.

Nakajima, K. & Arima, M. 1998. Melting experiments on hydrous low-K tholeiite: implications for the genesis of tonalitic crust in the Izu–Bonin–Mariana arc. *The Island Arc*, **7**, 359–373.

Plank, T. & Langmuir, C. H. 1988. An evaluation of the global variations in the major element chemistry of arc basalts. *Earth and Planetary Science Letters*, **90**, 349–370H.

Shukuno, H., Tamura, Y., Tani, K., Chang, Q., Suzuki, T. & Fiske, R. S. 2006. Origin of silicic magmas and the compositional gap at Sumisu submarine caldera, Izu–Bonin arc, Japan. *Journal of Volcanology and Geothermal Research* **156**, 187–216.

STEIGER, R. H. & JÄGER, E., 1977. Subcommision on Geochronology: convention on the use of decay constants in geo- and cosmochronology. *Earth and Planetary Science Letters*, **36**, 359–362.

STERN, R. J. 2010. The anatomy and ontogeny of modern intra-oceanic arc systems. *In*: KUSKY, T., MINGGUO, Z. & XIAO, W. (eds) *The Evolving Continents: Understanding Processes of Continental Growth*. Geological Society, London, Special Publications, **338**, 7–34.

STERN, R. J. & HARGROVE, U. S. 2003. The Anatahan Felsic Province in the Mariana Arc system. *EOS Transactions of the American Geophysical Union*, **84**, F1562 (abstract).

STERN, R. J., FOUCH, M. J. & KLEMPERER, S. 2003. An overview of the Izu–Bonin–Mariana subduction factory. *In*: EILER, J. & HIRSCHMANN, M. (eds) *Inside the Subduction Factory*, Geophysical Monograph, American Geophysical Union, Washington, DC, **138**, 175–222.

STERN, R. J., KOHUT, E. J., BLOOMER, S. H., LEYBOURNE, M., FOUCH, M. & VERVOORT, J. 2006. Subduction factory processes beneath the Guguan Cross-chain, Mariana Arc: no role for sediments, are serpentinites important? *Contributions to Mineralogy and Petrology*, **151**, 202–221.

STERN, R. J., TAMURA, Y. *ET AL.* 2008. Evolution of West Rota Volcano, an extinct submarine volcano in the Southern Mariana Arc: evidence from sea floor morphology, remotely operated vehicle observations and $^{40}Ar/^{39}Ar$ Geochronology. *The Island Arc*, **17**, 70–89.

STRAUB, S. M. 2008. Uniform processes of melt differentiation in the central Izu Bonin volcanic arc (NW Pacific). *In*: ANEN, C. & ZELLMER, G. F. (eds) *Dynamics of Crustal Magma Transfer, Storage and Differentiation*. Geological Society, London, Special Publications, **304**, 261–283.

SYRACUSE, E. M. & ABERS, G. A. 2006. Global compilations of variations in slab depth beneath arc volcanoes and implications. *Geochemistry, Geophysics, Geosystems*, **7**, doi: http://dx.doi.org/10.1029/2005GC001045

TAKAHASHI, N., KODAIRA, S., KLEMPERER, S., TATSUMI, Y., KANEDA, Y. & SUYEHIRO, K. 2007. Crustal structure and evolution of the Mariana intra-oceanic island arc. *Geology*, **35**, 203–206.

TAMURA, Y. & TATSUMI, Y. 2002. Remelting of an andesitic crust as a possible origin for rhyolitic magma in oceanic arcs: an example from the Izu–Bonin arc. *Journal of Petrology*, **43**, 1029–1047.

TAMURA, Y., TANI, K., ISHIZUKA, O., CHANG, Q., SHUKUNO, H. & FISKE, R. S. 2005. Are arc basalts dry, wet, or both? Evidence from the Sumisu Caldera volcano, Izu–Bonin Arc, Japan. *Journal of Petrology*, **46**, 1769–1803.

TAMURA, Y., ISHIZUKA, O. *ET AL.* 2010. Missing oligocene crust of the izu-bonin arc: consumed or rejuvenated during Collision? *Journal of Petrology*, **51**, 823–846.

TATSUMI, Y. & SUZUKI, T. 2009. Tholeiitic vs calc-alkalic differentiation and evolution of arc crust: constraints from melting experiments on a basalt from the Izu–Bonin–Mariana Arc. *Journal of Petrology*, **50**, 1575–1603.

WADE, J. A., PLANK, T. *ET AL.* 2005. The May 2003 eruption of Anatahan volcano, Mariana Islands: geochemical evolution of a silicic island-arc volcano. *Journal of Volcanology and Geothermal Research*, **146**, 139–170.

WOODHEAD, J. D. 1989. Geochemistry of the Mariana Arc (western Pacific). *Chemical Geology*, **76**, 1–24.

YORK, D. 1969. Least squares fitting of a straight line with correlated errors. *Earth and Planetary Science Letters*, **5**, 320–324.

Preferential eruption of andesitic magmas: Implications for volcanic magma fluxes at convergent margins

ADAM J. R. KENT

College of Earth, Ocean, and Atmospheric Sciences, Oregon State University, Corvallis, OR, 97331, USA (e-mail: adam.kent@geo.oregonstate.edu)

Abstract: Andesitic magmas (herein defined as having *c.* 55–65 wt% SiO_2) are abundant in convergent margin volcanoes. Although andesites and other intermediate magmas can plausibly be produced by a number of mechanisms, a lack of evidence for andesitic liquids relative to the abundance of andesite volcanic rocks, together with widespread evidence for hybridization and recharge, suggests that many erupted andesites form by mixing between relatively mafic magmas that ultimately derive from the mantle wedge, and more felsic magmas or mush zones derived from differentiation and crustal melting.

A model of *recharge filtering* accounts for the high abundance of andesitic volcanic rocks and reflects the simple idea that the processes that create hybridized andesitic magmas in the shallow crust can also initiate volcanic activity, resulting in preferential eruption of andesitic magmas relative to the parental magmas that mix to produce them. This occurs via mafic recharge – intrusion of a mafic magma into a silicic magma or mush within the shallow crust. The overall abundance and variability from volcano to volcano of andesitic volcanism at convergent margins also suggests important roles for hydrous mafic magma in promoting mixing and eruption during mafic recharge, and the crust for modulating the compositions of volcanic outputs.

Andesitic volcanic rocks represent a significant proportion of the global eruptive output of convergent margins (Fig. 1; e.g. Ewart 1976; Gill 1981), and are an important and characteristic component of subduction zone magmatism – particularly in continental subduction systems (e.g. Kuno 1968; Anderson 1976; Ewart 1976; Eichelberger 1978; Gill 1981; Annen *et al.* 2006; Hildreth 2007; Reubi & Blundy 2009). It has also been long recognized that the bulk composition of the continental crust is andesitic, suggesting a link between magmatism in orogenic settings and the growth and evolution of the continental crust. (e.g. Rudnick 1995). In addition, andesitic magmas are associated with important mineral deposits and andesite-dominated volcanoes pose significant hazards to human populations and infrastructure.

The recognition of the importance of understanding andesite petrogenesis is also matched by many decades of research and debate regarding the origin of andesitic magmas – particularly in the convergent margin settings in which they are so abundant. Despite this there remains considerable uncertainty regarding the processes by which andesitic magmas form, and which of these are the most important in convergent margin settings. Gill (1981) in his landmark book, together with many other studies (e.g. Kuno 1968; Anderson 1976; Eichelberger 1978; Hildreth & Moorbath 1988; Carmichael 2002; Annen *et al.* 2006; Reubi & Blundy 2009; Kent *et al.* 2010; Straub *et al.* 2011), has emphasized the range of mechanisms by which andesites and other intermediate composition volcanics may form. These include: derivation of andesitic liquids directly from the mantle via hydrous partial melting of the mantle wedge or subducted ocean crust; partial melting of underplated basalts within the lower crust or uppermost mantle; differentiation of basaltic precursors at the base of the crust or within the lower crust; assimilation of sialic crustal material by basaltic magma; and mixing between mafic magmas derived from the mantle with silicic magmas produced by differentiation or crustal melting. Hildreth & Moorbath (1988) and Annen *et al.* (2006) also argued for an important role for a combination of differentiation, assimilation and mixing processes within lower crustal zones to produce a range of intermediate liquid compositions, including andesites. Although examples of andesitic magmas exist that plausibly formed by all these the mechanisms, resolution of which process or processes are dominant, and why andesites are so prevalent in convergent margin settings, remains one of the primary questions in the field of igneous petrology.

This contribution focuses on the record of subduction zone magmatic processes preserved within volcanic sequences. While a detailed evaluation of all the mechanisms of andesite genesis listed above is beyond the scope of this paper, some insight can be gained from the abundances of the different magma compositions that occur in convergent margins (e.g. Ewart 1976; Gill 1981; Hildreth 2007; Reubi & Blundy 2009). One clear

From: Gómez-Tuena, A., Straub, S. M. & Zellmer, G. F. (eds) 2014. *Orogenic Andesites and Crustal Growth*. Geological Society, London, Special Publications, **385**, 257–280.
First published online September 12, 2013, http://dx.doi.org/10.1144/SP385.10

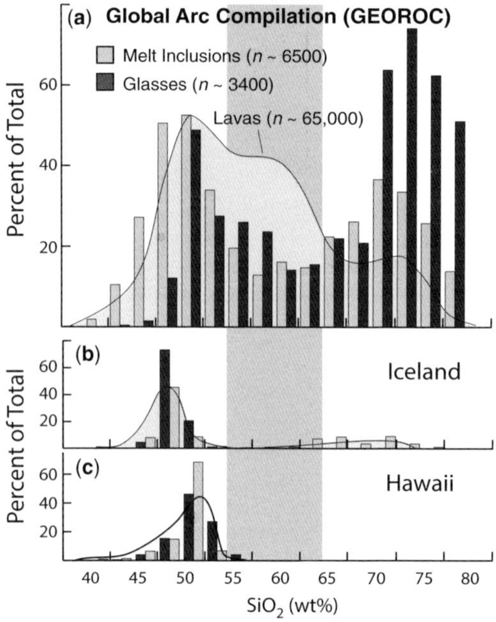

Fig. 1. (**a**) Global compilations of SiO_2 contents of whole rock, melt inclusion and glass (including pumice and matrix glass) from arc settings. Data are from GEOROC (georoc.mpch-mainz.gwdg.de/georoc/). (**b**), (**c**) The same comparison for Iceland and Hawaii. Data are from GEOROC and whole rock, glasses and melt inclusions consists of *c*. 3100, 700 and 450 and *c*. 6100, 3000 and 700 analyses for Iceland and Hawaii, respectively.

prediction of models that propose andesite genesis via differentiation or melting is that liquids of andesitic composition should constitute the precursors of the andesitic magmas that are eventually erupted. Examples of crystal-poor andesitic magmas that are interpreted to represent liquids do occur (e.g. Wallace & Carmichael 1994; Ownby *et al.* 2008; Schmidt & Grunder 2009); however, a significant majority of andesites from convergent margin settings are crystal rich (e.g. Ewart 1976; Gill 1981; Hildreth 2007), and Ewart (1976) estimated that *c*. 70% of island arc and continental arc lavas have between 20 vol% and 60 vol% phenocrysts. In such samples it is difficult to demonstrate *a priori* that the erupted magmas ever existed as a true igneous liquid. Note that, for simplicity herein, the term *andesitic* is used to denote intermediate volcanics with compositions of *c*. 55–65 wt% SiO_2 (a compositional range similar to that used by Gill 1981), although by currently accepted definition true andesites have SiO_2 contents that range from 57–63 wt% and basaltic andesites have SiO_2 between 52 wt% and 57 wt%).

Andesite genesis: the role of mixing

Any requirement for andesitic liquids to be the precursors for the andesitic volcanic rocks that erupt at subduction zones is also challenged by two broad sets of observations. The first is the presence of textures and other evidence that suggest that shallow mixing (at $<c.$10 km crustal depth) between magmas of diverse origins plays an important role in the constitution of many erupted andesitic magmas. In particular, the presence of disequilibrium phase assemblages, multiple melt compositions, mineral zoning and multiple populations of the same crystal phase that are not in equilibrium (e.g. Anderson 1976; Eichelberger 1978; Gill 1981; Gourgaud *et al.* 1989; Pallister *et al.* 1992*a*, 1996; Clynne 1999; Miller *et al.* 1999; Sato *et al.* 1999; Tepley *et al.* 1999; Patia 2004; Browne *et al.* 2006; Klemetti & Grunder 2008; Martel *et al.* 2006; Nakai *et al.* 2008; Salisbury *et al.* 2008; Hirotani *et al.* 2009; Humphreys *et al.* 2009, 2013; Kent *et al.* 2010; Dogan *et al.* 2011; Koleszar 2011; Coombs *et al.* 2013; Koleszar *et al.* 2012, Ruprecht *et al.* 2012; Ruprecht & Cooper 2012; Viccaro *et al.* 2012; Walker *et al.* 2012; Arce *et al.* 2013; Costa *et al.* 2013) reveal the importance of hybridization of crystal-bearing magmas (or crystal-rich mushes) of different bulk compositions. Observations of active volcanoes also suggest that mixing processes occur within the shallow crust ($<c.$10 km; e.g. Pallister *et al.* 1992*a*, 1996; Aspinall *et al.* 1998; Miller *et al.* 1999; Nishi *et al.* 1999; Elsworth *et al.* 2008; Foroozan *et al.* 2010). Although the role of mixing has been recognized for some time (e.g. Anderson 1976), increasing application of micro-analytical techniques is promoting a burgeoning awareness of the importance and extent of hybridization in many intermediate magmas (e.g. Tepley *et al.* 1999; Sato *et al.* 1999; Browne *et al.* 2006; Martel *et al.* 2006; Salisbury *et al.* 2008; Hirotani *et al.* 2009; Humphreys *et al.* 2009, 2013; Smith *et al.* 2009; Kent *et al.* 2010; Koleszar 2011; Koleszar *et al.* 2012; Ruprecht *et al.* 2012; Ruprecht & Cooper 2012; Viccaro *et al.* 2012; Costa *et al.* 2013).

Mineral chemistry, melt inclusions, mafic enclaves and other observations also suggest that in many cases erupted andesitic magmas are the result of mixing between parental magmas that have broadly mafic (basaltic-basaltic andesite) and felsic (dacitic-rhyolitic) compositions (e.g. Walker & Skelhorn 1966; Anderson 1976; Eichelberger 1978; Sakuyama 1978; Bacon & Metz 1984; Pallister *et al.* 1992*a*, 1996; Clynne 1999; Tepley *et al.* 1999; Patia 2004; Browne *et al.* 2006; Salisbury *et al.* 2008; Hirotani *et al.* 2009; Humphreys *et al.* 2009, 2010; Reubi & Blundy 2009; Christopher *et al.* 2010; Kent *et al.* 2010; Koleszar 2011;

Koleszar *et al.* 2012; Almeev *et al.* in press) rather than the result of mixing of multiple generations of more intermediate composition magmas – although this undoubtedly also occurs (e.g. Smith *et al.* 2009; Coombs *et al.* 2013). This equates to mixing between magmas that are only moderately evolved from basaltic compositions derived from the mantle wedge and silicic magmas derived from more extensive crustal differentiation and melting (e.g. Hildreth & Moorbath 1988; Annen *et al.* 2006; Reubi & Blundy 2009). Furthermore, geobarometry and geophysical observations (when mixing events are associated with eruptions) suggest that mixing typically occurs within the shallow crust (<10 km; e.g. Pallister *et al.* 1992*a*, 1996; Nakada *et al.* 1995; Mori *et al.* 1996; Aspinall *et al.* 1998; Barclay *et al.* 1998; Devine *et al.* 1998; Miller *et al.* 1999; Elsworth *et al.* 2008; Foroozan *et al.* 2010; Koleszar 2011; Coombs *et al.* 2013) and often occurs immediately prior to eruption (Pallister *et al.* 1992*a*, 1996; Nakamura 1995; Devine *et al.* 2003; Humphreys *et al.* 2009; Smith *et al.* 2009; Kent *et al.* 2010).

Secondly, amongst melt inclusion and matrix glasses' compositions from subduction zone volcanic rocks – which represent unambiguous samples of liquid compositions – there is relatively little direct evidence for andesitic liquids in proportions that appear to match the relatively high abundances of andesitic magmas in convergent margins. Reubi & Blundy (2009) illustrated this observation using a compilation of melt inclusion and magma compositions from selected convergent margins, although the dataset used was relatively small (*c.* 2500 melt inclusions) and thus potentially prone to being distorted by melt inclusion derived from a smaller number of individual lava samples. In light of this the author has re-examined this issue for a number of selected subduction zones using updated GEOROC (georoc.mpch-mainz.gwdg.de/georoc/) melt inclusion, glass and bulk rock compilations. This is shown in Figure 1 and includes more than 6000 melt inclusion analyses (and represents over 250 separate localities), more than 3400 glass analyses (primarily pumice, groundmass glass, matrix glass and pillow rim glasses) and almost 65 000 whole-rock analyses. The new compilation is consistent with the original findings of Reubi & Blundy (2009) and shows that andesitic liquids are relatively uncommon in subduction zone magmas relative to melts of more felsic and mafic composition. The lower apparent abundances of andesitic liquids preserved as glasses or melt inclusions also contrast with the relatively high apparent abundances of andesitic magmas (as reflected by the number of whole-rock analyses). This difference is dramatic: overall melt inclusions with andesitic compositions appear about one third less common than those <55 wt% SiO_2 (Fig. 2a), but in terms of bulk magma compositions andesitic and more mafic compositions are broadly sub-equal (each represents *c.* 40% of all magmas sampled). Other magmatic environments, such as the examples from Hawaii and Iceland shown in Figure 1b and c, show much closer correspondence between melt inclusion and erupted lava compositions. Further, although melt inclusion compositions can potentially be modified by a range of post-entrapment processes (e.g. Kent 2008), the observation that melt inclusion patterns are independent of host mineral (e.g. Reubi & Blundy 2009), and have compositions that broadly parallel those evident in other glasses (although melt inclusions have a higher proportion of mafic compositions; Fig. 2), suggests inclusion modification has had little effect.

Although these compilations are based on the number of published analyses and thus do not specifically reflect magma volumes, the large amount and geographical spread of the data suggest that they are a reasonably robust measure of the overall output of convergent margin systems. In addition, published estimates of eruptive volumes for arcs or arc segments, such as the Cascades and parts of the Mexican Arc (Fig. 2b; Hildreth 2007; Ownby *et al.* 2011), also suggest that erupted volumes of andesites are high relative to their apparent low abundance in melt inclusions or glasses, particularly where much of the volume of erupted material occurs in large volcanic edifices (Gill 1981; Hildreth 2007). Individual arc segments or smaller volcanic fields can show greater variation and can be dominated by more dispersed mafic volcanism (e.g. Priest 1990; Ownby *et al.* 2008). In addition a comparison between abundances based on the number of published estimates and volume estimates derived from Hildreth (2007) for the major Cascades volcanic centres shown in Figure 2b shows a high degree of consistency between these two datasets. Volume estimates that also include dispersed volcanism in the Cascades show higher proportions of mafic compositions relative to the equivalent number of published analyses (Fig. 2b), suggesting an overall bias against this dispersed mafic style of volcanism in available data compilations, but still show higher abundances of andesites relative to more felsic compositions.

Collectively the above evidence infers that, as noted by Reubi & Blundy (2009), there is a dearth of andesitic liquids in subduction zone settings, relative to their abundance as magmas. This in turn suggests that magma mixing and hybridization plays an important role in forming many andesitic magmas – as, unlike the other mechanisms for andesite genesis listed above, these do not require an andesite liquid that is the precursor to an andesite magma. This issue has been has been discussed in a number of previous studies (Anderson 1976; Eichelberger 1978; Eichelberger *et al.* 2006; Reubi & Blundy 2009; Kent *et al.* 2010) and, given the widespread evidence for magma mixing in andesitic systems, a case can be made for mixing being an important, and perhaps dominant, process in the formation of andesites in convergent margins (but see other papers in Gómez-Tuena *et al.* (2014) for alternative viewpoints). However, if this is the case there remains an important outstanding issue to resolve. Namely: *why* should andesitic volcanic rocks be so common in convergent margin environments relative to their apparent low abundances as liquids? Mafic and felsic magmas occur widely in other tectono-magmatic settings (as do andesitic or equivalent magmas, but in subordinate proportions, Gill 1981; e.g. Fig. 1b, c), but abundant andesitic volcanics occur only in convergent margins. In effect this requires that andesitic magmas erupt in significantly greater quantities than they are produced and preserved as liquids, and that andesitic magmas must be preferentially erupted with respect to the more mafic and felsic compositions that appear more frequently as liquid compositions. This contribution addresses these issues and explores the processes that control the formation and eruption of andesitic magmas, building on earlier studies (e.g. Eichelberger 1978; Marsh 1981; Eichelberger *et al.* 2006) and in particular expanding the treatment presented by Kent *et al.* (2010).

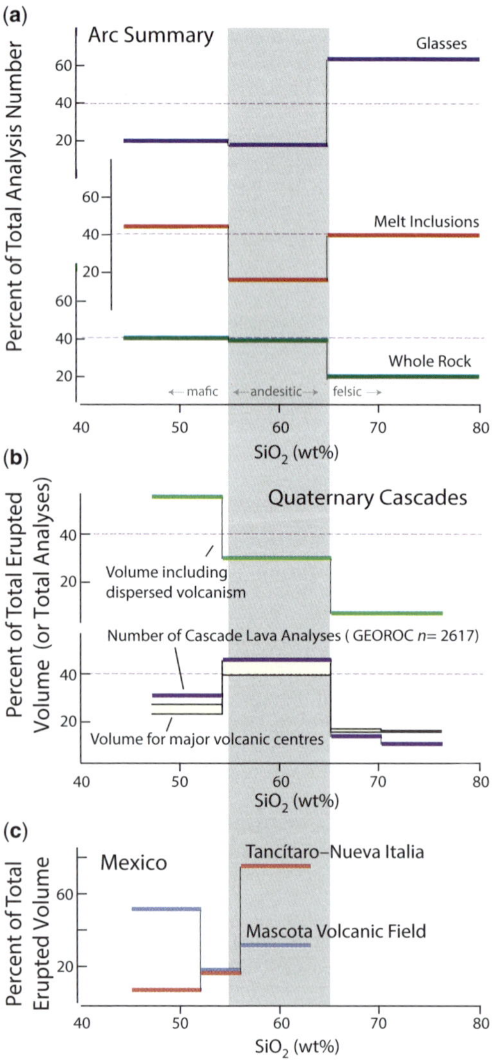

Fig. 2. (**a**) SiO_2 contents of whole rock, glasses and melt inclusions from analyses of arc samples binned into mafic (<55 wt%), andesitic (55–65 wt%) and felsic (>65 wt%) compositions. (**b**) Estimates of erupted volumes for major Cascade volcanic centres based on the work of Hildreth (2007). Lines are shown representing both the volcanic centres alone (boxes include the maximum and minimum estimates) and volcanic centres together with the dispersed and predominantly mafic volcanism (the single line reflects the estimated value). Compositional ranges used follow those of Hildreth (2007) and differ slightly from those used in the rest of this paper). The estimates of Hildreth (2007, table 6) were converted to volumes for each volcanic centre using the following rubric: Hildreth's estimates of A (Andesite) > B (Basalt) was taken to be equivalent to 67% andesite and 33% basalt; A ≫ B was taken as 90% andesite; the composition of dispersed magmatism was estimated to be 80% basalt and 20% andesite. Also shown is the total number of GEOROC analyses in each compositional range. (**c**) Erupted volume estimates for the Mascota volcanic field and the Tancítaro–Nueva Italia region in the Mexican Arc from Ownby *et al.* (2008, 2011). Compositional ranges used follow those in the cited sources.

Magma mobility within the crust

It has long been recognized that volcanic rocks represent a specialized subset of magmatic rocks – namely those that *are able to erupt* (Stolper & Walker 1980; Sparks *et al.* 1980; Marsh 1981). The physical properties of magmas that control mobility within the crust vary substantially and, across the wide spectrum of known igneous compositions, magmas exhibit considerable differences in their ability to be mobilized and to eventually erupt. For silicate liquids with SiO_2 contents between basalt and rhyolite, density varies by more than 25% and viscosity varies by at least eight orders of magnitude (e.g. Giordano *et al.* 2008; Takeuchi 2011). Larger variations in density and viscosity will also occur for magmas that are crystal or bubble bearing. As a result, any process of igneous differentiation will produce derivative magmas whose ability to be mobilized may be quite different from that of their parents. Given that mobility is required for magmas to reach the surface and erupt, there is likely to be a strong compositional bias in volcanic rocks towards those compositions that are able to erupt. In addition, magmas that are formed by processes that favour magma mobility will also be preferentially represented in the volcanic record. Collectively these effects are described as the *eruption filter*, and by definition the volcanic record consists only of magmas that have successfully passed through this filter. In earlier studies Stolper & Walker (1980) and Sparks *et al.* (1980) discussed the role of density filtering in producing the spectrum of erupted mid-ocean ridge basalt and other basaltic rocks. Marsh (1981) focused on the role of crystallinity and its effect on viscosity in terms of controlling the eruption of more evolved magmas, including andesites. More recently the concept of eruption filtering has received less attention.

This issue becomes all the more germane when it is considered that only a relatively small fraction of magmas present within subvolcanic magma storage zones are ever erupted. Although it is a difficult parameter to constrain precisely, current estimates place the ratio of extruded to intruded magmas in various crustal settings at *c.* 3–30% (Hildreth & Moorbath 1988; White *et al.* 2006). Given this, and the likelihood that the composition and mode of formation of magmas is likely to strongly influence their mobility within the crust, it is highly likely that the compositions of volcanic rocks alone provide an incomplete record of the range of magma compositions produced and present in a given crustal-scale magmatic system. Magmas that are more mobile, or that are formed by processes that also initiate eruptions, are likely to be over-represented, and those magmas that do erupt are probably strongly biased in favour of eruptible compositions.

A corollary of this conclusion is that conditions within the crust itself – such as the thermal and stress state, composition, density and strength – will influence magma mobility, and thus can exert a significant control on the compositions of rocks erupted in a given region or from a given magmatic system. As crustal conditions and lithology can also vary over short lateral length scales it is likely that the influence of the crust varies considerably – within individual subduction zone segments or even between neighbouring volcanoes. This effect is evident when looking at the variations in magma compositions of rocks erupted in volcanoes along a given arc, where considerable variations in the range and proportions of erupted compositions are apparent – even in adjacent volcanoes (e.g. Fig. 3; Hildreth & Moorbath 1988; Hildreth 2007; Klemetti & Grunder 2008; Kent *et al.* 2010). Within individual arc segments it is unlikely that the overall forcing factors that control melting during subduction (convergence rate, slab age, slab and wedge thermal structure) vary substantially, and are unlikely to drive major compositional differences between nearby volcanoes. Local crustal conditions, and the interplay between these and variations in magma flux, should thus play an important role in dictating the compositions of volcanic rocks erupted.

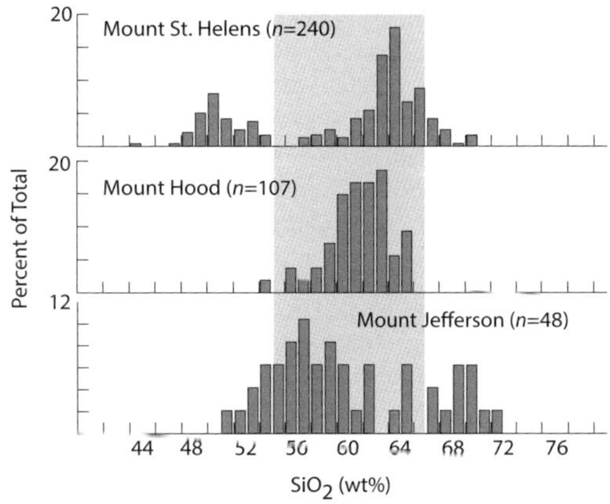

Fig. 3. Histograms showing the number of analysed compositions in the GEOROC database for Mount St Helens, Mount Hood, and Mount Jefferson. Data for Mount Hood also include data from Woods (2004), Kent *et al.* (2010) and Koleszar (2011).

Eruption filtering and the origin of andesitic magmas

Kent *et al.* (2010) argued that eruption filtering plays an important role in andesitic volcanism in convergent margins, and identified a critical role for mafic recharge, magma mixing and hybridization in andesitic volcanism. This idea originated from a study of Mount Hood, Oregon, and this andesitic volcano provides a compelling case for eruption filtering of this style.

Mount Hood – a case study of a compositionally restricted andesitic volcano controlled by mafic recharge

Mount Hood, Oregon, forms part of the Cascadia subduction zone, and represents a volcanic system whose behaviour appears fundamentally controlled by recharge and magma mixing. As such, it is useful here to briefly summarize the characteristics of the volcano and the magmas it erupts. The following section and Table 1 summarize a number of previous studies (e.g. Cribb & Barton 1997; Scott *et al.* 1997; Woods 2004; Darr 2006; Kent *et al.* 2010; Koleszar 2011; Eppich *et al.* 2012; Koleszar *et al.* 2012).

The current edifice of Mount Hood has been constructed over the last *c.* 500 ka, with the most recent growth phases occurring between 15 ka and12 ka (Polallie), *c.* 1500 years (Timberline) and *c.* 220 years (Old Maid) before present (Scott *et al.* 1997). In comparison with many other subduction volcanoes, Mount Hood exhibits a marked degree of homogeneity in the compositions of erupted magmas, and contrasts with other nearby Cascade volcanoes (Fig. 3). Some 95% of the lavas erupted over the life of the current edifice have SiO_2 contents that range between 58 wt% and 66 wt%, (Figs 3 & 4), and lie primarily within the andesite field (although some compositions extend to low SiO_2 dacite). Erupted lavas are crystal-rich and mineralogically similar, being dominated by plagioclase with lesser amounts of amphibole, pyroxene, oxides and olivine.

Mount Hood lavas exhibit a range of textural, mineralogical and composition features that suggest that magma mixing and hybridization play a dominant role in petrogenesis.

- The presence of distinct populations of plagioclase and amphibole – the dominant modal phases (Kent *et al.* 2010; Koleszar 2011; Koleszar *et al.* 2012). Kent *et al.* (2010) defined two populations of plagioclase on the basis of kinked crystal size distributions (CSD), textural form and chemical compositions (Figs 5 & 6). Population 1 plagioclase are smaller, elongate or acicular in form, show little chemical zoning and have higher Fe, Ti, Sr contents and lower Sr/Ba and Sr/light rare earth elements (LREE) ratios consistent with derivation from a more mafic melt. Population 2 plagioclase are larger and more tabular, often exhibit complex internal zoning and have lower Fe, Mg, Ti, Sr/Ba and Sr/LREE, consistent with formation from more felsic melts. Zoning profiles, specifically thin An-rich rims (with higher Fe, Mg, Ti, Sr and Sr/Ba) on Population 2 plagioclase (Fig. 6), show that the two plagioclase populations were juxtaposed at a late stage of magmatic development, immediately prior to eruption. $^{230}Th-^{226}Ra$ disequilibria measurements also suggest that each plagioclase population has distinctly different ages (Eppich *et al.* 2012), with the larger Population 2 grains, having average ages of >4.5–5.5 ka, with cores that extend to older than 10 ka (the oldest resolvable age for the $^{230}Th-^{226}Ra$ system). The ages of smaller Population 1 plagioclase are more difficult to estimate but are <3 ka and likely close to the eruption ages. Two populations of amphibole are also apparent, with magnesiohornblende crystallized from mafic melts formed at 10–16 km depth and at temperatures of *c.* 900–1000 °C, and tschermakitic pargasites from more felsic melt at 3.5–7 km and *c.* 800–870 °C (data from Koleszar (2011) recalculated using the calibration of Ridolfi & Renzulli 2012).
- Quenched mafic enclaves are widespread and occur in almost all lavas. These typically have quenched groundmass textures, lower SiO_2 contents than their host lava (Fig. 4) and higher proportions of phenocrysts derived from mafic melt, although there is also evidence for substantial crystal exchange. Enclaves are viewed as partially hybridized samples of unmixed mafic melt left following recharge and mixing (e.g. Bacon 1986).
- Whole-rock compositions of Mount Hood lavas define linear trends on bivariate plots between major element oxides (Fig. 4), suggesting mixing between mafic and felsic magmas. The compositions of the endmember compositions that define these arrays have remained broadly constant through time (although there is greater diversity for the mafic magma in terms of trace element compositions; Woods 2004; Koleszar 2011). Plots of highly incompatible v. highly compatible trace elements (e.g. Rb v. Cr; Fig. 7) also show linear or fan-shaped trends rather than the strongly curved trends predicted by crystal fractionation.

In order to investigate the dynamics of magmatic systems dominated by recharge and mixing it is important to establish the compositions of the parental magmas involved (e.g. Sparks & Marshall 1986). However, this can also be difficult as, even with the simple binary mixing such as that observed at Mount Hood, it is easy to define a line along which parental compositions must lie – which is simply the extension of the mixing array outside the compositional space defined by mixed lava compositions – but more difficult to establish exactly where along these arrays the parental compositions lie. Previous studies on other systems have used mass balance approaches based on mineral chemistry, whole-rock analyses, glass inclusions and quenched mafic enclaves (e.g. Nakamura 1995; Holtz *et al.* 2005; Humphreys *et al.* 2013), although these approaches also have their limitations. At Mount Hood major element oxides are highly correlated with the proportion of Population 1 plagioclase (that derived from a mafic source)

Table 1. *Summary of erupted magmas and magma storage conditions for Mount Hood, Mount Unzen, Mont Pelée and Soufriére Hills volcanoes*

		Mount Hood	Mount Unzen	Mont Pelée	Soufriére Hills
Duration of andesitic-dominated eruptions		*c.* 500 ka*	*c.* 500 ka*	At least 2.5 ka and possibly up to 14 ka	*c.* 175 ka
Range of erupted magma compositions[†]		54–66 wt% SiO_2 (Timberline and Old Maid periods, *c.* 1500 and *c.* 220 years BP *c.* 62–64 wt% SiO_2)	58–67 wt% SiO_2 (1992–1995: 59–65 wt% SiO_2)	57–66 wt% SiO_2 (1902–1930: 58–66 wt% SiO_2)	58–68 wt% SiO_2 (1995–2007: 59–65 wt% SiO_2)
Range of mafic inclusion compositions		50–60 wt% SiO_2	51–59 wt% SiO_2	51–58 wt% SiO_2	51–55 wt% SiO_2
Dominant modal phases		Plagioclase, amphibole, orthopyroxene, clinopyroxene (Fe–Ti oxides, olivine, apatite)	Plagioclase, amphibole, biotite, clinopyroxene, orthopyroxene, quartz, magnetite (Fe–Ti oxides, olivine)	Plagioclase, orthopyroxene; clinopyroxene; amphibole, (Fe–Ti oxides, olivine)	Plagioclase, amphibole, orthopyroxene, Fe–Ti oxides
Crystallinity		*c.* 25–60 vol%	17–48 vol%	30–50 vol%	*c.* 40 vol%
Estimated resident magma composition (see text for method)		68–76 wt% SiO_2	72–77 wt% SiO_2	67–72 wt% SiO_2	65–70 wt% SiO_2
Estimated temperature and pressure from amphibole compositions[‡]	*Magnesio-hornblende*	100–200 MPa (3.5–7 km) 790–900 °C	110–160 MPa (3.9–5.6 km) 790–850 °C	100–180 MPa (3.5–6.3 km) *c.* 850 °C	100–160 MPa (3.5–5.6 km) 720–850 °C
	Tschermakitic pargasite	300–500 MPa (*c.* 10.5–17.5 km) 910–900 °C	300–400 MPa (10.5–14 km) 900–950 °C	340–500 MPa (12–17.5 km) 910–1050 °C	350–530 MPa (12.3–18.6 km) 910–950 °C
Geophysical estimates of magma residence depth	Shallow	1.8–8.8 km	5–8 km		5–7 km
	Deep		*c.* 11 km		12–17 km
Phase equilibria constraints on magma storage	Shallow			150–250 MPa (5.2–8.8 km)	50–100 MPa (1.8–3.5)
	Deep			*c.* 400 MPa (*c.* 14 km)	

See text for references unless noted.
*This is the life of the current edifice.
[†]Total range and most recent eruptions are shown. For Mount Hood this includes the two most recent prehistoric eruptive phases.
[‡]Estimates are from data summarized by Koleszar (2011) and calculated using the calibrations of Ridolfi & Renzulli (2012).

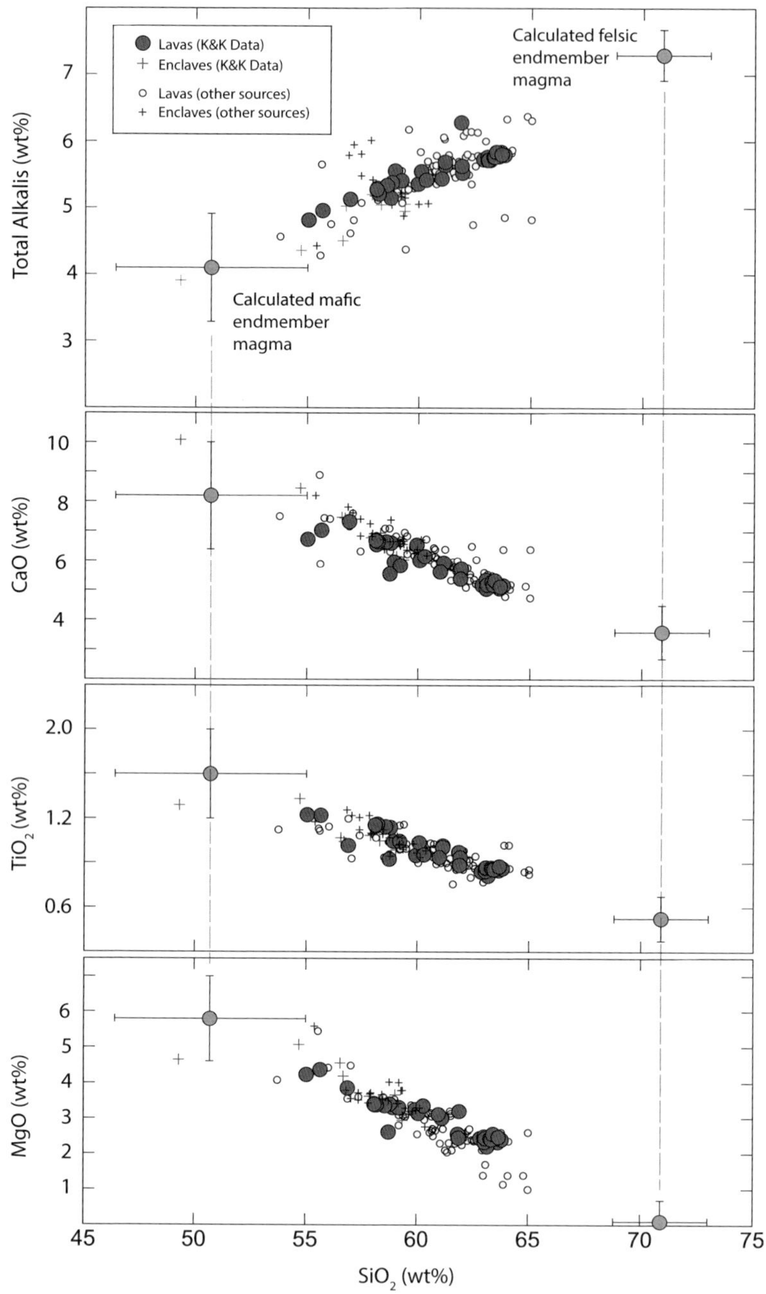

Fig. 4. Total alkalis, CaO, TiO_2 and MgO v. SiO_2 for Mount Hood lava and mafic enclave samples. Data from this study are reported in Kent *et al.* (2010) and Koleszar (2011) and are identified separately ('K&K Data') as all come from the same analytical facility; data from other sources are from GEOROC. Calculated mafic and felsic parental magma compositions and uncertainties are from Kent *et al.* (2010).

present in each lava (as estimated from CSD measurements). This is precisely what would be expected from mixing of two parental magmas of differing crystallinity and composition. Extension of these regressions to 0 and 100% of Population 1 plagioclase provides an estimate of the parental

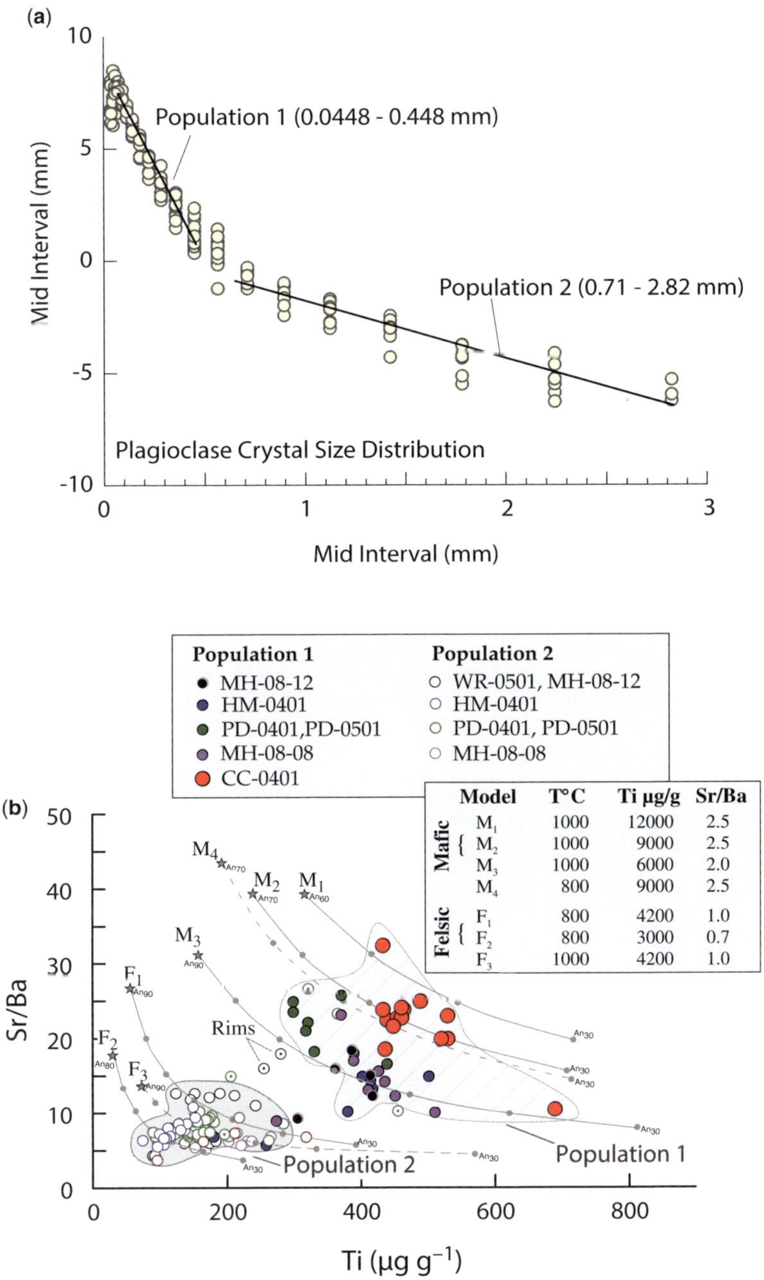

Fig. 5. Textural and compositional populations in plagioclase from Mount Hood lavas. (**a**) Crystal size distributions (CSD) for plagioclase from eight lavas ranging in age from *c.* 470 000 to 220 years (Kent *et al.* 2010). Populations 1 and 2 plagioclase are defined on the basis of crystal length as shown. Solid lines represent linear regressions for crystals from 0.0448–0.448 mm (Population 1) and 0.710–2.82 mm (Population 2). (**b**) Sr/Ba v. Ti ($\mu g\ g^{-1}$) measured in Populations 1 and 2 plagioclase crystals, modified from Kent *et al.* (2010). Also shown are curves representing equilibrium partitioning between plagioclase and a range of representative mafic and felsic melt compositions (see inset table). Anorthite contents of equilibrium plagioclase are shown starting at An_{30} (see Kent *et al.* 2010 for more details). Each point marked along the curves represents 10 mol. percent higher An content. Open circle symbols with dot in the middle represent rims to Population 2 plagioclase crystals. Data and further sample details are in Kent *et al.* (2010).

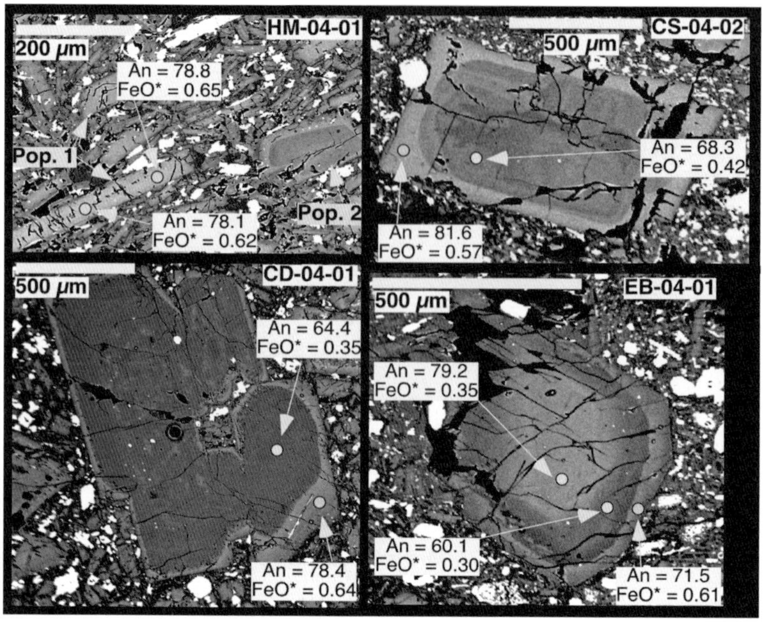

Fig. 6. Backscattered electron (BSE) photos of typical textures from four Mount Hood lava samples (see Kent *et al.* 2010 for sample details). Anorthite (An) and total iron (FeO*) contents are marked for selected points. Morphologies representing Populations 1 and 2 plagioclase are shown for HM-04-01. All other photos show Population 2 plagioclase.

magma compositions involved (Kent *et al.* 2010), and shows that these parental magmas have SiO_2 contents of 50 ± 4 and 72 ± 4 wt% SiO_2 (the uncertainties reflect the uncertainties in regressions). Note that all other lavas lie along similar trends in bivariate plots of major element

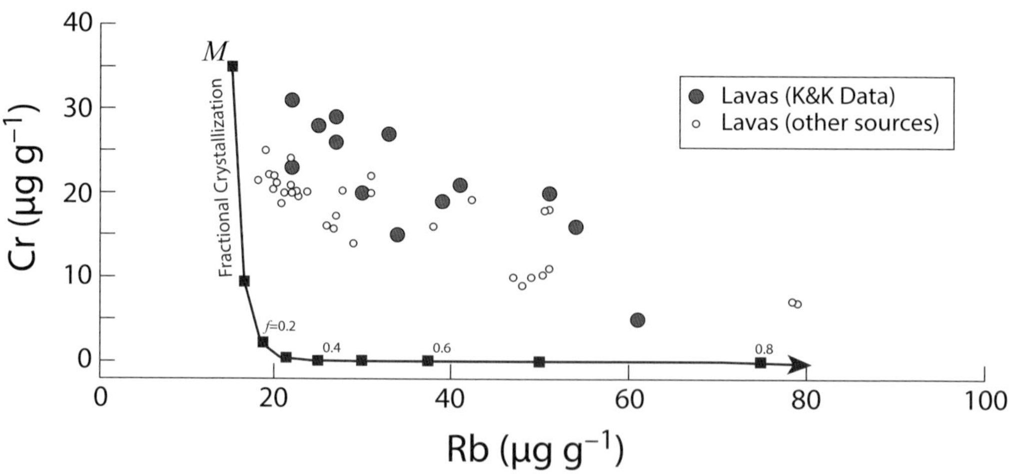

Fig. 7. Rb v. Cr for Mount Hood lavas. Data sources as per legend and Figure 4. Also shown is a calculated fractional crystallization trend of representative mafic melt (*M*) with and initial composition of 15 $\mu g\ g^{-1}$ Rb and 35 $\mu g\ g^{-1}$ Cr. Calculations used bulk mineral-melt partition coefficients of $D_{Rb} = 0.01$ and $D_{Cr} = 13$, calculated for crystallization of an assemblage consisting of 0.55 (number refers to modal mass fraction) plagioclase, 0.15 orthopyroxene, 0.05 clinopyroxene, 0.20 amphibole, and 0.05 titanomagnetite, using mineral-melt partition coefficient from Rollinson 1993). Crystal fraction values (*f*) are marked on the fractional crystallization trend.

composition (e.g. Fig. 4), suggesting that broadly similar parental magma compositions are common to all magmas erupted at Mount Hood.

The estimated compositions of parental magmas from Mount Hood are shown on the major element oxide v. silica plots in Figure 4. Importantly there is little or no overlap between the erupted magma compositions and those of the parental magmas. Mafic enclaves extend to more mafic compositions and are close to the estimated mafic parental magma, although only cumulate texture enclaves fall within the mafic parent magma range. Thus, although mafic and parental magmas demonstrably contribute to each of the erupted lavas at Mount Hood they appear to *never erupt by themselves*, and an obvious conclusion is that, although both mafic and felsic precursors are key participants in magma genesis, they are themselves *unable to erupt*.

Another feature of Mount Hood lavas that is important for the interpretation of the petrogenesis are mineral zoning profiles that reveal the timing of magma mixing events. This is most evident in the larger Population 2 plagioclase crystals that exhibit narrow (<*c.* 50–100 μm) rims of plagioclase with higher anorthite contents (*c.* An_{50-70}) and higher Mg, Fe, Ti, Sr and Sr/Ba (Figs 5 & 6). These rims have compositions that approach those of the smaller Population 1 plagioclase that derive from the mafic parental melt, and are the result of plagioclase growth following mixing, whereas the interior of Population 2 plagioclase grew within the felsic parental magma. The compositions that are evident as rims are not observed within the interior of Population 2 plagioclase grains, suggesting that the only time that mixing occurs is immediately prior to eruption – and thus strongly implies that mixing and eruption are linked processes. Diffusion of Mg across the high An rims on Population 1 crystals can be used to estimate the time at which rims remained at high temperature prior to eruption and quenching, and this timescale ranges between a few days to a few weeks (Kent *et al.* 2010).

Mixing proportions

One additional feature of the homogeneity of Mount Hood magmas that must be explained is the broad constancy of the proportions with which mafic and felsic parental melts are combined to produce the erupted andesitic magmas. A homogeneous series of magmas will not necessarily result from mixing processes alone, as theoretically any composition along the length of the mixing array can be produced. Thus, there must be other mechanisms that restrict the possible range of mixing proportions (Eichelberger *et al.* 2006). From the estimated parental magma compositions at Mount Hood it is possible to calculate the proportion of felsic magma added to mafic (f_{felsic}) to produce each magma composition. As shown in Figure 8 these are quite restricted; over 90% of the erupted lavas have f_{felsic} values that lie within 0.5 ± 0.2 (calculated using SiO_2 contents, although mixing proportions estimated using other major elements are similar). Thus, not only is it apparently difficult for the endmember parental mafic and silicic magmas to erupt at Mount Hood, but it also appears that mixed magmas with relatively small admixtures (<20%) of either parental melt either do not form or, if they do, are also unable to erupt.

The apparent restriction of mixing proportions probably reflects, at least in part, the requirement that any two magmas must have similar viscosities after thermal equilibration in order to promote efficient mixing and hybridization (e.g. Sparks & Marshall 1986; Koyaguchi & Blake 1989; Jellinek *et al.* 1999; Jellinek & Kerr 1999). Sparks & Marshall (1986) suggest that there is a 'window' of mixing proportions, controlled by the temperature and crystallinity of the mixing magmas, where hybridization is strongly favoured. Application of the approach set out by Sparks & Marshall (1986) to Mount Hood using the calculated parental mafic and silicic magma compositions (Fig. 8) suggests that the mixing proportions where the thermally equilibrated parental magmas will have similar viscosities (and where the silicic parent is crystal-rich) closely match the peak in the observed mixing proportions. Thus the ability of magmas to mix and hybridize during recharge probably exerts an important influence on the composition of hybrid magmas erupted at Mount Hood and other andesitic volcanoes.

Furthermore, the observation that only hybridized magmas erupt at Mount Hood suggests that there is also a link between magma mobility and efficient magma mixing. One simple option, as noted by Kent *et al.* (2010) is that the mixed magmas occupy a 'goldilocks' zone in terms of physical properties that allow them to be sufficiently mobile to erupt. Addition of small amounts of mafic material to felsic magmas, or vice versa, may not sufficiently modify viscosity or density to surmount the same barriers to eruption that restrict the parental magmas. Several studies (Ruprecht & Bachmann 2010; Takeuchi 2011; Koleszar *et al.* 2012) have highlighted the role of recharge by hotter mafic magmas in reducing the viscosity of cooler crystal-rich magma stored at shallow crustal levels. For Mount Hood the estimated thermal effects of recharge reduce the viscosity of silicic parental magma by a factor of four (Koleszar *et al.* 2012). In this instance the effectiveness of recharge in reducing the viscosity may also play an important

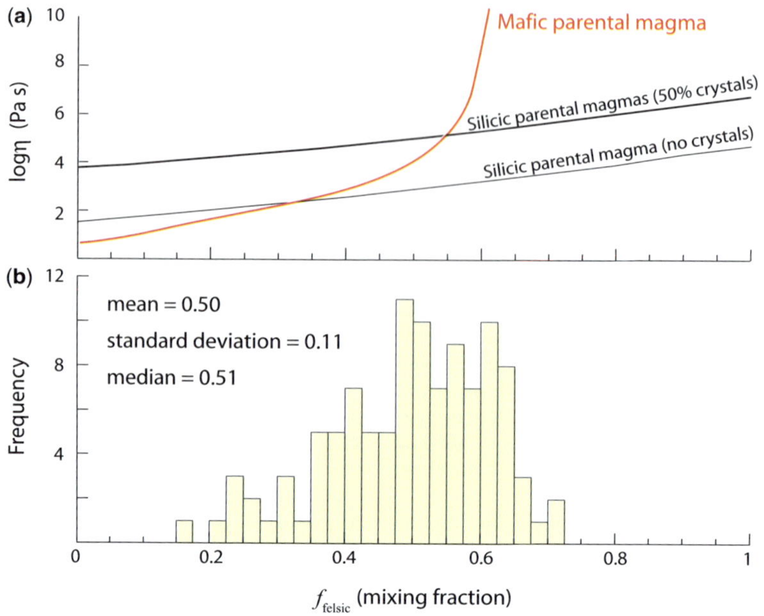

Fig. 8. (**a**) Calculated viscosity (log η) for mafic and felsic parental magma compositions v. f_{felsic}, the proportion of felsic parental magma present in mixed magmas. Curves represent the viscosities after thermal equilibration but prior to hybridization and were calculated following the method outlined in Sparks & Marshall (1986). The approach of Giordano *et al.* (2008) was used to estimate the silicate liquid viscosities and the MELTS algorithm used to estimate the crystallinity of the mafic parent magma during cooling below a liquidus temperature of 1144 °C. Calculations were performed for 50% and zero crystals for the felsic parent composition. (**b**) Calculated mixing proportions of felsic and mafic parent magmas contributing to erupted Mount Hood magma compositions, based on SiO_2 contents and the estimated parental magma compositions from Kent *et al.* (2010).

role in promoting effusive styles of eruption (Ruprecht & Bachmann 2010; Koleszar *et al.* 2012).

Other examples of compositionally restricted andesitic volcanoes

Several other well-characterized intermediate volcanic systems are remarkably similar to Mount Hood, in that they produce compositionally-restricted andesitic magmas over significant periods (several thousand years or more), but erupt magmas with clear evidence of mixing between mafic and felsic parental magmas during mafic recharge. Collectively, these volcanoes are described here as *compositionally restricted andesitic volcanoes*. Three volcanoes of this type – Mount Unzen, Mont Pelée and Soufriere Hills – are particularly well known, partly due to historic or recent eruptive activity.

The compositional range of lavas and mafic enclaves from Mount Unzen, Soufriére Hills and Mont Pelée are shown in in Figure 9 and summarized in Table 1. In these three volcanoes the long-term eruptive output is dominated by andesitic magmas (e.g. Gourgaud *et al.* 1989; Smith & Roobol 1990; Murphy *et al.* 1998, 2000; Nakada & Motomura 1999; Venezky & Rutherford 1999; Pichavant *et al.* 2002; Zellmer *et al.* 2003; Sugimoto *et al.* 2005; Davidson & Wilson 2011), with SiO_2 contents typically between 58 wt% and 66 wt%. Most erupted magmas also contain abundant quenched mafic enclaves with compositions that are more mafic than their host magmas (50–59 wt%, Table 1, Fig. 9 (Gourgaud *et al.* 1989; Smith & Roobol 1990; Nakada *et al.* 1999; Murphy *et al.* 2000; Browne *et al.* 2006)). Erupted magmas lie on linear compositional arrays and also contain multiple mineral populations, consistent with magma mixing (e.g. Fig. 9; Gourgaud *et al.* 1989; Smith & Roobol 1990; Nakada *et al.* 1999, Sato *et al.* 1999; Venezky & Rutherford 1999; Browne *et al.* 2006; Martel *et al.* 2006; Humphreys *et al.* 2009, 2010; Koleszar 2011). Zoning patterns in plagioclase, oxides and other minerals also show the late influence of mafic recharge as an eruption trigger (Devine *et al.* 2003, Nakamura 1995; Murphy *et al.* 2000; Stewart & Fowler 2001; Browne *et al.* 2006; Humphreys *et al.* 2009). Estimates of the proportions and compositions of the

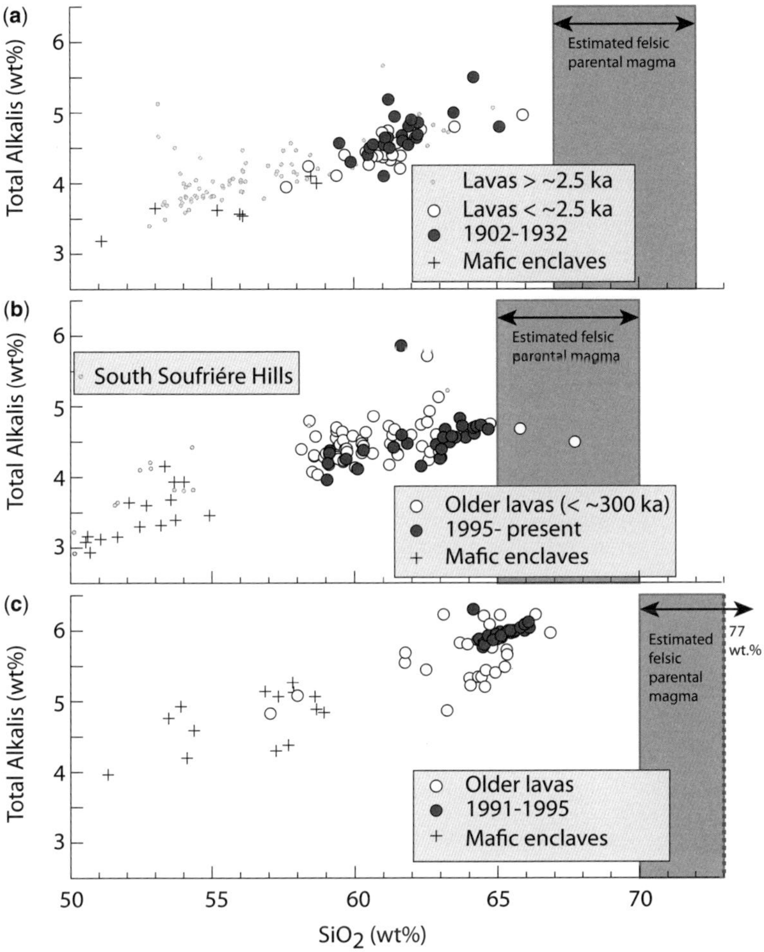

Fig. 9. Total alkali v. SiO_2 for three other compositionally restricted andesitic volcanoes. (**a**) Mont Pelée, Martinique (data sources: Gourgaud *et al.* 1989; Smith & Roobol 1990; Pichavant *et al.* 2002). Lavas from the most recent phase of volcanism (younger than *c.* 2.5 ka; Smith & Roobol 1990) represent a series of crystal-rich andesites. Older lavas are more variable and more mafic (note that Gourgaud *et al.* (1989) and Pichavant *et al.* (2002) consider that the homogeneous andesitic eruptions commenced at *c.* 14 ka.) (**b**) Soufriére Hills, Montserrat (data sources: Murphy *et al.* 1998, 2000; Zellmer *et al.* 2003). (**c**) Mount Unzen, Japan (data from GEOROC). In all plots the range of SiO_2 estimated for felsic parental melts following the approach in the text is shown by a grey field. For (c) the composition of this field extends off scale to 77 wt% SiO_2.

magmas involved in mixing vary (e.g. Nakamura 1995; Murphy *et al.* 2000; Pichavant *et al.* 2002; Devine *et al.* 2003; Holtz *et al.* 2005; Sugimoto *et al.* 2005; Humphreys *et al.* 2013). However, recognition of the ubiquitous presence of microlites and microphenocrysts derived from the recharging mafic magma suggests that the proportion of mafic magma involved may be greater than previously considered (Browne *et al.* 2006; Martel *et al.* 2009; Humphreys *et al.* 2009, 2013), and that the resident magma is also more evolved in composition. A rough estimate for the composition of resident magmas can be made by assuming that mixing to produce the erupted magmas occurs in broadly 50:50 proportions of recharge and resident magma, as suggested by the dynamics of mixing between mafic and felsic magmas (e.g. Sparks & Marshall 1986), and as occurs at Mount Hood (see discussion above; Fig. 8). For Mont Pelée, Soufriére Hills and Mount Unzen resident magma compositions can be estimated by assuming $f_{felsic} = 0.5$–0.6, by using the composition of mafic enclaves (51–55 wt% SiO_2) as the composition of the mafic recharging magma, and the median values of 61,

62 and 65 wt% SiO_2 of erupted lavas from the most recent eruptive episodes (Fig. 9). This approach suggests bulk rock SiO_2 contents of the resident magmas of 67–72, 65–70 and 72–77 wt% for Mont Pelée, Soufriére Hills and Mount Unzen, respectively (also see Table 1). These ranges overlap with the composition of the felsic parental magma given above for Mount Hood, and suggest that resident magmas are relatively silicic in all these systems (ranging overall from dacite to rhyolite) and rarely, if ever, erupt directly. Our estimates are more SiO_2-rich than previous studies that suggested that resident magmas were broadly similar to the composition of erupted andesites (e.g. Nakamura 1995; Murphy *et al.* 1998, 2000; Pichavant *et al.* 2002; Devine *et al.* 2003) – although many of these estimates were made prior to the realization of the extent of mafic material present within mixed magmas. For example, recent work by Humphreys *et al.* (2013) suggests that there is *c.* 6% of ‘cryptic’ mafic-derived crystalline material in Soufriére Hills andesite (not including those in quenched mafic enclaves). Although it is difficult to estimate the amount of mafic liquid that must accompany these crystals, it is likely significant. Using the typical crystallinities observed in arc basalts of 20–30% (Ewart 1976) suggests that the composition of the resident magma (calculated using the same compositions for mafic enclaves and andesites used above) has SiO_2 contents between 65 wt% and 68 wt%.

Amphibole barometry also reveals remarkably similarities in magmatic architecture between Mont Pelée, Soufriére Hills, Mount Unzen and Mount Hood. Two compositionally-distinct populations of amphibole occur in erupted magmas from all these volcanoes (Table 1; Fig. 10; Koleszar 2011): (1) tschermakitic pargasite which crystallized from more mafic magma (molar Al/Si_{melt} *c.* 0.3–0.4) at 900–990 °C and 300–500 MPa and (2) magnesiohornblende which crystallized from 770 to 870 °C and <200 MPa from a more silicic host magma (molar Al/Si_{melt} *c.* 0.15–0.25). Note that all amphibole classifications given herein use the nomenclature scheme of Leake *et al.* (1997) and Ridolfi *et al.* (2010), and calculation of pressure, temperature and molar Al/Si_{melt} follows the methods given in Koleszar (2011) and Ridolfi & Renzulli (2012). Estimates of the depth of magma storage from phase equilibria, gas and geophysical observations (e.g. Nakada *et al.* 1995; Aspinall *et al.* 1998; Barclay *et al.* 1998; Devine *et al.* 1998; Nishi *et al.* 1999; Pichavant *et al.* 2002; Elsworth *et al.* 2008; Christopher *et al.* 2010; Foroozan *et al.* 2010) overlap with pressures calculated from amphibole (Table 1), showing that in all these volcanoes hotter mafic magmas ascend from deeper sources to mix with silicic resident magma within the shallow crust. The depth of shallow magma storage of the felsic magma also overlaps with the *c.* 4–10 km depth estimated by Annen *et al.* (2006) for viscous stalling of the ascent of silicic melts derived from within the lower and middle crust.

It is also important to note that this basic architecture evident at Mount Hood and other compositionally restricted andesite volcanoes appears to be a consistent global feature of subduction zone volcanic systems. A compilation of amphibole compositions of intermediate lavas from convergent margin reveals that calculated pressures, temperature and co-existing liquid compositions (using the method of Ridolfi & Renzulli 2012) are also strongly bimodal (e.g. Fig. 10c). One population of amphibole forms between *c.* 100 and 200 MPa, at 770–850 °C and from relatively felsic liquids, the second population formed at 300–425 MPa and 900–970 °C, and from more mafic liquids. These closely mirror the amphibole populations observed at Mount Hood and the other andesitic volcanoes discussed above (e.g. Fig. 10c). The widespread presence of these bimodal amphibole populations emphasizes the global importance of mixing between deep and hot mafic magmas and shallowly stored felsic magmas in the production of intermediate magmas, and is further evidence for the important role of mixing processes in the formation of many andesitic volcanic rocks.

Other examples of compositionally restricted intermediate volcanoes that show evidence for an important role for hybridization and mixing include Mount Dutton, Alaska (Miller *et al.* 1999), the Plat Plays complex in the Lesser Antilles (Halama *et al.* 2006) and Volcán Aucanchilcha, Chile (Klemetti & Grunder 2008; Walker *et al.* 2012).

The recharge filtering model for andesitic volcanoes

Any model for magma genesis at Mount Hood and other compositionally restricted andesitic volcanoes must explain the following observations: (1) the broad homogeneity of magma compositions through time; (2) the macro- and microscopic features which indicate that magma mixing between mafic and felsic magmas plays a key role in magma genesis, and that these parental magmas seem to rarely erupt without mixing; (3) the observation that magma mixing often occurs immediately prior to eruption, which is initiated by episodes of mafic recharge.

Kent *et al.* (2010) proposed a model to account for the behaviour of Mount Hood via a process they termed *recharge filtering*. This is a reference

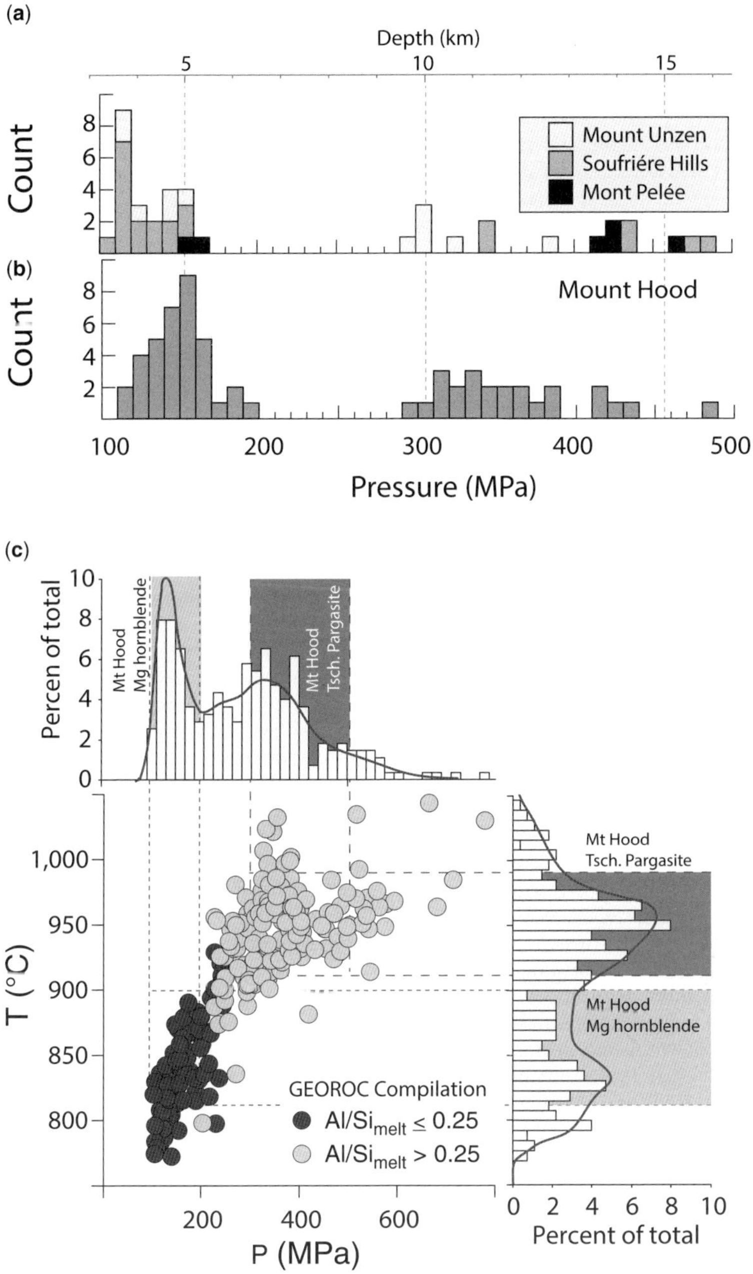

Fig. 10. Summary of crystallization pressures calculated from amphibole for (**a**) Mount Unzen, Mont Pelée and Soufriére Hills, (**b**) Mount Hood and (**c**) a global summary of pressure, temperatures calculated from amphiboles from convergent margin volcanoes from GEOROC. Samples have been grouped according to calculated Al/Si_{melt} values, with Al/Si_{melt} <0.25 corresponding to more silicic liquids. Pressure, temperature and Al/Si_{melt} estimates have been calculated using the procedure of Ridolfi & Renzulli (2012). Amphibole compositions from GEOROC were filtered to include only analyses from the cores of amphibole phenocrysts from subduction-related volcanic rocks. The global compilation includes data from 300 different amphiboles taken from 15 arcs and more than 50 different volcanic centres. Fields showing the range of pressures calculated for Mount Hood amphiboles are also shown for reference. For (a) and (b) all amphibole crystallized at pressures greater than 250 MPa are tschermakitic pargasite, all those at lower pressures are magnesiohornblende, see Koleszar (2011) for more details.

both to the general phenomenon of eruption filtering, and the recognition that at Mount Hood, as with many other andesitic volcanoes, it is mafic recharge that is the driving force for initiating eruptions.

The observation that mafic and silicic parental magmas involved in magma genesis at compositionally-restricted andesitic volcanoes appear unable to erupt without mixing suggests that there are significant barriers to the movement and eruption of these parental magmas within the shallow crust. This is readily understood in terms of the physical properties expected for these magmas, but is also probably heavily influenced by the nature of the crust itself. Felsic magmas are typically cooler and difficult to mobilize due to high crystallinities and viscosities (e.g. Marsh 1981; Rubin 1995; Takeuchi 2011; Fig. 8), and mafic magmas are restricted by their higher densities relative to many crustal lithologies and by difficulty in propagating through (or near) overpressured magma reservoirs (e.g. Wiebe 1994; Pinel & Jaupart 2000; Eichelberger *et al.* 2006; Hildreth 2007; Karlstrom *et al.* 2009). Crustal limits on magma mobility also likely reflect regional and local crustal stress fields, basement lithology, thermal state and strength of the crust and the presence of existing magma reservoirs and the volcanic edifice. These factors are also likely to interact as the changes in magmatic overpressure and the size of the edifice influence local stress conditions (e.g. Pinel & Jaupart 2000; Karlstrom *et al.* 2009; Kenedi *et al.* 2010).

In volcanic systems where crustal and other conditions reduce magma mobility and limit the ability of magmas to erupt, mafic recharge appears to be a common means by which eruptions can initiate. Mafic recharge is widely recognized as a powerful means of mobilizing magmas and initiating eruptions, and has been documented in a considerable number of volcanic systems (e.g. Sparks *et al.* 1977; Eichelberger 1980; Huppert *et al.* 1982; Pallister *et al.* 1992*a*, 1996). During recharge increases in magma reservoir overpressure (required to propagate a fracture to the surface) and magma mobility (required to allow magma to move along the fracture or conduit without freezing) results from thermal and volatile exchanges between mixing magmas, changes in volatile solubility, addition of material to a fixed-volume magma chamber and changes in physical properties of the new mixed magma (e.g. Sparks *et al.* 1977; Eichelberger 1978, 1980; Huppert *et al.* 1982; Snyder & Tait 1996; Ruprecht *et al.* 2008; Ruprecht & Bachmann 2010; Burgisser & Bergantz 2011; Takeuchi 2011; Koleszar *et al.* 2012).

In addition to initiating eruptions, mafic recharge is also often associated with rapid and intense hybridization between the mafic recharging magma and the magma or crystal-rich mush (typically more felsic in composition) residing in shallow storage zones. Petrographic, experimental and numerical studies demonstrate that vigorous convective overturn and other dynamic processes associated with recharge and magma migration can produce highly hybridized magmas (Eichelberger 1980; Huppert *et al.* 1982; Sparks & Marshall 1986; Snyder & Tait 1996; Jellinek & Kerr 1999; Eichelberger *et al.* 2006; Ruprecht *et al.* 2008; Burgisser & Bergantz 2011). Studies of mineral compositions in many andesitic systems suggest that the timescales of mixing and hybridization following recharge are rapid (e.g. Nakamura 1995; Venezky & Rutherford 1999; Devine *et al.* 2003; Browne *et al.* 2006; Martel *et al.* 2006; Salisbury *et al.* 2008; Humphreys *et al.* 2009, 2010; Kent *et al.* 2010; Koleszar 2011; Ruprecht & Cooper 2012). Less complete hybridization can also result in magma mingling textures and the presence of mafic enclaves, which are also common in andesitic magmas (e.g. Bacon 1986; Sparks & Marshall 1986).

Where recharge involves mafic magma mixing with a more silicic composition, efficient hybridization must also produce a bulk magma with a composition that is intermediate between the two, producing mixed magmas of intermediate composition (e.g. Walker & Skelhorn 1966; Eichelberger *et al.* 2006). Thus an important link exists between the increased probability of eruption via mafic recharge and the production and eruption of mixed andesitic magmas. Kent *et al.* (2010) termed this effect *recharge filtering* as it results in filtering of erupted magmas in favour of the mixed andesitic compositions over the original more diverse parental magma compositions. As a result recharge filtering involves the preferential production *and* eruption of andesitic magmas.

A recharge filtering explanation for compositionally restricted andesitic volcanoes

The recharge filtering model provides a straightforward answer to the question of how compositionally-restricted andesitic volcanoes form – in that these are systems where crustal or other conditions are highly restrictive for mobility and eruption of more mafic and silicic parental melts. As a result the only means by which eruption can occur is via mafic recharge, which results only in eruption of hybridized andesites, effectively filtering the volcanic record in favour of these compositions.

Eichelberger (1978) argued that compressional settings common in convergent margins also favour andesite production by promoting recharge and, although this may be broadly true, compositionally-restricted andesitic volcanoes appear to occur in a range of local stress regimes, including extensional (Unzen) and transtensional (Soufriére Hills). Mount Hood also lies within a local extensional zone, close to the propagating tip of the High Cascade graben, although compressional faults related to the northward movement of the Oregon coast block also occur in the vicinity of the edifice (Scott *et al.* 1997). This variability underscores the likelihood that the ultimate combinations of crustal conditions that promote or retard magma mobility are probably complex.

Parameters, such as integrated magma flux and the age and maturity of a given magma plumbing system, are also influential as they control thermal conditions within the shallow crust (e.g. Annen *et al.* 2006; Klemetti & Grunder 2008; Gregg *et al.* 2012), and establishment of silicic magma chambers in the shallow crust are likely to be particularly important for retarding mobility of mafic magmas (e.g. Wiebe 1994; Eichelberger *et al.* 2006; Karlstrom *et al.* 2009). It is probable that magma plumbing systems with relatively low and/or sporadic magma fluxes are more likely to be restricted by recharge filtering, as in these cases it is conditions within the crust that will have the primary influence on magma mobility. Smaller magma storage zones would also be able to react faster following recharge, reducing the time between recharge and eruption (e.g. Miller *et al.* 1999; Ruprecht & Wörner 2007). Where magma fluxes and magma accumulation rates are higher, heat flow from magma to the surrounding crust can strongly influence the temperature and viscosity of wallrocks (e.g. Annen *et al.* 2006; Lipman 2007; Gregg *et al.* 2012) and, for larger systems, the triggering of eruptions can occur via thermal weakening and collapse of the crust overlying the magma reservoir (e.g. Gregg *et al.* 2012). In addition, in some systems (e.g. Mount Pinatubo) it appears that mafic recharge results in mixing to produce andesites, but conditions within the shallow magma reservoir are such that the resident magma may also be mobilized and erupt in significant quantities (e.g. Pallister *et al.* 1992*a*, 1996; Eichelberger *et al.* 2006).

Previous explanations for compositionally restricted andesitic volcanism

The occurrence of volcanoes that produce consistent intermediate magma compositions through time has been noted previously in the literature (e.g. Pichavant *et al.* 2002; Annen *et al.* 2008; Klemetti & Grunder 2008). In the most detailed studies conducted to date, Pichavant *et al.* (2002) and Annen *et al.* (2008) argued that the relative constancy of andesitic compositions erupted by Mont Pelée over the last 13 000 years suggests that the physical state of magma beneath Mont Pelée has remained largely constant over this time, with temperature maintained by periodic input of hotter mafic material. Annen *et al.* (2008) used thermal models to estimate the magma flux required to maintain temperatures above 875 °C (as estimated from experimental petrology; Pichavant *et al.* 2002) at $4–5 \times 10^{-4}$ km^3 a^{-1}. Although frequent intrusions of mafic magma could maintain constant temperatures, it is unlikely that this represents a general model to explain compositionally-restricted andesitic volcanoes. In many andesitic systems, including Mont Pelée, the restriction of crystalline material derived from mafic magmas to the rims of larger plagioclase phenocrysts and microlite-sized crystals (e.g. Eichelberger 1978; Gourgaud *et al.* 1989; Browne *et al.* 2006; Martel *et al.* 2006; Humphreys *et al.* 2009, 2013; Kent *et al.* 2010), as well as the limited diffusional equilibration (Nakamura *et al.* 1995; Venezky & Rutherford 1999; Devine *et al.* 2003; Kent *et al.* 2010), show that interaction with mafic melts occurs immediately prior to eruption, rather than semi-continuously throughout magmatic evolution (although some systems do show evidence for periodic influx of mafic material with only the final interaction being sufficient to initiate eruption; e.g. Ruprecht & Wörner 2007; Smith *et al.* 2009). The presence of microlite-sized plagioclase derived from mafic magma in many andesites is also inconsistent with long periods of storage after mafic intrusion, as it would be expected that textural and chemical equilibration would modify these small crystals.

Recharge filtering in other volcanoes

Recharge filtering is also unlikely to occur solely in compositionally restricted andesitic systems. The petrographic features evident at Mount Hood and other andesite-dominated volcanoes described above are not unique – and, in fact, features such as the presence of disequilibrium mineral assemblages, reverse zoning patterns and mineral dissolution, and quenched mafic enclaves are often considered defining features of intermediate magmas from convergent margins (e.g. Eichelberger 1978; Gill 1981). Petrological and other evidence that demonstrates the important role of mafic recharge in the formation of andesites and other intermediate magmas in convergent margins is also widespread (e.g. Anderson 1976; Eichelberger 1978; Gill 1981). Thus, even where volcanoes might be hosted in crustal settings that permit eruption of more diverse magma types, the strong link between mafic recharge, hybridization and initiation of eruption is still likely to result in preferential eruption of andesitic magmas. As a result many other convergent margin volcanoes show high relative abundances of andesites amongst their eruptive output (e.g. Ewart 1976; Gill 1981; Hildreth & Moorbath 1988; Hildreth 2007), even where more diverse compositions also occur. In these volcanoes the high overall abundance of andesitic magmas is consistent with recharge filtering and is not well explained by either the volumes of andesitic magma expected from progressive crystal fractionation of a basaltic parent – where andesitic magmas should be less abundant than more mafic magmas as they require greater extents of crystal fractionation (e.g. Marsh 1981), or from melting of crustal lithologies, which produce more silicic magma compositions for most crustal lithologies at reasonable degrees of melting (e.g. Sisson *et al.* 2005).

One interesting example is Mount St Helens, located *c.* 100 km to the NW of Mount Hood. Over the previous 300 000 years Mount St Helens has erupted a diverse range of magma compositions from basalts through to low silica rhyolite (Fig. 3; *c.* 48–70 wt% SiO_2; Clynne *et al.* 2008). In particular the significant proportion of lavas with SiO_2 <55 wt% suggests that at times mafic magmas have been able to erupt with little or no mixing with evolved melts. In addition, some magmas such as the 1980–1986 and 2004–2008 dacites (*c.* 65 wt% SiO_2; Pallister *et al.* 1992*b*, 2008) show little petrological evidence for mafic recharge, appearing to derive directly from melting of basaltic protolith within the lower crust (e.g. Smith & Leeman 1993). These recent eruptions also appear to have been initiated by mechanisms other than mafic recharge (e.g. Kent *et al.* 2007; Pallister *et al.* 2008). Despite this, significant volumes of andesites also occur within the eruptive output of the volcano, and these are considered to result from hybridizing between mafic and dacitic magmas (Pallister *et al.* 1992*b*, 2008; Smith & Leeman 1993; Clynne *et al.* 2008). Petrological evidence indicative of mafic recharge and hybridization is also relatively common (e.g. Clynne *et al.* 2008). Thus in this system it is likely that recharge filtering plays an important, but not dominant role.

Formation of andesites in convergent margins

The idea that recharge filtering results in preferential eruption of andesitic magmas also provides an explanation for the overall high abundance of andesitic magmas in convergent margin settings relative to other tectono-magmatic environments. Although andesitic magmas and their equivalents occur in other settings, they are generally subordinate to other magma types (e.g. Fig. 1b, c). Gill (1981) suggested that two conditions are required for orogenic andesites to predominate: (i) the presence of subduction-modified asthenosphere (but not necessarily active subduction) to produce parental magmas, and (ii) compressional stress or thick sialic crust to promote differentiation of mantle-derived magmas. Eichelberger (1978) also emphasized the role of compressional environments for andesite formation, but in this case argued that this promoted a dependence on mixing to produce andesites.

If it is accepted that mafic recharge and mixing drive production of a significant proportion of andesitic magmas in convergent margins and also promote their preferential eruption, then the high abundances of water and other volatiles in the basaltic melts that form in subduction zones are also likely to play a decisive role in andesite genesis. Petrological evidence, together with experimental and numerical studies, highlights the important role of density instabilities driven by vapour exsolution within an underplated mafic recharge magma to drive convective overturn associated with mafic recharge events (e.g. Eichelberger 1978, 1980; Huppert *et al.* 1982; Bacon 1986; Ruprecht *et al.* 2008; Burgisser & Bergantz 2011). Hydrous mafic magmas are key to this process as the higher volatile abundances are more liable to promote the density-driven overturn and subsequent mixing required to produce and erupt mixed andesitic magmas during recharge events. The volatile content of the silicic parental magmas is probably less important, and the melt component of silicic magmas produced in a range of tectonic settings is typically saturated or close to saturated with vapour at shallow pressures, due to partial melting of hydrous crustal lithologies or extensive differentiation involving anhydrous phases. Importantly, it is only in subduction zone environments where volatile-rich mafic magmas predominate (e.g. Wallace 2005) and where significant volumes of andesitic magmas occur. The likelihood of andesite production and eruption is correspondingly reduced in environments, such as intraplate volcanic fields, ocean islands and mid-ocean ridges, where mafic magmas are less hydrous.

Conclusions and outstanding issues

The global record of volcanism represents a tremendous resource for geologists, geochemists and others who seek to understand the inner workings of the Earth. Melting of the crust and mantle is strongly correlated with plate tectonic and other deep Earth processes. Moreover, as magma generation occurs deep within the crust and mantle, any direct observations of the processes that create magmas, and direct samples of the materials that undergo melting, are difficult or impossible to obtain. Volcanic rocks represent the endpoint of these processes and the material is conveniently transported to the surface of the Earth, where sampling and observation are possible. As a result the physical state and the mineralogical, chemical and isotopic compositions of volcanic rocks represent one of the primary observational datasets used to constrain the inner workings of the Earth. However, the volcanic record should not be considered to simply represent an unbiased sampling of magmas produced within a given setting, and it must always be remembered that volcanic rocks represent a special set of magmas that are capable of being erupted.

In this contribution I have promoted the idea that the high proportions of andesitic magmas erupted at convergent margin volcanoes need not directly reflect production of large volumes of andesitic melt produced by igneous differentiation or melting processes associated with subduction. Instead, many erupted andesitic magmas result from preferential eruption of andesitic magmas formed by mafic recharge and hybridization processes within the shallow crust. Although this does not rule out other modes of andesite formation, the process of preferential eruption via recharge filtering is likely to play an important, and possibly dominant role in controlling the high abundances of andesitic magma erupted from convergent margin volcanoes. The hydrous nature of mantle-derived magmas in subduction zones also favours andesite formation relative to other tectono-magmatic settings as this promotes the vigorous convection and overturn following mafic recharge required to initiate eruptions and to produce hybridized magmas. Andesite formation is also promoted by crustal conditions that limit the ability of more mafic and felsic parental magmas to erupt, and thus an important corollary of the recharge filtering concept is that local conditions within the crust, probably in concert with magma fluxes, play an important role in controlling the range and compositional proportions of magmas erupted at a given volcanic system. Volcanoes that erupt compositionally-restricted andesitic sequences through time probably reflect crustal settings where parental magmas are restricted from erupting by crustal density and viscosity filters or other factors and, thus, where recharge is the dominant form of eruption initiation. For volcanoes in crustal settings where a broader range of compositions erupt, the common prevalence of andesitic magmas formed by recharge and hybridization suggest that recharge filtering is still an important process. In addition, as crustal conditions such as lithology and stress field may change over short lateral length scales, it is likely that the role of recharge filtering varies considerably between neighbouring volcanoes, providing an explanation for the common juxtaposition of compositionally-restricted volcanoes with those that erupt more variable compositions and for the general variations observed in erupted compositions that occur within individual arc segments. It is also possible to envisage changes in crustal conditions and regional or local stress fields driven by far-field tectonic processes, by development of the volcanic plumbing system or by growth and collapse of the volcanic edifice. Thus, the degree to which eruption filtering controls erupted compositions at an individual volcano may also change, sometimes abruptly.

Another implication of the recharge filtering model is that in many convergent margins the rocks that constitute the volcanic record may be an unreliable guide to the magmatic outputs of the crustal magmatic systems involved. Andesitic and other intermediate volcanics that are formed from mixing and hybridization will be over-represented relative to the parental magmas that mix to form them. This paper has not focused on the plutonic record of convergent margin magmatism, and overall reconciliation of the volcanic and plutonic records of convergent margin magmatism remains an area of active debate (e.g. Bachmann *et al.* 2007; Lipman 2007). Recharge filtering and other forms of eruption filtering will result in significant differences between the volcanic and plutonic record of magma fluxes in convergent margins, and complicate the comparison of the plutonic and volcanic records. In addition – although the role of mixing and hybridization in the formation of magmas that occur in large batholiths in exhumed subduction terranes is often less evident than in many volcanic rocks – macroscopic and other indications of magma mixing in plutonic rocks are common, and mixing probably plays a prominent role in the formation of many intermediate magmas that occur in convergent margin batholiths (e.g. Sisson *et al.* 1996; Wiebe 1994; Reubi & Blundy 2009). Thus, it is also possible that mixing and magma mobility are also linked in the construction of batholiths in convergent margins.

Several other important avenues related to this topic await further study. One of the most important is the development and application of techniques to identify the compositions and physical state of the parental magmas prior to mafic recharge. Modern micro-analytical techniques provide unprecedented access to the compositions of minerals and melts (via melt inclusions and matrix glasses) within hybridized magmas. Although these studies are driving an increasing appreciation of the role of mixing and hybridization in andesite genesis, there are still limitations. In particular, it remains difficult to estimate the compositions and the physical conditions of the parental magmas that are involved in hybridization. These parameters are critical for understanding the dynamics of recharge, mixing and eruption, but are also difficult to estimate from mineral or melt inclusion chemistry alone. Mineral compositions can be inverted to estimate the compositions of the liquid from which they crystallize, but it can be difficult to determine the crystallinity of the magma involved, which is key to understanding viscosity. Mineral thermometry can also be used to estimate temperatures of various minerals or mineral pairs, but these reflect the conditions at the time of crystallization of the mineral or most recent re-equilibration. For the case of a felsic magma or mush present within the shallow crust, significant cooling to solidus or subsolidus conditions may occur after mineral formation and prior to recharge, and thus the state of the magma at the time of remobilization may differ substantially from that inferred by mineral equilibria.

One way to address these issues is to combine multiple lines of inquiry. For example recent studies combine textural quantification using crystal size distributions or other approach with mineral and bulk rock geochemistry (e.g. Salisbury *et al.* 2008; Kent *et al.* 2010). Kent *et al.* (2010) combined CSD, mineral and bulk rock chemistry to estimate parental magma compositions at Mount Hood. Martel *et al.* (2006), Salisbury *et al.* (2008) and Humphreys *et al.* (2013) have also demonstrated novel approaches for identifying crystalline contributions from mafic magma in hybrid lavas.

Further work is also needed to illuminate the dynamics of magma mixing and hybridization during mafic recharge and eruption. A basic understanding of these processes is already known from numerical, analogue and other approaches, but micro-analytical and related studies are increasingly revealing the high degree to which andesitic magmas are hybridized, with mineral components derived from parental magmas intimately juxtaposed. In particular mafic magma-derived microlite-size crystals are now known to occur throughout many andesites, but important questions remain as to whether mixing occurs via hybridization of two magmas (e.g. Sparks & Marshall 1986) or disaggregation and shear of quenched mafic enclaves (e.g. Clynne 1999; Humphreys *et al.* 2009). Timing constraints also suggest that such efficient mixing is also extremely rapid – occurring within days to weeks prior to eruption (e.g. Nakamura 1995; Venezky & Rutherford 1999; Devine *et al.* 2003; Browne *et al.* 2006; Salisbury *et al.* 2008; Kent *et al.* 2010; Ruprecht & Cooper 2012) and, although models that describe such mixing are now being developed (e.g. Huber *et al.* 2009; Burgisser & Bergantz 2011), understanding this rapid timescale remains a challenge. Finally, there is also the puzzling observation that for long-lived eruptions mixing timescales appear to be shorter than the duration of the eruption (Nakamura 1995; Venezky & Rutherford 1999; Devine *et al.* 2003), suggesting that, despite the compositional homogeneity of individual eruptive units, mixing is a continual process throughout the eruptions.

The author thanks many colleagues for discussions regarding the origin of andesites, and especially A. Koleszar, C. Darr, K. Cooper, G. Eppich, W. Scott and C. Gardner for collaboration on studies of Mount Hood. Members of the Oregon State University VIPER group also contributed to lively discussions on andesites and other topics. Comments from R. Lange and an anonymous reviewer, together with comments and editorial suggestions from A. Gómez-Tuena, improved the manuscript significantly. This work was supported by National Science Foundation grant EAR 1144555.

References

Almeev, R. R., Ariskin, A. A., Kimura, J.-I. & Barmina, G. in press. The role of polybaric crystallization in genesis of andesitic magmas: phase equilibria simulations of the Bezymianny volcanic subseries. *Journal of Volcanology and Geothermal Research*. http://dx.doi.org/10.1016/j.jvolgeores.2013.01.004

Anderson, A. T. 1976. Magma mixing: Petrological process and volcanological tool. *Journal of Volcanology and Geothermal Research*, **1**, 3–33.

Annen, C., Blundy, J. D. & Sparks, R. S. J. 2006. The genesis of intermediate and silicic magmas in deep crustal hot zones. *Journal of Petrology*, **47**, 505–539.

Annen, C., Pichavant, M., Bachmann, O. & Burgisser, A. 2008. Conditions for the growth of a long-lived shallow crustal magma chamber below Mount Pelee volcano (Martinique, Lesser Antilles Arc). *Journal of Geophysical Research – Solid Earth*, **113**, http://dx.doi.org/10.1029/2007jb005049

Arce, J. L., Gardner, J. E. & Macías, J. L. 2013. Pre-eruptive conditions of dacitic magma erupted during the 21.7 ka Plinian event at Nevado de Toluca volcano, Central Mexico. *Journal of Volcanology and Geothermal Research*, **249**, 49–65.

Aspinall, W. P., Miller, A. D., Lynch, L. L., Latchman, J. L., Stewart, R. C., White, R. A. & Power, J. A. 1998. Soufriere Hills eruption, Montserrat, 1995–1997: volcanic earthquake locations and fault plane solutions. *Geophysical Research Letters*, **25**, 3397–3400.

Bachmann, O., Miller, C. F. & De Silva, S. L. 2007. The volcanic-plutonic connection as a stage for understanding crustal magmatism. *Journal of Volcanology and Geothermal Research*, **167**, 1–23.

BACON, C. R. 1986. Magmatic inclusions in silicic and intermediate volcanic rocks. *Journal of Geophysical Research – Solid Earth and Planets*, **91**, 6091–6112.

BACON, C. R. & METZ, J. 1984. Magmatic inclusions in rhyolites, contaminated basalts, and compositional zonation beneath the Coso Volcanic Field, California. *Contributions to Mineralogy and Petrology*, **85**, 346–365.

BARCLAY, J., RUTHERFORD, M. J., CARROLL, M. R., MURPHY, M. D., DEVINE, J. D., GARDNER, J. & SPARKS, R. S. J. 1998. Experimental phase equilibria constraints on pre-eruptive storage conditions of the Soufriere Hills magma. *Geophysical Research Letters*, **25**, 3437–3440.

BROWNE, B. L., EICHELBERGER, J. C., PATINO, L. C., VOGEL, T. A., UTO, K. & HOSHIZUMI, H. 2006. Magma mingling as indicated by texture and Sr/Ba ratios of plagioclase phenocrysts from Unzen volcano, SW Japan. *Journal of Volcanology and Geothermal Research*, **154**, 103–116.

BURGISSER, A. & BERGANTZ, G. W. 2011. A rapid mechanism to remobilize and homogenize highly crystalline magma bodies. *Nature*, **471**, 212–297.

CARMICHAEL, I. S. E. 2002. The andesite aqueduct: perspectives on the evolution of intermediate magmatism in west-central (105–99 degrees W) Mexico. *Contributions to Mineralogy and Petrology*, **143**, 641–663.

CHRISTOPHER, T., EDMONDS, M., HUMPHREYS, M. C. S. & HERD, R. A. 2010. Volcanic gas emissions from Soufriere Hills Volcano, Montserrat 1995–2009, with implications for mafic magma supply and degassing. *Geophysical Research Letters*, **37**, http://dx.doi.org/10.1029/2009GL041325

CLYNNE, M. A. 1999. A complex magma mixing origin for rocks erupted in 1915, Lassen Peak, California. *Journal of Petrology*, **40**, 105–132.

CLYNNE, M. A., CALVERT, A. T., WOLFE, E. W., EVARTS, R. C., FLECK, R. J. & LANPHERE, M. A. 2008. The pleistocene eruptive history of Mount St. Helens, Washington – from 300 000 to 12 800 years ago. *In*: SHERROD, D. R., SCOTT, W. E. & STAUFFER, P. H. (eds) *A Volcano Rekindled; the Renewed Eruption of Mount St. Helens, 2004–2006*. United Stated Geological Survey, Virginia, Professional Paper 1750.

COOMBS, M. L., SISSON, T. W. *ET AL.* 2013. Andesites of the 2009 eruption of Redoubt Volcano, Alaska. *Journal of Volcanology and Geothermal Research*, **259**, 349–372.

COSTA, F., ANDREASTUTI, S., BOUVET DE MAISONNEUVE, C. & PALLISTER, J. S. 2013. Petrological insights into the storage conditions and magmatic processes that yielded the centennial 2010 Merapi explosive eruption. *Journal of Volcanology and Geothermal Research*, **261**, 209–235.

CRIBB, J. W. & BARTON, M. 1997. Significance of crustal and source region processes on the evolution of compositionally similar calc-alkaline lavas, Mt Hood, Oregon. *Journal of Volcanology and Geothermal Research*, **76**, 229–249.

DARR, C. M. 2006. *Magma Chamber Processes Over the Past 475 000 years at Mount Hood, Oregon: Insights from Crystal Zoning and Crystal Size Distribution Studies*. MS thesis, Oregon State University.

DAVIDSON, J. & WILSON, M. 2011. Differentiation and source processes at Mt Pelee and the Quill; active volcanoes in the lesser antilles arc. *Journal of Petrology*, **52**, 1493–1531.

DEVINE, J. D., MURPHY, M. D. *ET AL.* 1998. Petrologic evidence for pre-eruptive pressure-temperature conditions, and recent reheating, of andesitic magma erupting at the Soufriere Hills Volcano, Montserrat, WI. *Geophysical Research Letters*, **25**, 3669–3672.

DEVINE, J. D., RUTHERFORD, M. J., NORTON, G. E. & YOUNG, S. R. 2003. Magma storage region processes inferred from geochemistry of Fe-Ti oxides in andesitic magma, Soufriere Hills Volcano, Montserrat, WI. *Journal of Petrology*, **44**, 1375–1400.

DOGAN, A. U., DOGAN, M., PEATE, D. W. & DOGRUEL, Z. 2011. Textural and mineralogical diversity of compositionally homogeneous dacites from the summit of Mt Erciyes, Central Anatolia, Turkey. *Lithos*, **127**, 387–400.

EICHELBERGER, J. C. 1978. Andesitic volcanism and crustal evolution. *Nature*, **275**, 21–27.

EICHELBERGER, J. C. 1980. Vesiculation of mafic magma during replenishment of silicic magma reservoirs. *Nature*, **288**, 446–450.

EICHELBERGER, J. C., IZBEKOV, P. E. & BROWNE, B. L. 2006. Bulk chemical trends at arc volcanoes are not liquid lines of descent. *Lithos*, **87**, 135–154.

ELSWORTH, D., MATTIOLI, G., TARON, J., VOIGHT, B. & HERD, R. 2008. Implications of magma transfer between multiple reservoirs on eruption cycling. *Science*, **322**, 246–248.

EPPICH, G. R., COOPER, K. M., KENT, A. J. R. & KOLESZAR, A. 2012. Constraints on crystal storage timescales in mixed magmas: Uranium-series disequilibria in plagioclase from Holocene magmas at Mount Hood, Oregon. *Earth and Planetary Science Letters*, **317**, 319–330.

EWART, A. 1976. Mineralogy and chemistry of modern Orogenic Lavas – some statistics and implications. *Earth and Planetary Science Letters*, **31**, 417–432.

FOROOZAN, R., ELSWORTH, D., VOIGHT, B. & MATTIOLI, G. S. 2010. Dual reservoir structure at Soufriere Hills Volcano inferred from continuous GPS observations and heterogeneous elastic modeling. *Geophysical Research Letters*, **37**, http://dx.doi.org/10.1029/2010GL042511

GILL, J. B. 1981. *Orogenic Andesites and Plate Tectonics*. Springer, Berlin.

GIORDANO, D., RUSSELL, J. K. & DINGWELL, D. B. 2008. Viscosity of magmatic liquids: a model. *Earth and Planetary Science Letters*, **271**, 123–134.

GÓMEZ-TUENA, A., STRAUB, S. M. & ZELLMER, G. F. (eds) 2014. *Orogenic Andesites and Crustal Growth*. Geological Society, London, Special Publications, **385**.

GOURGAUD, A., FICHAUT, M. & JORON, J. L. 1989. Magmatology of Mt Pelee (Martinique, Fwi).1. Magma mixing and triggering of the 1902 and 1929 Peleean Nuees-Ardentes. *Journal of Volcanology and Geothermal Research*, **38**, 143–169.

GREGG, P. M., DE SILVA, S. L., GROSFILS, E. B. & PARMIGIANI, J. P. 2012. Catastrophic caldera-forming eruptions: thermomechanics and implications for eruption triggering and maximum caldera dimensions on Earth. *Journal of Volcanology and Geothermal Research*, **241–242**, 1–12.

HALAMA, R., BOUDON, G., VILLEMANT, B., JORON, J.-L., LE FRIANT, A. & KOMOROWSKI, J.-C. 2006. Pre-eruptive crystallization conditions of mafic and silicic magmas at the Plat Pays volcanic complex, Dominica (Lesser Antilles). *Journal of Volcanology and Geothermal Research*, **153**, 200–220.

HILDRETH, W. 2007. *Quaternary magmatism in the Cascades – geologic perspectives*. Professional Papers of the United States Geological Survey, **1744**, 14.

HILDRETH, W. & MOORBATH, S. 1988. Crustal contributions to arc magmatism in the Andes of Central Chile. *Contributions to Mineralogy and Petrology*, **98**, 455–489.

HIROTANI, S., BAN, M. & NAKAGAWA, M. 2009. Petrogenesis of mafic and associated silicic end-member magmas for calc-alkaline mixed rocks in the Shirataka volcano, NE Japan. *Contributions to Mineralogy and Petrology*, **157**, 709–734.

HOLTZ, F., SATO, H., LEWIS, J., BEHRENS, H. & NAKADA, S. 2005. Experimental petrology of the 1991–1995 Unzen dacite, Japan. Part I: phase relations, phase composition and pre-eruptive conditions. *Journal of Petrology*, **46**, 319–337.

HUBER, C., BACHMANN, O. & MANGA, M. 2009. Homogenization processes in silicic magma chambers by stirring and mushification (latent heat buffering). *Earth and Planetary Science Letters*, **283**, 38–47.

HUMPHREYS, M. C. S., CHRISTOPHER, T. & HARDS, V. 2009. Microlite transfer by disaggregation of mafic inclusions following magma mixing at Soufriere Hills volcano, Montserrat. *Contributions to Mineralogy and Petrology*, **157**, 609–624.

HUMPHREYS, M. C. S., EDMONDS, M., CHRISTOPHER, T. & HARDS, V. 2010. Magma hybridisation and diffusive exchange recorded in heterogeneous glasses from Soufriere Hills Volcano, Montserrat. *Geophysical Research Letters*, **37**, http://dx.doi.org/10.1029/2009GL041926

HUMPHREYS, M. C. S., EDMONDS, M., PLAIL, M., BARCLAY, J., PARKES, D. & CHRISTOPHER, T. 2013. A new method to quantify the real supply of mafic components to a hybrid andesite. *Contributions to Mineralogy and Petrology*, **165**, 191–215.

HUPPERT, H. E., SPARKS, R. S. J. & TURNER, J. S. 1982. Effects of volatiles on mixing in calc-alkaline magma systems. *Nature*, **297**, 554–557.

JELLINEK, A. M. & KERR, R. C. 1999. Mixing and compositional stratification produced by natural convection 2. Applications to the differentiation of basaltic and silicic magma chambers and komatiite lava flows. *Journal of Geophysical Research – Solid Earth*, **104**, 7203–7218.

JELLINEK, A. M., KERR, R. C. & GRIFFITHS, R. W. 1999. Mixing and compositional stratification produced by natural convection 1. Experiments and their application to Earth's core and mantle. *Journal of Geophysical Research – Solid Earth*, **104**, 7183–7201.

KARLSTROM, L., DUFEK, J. & MANGA, M. 2009. Organization of volcanic plumbing through magmatic lensing by magma chambers and volcanic loads. *Journal of Geophysical Research – Solid Earth*, **114**, http://dx.doi.org/10.1029/2009JB006339

KENEDI, C. L., SPARKS, R. S. J. *ET AL*. 2010. Contrasts in morphology and deformation offshore Montserrat: new insights from the SEA-CALIPSO marine cruise data. *Geophysical Research Letters*, **37**, http://dx.doi.org/10.1029/2010GL043925

KENT, A. J. R. 2008. Melt inclusions in basaltic and related volcanic Rocks. *Minerals, Inclusions and Volcanic Processes*, **69**, 273–331.

KENT, A. J. R., BLUNDY, J. *ET AL*. 2007. Vapour transfer prior to the October 2004 eruption of Mount St. Helens, Washington. *Geology*, **35**, 231–234.

KENT, A. J. R., DARR, C., KOLESZAR, A. M., SALISBURY, M. J. & COOPER, K. M. 2010. Preferential eruption of andesitic magmas through recharge filtering. *Nature Geoscience*, **3**, 631–636.

KLEMETTI, E. W. & GRUNDER, A. L. 2008. Volcanic evolution of Volcan Aucanquilcha: a long-lived dacite volcano in the Central Andes of northern Chile. *Bulletin of Volcanology*, **70**, 633–650.

KOLESZAR, A. 2011. *Controls on Eruption style and Magma Compositions at Mount Hood, Oregon*. PhD thesis, Oregon State University.

KOLESZAR, A. M., KENT, A. J. R., WALLACE, P. J. & SCOTT, W. E. 2012. Controls on long-term low explosivity at andesitic arc volcanoes: Insights from Mount Hood, Oregon. *Journal of Volcanology and Geothermal Research*, **219**, 1–14.

KOYAGUCHI, T. & BLAKE, S. 1989. The dynamics of magma mixing in a rising magma batch. *Bulletin of Volcanology*, **52**, 127–137.

KUNO, H. 1968. Origin of andesite and its bearing on the island arc structure. *Bulletin of Volcanology*, **32**, 141–176.

LEAKE, B. E., Woodley, A. R. *ET AL*. 1997. Nomenclature of amphiboles: report of the subcommittee on amphiboles of the International Mineralogical Association, commission on new minerals and mineral names. *American Mineralogist*, **82**, 1019–1037.

LIPMAN, P. W. 2007. Incremental assembly and prolonged consolidation of Cordilleran magma chambers: evidence from the Southern Rocky Mountain volcanic field. *Geosphere*, **3**, 42–70.

MARSH, B. D. 1981. On the crystallinity, probability of occurrence, and rheology of lava and magma. *Contributions to Mineralogy and Petrology*, **78**, 85–98.

MARTEL, C., ALI, A. R., POUSSINEAU, S., GOURGAUD, A. & PICHAVANT, M. 2006. Basalt-inherited microlites in silicic magmas: evidence from Mount Pelee (Martinique, French West Indies). *Geology*, **34**, 905–908.

MILLER, T. P., CHERTKOFF, D. G., EICHELBERGER, J. C. & COOMBS, M. L. 1999. Mount Dutton volcano, Alaska: Aleutian arc analog to Unzen volcano, Japan. *Journal of Volcanology and Geothermal Research*, **89**, 275–301.

MORI, J., EBERHART-PHILLIPS, D. & HARLOW, D. H. 1996. Three-dimensional velocity structure at Mount Pinatubo: Resolving magma bodies and earthquake hypocenters. *In*: NEWHALL, C. G. & PUNONGBAYAN, R. S. (eds) *Fire and Mud: Eruptions and Lahars of Mount Pinatubo, Philippines*. University of Washington Press, Seattle, 371–382.

MURPHY, M. D., SPARKS, R. S. J. *ET AL*. 1998. The role of magma mixing in triggering the current eruption at the Soufriere Hills volcano, Montserrat, West Indies. *Geophysical Research Letters*, **25**, 3433–3436.

MURPHY, M. D., SPARKS, R. S. J., BARCLAY, J., CARROLL, M. R. & BREWER, T. S. 2000. Remobilization of andesite magma by intrusion of mafic magma at the Soufriere Hills Volcano, Montserrat, West Indies. *Journal of Petrology*, **41**, 21–42.

NAKADA, S. & MOTOMURA, Y. 1999. Petrology of the 1991–1995 eruption at Unzen: effusion pulsation and groundmass crystallization. *Journal of Volcanology and Geothermal Research*, **89**, 173–196.

NAKADA, S., MOTOMURA, Y. & SHIMIZU, H. 1995. Manner of magma ascent at Unzen volcano (Japan). *Geophysical Research Letters*, **22**, 567–570.

NAKAI, S., MAEDA, Y. & NAKADA, S. 2008. Common origin of plagioclase in the last three eruptions of Unzen volcano, Japan. *Journal of Volcanology and Geothermal Research*, **175**, 181–188.

NAKAMURA, M. 1995. Continuous mixing of crystal mush and replenished magma in the ongoing Unzen eruption. *Geology*, **23**, 807–810.

NISHI, K., ONO, H. & MORI, H. 1999. Global positioning system measurements of ground deformation caused by magma intrusion and lava discharge: the 1990–1995 eruption at Unzendake volcano, Kyushu, Japan. *Journal of Volcanology and Geothermal Research*, **89**, 23–54.

OWNBY, S. E., LANGE, R. A. & HALL, C. M. 2008. The eruptive history of the Mascota volcanic field, western Mexico: age and volume constraints on the origin of andesite among a diverse suite of lamprophyric and calc-alkaline lavas. *Journal of Volcanology and Geothermal Research*, **177**, 1077–1091.

OWNBY, S. E., LANGE, R. A., HALL, C. M. & DELGADO-GRANADOS, H. 2011. Origin of andesite in the deep crust and eruption rates in the Tancitaro–Nueva Italia region of the central Mexican arc. *Geological Society of America Bulletin*, **123**, 274–294.

PALLISTER, J. S., HOBLITT, R. P., CRANDELL, D. R. & MULLINEAUX, D. R. 1992*a*. Mount St Helens a decade after the 1980 eruptions: magmatic models, chemical cycles, and a revised hazards assessment. *Bulletin of Volcanology*, **54**, 126–146.

PALLISTER, J. S., HOBLITT, R. P. & REYES, A. G. 1992*b*. A basalt trigger for the 1991 eruptions of Pinatubo volcano. *Nature*, **356**, 426–428.

PALLISTER, J. S., HOBLITT, R. P., MEEKER, G. P., KNIGHT, R. J. & SIEMS, D. F. 1996. Magma Mixing at Mount Pinatubo: petrographic and chemical evidence from the 1991 deposits. *In*: NEWHALL, C. G. & PUNONGBAYAN, R. S. (eds) *Fire and Mud: Eruptions and Lahars of Mount Pinatubo, Phhillipines*. University of Washington Press, Seattle, 687–732.

PALLISTER, J. S., THORNBER, C. R., CASHMAN, K. V., CLYNNE, M. A., LOWERS, H. A., MANDEVILLE, C. W. & MEEKER, G. P. 2008. Petrology of the 2004–2006 Mount St Helens lava dome – implications for magmatic plumbing and eruption triggering. *In*: SHERROD, D. R., SCOTT, W. E. & STAUFFER, P. H. (eds) *A Volcano Rekindled: The Renewed Eruption of Mount St Helens, 2004–2006*. United Stated Geological Survey, Virginia, Professional Paper 1750, 647–702.

PATIA, H. 2004. *Petrology and Geochemistry of the Recent Eruption History at Rabaul Caldera, Papua New Guinea: Implications for Magmatic Processes and Recurring Volcanic Activity*. Australian National University, Canberra, Australia.

PICHAVANT, M., MARTEL, C., BOURDIER, J. L. & SCAILLET, B. 2002. Physical conditions, structure, and dynamics of a zoned magma chamber: Mount Pelee (Martinique, Lesser Antilles Arc). *Journal of Geophysical Research – Solid Earth*, **107**, http://dx.doi.org/10.1029/2001JB000315

PINEL, V. & JAUPART, C. 2000. The effect of edifice load on magma ascent beneath a volcano. *Philosophical Transactions of the Royal Society of Edinburgh*, **358**, 1515–1532.

PRIEST, G. R. 1990. Volcanic and tectonic evolution of the cascade volcanic arc, central Oregon. *Journal of Geophysical Research-Solid Earth and Planets*, **95**, 19 583–19 599.

REUBI, O. & BLUNDY, J. 2009. A dearth of intermediate melts at subduction zone volcanoes and the petrogenesis of arc andesites. *Nature*, **461**, 1269–1273.

RIDOLFI, F. & RENZULLI, A. 2012. Calcic amphiboles in calc-alkaline and alkaline magmas: thermobarometric and chemometric empirical equations valid up to 1130A degrees C and 2.2 GPa. *Contributions to Mineralogy and Petrology*, **163**, 877–895.

RIDOLFI, F., RENZULLI, A. & PUERINI, M. 2010. Stability and chemical equilibrium of amphibole in calc-alkaline magmas: an overview, new thermobarometric formulations and application to subduction-related volcanoes. *Contributions to Mineralogy and Petrology*, **160**, 45–66.

ROLLINSON, H. 1993. *Using Geochemical Data: Evaluation, Presentation, Interpretation*. Longman, Essex.

RUBIN, A. M. 1995. Getting granite dikes out of the source region. *Journal of Geophysical Research – Solid Earth*, **100**, 5911–5929.

RUDNICK, R. L. 1995. Making continental-crust. *Nature*, **378**, 571–578.

RUPRECHT, P. & BACHMANN, O. 2010. Pre-eruptive reheating during magma mixing at Quizapu volcano and the implications for the explosiveness of silicic arc volcanoes. *Geology*, **38**, 919–922.

RUPRECHT, P. & COOPER, K. M. 2012. Integrating the uranium-series and elemental diffusion geochronometers in mixed magmas from Volcan Quizapu, Central Chile. *Journal of Petrology*, **53**, 841–871.

RUPRECHT, P. & WÖRNER, G. 2007. Variable regimes in magma systems documented in palgioclase zoning patterns: El Misti stratovolcano and Andahua monogenetic cones. *Journal of Volcanology and Geothermal Research*, **165**, 142–162.

RUPRECHT, P., BERGANTZ, G. W. & DUFEK, J. 2008. Modeling of gas-driven magmatic overturn: tracking of phenocryst dispersal and gathering during magma mixing. *Geochemistry Geophysics Geosystems*, **9**, http://dx.doi.org/10.1029/2008GC002022

RUPRECHT, P., BERGANTZ, G. W., COOPER, K. M. & HILDRETH, W. 2012. The crustal magma storage system of volcan Quizapu, Chile, and the effects of magma mixing on magma diversity. *Journal of Petrology*, **53**, 801–840.

SAKUYAMA, M. 1978. Evidence of magma mixing: petrological study of Shiroumaoike calc-alkalilne andesite volcano. *Journal of Volcanology and Geothermal Research*, **5**, 179–208.

SALISBURY, M. J., BOHRSON, W. A., CLYNNE, M. A., RAMOS, F. C. & HOSKIN, P. 2008. Origin of the 1915

Lassen Peak eruption by magma mixing: evidence for formation of chemically distinct plagioclase populations from crystal size distributions and in situ chemical data. *Journal of Petrology*, **49**, 1755–1780.

Sato, H., Nakada, S., Fujii, T., Nakamura, M. & Suzuki-Kamata, K. 1999. Groundmass pargasite in the 1991–1995 dacite of Unzen volcano: phase stability experiments and volcanological implications. *Journal of Volcanology and Geothermal Research*, **89**, 197–212.

Schmidt, M. E. & Grunder, A. L. 2009. The evolution of North Sister: a volcano shaped by extension and ice in the central Oregon Cascade Arc. *Geological Society of America Bulletin*, **121**, 643–662.

Scott, W. E., Gardner, C. A., Pierson, T. C., Schilling, S. P., Costa, J. E., Vallance, J. W. & Major, J. J. 1997. *Geologic history of Mount Hood Volcano, Oregon: a field-trip guidebook*. United States Geological Survey Open File Report, **97**.

Sisson, T. W., Grove, T. L. & Coleman, D. S. 1996. Hornblende gabbro sill complex at Onion Valley, California, and a mixing origin for the Sierra Nevada batholith. *Contributions to Mineralogy and Petrology*, **126**, 81–108.

Sisson, T. W., Ratajeski, K., Hankins, W. B. & Glazner, A. F. 2005. Voluminous granitic magmas from common basaltic sources. *Contributions to Mineralogy and Petrology*, **148**, 635–661.

Smith, D. R. & Leeman, W. P. 1993. The origin of Mount St Helens andesites. *Journal of Volcanology and Geothermal Research*, **55**, 271–303.

Smith, A. L. & Roobol, M. J. 1990. *Mt Pelée, Martinique: a study of an active island arc volcano*. Geological Society of America Memoirs, Boulder, **175**.

Smith, V. C., Blundy, J. D. & Arce, J. L. 2009. A temporal record of magma accumulation and evolution beneath Nevado de Toluca, Mexico, preserved in Plagioclase Phenocrysts. *Journal of Petrology*, **50**, 405–426.

Snyder, D. & Tait, S. 1996. Magma mixing by convective entrainment. *Nature*, **379**, 529–531.

Sparks, R. S. J. & Marshall, L. A. 1986. Thermal and mechanical constraints on mixing between mafic and silicic magmas. *Journal of Volcanology and Geothermal Research*, **29**, 99–124.

Sparks, S. R. J., Sigurdsson, H. & Wilson, L. 1977. Magma mixing – mechanism for triggering acid explosive eruptions. *Nature*, **267**, 315–318.

Sparks, R. S. J., Meyer, P. & Sigurdsson, H. 1980. Density variation amongst mid-ocean ridge basalts – implications for magma mixing and the scarcity of primitive lavas. *Earth and Planetary Science Letters*, **46**, 419–430.

Stewart, M. L. & Fowler, A. D. 2001. The nature and occurrence of discrete zoning in plagioclase from recently erupted andesitic volcanic rocks, Montserrat. *Journal of Volcanology and Geothermal Research*, **106**, 243–253.

Stolper, E. & Walker, D. 1980. Melt density and the average composition of basalt. *Contributions to Mineralogy and Petrology*, **74**, 7–12.

Straub, S. M., Gomez-Tuena, A., Stuart, F. M., Zellmer, G. F., Espinasa-Perena, R., Cai, Y. & Iizuka, Y. 2011. Formation of hybrid arc andesites beneath thick continental crust. *Earth and Planetary Science Letters*, **303**, 337–347.

Sugimoto, T., Ishibashi, H., Wakamatsu, S. & Yanagi, T. 2005. Petrologic evolution of Pre-Unzen and Unzen magma chambers beneath the Shimabara Peninsula, Kyushu, Japan: evidence from petrography and bulk rock chemistry. *Geochemical Journal*, **39**, 241–256.

Takeuchi, S. 2011. Preeruptive magma viscosity: an important measure of magma eruptibility. *Journal of Geophysical Research – Solid Earth*, **116**, http://dx.doi.org/10.1029/2011JB008243

Tepley, F. J., Davidson, J. P. & Clynne, M. A. 1999. Magmatic interactions as recorded in plagioclase phenocrysts of Chaos Crags, Lassen Volcanic Center, California. *Journal of Petrology*, **40**, 787–806.

Venezky, D. Y. & Rutherford, M. J. 1999. Petrology and Fe–Ti oxide reequilibration of the 1991 Mount Unzen mixed magma. *Journal of Volcanology and Geothermal Research*, **89**, 213–230.

Viccaro, M., Giuffrida, M., Nicotra, E. & Ozerov, A. Y. 2012. Magma storage, ascent and recharge history prior to the 1991 eruption at Avachinsky volcano, Kamchatka, Russia: inferences on the plumbing system geometry. *Lithos*, **140**, 11–24.

Walker, B. A. J., Klemetti, E. W., Grunder, A. L., Dilles, J. H., Tepley, F. J. & Giles, D. 2012. Crystal reaming during the assembly, maturation, and waning of an eleven-million-year crustal magma cycle: thermobarometry of the Aucanquilcha Volcanic Cluster. *Contributions to Mineralogy and Petrology*, **165**, 663–682.

Walker, G. P. L. & Skelhorn, R. R. 1966. Some associations of acid and basic ignbeous rocks. *Earth Science Reviews*, **2**, 93–109.

Wallace, P. J. 2005. Volatiles in subduction zone magmas: concentrations and fluxes based on melt inclusion and volcanic gas data. *Journal of Volcanology and Geothermal Research*, **140**, 217–240.

Wallace, P. J. & Carmichael, I. S. E. 1994. Petrology of Volcan-Tequila, Jalisco, Mexico – disequilibrium phenocryst assemblages and evolution of the subvolcanic magma system. *Contributions to Mineralogy and Petrology*, **117**, 345–361.

White, S. M., Crisp, J. A. & Spera, F. J. 2006. Long-term volumetric eruption rates and magma budgets. *Geochemistry Geophysics Geosystems*, **7**, http://dx.doi.org/10.1029/2005GC001002

Wiebe, R. A. 1994. Silicic magma chambers as traps for basaltic magmas – the Cadillac Mountain intrusive complex, Mount Desert Island, Maine. *Journal of Geology*, **102**, 423–437.

Woods, M. M. 2004. *Compositional and Mineralogical Relationships between Mafic Inclusions and Host Lavas as key to andesite petrogenesis at Mount Hood Volcano, Oregon*. MSc thesis, Portland State University.

Zellmer, G. F., Hawkesworth, C. J., Sparks, R. S. J., Thomas, L. E., Harford, C. L., Brewer, T. S. & Loughlin, S. C. 2003. Geochemical evolution of the Soufriere Hills volcano, Montserrat, Lesser Antilles volcanic arc. *Journal of Petrology*, **44**, 1349–1374.

The Strawberry Volcanics: generation of 'orogenic' andesites from tholeiite within an intra-continental volcanic suite centred on the Columbia River flood basalt province, USA

ARRON STEINER & MARTIN J. STRECK*

Department of Geology, Portland State University, PO Box 751, Portland, OR 97207-0751, USA

**Corresponding author (e-mail: streckm@pdx.edu)*

Abstract: The widely distributed, mid-Miocene lavas of the Strawberry Volcanics of NE Oregon are compositionally diverse, ranging from basalt to rhyolite. They are composed mainly of calc-alkaline and mildly tholeiitic basaltic andesite and andesite. Ar–Ar dating and stratigraphic relationships indicate that the volcanic field was active from >16 Ma to *c.* 12 Ma ago and thus is coeval for the first 1–2 Ma with strongly tholeiitic flood basalts of the Columbia River Province that encircle the Strawberry Volcanics. Tholeiitic and calc-alkaline compositions develop subtle but noticeable differences towards higher silica contents. At silica contents of <55 wt% SiO_2, calc-alkaline and tholeiitic lavas are essentially indistinguishable. Trace element constraints among Strawberry Volcanics and crustal rocks indicate that open-system processes, such as assimilation or magma mixing, are responsible for evolution along a calc-alkaline trend leading to 'orogenic' andesites from tholeiite. Exclusively tholeiitic basalts carrying evidence for a metasomatized mantle source erupted during the mid-Miocene of eastern Oregon. Consequently, tholeiite imparted the 'subduction signals' and crustal processing generated the calc-alkaline character to end up with compositions of typical 'orogenic' andesites at the Strawberry Volcanics. No primitive calc-alkaline basalt from the mantle is needed as parental magma here and possibly at other similar intra-continental calc-alkaline suites.

Magmas of calc-alkaline affinity (see Arculus 2003) are the typical magmas of subduction zone settings, yet they occur in other tectonic regimes. Because of their prevalence at active plate margins, the generation of the calc-alkaline trend has been closely tied to processes that occur as a consequence of the subduction of hydrated oceanic lithosphere. Key parameters that drive magmas toward calc-alkaline evolution are higher water content and higher degree of oxidation, which researchers have pointed out for many years (e.g. Hora *et al.* 2009; Zimmer *et al.* 2010 and references therein). At subduction zones, these two conditions likely exist at primary magma generation in the mantle wedge, although this has been questioned recently (Lee *et al.* 2005). In regions without active subduction (i.e. intra-continental or anorogenic settings), explanation of magmas with calc-alkaline character is more challenging, as there is no dehydrating slab that can introduce fluids to the overlying mantle. However, in such settings, partial melting of mantle that was previously affected by subduction zone magmatism during an earlier stage is thought to be the principal control (e.g. Hawkesworth *et al.* 1995; Hooper *et al.* 1995; Christiansen & McCurry 2008). Thus, even in such non-subduction settings, most workers would argue that primitive calc-alkaline basalts originating from the melting of a subduction modified (i.e. metasomatized) mantle are required in order to generate the calc-alkaline trend with andesite as their intermediate magmas. In this study, we address this issue and present data on generating voluminous calc-alkaline 'orogenic' andesites from tholeiitic parents instead of primitive mafic parents that already are calc-alkaline. Our case study is centred on a volcanic field located in NE Oregon, USA, called the Strawberry Volcanics, which consists of a diverse suite of calc-alkaline and mildly tholeiitic lavas (Figs 1 & 2). The Strawberry Volcanics are coeval with flood basalts of the Columbia River Basalt Group (CRBG) that were issued from vents geographically encircling the Strawberry Volcanics (Robyn 1977, this study), documenting that the prevalent magmas from the mantle during activity of the Strawberry Volcanics were tholeiitic. Our primary focus in this contribution is to document the distribution, age and composition of mafic to intermediate lavas of the Strawberry Volcanics, followed by a discussion of likely petrogenetic causes.

Methods

Major and selected trace elements were analysed using an X-ray fluorescence spectrometer (XRF) at Washington State University (WSU). Sample preparation followed the analytical procedures of the

From: GÓMEZ-TUENA, A., STRAUB, S. M. & ZELLMER, G. F. (eds) 2014. *Orogenic Andesites and Crustal Growth*. Geological Society, London, Special Publications, **385**, 281–302.
First published online August 6, 2013, http://dx.doi.org/10.1144/SP385.12

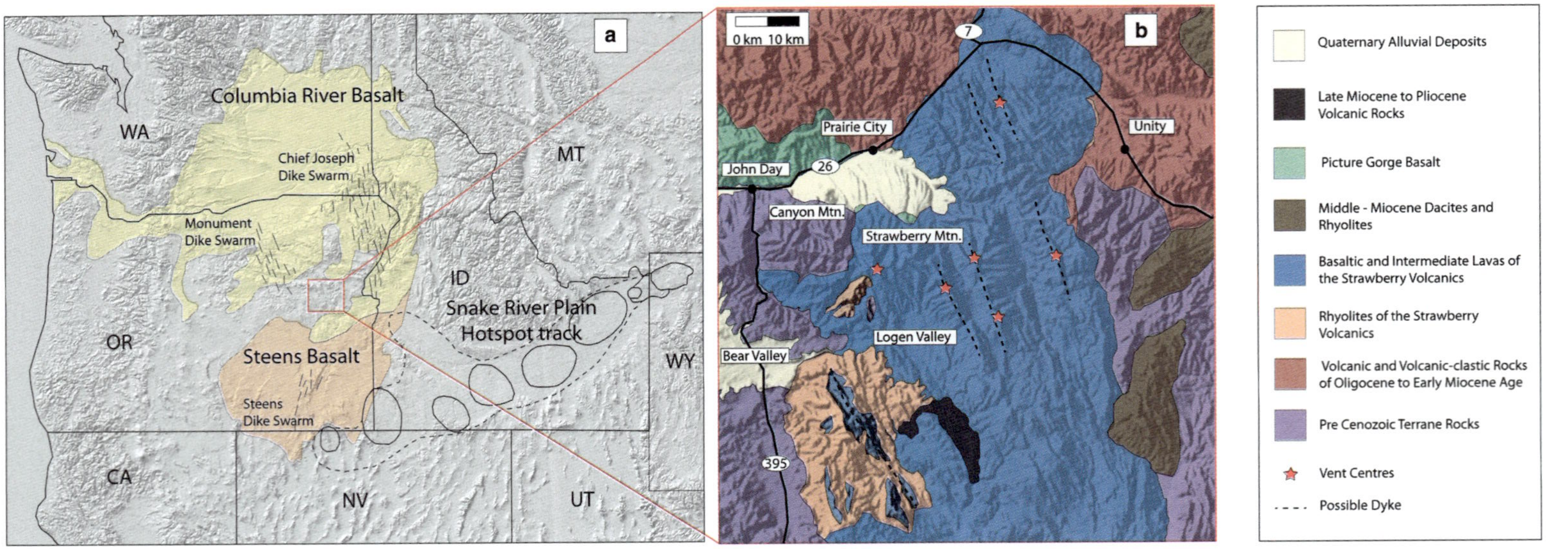

Fig. 1. (**a**) Regional map showing the location of Columbia River Basalt Group (CRBG) lavas including Steens Basalt in relation to study area (red square) with the Strawberry Volcanics. The solid circles are presumed rhyolitic centres associated with the Yellowstone hotspot (modified after Camp & Ross 2004). WA, Washington; OR, Oregon; CA, California; ID, Idaho; NV, Nevada; MT, Montana; WY, Wyoming; UT, Utah. (**b**) Simplified geological map from 44° to 44°30′ latitude and from 118° to 119° longitude of the Strawberry Volcanics and surrounding units. The geological units surrounding the Strawberry Volcanics are based on Thayer (1956), Crowley (1960), Brown & Thayer (1966*a*, *b*, 1977), Thayer *et al.* (1967), Greene *et al.* (1972), Robyn (1977), Brooks & Ferns (1979), Ferns *et al.* (1983) and Mullen (1983).

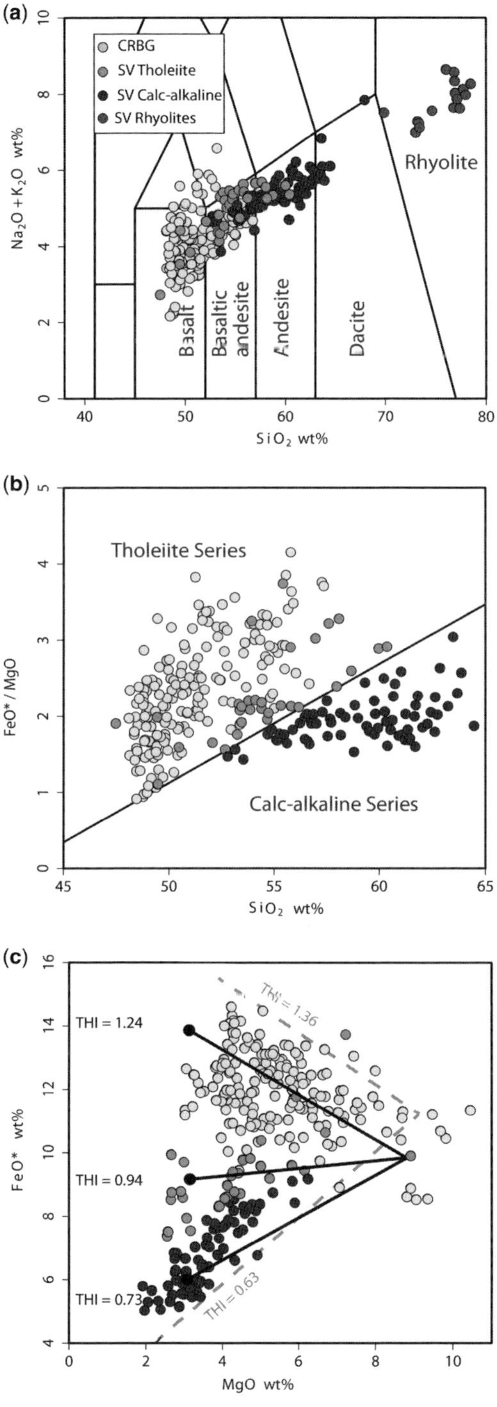

Fig. 2. Compositional spread of lavas of the Strawberry Volcanics (this study) and of the Columbia River Basalt Group (CRBG) (Wolff *et al.* 2008). (**a**) Total alkali–silica diagram (Le Bas *et al.* 1986). (**b**) FeO*/MgO v. SiO_2 diagram with tholeiitic/calc-alkaline fields as based on Miyashiro (1974). (**c**) FeO* v. MgO wt% diagram. Overall, FeO* wt% decreases with decreasing MgO wt% in the tholeiitic and calc-alkaline lavas of the Strawberry Volcanics, yet the tholeiitic lavas decrease less than the calc-alkaline. The most mafic tholeiitic lavas of the Strawberry Volcanics overlap considerably with the CRBG and transition towards calc-alkaline compositions as MgO wt% decreases. Calc-alkaline lavas of the Strawberry Volcanics have a Tholeiitic Index (THI) of 0.73 while tholeiitic basalt and intermediate lavas have a THI of 0.94. The tholeiitic index (THI) was developed by Zimmer *et al.* (2010) and is defined in the following way: THI = $Fe_{4.0}/Fe_{8.0}$ where $Fe_{4.0}$ is the average FeO* wt% concentration of samples with 4 ± 1 wt% MgO, and $Fe_{8.0}$ is the average FeO* wt% concentration at 8 ± 1 wt% MgO thus, samples with THI > 1 are tholeiitic and samples with THI < 1 are calc-alkaline. The three solid lines correspond with calculated THI values for the various suites, that is calc-alkaline and tholeiitic suites of the Strawberry Volcanic and the tholeiitic lavas of the CRBG if we use the composition of a primitive basalt of the Strawberry Volcanics for $Fe_{8.0}$. Alternatively, if we use an initially higher FeO* at 8 wt% MgO, this will lead to a reduced THI value for the Strawberry lavas (dashed grey line) and consequently to a stronger calc-alkaline signature.

WSU geochemical laboratory. Beads were fused at 1000 °C using a 2:1 dilithium tetraborate ($Li_2B_4O_7$) flux to powered sample. Selected trace elements and rare earth elements (REEs) were analysed using inductively-coupled plasma mass spectrometry (ICP-MS) at WSU. Samples were analysed following standard analytical procedures at the geochemical laboratory at WSU: samples are ground into fine powder and mixed with an equal amount of dilithium tetraborate flux and fused at 1000 °C. The beads are then re-ground to a fine-grained powder and dissolved for final analysis.

$^{40}Ar/^{39}Ar$ analyses were conducted on fresh groundmass samples which were handpicked after crushing and sieving whole-rock samples at the Noble Gas Mass Spectrometry Laboratory at Oregon State University. All samples were washed in a solution of diluted HCl (<5%) for approximately 10 min to remove any secondary carbonation. All samples with small quantities of mineral monitor FCT-3 biotite (28.030 ± 0.003 Ma ago, Renne *et al.* 1998) were loaded in quartz vials and irradiated at the OSU TRIGA research reactor. $^{40}Ar/^{39}Ar$ analysis was performed on samples using an MAP 215–50 gas mass spectrometer and the incremental heating method. Starting temperature of 400 °C was followed by 100 °C incremental increases until significant ^{39}Ar was released. Thereafter, temperature was increased in 50–100 °C steps until reaching 1400 °C. The isotopic data were reduced and age calculations were made using the ArArCALC software provided by Koppers (2002) using the currently accepted ^{40}K decay constant

(Steiger & Jäger 1977). Further details of the analytical procedures are described in Duncan & Keller (2004) and on the laboratory website (http://www.coas.oregonstate.edu/research/mgg/chronology.html).

Five samples were chosen to represent the large area of the Strawberry Volcanics. One sample from each cardinal direction (north, east, south and west), along with a central location, was chosen for $^{40}Ar/^{39}Ar$ analysis (Table 1). All samples were basaltic andesite to andesite in composition (53–58 wt% SiO_2) and field studies indicate that they represent the upper stratigraphy of the Strawberry Volcanics.

Results

The lavas of the Strawberry Volcanics are distributed over 3600 km^2 and make up a diverse volcanic suite which spans the range of compositions from basalt to rhyolite (Figs 1 & 2). Below, we present our current data on field relationships of the Strawberry Volcanics, when and for how long volcanic activity occurred, and the erupted magma types.

Stratigraphy and new ages of the Strawberry Volcanics

Field relationships. The Strawberry Volcanics were named after Strawberry Mountain located along the southeastern margin of the John Day Valley in NE Oregon (Fig. 1) (Thayer 1957). There, uplift and glaciation produced the largest continuously exposed sections consisting of 30 or more, mostly conformably overlying lava flows (Figs 3 & 4). Lava flow sections can reach thicknesses of approximately 1000 m. Where large continuous sections are exposed, thickness of individual lava flows varies dramatically from the lower to upper sections. Near the base of the Strawberry Volcanics lava thickness can reach >40 m while the upper sections thicknesses decrease to no greater than 15 m (Fig. 3). Independent of thickness, individual lava flow thickness appears nearly constant for the length of exposed section, which can reach several kilometres along fault blocks (Fig. 3). At Strawberry Mountain, andesitic lava flows dominate and are typically phenocryst-poor (<3%) to essentially aphyric. The groundmass includes microphenocrysts of plagioclase and pyroxene. Phenocrysts occasionally occur as crystal clots instead of single crystals of 1–3 mm dominated by plagioclase, orthopyroxene and minor clinopyroxene. Xenoliths >3 cm of the same mineral assemblage with plutonic textures have also been observed. Strawberry Volcanics are emplaced on top of pre-Tertiary accreted terranes of the Blue Mountain Province (Thayer 1957), Mesozoic plutonic rocks, and older Tertiary volcanic rocks thought to be mostly Oligocene in age (Thayer *et al.* 1967). More specifically, Strawberry Volcanics overlie Oligocene volcanic rocks to the north and east of Strawberry Mountain, and Palaeozoic marine sediment (Izee terrane) and ophiolitic sequences (Baker terrane) to the south and west (Fig. 1b). Oligocene volcanic rocks are typically phenocryst-rich and display large phenocrysts (>0.5–3 mm). This lithological contrast has been used to differentiate Strawberry Volcanics from older intermediate volcanic rocks. Silica undersaturated alkali basaltic lavas erupted within the area of this study and overlie Strawberry Volcanics in the southern portion and are thought to be late Miocene/Pliocene in age (Fig. 1b). In addition, several regional ignimbrites reach into the area. The 9.7 Ma Devine Canyon Tuff and the 7.1 Ma Rattlesnake Tuff (cf. Greene 1973; Streck & Grunder 1995; Jordan *et al.* 2004) overlie the Strawberry Volcanics. The 15.9–15.4 Ma Dinner Creek Tuff (Streck *et al.* 2011) is also mid-Miocene and clear stratigraphic relationships are found in only one area where the tuff is intercalated between thick Strawberry Volcanics lava flows (Fig. 4). Lavas with characteristic Strawberry Volcanics lithology have a wide distribution and are at least distributed over an area of 40 × 90 km centred 20 km SW of Strawberry Mountain proper (Fig. 1b). The more mafic flows locally exhibit inflated pahoehoe and a'a features. Lavas with characteristics similar to Strawberry Volcanics occur south of 44 °N and north of 44°30′N (Fig. 1b), but our investigation does not go past these two limits. Massive rhyolites (*c.* 300 m) are exposed mainly along the western flank (Fig. 1b) and underlie the intermediate composition lavas. Because of the proximity to Strawberry Mountain proper, we group these rhyolites with the Strawberry Volcanics, as was done by earlier investigators for the northernmost rhyolites (Thayer *et al.* 1967; Robyn 1977). These rhyolites are phenocryst-poor (≤1%) and range from obsidian to devitrified flows. We previously named them Strawberry Rhyolites (Steiner & Streck 2011) whereas Thayer *et al.* (1967) and Robyn (1977) referred to them as the Wildcat Rhyolite Member of the Strawberry Volcanics. Other rhyolites that also appear slightly older than intermediate lavas of the Strawberry Volcanics crop out along the NE near the town of Unity (Fig. 1b). This location is where a series of other rhyolite occurrences are found nearby as well (Fig. 1b). These rhyolites are more phenocryst-rich (>10%) and include mineral phases of plagioclase, quartz, biotite and occasional amphibole. Mafic lavas of the Strawberry Volcanics occur sporadically throughout the area, but are concentrated in the NE where mostly basaltic andesites have been called Slide Creek

Table 1. *Incremental heating age determinations*

Sample	Location	Rock type*	Material dated[†]	Plateau Age (Ma)	Plateau ± (2σ)	Steps Plateau	^{39}Ar (%)	Plateau MSWD[‡]	Isochron Age (Ma)	Isochron ± (2σ)	$^{40}Ar/^{39}Ar$ Intercept	± (2σ)	Isochron MSWD[‡]
AS-SV-156	44°2′38.60″N, 118°45′49.50″W	A	GM	14.78	0.13	6	87.43	0.63	14.74	0.25	296.24	4.45	0.76
AS-SV-230	44°18′50.39″N, 118°17′16.61″W	BA	GM	13.45	0.24	5	70.78	2.25	14.01	0.43	288.61	5.50	0.99
AS-SV-82	44°15′55.50″N, 118°22′57.40″W	A	GM	14.12	0.26	–	88.91	4.34	13.94	0.29	301.10	41.01	2.76
AS-SV-14	44°37′13.70″N, 118°28′13.20″W	BA	GM	13.68	0.16	9	85.41	1.04	13.84	0.35	293.20	4.26	1.03
AS-SV-192	44°17′05.30″N, 118°41′6.60″W	A	GM	12.44	0.12	9	98.84	0.19	12.40	0.13	295.60	53.64	0.26

*A, andesite; BA, basaltic andesite.
[†]GM, groundmass.
[‡]MSWD, mean square weighted deviation.
J = 0.0014610 ± 0.0000056.
FCT-3 = 28.030 ± 0.003 Ma.

Fig. 3. (**a**) Section of >1000 m in thickness of lava flows of the Strawberry Volcanics exposed in the glaciated wall of Strawberry Mountain (44°17′44″N, 118°40′41″W); (**b**) NNW–SSE-trending dyke of the Strawberry Volcanics (44°16′54″N, 118°31′18″W) that follows a regional dyke trend; (**c**) close up of a Strawberry Volcanics dyke cross-cutting other volcanic rocks (44°22′04″N, 118°26′30″W).

Basalts (Robyn 1977; Goles *et al.* 1989), and this is also the area where lava flows of the Strawberry Volcanics interfinger with flows of the CRBG (Picture Gorge Member) (Thayer 1957). These basalts have an ophitic texture dominated by plagioclase, clinopyroxene, orthopyroxene and olivine.

Eruption sites have locally been identified either by the accumulation of near-vent pyroclastic material or by intrusive rocks and contacts. The most prominent occurs at Strawberry Mountain where an elongated intrusion >500 m wide cross-cuts more than 1000 m of lava flows (Fig. 1b). The intrusions are fine-grained with interlocking crystal boundaries. Similar but smaller vent structures appear throughout the area (Fig. 1b). In addition, numerous NW–SE-trending dykes accompany these volcanic intrusions (Figs 1b, 3b, c).

^{40}Ar/^{39}Ar ages of Strawberry Volcanics and stratigraphic relationships to mid-Miocene flood basalts and other regional rock units. New ^{40}Ar/^{39}Ar ages indicate calc-alkaline and tholeiitic basaltic andesites to andesites of the Strawberry Volcanics erupted from 14.78 ± 0.13 Ma (2σ) to 12.44 ± 0.12 Ma (2σ) ago (Table 1, Fig. 5). This demonstrates magmatism of the Strawberry Volcanics was ongoing during the eruptions of the Wanapum Basalt (15–14.5 Ma ago) and the Saddle Mountain Basalts (13–6 Ma ago) of the CRBG (Table 1, Fig. 5). We expect to extend this range to older ages as our last field work revealed a regional stratigraphic marker bed intercalated with the Strawberry Volcanics. This maker bed (Fig. 4) consists of two ignimbrites that we tentatively correlate with the 15.9–15.4 Ma Dinner Creek Tuff (Streck *et al.* 2011). The sample SW of Strawberry Mountain, AS-SV-156, gave the oldest date of 14.78 ± 0.13 Ma (2σ) while to the SE, AS-SV-82 yielded the second oldest date of 14.12 ± 0.26 Ma (2σ) (Table 1). The sample to the north of Strawberry Mountain (the Slide Creek Basalt member) yielded an age of 13.68 ± 0.16 Ma (2σ) and the sample to the east yielded an age of 13.45 ± 0.24 Ma (2σ) (Table 1). The sample directly from Strawberry Mountain (AS-SV-192) is a micronorite volcanic neck/plug with the youngest age of 12.44 ± 0.12 Ma (2σ) (Table 1). Previously, the eruption period of the Strawberry Volcanics was bracketed by K–Ar ages of 19.8 ± 0.38 Ma (2σ) and 12.4 ± 0.41 Ma (2σ) (Robyn 1977). Robyn (1977) considered dacites that were dated at 19.8 ± 0.38 Ma (2σ) and 18.1 ± 0.50 Ma (2σ) to be part of the Strawberry Volcanics. These dacites underlie the Strawberry Volcanics and we argue that they are not likely part of them because they are texturally and compositionally distinct. Other reported ages (Robyn 1977) are 14.9 ± 0.85 Ma (2σ) on a basalt and 12.4 ± 0.41 (2σ) Ma on basaltic andesite. Both of these ages correspond closely with our new Ar–Ar ages.

Geochemistry

Based on the discrimination scheme by Miyashiro (1974) and Arculus (2003), mafic to intermediate magmas of the Strawberry Volcanics are both

Fig. 4. 8 m thick rhyolitic ash-flow tuffs intercalated with lava flows of the Strawberry Volcanics (44°18′52″N, 118°41′32″W). These tuffs are correlated with the Dinner Creek Tuff which erupted between 15.9–15.4 Ma ago (Streck *et al.* 2011). This stratigraphic evidence indicates that activity of the Strawberry Volcanics likely started at ≥16 Ma ago.

tholeiitic and calc-alkaline (Fig. 2). All basalts of the Strawberry Volcanics are tholeiitic, whereas lavas of intermediate compositions are either tholeiitic or calc-alkaline (Fig. 2). As compositions become more evolved, the range in tholeiitic v. calc-alkaline character increases with transitional samples in between (Fig. 2). Continuous ranges from tholeiite to calc-alkaline similar to those of the Strawberry Volcanics have been observed in other magma series (e.g. Arculus 2003; Brueseke & Hart 2009; Hora *et al.* 2009).

Tholeiitic lavas of the Strawberry Volcanics. The tholeiitic lavas of the Strawberry Volcanics comprise compositions ranging from basalt (49–50 wt% SiO_2) to andesite (57–60 wt% SiO_2) (Table 2). All tholeiitic lavas have high Al_2O_3, ranging from 16–17 wt%. MgO decreases from 8.9–5.9 wt% in the basalts to 2.5 wt% in the andesites. As MgO decreases FeO* decreases as well, yet not as fast, and it is more variable. FeO* values range from 9.9–13.7 wt% in basalts to 7.4 wt% in andesites (Table 2). Consequently, the range in terms of FeO*/MgO generally becomes wider towards higher SiO_2 contents (Fig. 2b). In a total alkalis v. SiO_2 diagram, tholeiitic lavas plot on or very near the subalkaline/alkaline divide (Fig. 2a).

Select incompatible element ranges are as follows: Rb (4–37 ppm), Ba (408–929 ppm), La (18–34 ppm), Th (1.2–3.5 ppm), U (0.33–2.10 ppm) and K_2O (1.80–0.45 wt%), with the basalts being the least enriched (Table 2). The high field strength (HFS) trace elements (Nb, Ta, Zr, Hf) display near-constant or slightly decreasing concentrations towards andesitic compositions (Fig. 6). There are no Eu anomalies. REE patterns have negative slopes with $(La/Yb)_N$ values ranging from 4.9 to 7.4, $(La/Sm)_N$ from 2.1 to 2.9 and $(Gd/Yb)_N$ from 2.1 to 1.7 (Fig. 7). Incompatible trace element concentrations of the tholeiitic lavas of the Strawberry Volcanics indicate elevated concentrations at 20 to 100 times primitive mantle with the most incompatible elements being most enriched (Table 2, Figs 8 & 9). Normalized incompatible trace element patterns indicate distinct spikes at Ba, Pb, and in some samples Sr and K (Figs 8 & 9). Ta and Nb have small troughs relative to neighbouring elements.

The calc-alkaline suite of the Strawberry Volcanics. The calc-alkaline lavas (excluding rhyolites) mostly consist of basaltic andesite and andesite (53–63 wt% SiO_2). Only a few dacitic lavas and no calc-alkaline basalt have been found (Fig. 2). All calc-alkaline compositions plot in the sub-alkaline field and contain generally less total alkalis than tholeiitic samples. Al_2O_3 is similar to tholeiitic lavas with a range of 16.0–17.9 wt%. MgO and FeO* decrease with increased silica ranging from 6.2 to 2.0 wt% and 9.2 to 5.2 wt%, respectively.

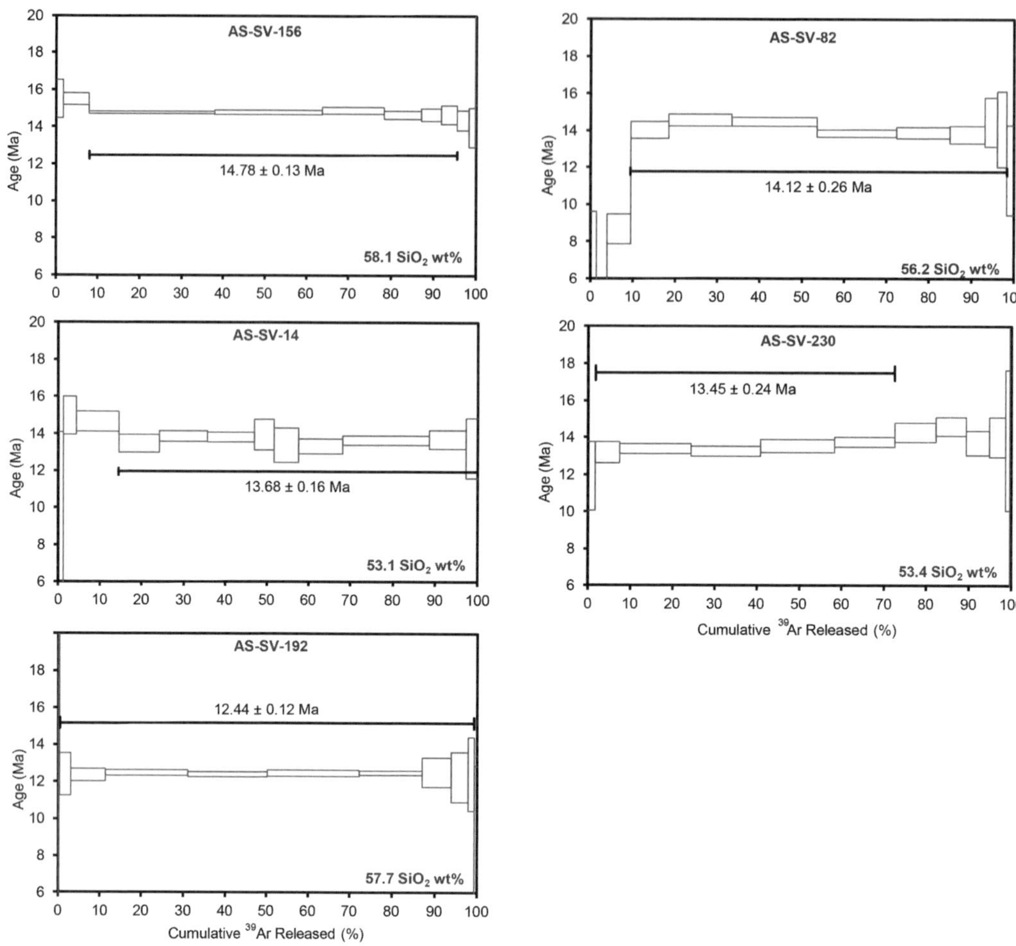

Fig. 5. Results of $^{40}Ar/^{39}Ar$ dating experiments indicating that the Strawberry Volcanics were at least active between 14.78 ± 0.13 Ma and 12.44 ± 0.12 Ma ago, see text for discussion. SiO_2 content of dated sample is given in right-hand corner.

FeO*/MgO ratios display no increase to a slight increase and trend horizontally from the tholeiitic/calc-alkaline dividing line towards higher SiO_2 to straddling it (Table 2, Fig. 2).

Normalized incompatible trace element patterns indicate similar enrichment levels and patterns as tholeiitic samples (Table 2, Figs 6 & 8). Select element ranges for samples of the calc-alkaline suite are: Rb (10.8–42.0 ppm), Ba (550–1124 ppm), Th (0.8–4.0 ppm), U (0.3–2.3), and K_2O wt% (0.9–2.8) (Table 2). Typical incompatible elements during evolution from basaltic andesite to andesite behave variably within the calc-alkaline suite of the Strawberry Volcanics. Incompatible elements that do increase include Rb, Ba, Th, U; those that are nearly constant include Zr, Hf; and others decrease, including Nb, Ta. Another noteworthy feature is that Sr also stays constant. $(La/Yb)_N$ values of the calc-alkaline lavas range from 5.6 to 9.0, $(La/Sm)_N$ from 2.4 to 3.5 and $(Gd/Yb)_N$ from 1.5 to 1.9 (Fig. 7). Some of the more evolved samples indicate small negative Eu anomalies.

Compositional difference between tholeiitic and calc-alkaline suite. The calc-alkaline suite strongly overlaps with the tholeiitic suite in regards to most major elements (Table 2). Both suites are medium to high-K lavas (Le Bas *et al.* 1986) and normalized trace element patterns are comparable and similar to the most Mg-rich basalt (Fig. 8). Despite these similarities, there are distinct compositional differences – albeit some are quite small – between calc-alkaline and tholeiitic samples. Differences become more notable the more evolved the samples are and the greater their relative difference in FeO*/MgO (Figs 8 & 9). Tholeiitic samples are generally lower in the large ion lithophile elements (LILE) (e.g. Rb, Ba, Th, U and K) and tend to be higher in high field strength elements (HFSE) (Figs 6, 8 & 9). There is also a difference in the slope of the REE pattern. The slope of light rare earth elements (LREE)–middle rare earth elements (MREE) is generally steeper and the slope of MREE–heavy rare earth elements (HREE) shallower in calc-alkaline samples than in tholeiitic samples (Fig. 7). And finally, Ti and P are lower in calc-alkaline than in tholeiitic samples (Table 2, Fig. 8).

The range in FeO*/MgO at any given SiO_2 concentration among lavas of the Strawberry Volcanics allows us to investigate whether the mentioned differences change progressively from samples with the highest to lowest tholeiitic index. Using a series of samples at 55–62 wt% SiO_2 and comparing normalized trace element diagrams illustrates how observed changes correlate with the tholeiitic index at a given SiO_2 (Fig. 9).

Discussion

Field interpretations

One question to answer is did lavas of the Strawberry Volcanics erupt from central vent volcanoes with a few localized general eruption sites, or did eruptions occur from numerous vents widely distributed, constituting a volcanic field without prominent topographic edifices? Evidence that bears on this question includes field data on thickness distribution of lava flows, on location of venting sites, and distribution of dykes. Erosional levels as encountered within the area of the Strawberry Volcanics are likely to be near (<1 km) the original surface as judged from intrusive contact relationships and volcanic textures of intrusive magmas cooled in a subvolcanic environment. The exception may be recorded at Strawberry Mountain where rocks do begin to show interlocking plutonic textures. With this in mind, we have no evidence for the existence of a major volcanic edifice rising 1000–2000 m above its base anywhere in the area as could have been expected given the intermediate magma composition which is typical for volcanic arcs. In contrast, equal thicknesses of single lava flows over considerable distances argue that original landscape was relatively flat for lava flows to spread out evenly with or without inflation. Dyke patterns follow a regional NNW–SSE trend as seen, for example, by the nearby Monument dyke swarms that gave rise to the Picture Gorge Basalt (Fig. 1) (Wilcox & Fisher 1966) instead of radial arrangements characteristic for large composite volcanoes. These data point to the interpretation that a volcanic field with topographically subdued volcanic edifices in the form of fissure eruptions or small central vent volcanoes produced the lavas of the Strawberry Volcanics. A likely analogue to the Strawberry Volcanics is found in the Harney Basin of the eastern High Lava Plains, eastern Oregon, where intermediate composition magmas erupted during the late Miocene from small, mostly low-relief venting sites, as identified by agglutinates and other pyroclastic material, to form thin lava flows (Streck & Grunder 2012).

The eruptions of lavas of the Strawberry Volcanics were not only coeval with the eruptions of the Wanapum member (15–14.5 Ma ago) but likely also with earlier units of the CRBG (Camp & Hooper 1981; Baksi 1989), because our oldest age of 14.8 Ma (Table 1) does not correspond with the activity onset of the Strawberry Volcanics. Onset must have been considerably earlier as approximately 850 m of lavas sit underneath the lava flow that yielded the oldest date reported here. The intercalated ash-flow tuffs (Fig. 4), that are correlated with the 15.4–15.9 Ma Dinner Creek Tuff (Streck *et al.* 2011), are found in the lower section, indicating some lavas flows of the Strawberry Volcanics are even older than the Dinner Creek Tuff. Therefore, it is reasonable to infer that activity started around 16 Ma ago, if not before. The shallow intrusive andesite plug of Strawberry Mountain dated at 12.44 ± 0.12 Ma is probably close to the terminal activity of the Strawberry Volcanics (Table 1).

Rhyolite lava flows subordinate in volume to basaltic and andesitic lava flows mainly crop out in the west and south of the area (SW of Strawberry Mountain) (Fig. 1b). Field evidence suggests a stratigraphic level of the rhyolites below basalts and andesites, and thus rhyolites constitute the first erupted phase of the Strawberry Volcanics. One date of 17.3 ± 0.36 Ma (Robyn 1977) places these rhyolites at the very onset of plume-head-related rhyolites associated with the CRBG (Pierce & Morgan 1992; Coble & Mahood 2012). There is no evidence for a significant time hiatus, such as palaeosol development, scouring or reworking between the emplacement of rhyolite lavas and the basalt to intermediate flows of the Strawberry Volcanics. This suggests earliest eruptions of basalts and andesites closely followed those of rhyolite, which in turn also suggests an onset of basaltic to andesitic volcanism of the Strawberry Volcanics at ≥16 Ma ago.

In summary, basaltic to andesitic magmas of the Strawberry Volcanics erupted over a period of 3–4 Ma from ≥16 to 12 Ma ago, and for the first 1–2 Ma were coeval with the eruption of flood

Table 2. *Representative analyses of basaltic to andesitic Strawberry Volcanics*

	AS-SV-171	AS-SV-11	AS-SV-217	AS-SV-14	AS-SV-230	AS-SV-82	AS-SV-192	AS-SV-159c	AS-SV-156	AS-SV-182	AS-SV-206	AS-SV-231
*CA/TH	0.94	0.75	0.88	1.03	0.87	1.1	1.18	0.72	0.73	0.94	1.59	1.84
wt% normalized												
SiO_2	49.48	50.47	52.06	53.13	53.43	56.23	57.66	57.59	58.08	60.36	61.35	61.70
TiO_2	1.85	1.86	1.60	1.57	1.81	1.42	1.20	1.77	1.68	1.15	0.81	0.69
Al_2O_3	16.10	16.17	17.09	16.45	16.82	16.59	16.98	16.77	16.31	16.10	16.81	16.78
FeO*	9.91	10.65	9.61	9.11	9.59	8.19	7.66	8.53	8.76	7.52	5.73	5.47
MnO	0.16	0.19	0.167	0.16	0.16	0.16	0.14	0.133	0.12	0.098	0.114	0.108
MgO	8.91	6.69	5.82	5.83	5.00	4.29	3.90	2.65	2.67	2.58	3.14	3.42
CaO	8.74	9.52	8.41	8.51	8.26	7.28	6.92	6.42	6.47	6.44	6.20	6.22
Na_2O	3.34	3.09	3.27	3.44	3.27	3.75	3.72	4.15	3.51	3.72	3.29	3.57
K_2O	1.09	0.76	1.394	1.22	1.08	1.58	1.34	1.541	1.75	1.645	2.209	1.837
P_2O5	0.42	0.59	0.581	0.56	0.60	0.51	0.48	0.464	0.64	0.394	0.346	0.214
ppm												
Ni	188	69	91.20	96	53	66	37	6.40	14	9.60	30.90	43.40
Cr	270	167	101.80	214	82	90	62	19.40	29	29.10	53.10	68.50
Sc	24	29	23.26	24	24	21	19	23.26	22	21.54	16.21	16.50
V	222	265	232.50	207	217	183	153	195.40	209	192.60	129.20	132
Ba	408	431	576.77	550	578	746	671	808.70	875	780.57	818.39	828.31
Rb	18	9	12.83	15	13	21	17	25.39	31	30.68	27.74	32.21
Sr	799	527	564.36	560	498	522	520	435.22	483	414.28	505.90	435.63
Zr	126	119	143.46	154	172	185	169	194.67	165	161.50	156.21	145.01
Y	21	26	24.58	26	30	27	26	32.53	30	29.81	21.98	18.96
Nb	16	12	10.64	16	15	15	13	14.45	11	9.33	7.77	6.46
Ga	19	18	19.30	19	20	20	19	21.10	19	17.10	17.70	17.90
Cu	64	50	71.60	56	33	52	39	24	22	21	46.20	47.10
Zn	96	102	101.50	96	107	102	94	108.40	102	87.90	77.60	65.40

Pb	3.10	3.50	5.15	4.80	5.40	6.70	6.10	8.03	7.00	6.89	7.66	7.51
U	0.60	0.40	0.43	0.60	0.60	0.70	0.70	1.13	1.00	1.06	0.98	0.96
Th	1.70	1.20	1.22	1.60	2.00	2.10	1.80	3.46	2.80	2.73	2.89	2.89
Cs	0.44	0.18	0.19	0.31	0.33	0.37	0.25	1.40	1.68	1.75	0.80	0.50
Hf	3.15	2.85	3.41	3.62	4.01	4.44	4.09	4.91	4.18	4.02	3.86	3.63
Ta	1.00	0.68	0.60	0.94	0.89	0.87	0.77	0.94	0.70	0.58	0.47	0.42
La	18.80	19.60	22.22	23.10	27.10	27.90	24.60	24.21	23.70	21.21	22.66	19.88
Ce	41.20	42.60	45.59	49.00	55.40	56.50	51.10	48.55	48.10	41.84	45.70	38.46
Pr	5.44	5.67	5.94	6.40	7.40	7.13	6.51	6.72	6.69	5.81	5.70	4.63
Nd	23.10	24.50	24.90	26.60	30.70	28.80	26.80	28.51	28.10	24.04	22.34	18.18
Sm	5.22	5.61	5.46	5.94	6.83	6.29	5.67	6.58	6.47	5.51	4.79	3.90
Eu	1.72	1.90	1.75	1.84	2.12	1.87	1.77	2.04	1.90	1.57	1.32	1.09
Gd	4.82	5.46	5.08	5.56	6.39	5.78	5.20	6.62	6.23	5.47	4.38	3.60
Tb	0.74	0.84	0.82	0.86	1.02	0.89	0.83	1.04	0.99	0.91	0.69	0.58
Dy	4.37	5.06	4.69	5.07	5.87	5.24	5.01	6.36	5.89	5.59	4.18	3.48
Ho	0.85	1.01	0.95	1	1.19	1.05	1.01	1.26	1.17	1.15	0.84	0.72
Er	2.20	2.68	2.53	2.60	3.09	2.76	2.66	3.36	3.09	3.20	2.28	1.95
Tm	0.30	0.38	0.36	0.37	0.43	0.40	0.38	0.47	0.44	0.47	0.33	0.28
Yb	1.82	2.29	2.22	2.31	2.68	2.46	2.40	2.92	2.72	2.93	2.10	1.86
Lu	0.27	0.35	0.35	0.37	0.42	0.41	0.38	0.46	0.42	0.46	0.34	0.30

*CA/TH < 1 = tholeiitic, CA/TH > 1 = calc-alkaline (cf. Hora *et al* 2009).

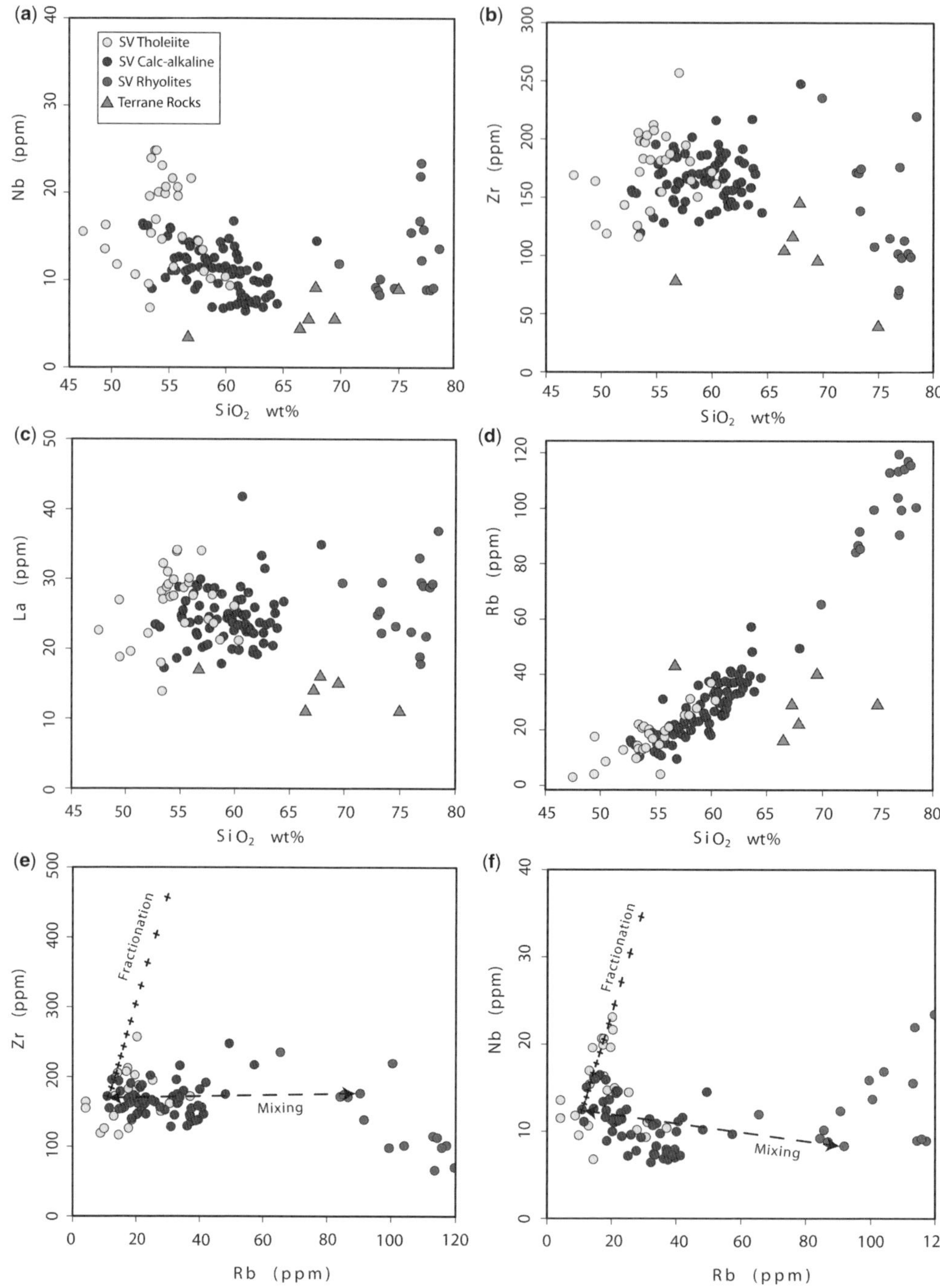

Fig. 6. Variation diagrams of Nb (**a**), Zr (**b**), La (**c**), and Rb v. SiO_2 (**d**). Observed trends among incompatible elements towards increased silica rule out fractional crystallization as main process to generate calc-alkaline andesites. Diagrams (**e**) and (**f**) include mixing and fractionation models. Fractionation models are based on partition coefficients for Rb of 0.071, 0.031, 0.02, for Zr of 0.048, 0.10, 0.18 and for Nb of 0.01, 0.15 and 0.005 for plagioclase, clinopyroxene and othropyroxene, respectively. The modal proportions used are based on observed mineral assemblages and are as follows; 68% plagioclase, 17% clinopyroxene and 15% orthopyroxene. The composition of the liquid has been calculated in intervals of 5%, as illustrated by + marks.

basalts of the Wanapum, Picture Gorge, and Grande Ronde members of the CRBG, and possibly even with Imnaha and Steens Basalt (Camp & Hooper 1981; Baksi 1989). The volcanic suite of the Strawberry Volcanics is surrounded by the voluminous (230 000 km^3) exclusively tholeiitic flood basalts and basaltic andesites of the CRBG (Fig. 1a). The upshot of the Strawberry Volcanics being in the centre of and coeval to CRBG magmatism is that the melting anomaly responsible for the mid-Miocene Columbia River flood basalt province likely also caused magmatism at the Strawberry Volcanics, yet at the Strawberry Volcanics we observe calc-alkaline and mildly tholeiitic volcanism contrasting with CRBG magmas (Fig. 2).

Common parent for calc-alkaline and tholeiitic magmas

The calc-alkaline and tholeiitic rock suite of the Strawberry Volcanics and the tholeiites of the CRBG all converge towards higher Mg concentrations to a composition with about 8 wt% MgO (Fig. 2c). This suggests the same parental basalt composition in terms of Fe/Mg ratio can lead to vastly different trajectories in a FeO* v. MgO diagram, leading to both tholeiitic and calc-alkaline compositions. This is not only observed for eastern Oregon during the mid-Miocene but has also been documented for a late Miocene/Pliocene basaltic suite from farther SSW along the High Lava Plains (Streck & Grunder 2012). There, an evolved high-alumina olivine tholeiite gave rise to a strongly calc-alkaline suite – just like the calc-alkaline Strawberry Volcanics – as well as a ferrotrachytic suite (Streck & Grunder 2012). Furthermore, there is trace element evidence that tholeiitic and calc-alkaline magmas of the Strawberry Volcanics developed from the same tholeiitic parent. Incompatible trace element and REE patterns among tholeiitic and calc-alkaline basaltic andesites ($\leq$53 wt% SiO_2) are identical suggesting that both have a common parent (Fig. 8c). The mentioned subtle differences between tholeiitic and calc-alkaline compositions become significant only at higher silica contents (Fig. 8a) and are discussed in the next section but compositions are essentially indistinguishable at lower silica contents irrespectively of whether they are tholeiitic or calc-alkaline in character. Some variations among Strawberry Volcanics basalts do exist, but our current data also include compositions that are compatible with being parental to basaltic andesites, again judged from nearly identical trace element concentrations and REE patterns (Table 2, Fig. 8c). Furthermore, tholeiitic basalts of the Strawberry Volcanics and lavas of the CRBG lavas overlap in some trace-element ratios (Fig. 10) and are otherwise similar in their incompatible trace element patterns (Fig. 11). This is testimony that Strawberry basalts are derived from a more or less similar mantle source as the flood basalts of the CRBG. The fact that all CRBG magmas and all basalts of the Strawberry Volcanics are tholeiitic indicates that this mantle source apparently gave rise to only tholeiitic magmas during the mid-Miocene in eastern Oregon. This is further supported by the fact that neither our study to date nor any other published data have revealed the existence of any calc-alkaline basalt among tholeiites. The mantle source for the tholeiites was metasomatized by subduction processes at some earlier point in time because all CRBG and Strawberry tholeiites show clear geochemical signs in the form of elevated concentration levels of, for example Ba, Pb, Sr and lower Nb, Ta relative to similarly incompatible elements (Figs 10 & 11) (Carlson 1984; Brandon *et al.* 1993; Wolff *et al.* 2008). It is such tholeiitic magma that is parental to the compositional intermediate, tholeiitic as well as the calc-alkaline suites of the Strawberry Volcanics, as based on geological and geochemical arguments. There is no evidence for, and no need to call upon, an unseen calc-alkaline basalt magma from which calc-alkaline lavas evolved.

Compositional changes and petrogenesis associated with calc-alkaline and tholeiitic evolution

As Strawberry Volcanics magmas become more silica rich and more evolved, their compositional diversity increases in terms of being less or more strongly calc-alkaline. Here we evaluate what compositional changes take place along with changes in the calc-alkalinity index of the magmas, and use this in turn as constraints for the processes involved. Detailed modelling of these trends is beyond the scope of this paper, but will be the subject of subsequent work. The main changes between the strongest calc-alkaline and tholeiitic andesite of the Strawberry Volcanics is that tholeiitic andesites tend to be slightly richer in REE, P, Ti, Ta and Nb, (Figs 7–9, Table 2). All other elements are practically indistinguishable. Also, calc-alkaline REE patterns have in general a steeper LREE–MREE (higher $(La/Sm)_N$ ratios) part and a shallower MREE–HREE portion (i.e. lower $(Gd/Yb)_N$ ratios) (Fig. 7). If these differences were to be explained by fractional crystallization processes, then observed compositional systematics would be consistent with fractionating liquidus minerals for the calc-alkaline magmas that include more Ti–Fe oxides, apatite and silicates that preferentially uptake the MREE, such as amphibole compared to tholeiitic magmas. However, near-constant incompatible element concentrations going from basalt to silicic andesites, as observed, for example, for Zr and La (Fig. 6b, c), are inconsistent with a dominant control by fractional crystallization, except for some tholeiitic magmas (Fig. 6e). Even more striking are the generally decreasing Nb concentrations towards calc-alkaline andesite (Fig. 6a) that violate any reasonable fractional crystallization scenario (Fig. 6f). Consequently, evolution towards calc-alkaline andesite must therefore be controlled by open-system processes either in the form of crustal contamination, magma-mixing processes or a combination thereof. Contaminant or magma for magma mixing must be silicic in order to generate the more silicic compositions of the Strawberry Volcanics. Some of the Strawberry rhyolites have compositions that make them suitable for magma mixing as they have concentrations of incompatible elements as low, lower, or only slightly higher than basalt, which means concentrations around 8–9, 110, and 28 ppm for the key elements Nb, Zr, and La, respectively (Fig. 6, Table 3). Similarly, our preliminary data on composition of crustal rocks also indicate low Nb, Zr and La, and thus they are permissive as contaminants in the case of silicic bulk composition (>65 wt% SiO_2) or suitable as source rocks to generate permissive silicic partial melts when crustal rocks are more mafic (Fig. 6, Table 3).

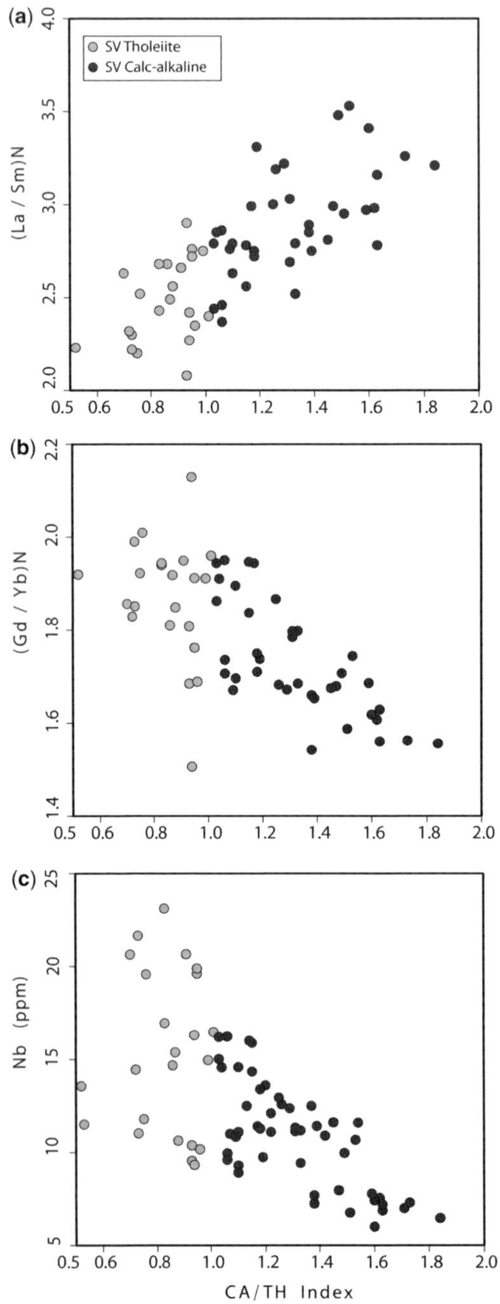

Implications for other intra-continental settings with calc-alkaline suites

Calc-alkaline suites of intra-continental settings are numerous. Examples from Oregon include the late Miocene Harney Basin suite (Streck & Grunder 2012), other mid-Miocene suites of the Powder River Volcanics (Hooper *et al.* 1995; Ferns & McClaughry 2013) and of the Owyhee volcanic field (Camp *et al.* 2003; Ferns & McClaughry 2013), Oligocene suites of the Steens Mountain Volcanics (Langer 1991) and Tower Mountain volcanic complex (Ferns *et al.* 2001). They are equally abundant throughout the western US, for example in the Basin and Range province (e.g. Hawkesworth *et al.* 1995) and elsewhere (Zielinski & Lipman 1976; Ewart *et al.* 1992; Lipman 2000; Kürkcüoglu *et al.* 2001, Harangi *et al.* 2007).

Fig. 7. Ratios of $(La/Sm)_N$ and $(Gd/Yb)_N$, and concentrations of Nb plotted v. calc-alkaline/tholeiite (CA/TH) index of Hora *et al.* (2009) determined for each sample. CA/TH index = (wt% SiO_2 −42.8)/(6.4 × FeO^*/MgO). A CA/TH index of <1 indicates tholeiitic character and of >1 calc-alkaline character. The CA/TH index is used here because it permits assigning a value of calc-alkaline/tholeiitic character to individual samples, as opposed to the THI index which describes the trend within a suite of samples. On average, tholeiitic lavas of the Strawberry Volcanics have a flatter LREE $(La/Sm)_N$ slope but a steeper HREE $(Gd/Yb)_N$ slope than calc-alkaline lavas. Nb concentrations are higher in the tholeiitic lavas and decrease with increasing calc-alkalinity.

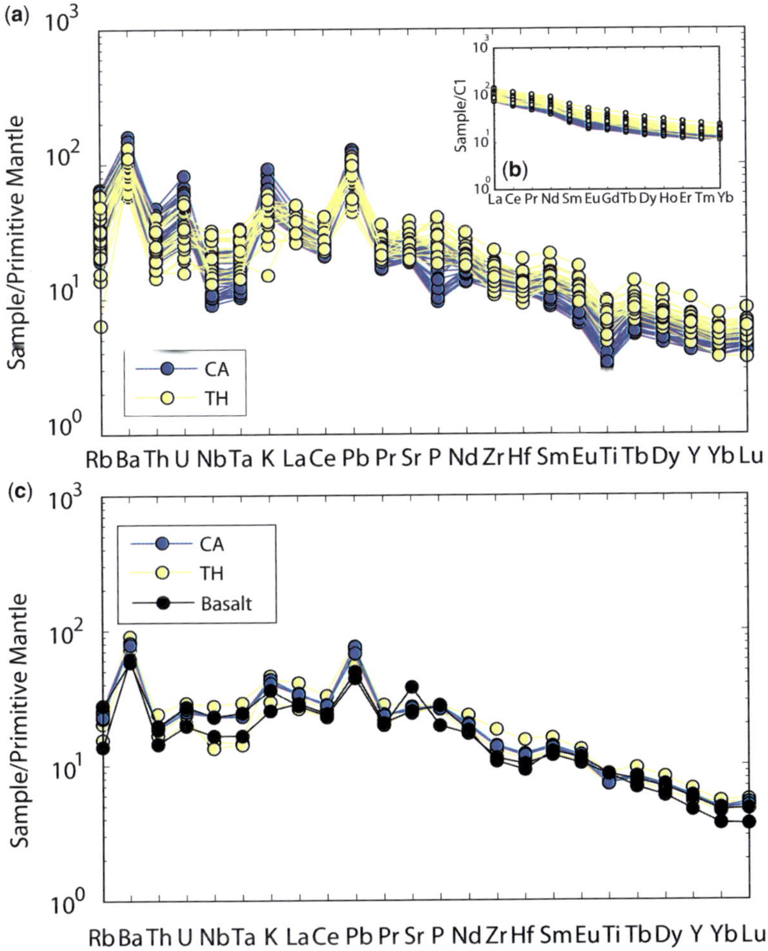

Fig. 8. Primitive mantle-normalized incompatible element diagram (**a**) and chondrite-normalized REE diagram (**b**) of select andesites (<62 wt% SiO_2) of Strawberry Volcanics showing similarities as well as difference towards more evolved compositions. Normalization values are taken from Sun & McDonough (1989) for (a) and McDonough & Sun (1995) for (b). Diagram (**c**) is a primitive mantle-normalized incompatible element diagram of selected basalts and basaltic andesites with compositions of ≤53 wt% SiO_2 showing similarity of both tholeiitic (yellow) and calc-alkaline (blue) basaltic andesites and tholeiitic basalts.

Some of the calc-alkaline centres that occur in an intra-continental setting have been interpreted to be due to very shallow subduction that caused volcanic-arc-style volcanism much further inward than is known from modern volcanic arc settings. This was championed by Lipman *et al.* (1972) for the Oligocene San Juan volcanic field. In fact, the occurrence of calc-alkaline suites of older age, where the original geotectonic setting is unsure, is traditionally used to argue for a subduction setting. Even in cases of calc-alkaline suites of young age, where a non-subduction setting is unquestionable, a common precondition for the generation of calc-alkaline magmas is a mantle that has previously been metasomatized by earlier subduction processes from which first calc-alkaline basalt is generated (Hawkesworth *et al.* 1995; Hooper *et al.* 1995; Harangi *et al.* 2007). It is here where the record of the Strawberry Volcanics has implications. Calc-alkaline rocks of the Strawberry Volcanics indeed show evidence of this metasomatized mantle, but this signal was most likely imparted by tholeiites and not by calc-alkaline basalt. Instead, as we discussed, open-system processes in the form of contamination and mixing facilitated the development along calc-alkaline evolution with the production of arc-like orogenic andesites from ambient tholeiitic magmas delivered from the mantle. The

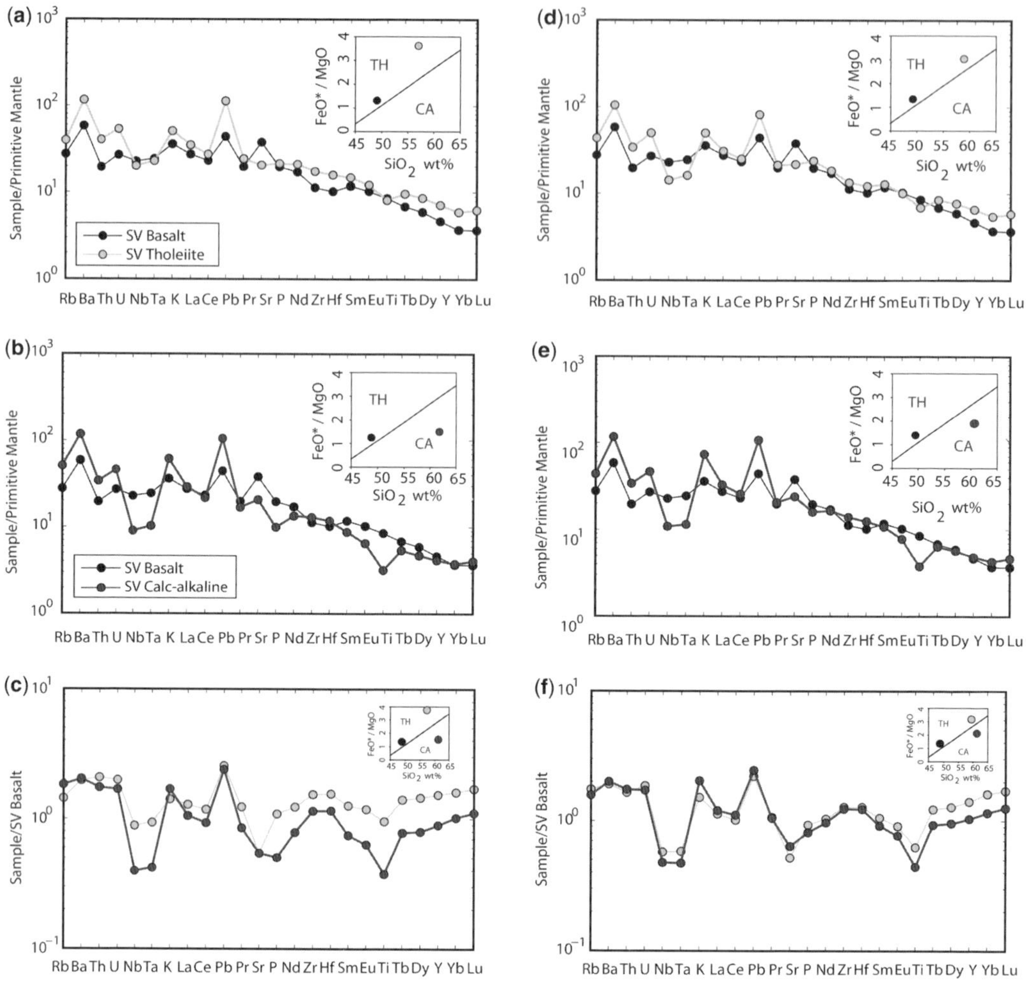

Fig. 9. Primitive mantle-normalized (Sun & McDonough 1989) incompatible element diagrams of select samples of Strawberry Volcanics comparing primitive tholeiite (AS-SV-171) to single intermediate composition members of the tholeiitic (**a** and **d**) and calc-alkaline (**b** and **e**) suites within a narrow silica range (57–61 wt%). (a) and (b) compare samples that are the most tholeiitic and the most calc-alkaline and (d) and (e) compare samples that are tholeiitic and calc-alkaline, but are closer in their FeO*/MgO ratios and straddle the tholeiitic/calc-alkaline divide. (c) and (f) show normalization diagrams using elemental concentrations of basalt (AS-SV-171) of the Strawberry Volcanics; (c) displays differences between andesites of (a) and (b), and (f) shows differences between the andesites of (d) and (e). Insets show position of samples relative to the tholeiitic/calc-alkaline divide of Miyashiro (1974). Differences in incompatible element compositions between tholeiitic and calc-alkaline samples are notable the more different samples are in terms of their FeO*/MgO ratio (or CA/TH index), but are less notable as samples converge in their FeO*/MgO ratio (see text for discussion).

follow-up question is how much of the other intracontinental calc-alkaline systems indeed require delivery of calc-alkaline basalt from the mantle that serves as parent to calc-alkaline intermediate lavas? Tholeiites carrying subduction signals, such as those of the mid-Miocene of eastern Oregon, are able to spawn calc-alkaline andesites during crustal processing, eliminating the need for a mantle-derived calc-alkaline basalt. Our example of the Strawberry Volcanics calls into question the general interpretation that a metasomatized mantle signal is imparted to the more evolved magmas by calc-alkaline basalts when the most mafic lavas are basaltic andesite or more silicic. In the case of the Strawberry Volcanics, calc-alkaline basaltic andesites have already undergone some crustal processing, despite the fact that some are fairly mafic, with MgO and Cr of around 6 wt% and 200 ppm,

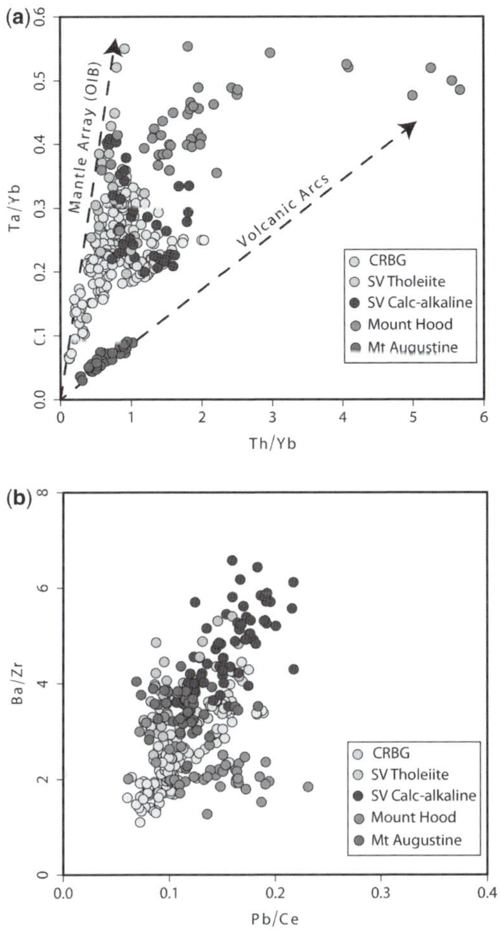

Fig. 10. Lavas of the Strawberry Volcanics relative to lavas of CRBG and typical continental arcs represented by Mt Hood and Mt Augustine: (**a**) Th/Yb v. Ta/Yb diagram; dashed arrows indicate typical volcanic arcs and a 'mantle array' (OIB) trajectories taken from Leeman *et al.* (1990); (**b**) Pb/Ce v. Ba/Zr diagram. Significant overlap between lavas of the Strawberry Volcanics and the CRBG and their compositional spread indicate derivation from a similar mantle source that has been previously metasomatized. Data for Mt Hood taken from Woods (2004) and for Mt Augustine from Steiner *et al.* (2012).

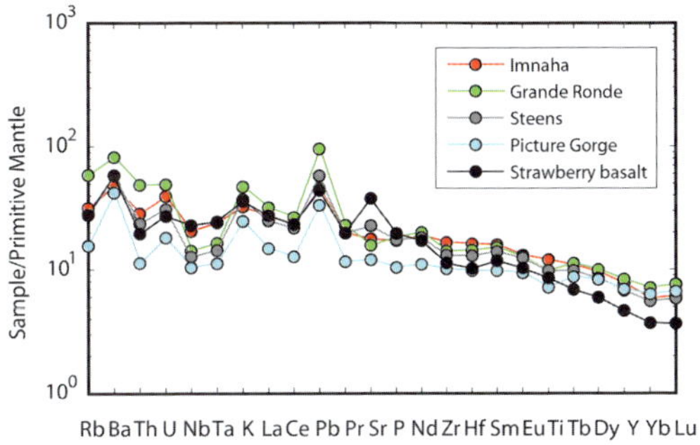

Fig. 11. Primitive mantle-normalized (Sun & McDonough 1989) incompatible element diagram comparing one basalt of the Strawberry Volcanics (black) with select lavas of the CRBG (Wolff *et al.* 2008) representing the following members: Imnaha Basalt (red), Grande Ronde Basalt (green), Steens Basalt (grey), and Picture Gorge Basalt (blue). The tholeiite of the Strawberry Volcanics is lightly depleted in HREE relative to CRBG lavas, but otherwise indicates a very similar pattern. Best match is found with Steens-type basalt.

respectively (Table 2). Such compositions are often the most mafic samples encountered at other intra-continental calc-alkaline systems (e.g. Zielinski & Lipman 1976; Askren *et al.* 1997). The up-shot of this is that, even though calc-alkaline basaltic andesites and andesites of intra-continental settings carry compositional evidence for being derived from a mantle that experienced some compositional overprint by subduction (e.g. high Ba/Zr), there it is no guarantee that parental mantle magmas were already calc-alkaline, as tholeiitic magmas can be carriers of subduction signals just as well – as our study demonstrates. This challenges the *a priori* interpretation that, when chemical subduction signals are observed in basaltic andesites and andesites, they are coming, indeed, from calc-alkaline basalt. Our study shows that crustal interactions and/or magma mixing can be solely responsible for developing a calc-alkaline trend from tholeiite and, therefore, merging the subduction signals with calc-alkaline character.

Conclusions

The Strawberry Volcanics is a diverse group of lavas ranging from basalt to rhyolite and from tholeiitic to calc-alkaline. The most primitive lavas of the Strawberry Volcanics are tholeiitic basalts, and they are also the parent to the intermediate calc-alkaline lavas. Calc-alkaline and tholeiitic basaltic andesites show identical enrichment and depletion patterns on trace element normalization diagrams that are similar to the ones of basalts, attesting to their common tholeiitic ancestry. Subtle differences show up as magmas evolve and acquire progressively stronger calc-alkaline or tholeiitic character. Decreasing, constant or increasing incompatible element abundances, along with increased silica contents among calc-alkaline compositions, are not reconcilable with fractional crystallization scenarios, but instead require open-system processes in

Table 3. *Geochemical data of rhyolites and basement rocks (accreted terranes)*

	Strawberry Mountain Rhyolites						Strawberry Mountain Basement Rocks				
	AS-SV-173	AS-SV-151	AS-SV-60	AS-SV-85	AS-SV-110a	AS-SV-144	SV-BASE-1	SV-BASE-2	SV-BASE-3	SV-BASE-7	SV-BASE-9
wt% normalized											
SiO_2	73.33	73.19	77.71	77.36	77.95	72.99	67.85	67.24	66.48	74.97	69.49
TiO_2	0.27	0.35	0.10	0.13	0.10	0.35	0.65	0.56	0.55	0.04	0.60
Al_2O_3	14.24	14.57	12.40	12.76	12.54	15.00	15.44	16.25	16.22	15.68	13.66
FeO*	2.02	2.23	0.85	1.39	0.74	2.29	3.67	3.61	3.48	0.97	5.58
MnO	0.05	0.05	0.03	0.02	0.01	0.04	0.07	0.07	0.08	0.05	0.05
MgO	0.57	0.23	0.08	0.06	0.03	0.25	2.46	1.90	2.18	0.14	3.12
CaO	2.21	1.99	0.68	0.63	0.63	2.03	3.86	3.95	4.48	1.61	3.37
Na_2O	3.28	3.86	3.76	3.29	3.50	3.53	4.66	4.73	5.30	5.08	2.64
K_2O	3.96	3.42	4.37	4.34	4.49	3.47	1.16	1.50	1.08	1.40	1.36
P_2O_5	0.08	0.10	0.02	0.02	0.02	0.05	0.17	0.19	0.15	0.07	0.14
ppm											
Ni	5	3	1	3	2	3	30	15	19	7	31
Cr	7	5	4	2	3	5	46	24	25	4	52
Sc	6	6	4	4	4	7	9	9	9	1	15
V	27	30	2	8	3	25	78	74	77	5	156
Ba	1192	1485	1445	1587	1417	1450	662	721	538	577	843
Rb	92	87	117	114	116	84	22	29	16	29	40
Sr	198	231	58	65	58	258	536	526	537	389	403
Zr	139	171	102	113	99	172	145	116	104	39	95
Y	25	18	23	20	18	19	16	13	16	16	21
Nb	8	9	9	9	9	9	9	5	4	9	5

Ga	16	16	14	13	15	15	17	19	18	20	15
Cu	6	7	1	2	2	3	5	10	31	5	81
Zn	42	49	25	19	20	43	41	57	172	49	71
Pb	13.3	13.9	16.1	17.7	17.6	13.7	8.9	8.2	12.2	10.8	4.8
U	4.7	3.7	4.6	4.1	4.2	3.4	0	0.3	0.0	1.0	1.8
Th	8.7	7.7	10.5	9.4	10.2	7.3	3.5	3.6	4.5	2.1	4.6
Cs	4.42	4.20	5.53	–	–	4.32	–	–	–	–	–
Hf	3.99	4.59	3.66	–	–	4.59	–	–	–	–	–
Ta	0.83	0.74	0.86	–	–	0.74	–	–	–	–	–
La	22.3	25.4	28.8	21.8	29.3	24.9	16.1	13.5	11.2	10.5	15.3
Ce	43.6	46.2	54.5	41.3	43.0	43.5	32.7	27.3	25.8	23.9	25.3
Pr	5.16	5.36	6.04	–	–	5.14	–	–	–	–	–
Nd	19.2	19.6	21.2	17.2	19.9	18.8	14.7	13.7	12.9	14.2	14.4
Sm	4.32	3.82	4.10	–	–	3.69	–	–	–	–	–
Eu	0.77	0.88	0.46	–	–	0.77	–	–	–	–	–
Gd	3.98	3.35	3.64	–	–	3.24	–	–	–	–	–
Tb	0.68	0.54	0.64	–	–	0.54	–	–	–	–	–
Dy	4.28	3.26	3.98	–	–	3.33	–	–	–	–	–
Ho	0.89	0.67	0.84	–	–	0.68	–	–	–	–	–
Er	2.55	1.88	2.42	–	–	1.95	–	–	–	–	–
Tm	0.39	0.29	0.37	–	–	0.30	–	–	–	–	–
Yb	2.61	1.93	2.49	–	–	2.02	–	–	–	–	–
Lu	0.42	0.32	0.41	–	–	0.33	–	–	–	–	–

the form of crustal contamination and magma mixing, and are the main driver for developing a calc-alkaline suite. Preliminary evaluations indicate that observed rhyolites and nearby exposed crustal rocks have compositions suitable for magma mixing and/or contamination. Crustal interactions appear to be important for the early stage of the calc-alkaline liquid line in order to diverge from the regional tholeiitic trend.

Our study holds implications for intra-continental calc-alkaline suites where relationships to tholeiitic magma are equivocal or do not exist. When calc-alkaline basaltic andesites and andesites of intra-continental settings carry compositional evidence of being derived from a mantle influenced by subduction, then this signal may also have been delivered by tholeiitic magmas that subsequently evolved a calc-alkaline suite during open system processes in the crust. Therefore, the notion that chemical subduction signals, as observed in evolved mafic magmas, must come from calc-alkaline basalt, has to be re-evaluated. Subsequent crustal processing can be solely responsible for developing a calc-alkaline trend including 'orogenic' andesites and, therefore, merging the subduction signals and calc-alkaline character – just as observed at volcanic arcs.

This work has been supported by NSF-EAR grant #1220676 and PSU Faculty Enhancement grant to Streck. The authors are grateful to M. Ferns for discussions of eastern Oregon geology. The authors thank E. Christiansen and E. Gazel for their comments which considerably helped to improve this paper. The authors also thank A. Gómez-Tuena for his editorial handling and additional comments.

References

ARCULUS, R. J. 2003. Use and abuse of the terms calc-alkaline and calc-alkalic. *Journal of Petrology*, **44**, 929–935.

ASKREN, D. R., RODEN, M. F. & WHITNEY, J. A. 1997. Petrogenesis of Tertiary andesite lava flows interlayered with large-volume felsic ash-flow tuffs of the western USA. *Journal of Petrology*, **38**, 1021–1046.

BAKSI, A. K. 1989. Reevaluation of the timing of extrusion of Imnaha, Picture Gorge, and Grande Ronde Basalts, Columbia River Basalt Group. *In*: REIDEL, S. P. & HOOPER, P. R. (eds) *Volcanism and Tectonism in the Columbia River Flood-Basalt Province*. Geological Society of America Special Papers, **239**, 105–112.

BRANDON, A. D., HOOPER, P. R., GOLES, G. G. & LAMBERT, R. ST. J. 1993. Evaluating crustal contamination in continental basalts: the isotopic composition of the Picture Gorge Basalt of the Columbia River Basalt Group. *Contributions to Mineralogy and Petrology*, **114**, 452–464.

BROOKS, H. C. & FERNS, M. L. 1979. *Geologic map of the Bull Run Rock quadrangle, Oregon*. Oregon Department of Geology and Mineral Industries, Portland, OR. Open-File Report **O-79–6**, scale 1:24 000.

BROWN, C. E. & THAYER, T. P. 1966*a*. *Geologic map of the Canyon City quadrangle, northeastern Oregon*. United States Geological Survey, Washington, DC, Miscellaneous Investigations Map 447, scale 1:250 000.

BROWN, C. E. & THAYER, T. P. 1966*b*. *Geologic map of the Mount Vernon quadrangle, Grant County*. United States Geological Survey, Washington, DC, Geologic Quadrangle Map GQ-548, scale 1:62 500.

BROWN, C. E. & THAYER, T. P. 1977. *Geologic map of pre-Tertiary rocks in the eastern Aldrich Mountains and adjacent areas to the south, Grant County, Oregon*. United States Geological Survey, Reston, VA, Miscellaneous Investigations Map 1021, scale 1:62 500.

BRUESEKE, M. E. & HART, W. K. 2009. Intermediate composition magma production in an intracontinental setting: unusual andesites and dacites of the mid-Miocene Santa Rosa–Calico volcanic field, Northern Nevada. *Journal of Volcanology and Geothermal Research*, **188**, 197–213.

CAMP, V. E. & HOOPER, P. R. 1981. Geologic studies of the Columbia Plateau: Part I. Late Cenozoic evolution of the southeast part of the Columbia River Basalt Province. *Geological Society of America Bulletin*, **92**, 659–668.

CAMP, V. E. & ROSS, M. E. 2004. Mantle dynamics and genesis of mafic magmatism in the intermontane Pacific Northwest. *Journal of Geophysical Research*, **109**, B08204, http://dx.doi.org/10.1029/2003JB002838.

CAMP, V. E., ROSS, M. E. & HANSON, W. E. 2003. Genesis of flood basalts and Basin and Range volcanic rocks from Steens Mountain to the Malheur River Gorge, Oregon. *Geological Society of America Bulletin*, **115**, 105–128.

CARLSON, R. W. 1984. Isotopic constraints on Columbia River flood basalt genesis and the nature of the subcontinental mantle. *Geochimica et Cosmochimica Acta*, **48**, 2357–2372.

CHRISTIANSEN, E. H. & MCCURRY, M. 2008. Contrasting origins of Cenozoic silicic volcanic rocks from the western Cordillera of the United States. *Bulletin of Volcanology*, **70**, 251–267.

COBLE, M. A. & MAHOOD, G. A. 2012. Initial impingement of the Yellowstone plume located by widespread silicic volcanism contemporaneous with Columbia River flood basalts. *Geology*, **40**, 655–658.

CROWLEY, K. C. 1960. *Geology of the Seneca–Silvies area, Grant County, Oregon*. Unpublished thesis, University of Oregon, Eugene, map scale 1:55 000.

DUNCAN, R. A. & KELLER, R. A. 2004. Radiometric ages for basement rocks from the Emperor Seamounts, ODP Leg 197. *Geochemistry Geophysics Geosystems*, **5**, Q08L03, http://dx.doi.org/10.1029/2004GC000704.

EWART, A., SCHÖN, R. W. & CHAPPELL, B. W. 1992. The Cretaceous volcanic-plutonic province of the central Queensland (Australia) coast – a rift-related 'calc-alkaline' province. *Transactions of the Royal Society of Edinburgh*, **83**, 327–346.

FERNS, M. L. & MCCLAUGHRY, J. D. 2013. Stratigraphy and volcanic evolution of the middle Miocene to Pliocene La Grande–Owyhee eruptive axis in eastern

Oregon. *In*: REIDEL, S. & TOLAN, T. (eds) *The Columbia River Basalt Group*. Geological Society of America Special Papers, in press.

FERNS, M. L., BROOKS, H. C. & AVERY, D. G. 1983. *Geology and gold deposits of the Greenhorn quadrangle, Baker and Grant Counties, Oregon*. Oregon Department of Geology and Mineral Industries, Portland, OR, GMS-28, scale 1:24 000.

FERNS, M. L., MADIN, I. P. & TAUBENECK, W. H. 2001. *Reconnaissance Geologic Map of the La Grande 30′ × 60′ quadrangle, Baker, Grant, Umatilla, and Union Counties, Oregon*. Oregon Department of Geology and Mineral Industries. Reconnaissance Map, RMS-1, scale 1:100 000.

GOLES, G. G., BRANDON, A. D. & LAMBERT, R. ST J. 1989. Miocene basalts of the Blue Mountains province in Oregon: Part 2, Sr isotopic ratios and trace element features of little-known Miocene basalts of central and eastern Oregon. *In*: REIDEL, S. P. & HOOPER, P. R. (eds) *Volcanism and Tectonism in the Columbia River Flood Basalt Province*. Geological Society of America Special Papers, **239**, 357–365.

GREENE, R. C. 1973. Petrology of the welded tuff of Devine Canyon, southeastern Oregon. *United States Geological Survey, Professional Paper*, **797**, 1–26.

GREENE, R. C., WALKER, G. W. & CORCORAN, R. E. 1972. *Geologic map of the Burns quadrangle, Oregon*. United States Geological Survey, Interpretive Map I-680, scale 1:250 000.

HARANGI, S., DOWNES, H., THIRLWALL, M. & GMÉLING, K. 2007. Geochemistry, Petrogenesis and Geodynamic Relationships of Miocene Calc-alkaline Volcanic Rocks in the Western Carpathian Arc, Eastern Central Europe. *Journal of Petrology*, **48**, 2261–2287.

HAWKESWORTH, C., TURNER, S., GALLAGHER, K., HUNTER, A., BRADSHAW, T. & ROGERS, N. 1995. Calc-alkaline magmatism, lithospheric thinning and extension in the Basin and Range. *Journal of Geophysical Research*, **100**, 10 271–10 286.

HOOPER, P. R., BAILEY, D. G. & MCCARLEY HOLDER, G. A. 1995. Tertiary calc-alkaline magmatism associated with lithospheric extension in the Pacific Northwest. *Journal of Geophysical Research*, **100**, 10303–10319.

HOOPER, P. R., BINGER, G. B. & LEE, K. R. 2002. Ages of the Steens and Columbia River flood basalts and their relationship to extension-related calc-alkalic volcanism in eastern Oregon. *Geological Society of America, Bulletin*, **114**, 43–50.

HORA, J. M., SINGER, B. S., WÖRNER, G., BEARD, B. L., JICHA, B. R. & JOHNSON, C. M. 2009. Shallow and deep crustal control on differentiation of calc-alkaline and tholeiitic magma. *Earth and Planetary Science Letters*, **285**, 75–86.

JORDAN, B. T., GRUNDER, A. L., DUNCAM, R. A. & DEINO, A. L. 2004. Geochronology of age-progressive volcanism of the Oregon High Lava Plains: implications for the plume interpretation of Yellowstone. *Journal of Geophysical Research*, **109**, B10202, http://dx.doi.org/10.1029/2003JB002776

KOPPERS, A. A. P. 2002. ArArCALC – software for Ar-40/Ar-39 age calculations. *Computers & Geosciences*, **28**, 605–619.

KÜRKCÜOGLU, B. S., SEN, E., TEMEL, A., AYDAR, E. & GOURGAUD, A. 2001. Trace-element modeling and source constraints for tholeiitic and calc-alkaline basalts from a depleted ashenospheric mantle source, Mt. Erciyes Stratovolcano, Turkey. *International Geology Review*, **43**, 508–522.

LANGER, V. 1991. *Geology and petrologic evolution of the silicic to intermediate volcanic rocks underneath Steens Mountain basalt, southeastern Oregon*. Unpublished MSc thesis, Oregon State University.

LE BAS, M. J., LE MAITRE, R. W., STRECKEISEN, A. & ZANETTIN, B. 1986. A chemical classification of volcanic rocks based on the total alkali–silica diagram. *Journal of Petrology*, **27**, 745–750.

LEE, C.-T. A., LEEMAN, W. P., CANIL, D. & LI, Z. X. A. 2005. Similar V/Sc systematics in MORB and arc basalts: Implications for the oxygen fugacities of their mantle source regions. *Journal of Petrology*, **46**, 2313–2336.

LEEMAN, W. P., SMITH, D. R., HILDRETH, W., PALACZ, Z. & ROGERS, N. 1990. Compositional diversity of Late Cenozoic basalts in a transect across the southern Washington Cascades: implications for subduction zone magmatism. *Journal of Geophysical Research*, **95**, 19561–19582.

LIPMAN, P. W. 2000. The central San Juan caldera cluster: regional volcanic framework. *In*: BETHKE, P. M. & HAY, R. L. (eds) *Ancient Lake Creede: Its Volcano-Tectonic Setting, History of Sedimentation, and Relation of Mineralization in the Creede Mining District*. Geological Society of America Special Papers, **346**, 9–69.

LIPMAN, P. W., PROSTKA, H. J. & CHRISTIANSEN, R. L. 1972. Cenozoic volcanism and plate-tectonic evolution of the Western United States. I. Early and Middle Cenozoic. *Royal Society of London Philosophical Transactions*, **271**, 217–248.

MCDONOUGH, W. F. & SUN, S. S. 1995. The composition of the Earth. *Chemical Geology*, **120**, 223–253.

MIYASHIRO, A. 1974. Volcanic rock series in island arcs and active continental margins. *American Journal of Science*, **274**, 321–355.

MULLEN, E. D. 1983. *Petrology and regional setting of peridotite and gabbro of the Canyon Mountain complex, northeast Oregon*. PhD thesis, University of Oregon, map scale 1:12 000.

PIERCE, K. L. & MORGAN, L. A. 1992. The track of the Yellowstone Hotspot: volcanism, faulting, and uplift. *In*: LINK, P. K., KUNTZ, M. A. & PLATT, L. P. (eds) *Regional geology of eastern Idaho and western Wyoming*. Geological Society of America, Memoirs, **179**, 1–52.

RENNE, P. R., SWISHER, C. C., DEINO, A. L., KARNER, D. B., OWENS, T. L. & DEPAOLO, D. J. 1998. Intercalibration of standards, absolute ages and uncertainties in $^{40}Ar/^{39}Ar$ dating. *Chemical Geology*, **145**, 117–152.

ROBYN, T. L. 1977. *Geology and petrology of the Strawberry Mountain volcanic series, Central Oregon*. PhD thesis, University of Oregon.

STEIGER, R. H. & JÄGER, E. 1977. Subcommision on geochronology: convention on the use of decay constants in geo- and cosmochronology. *Earth and Planetary Science Letters*, **36**, 359–362.

STEINER, A. R. & STRECK, M. J. 2011. Strawberry Rhyolites, Oregon: Northwestern extent of mid-Miocene flood basalt related rhyolites of the Pacific Northwest.

Abstract T51H-2473, presented at the 2011 Fall Meeting, American Geophysical Union, San Francisco, CA, December 5–9.

Steiner, A. R., Browne, B. L. & Nye, C. J. 2012. Quenched mafic inclusions in ≤2200 years B.P. deposits at Augustine Volcano, Alaskal. *International Geology Review*, **54**, 1241–1270, http://dx.doi.org/10.1080/00206814.2011.636641

Streck, M. J. & Grunder, A. L. 1995. Crystallization and welding variations in a widespread ignimbrite sheet: the Rattlesnake Tuff, eastern Oregon. *Bulletin of Volcanology*, **57**, 151–169.

Streck, M. J. & Grunder, A. L. 2012. Temporal and crustal effects on differentiation of tholeiite to calcalkaline and ferro-trachytic suites, High Lava Plains, Oregon, USA. *Geochemical Geophysical Geosystems*, **13**, Q0AN02, http://dx.doi.org/10.1029/2012GC004237

Streck, M. J., Ferns, M. & Ricker, C. 2011. The Dinner Creek Tuff and other mid-Miocene rhyolites at the magmatic focal zone of the Columbia River Basalt group. *Geological Society of America, Abstracts with Programs*, **43**, 5.

Sun, S.-S. & McDonough, W. F. 1989. Chemical and isotopic systematics of oceanic basalts: implications for mantle composition and processes. *In*: Saunders, A. D. & Norry, M. J. (eds) *Magmatism in the Ocean Basin*. Geological Society, London, Special Publications, **42**, 313–345.

Thayer, T. P. 1956. *Preliminary geologic map of the John Day quadrangle, Oregon*. United States Geological Survey, Washington, DC, Miscellaneous Field Studies Map MF-51, scale 1:62 500.

Thayer, T. P. 1957. Some relations of later Tertiary volcanology and structure in eastern Oregon. *International Geological Congress 20th*, Mexico, Sec. 1, Volcanologia del Cenozoico, 1, 231.

Thayer, T. P., Brown, C. E. & Hay, R. L. 1967. *Preliminary geologic map of the Prairie City quadrangle, Grant County, Oregon*. United States Geological Survey, Washington, DC, Open File Map, scale 1:62 500.

Wilcox, R. E. & Fisher, R. V. 1966. *Geologic map of the Monument quadrangle, Grant County, Oregon*. United States Geological Survey, Washington, DC, Geology Quadrangle Map GQ-541.

Wolff, J. A., Ramos, F. C., Hart, G. L., Patterson, J. D. & Brandon, A. D. 2008. Columbia River flood basalts from a centralized crustal magmatic system. *Nature geoscience*, **1**, 177–180.

Woods, M. 2004. *Compositional and mineralogical relationships between mafic inclusions and host lava as key to andesite petrogenesis at Mount Hood volcano, Oregon*. MS thesis, Portland State University.

Zielinski, R. A. & Lipman, P. W. 1976. Trace-element variations at Summer Coon volcano, San Juan Mountains, Colorado, and the origin of continental-interior andesite. *Geological Society of America Bulletin*, **87**, 1477–1485.

Zimmer, M. M., Plank, T. *et al.* 2010. The role of water in generating the calc-alkaline trend: new volatile data for aleutian magmas and a new tholeiitic index. *Journal of Petrology*, **51**, 2411–2444.

Magma sources and tectonic setting of Central Andean andesites (25.5–28°S) related to crustal thickening, forearc subduction erosion and delamination

SUZANNE MAHLBURG KAY[1]*, CONSTANTINO MPODOZIS[2] & MOYRA GARDEWEG[3]

[1]*Department of Earth and Atmospheric Sciences, Cornell University, Snee Hall, Ithaca, NY 14853, USA*

[2]*Antofagasta Minerals, Apoquindo 4001, Santiago, Chile*

[3]*AURUM Consultores, Del Inca 4446, suite 701, Los Condes, Santiago, Chile*

**Corresponding author (e-mail: smk16@cornell.edu)*

Abstract: The Central Andean margin, where the name andesite originated, is the type locality for arc andesites erupted through thick continental crust. The <25 Ma mid to high K_2O andesites erupted from 25.5°S to 28.2°S exhibit a large variation in trace element and isotopic ratios, reflecting formation over an evolving slab, a crust thickening to 65–75 km and a frontal arc that migrated *c.* 45 km eastward at 8–3 Ma. Andesites at 28–26.8°S have the most variable and extreme heavy rare earth element (REE), high field strength element (HFSE) and Ba/La ratios and wt% Na_2O, with the highest values in those erupted as the frontal arc migrated and the slab shallowed to the south. The required garnet-bearing, feldspar-free residue is generated in both the thick crust and the mantle wedge, into which crust was injected in a peak of forearc subduction erosion as the arc migrated. Andesites at 25.5–26.8°S, east of the Puna plateau under which the slab shallowed at 18–7 Ma and then steepened as lithospheric delamination occurred, generally lack extreme REE and HFSE ratios. Their upper crust-like features reflect eruption in a mixed stress regime and incorporation of westward-flowing radiogenic crust from a region of extensive deep crustal melting to the east.

The Central Andes represent the type example of a continental margin arc built on thick crust, and their study provides an excellent opportunity to examine processes linking continental arc magmatism, crustal evolution and tectonics. In this paper, we look at the latest Oligocene to Holocene Andean arc volcanism at the latitude of the modern southern Central Volcanic Zone (CVZ) from near 25°S to 27°45′S (69°15′W to 68°15′W), a region which is just north of the currently amagmatic Chilean–Pampean flatslab region (Fig. 1). The time period discussed incorporates the events and processes that occurred as the volcanic front evolved in the Maricunga Belt arc in Chile from 26 to 6 Ma (e.g. Kay *et al.* 1994*a*; Mpodozis *et al.* 1995), andesitic volcanism spread widely into the back-arc after 16 Ma, and the arc front migrated to its current position in the CVZ along the Argentine border between *c.* 8 and 3 Ma. Over the same time, the crust thickened substantially, reaching its modern thickness of 60–70 km (McGlashan *et al.* 2008; Bianchi *et al.* 2013), the Puna plateau to the east was uplifted, and the Chilean–Pampean flatslab region developed as arc volcanism terminated south of 28°S. These events are contemporaneous with the southward passage of the subducting aseismic Juan Fernandez ridge on the Nazca plate through this region (Yañéz *et al.* 2001). The subduction of the ridge has been suggested to correlate with an episode of shallowing of the Nazca plate under the southern Puna, which preceded the steepening that facilitated delamination of the lowermost crust and lithosphere after 7 Ma (e.g. Kay *et al.* 1994*a*, *b*, 1999; Kay & Coira 2009; Bianchi *et al.* 2013) and the tectonic removal of *c.* 40 to 50 km of forearc crust by forearc subduction erosion between *c.* 8 Ma and 3 Ma as the frontal arc migrated eastward (Kay & Mpodozis 2002; Goss & Kay 2009; Goss *et al.* 2013).

In this paper, we review the chronology of Neogene volcanic events between 28°S and 25°S and examine the changing major and trace element signatures, and Nd, Sr and Pb isotopic ratios of the 26 to 0 Ma arc magmas that erupted in the arc and proximal back-arc. The changing magmatic source regions of these magmas are interpreted in view of their evolving tectonic setting and relative to geochemical, petrological and isotopic changes. The analysis shows how the magmas evolved at different lithospheric levels as crustal thickness changed in a fluctuating stress regime, the mantle wedge adjusted to shallowing and steepening of the

From: Gómez-Tuena, A., Straub, S. M. & Zellmer, G. F. (eds) 2014. *Orogenic Andesites and Crustal Growth*. Geological Society, London, Special Publications, **385**, 303–334.
First published online August 7, 2013, http://dx.doi.org/10.1144/SP385.11

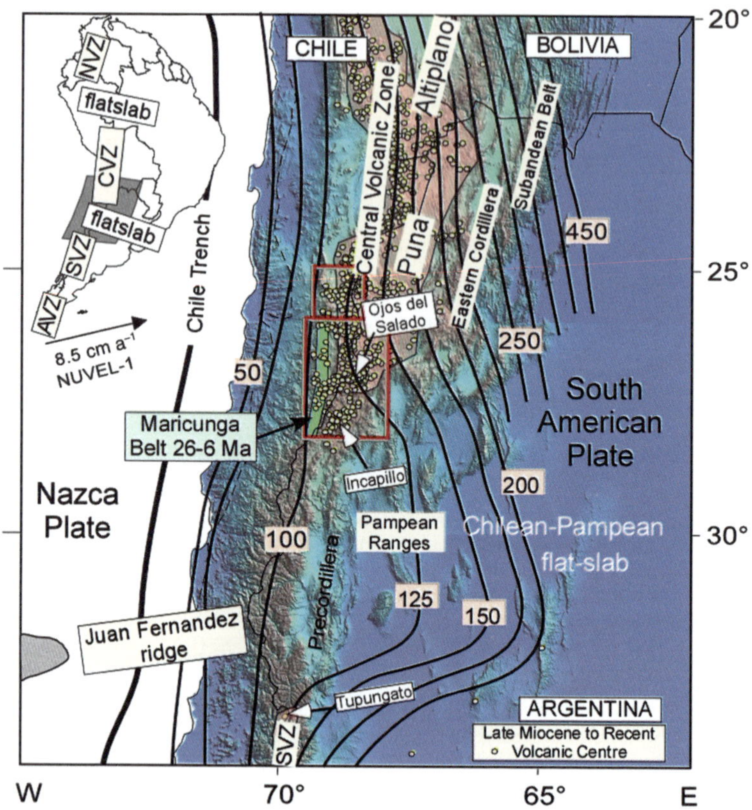

Fig. 1. Satellite Radar Topography Mission (SRTM) image showing the main morphotectonic features of the Central Andes. The Nazca plate convergence direction is the integrated NUVEL-1 rate from DeMets *et al.* (1990). Dark solid lines show contours of the subducting Nazca plate with depths in kilometers to the Wadati–Benioff zone as modified by Mulcahy *et al.* (2010) and Mulcahy (2012) from Cahill & Isacks (1992). Small circles show late Miocene to Recent volcanic centres of the Andean Central (CVZ) and Southern (SVZ) Volcanic Zones. Green box shows position of the early to late Miocene Maricunga Belt volcanic arc (Kay *et al.* 1994*a*; Mpodozis *et al.* 1995). White tipped arrows point to the southern CVZ Pleistocene to Recent Ojos del Salado volcanic and Incapillo caldera complex regions and the young SVZ Tupungato volcanic centre. The southern red box shows the region shown in detail in Figure 2; the northern red box shows the region of the volcanic centres with analyses in Trumbull *et al.* (1999), Siebel *et al.* (2001) and Schnurr *et al.* (2007).

subducting plate, forearc crust was introduced into the mantle wedge as the frontal arc migrated at 8–3 Ma, and in the aftermath of lower continental lithosphere removal in the back-arc.

Geological and tectonic setting

The largely Neogene (26–0 Ma) volcanic rocks discussed here erupted along the western margin and just south of the Puna plateau near the southern termination of the active CVZ arc and the northern limit of the modern Chilean–Pampean flat-slab (Fig. 1). To the north of this region, the Nazca plate is subducting at an angle of *c.* 30°, whereas to the south, a near-horizontal flat-slab segment forms at a depth of 100 km (Cahill & Isacks 1992; Mulcahy 2012; Fig. 1). The region is located north of the area where the contours of the Wadati–Benioff zone bend eastward under the Incapillo volcanic complex (Goss *et al.* 2009, 2011), which is the southernmost volcanic centre with Pleistocene activity until reaching the Tupungato volcano in the northernmost Southern Volcanic Zone (SVZ) at 33°S (Fig. 1). The Wadati–Benioff zone is currently at *c.* 105 km under the 26–6 Ma extinct Maricunga Belt arc and at *c.* 120–125 km under the active southern CVZ to the east (Mulcahy 2012). Based on depth phase precursors (McGlashan *et al.* 2008) and receiver functions (Bianchi *et al.* 2013), modern crustal thicknesses reach *c.* 60–70 km across the region.

Neogene arc volcanism began in the region when the late Oligocene break-up of the Farallon plate led to the onset of rapid, near-normal convergence between the Nazca and South American plates (e.g. Pardo-Casas & Molnar 1987; Somoza 1998). The convergence rate slowed through the Miocene as the Puna plateau uplifted, and the Nazca plate has been subducting at an average rate of 84 mm a^{-1} over the last 3 Ma at an angle of 075°N (DeMets *et al.* 1990). The volcanism that initiated in the Maricunga belt arc (Figs 1 & 2a) at *c.* 26 Ma persisted until *c.* 6 Ma when the arc front was displaced eastward to the present location in the southern CVZ (e.g. Kay & Mpodozis 2002).

The evolution of volcanic activity in the region discussed in this paper is considered relative to five zones. The first, and westernmost, is the *c.* 26–6 Ma Maricunga Belt arc; the second is the *c.* 26–14 Ma Maricunga Belt proximal back-arc; the third is the *c.* 10–3 Ma Maricunga Belt distal back-arc and the post Maricunga-belt syn-migration volcanic front; the fourth is the <3 Ma CVZ arc, some 40–50 km east of the extinct Maricunga Belt arc; and the fifth includes the 25–0 Ma arc and back centres in the region located further north between about 25°S and 26.4°S. Overviews of the geology, volcanology and geochemistry of these zones can be found in various sources including González Ferrán *et al.* (1985), Baker *et al.* (1987), Kay *et al.* (1994*a*, 2008), Mpodozis *et al.* (1995, 1996, 1997), Gardeweg *et al.* (1997, 1998, 2000), Trumbull *et al.* (1999), Kay & Mpodozis (2001, 2002), Clavero *et al.* (in press), Schnurr *et al.* (2007), Kay & Coira (2009) and Goss *et al.* (2009, 2011, 2013). The principal volcanic centres and geographical features of the region are shown in Figures 1 and 2 and summarized in Table 1.

Maricunga Belt arc (26–6 Ma)

The *c.* 200 km long late Oligocene to late Miocene Maricunga Belt is located some 40–50 km west of the <3 Ma CVZ front. Magmatic activity began there at *c.* 26 Ma and lasted until *c.* 6 Ma, when the volcanic front was displaced eastward. Details on the geology and volcanic centres of the Maricunga Belt between 26°S and 28°S can be found in papers by Davidson & Mpodozis (1991), Mercado (1982), Vila & Sillitoe (1991), Walker *et al.* (1991), Sillitoe (1991), Kay *et al.* (1991, 1994*a*, 1999), Cornejo *et al.* (1993, 1998), McKee *et al.* (1994), Mpodozis *et al.* (1995, in press) and Tomlinson *et al.* (1999) among others.

The evolution of the Maricunga Belt can be viewed in three main stages. In the first, *c.* 26–20 Ma volcanic complexes (green shades in Fig. 2) erupted over a moderately dipping subduction zone through a *c.* 45 km thick crust in the north and a *c.* 35–40 km thick crust in the south (see Kay & Mpodozis 2002). These centres include the large andesitic to dacitic Cerros Bravos stratovolcanic complex in the north and smaller clusters of dome complexes to the south. Many are hydrothermally altered and they host widespread epithermal and porphyry style gold–silver mineralization (e.g. La Coipa, Pantanillo, Refugio; e.g. Vila & Sillitoe 1991; Mpodozis *et al.* 1995; Muntean & Einaudi 2001). This stage terminated with a pulse of contractional deformation and a near volcanic-lull from *c.* 20 to 17 Ma. The next stage, in the middle Miocene (16–12 Ma) was marked by the eruption of voluminous stratovolcanic complexes which, from north to south, include the Doña Inés, Ojos del Maricunga, Santa Rosa, Lagunillas, La Laguna, Jotabeche Norte and Cadillal-Yeguas Heladas centres, some of which have associated dacitic ignimbrites. This stage ended with the emplacement of 12–10 Ma amphibole-bearing mineralized Au–Cu porphyries (e.g. Marte, Lobo, Casale-Alderaban; Vila & Sillitoe 1991; Mpodozis *et al.* 1995). The final stage was marked by a radical change, with volcanic activity first concentrated in the large *c.* 11–7 Ma Copiapó-Azufre (6300 m) dacitic complex and then terminating at *c.* 7 to 5 Ma with the eruption of rhyodacitic ignimbrites from the Jotabeche caldera (5800 m) and the small glassy adakitic Pircas Negras andesitic flows in the south (e.g. Goss *et al.* 2013). The crust under the Maricunga Belt is considered to have reached a thickness of over 50 km by *c.* 10 Ma (Kay *et al.* 1994*a*).

Maricunga Belt proximal back-arc (26–14 Ma)

Intermittent volcanic activity (pink-shaded colours in Fig. 2) also occurred in the proximal back-arc of the Maricunga belt from 26 to 14 Ma. The early tectonic and volcanic history is recorded in the sedimentary and volcaniclastic sequences in the Pedernales–Maricunga basin and the Cordillera Claudio Gay (26–27.2°S), where abundant reworked dacitic pyroclastic deposits occur in the Claudio Gay Sequence and scattered dacitic dome complexes with K/Ar ages of 23–24 Ma are found (Mpodozis & Clavero 2002; Clavero *et al.* in press). The Claudio Gay Sequence includes the tholeiitic Segerstrom mafic flows with K/Ar ages from 25.0 ± 0.9 Ma to 23.4 ± 1.5 Ma (Mpodozis *et al.* 1995, 1997). Importantly, the Segerstrom lavas include rare primitive olivine basalts, which are also present further south near 30°S over the modern Chilean–Pampean flat-slab region (*c.* 22 Ma Máquinas flows; Kay *et al.* 1991). In the same period, *c.* 23–22 Ma dacitic ignimbrites and 21–18 Ma mafic andesitic to andesitic flows equivalent to those in the Tilito and Escabroso

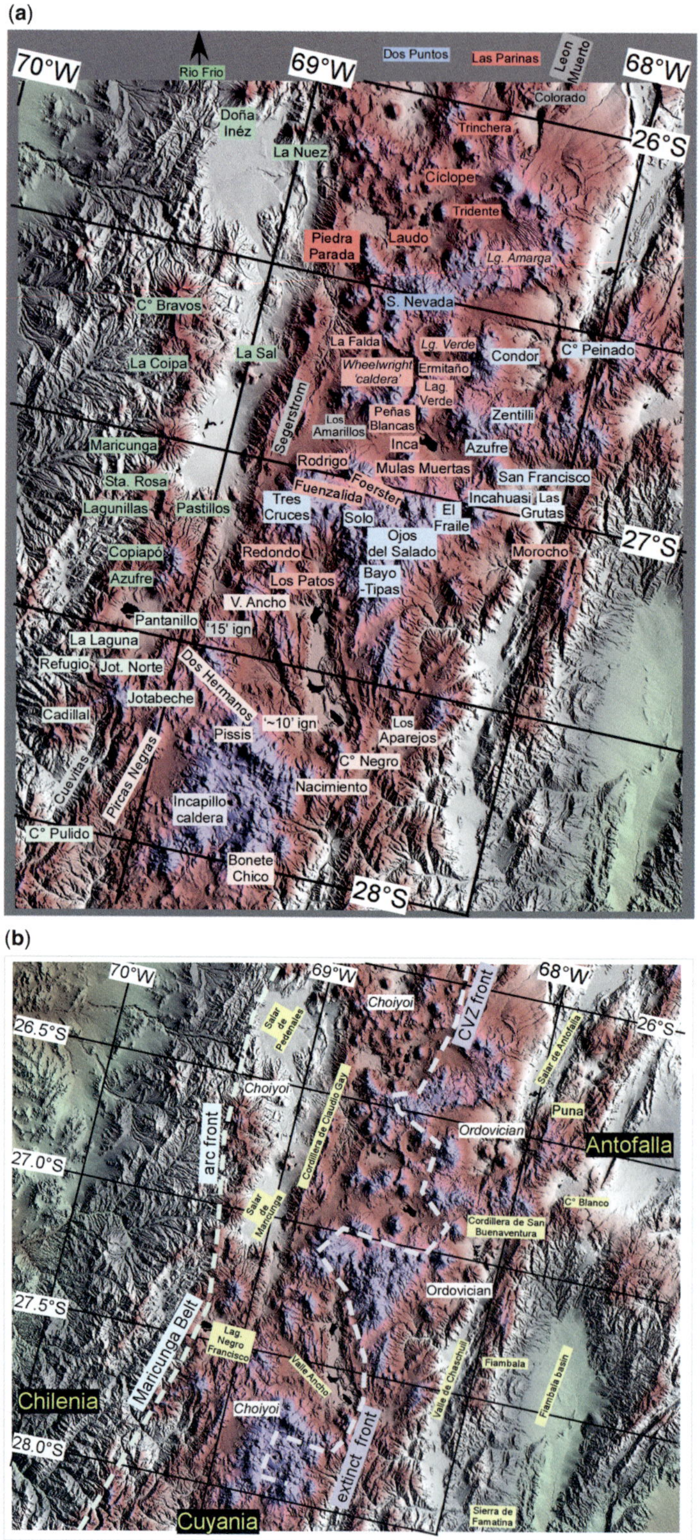
(a)
70°W
69°W
68°W
26°S
27°S
28°S
Rio Frio
Dos Puntos
Las Parinas
Leon Muerto
Colorado
Doña Inéz
La Nuez
Trinchera
Ciclope
Tridente
Piedra Parada
Laudo
Lg. Amarga
S. Nevada
C° Bravos
La Sal
La Falda
Lg. Verde
Condor
C° Peinado
La Coipa
Wheelwright 'caldera'
Ermitaño
Lag. Verde
Segerstrom
Los Amarillos
Peñas Blancas
Zentilli
Maricunga
Inca
Azufre
Rodrigo
Mulas Muertas
San Francisco
Sta. Rosa
Foerster
Fuenzalida
Incahuasi
Las Grutas
Lagunillas
Pastillos
Tres Cruces
Solo
El Fraile
Ojos del Salado
Morocho
Copiapó
Redondo
Azufre
Los Patos
Bayo -Tipas
V. Ancho
Pantanillo
'15' ign
La Laguna
Refugio
Jot. Norte
Dos Hermanos
Jotabeche
Cadillal
'~10' ign
Pissis
Los Aparejos
C° Negro
Cuevitas
Pircas Negras
Nacimiento
Incapillo caldera
C° Pulido
Bonete Chico
(b)
70°W
69°W
68°W
26°S
26.5°S
27.0°S
27.5°S
28.0°S
Choiyoi
Salar de Pedernales
CVZ front
Salar de Antofalla
Choiyoi
Cordillera de Claudio Gay
Puna
arc front
Ordovician
Antofalla
Salar de Maricunga
C° Blanco
Cordillera de San Buenaventura
Ordovician
Maricunga Belt
Lag. Negro Francisco
Valle Ancho
Valle de Chaschuil
Fiambala
Fiambala basin
Chilenia
Choiyoi
extinct front
Sierra de Famatina
Cuyania

Formations in the El Indio region at 29–30°S (e.g. Kay *et al.* 1991, 1999; Bissig *et al.* 2001) erupted in the proximal back-arc near 27.5–28°S (Cuevitas region in Fig. 2b; see Goss & Kay 2009; Goss *et al.* 2013). This period, which is generally considered to have been in an extensional tectonic regime (e.g. Mpodozis & Clavero 2002), ended with east-verging high-angle reverse faults uplifting the Late Palaeozoic basement in the northern Cordillera Claudio Gay, and as 20–19 Ma dome complexes and extensive block and ash flows (Sierra de Aliste sequence) erupted to the east of the Pedernales and Maricunga basins (Mpodozis & Clavero 2002). The uplift that initiated at this time continued into the mid-Miocene, as recorded by thick syntectonic gravel sequences interbedded with distal 15–16 Ma ignimbrites (Rio Lamas sequence) south of the Cordillera Claudio Gay (Fig. 2). The magmatism of this period ended with the emplacement of the *c.* 14 Ma Cerros Amarillos domes, east of the Cordillera Claudio Gay, and the eruption of *c.* 15–14 Ma dacitic ignimbrites in the western Valle Ancho (27.4°S, Fig. 2).

*Late Maricunga back-arc and migrating volcanic arc (*c. *8–4 Ma)*

From 8 to 3.5 Ma, important volcanic eruptions occurred between the ebbing and extinct Maricunga Belt arc front and the future CVZ arc front (pink-shaded colours in Fig. 2; e.g. Mpodozis *et al.* 1996, 1997; Kay & Mpodozis 2001, 2002; Kay *et al.* 2006, 2008; Goss & Kay 2009; Mpodozis & Kay 2009; Goss *et al.* 2013). The immediate precursors were 11–9 Ma dacitic domes, porphyries and ignimbrite flows and *c.* 9.5–9.1 Ma mafic andesitic flows that erupted along the NNE-trending Los Aparejos fault in the Valle Ancho. These eruptions were followed by *c.* 7.7 to 6.6 Ma mafic andesitic to andesitic flows erupted along NW- and NNW-trending faults, east of the *c.* 7.7 Ma Dos Hermanos volcano. Isolated andesitic to dacitic lavas of similar age erupted near *c.* 26.5–27.5°S at the Laguna Verde (*c.* 8.9 Ma), Los Patos *c.* (7.6 Ma) and Cordon Foerster (*c.* 7.8–7.9 Ma) stratovolcanoes and the *c.* 7.6 Ma Morocho centre to the east (Rubiolo *et al.* 2000). The distinctive *c.* 8–3 Ma adakitic Pircas Negras andesitic lavas (Kay *et al.* 1991; Mpodozis *et al.* 1995; Goss *et al.* 2013) erupted at isolated, small centres east of the southern Maricunga belt as andesitic to dacitic lavas were emplaced along the Robertson Escarpment (*c.* 6.5 Ma) and at the Wheelwright (*c.* 5.9–5.0 Ma) and Mulas Muertas (*c.* 5.0–5.1 Ma) centres near 27°S (Fig. 1).

The construction of large andesitic to dacitic volcanic centres began at *c.* 6.5–3.5 Ma with those in the south, including the Nevado Pissis (*c.* 6.6–6.2 Ma), Cerro Nacimiento de Jagüe (*c.* 6–3.5 Ma), Cerro Negro (*c.* 5.5–4.5), Bonete Chico, Bonete Grande and Fandango centres. Further north, dacitic lavas and domes created parts of the La Barda Complex (*c.* 4.9 Ma) and Mulas Muertas centres (*c.* 5.6 Ma), and the 5.8 Ma Fuenzalida flow banded rhyolite dome was emplaced. Subsequent widespread activity from 4.5 to 3 Ma includes the eruption of andesitic to dacitic lavas at the Los Patos stratovolcano (*c.* 4.5 Ma), a series of eruptions near Nevado de Tres Cruces (*c.* 4.5–4.0 Ma; Gardeweg *et al.* 1997), including the 4.4 Ma adakitic Rodrigo andesitic cone, and the andesitic lavas erupted at Peñas Blancas (*c.* 4.2 Ma), Ermitaño (*c.* 3.7–3.6 Ma) and a centre north of the Laguna Verde volcano (*c.* 3.9 Ma) in the 15 km wide topographic ring called the Wheelwright Caldera (Baker *et al.* 1987; Clavero *et al.* 2000). A change in this regime is signalled by the eruption of the 4 to 3.7 Ma Laguna Verde and Laguna Amarga rhyolitic ignimbrites.

CVZ arc from 26.4° to 28°S (<3.5 Ma)

From 3.5 to 2 Ma, volcanism was concentrated in the Ojos del Salado and Nevado de Tres Cruces complexes in the north and in the Incapillo region to the south (Figs 1 & 2b). Among the oldest volcanic rocks emplaced at this time are thick dacitic

Fig. 2. Satellite Radar Topography Mission (SRTM) images showing centres and localities discussed in this paper. Colours on images correspond to elevation with highest in blue (*c.* 5500–6800 m) and red (darkest is *c.* 4500 m), intermediate in white (clearest white is *c.* 3800 m) and grey (darkest is *c.* 2500 m) and lowest in green and brown. (**a**) Names of volcanic centres are grouped by colour of labels. Green: 26–7 Ma Maricunga Belt arc centres; grey: 24–15 Ma Maricunga Belt back-arc centres; red or rose: *c.* 9–4 Ma syn-migration centres; blue: <3 Ma post-migration southern Central Volcanic Zone arc. Centres are put in three geographic regions by colour shades of labels. Descriptions for southern (light shades) and central (intermediate shades) centres are in González Ferrán *et al.* (1985), Baker *et al.* (1987), Mpodozis *et al.* (1995, 1996, 1997), Kay *et al.* (1991, 1994*a*, 1999, 2008) and Goss *et al.* (2011, 2013) and references therein. Descriptions for centres to the north (darkest shades) in Trumbull *et al.* (1999), Siebel *et al.* (2001), Mpodozis & Clavero (2002) and Schnurr *et al.* (2007) and references therein. (**b**) Image shows region in rectangles (thick lines) in Figure 1. Maricunga Belt arc front is indicated as light green dashed line and CVZ front as a light blue dashed line. Names of major geographical features mentioned in text are shown with yellow backgrounds, representative regions of Ordovician and Permo-Triassic basement are shown by white labels and names of Palaeozoic terranes are in yellow letters with black backgrounds.

Table 1. *Summary of Central Andean geological and magmatic history from 25 to 0 Ma between* c. *28.2°S and 25.5°S latitude*

South			to		North
Early Miocene Maricunga Belt arc					
c. 25–20 Ma	Cerro Pulido, Chacay andesites	Au (Cu) porphyries	Cerros Bravos (La Coipa) dacitic stratovolcano and dome complex		
	Cantarito ignimbrites	(Refugio, Caspiche)			Río Frío ignimbrites
Proximal Maricunga Belt Back-arc					
c. 23–22 Ma	Tilito ignimbrites		Claudio Gay Sequence (pyroclastic and sedimentary rocks)		
		Valle Ancho porphyries	*c.* 25–23 Ma Segerstrom basalts to mafic andesites		
c. 21–18 Ma	Escabroso mafic andesites				*c.* 20 Ma León Muerto mafic flows
			Sierra de Aliste Sequence (20–19 Ma)		Colorado andesites
c. 19–17 Compression and near volcanic lull					
c. 16–9 Ma Maricunga Arc andesitic centres (many with hornblende)					
c. 16–13 Ma Arc front andestic to dacitic stratovolcanoes					
Cadillal,Yeguas Heladas, Jotabeche Norte, La Laguna centres					Northern centres
		Pastillos, Pastillitos, Santa Rosa, Lagunillas	Ojos del Maricunga	Doña Inés	Chaco
		c. 12–10 Ma Au (Cu) Porphyries			(Cerro Blanco, Cerro Azufre, etc)
		c. 11–10 Ma Copiapó (Azufre) dacitic ignimbrites and domes			*c.* 13–11 Ma ignimbrites
c. 16–9 Ma Proximal Maricunga Belt arc back-arc					
			c. 20–19 Ma Sierra de Aliste sequence Vitrophyres		
	c. 15–14 Ma Valle Ancho dacitic Ignimbrites		*c.* 14 Ma Rio Lamas sequence		
			c. 14 Ma Cerros Amarillos domes		*c.* 14–12 Ma La Falda ignimbrite-dome-porphyries
	c. 10–9 Ma Valle Ancho Ignimbrites, Aparejos mafic lavas		Pedernales ignimbrites		
		c. 9 Ma Laguna Verde (Wheelwright complex)			Tridente and northern centres

c. 8–4.5 Ma Dying Maricunga Belt arc, eastward migrating arc and back-arc					
Dying Maricunga Belt arc	*c.* 7–5 Ma Jotabeche rhyodacitic ignimbrites		*c.* 8–7 Ma Copiapó dacites		
Migrating arc and back-arc *c.* 8–5 Ma mafic andesitic to dacite					Northern Centres
	Dos Hermanos	Los Patos	Foerster	Morocho (Las Grutas)	Aguas Blancas, Dos Craters, Juncalito
	Pircas Negras andesites		Wheelwright Complex Centres, Juncalito		Colorado Plomizo, Sierra de Gorbea
		Valle Ancho mafic lavas	Mulas Muertas	*c.* 4.4 Ma Rodrigo andesites	
6–5 Ma rhyodacitic to rhyolitic ignimbrites			*c.* 6–5 Ma Fuenzalida, Redondo		*c.* 6–5 Ma Parinas, Ciclope, Trinchera, León Muerto ignimbrites
c. 4–3 Early Central Volcanic Zone arc (primarily andesite and dacite)					
'Pissis-Bonete' stratovolcano-dome complexes; Pircas Negras (Cerro Nacimiento de Jagüe, Nevado Pissis, Bonete Chico, Bonete Grande, Fandango)					
Incapillo domes		Los Patos	>3–2 Ma Nevado Tres Cruces – Ojos del Salado – Las Grutas region		
		c. 3.3 Ma Inca mafic lavas	4–3.8 Ma Laguna Amarga-Laguna Verde ignimbrites		
<3 Ma Central Volcanic Zone arc					
Incapillo dacitic Ignimbrites		Tipas (Bayo)	Ojos del Salado-Nevado Tres Cruces- El Fraile domes		Sierra Nevada
<3 Ma Volcanic complexes		Peinado, San Francisco, Incahuasi, Zentilli, Falso Azufre Las Grutas (Cordillera de San Buenaventura domes)			Volcán Laguna, Azufrera, Los Cuyanos

3.4–3.5 Ma flows under the main younger Ojos del Salado volcanic complex, mafic andesite flows at the El Muertito (3.2 Ma) and del Inca (3.3 Ma) volcanoes and 3–2 Ma dacitic lavas near Nevado de Tres Cruces (Gardeweg *et al.* 1997, 2000). Other dacitic to rhyolitic activity occurred to the south at the Cerro Bayo–Tipas complex (2.9–1.2 Ma) and minor flows and domes (*c.* 2.4–1.7 Ma) were erupted in the Valle Ancho region.

The most recent volcanic activity of the active southern CVZ arc is concentrated at the major centres around the Ojos del Salado complex. The main volcanic edifices of this period, shown in Figure 2, include Sierra Nevada near 26.5°S, El Condor and Cerro Peinado near 26.6°S, Falso Azufre and San Francisco near 26.8°S and a *c.* 50 km long east–west-trending line of major centres just south of 27°S that includes Nevado Tres Cruces (6330 m), Solo (6190 m), Ojos del Salado (6880 m), El Muerto (6420 m), Frailes (>6000 m) and Incahuasi (6610 m). Other smaller centres include 2.3 to <1 Ma domes and silicic flows. Among these is the chain of domes along the ENE-trending faults in the Cordillera de San Buenaventura (Seggiaro *et al.* 2002; Kay *et al.* 2006). To the south are the isolated Bayo (also called Tipas) centre near 27.3°S and the Incapillo ignimbrite-dome complex near 27.8°S, which is notable for an extensive NW–SE-orientated dome field and an ignimbrite with an Ar/Ar age of 0.5 Ma (Goss *et al.* 2009, 2011). These centres, whose K/Ar and Ar/Ar ages are mostly less than 2 Ma (Zentilli 1974; González Ferrán *et al.* 1985; Mpodozis *et al.* 1996, 1997; Gardeweg *et al.* 1997, 1998, 2000; Clavero *et al.* 2000; Seggiaro *et al.* 2002; Kay *et al.* 2006, 2008; Schnurr *et al.* 2007; Goss *et al.* 2009; Mpodozis & Kay 2009) constitute more than 50% of the volcanic volume in this zone.

The western centres in the west–east-trending Ojos del Salado line (Nevado Tres Cruces, Cerro Solo, Ojos del Salado, El Muerto, El Fraile) are multiple dacitic domes, lava domes and flow complexes. The westernmost Nevado Tres Cruces complex has three large dacitic cones located along a NNW-trending fault, the southernmost of which likely erupted the 100 m thick Tres Cruces ignimbrite (0.067 Ma, Ar/Ar age in Gardeweg *et al.* 2000). The very large Cerro Solo (1.3 Ma) and El Fraile dacitic domes flank the active Ojos del Salado volcano (the highest on Earth), which is composed of multiple dacitic domes, lava-domes and short lava flows erupted from centres clustered in an 13 × 12 km area. The oldest part of the young Ojos del Salado complex is early Pleistocene in age, as shown by K/Ar age of 1.53 ± 0.13 Ma and 1.08 ± 0.09 Ma, with lavas and domes near the summit yielding a younger Ar/Ar age of 0.34 ± 0.19 Ma. The large Bayo-Tipas, San Francisco, Falso Azufre, Incahausi, Condor and Cerro Peinado volcanic centres are typical Andean stratovolcanoes dominated by pyroxene–hornblende andesites containing scarce biotite and rare olivine xenocrysts. The Incahuasi and San Francisco centres are notable for having olivine-bearing basaltic andesitic monogenetic parasitic cones.

25–0 Ma volcanism in the northern arc and back-arc from 25°S to 26.4°S

The northern zone includes the arc and back-arc volcanic centres located north of the Maricunga belt proper (Figs 1 & 2), which have been mapped, dated and discussed by Naranjo & Cornejo (1993), Mpodozis *et al.* (1995), Cornejo & Mpodozis (1996), Cornejo *et al.* (2009) and Clavero *et al.* (2000, in press). Additional ages and sketch maps with information on centres considered in this paper can be found in Trumbull *et al.* (1999), Siebel *et al.* (2001) and Schnurr *et al.* (2007).

A decreasing age progression of volcanic centres towards the east is not as obvious in this zone, as early Miocene eruptions were widespread. Large early Miocene eruptions produced the extensive Rio Frio–Llano Las Vicuñas dacitic to rhyodacitic ignimbrites in the west, whose ages are mostly near 18 Ma (Naranjo & Cornejo 1993; Cornejo & Mpodozis 1996; Cornejo *et al.* 2009), but can be up to 25.9 Ma (see Siebel *et al.* 2001). In the same period, 19.9–19.6 Ma back-arc basaltic andesites and andesites erupted from centres at Léon Muerto near 25°57′S and Colorado near 25°45′S. Following a volcanic lull that was nearly contemporaneous with that in the Maricunga Belt further south, widespread middle- to early-late Miocene volcanism resumed. Among centres erupting in or near the frontal arc were Chaco (17–15.6 Ma) near 25°30′S, Puntaguido de la Isla (16.4 Ma) near 25°58′S, Cerro Blanco (15.7 Ma) near 25°30′S, Cerro del Azufre (11.2–19.7 Ma) near 25°20′S and Piedra Parada (9.5 Ma) near 26°25′S. One of the back-arc centres is Tridente (11.7 Ma) near 25°15′S. These generally andesitic eruptions were accompanied by silicic eruptions like those that produced the 13.0 Ma Plato Superior (26°28′S), 11.0 Ma Azufre (25°21′S) and 10.7 Ma Pedernales (26°21′S) ignimbrites (Siebel *et al.* 2001).

In the late Miocene and Pliocene, volcanic centres erupted to the east at the same time as the ebbing Maricunga Belt arc and syn-migration centres erupted to the south. Among the lavas erupted are the 8.3 Ma Aguas Blancas (25°20′S, 68°30′W), 7.9 Ma Dos Craters (25°36′S, 68°36′W), 5.7 Ma Colorado Plomizo (26°26′S, 68°39′W) and *c.* 5 Ma Sierra de Gorbea (25°30′S, 68°45′W) andesites (Trumbull *et al.* 1999). Contemporaneous silicic rocks include the 6.1 Ma Juncalito (26°31′S), 5.9 Ma Cíclope (26°13′S), 5.1 Ma Trinchera (26°4′S), 5.1 Ma Parinas (25°59′S) and 5.1 Ma Léon Muerto (25°52′S) ignimbrites. Among the youngest activity in this zone is that at the <2 Ma Bayo (25°25′S) and Dos Puntos (25°58′S) and the <1 Ma Cordon del Asufre (25°19′'S) and Lastaria (25°7′S) CVZ centres in the north and the 3.1 Ma Cerros Colorados (25°9′S) and 1.2–1 Ma Volcán Laguna (26°10′S), Azufrera Cuyanos (26°25′S) and Sierra Nevada (26°24′S) CVZ centres to the south.

Whole-rock major, trace element and isotopic analyses

The major and trace element analyses and Sr, Nd and Pb isotopic ratios plotted in Figures 3–12 are from McNutt *et al.* (1975), González Ferrán *et al.* (1985), Baker *et al.* (1987), Kay *et al.* (1991, 1994*b*, 1999, 2006, 2010), Walker *et al.* (1991), Moscoso *et al.* (1993), McKee *et al.* (1994), Kay & Abbruzzi (1996), Mpodozis *et al.* (1995, 1996, 1997), Trumbull *et al.* (1999), Siebel *et al.* (2001), Schnurr *et al.* (2007), Goss & Kay (2009), Goss *et al.* (2009, 2011, 2013) and Mpodozis & Kay (2009) and our unpublished data. In the case of the unpublished data, major elements were either measured on glass beads made from rock powders in an Ar atmosphere using the JEOL 733 electron microprobe (EMPA) in the Cornell Center for Materials Research and primary standards from the Smithsonian Institute, or in the geochemical facilities of the SERNAGEOMIN in Santiago, Chile. In all cases, ignimbrite analyses were carried out on pumices or lithic-free matrix samples. All samples were analysed for Na, La, Ce, Nd, Sm, Eu, Tb, Yb, Lu, Sr, Ba, Cs, U, Th, Hf, Sc, Ta, Cr, Ni and Co by Instrumental Neutron Activation (INAA) at Cornell University with analyses of other elements by atomic absorption and other techniques at the SERNAGEOMIN. $^{87}Sr/^{86}Sr$, $^{143}Nd/^{144}Nd$ and Pb isotopic ratios were measured on a VG Sector 54 Thermal Ionization mass spectrometer in the Keck Isotope Laboratory at Cornell. Multiple run values for standards are 0.7102957 for $^{87}Sr/^{86}Sr$ on NBS987 and 0.5121436 and 0.5118496 for $^{143}Nd/^{144}Nd$ on Ames and La Jolla, respectively. Pb isotopic analyses are relative to NBS standard values of $^{206}Pb/^{204}Pb = 16.937$, $^{207}Pb/^{204}Pb = 15.49$ and $^{208}Pb/^{204}Pb = 36.695$. Additional details on the methods can be found in Goss *et al.* (2013).

Major and trace element data

The major and trace element contents of analysed samples are summarized in plots of wt% K_2O v. SiO_2 (Fig. 3), wt% Na_2O v. La/Yb (Fig. 4), wt% SiO_2 v. Sm/Yb (Fig. 5), wt% SiO_2 v. La/Ta (Fig. 6) and Ba/La v. Th/La ratios (Fig. 7). From the many elemental and ratio indicators possible, wt% SiO_2 and K_2O were chosen as measures of rock type and differentiation; wt% Na_2O as an index of source rock composition, residual plagioclase and, in some sense, melting percentage (when Na is an incompatible element); La/Yb and Sm Yb ratios as measures of rare earth element (REE) pattern steepness, with the Sm/Yb ratios particularly sensitive to high pressure residual phases like amphibole and especially garnet, La/Ta ratios as measures of relative high field strength element (HFSE) depletion to the REE and general indicators of subduction zone character; Ba/La ratios as potential measures of slab fluids; and Th/La and Ba/Th ratios as potential indicators of crustal contaminants including melted subducted sediments. La/Ta and Ba/La ratios and REE ratios can also be modified by contaminants from the underlying crust or crust introduced into the mantle by forearc subduction erosion, sediment subduction or continental lithospheric delamination as is further discussed below.

The samples are grouped into five time periods on the plots: (1) 25–18 Ma early Maricunga arc prior to major contraction; (2) 16–9 Ma premigration post-contraction; (3) 8–5 Ma terminal Maricunga arc and early syn-migration; (4) 4–3 Ma late syn-migration; and (5) <3 Ma post arc migration CVZ stage. The fields on the plots follow the colour shading shown in Figure 2b with Maricunga Belt arc fields in green, early Maricunga back-arc fields in grey, <17 Ma back-arc and syn-migration fields in red and CVZ arc fields in blue. The lightest colour shades are for centres in the south, intermediate shades for centres in the central region and darker shades for centres further north, with the approximate boundaries being near the Valle Ancho at 27.5°S and the Wheelwright caldera at 26.75°S. Fields for samples from the Maricunga Belt and CVZ front arcs are striped. Field sizes reflect compositional diversity and not volume and can be biased by sampling density. A summary of the centres plotted on the graphs is shown relative to age and relative location from south to north in Table 1.

Wt% K_2O v. wt% SiO_2. The plots in Figure 3 show that most of the analysed samples fall in the medium to high K_2O fields for arc volcanic rocks with the majority plotting near or within the high K_2O field. The samples that fall in or near the medium K_2O field are generally from the early to mid-Miocene Maricunga Belt and/or from south of the Valle Ancho (Fig. 2). The overall range of wt% SiO_2 for all samples covers the spectrum from

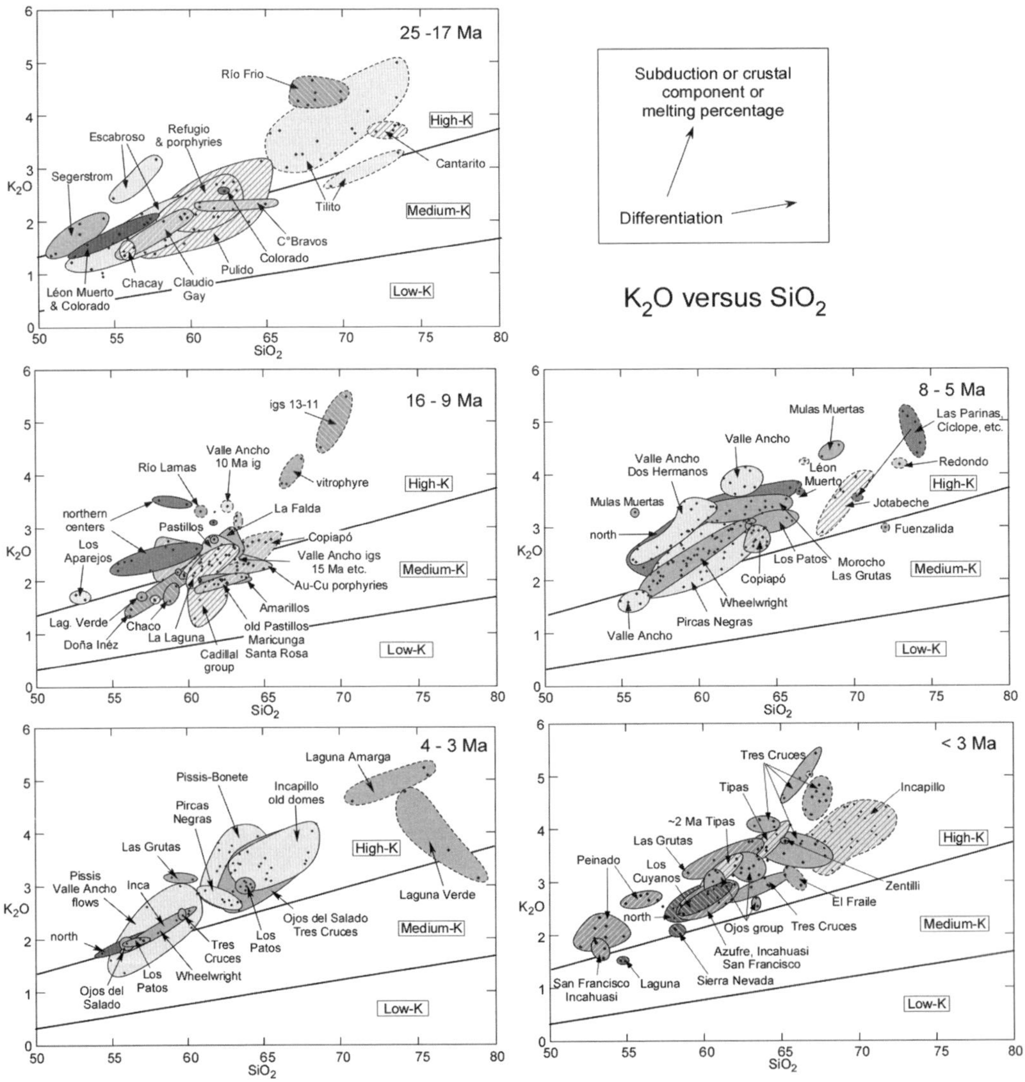

Fig. 3. Plot of wt% K_2O v. wt% SiO_2 with boundaries for low-K, medium-K and high-K andesites as in Mamani *et al.* (2010). Samples are grouped by age with colours and shades of fields and names keyed to centres listed in Table 1 and shown in Figure 2. Striped patterns are for frontal arc groups. Dashed fields are for ignimbrites and dome complexes. Points are analyses; data are from sources listed in text.

basalt to rhyolite with many of the most silicic and most mafic samples being of early Miocene (15–17 Ma) or latest Pliocene to Pleistocene (<4 Ma) age. Most of the 16–3 Ma samples plot between 55 and 67 wt% SiO_2 with the 16–9 Ma samples generally falling near the medium-high K_2O boundary and the <8 Ma samples largely being in the high K_2O field. The plots also show that samples from the west generally have lower wt% K_2O than those from further east, as seen by comparing Maricunga belt 25–18 Ma Pulido–Chacay and 8–5 Ma Copiapó–Jotabeche samples (striped-green-shaded fields) with similar age and younger samples from further east. However, there are exceptions as the frontal arc early Miocene Río Frío ignimbrites in the north have generally higher K_2O contents than the overlapping age proximal back-arc Tilito ignimbrites further south, and <3 Ma Tres Cruces volcanic rocks having higher wt% K_2O than contemporaneous CVZ volcanic rocks to the east.

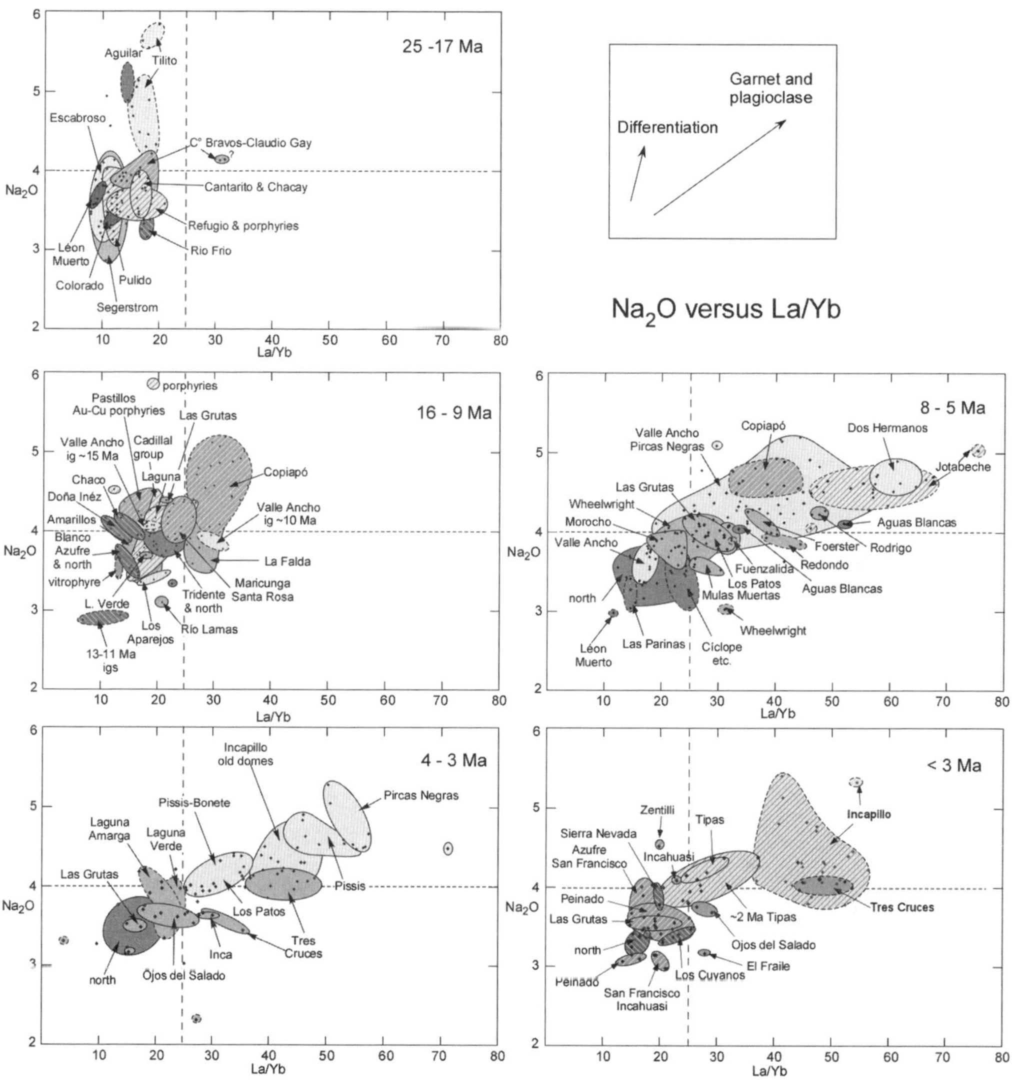

Fig. 4. Plot of wt% Na_2O v. La/Yb ratios for the same sample groups in Figure 3. Fields and data sources are as in Figure 3. Note extreme range of La/Yb ratios at 8–5 Ma compared to other periods and increase in ratios after 8 Ma.

Wt% Na_2O v. La/Yb ratio. The plots in Figure 4 show that wt% Na_2O and La/Yb ratios span a broad range with Na_2O from 2.9 to 5.3 wt% and La/Yb from 9 to *c.* 80. The higher wt% Na_2O (4–5.2 wt%) values are prominent in samples with La/Yb ratios above 25, but also occur in 25–17 Ma ignimbrites (Tilito, Aquilar) and in 16–12 Ma arc volcanic rocks in and near the southern and central Maricunga Belt arc. Overall, La/Yb ratios in samples from south of *c.* 26.8°S are low in all 25–17 Ma volcanic rocks (<22; exceptions are a clast of uncertain age from the Claudio Gay Sequence and a few porphyries), increase in 16–9 Ma Maricunga Belt arc volcanic rocks with ratios reaching *c.* 35 in *c.* 12–9 Ma Copiapó dacites, and peak in the *c.* 8–6 Ma Jotabeche ignimbrites erupted in the last stages of the Maricunga Belt arc and the 9–3 Ma Dos Hermanos and Pircas Negras andesites erupted as the arc front migrated (up to *c.* 70). From a broader perspective, the wt% Na_2O and La/Yb ratios in the Pircas Negras and Dos Hermanos andesites are among the highest in the Central Andes (see compilation in Mamani *et al.* 2010). High La/Yb ratios and wt% Na_2O values also occur in the <9 Ma Maricunga Belt arc Copiapó ignimbrites, some 8–3 Ma

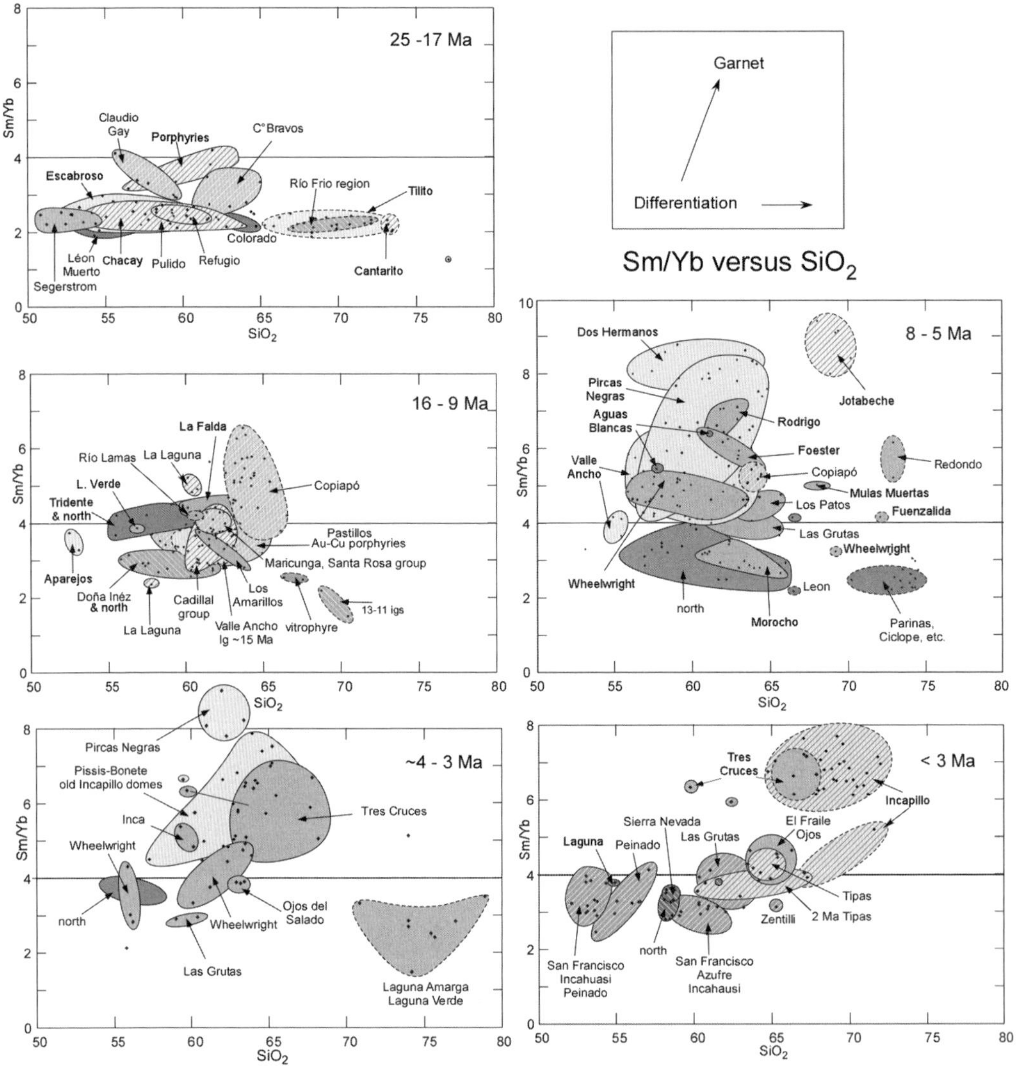

Fig. 5. Plot of Sm/Yb ratio v. wt% SiO_2 for same sample groups as in Figure 3. Note variability of Sm/Yb ratios in time and space at a similar wt% SiO_2 and the dramatic increase in values after 8 Ma.

Valle Ancho and Pissis-Bonete region andesites and dacites and in <3 Ma CVZ Incapillo caldera and Tres Cruces region samples. In contrast, wt% Na_2O and La/Yb ratios in the <3 Ma CVZ arc rocks approach those in the early Miocene Maricunga Belt arc samples. The picture is quite distinct to the north of *c.* 26.8°S, where samples of all ages have $Na_2O < 4.1$ wt% and La/Yb ratios are all between 10 and 26 except for two *c.* 8 Ma Aguas Blanca andesites (*c.* 4.1 wt% Na_2O; La/Yb = 34 & 52).

Sm/Yb ratio v. wt% SiO_2. The plots in Figure 5 complement those in Figure 4 in adding wt% SiO_2 as a differentiation index, and in keying in on heavy REE behaviour through Sm/Yb ratios. Importantly, Sm/Yb ratios over *c.* 4 generally signal Yb retention in residual garnet formed at high pressure. In parallel with La/Yb ratios of <25, most Sm/Yb ratios of <4 occur in 25–17 Ma and <4 Ma samples from south of *c.* 26.8°S as well as in samples from north of 26°S (except *c.* 8 Ma Aguas Blancas andesites with Sm/Yb = 5.5 and 6.4). Sm/Yb ratios of <4 are also found in *c.* 8 Morocho back-arc centre andesites and late Miocene Valle Ancho basaltic andesites. In parallel with the occurrence of high La/Yb

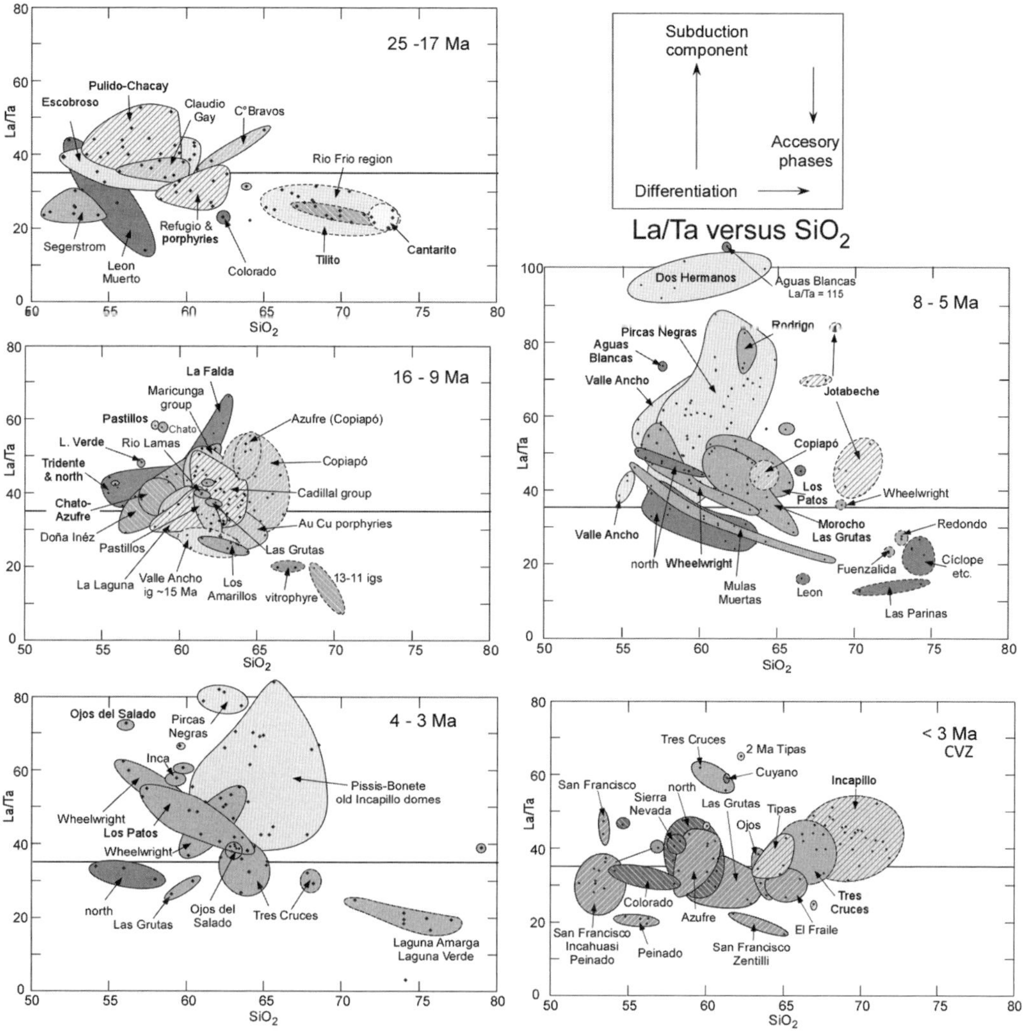

Fig. 6. Plot of La/Ta ratio v. wt% SiO_2 for same sample groups as in Figure 3. Note variability of La/Ta ratios in time and space at a similar wt% SiO_2 and extreme values in samples from the south between 8 and 3 Ma.

ratios, Sm/Yb ratios over 5 and up to 9.5 occur in 11–6 Ma Maricunga Belt Copiapó and Jotabeche arc samples, *c.* 8 Ma Dos Hermanos and *c.* 8–3 Ma Pircas Negras volcanic rocks erupted in the migrating arc and <3 Ma CVZ Tres Cruces and Incapillo samples (5.5–8). Another observation is that similar age andesitic to rhyolitic samples erupted in the same region generally have parallel high or low Sm/Yb ratios as seen by comparing the ratios in the 8–6 Ma Dos Hermanos andesites and Jotabeche rhyodacites (7.5–9.5), <6 Ma Pissis-Bonete andesites and Incapillo ignimbrites (up to 8), Wheelwright andesites and Fuenzulida rhyodacites (up to 5.5), CVZ andesites and Laguna Amarga-Verde ignimbrites (<4.2), and the less than 6 Ma andesites and ignimbrites in the north (<4).

La/Ta ratio v. wt% SiO_2. The plots in Figure 6 use La/Ta ratios as a proxy for arc-like HFSE depletion with ratios of >35 being typical of Andean SVZ arc lavas (Hickey *et al.* 1986). Overall La/Ta ratios range from *c.* 10 in early Miocene proximal back-arc mafic lavas to over 100 in *c.* 8 Ma Dos Hermanos and Aguas Blancas andesites erupted in the migrating arc. In general, La/Ta ratios in andesites and dacites south of *c.* 26.8°S are from *c.* 25 to 55 in 25–9 Ma Maricunga Belt arc samples, from 25 to over 100 in

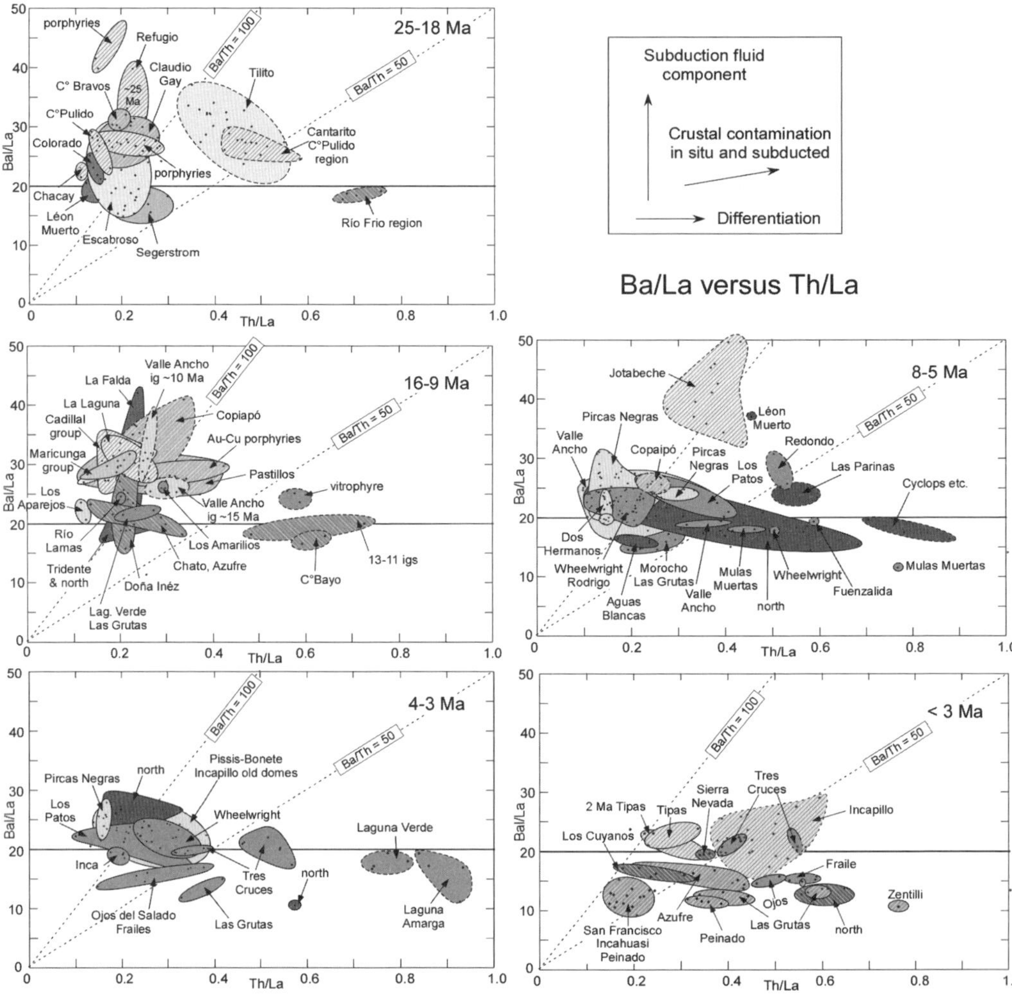

Fig. 7. Plot of Ba/La ratio vs Th/La ratios for same sample groups as in Figure 3. Note decrease in Ba/La ratios with time and high Ba/Th and Th/La ratios in northern and younger samples.

samples from the eastward-migrating arc and from *c.* 15 to 50 in <3 Ma CVZ arc samples. Most andesitic and dacitic samples north of *c.* 26.8°S have La/Ta ratios of <45. The highest La/Ta ratios (65–115), like the highest Sm/Yb ratios, are in the *c.* 9–3 Ma volcanic rocks erupted as the arc front migrated eastward (e.g. Pircas Negras, Dos Hermanos, Rodrigo, Pissis-Bonete region and Aguas Blancas groups) with La/Ta ratios of over 100 being among the highest in Andean magmatic rocks (Goss & Kay 2009). Lower La/Ta ratios in rhyodacitic to rhyolitic samples can reflect removal of REE- and HFSE-bearing phases, in the same vein that Rb, Sr and Ba trends in CVZ and Puna silicic samples were argued to reflect removal of accessory phases by Schnurr *et al.* (2007).

Ba/La v. Th/La ratios. Figure 7 shows Ba/La ratios, which are >20 in most arc rocks along with Th/La and Ba/Th ratios, which generally increase with crustal assimilation.

The plots show that 25–7 Ma Maricunga Belt arc samples generally have typical Andean frontal arc-like Ba/La ratios (20–40) with ratios over 40 in some hornblende porphyries and up to 55 in the *c.* 8–6 Ma Jotabeche rhyodacites. Ba/La ratios <20 are common in the 25–18 Ma proximal back-arc mafic lavas (Escabroso, Segerstrom and Léon Muerto). Overall, the lowest Ba/La ratios among 25–9 Ma arc region samples tend be in the north, where ratios of <20 occur. The *c.* 8–3 Ma andesites and dacites erupted as the arc migrated have Ba/La ratios from *c.* 30 to 15 as do <3 Ma

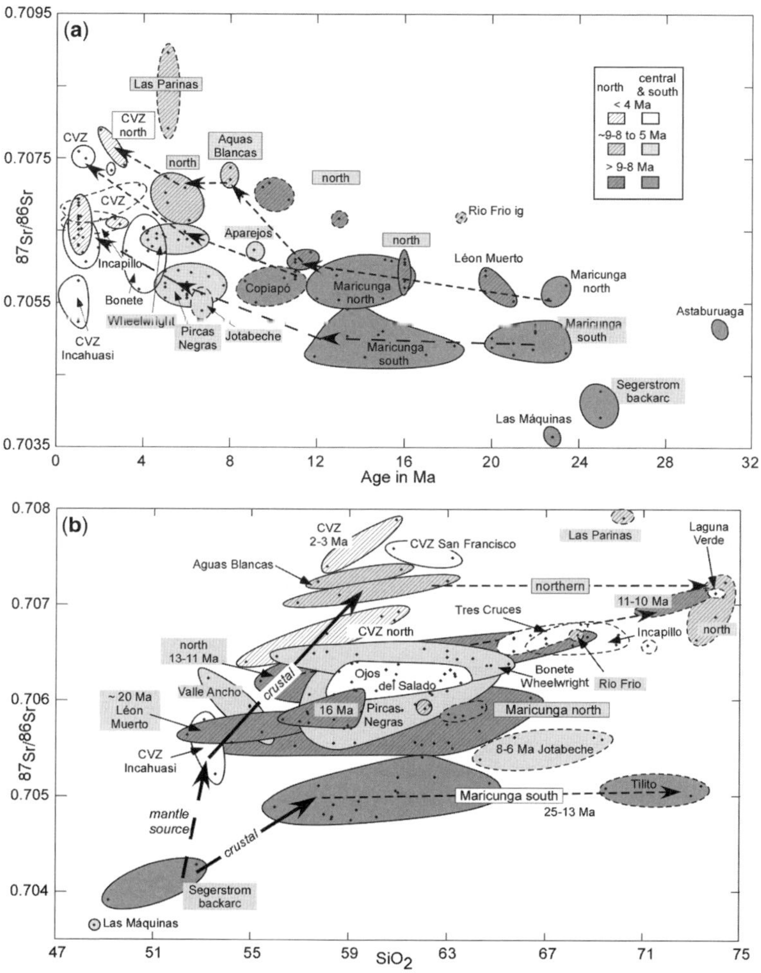

Fig. 8. Plot of $^{87}Sr/^{86}Sr$ ratio v. (**a**) age and (**b**) wt% SiO_2. Fields for samples are keyed to age with those from the northern region in striped patterns. Northern region data are from Trumbull *et al.* (1999), Siebel *et al.* (2001) and Schnurr *et al.* (2007). Other data are from McNutt *et al.* (1975), Baker *et al.* (1987), Kay *et al.* (1991, 1994*a*, 1999), Walker *et al.* (1991), Moscoso *et al.* (1993), McKee *et al.* (1994), Kay & Abbruzzi (1996), Mpodozis *et al.* (1996) and Goss *et al.* (2011, 2013). Thick dashed arrow in (b) is inferred to primarily indicate mantle source enrichment; thick undashed arrows primarily indicate *in situ* crustal contamination. Fine dashed arrows show evolution from andesite to rhyodacite where additional crustal contamination is considered to be limited.

Incapillo volcanic rocks in the south. All of the <3.5 Ma central and northern region volcanic rocks are notable for having Ba/La ratios under 25, with ratios in the <3 Ma samples being between 10 and 20.

Further comparisons show that 25–12 Ma Maricunga Belt andesites and dacites have lower Th/La (most <0.3) and higher Ba/Th ratios than andesites and dacites erupted in the Maricunga Belt arc or as the frontal arc migrated. Overall, the highest Th/La (up to 0.7) and lowest Ba/Th ratios among andesites and dacites occur in 8–5 Ma samples from the north and <4 Ma CVZ samples. Among rhyodacites and rhyolites, the lowest Th/La ratios are in Maricunga Belt ignimbrites (Tilito, Copiapó, Jotabeche) and the highest in ignimbrites to the north (*c.* 18 Ma Rio Frio, *c.* 4 Ma Laguna Amarga, Laguna Verde) and the CVZ Zentilli dome in the Cordillera San Buenaventura.

Isotopic data

The Sr, Nd and Pb isotopic ratios of analysed samples are plotted by age and region in Figures

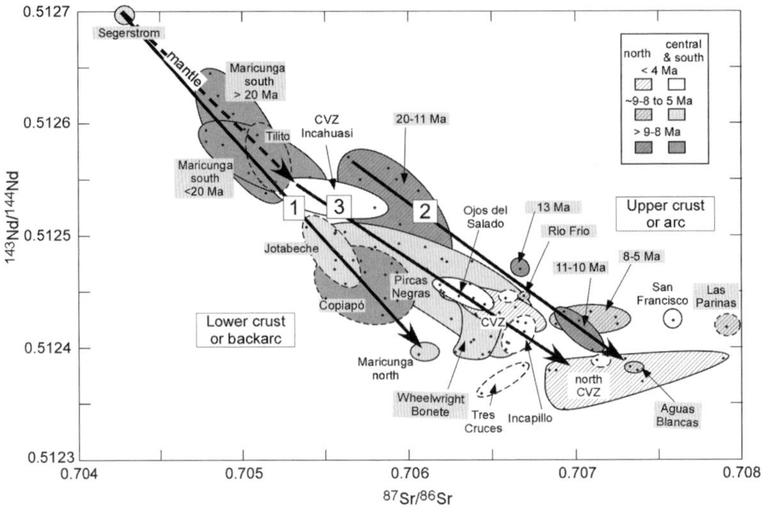

Fig. 9. Plot of $^{143}Nd/^{144}Nd$ v. $^{87}Sr/^{86}Sr$ ratios for same sample groups as in Figure 8. Dashed arrow shows inferred mantle source enrichment path; and numbered solid line arrows show three inferred general crustal contamination paths discussed in text.

8–11 with samples from north of 26°S in striped fields and those from further south in unstriped fields. The age ranges were chosen to correspond to the most active stage of the Maricunga Belt arc from *c.* 25 to *c.* 9–8 Ma (pink, darkest in PDF), the principal time of frontal arc migration from *c.* 9–8 Ma to *c.* 5–4 Ma (blue in PDF) and the formation and stabilization time of the modern CVZ arc (white). As for the trace element data, field sizes reflect compositional diversity and not volume, and can be biased by sampling density.

$^{87}Sr/^{86}Sr$ ratios. The plots in Figure 8 demonstrate that $^{87}Sr/^{86}Sr$ ratios span a range from *c.* 0.7035 to 0.7085 and show a general increase in both maximum and minimum ratios with age (Fig. 8a) and a wide variability at similar wt% SiO_2 (Fig. 8b). The lowest ratios occur in mafic lavas (<54% SiO_2) with high ppm Cr and Ni, particularly the >20 Ma back arc Segerstrom and Escabroso basalts and the <2 Ma mafic CVZ lavas (Fig. 2). The highest ratios are generally in silicic ignimbrites, with the more northerly samples tending to have the higher ratios. The south-to-north variation in the Maricunga Belt noted by McKee *et al.* (1994) and Kay *et al.* (1994*a*) is evidenced by samples from the southern Maricunga Belt 26–20 Ma (Pantanillo, Escabroso, Tilito) and *c.* 16–12 Ma (Laguna–Jotabeche Norte, Cadillal–Yeguas Heladas) centres having lower $^{87}Sr/^{86}Sr$ ratios than Maricunga Belt samples from the *c.* 23 Ma (La Coipa) and *c.* 16–10 Ma (Maricunga–Santa Rosa, Pastillos, 11–7 Ma (Copiapó–Azufre) and other arc centres further north. Another feature is that as the Maricunga Belt shut off at 8–6 Ma and syn-migration volcanic lavas erupted further east, $^{87}Sr/^{86}Sr$ ratios increased eastward as shown by comparing ratios of Maricunga Belt Jotabeche samples with those of syn-migration Pircas Negra and Bonete-Wheelwright region samples (Fig. 8; see Goss *et al.* 2013). A similar trend towards the east occurs in the northern zone (Trumbull *et al.* 1999). Other notable features on Figure 8b are the increase in $^{87}Sr/^{86}Sr$ ratio from the *c.* 24 Ma Segerstrom basalts to the <2 Ma CVZ Incahuasi mafic lavas (Kay *et al.* 1999), the relatively constant values from andesitic to rhyolitic samples at the same age and the broad range of $^{87}Sr/^{86}Sr$ ratios at similar wt% SiO_2 in samples with different ages.

$^{87}Sr/^{86}Sr$ v. $^{143}Nd/^{144}Nd$ ratios. Figure 9 shows a plot for $^{87}Sr/^{86}Sr$ v. $^{143}Nd/^{144}Nd$ ratios on which the most primitive Segerstrom and Máquinas basalts, respectively, plot off-scale at 0.7039 and 0.51284 and at 0.70365 and 0.51287. The dashed arrow shows the shift in from a somewhat-evolved Segerstrom mafic lava to the <2 Ma CVZ mafic lavas (Kay *et al.* 1999). Three important generalized trends, which are considered in the discussion section, are shown by the black arrows. The first, for the *c.* 26–8 Ma Maricunga Belt samples, runs from the back-arc Segerstrom basalts to the lowest $^{87}Sr/^{86}Sr$ ratios at a given $^{143}Nd/^{144}Nd$ ratio among the three trends. The second runs from the northern back-arc 20 Ma Léon Muerto mafic lavas to the *c.* 8 Ma Aguas Blancas andesites,

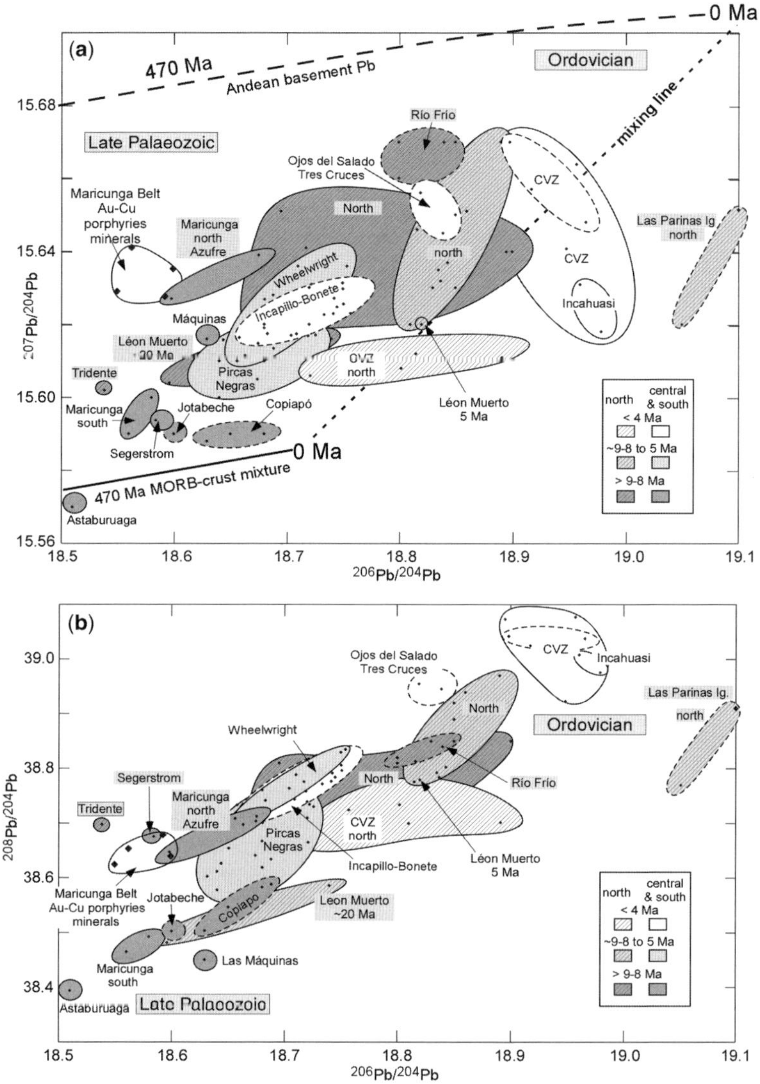

Fig. 10. Plot of ratios of (**a**) $^{207}Pb/^{204}Pb$ and (**b**) $^{208}Pb/^{204}Pb$ v. $^{206}Pb/^{204}Pb$. Patterns and data sources are as in Figure 8. Following Lucassen *et al.* (2007), in (a) dashed black line is a modeled Pb growth curve for 470 Ma Central Andean basement, solid line shows radiogenic growth of a mixture of 470 Ma Pacific MORB-like mantle with a Mu of 8.1 and a Andean crustal like component and short dashed line shows mixing between the two at 0 Ma. Labels indicate general regions of contaminants from the dominantly Ordovician and Late Palaeozoic modified basement. See discussion of Figure in text.

and shows the highest $^{87}Sr/^{86}Sr$ ratios at a given $^{143}Nd/^{144}Nd$ ratio. The third runs from the <2 Ma Incahuasi mafic lavas to the more silicic CVZ samples, and shows intermediate values of $^{87}Sr/^{86}Sr$ ratios at a given $^{143}Nd/^{144}Nd$ ratio.

Pb isotopic ratios. The plots in Figure 10 show the trends for $^{207}Pb/^{204}Pb$ and $^{208}Pb/^{204}Pb$ v. $^{206}Pb/^{204}Pb$ ratios along with analyses of galena, pyrite and enargite in the *c.* 10 Ma Marte and Alderaban porphyries in the Maricunga Belt from Moscoso *et al.* (1993) and an analysis of a *c.* 35 Ma Astaburuaga Group andesite from the southern Maricunga Belt. The analyses can be divided into low and high $^{206}Pb/^{204}Pb$ groups. The lower $^{206}Pb/^{204}Pb$ group includes the 18–11 Ma Maricunga Belt samples (Laguna, Jotabeche Norte, Copiapó), which are characterized by

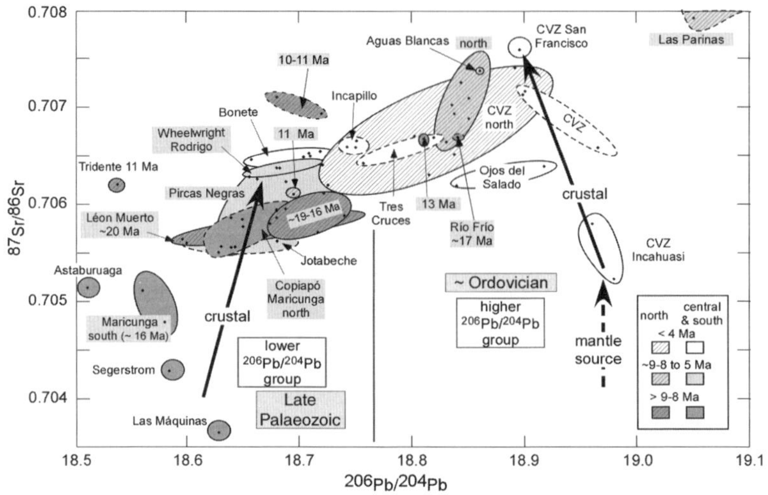

Fig. 11. Plot of $^{87}Sr/^{86}Sr$ v. $^{206}Pb/^{204}Pb$ ratios with patterns and data sources as in Figure 8. The dashed arrow shows the mantle source enrichment path and the solid arrows show general the crustal enrichment paths discussed in text. The vertical line separates the high and low $^{206}Pb/^{204}Pb$ groups discussed in the text, which generally correspond to regions where contaminants are, respectively, influenced by Ordovician and Late Palaeozoic modified basement.

relatively low $^{207}Pb/^{204}Pb$ and $^{208}Pb/^{204}Pb$ ratios, and the 24 Ma Máquinas, 22 Ma Segerstrom and 20 Ma Léon Muerto back-arc mafic lavas, which have variable $^{207}Pb/^{204}Pb$ and $^{208}Pb/^{204}Pb$ ratios. A *c.* 12 Ma Maricunga Belt Azufre volcano sample with higher $^{207}Pb/^{204}Pb$ and $^{208}Pb/^{204}Pb$ ratios plots near the Au–Cu mineralized porphyry minerals and on the low $^{206}Pb/^{204}Pb$ side of the 8–5 Ma Pircas Negras, Valle Ancho and Dos Hermanos samples. The 8–4 Ma syn-migration Wheelwright and Bonete region and <2 Ma Incapillo samples plot in the higher $^{206}Pb/^{204}Pb$ part of this group. The higher $^{206}Pb/^{204}Pb$ group includes samples from the <4 Ma CVZ Ojos del Salado region (Incahuasi, Ojos del Salado, Laguna Verde, Blanco Ignimbrite) and most of the <9 Ma volcanic rocks from the northern zone. The highest $^{206}Pb/^{204}Pb$ ratio occurs in the 6–5 Ma Parinas rhyolitic ignimbrite located in the CVZ region axis and is on a trend to higher $^{206}Pb/^{204}Pb$ ratios that continues into Argentina. The *c.* 5 Ma ignimbrite outcropping near Cerro Léon Muerto plots with this group. The early Miocene Rio Frio rhyolitic ignimbrites fall on the low $^{206}Pb/^{204}Pb$ side at high $^{207}Pb/^{204}Pb$ and $^{208}Pb/^{204}Pb$ ratios.

Sr and Pb isotopic ratios. The plot in Figure 11 shows relations between $^{206}Pb/^{204}Pb$ and $^{87}Sr/^{86}Sr$ ratios and includes black arrows indicating generalized trends from mafic to silicic samples and a dashed arrow showing the change in $^{87}Sr/^{86}Sr$ ratios from the *c.* 24–22 Ma Segerstrom/Máquinas olivine basalts to the CVZ Incahuasi mafic lavas. A trend of increasing $^{87}Sr/^{86}Sr$ with little change in $^{206}Pb/^{204}Pb$ is clearly seen from the Segerstrom/Máquinas lavas to the *c.* 16 Ma southern Maricunga Belt to the *c.* 12–6 Ma Copiapó–Jotabeche samples. $^{87}Sr/^{86}Sr$ ratios continue to increase in the 8–5 Ma syn-migration Pircas Negras and Wheelwright lavas and reach a maximum in <6 Ma Bonete and CVZ Incapillo samples. The *c.* 20 Ma Léon Muerto and 19–16 Ma northern zone samples plot near the Copiapó field, the 11 Ma Tridente andesite at lower $^{206}Pb/^{204}Pb$ and the 11–10 Ma northern ignimbrites at higher $^{87}Sr/^{86}Sr$. On the higher $^{206}Pb/^{204}Pb$ side are the <9 Ma northern centres and the <4 Ma CVZ samples with the more silicic samples tending to plot at the higher $^{87}Sr/^{86}Sr$ ratios. In general, the samples with lower $^{206}Pb/^{204}Pb$ ratios are from areas where Late Palaeozoic granitoids and acid volcanic rocks crop out whereas those with higher $^{206}Pb/^{204}Pb$ ratios are from areas where Ordovician rocks occur (see discussion below).

Discussion

Among the important issues in understanding the sources of Neogene magmas in the Andes are the origin and proportions of diverse crustal and mantle components, how these components change in space and time and the processes that control

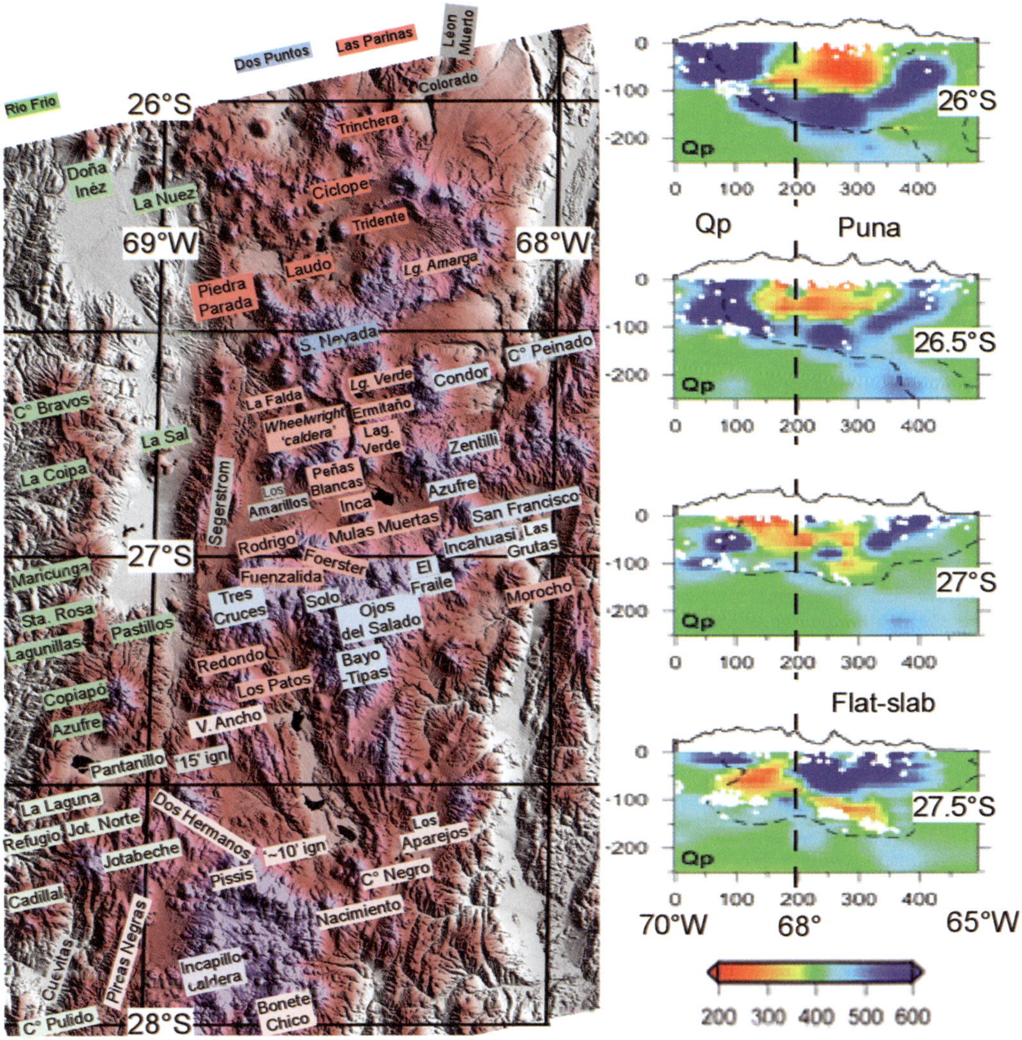

Fig. 12. Satellite Radar Topography Mission (SRTM) image from Figure 2a shown with east–west trending Qp seismic attenuation tomographic profiles from Liang *et al.* (2011; pers. comm. 2013) across the region at the latitude indicated. Colour scale for dimensionless quality factor (Qp) in tomographic image is shown at base. White points are earthquake locations and dashed lines enclose the region of greatest data confidence in the image. Red colours indicating the regions of highest attenuation are interpreted as zones of melts or fluids and blue areas showing low attenuation generally correspond to the subducting slab or regions of cool continental lithosphere. Note the regions of highest attenuation (red) are in the Puna east of the northern centres, whereas the region east of the southernmost centres shows low attenuation (blue).

these changes. Below, we use the trace element and isotopic trends in Figures 3–11 to discuss differences in the pre-existing lithosphere, temporal changes in the underlying crust related to crustal thickening and flow, introduction of continental lithosphere into the mantle source by fore-arc subduction erosion or delamination and changes in subduction zone parameters. We show that all of these factors were involved to varying degrees through time and space.

Crustal contamination and mantle sources: spatial variations

The variable trace element and isotopic signatures of the Maricunga Belt to CVZ arc and back-arc volcanic rocks appear to reflect the nature of their mantle source, the amount of crustal contamination and types of crustal and lithospheric components. In examining the nature of crustal and lithospheric mantle sources, a key issue to consider is the state of the local Andean crust and lithosphere in the Oligocene to Recent. In a review of zircon ages and εNd and εHf model ages, Bahlburg *et al.* (2009) suggest that much of the central Andean crust formed at 1.7–1.2 Ga with little juvenile magmatic addition after 800 Ma. However, the area under discussion has been suggested to be underlain by several Palaeozoic terranes (see Fig. 2b), with Mesoproterozoic Grenville-age continental basements that have been reworked in Palaeozoic magmatic and collisional events (e.g. Ramos *et al.* 2010; see Fig. 2b). The first, the Antofalla terrane, is suggested to underlie the southern Altiplano–Puna plateau and most of the Central Andean western Cordillera and slope north of 27°S. According to Ramos (2008), the Antofalla terrane was part of the Pampia terrane that was sutured to Gondwana during the Grenville orogeny, then rifted away to the west in the earliest Palaeozoic and was then reattached to Gondwana during the Ordovician Ocloyic orogeny. The second, the Cuyania terrane, is essentially located in Argentina south of the latitude of the Incapillo caldera (28°S). This terrane, which is also considered to have been sutured to Gondwana in the Ordovician, is thought to be an exotic terrane that came from Laurentia (Thomas & Astini 2003). The third, the Chilenia terrane, which is said to occur west of the Cuyania terrane and south of the Antofalla terrane, is proposed to have been sutured to Gondwana around 390 Ma (e.g. Willner *et al.* 2011). Parts of all of these terranes were subsequently modified in the Carboniferous–Triassic magmatic episode that produced the extensive Choiyoi granite–rhyolite province (e.g. Mpodozis & Kay 1992). The parts of these terranes west of the longitude of the Maricunga Belt were further reworked by Mesozoic to Palaeogene arc magmatism (e.g. Lucassen *et al.* 2007). This history has resulted in average pre-Neogene crustal compositions that are more highly radiogenic in Sr and Pb and less radiogenic in Nd in the east (Argentina) than in the west (Chile), as shown by the compilation of Lucassen *et al.* (2001).

$^{87}Sr/^{86}Sr$, $^{143}Nd/^{144}Nd$, and Pb ratios are useful in discriminating crustal- and mantle-derived components, with Pb ratios being the best for providing information on central Andean basement domains (e.g. Wörner *et al.* 1992; Aitcheson *et al.* 1995). A $^{206}Pb/^{204}Pb$ isotopic map in Mamani *et al.* (2008) that includes the area discussed here shows a general gradient, with values of >18.9 near the CVZ arc to values as low as 18.5 to the west. This range encompasses that of the high and low $^{206}Pb/^{204}Pb$ groups in Figures 10 and 11. The pattern can be explained as a contrast between regions last reworked in the Ordovician (18.8–19.1) to the east and in the Permo-Triassic (18.5–18.75) to the west with most of the region falling in the latter group. Importantly, the Pb isotopic ratios of the sulphide minerals in the Maricunga Belt mineralized (Au–Cu) porphyries, which are interpreted as mixtures of mantle and crust modified by late Palaeozoic magmatism (Moscoso *et al.* 1993), approach those in the 24 Ma Segerstrom and *c.* 20 Ma León Muerto mafic lavas and the 12–11 Ma Tridente and Azufre andesites. A tendency for isotopic ratios in the Pircas Negras, Incapillo, Ojos del Salado–Tres Cruces samples and <9 Ma northern zone lavas to have intermediate $^{206}Pb/^{204}Pb$ ratios is consistent with their eruption on the margins of the region characterized by the occurrence of widespread upper Palaeozoic granites and rhyolites. The higher $^{207}Pb/^{204}Pb$ and lower $^{208}Pb/^{204}Pb$ and $^{206}Pb/^{204}Pb$ ratios of the Rio Frio ignimbrites likely reflect the basement to the west.

One way to explain the range of Pb isotopic ratios in the region is that higher $^{206}Pb/^{204}Pb$ ratios at a given Mu ($^{238}U/^{204}Pb$) indicate generation in an older lithosphere, and incorporation of old radiogenic crust influences $^{207}Pb/^{204}Pb$ and $^{208}Pb/^{204}Pb$ ratios. Figure 10a shows a reference model from Lucassen *et al.* (2007), which has a 0 Ma mixing line for a radiogenically corrected mixture of (1) 470 Ma Andean basement (based on average early Palaeozoic basement from 12°S to 27°S in Lucassen *et al.* 2002) and (2) a 470 Ma enriched lithosphere consisting of Pacific-like mid-ocean ridge basalt (MORB) with a Mu of 8.1 and a crustal-like enrichment, which started with $^{206}Pb/^{204}Pb$ and $^{207}Pb/^{204}Pb$ ratios of *c.* 18 and *c.* 15.525 at 470 Ma. Importantly, the Pb isotopic ratios of the *c.* 22 Ma Segerstrom basaltic lavas (Kay *et al.* 1999) plot near the 0 Ma point, consistent with this composition being a possible proxy for enriched modern lithosphere. The mixing line from this 0 Ma point to the 470 Ma Andean basement passes through the fields of the <4 Ma CVZ arc rocks. Higher $^{206}Pb/^{204}Pb$ ratios in these <4 Ma CVZ volcanic rocks than in the >8 Ma southern and central Maricunga Belt arc and proximal back-arc volcanic rocks are consistent with a contrast between a lithosphere last reworked in the Ordovician (higher $^{206}Pb/^{204}Pb$) and one reworked in and after the Permo-Triassic (lower $^{206}Pb/^{204}Pb$). Higher $^{207}Pb/^{204}Pb$ ratios in central CVZ ignimbrites than in the CVZ Incahuasi mafic lavas fit with contamination of mafic lavas by old crust. The relatively intermediate $^{206}Pb/^{204}Pb$ and $^{207}Pb/^{204}Pb$ ratios in *c.* northern CVZ region volcanic rocks can reflect lithospheric and crustal reworking in both the Permo-Triassic and Ordovician.

Crustal contamination and mantle sources: temporal considerations

$^{87}Sr/^{86}Sr$ and $^{143}Nd/^{144}Nd$ ratios provide another view of mixing of crustal and mantle components. Most workers agree that even the most mafic Central Andean lavas generally contain at least 20% crustal component (e.g. Aitcheson & Forrest 1994; Kay *et al.* 1994*a*, *b*) and that $^{87}Sr/^{86}Sr$ and $^{143}Nd/^{144}Nd$ ratios are largely established by the time that Central Andean magmas evolve to andesitic composition (e.g. Davidson *et al.* 1991; Schnurr *et al.* 2007). This effect can be seen in Figure 8b, where most isotopic ratios change little after reaching *c.* 56 wt% SiO_2, which is consistent with less effective assimilation at lower temperatures at shallower levels where evolution to higher wt% SiO_2 occurs. The increases of $^{87}Sr/^{86}Sr$ ratios from the *c.* 24 Ma Segerstrom basalts to the Miocene Maricunga Belt silicic rocks and from the Pleistocene Incahuasi mafic lavas to the CVZ andesites and dacites occur independently of $^{206}Pb/^{204}Pb$ ratios, which require different controls on their isotopic evolution. Schnurr *et al.* (2007) argued for different $^{87}Sr/^{86}Sr$ ratios in distinct assimilation–fractional (AFC) paths for Chilean and Argentine Neogene andesites with those in Chile evolving from *c.* 0.7055 to *c.* 0.707 and those in Argentina from *c.* 0.7055 to *c.* 0.709 as Rb/Sr and Eu/Eu* ratios increased. Distinct AFC paths are also needed to explain lower $^{87}Sr/^{86}Sr$ ratios in lavas from the southern and northern Maricunga Belt (Fig. 8a, b; Kay *et al.* 1994*b*; McKee *et al.* 1994).

The origin of the increase in $^{87}Sr/^{86}Sr$ ratios after 10–9 Ma (Fig. 8a) across the area is a continuing problem as to whether the control is spatial or temporal or both, as higher $^{87}Sr/^{86}Sr$ ratios generally occur in younger volcanic rocks erupted in the east. In the northern zone, Trumbull *et al.* (1999) argued that higher ratios in <8 Ma andesites (>0.707) than in 20–6 Ma andesites (<0.706) (Figs 8 & 9) reflect a temporal change caused by more crust being incorporated into the magmas as they passed through a crust thickened at *c.* 8 Ma. The evidence for crustal thickening at this time is based on increases in La/Yb and Sm/Yb ratios in <9 Ma andesites. A similar pattern is seen in the ignimbrites, with those erupted after 6 Ma having higher $^{87}Sr/^{86}Sr$ ratios. This increase in $^{87}Sr/^{86}Sr$ ratios near 9 Ma is clearly visible in Figure 8a and in Figure 8b, where the field flattens in the andesite to rhyolitic range.

Supporting evidence for temporal isotopic changes in the northern zone comes from increases in all three Pb isotopic ratios in andesites erupted in the Léon Muerto region between *c.* 20 and *c.* 5 Ma, and higher $^{207}Pb/^{204}Pb$ and $^{208}Pb/^{204}Pb$ ratios at similar wt% SiO_2 in 8–5 Ma than in 20–16 Ma andesites in the Salar de la Isla region to the west. Trumbull *et al.* (1999) further argued for crustal thickening near *c.* 8 Ma based on relative increases in Rb, Cs and Th ppm concentrations in andesites that they attributed to an acceleration of crustal assimilation in a thicker crust. Walker *et al.* (1991) argued for a similar change in crustal assimilation in northern Maricunga Belt magmas based on increases in Rb and Ba (ppm) concentrations and $^{87}Sr/^{86}Sr$ ratios.

Arguments for a temporal change in the southern Maricunga Belt region at a similar time come from increases in $^{87}Sr/^{86}Sr$ ratios with decreasing age. These increases can be seen by comparing Maricunga Belt 24–10 Ma andesites to syn-migration 8–3 Ma Pircas Negras lavas and <6 Ma Bonete–Incapillo andesites to the east (Kay *et al.* 1991, 1994*b*; *Goss et al.* 2013), and by comparing 25–10 Ma central Maricunga Belt andesites to 9–3 Ma syn-migration Wheelwright region and <3 Ma CVZ Ojos del Salado region andesites (Mpodozis *et al.* 1996; Figs 8a, b). Parallel changes at higher wt% SiO_2 in the southern Maricunga Belt can be shown by comparing *c.* 22 Ma Tilito ignimbrites to *c.* 8–6 Ma Jotabeche and <2 Ma Incapillo dacites and rhyolites, and by comparing >10 Ma central Maricunga Belt Copiapó ignimbrites to Tres Cruces and CVZ Ojos del Salado region dacites. Kay *et al.* (1999), Kay & Mpodozis (2002), Goss & Kay (2009) and Goss *et al.* (2013) all argue that these changes involve interaction of magmas with thickened *in situ* crust, as well with crust recycled into the mantle source by fore-arc subduction erosion as the arc front was displaced eastward.

More insight into these processes comes from the plot of Sr v. Nd isotopes (Fig. 9), in which an additional factor is that $^{143}Nd/^{144}Nd$ ratios are higher at a given $^{87}Sr/^{86}Sr$ ratios in the north than in the south. One explanation for lower $^{143}Nd/^{144}Nd$ ratios in the south is the involvement of a lower crustal component, consistent with higher La/Yb and Sm/Yb ratios in the southern samples (Figs 4 & 5). Goss & Kay (2009) interpreted the *c.* 8 Ma Dos Hermanos and <4 Ma Pircas Negras andesites with the steepest REE patterns and highest Nb/Ta ratios as dominantly having a crustal component that was in equilibrium with basaltic eclogite, and the 8–4 Ma Pircas Negras andesites with high ppm Cr and Ni and lower Nb/Ta ratios as having an additional crustal component introduced into the mantle wedge by fore-arc subduction erosion. Other arguments for changes in crustal involvement at this time include isotopic enrichment at the same wt% SiO_2 (Fig. 8b; e.g. Kay *et al.* 1999; Goss *et al.* 2013) and a general increase in Th/La ratios after 9 Ma (Fig. 7).

The best evidence for temporal changes in the mantle source come from comparisons of the most mafic lavas within a zone as they provide the clearest indication of the evolution of the mantle wedge. The strongest argument for a mantle source change comes from comparing the *c.* 24 Ma Segerstrom basalts to the <2 Ma Incahuasi mafic lavas as they all have high MgO (>9 wt%), Cr (>700 ppm) and Ni (>200 ppm) and contain mantle-equilibrated FO88 olivine phenocrysts (e.g. Kay *et al.* 1999). Mass balance considerations show that the high ppm Sr (>600) in these lavas permit little of their $^{87}Sr/^{86}Sr$ signature to be derived from mixing with low Sr (<50 ppm), high $^{87}Sr/^{86}Sr$ crust and their primitive Mg, Ni and Cr characteristics allow little overall crustal contamination. Kay *et al.* (1999) and Kay (2006) argued that the influence of crustal contaminants in the Incahuasi basaltic andesites can largely be removed by subtracting out *c.* 20% of a siliceous melt with a composition like that of the nearby *c.* 4 Ma Laguna Verde ignimbrites. The isotopic differences between the 24 Ma Segerstrom basalt and the contamination corrected Incahuasi basalt show a striking change to a more 'enriched' mantle source by 2 Ma, consistent with the underlying mantle being modified by the addition of crustal components, which probably came from crust removed by fore-arc subduction erosion to the west (Kay *et al.* 1999; Kay & Mpodozis 2002; Goss & Kay 2009; Goss *et al.* 2013) and delamination in the east (e.g. Kay *et al.* 1994*b*, 2011; Risse *et al.* 2013). Goss *et al.* (2013) argue that the most likely time for addition of subducted eroded crust is as the frontal arc migrated eastward, in parallel with $^{87}Sr/^{86}Sr$ ratios increasing after 9 Ma in the Pircas Negras andesites (Figs 8–11).

Other indicators of temporal and spatial variability are changes in major and trace element patterns. Among the distinctive changes are the appearance of high wt% Na_2O along with high La/Yb and Sm/Yb ratios (Figs 4 & 5) in the ebbing Maricunga Belt *c.* 8–6 Ma Jotabeche rhyodacites, the syn-migration plagioclase–phenocryst free *c.* 8–3 Ma Pircas Negras andesites and the <4 Ma Incapillo region volcanic rocks, which all erupted over or close to the developing bend in the subducting plate on the northern edge of the Chilean–Pampean flat-slab (e.g. Goss *et al.* 2013). The Tres Cruces region rhyodacites in the westernmost CVZ to the north resemble the Incapillo ignimbrites. These high La/Yb and Sm/Yb ratios require sequestration of heavy REEs by garnet at high pressures in the lower crust or mantle wedge, consistent with feldspar instability signalled by high wt% Na_2O.

Other indicators of change are that La/Ta (e.g. >35; Fig. 6) and Ba/La (e.g. >20, Fig. 7) ratios in the 25–9 Ma Maricunga Belt volcanic rocks are much more arc-like than those in <3 Ma CVZ volcanic rocks, which, with the exception of the westernmost Tres Cruces and southernmost Incapillo region volcanic rocks, have La/Ta (18–40) and Ba/La (10–25) ratios extending below values typically found in frontal arc volcanic rocks. In general, arc-like La/Ta ratios are attributed to retention of Ta in oxide phases in the mantle wedge and/or thick crust, with arc-like Ba/La ratios being attributed to mantle fluid contributions from dewatering of subducted sediments (e.g. Elliot *et al.* 1997) and serpentine breakdown in the slab (e.g. Rüpke *et al.* 2004). The tendency for the contrasting La/Ta and Ba/La ratios in the Maricunga Belt and CVZ volcanic rocks to occur in the most mafic lavas (basaltic andesitic to mafic andesite; Fig. 6) supports these ratios reflecting processes in the mantle wedge and/or deep crust.

At least partial support for these contrasting La/Ta and Ba/La ratios reflecting a temporal change comes from a striking temporal trend in the southern and central Maricunga Belt region, which starts with early (Cerro Pulido and Chacay) and middle (Cadillal and Ojos de Maricunga groups) Miocene hornblende-bearing andesites having arc-like La/Ta (up to 55; Fig. 6) and Ba/La ratios (up to 40; Fig. 7). These ratios begin to change in the ebbing Maricunga belt *c.* 8–6 Ma Jotabeche rhyodacites, which have arc-like Ba/La ratios (up to 52) at more variable La/Ta ratios (35–80), and then change dramatically in the pre- and syn-arc-migration 8–3 Ma Dos Hermanos, Pircas Negras and Rodrigo lavas, which have Ba/La ratios of 15–30 at La/Ta ratios that can reach over 100. These extreme La/Ta ratios are interpreted to reflect residual HFSE-element phases (amphibole and rutile) in the mantle wedge and crust (Goss & Kay 2009), with the generally low Ba/La ratios interpreted to reflect less hydrous conditions in a narrowing mantle wedge above a shallowing slab under the migrating arc front (Goss *et al.* 2011, 2013).

A similar pattern does not occur in Miocene and Pliocene volcanic rocks in the north, where La/Ta ratios are mostly <40 and Ba/La ratios <25 (Figs 6 & 7), and only the *c.* 8 Ma Aguas Blancas andesites (La/Ta = 75 and 120) have La/Ta ratios > 50. The striking correlation between high Sm/Yb and La/Ta ratios (Figs 5 & 6) suggests a link in time and space in their creation, with the conditions that produced them being more commonly met in the south than the north, as discussed in more detail below.

A general correlation between higher Th/La and Ba/Th and $^{87}Sr/^{86}Sr$ ratios and lower Ba/La ratios (Figs 7 & 8) raises the question of low Ba/La ratios being related to processes within the crust. In the case of the northern region, Trumbull *et al.* (1999) considered that there was little to no change in slab fluid or crustal influence on Ba over the last 25 Ma, based on a flat trend in Ba/Zr ratios, whereas an increase in Th/Zr ratios after *c.* 8 Ma was attributed to an increase in Th by crustal contamination in a thickened crust. A similarity in Th/La and Ba/Th ratios in the Ojos del Salado CVZ region (e.g. Cerro Peinado, San Francisco, Incahuasi) and Maricunga Belt volcanic rocks at their, respectively, lower and higher high Ba/La ratios provides support for their Ba/La ratios being controlled by different factors from their Th/La and Ba/Th ratios.

The fact that the most mafic lavas from the CVZ (Incahuasi lavas) with very high Cr, Ni and Mg concentrations show among the lowest Ba/La ratios is consistent with a small contribution of fluids from the slab, which is now at a depth of *c.* 120 km under the Incahuasi stratovolcano (Mulcahy *et al.* 2010). Another relevant point is that the Peru–Chile trench to the west became nearly devoid of terrestrially derived sediments near 10 Ma (e.g. Nalpas *et al.* 2008), thus reducing a source of potential fluid input to the mantle wedge. Further, in analogy with the lack of fracturing and implied serpentinization of the subducting plate where the Cocos ridge is subducting along the Central American trench (Ranero *et al.* 2003; Rüpke *et al.* 2004), the subduction of the Juan Fernandez ridge would appear not to favour serpentinization of the down-going Nazca plate, again diminishing the potential for slab fluid input. Finally, as elaborated below, proximity of a hot mantle wedge and crust under the Puna to the east would particularly promote shallow dehydration of the slab in the north, and lead to low Ba/La ratios in that region. As such, decreasing Ba/La ratios over time in the region can be correlated with decreasing amounts of slab fluids.

Causes of spatial and temporal changes

The Neogene magmas discussed here erupted in a complex tectonic setting where crustal thickening, plateau uplift and frontal arc migration were all occurring over a changing Benioff zone geometry. To a first order, there appears to be a correlation between the most extreme trace element variations (Figs 3–7) and regions where the greatest changes in overall plate geometry occurred.

The extreme Sm/Yb ratios and high wt% Na_2O in some of these lavas, which are virtually unique in the Andes, require a high-pressure garnet-bearing residual assemblage. It is clear that these features do not strictly correlate with crustal thickness (e.g. Kay & Mpodozis 2002; Mamani *et al.* 2010), with an example coming from the CVZ Tres Cruces region volcanic rocks having higher Sm/Yb ratios than nearby Ojos del Salado region centres that erupted at nearly the same time through a similar thickness crust. At the same time, the appearance of Sm/Yb ratios over 5 in Maricunga Belt Copiapó region volcanic rocks at *c.* 11–10 Ma (Fig. 5) does appear to signal the formation of a thick crustal keel under the region (e.g. Kay *et al.* 1994*a*; Kay & Mpodozis 2002). Additional factors shaping the final chemical signatures of these volcanic rocks are regional stresses that influence magma entrapment times at various depths, and additions of eclogitic continental crust into the mantle wedge through fore-arc subduction erosion at times of arc migration, and delamination. The observation that, except in a few cases, like the *c.* 8 Ma Aguas Blancas andesites, samples with extreme Sm/Yb and La/Ta ratios come from south of or near the southern boundary of the Puna plateau (Fig. 2b) also shows there is no simple correlation of these extreme ratios with frontal arc migration, which occurred all along the region. A final question is why the topographically high and massive Bonete–Incapillo, Tres Cruces–Ojos del Salado and Sierra Nevada dome–ignimbrite-lava complexes, which all formed after the arc began to migrate at *c.* 8 Ma, are concentrated south of and near the southern boundary of the Puna plateau (Fig. 2b).

Clues to difference in processes from north (*c.* 25°S) to south (*c.* 28°S) come from seismic attenuation tomographic images in Figure 12, from Liang *et al.* (2011; pers. comm. 2013) that show not only the modern situation, but also reflect the cumulative history of the area. These images show a marked increase in attenuation (lower Qp) in the back-arc crust and upper mantle east of 68°W and north of 26.5°S, where the broad part of the Puna plateau begins, and a diminishing attenuation (higher Qp) in the back-arc towards the south. Attenuation is also notably high in the upper crust west of 68°W below the Sierra Nevada and Cerro Peinado regions, noticeable below the CVZ Tres Cruces and Ojos del Salado regions and subtle below the Cerro Bayo–Tipas region in good correspondence with regions of potentially active volcanism (Fig. 2). High attenuation (lower Qp) west of 68°W is also seen above the slab from at least 25.5–27.5°S and occurs at a depth of *c.* 100–150 km above the slab in the 27.5°S profile.

Interestingly, magmas erupted north of 26°S, west of where strong attenuation is now seen under the Puna show increases in $^{87}Sr/^{86}Sr$ ratios at relatively high $^{143}Nd/^{144}Nd$ ratios (Fig. 9) after *c.* 8 Ma that, together with trace element changes, are consistent with upper crustal type contaminants (Fig. 7) that could be due to enhanced westward deep crustal flow from under the Puna. Such crustal flow can also help to explain why, with the exception of the *c.* 8 Ma Aguas Blancas andesites, increases in Sm/Yb ratios after 8 Ma (Trumbull

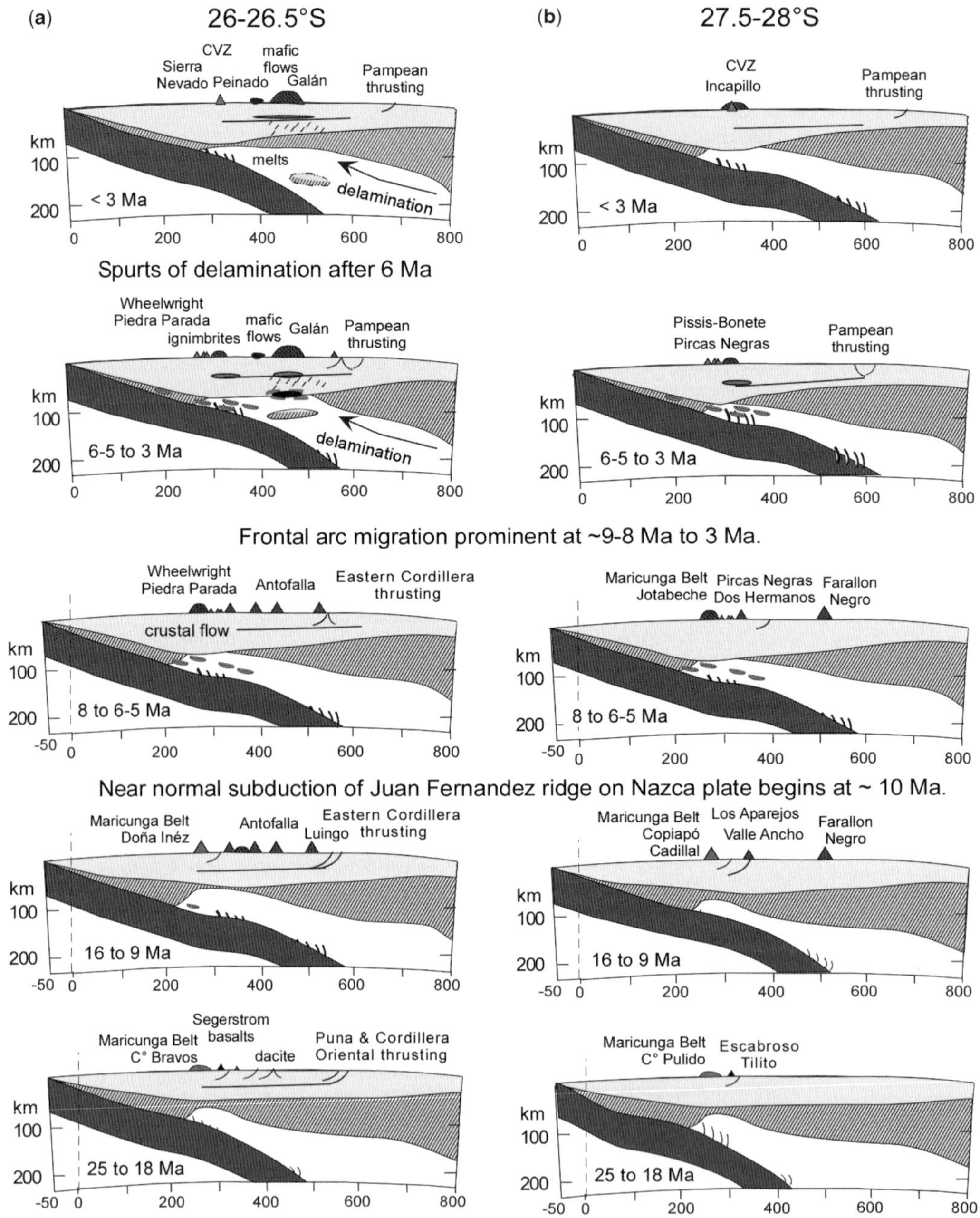

Fig. 13. Lithospheric cartoons showing generalized transects near 26–26.5°S and 27.5–28°S at 25–18 Ma, 16–9 Ma, 8–5 Ma, 5–3 Ma and less than <3 Ma, corresponding to time frames in major and trace element plots in Figures 3–7. Framework for 26–26.5°S section is modified from Kay & Coira (2009). The shapes of the subducting slab (blue) are based on modern profiles at various latitudes of the central Andean Benioff zone presented by Cahill & Isacks (1992). The vertical dashed lines in the 25–6 Ma sections show the amount of eastward displacement of the trench required if frontal arc migration is tied to fore-arc subduction erosion. (**a**) Temporal sections at 26–26.5°S show expansion of magmatism to the east at 16–9 Ma argued to be tied to shallowing of the Nazca plate and <6.8 Ma mafic lavas and Cerro Galan caldera ignimbrites argued to reflect continental lithospheric delamination, decompression melting of the mantle and partial melting of the crust as the formerly shallow slab steepened. (**b**) Temporal sections at 27.5–28°S show expansion of magmatism to the east at 16–9 Ma over shallowing Nazca plate, followed by more intense shallowing as

et al. 1999) do not reach the high levels seen further south (Fig. 5). Enhanced crustal and mantle melting under the Puna at this time is also consistent with the beginning of the steepening of a formerly shallow subducting slab and the start of lower crustal and mantle lithospheric delamination, as proposed by Kay & Coira (2009; see Fig. 13a). A hot mantle wedge can also contribute to low fluid release from the slab, leading to the low Ba/La ratios seen in the northern zone lavas (Fig. 7). In contrast, Sm/Yb ratios up to 9 (Fig. 5) occur in similar age magmas that erupted south of 26.75°S as Maricunga belt arc volcanism ebbed and the arc front migrated eastward. These high ratios, coupled with increases in $^{87}Sr/^{86}Sr$ ratios at lower $^{143}Nd/^{144}Nd$ ratios (Fig. 9), suggest the involvement of lower crustal contaminants. Significantly, these eruptions occurred west and south of where the broad part of the Puna ends, and east of where the high attenuation (low Qp) zone decays southward.

Lithospheric-scale models for magma evolution in the southern CVZ

Contrasting lithospheric-scale cross-sections for the evolution of the Maricunga Belt and CVZ arc region volcanic rocks are presented in the transects in Figure 13 through the southern Puna near 26°S and the northernmost flat-slab region near 27.5°S. These sections, which start in the early Miocene with a relatively steep slab geometry and end with the modern slab geometries at 26° and 27.5°S, are discussed below relative to the magmatic evolution in each time frame.

Early Miocene. Arc volcanic activity from 25–18 Ma was concentrated in the Maricunga Belt, with eruption of andesites and dacites in the arc, and basaltic andesites and dacites in the proximal back-arc over a moderately steeply dipping subduction zone. The emplacement of small back-arc dacitic dome and ignimbrite complexes (e.g. Coira *et al.* 1993) and the inversion of high angle normal faults in the Eastern Cordillera (e.g. Deeken *et al.* 2006) suggest a slightly shallower slab dip existed under the northern transect. This picture is consistent with the eruption of medium- to high-K arc magmas (Fig. 3), with typical frontal-arc Na_2O wt% and Ta/La and Ba/La ratios (Figs 4, 6 & 7) from widely dispersed centres along the Maricunga Belt arc. The relatively low Sm/Yb and La/Yb ratios (Figs 4 & 5) in these volcanic rocks are consistent with the magmas passing through a near normal thickness crust. The voluminous arc (Rio Frio) and proximal back-arc (Tilito) dacitic to rhyodacitic ignimbrites, small back-arc olivine basalts and proximal sedimentary basins reflect extensional conditions that facilitated magma transit through the crust. Isotopic variations along the arc are consistent with a more radiogenic basement in the north than in the south (Figs 8–11).

Middle to late Miocene. Magmatism continued at 16–9 Ma as andesitic to dacitic stratovolcanic activity reached a peak in the Maricunga Belt arc and predominantly medium- to borderline high-K andesites and low-Si dacites (Fig. 3) with arc-like La/Ta and Ba/La ratios (Figs 4 & 5) erupted. Volcanism also spread widely into the back-arc at this time (see Kay & Coira 2009). In the northern transect, the eastward spread across the Puna is represented by the 14–8 Ma Antofalla andesitic to dacitic stratovolcanic group (e.g. Coira *et al.* 1993) and the 12–9 Ma Luingo dacitic ignimbrites (Guzmán & Petrinovic 2010), which extend as far east as 66.5°W and the initial eruption of the far back-arc Farallon Negro by 9.5 Ma (Sasso & Clark 1998; Harris *et al.* 2006). In the southern transect, the eastward spread is shown by the proximal 15–9 Ma andesitic to dacitic centres in the Valle Ancho. The presence of arc-like HFSE depletion in back-arc lavas in both transects is consistent with shallowing of the underlying slab (Kay *et al.* 1999; Kay & Coira 2009) in response to the oblique subduction of the Juan Fernandez ridge on the Nazca plate (Yañéz *et al.* 2001). Lower Ba/La ratios in the northern transect arc volcanic rocks (Figs 6 & 7) are consistent with a shallowing slab losing the heat necessary to release slab-related fluids.

The formation of the Copiapó volcanic centre at *c.* 11–10 Ma marks the initial ebbing of Maricunga Belt volcanism as near perpendicular subduction of the Juan Fernandez ridge began to the south (Yañéz *et al.* 2001), and the sharp bend in the subducting slab bounding the northern margin of the Chilean–Pampean flat-slab region began to form beneath the southern transect (Fig. 1). The higher Sm/Yb and La/Yb ratios and wt% Na_2O (Figs 4 & 5) in the Copiapó volcanic rocks can be interpreted as reflecting crustal thickening (Kay

Fig. 13. (*Continued*) the Juan Fernandez ridge begins to subduct at a near-normal angle on the Nazca plate after 10 Ma. Signatures of <10 Ma are argued to be distinct from those in section at 26–26.5°S as crustal and mantle flow from the east has little influence and erupted magmas retains signatures of crust transported into the mantle by fore-arc subduction erosion, as discussed by Kay & Mpodozis (2002), Goss & Kay (2009) and Goss *et al.* (2013). See further discussion of both sequences in the text.

et al. 1994*a*, *b*) associated with contractional deformation in the thrust belts to the east (e.g. Deeken *et al.* 2006).

Latest Miocene to Pliocene. The picture changed as Maricunga Belt arc volcanism ebbed and finally ceased at 8–6 Ma as the arc front was displaced *c.* 40–50 km to the east. This is the period of the syn-migration volcanism that occurred between the old Maricunga Belt arc and future CVZ arc fronts.

In the northern transect, Kay & Coira (2009) interpret this period as including both the peak of Nazca plate shallowing under the back-arc and the initiation of the slab steepening, which facilitated lithospheric delamination beneath the Puna and resulted in the eruptions of mafic lavas along faults with strike–slip and normal components (Kay *et al.* 1994*b*; Risse *et al.* 2008) and the giant 6–4 Ma Toconquis dacitic ignimbrites (Sparks *et al.* 1985; Folkes *et al.* 2011). The crustal and mantle melting that is inferred to have begun at this time is considered to still be reflected in the regions of high attenuation (low Qp) in the tomographic image in Figure 12. The appearance of high Sm/Yb ratios in $<$ 9 Ma volcanic rocks in Chile and in the Puna Antofalla centre indicate that a thick crust was present under both the arc and back-arc by this time (Kraemer *et al.* 1999; Trumbull *et al.* 1999; Richards *et al.* 2006).

The situation is different in the southern transect, where eruption of the far back-arc *c.* 8–6 Ma Farallon Negro high-K volcanic rocks (e.g. Harris *et al.* 2006) followed by cessation of back-arc volcanism and initial uplift of the Sierras Pampeanas by *c.* 6 Ma (e.g. Sobel & Strecker 2003) signal continued flattening of the subducting slab. The depth to the Benioff zone at the time of the eruption of the Farallon Negro is inferred to have been *c.* 180–200 km, based on an analogy with the Pocho volcanic centre over the modern flab-slab at *c.* 32°S (see Fig. 1, Kay & Gordillo 1994). In the arc, the termination of Maricunga Belt volcanism by 6 Ma and subsequent migration of the arc front to the east are considered to coincide with a pulse of removal and recycling of fore-arc crust into the mantle wedge.

It is in this period that the geochemistry of the volcanic rocks in the southern transect diverges most from those in the northern transect, as shown by their very high La/Ta, La/Yb and Sm/Yb ratios (Figs 4–6). These high ratios require residual garnet, consistent with stalling and contamination of magmas in thickened crust, incorporation of melts of subducted eroded crust from the mantle wedge or a combination both of the processes. The main argument in favour of introducing fore-arc crust into the mantle wedge comes from the chemistry of the glassy plagioclase–phenocryst-free Pircas Negras andesites that erupted over the developing northern bend of the Chilean–Pampean flat-slab during the height of arc migration (see Goss *et al.* 2013). These lavas are marked by high Cr (to 250 ppm) and Ni (to 60 ppm) concentrations, extreme Sm/Yb ratios (Fig. 5), superchondritic Nb/Ta ratios and higher $^{87}Sr/^{86}Sr$ ratios at similar wt% SiO_2 than older Maricunga Belt lavas (Fig. 8). The *in situ* assimilation of garnet-bearing crust alone provides no ready explanation for their high ppm Cr and Ni or the transient nature of their very steep REE patterns and La/Ta ratios. Goss & Kay (2009) and Goss *et al.* (2013) show that these features can be modelled by interacting mantle-derived magmas with fore-arc crustal components in the mantle wedge, and that their high La/Ta and low Ba/La ratios can be explained by titanite retention and low water contents. The Tres Cruces (e.g. Redondo, Foerster, Rodrigo) and Wheelwright region lavas further north, which also erupted as the arc front migrated, have Sm/Yb and La/Ta ratios in the lower range of the Pircas Negras lavas.

In contrast, the northern transect magmas erupted west of the Puna have less pronounced residual garnet signatures, permitting little stalling of magmas at depth in the thick crust. This environment is consistent with a more extensional regime linked to steepening of the underlying slab. Upper crustal-like high $^{87}Sr/^{86}Sr$ ratios at a given $^{143}Nd/^{144}Nd$ ratio (Fig. 9), like $^{207}Pb/^{208}Pb$ and $^{206}Pb/^{204}Pb$ ratios and trace element features in these northern volcanic rocks, can be interpreted as reflecting components from the *in situ* underlying crust as well as crustal flowing from the east. This interpretation is based on the expectation that *in situ* crust contaminating these lavas should be like those to the south, whereas more radiogenic crust inflowing from the east can explain their more enriched isotopic signatures (Figs 9–11). The easiest way to inject this crust is by westward flow of partially molten crust. Evidence for extensive crustal melting at this time comes from eruption of the rhyodacitic 6–4 Ma Las Parinas, Cíclope, Laguna Amarga and Laguna Verde ignimbrites in the CVZ region, and the voluminous Toconquis ignimbrites and high temperature mafic magmas in the Puna to the east.

At the same time, the *c.* 8 Ma Aguas Blancas andesites that erupted as the arc was migrating and the slab started to steepen are distinct from other lavas in the region. As with the Pircas Negras lavas to the south, their very high Sm/Yb and La/Ta ratios at moderately high Cr (40–50 ppm) and Ni (*c.* 20 ppm) contents could be interpreted as reflecting fore-arc crust incorporated into the mantle wedge at this time.

The 6–3 Ma period is also the time when the construction of the large dome complexes with elevations of 5000–6800 m (purple regions in Fig. 2) began at Pissis, Bonete Grande and Bonete Chico in the south, and at Ojos del Salado, Tres Cruces and Sierra Nevada further north. The largest of these massive complexes are in the south above where the slab was bending into the Chilean–Pampean flat-slab and the mantle wedge was thinning. These conditions provide little opportunity for westward crustal flow allowing these complexes to reflect build-up of magmatic rocks and crustal material that was unable to spread out to the east. A further question in the southern transect is if the high attenuation zone (low Qp) above the slab at depth in the back-arc (near 27.57°S, Fig. 12) could be fore-arc lithosphere removed by fore-arc subduction erosion that is now melting in the mantle wedge.

Latest Pliocene to Recent. The last period depicted in Figure 13 shows the setting of the CVZ arc after the end of arc front migration. A strong contrast exists between the two transects by this time as the subduction zone has stabilized into a steeper mode in the north and a flat-slab mode in the south. The subducting slab under the northern transect has a generally similar dip to that envisioned under the early Miocene Maricunga Belt arc. However, the overall tectonic setting is distinct as the underlying crust is much thicker (*c.* 5–70 km) and the immediate past history involved slab steepening as ongoing back-arc crustal and mantle melting produced mafic lavas and the giant *c.* 2 Ma Cerro Galan ignimbrite (e.g. Kay *et al.* 2011), in accord with high attenuation (low Qp) zones in the tomographic images (Fig. 13; Bianchi *et al.* 2013). Overlapping wt% Na_2O and La/Yb and Sm/Yb ratios of CVZ and early Maricunga Belt volcanic rocks (Figs 4–5) reflect local extension that allowed rapid transit of mantle-derived magmas along faults (Baldwin & Marrett 2004). The higher wt% K_2O (Fig. 3), Ba/Th and Th/La ratios (Fig. 7) and $^{206}Pb/^{204}Pb$ (Figs 10 & 11) and $^{87}Sr/^{86}Sr$ ratios at higher $^{143}Nd/^{144}Nd$ ratios (Figs 8 & 9) in the CVZ volcanic rocks are consistent with contaminants from a thicker crust reshaped by Ordovician as well as Permo-Triassic events relative to the early Maricunga Belt volcanic rocks. The CVZ volcanic rocks are also related to a mantle wedge that was enriched and modified by Miocene to Pliocene fore-arc subduction erosion and lithospheric delamination. Both a hotter mantle wedge to the east and a dearth of terriginous subducted sediments help to account for lower Ba/La (Fig. 7) ratios in the CVZ lavas.

The Incapillo dome and ignimbrite complex in the southern transect was the last of the CVZ volcanic centres to form before the subducting slab became too shallow to the south and east to support arc magmatism (Goss *et al.* 2009, 2011). The erupted dacitic and rhyodacitic magmas have higher Na_2O and La/Yb and Sm/Yb ratios than those from CVZ centres to the north, reflecting magma entrapment and evolution at depth in a *c.* 65–70 km thick crust in a contractional centre over a shallow slab. Higher Ba/La and La/Ta ratios than in CVZ rocks to the north could reflect magma evolution over a shallower slab closer to the trench and to the west of a thinner back-arc mantle wedge.

Conclusions

Neogene medium to high-K andesites and associated arc rocks in the southern Central Volcanic Zone (25.5–28.2°S) and on the northern margin of the Chilean–Pampean flat-slab exhibit some of the largest variations in REE, HFSE and large ion lithophile elements (LILE) signatures and Sr, Nd and Pb isotopic ratios seen in the Central Andes. These variations reflect their eruption in a region with a changing subducting zone geometry, a crust in the process of thickening to 65–75 km, an uplifting arc, an arc front being displaced *c.* 40–50 km eastward from the Maricunga Belt to the CVZ at *c.* 8–3 Ma and an uplifting plateau in the northern part and the developing Chilean–Pampean flat-slab to the south. Some key conclusions are:

(1) Arc magmas erupted across the region commonly show trends toward steeper REE patterns, enriched isotopic signatures and more lower or upper crustal-like signatures as mantle-derived magmas are contaminated in a thickening or thick crust. The trends are well expressed in the calc-alkaline medium to high-K_2O andesitic to dacitic rocks of the 26–6 Ma Maricunga Belt, that show a temporal trend from low pressure pyroxene-dominated to high pressure garnet-bearing, plagioclase-free residual mineral assemblages, as expressed by steepening heavy REE patterns and increasing wt% Na_2O.

(2) Chemical and isotopic signatures of arc magmas erupted between *c.* 28.2 and 26.7°S, east of where the Nazca plate shallowed on the northern edge of the modern Chilean–Pampean flat-slab after 10 Ma, show more extreme trace element variability than those erupted further to the north at 26.7–25.5°S, west of the Puna plateau. The extreme and variable heavy REE slopes (Sm/Yb = 2–9), wt% Na_2O (3–5.5), HFSE depletion (La/Ta = 15–110) and Ba/La (15–55) ratios in the 8–3 Ma Pircas Negras lavas on the margin of the developing flat-slab can be explained by contaminants from the thick

underlying crust as well as by partial melts of crust incorporated into the mantle wedge by fore-arc subduction erosion (Goss *et al.* 2013). Although fore-arc subduction erosion likely accompanied the *c.* 40–50 km eastward migration of the arc front along the entire region, the chemical signatures of the volcanic rocks are distinct in the north, where there was extensive mid to deep crustal flow to the west above a hot mantle, from the south, where the mantle wedge was cooling above a shallowing slab.

(3) Magmas erupted through the thick arc crust north of 26.7°S to the west of where the Nazca slab is argued to have shallowed at *c.* 18–7 Ma and then resteepened as the Puna plateau rose and the lower crust and lithospheric mantle delaminated (Kay & Coira 2009), generally have lower Sm/Yb (2–4) and La/Ta (15–45) ratios and wt% Na_2O (3–4) than those to the south. A temporal trend towards more upper crustal-like trace element and isotopic signatures in these volcanic rocks is attributed to a stress and melting regime under the Puna that facilitated mid to deep crustal flow to the west.

(4) Overlapping wt% Na_2O and La/Yb and Sm/Yb ratios in early Miocene Maricunga Belt and <3 Ma CVZ arc volcanic rocks, which erupted through a thicker crust 40–50 km to the east, reflect local extension that allowed rapid transit of mantle-derived magmas along faults. The higher wt% K_2O and Ba/Th, Th/La, $^{206}Pb/^{204}Pb$ and $^{87}Sr/^{86}Sr$ ratios at lower $^{143}Nd/^{144}Nd$ ratios in the CVZ rocks reflect contamination of these magmas in a thicker crust and derivation of mantle-derived magmas from a wedge enriched in lithospheric components by fore-arc subduction erosion to the west and delamination to the east. Lower Ba/La ratios in the CVZ than the early Miocene lavas, at least partially, reflect a terriginous sediment-free Chile trench after 10 Ma, a hotter back-arc mantle to the east and, possibly, little serpentinization of the down-going slab in the absence of prominent bend fault in parallel with the subducting Nazca ridge acting like the Cocos ridge in the Middle American trench (Ranero *et al.* 2003; Rüpke *et al.* 2004).

(5) Pb isotopic ratios throughout the region correlate well with crustal domains reflecting early Palaeozoic and Permo-Triassic reworking of the crust, whereas a wide range of $^{87}Sr/^{86}Sr$ and $^{143}Nd/^{144}Nd$ ratios at similar wt% SiO_2 at different times and places reflect varying amounts and types of *in situ* crustal contaminants as well as mantle source variability related to recycling of continental lithosphere by fore-arc subduction erosion and delamination.

As a final comment, Gómez-Tuena *et al.* (2007) and Straub *et al.* (2008, 2011) propose a general model for arc andesite formation in which silicic components from the subducting slab infiltrate the sub-arc peridotitic mantle creating segregations of Si-deficient and Si-excess reaction pyroxenites that then partially melt to produce high-Mg# melts. These basaltic to dacitic melts are then argued to mix with subordinate amounts of peridotite melt and with each other to produce a broad range of high-Mg# (70) melts, which then undergo moderate fractional crystallization, recharge melting and mixing in the upper crust to produce the basaltic, andesitic and dacitic magmas that finally erupt. In contrast, we suggest that the more silicic components introduced into the mantle wedge in the Central Andes come from partial melts of crust introduced into the mantle by delamination and fore-arc subduction erosion. We consider that partial melts of hybridized garnet pyroxenites mixed with those of peridotite produce high Mg# Pircas Negras-type magmas, whose formation is closely tied to the time of frontal arc migration. Contamination of mantle-derived basaltic composition melts in the deep crust is considered also to play an important role in production of evolved Central Andean magmas.

We thank R. Kay for stimulating discussions and participation in collecting the INAA data; P. Cornejo, J. Clavero, A. Goss, A. Tittler, B. Coira, D. Rubiolo, P. Caffe and A. Folguera for valuable contributions in the field and discussions; the Chilean SERNAGEOMIN for K/Ar ages and some of the chemical data, and A. Diaz and J. Lemp for professional driving skills that made sampling of extremely difficult localities possible. P. Caffe and A. Gómez-Tuena are thanked for their constructive comments on an earlier version of this manuscript. Financial support variably came from Chilean FONDECYT grants 14992, 1950025 and 1990009, SERNAGEOMIN and US National Science Foundation EAR Grants 92-05042, 95-08023, 01-26000 and 05-38112.

References

AITCHESON, S. J. & FORREST, A. H. 1994. Quantification of crustal contamination in open magmatic systems. *Journal of Petrology*, **35**, 461–488.

AITCHESON, S. J., HARMON, R. S. ET AL. 1995. Pb isotopes define basement domains of the Altiplano, Central Andes. *Geology*, **23**, 555–558.

BALDWIN, A. K. & MARRETT, R. A. 2004. Evidence for a releasing bend at the southern margin of the Puna plateau, Argentine Andes. *Geological Society of America Abstracts with Programs*, **36**, 50.

BAKER, P. E., GONZÁLEZ FERRÁN, O. & REX, D. C. 1987. Geology and geochemistry of the Ojos del Salado volcanic region, Chile. *Journal of the Geological Society, London*, **144**, 85–96.

BAHLBURG, H., VERVOORT, J. D., DUFRANE, S., BOCK, B., AUGUSTSSON, C. & REIMANN, C. 2009. Timing of crust formation and recycling in accretionary orogens: insights learned from the western margin of South America. *Earth-Science Reviews*, **97**, 215–241.

BIANCHI, M., HEIT, B. ET AL. 2013. Teleseismic tomography of the southern Puna plateau in Argentina and adjacent regions. *Tectonophysics*, **586**, 65–83.

BISSIG, T., CLARK, A. H., LEE, J. K. W. & HEATHER, K. B. 2001. The Cenozoic history of volcanism and hydrothermal alteration in the Central Andean flat-slab region: New ^{40}Ar–^{39}Ar constraints from the El Indio-Pascua Au (-Ag, Cu) belt, 29°20–30°30'S. *International Geology Review*, **43**, 312–340.

CAHILL, T. & ISACKS, B. L. 1992. Seismicity and shape of the subducted Nazca Plate. *Journal of Geophysical Research*, **97**, 17 503–17 529.

CLAVERO, J., MPODOZIS, C. & GARDEWEG, M. 2000. La caldera Wheelwright, una estructura volcánica circular en la zona sur de la cadena volcánica los Andes centrales (región de Atacama, Chile). *In*: *IX Congreso Geológico Chileno Actas,* Puerto Varas, 13 July–4 August, **2**, 274.

CLAVERO, J., MPODOZIS, C., GARDEWEG, M. & VALENZUELA, M. in press. *Cartas Salar de Wheelwright (esc. 1:100 000).* Servicio Nacional de Geología y Minería Carta Geológica de Chile Santiago.

COIRA, B., KAY, S. M. & VIRAMONTE, J. 1993. Upper Cenozoic magmatic evolution of the Argentine Puna – a model for changing subduction geometry. *International Geology Review*, **35**, 677–720.

CORNEJO, P. & MPODOZIS, C. 1996. *Geología de la región de Sierra Exploradora (25°–26° Lat. S).* Registered Report, IR-96-09, 2 vols, 8 maps, scale 1:50,000. Servicio Nacional de Geología y Minería, Santiago, Chile.

CORNEJO, P., MPODOZIS, C., RAMÍREZ, C. & TOMLINSON, A. 1993. *Estudio geológico de la región de Potrerillos y El Salvador (26°–27° lat. S).* Registered Report, IR-93-01, 2 volumes, 12 maps 1:50 000 scale. Servicio Nacional de Geología y Minería, Santiago.

CORNEJO, P., MPODOZIS, C. & TOMLINSON, A. 1998. *Hoja Salar de Maricunga (scale 1:100 000).* Servicio Nacional de Geología y Minería Carta Geológica de Chile 7, Santiago.

CORNEJO, P., MPODOZIS, C., RIVERA, O. & MATTHEWS, S. 2009. *Carta Exploradora (1:100.000), Regiones de Antofagasta y Atacama.* Carta Geológica de Chile, **119**, 1–99. Servicio Nacional de Geología y Minería, Santiago, Chile.

DAVIDSON, J. & MPODOZIS, C. 1991. Regional geologic setting of epithermal gold deposits, Chile. *Economic Geology*, **86**, 1174–1186.

DAVIDSON, J., HARMON, R. & WÖRNER, G. 1991. The source of central Andean magmas: some considerations. *In*: HARMON, R. S. & RAPELA, C. W. (eds) *Andean Magmatism and Its Tectonic Setting*. Geological Society of America Special Papers, **265**, 233–243, http://dx.doi.org/10.1130/SPE265-p233

DEEKEN, A., SOBEL, E. R., COUTAND, I., HASCHKE, M., RILLER, U. & STRECKER, M. R. 2006. Development of the southern Eastern Cordillera, NW Argentina, constrained by apatite fission track thermochronology: from Early Cretaceous extension to middle Miocene shortening. *Tectonics*, **25**, TC6003, http://dx.doi.org/10.1029/2005TC001894

DEMETS, C., GORDON, R. G., ARGUS, D. F. & STEIN, S. 1990. Current plate motions. *Geophysical Journal International*, **101**, 425–478.

ELLIOT, T., PLANK, T., ZINDLER, A., WHITE, W. & BOURDON, B. 1997. Element transport from slab to volcanic front in the Mariana arc. *Journal of Geophysical Research*, **102**, 14991–15019.

FOLKES, C. B., WRIGHT, H. M., CAS, R. A. F., DESILVA, S. L., LESTI, C. & VIRAMONTE, J. G. 2011. A re-appraisal of the stratigraphy and volcanology of the Cerro Galán volcanic system, NW Argentina. *Bulletin of Volcanology*, **73**, 1427–1454.

GARDEWEG, M., MPODOZIS, C., CLAVERO, J. & CUITIÑO, L. 1997. *Mapa Geológico de la Hoja Nevado Ojos de Salado, Región de Atacama, escala 1:100 000.* Servicio Nacional de Geología y Minería Carta Geológica de Chile, Santiago.

GARDEWEG, M., MPODOZIS, C., CLAVERO, J. & CUITIÑO, L. 1998. The Ojos del Salado complex; the highest active volcano in the world. *Abstracts on Magmatic Diversity: Volcanoes and their roots*, IAVCEI, 21, Cape Town.

GARDEWEG, M., CLAVERO, J., MPODOZIS, C., PÉREZ DE ARCE, C. & VILLENEUVE, M. 2000. El Macizo Tres Cruces: un complejo volcánico longevo y potencialmente activo en la Alta Cordillera de Copiapó, Chile. *In*: *IX Congreso Geológico Chileno Actas*, Puerto Varas, 13 July–4 August, **2**, 291–295.

GÓMEZ-TUENA, A., LANGMUIR, C. H., GOLDSTEIN, S. L., STRAUB, S. M. & ORTEGA-GUTIERREZ, F. 2007. Geochemical evidence for slab melting in the Trans-Mexican Volcanic Belt. *Journal of Petrology*, **48**, 537–562.

GONZÁLEZ FERRÁN, O., BAKER, P. E. & REX, D. C. 1985. Tectonic–volcanic discontinuity at latitude 27° South, Andean Range, associated with Nazca plate subduction. *Tectonophysics*, **112**, 423–441.

GOSS, A. R. & KAY, S. M. 2009. Extreme high field strength element (HFSE) depletion and near-chondritic Nb/Ta ratios in Central Andean adakite-like lavas (~27°S, ~68°W). *Earth and Planetary Science Letters*, **270**, 97–109.

GOSS, A. R., KAY, S. M., MPODOZIS, C. & SINGER, B. S. 2009. The Incapillo Caldera and dome complex (~28°S): a stranded magma chamber over a dying Andean arc. *Journal of Volcanology and Geothermal Research*, **184**, 384–404.

GOSS, A. R., KAY, S. M. & MPODOZIS, C. 2011. The geochemistry of a dying continental arc: the Incapillo Caldera and Dome Complex of the southernmost Central Andean Volcanic Zone (~28°S). *Contributions to Mineralogy and Petrology*, **161**, 101–128.

GOSS, A. R., KAY, S. M. & MPODOZIS, C. 2013. Andean adakites from the northern edge of the Chilean–Pampean flat-slab (27–28.5°S) associated with frontal arc migration and forearc subduction erosion. *Journal of Petrology*, in press.

GUZMÁN, S. & PETRINOVIC, I. 2010. The Luingo caldera: the south-easternmost collapse caldera in the

Altiplano-Puna plateau, NW Argentina. *Journal of Volcanology and Geothermal Research*, **194**, 174–188.

Harris, A. C., Dunlap, W. J. *et al.* 2006. Multimillion year thermal history of a porphyry copper deposit: application of U–Pb, 40Ar/39Ar and (U–Th)/He chronometers, Bajo de la Alumbrera copper–gold deposit, Argentina. *Miner Deposita*, **43**, 295–314.

Hickey, R. L., Frey, F. A., Gerlach, D. C. & Lopez Escobar, L. 1986. Multiple sources for basaltic arc rocks from the southern volcanic zone of the Andes (34°–41°S): trace element and isotopic evidence for contributions from subducted oceanic crust, mantle and continental crust. *Journal of Geophysical Research*, **91**, 5963–5983.

Kay, S. M. 2006. Chemical and isotopic evidence for modification of the Central Andean arc mantle by crust removed by forearc subduction erosion. *Geochimica et Cosmochimica Acta*, **170**, A385–A385.

Kay, S. M. & Abbruzzi, J. M. 1996. Magmatic evidence for Neogene lithospheric evolution of the central Andean 'flat-slab' between 30° and 32°S. *Tectonophysics*, **259**, 15–28.

Kay, S. M. & Coira, B. L. 2009. Shallowing and steepening subduction zones, continental lithospheric loss, magmatism, and crustal flow under the Central Andean Altiplano–Puna Plateau. *In*: Kay, S. M., Ramos, V. A. & Dickinson, W. R. (eds) *Backbone of the Americas: Plateau Uplift, Shallow Subduction, and Ridge Collision*. Geological Society of America, Memoirs, Boulder, **204**, 229–259.

Kay, S. M. & Gordillo, C. E. 1994. Pocho volcanic rocks and the melting of depleted continental lithosphere above a shallowly dipping subduction zone in the Central Andes. *Contributions to Mineralogy and Petrology*, **117**, 25–44.

Kay, S. M. & Mpodozis, C. 2001. Central Andean ore deposits linked to evolving shallow subduction systems and thickening crust. *GSA Today*, **11**, 4–9.

Kay, S. M. & Mpodozis, C. 2002. Magmatism as a probe to the Neogene shallowing of the Nazca plate beneath the modern Chilean flatslab. *Journal of South American Earth Science*, **15**, 39–59.

Kay, S. M., Mpodozis, C., Ramos, V. A. & Munizaga, F. 1991. Magma source variations for mid-late Tertiary magmatic rocks associated with a shallowing subduction zone and a thickening crust in the central Andes (28 to 33°S). *In*: Harmon, R. S. & Rapela, C. W. (eds) *Andean Magmatism and its Tectonic Setting*. Geological Society of America Special Papers, **265**, 113–137.

Kay, S. M., Coira, B. & Viramonte, J. 1994*a*. Young mafic back-arc volcanic rocks as indicators of continental lithospheric delamination beneath the Argentine Puna plateau, Central Andes. *Journal of Geophysical Research*, **99**, 24 323–24 339.

Kay, S. M., Mpodozis, C., Tittler, A. & Cornejo, P. 1994*b*. Tertiary magmatic evolution of the Maricunga mineral belt in Chile. *International Geology Review*, **36**, 1079–1112.

Kay, S. M., Mpodozis, C. & Coira, B. 1999. Magmatism, tectonism and mineral deposits of the Central Andes (22°–33°S latitude). *In*: Skinner, B. J. (ed.) *Geology and Ore Deposits of the Central Andes*. Society of Economic Geology, Special Publications, Littleton, CO, **7**, 27–59.

Kay, S. M., Coira, B. & Mpodozis, C. 2006. Late Neogene volcanism in the Cerro Blanco region of the Puna Austral Argentina (*c.* 26.5°S, *c.* 67.5°W). *In*: *XI Congreso Geológico Chileno Actas*, Antofagasta, 7–11 August, **2**, 499–502.

Kay, S. M., Coira, B. & Mpodozis, C. 2008. Field trip guide: Neogene evolution of the central Andean Puna plateau and southern Central Volcanic Zone. *In*: Kay, S. M. & Ramos, V. A. (eds) *Field Trip Guides to the Backbone of the Americas in the Southern and Central Andes: Ridge Collision, Shallow Subduction, and Plateau Uplift*. Geological Society of America, Field Guide, Boulder, **13**, 119–183.

Kay, S. M., Coira, B. L., Caffe, P. J. & Chen, C.-H. 2010. Geochemical diversity, crustal and mantle sources and evolution of the Neogene Puna plateau ignimbrites of the Central Andes. *Journal of Volcanology and Geothermal Research*, **198**, 81–111.

Kay, S. M., Coira, B., Wörner, G., Kay, R. W. & Singer, B. S. 2011. Geochemical, isotopic and single crystal $^{40}Ar/^{39}Ar$ age constraints on the evolution of the Cerro Galán ignimbrites. *Bulletin of Volcanology*, **73**, 1487–1511.

Kraemer, B., Adelmann, D. *et al.* 1999. Incorporation of the Paleogene foreland into the Neogene Puna plateau: the Salar de Antofalla area, NW Argentina. *Journal of South American Earth Sciences*, **12**, 157–182.

Liang, X., Sandvol, E. A., Brown, L. D., Heit, B., Yuan, X., Mulcahy, P. & Kay, S. M. 2011. The Q images of southern Puna plateau from body wave attenuation tomography. Abstract S51C-2226, presented at the Fall Meeting, American Geophysical Union, San Francisco, 5–9 December 2011.

Lucassen, F., Becchio, R. *et al.* 2001. Composition and density model of the continental crust at an active continental margin: Central Andes between 21° and 27°S. *Tectonophysics*, **341**, 195–223.

Lucassen, F., Harmon, R., Franz, G., Romer, R. L., Becchio, R. & Siebel, W. 2002. Lead evolution of the Pre-Mesozoic crust in the Central Andes (18°–27°): progressive homogenisation of Pb. *Chemical Geology*, **186**, 183–197.

Lucassen, F., Franze Romer, R. G., Schultz, F., Dulski, P. & Wemmer, K. 2007. Pre-Cenozoic intraplate magmatism along the Central Andes (17–34°S): Composition of the mantle at an active margin. *Lithos*, **99**, 312–338.

Mamani, M., Tassara, A. & Wörner, G. 2008. Composition and structural control of crustal domains in the Central Andes. *Geochemistry, Geophysics, Geosystems*, **9**, http://dx.doi.org/10.1029/2007GC001925

Mamani, M., Wörner, G. & Sempere, T. 2010. Geochemical variations in igneous rocks of the Central Andean orocline (13°S to 18°S): tracing crustal thickening and magma generation through time and space. *Geological Society of America Bulletin*, **122**, 162–182.

Mcglashan, N., Brown, L. D. & Kay, S. M. 2008. Crustal thicknesses in the Central Andes from teleseismically recorded depth phase precursors. *Geophysical Journal International*, **175**, 1013–1022.

McKee, E. H., Robinson, A. C., Rybuta, J. J., Cuitini, L. & Moscoso, R. E. 1994. Age and Sr isotopic composition of volcanic rocks in the Maricunga Belt, Chile: implications for magma sources. *Journal of South American Earth Sciences*, **7**, 167–177.

McNutt, R. R., Crocket, J. H., Clark, A. H., Caelles, J. C., Farrar, E., Haynes, S. J. & Zentilli, M. 1975. Initial $^{87}Sr/^{86}Sr$ ratios of plutonic and volcanic rocks of the Central Andes between latitudes 26° and 29° South. *Earth and Planetary Science Letters*, **27**, 305–13.

Mercado, M. 1982. *Hoja Laguna del Negro Francisco, Región de Atacama*. Servicio Nacional de Geología y Mineria Carta Geológica de Chile, 56, Santiago.

Moscoso, R., Maksaev, V. et al. 1993. El complejo vol cánico Cerro Bravos, región de Maricunga, Chile: geología, alteración hidrotermal y mineralización. *In*: *Investigaciones de Metales Preciosos en el complejo volcánico neógeno-cuaternario de los Andes Centrales*. Bolivia Servicio Geológico, Banco Interamericano de Desarrollo, 131–165.

Mpodozis, C. & Clavero, J. 2002. Tertiary tectonic evolution of the southwestern edge of the Puna Plateau: Cordillera Claudio Gay (26°–27°S). *In*: *Proceedings of 5th International Symposium on Andean Geodynamics*, Toulouse, 16–18 September, IRD, Paris, 445–448.

Mpodozis, C. & Kay, S. M. 1992. Late Paleozoic to Triassic evolution of the Gondwana margin: Evidence from Chilean Frontal Cordilleran batholiths (28ºS to 31ºS). *Geological Society of America Bulletin*, **104**, 999–1014.

Mpodozis, C. & Kay, S. M. 2009. Evolution of <10 Ma Valle Ancho region lavas, southern end of the Central Andean Volcanic Zone (*c*. 27.5°S). *In*: *XII Congreso Geológico Chileno Actas*, Santiago, 22–26 November, S7–019.

Mpodozis, C., Cornejo, P., Kay, S. M. & Tittler, A. 1995. La Franja de Maricunga: síntesis de la evolución del frente volcánico oligoceno-mioceno de la zona sur de los Andes Centrales. *Revista Geológica de Chile*, **22**, 273–314.

Mpodozis, C., Kay, S. M., Gardeweg, M. & Coira, B. 1996. Geología de la región de Ojos del Salado (Andes centrales, 27°S): Implicancias de la migración hacia el este del frente volcánico Cenozoico Superior. *In*: *XIII Congreso Geológico Argentino Actas*, Buenos Aires, 13–18 October, **3**, 539–554.

Mpodozis, C., Kay, S. M., Gardeweg, M. & Coira, B. 1997. Geología de la región de Valle-Ancho-Laguna Verde (Catamarca, Argentina): Una ventana al basamento del extremo sur de la zona volcánica de los Andes Centrales. *In*: *XIII Congreso Geológico Chileno Actas*, Antofagasta, 13–17 October, 1689–1693.

Mpodozis, C., Iriarte, S., Gardeweg, M. & Valenzuela, M. in press. *Carta Laguna del Negro Francisco (1:100.00)*. Servicio Nacional de Geología y Minería Carta Geológica de Chile, Santiago.

Mulcahy, P. 2012. *The southern Puna seismic experiment: seismicity and the morphology of the subduction zone*. MS thesis, Cornell University.

Mulcahy, P., Chen, C. et al. 2010. The SOUTHERN PUNA seismic experiment: shape of the subducting Nazca Plate, areas of concentrated mantle and crustal earthquakes, and crustal focal mechanisms. *Eos Transactions of the AGU*, **91**, Fall Meeting Suppl., T11A-2050.

Muntean, J. & Einaudi, M. 2001. Porphyry epithermal transition, Maricunga Belt, northern Chile. *Economic Geology*, **96**, 743–772.

Nalpas, T., Dabard, M.-P., Ruffet, G., Vernon, A., Mpodozis, C., Loi, A. & Hérail, G. 2008. Sedimentation and preservation of the Miocene Atacama gravels in the Pedernales–Chañaral Area, Northern Chile: Climatic or tectonic control? *Tectonophysics*, **459**, 161–173.

Naranjo, J. A. & Cornejo, P. 1993. *Hoja Sala de la Isla (1:250 000)*. Servicio Nacional de Geología y Minería Carta Geológica de Chile, Santiago.

Pardo-Casas, F. & Molnar, P. 1987. Relative motion of the Nazca (Farallon) and South American plates since Late Cretaceous time. *Tectonics*, **6**, 233–248.

Ramos, V. A. 2008. The basement of the Central Andes: the Arequipa and related terranes. *Annual Review Earth and Planetary Sciences*, **36**, 289–324.

Ramos, V. A., Vujovich, G., Martino, R. & Otamendi, J. 2010. Pampia: A large cratonic block missing in the Rodinia supercontinent. *Journal of Geodynamics*, **50**, 243–255.

Ranero, C. R., Morgan, J. P., McIntosh, K. & Reichert, C. 2003. Bending related faulting and mantle serpentinization at the Middle America trench. *Nature*, **425**, 367–373.

Richards, J. P., Ulrich, T. & Kerrick, R. 2006. The late Miocene Quaternary Antofalla volcanic complex, southern Puna, NW Argentina: Protracted history, diverse petrology and economic potential. *Journal of Volcanology and Geothermal Research*, **152**, 197–239.

Risse, A., Trumbull, R. B., Coira, B., Kay, S. M. & Van Den Bogaard, P. 2008. $^{40}Ar/^{39}Ar$ geochronology of basaltic volcanism in the back-arc region of the southern Puna plateau, Argentina. *Journal of South America Geology*, **26**, 1–15.

Risse, A., Trumbull, R. B., Kay, S. M., Coira, B. & Romer, R. L. 2013. Multi-stage evolution of late Neogene mantle derived magmas from the Central Andes back arc in the southern Puna Plateau of Argentina. *Journal of Petrology*, in press.

Rüpke, L. H., Morgan, J. P. & Connolly, J. A. D. 2004. Serpentine and the subduction zone water cycle. *Earth and Planetary Science Letters*, **223**, 17–34, http://dx.doi.org/10.1016/j.epsl.2004.04.018

Rubiolo, D., Hickson, C., Busteros, A., Villenueve, M. & Edwards, B. 2000. Petrología del Cerro Morocho: volcan del mioceno superior en ambiente glacial? (Valle de Chaschuil, Catamarca, Argentina). *In*: *IX Congreso Geológico Chileno Actas*, Puerto Varas, 13 July–4 August, **2**, 329–333.

Sasso, A. & Clark, A. H. 1998. The Farallon Negro Group, northwest Argentina: magmatic, hydrothermal and tectonic evolution and implications for Cu–Au metallogeny in the Andean back-arc. *Society of Economic Geology Newsletter*, **34**, 1–18.

Schnurr, W. B. W., Trumbull, R. B., Clavero, J., Hahne, K., Siebel, W. & Gardeweg, M. 2007. Twenty-five million years of silicic volcanism in the southern central volcanic zone of the Andes.

Geochemistry and magma genesis of ignimbrites from 25° to 27°S, 67° to 72°W. *Journal of Volcanology and Geothermal Research*, **266**, 27–46.

Seggiaro, R., Hongn, F., Folguera, A. & Clavero, J. 2002. *Mapa geológico de la Hoja 2769 II, Paso de San Francisco, Provincia de Catamarca, scale 1:250 000.* Servicio Geológico Minero Argentino Boletín, **294**, Buenos Aires.

Siebel, W., Schnurr, W., Hahne, K., Kraemer, B., Trumbull, R. B., Van Den Bogaard, P. & Emmermann, R. 2001. Geochemistry and isotope systematics of small- to medium-volume Neogene–Quaternary ignimbrites in the southern central Andes: evidence for derivation from andesitic magma sources. *Chemical Geology*, **171**, 213–237.

Sillitoe, R. H. 1991. Gold rich porphyry systems in the Maricunga Belt, northern Chile. *Economic Geology*, **86**, 1238–1260.

Sobel, E. & Strecker, M. R. 2003. Uplift, exhumation and precipitation: tectonic and climatic control of late Cenozoic landscape evolution in the northern Sierras Pampeanas, Argentina. *Basin Research*, **15**, 431–451, http://dx.doi.org/10.1046/j.1365-2117.2003.00214.x

Somoza, R. 1998. Updated Nazca (Farallon)–South America relative motions during the last 40 my: implications for mountain building in the Central Andean region. *Journal of South American Earth Sciences*, **11**, 211–215.

Sparks, R. S. J., Francis, P. W., Hamer, R. D., Pankhurst, R. J., O'Callaghan, L. L., Thorpe, R. S. & Page, R. S. 1985. Ignimbrites of the Cerro Galán Caldera, NW Argentina. *Journal of Volcanology and Geothermal Research*, **24**, 205–248.

Straub, S. M., Lagatta, A. B., Martin-Del Pozzo, A. L. & Langmuir, C. H. 2008. Evidence from high-Ni olivines for a hybridized peridotite/pyroxenite source for orogenic andesites from the central Mexican Volcanic Belt. *Geochemistry Geophysics Geosystems*, **9**, http://dx.doi.org/10.1029/2007GC001583

Straub, S. M., Gomez-Tuena, A., Stuart, F. M., Zellmer, G. F., Espinasa-Perena, R., Cai, Y. & Iizuka, Y. 2011. Formation of hybrid arc andesites beneath thick continental crust. *Earth and Planetary Science Letters*, **303**, 337–347

Thomas, W. & Astini, R. 2003. Ordovician accretion of the Argentine Precordillera terrane to Gondwana: a review. *Journal of South American Earth Sciences*, **16**, 67–79.

Tomlinson, A., Cornejo, P. & Mpodozis, C. 1999. *Hoja Potrerillos, Región de Atacama scale 1:100 000.* Servicio Nacional de Geología y Minería Carta Geológica de Chile, 14, Santiago.

Trumbull, R. B., Wittenbrink, R., Hahne, K., Emmermann, R., BüSch, W., Gerstenberger, H. & Siebel, W. 1999. Evidence for late Miocene to Recent contamination of arc andesites by crustal melts in the Chilean Andes (25–26°S) and its geodynamic implications. *Journal of South American Earth Sciences*, **12**, 135–155.

Vila, T. & Sillitoe, R. H. 1991. Gold-rich porphyry systems in the Maricunga belt, northern Chile. *Economic Geology*, **86**, 1238–1260.

Walker, J., Moulds, T., Zentilli, M. & Feigenson, M. D. 1991. Spatial and temporal variations in volcanics of the Andean Central Volcanic Zone (26° to 28°S). *In*: Harmon, R. S. & Rapela, C. W. (eds) *Andean Magmatism and its Tectonic Setting*. Geological Society of America Special Papers, **265**, 139–155.

Willner, A. P., Gerdes, A., Massonne, H.-J., Schmidt, A., Sudo, M., Thomson, S. N. & Vujovich, G. 2011. The geodynamics of collision of a microplate (Chilenia) in Devonian times deduced by the pressure–temperature–time evolution within part of a collisional belt (Guarguaraz Complex, W-Argentina). *Contributions to Mineralogy and Petrology*, **62**, 303–327.

Wörner, G., Moorbath, S. & Harmon, R. S. 1992. Andean Cenozoic volcanic centers reflect basement isotopic domains. *Geology*, **20**, 1103–1106.

Yañéz, G. A., Ramiro, C. R., Von Huene, R. & Diaz, J. 2001. Magnetic anomaly interpretation across the southern central Andes (32°–34°S): the role of the Juan Fernández Ridge in the late Tertiary evolution of the margin. *Journal of Geophysical Research*, **106**, 6325–6345.

Zentilli, M. 1974. *Geological evolution and metallogenetic relationships in the Andes of northern Chile between 26° and 29° south.* PhD thesis, Queen's University, Kingston, Ontario.

Evolution of late Cenozoic magmatism and the crust–mantle structure in the NE Japan Arc

TAKEYOSHI YOSHIDA[1]*, JUN-ICHI KIMURA[2], RYOICHI YAMADA[1], VALERIO ACOCELLA[3], HIROSHI SATO[4], DAPENG ZHAO[5], JUNICHI NAKAJIMA[5], AKIRA HASEGAWA[5], TOMOMI OKADA[5], SATORU HONDA[4], MASAHIRO ISHIKAWA[6], OKY DICKY ARDIANSYAH PRIMA[7], TAKESHI KUDO[8], BUNICHIRO SHIBAZAKI[9], AKIKO TANAKA[10] & TOSHIFUMI IMAIZUMI[1]

[1]*Institute of Earth Sciences, Graduate School of Science, Tohoku University, Sendai 980-8578, Japan*

[2]*Institute for Research on Earth Evolution, Japan Agency for Marine-Earth Science and Technology, Yokosuka 237-0061, Japan*

[3]*Dipartimento Scienze Geologiche, Universita Roma TRE, Rome 00146, Italy*

[4]*Earthquake Research Institute, University of Tokyo, Tokyo 113-0032, Japan*

[5]*Research Center for Prediction of Earthquakes and Volcanic Eruptions, Tohoku University, Sendai 980-8578, Japan*

[6]*Graduate School of Environment and Information Sciences, Yokohama National University, Yokohama 240-8501, Japan*

[7]*Faculty of Software and Information Science, Iwate Prefectural University, Iwate 020-0193, Japan*

[8]*College of Engineering, Chubu University, Kasugai 487-8501, Japan*

[9]*International Institute for Seismology and Earthquake Engineering, Building Research Institute, Tsukuba 305-0802, Japan*

[10]*Geological Survey of Japan, AIST, Tsukuba 305-8567, Japan*

**Corresponding author (e-mail: tyoshida@m.tohoku.ac.jp)*

Abstract: We review the evolution of late Cenozoic magmatism in the NE Japan arc, and examine the relationship between the magmatism and the crust–mantle structure. Recent studies reveal secular changes in the mode of magmatic activity, the magma plumbing system, erupted volumes and magmatic composition associated with the evolution of crust–mantle structures related to the tectonic evolution of the arc. The evolution of Cenozoic magmatism in the arc can be divided into three periods: the continental margin (66–21 Ma), the back-arc basin (21–13.5 Ma) and the island-arc period (13.5–0 Ma). Magmatic evolution in the back-arc basin and the island-arc periods appears to be related to the 2D to 3D change in the convection pattern of the mantle wedge related to the asthenosphere upwelling and subsequent cooling of the mantle. Geodynamic changes in the mantle caused back-arc basin basalt eruptions during the back-arc basin opening (basalt phase) followed by crustal heating and re-melting, which generated many felsic plutons and calderas (rhyolite/granite phase) in the early stage of the island-arc period. This was followed by crustal cooling and strong compression, which ensured vent connections and mixing between deeper mafic and shallower felsic magmas, erupting large volumes of Quaternary andesites (andesite phase).

The NE Japan arc is a typical island arc associated with a cold subduction zone. It is one of the most well-known island arcs on Earth and provides a useful model for our understanding of how subduction zones produce magmas, including andesites. The NE Japan arc is related to the westward

From: GÓMEZ-TUENA, A., STRAUB, S. M. & ZELLMER, G. F. (eds) 2014. *Orogenic Andesites and Crustal Growth.* Geological Society, London, Special Publications, **385**, 335–387.
First published online September 24, 2013, http://dx.doi.org/10.1144/SP385.15

subduction of the Pacific plate beneath the Eurasian plate at a rate of 8–9 cm a^{-1} with a relatively shallow dip angle of about 30° (DeMets 1992). Integrated study of geology, petrology and geophysics is valuable for revealing processes such as the formation of crust–mantle structures, the origin of the magmas and the geodynamic evolution of the subduction zone system (Iwamori & Zhao 2000; Stern 2002; van Keken 2003; Sato *et al.* 2004; Hasegawa *et al.* 2005; Yoshida *et al.* 2005; Kimura & Yoshida 2006; Yoshida 2009).

Secular changes in the late Cenozoic NE Japan magmatic arc are key factors involved in the arc's evolution, and changes in (1) the mode of magmatic activity, (2) the magma plumbing system, (3) the erupted volume and (4) the magmatic composition have all been investigated in relation to the tectonic history of the arc (Sato 1994; Yoshida *et al.* 1995; Umeda *et al.* 1999; Yoshida 2001, 2009; Yamada & Yoshida 2011; Yamada *et al.* 2012). The evolution of late Cenozoic magmatism in the NE Japan arc can be divided into three major periods – (1) the continental margin (66–21 Ma), (2) the back-arc basin (21–13.5 Ma) and (3) the island-arc (13.5–0 Ma) periods (Ohguchi *et al.* 1989; Yoshida *et al.* 1995; Yoshida 2001) – all of which are related to the development of magmatism in the subduction zone, including the back-arc basin opening (Sato 1994; Yoshida *et al.* 1995; Yamada *et al.* 2012). The continental margin period is characterized by mafic to felsic subaerial volcanism, with thick ignimbrites in the continental rift systems. The back-arc basin period is characterized by submarine bimodal volcanism under extensional tectonics, and the island-arc period is characterized by felsic caldera-forming events followed by andesitic stratovolcano-forming volcanism under neutral to compressional stress fields (Ohguchi *et al.* 1989; Sato & Yoshida 1993; Sato 1994; Yoshida 2001).

The characteristic magma of subduction zones is andesite, although it is usually associated with basaltic to rhyolitic magmas (Gill 1981). Calc-alkaline andesites are typical, particularly in mature arcs constructed on a thick crust (Kuno 1968, and references therein). The mafic magmas derived from the mantle wedge may be underplated and stagnating around the Moho beneath the thick crust, especially under horizontal compression. Differentiated magmas can stagnate at levels of neutral buoyancy in the crust (Ryan 1987; Takada 1989, and references therein). There are many models for andesite genesis in the subduction zone of Japan, including differentiation by fractional crystallization of mafic magmas (low-K tholeiite from Adatara volcano: Fujinawa 1988), magma mixing/co-mingling between mafic and felsic end-members (low-K andesite: Toya *et al.* 2005; medium-K andesite: Ban & Yamamoto 2002; Hirotani & Ban 2006; high-K andesite: Inoue & Ban 1996), assimilation and fractional crystallization (DePaolo 1985) or crustal contamination (Kobayashi & Nakamura 2001; Hirotani *et al.* 2009), re-melting of lower-crustal hydrous mafic rocks (Takahashi 1986; Kimura *et al.* 2001, 2002) and melting of the mantle wedge or slab (high-magnesian andesite and adakite: Hanyu *et al.* 2006; Yamamoto & Hoang 2009; Hoang *et al.* 2009). Possible processes which could produce the felsic end-member of mixed magmas are: the fractional crystallization from mafic end-member magmas (Kanisawa & Yoshida 1989; Miyagi *et al.* 2012); the melt extraction or re-melting of partially or fully crystallized precursory mafic magmas in the upper to middle crust (Hildreth & Wilson 2007; Hirotani *et al.* 2009; Miyagi *et al.* 2012); the partial re-melting of mafic end-member material solidified to become hornblende gabbro or amphibolite (Feeley *et al.* 1998; Hansen *et al.* 2002; Toya *et al.* 2005); and the partial re-melting of crustal basement rocks (Shuto *et al.* 2006, 2008). All of these crustal processes forming calc-alkaline andesites work effectively in mature arcs on thick crust.

To understand the origin of island-arc andesites, it is very important to know in detail the structure of the magma plumbing system from the upper mantle through the entire crust. Moreover, the observation that andesite-dominated volcanism may occur only at specific stages of the evolution of a subduction zone justifies a comprehensive study of the evolutional history of the NE Japan arc and the crust–mantle structures throughout its evolution. The purpose of this paper is to review the evolutionary history of late Cenozoic magmatism in the arc, including its thermal and tectonic evolution, and to examine the relationship between magmatism and the crust–mantle structures at each evolutionary stage.

Tectonic and magmatic evolution in the NE Japan arc

Tectonic outline

The NE Japan arc consists of multiple tectonic segments separated by topographic depressions (Fig. 1a). Horst and graben-like structures developed in the Early to Middle Miocene, simultaneously with the opening of the back-arc basin of the Japan Sea (Okamura *et al.* 1995; Yoshida *et al.* 2005; Yamada & Yoshida 2011). Detailed seismological profiles of the arc enable the examination of the deep structures in the island-arc crust. From these profiles, two rift systems, which correspond to the surface topography and the geology, can be recognized (Fig. 1b: Sato *et al.* 2004); the

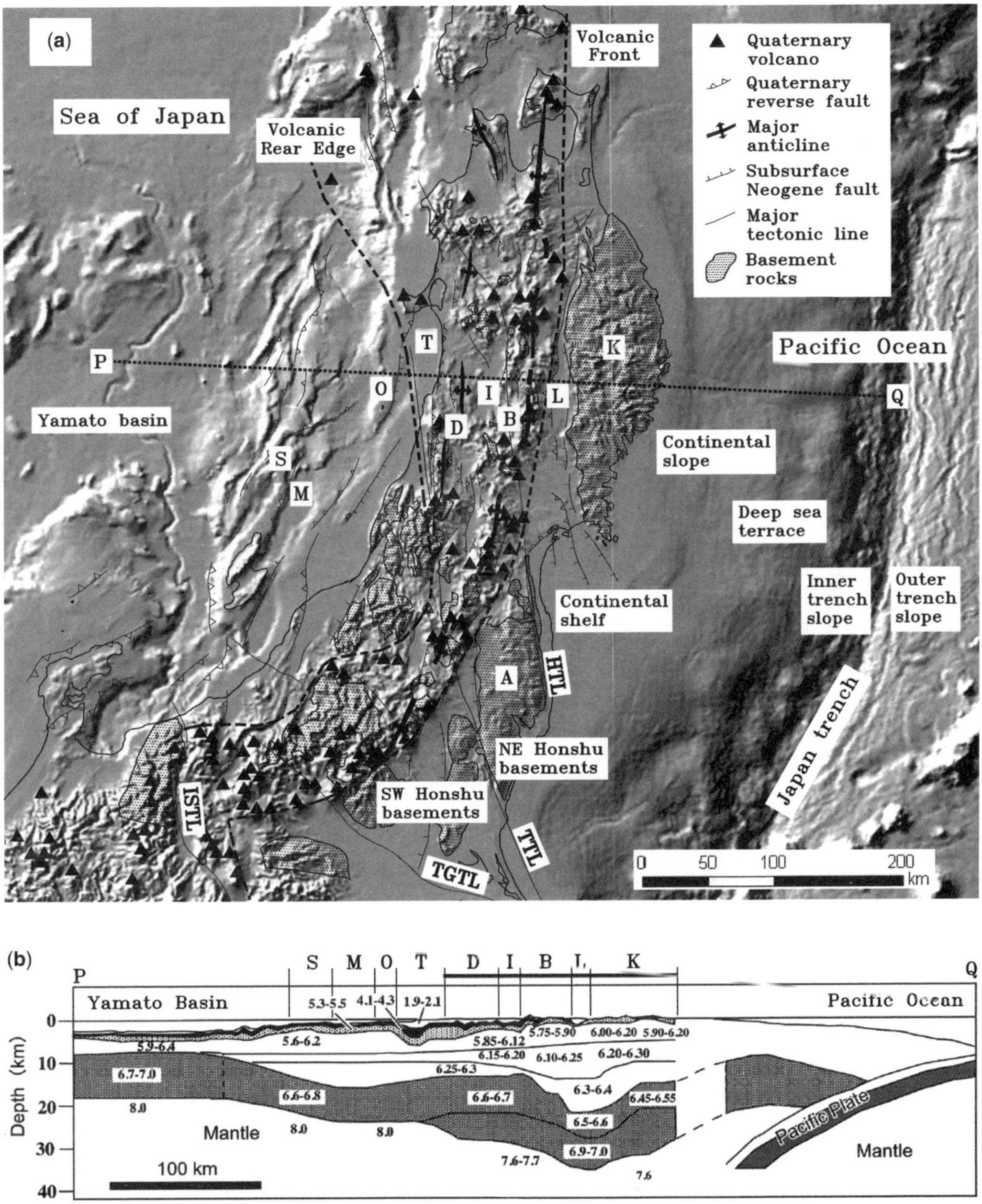

Fig. 1. (**a**) Digital elevation model showing tectonics and the distribution of Quaternary volcanoes in the NE Japan arc (modified from Sato 1994 and Yoshida *et al.* 2005). (**b**) Seismic V_p structure across the NE Japan arc (Iwasaki *et al.* 2001; Nishisaka *et al.* 2001; Takahashi *et al.* 2004; Sato *et al.* 2004). Numbers in (b) correspond to km s^{-1}. The two dashed lines in (a) show the western and eastern edges of Quaternary volcanism; the eastern edge is the present volcanic front, and the western edge is the (Quaternary) volcanic rear edge (Yoshida *et al.* 2005). Tanakura tectonic line (TTL) is the main boundary fault between the basement of NE and SW Honshu. The Hatagawa tectonic line (HTL) divides the basement of NE Honshu into the Abukuma Belt and the Kitakami Belt. Other symbols are explained in the upper-right box. K, Kitakami Mountains (Massif); A, Abukuma Mountains; L, Kitakami River Valley; B, Ou Backbone Range; I, Intermountain Basins; D, Dewa hills; T, Tobishima Basin (Akita–Niigata Basin); O, Oga–Awashima fault zone (Tobishima–Funakawa uplifted zone); M, Mogami Trough; S, Sado Ridge; TGTL, Tonegawa tectonic line; ISTL, Itoigawa Shizuoka tectonic line.

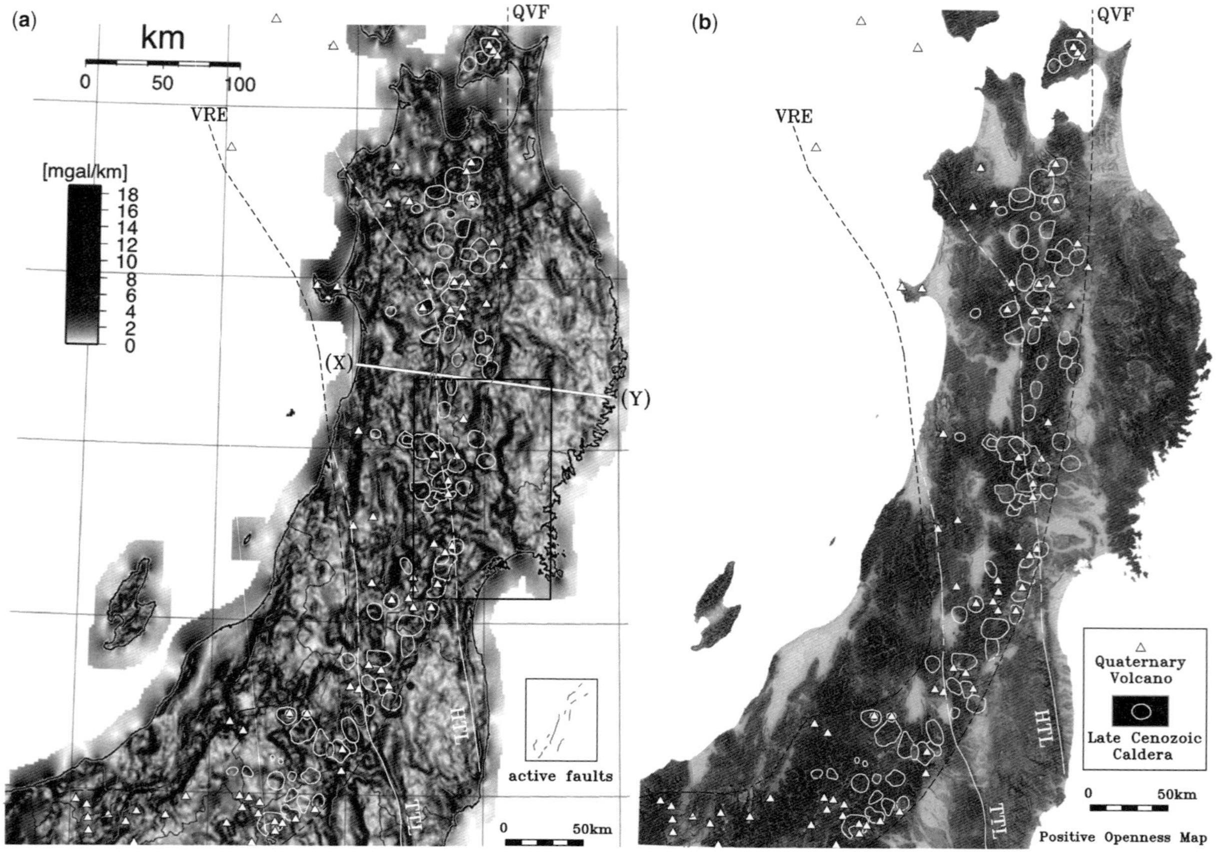

Fig. 2. (**a**) Horizontal gradient distribution of short-wavelength components of Bouguer gravity anomalies (Kudo *et al.* 2010; Yamamoto *et al.* 2011) and the distribution of late Cenozoic calderas (white circles) and Quaternary volcanoes (white triangles) in the NE Honshu arc. Topographic density is assumed to be 2.67 g cm^{-3}. The distribution of late Cenozoic large calderas corresponds strongly with the distribution of gravity depressions, and it is likely that the density structure of these calderas is a major cause of the gravity undulations of this region (Kudo *et al.* 2010). Section X–Y and the rectangle, show the study section from Figure 14a, b, and the study area in Figure 14c, respectively. (**b**) Distribution of the late Cenozoic large calderas and Quaternary volcanoes superimposed on a map of positive openness which is one of the thematic maps derived from the 50 m grid-size digital elevation models (Yokoyama *et al.* 1999; Prima *et al.* 2006). Adapted from Yoshida *et al.* (2005). QVF, Quaternary volcanic front; VRE, Quaternary volcanic rear edge; TTL, Tanakura tectonic line; HTL, Hatagawa tectonic line. Short lines represent active faults (Active Fault Research Group 1991).

Yamato Basin rift and the Northern Honshu rift. The Yamato Basin rift forms half-graben, bounded by westward-dipping Miocene normal faults. The thickness of the arc crust decreases westward owing to early Miocene crustal stretching, and is related to the opening of the Japan Sea. The Northern Honshu rift is younger (Middle Miocene), and formed a deep sedimentary basin by simple shear. Basalts were erupted along the rift axis associated with a thinning of the upper crust (Yagi *et al.* 2001; Sato *et al.* 2004). Very active felsic volcanism, which created more than 80 calderas concentrated along the topographic high of the Ou Backbone range (Ito *et al.* 1989; Sato & Yoshida 1993; Sato 1994; Yoshida *et al.* 1999*a*), occurred at the main stage of the island-arc period in NE Honshu (Fig. 2). During the late Miocene to late Pliocene, the Ou Backbone range was characterized by widespread north–south-trending dextral faults and NE–SW-trending normal faults (Acocella *et al.* 2008). During the Quaternary, the tectonic setting was characterized by a strong east–west compression (i.e. Sato 1994), and more andesites erupted from along-arc and east–west-aligned stratovolcanoes.

Temporal evolution of the NE Japan magmatic arc

Tectono-magmatic periods and spatio-temporal variation of magmatism. The three major evolutionary periods of the NE Japan arc can be further subdivided into 13 stages (Table 1 and Fig. 3) based on modes of volcanism, tectonic events and the inferred regional stress field (Ohguchi *et al.* 1989; Sato 1994; Yoshida *et al.* 1995; Nakajima *et al.* 2006*c*; Yoshida 2001, 2009). Figure 3 shows the migration of the position of the volcanic front in the NE Japan arc over the past 60 myr, where the age of each volcanic event is plotted against the distance from the Quaternary volcanic front (QVF). The volcanic front was located deeper in the back-arc in the time period immediately before the back-arc basin spreading, and it had quickly migrated to the trench side at a position approximately 40 km trench-ward of the QVF until the end of Stage 7. Through Stages 8–13 the volcanic front migrated back to the rear-arc and had settled in its present QVF position by Stage 13. It is clear that the rapid trench-ward migration of volcanism was coeval to the deepening of the back-arc basins under strong extension (Fig. 3), and it is possible that the formation of rift graben may have activated the back-arc basin volcanism.

The late Cenozoic NE Japan arc is subdivided into eight tectonic segments (Fig. 4a), based on gravity anomalies, volcanic stratigraphy and the age of volcanic rocks (Yamada & Yoshida 2011). Figure 4b shows the East–West cross-section through the forearc to the back-arc basin of the NE Japan arc, and Figure 5 shows the volcanic stratigraphy over the past 30 myr (Yamada & Yoshida 2011). Continental margin volcanism first occurred at *c.* 35 Ma in the western continental rift zone (Kano *et al.* 2007), and volcanism during this stage was dominated by subaerial andesite activity. Back-arc basin volcanism was active in the Yamato Basin between 21 and 17 Ma (Kaneoka *et al.* 1992), followed by rift volcanism in the Aosawa Rift at 17–15 Ma (Yagi *et al.* 2001), and in the Kuroko Rift at 15–13.5 Ma (Yamada & Yoshida 2011).

The first half of the back-arc basin period was characterized by the effusion of large amounts of seafloor basalt during the opening of the Yamato Basin, whereas the second half was characterized by spatially varying amounts of graben-fill basalt

Table 1. *The 13 stages of the NE Japan arc evolution*

(1) Continental margin period (66–21 Ma)	
Stage 1 (66–49 Ma)	Acidic to intermediate igneous activity
Stage 2 (49–35 Ma)	Hiatus with no volcanic activity.
Stage 3 (35–27 Ma)	Alkaline volcanism of within-plate type in the back-arc side
Stage 4 (27–21 Ma)	Calc-alkaline volcanism with alkaline rhyolites in the back-arc side
(2) Back-arc basin period (21–13.5 Ma)	
Stage 5 (21–17 Ma)	Yamato basin spreading
Stage 6 (17–15 Ma)	Early Northern Honshu rift (Aosawa rift) stage
Stage 7 (15–13.5 Ma)	Later Northern Honshu rift (Kuroko rift) stage
(3) Island-arc period (13.5–0 Ma)	
Stage 8 (13.5–10 Ma)	Oceanic island arc under neutral stress field
Stage 9 (10–8 Ma)	Oceanic island arc under weak compressional stress field
Stage 10 (8–5.3 Ma)	Early caldera-dominated arc (NE–SW compression)
Stage 11 (5.3–3.5 Ma)	Late caldera-dominated arc (NE–SW compression)
Stage 12 (3.5–1.0 Ma)	Transitional stage to strong compressional stress field
Stage 13 (1.0–0 Ma)	Andesitic stratovolcanoes (east–west strong compression)

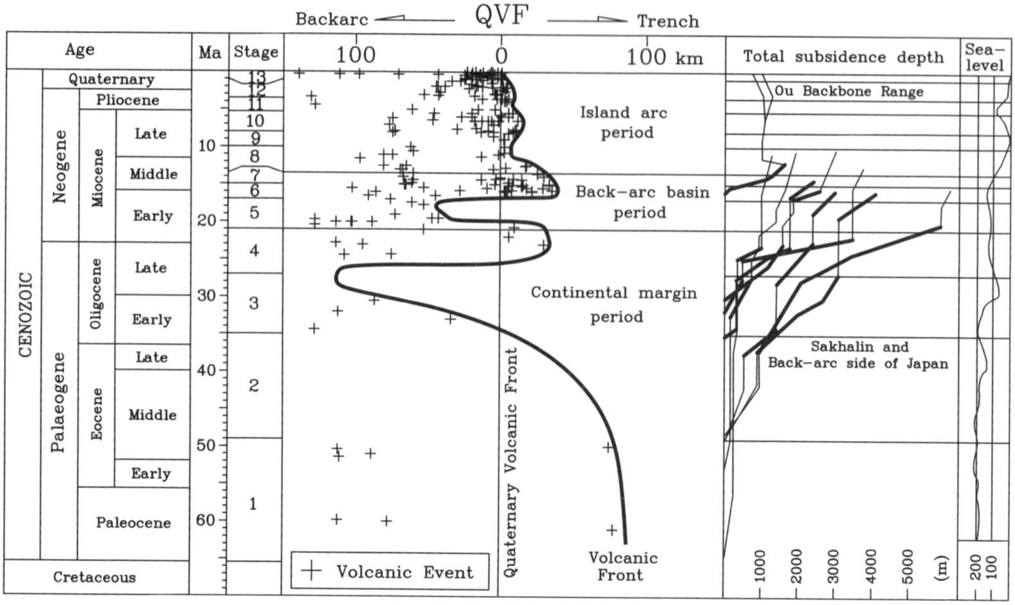

Fig. 3. Migration of the volcanic front over the past 60 myr in the NE Japan arc (modified from Yoshida *et al.* 1995; Yoshida 2001, 2009) and subsidence curves for sedimentary basins in the back-arc side of Japan, Sakhalin (Kano *et al.* 2007), and the Ou Backbone Range (Nakajima *et al.* 2006*c*). The three major periods (continental margin, back-arc basin and island arc), subdivided into 13 stages of igneous activity (Ohguchi *et al.* 1989; Sato 1994; Yoshida *et al.* 1995; Nakajima *et al.* 2006*c*; Yoshida 2001, 2009), are defined in the text and compared with the eustatic curve proposed by Haq *et al.* (1988). Some subsidence curves show a step pattern with flat thin lines and steep thick lines. The steep subsidence phase corresponds to the spreading phase of back-arc basin volcanism (Stages 4 and 6) under strong extension. QVF as in Figure 2.

and rhyolite effusions, as part of the bimodal volcanism in the Northern Honshu rift (Yagi *et al.* 2001; Yamada & Yoshida 2011). Rhyolitic magmas are mainly distributed in the forearc side, which is underlain by the inland continental margin crust (Fig. 4b). This activity in the back-arc basins was closely followed by a volcanic quiescence between 14 and 12 Ma. After the Japan Sea stopped opening at about 13.5 Ma, island-arc volcanism occurred in the eastern rift zones. Volcanism in the island-arc period is mainly characterized by caldera-forming felsic volcanism between 8 and 1.7 Ma (Yoshida *et al.* 1999*a*; Yamada & Yoshida 2002, 2011), which was accompanied by subordinate basaltic to andesitic volcanism (Acocella *et al.* 2008). After 1.7–1.0 Ma, many andesitic stratovolcanoes formed under strong compression (Umeda *et al.* 1999).

Erupted volume and regional stress field. The erupted magma volume reflects both the magma production rate in the mantle–crust section and the regional stress field (Umeda *et al.* 1999; Acocella *et al.* 2008; Yamada & Yoshida 2011). Compiled results (Fig. 6) show that erupted volumes of major volcanism changed from 4600 km^3 of basalts (per 1 myr in an arc length of 200 km) in the early back-arc basin period (21–19 Ma) to a volume of 2000–1000 km^3 from bimodal submarine volcanism in the late back-arc basin period (19–13.5 Ma), then to 200 km^3 from felsic pyroclastic eruptions in the early island-arc period (13.5–3.5 Ma), and finally to 500 km^3 of andesites in the latest island-arc period.

Changes in the erupted volume of magma along the volcanic front in the NE Japan arc since 2.0 Ma were estimated by Umeda *et al.* (1999) (Fig. 6b). Three stages are evident in the distinct types of volcanism, volume of eruptive magma and the locations of eruptive centres (Fig. 6c: Umeda *et al.* 1999): (1) felsic volcanism associated with large calderas (2.0–1.0 Ma); (2) felsic volcanism terminating at around 1.0 Ma followed by andesitic stratovolcanoes that became the dominant form of volcanism (1.0–0.6 Ma); (3) a continuation of andesite-dominated volcanism with a return of subordinate felsic volcanism. The eruptive rate increased drastically after 0.6 Ma (Fig. 6a, b). The change of eruption style from felsic pyroclastic flows associated with caldera formation, to andesitic stratovolcanism along the volcanic front after 2 Ma, was synchronous with the tectonic change from a

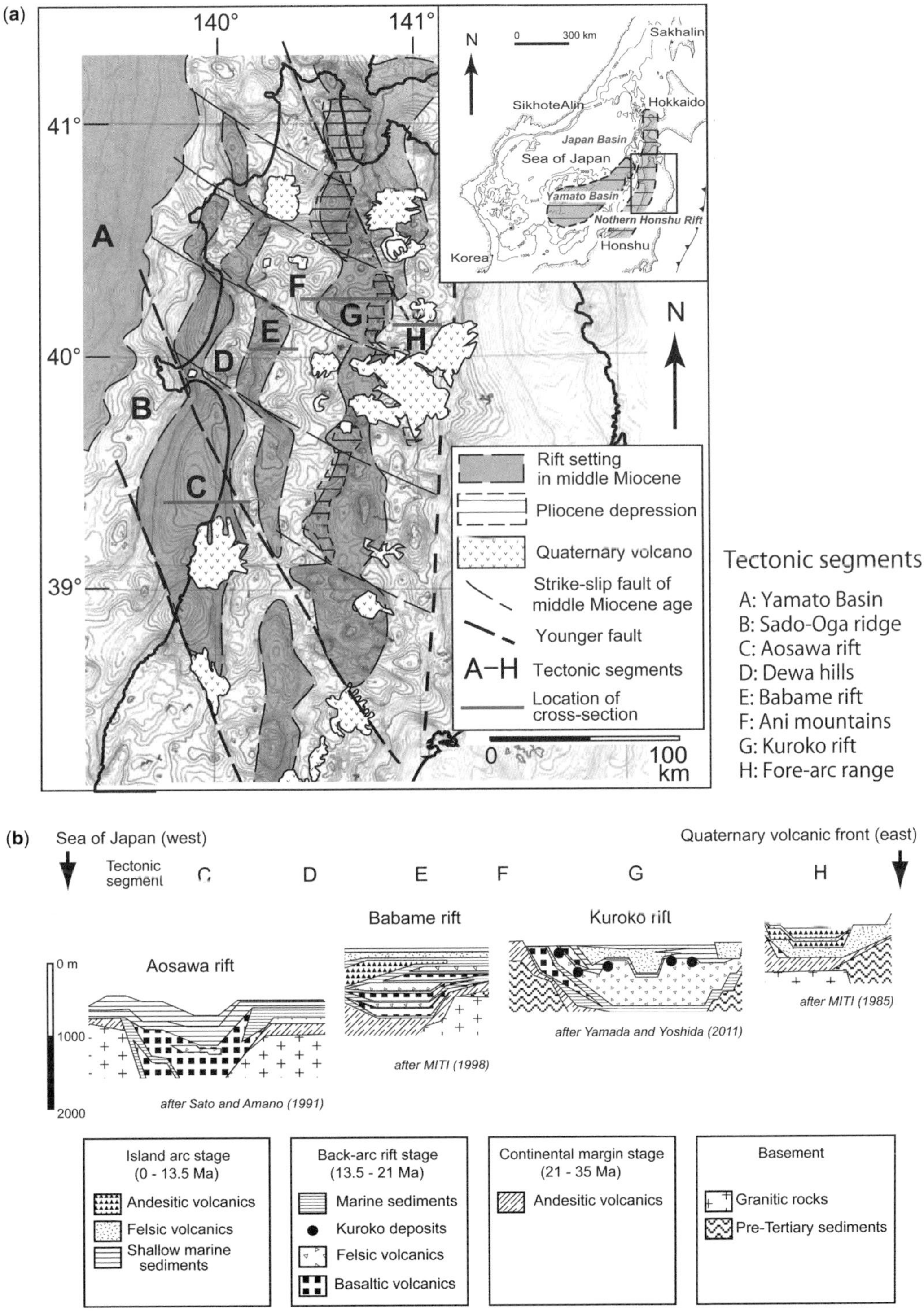

Fig. 4. (**a**) Tectonic framework (after Yamada & Yoshida 2011), and (**b**) east-to-west traverse profile (MITI 1985, 1998; Sato & Amano 1991; Yamada & Yoshida 2011) from the back-arc to the volcanic front of the NE Japan arc.

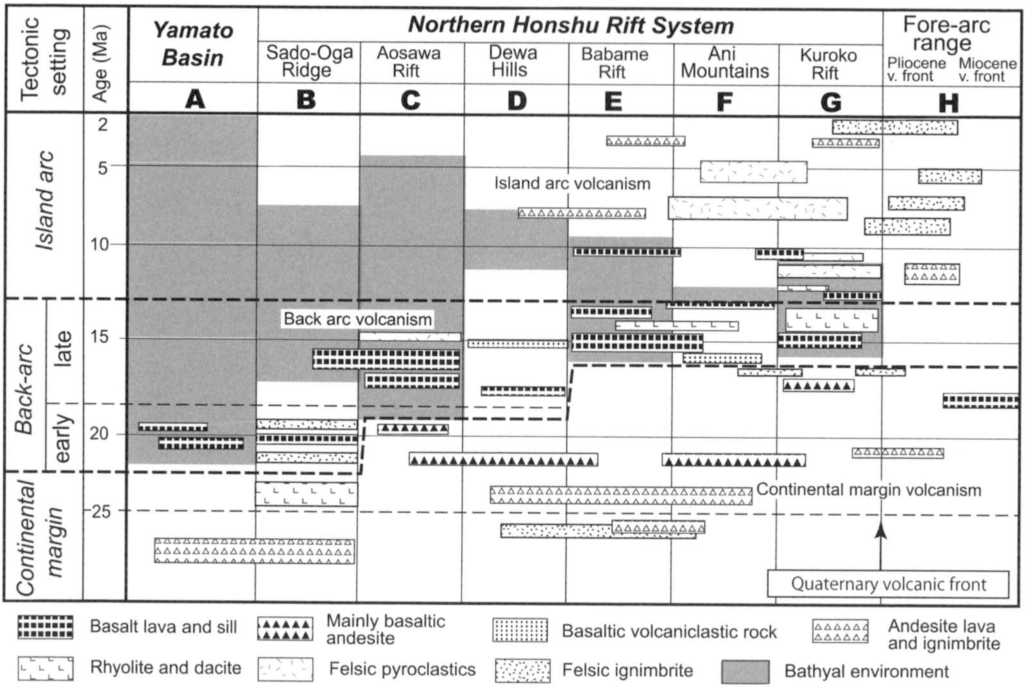

Fig. 5. Late Cenozoic volcanic stratigraphy along a transect from the Yamato Basin to the forearc range (Fig. 4a). Synthesis from radiometric ages of volcanic rocks, foraminiferal and micro-fossil ages in the intercalated sediments, and others (Yamada & Yoshida 2011 for detailed data source). A–H, Middle Miocene tectonic segments shown in Figure 4a.

NE–SW-oriented compression to an ENE–WSW- (east–west-) oriented compression (Umeda *et al.* 1999; Acocella *et al.* 2008).

Sato (1994) synthesized the regional tectonic stress field, basin development and crustal deformation of the NE Japan arc between the Oligocene and the Quaternary, based on dyke, vein and fault orientation data, and compiled regional geology (Takeuchi 1980; Sato *et al.* 1982; Tsunakawa 1986; Otsuki 1990; Yamamoto 1991). Dyke orientations are parallel to the horizontal compressive principal stress axis σ_{Hmax} (Anderson 1951; Nakamura 1977; Tsunakawa 1983). Results from veins and outcrop scale faults also highlight the regional tectonic stress field summarized in Figure 7 (Otsuki *et al.* 1977; Mimura 1979; Sato 1986, 1994; Oishi & Takahashi 1990; Otsuki 1990).

Temporal evolution of tectonic background

It is probable that the change from Stage 1 to Stage 2 (Fig. 3) was related to the change in the motion of the Pacific plate from NW to WNW at 50 Ma (Eocene plate reorganization: Patriat & Archache 1984; Gordon & Jurdy 1986; Maruyama & Seno 1986). Subsequent continental margin magmatism can be regarded as being due to the westward subduction of the Pacific Plate (e.g. Muller *et al.* 2008). The continental rifting which occurred deep in the back-arc region (Stages 3–4 in Fig. 3), before the opening of the Yamato Basin, is considered to have been a precursor of the back-arc basin opening (Kano *et al.* 2007, Nohda 2009). The late Eocene Monzen formation (Stage 3) contains alkaline to subalkaline within-plate type mafic to felsic volcanic rocks that erupted on land or in shallow water (Kano *et al.* 2011; Yamada *et al.* 2012). Sato (1994) showed that an extensional stress field became prevalent during the beginning of Stage 3, and major extension-related normal faulting (NNE to NE directed σ_{Hmax}) started at 25–20 Ma (Stage 4). This was followed by the back-arc opening of the Yamato and Aosawa Basins at 21–15 Ma (Stages 5–6). Normal faulting propagated from the present Japan Sea coast at 17–15 Ma (Stage 6) towards the forearc at 15–13.5 Ma (Stage 7), which followed the trench-ward migration of the volcanic field (Fig. 3). From 21 to 15 Ma (Stages 5–6), normal faults under NW- to NNW-directed σ_{Hmax} developed owing to the opening of the Japan Sea and the counter-clockwise rotation of NE Japan (Otofuji & Matsuda 1983,

1984; Otofuji *et al.* 1985; Nohda 2009). The formation of the Japan Sea would have begun in the central region of the rift zone (the Japan Basin and Yamato Basin) and rifting propagated subsequently to the east (Fig. 5). The clockwise rotation of SW Japan beginning in 16 Ma (Stage 6) produced a NW–SE-directed transtensional stress regime in the NE Japan arc at 15–13.5 Ma (Stage 7), as shown in Figure 7. Owing to the crustal stretching associated with pull-apart formation, the NE Japan back-arc subsided rapidly to middle bathyal depths (Sato 1994).

After the Japan Sea finished opening in about 13.5 Ma, a neutral stress regime prevailed between 13.5 and 10 Ma (Stage 8). The neutral to weak compressional stress regime with NE- to ENE-directed σ_{Hmax}, including both weak tension and compression, can be divided into two phases: 10–8 Ma (Stage 9) and 8–3.5 Ma (Stages 10–11). Lithospheric cooling led to thermal subsidence of the back-arc region at 13.5–10 Ma (Stage 8), and it is likely that the uplift of the axial zone of the volcanic arc was caused by igneous underplating during the back-arc basin period and the related crustal felsic magmatism at 13.5–8 Ma (Stage 9). The increase in the westward motion velocity of the Pacific plate at around 4 Ma (Pollitz 1986) produced stronger compression across the arc at 3.5–1 Ma (Stage 12), reactivating most of the Miocene normal faults, and uplifting the volcanic arc. The strong compressional stress regime since 3.5 Ma (Stage 12–13) is characterized by a σ_1-directed normal to the trench with a vertical σ_3 (Sato 1994). The greatest crustal shortening occurred within the volcanic arc at 1–0 Ma (Stage 13) in areas that were stretched the most during the Miocene, which implies tectonic inversion (Sato 1994).

Figure 8 shows the distribution and types of volcanoes in the NE Japan arc from the late Miocene to the Quaternary (Yoshida *et al.* 1999*a*; Acocella *et al.* 2008). The late Miocene–Pliocene and the Quaternary tectonic setting, estimated from fault analysis, are also shown in Figure 8 (adapted from Acocella *et al.* 2008). The Miocene–Pliocene deformation pattern is consistent with a NE–SW maximum compression, resulting from the NE–SW collision with the Kuril sliver, and the Quaternary deformation pattern is consistent with a WNW–ESE (east–west) maximum compression controlled by the motion of the Pacific plate (Acocella *et al.* 2008).

Crust–mantle structures and associated control of arc magmatism

Magmatism in the NE Japan arc is also controlled by the crust–mantle structures beneath the arc. Recent developments in seismological studies of the NE Japan arc have revealed heterogeneities, both in the mantle and in the arc crust. In this section, we summarize the heterogeneous structure of the arc crust and the underlying mantle, and then correlate this with the late Cenozoic magmatism.

Basement structures

Basement terranes. The distribution of volcanoes and the spatial variation in the composition of magma are closely related to the basement structures which exert a strong influence on the magma's composition (Kersting *et al.* 1996; Kimura & Yoshida 2006). Pre-Palaeogene basement rocks beneath the NE Japan arc (Geological Survey of Japan 1977) consist of Carboniferous to Jurassic metamorphic rocks of the Mino Belt, Cretaceous sedimentary rocks and granitoids of the Abukuma Belt, and an Ordovician to Cretaceous complex of the Kitakami Belt (Fig. 9a, b). The Mino and Abukuma Belts are in contact along the Tanakura Tectonic Line (TTL in Fig. 9a, b), which is the main boundary fault between the basement of NE and SW Honshu (Isozaki 1996). Another major fault, the Hatagawa Tectonic Line (HTL in Fig. 9a, b), separates the Abukuma and Kitakami Belts subparallel to the TTL.

The late Cenozoic basalts are isotopically enriched to the west of the TTL relative to those erupted to the east (Sato *et al.* 2007). The late Cenozoic rhylites to the west of the TTL are generally richer in alkalies than the associated mafic rocks, and those to the east are usually compositionally similar to the associated mafic rocks (Shuto *et al.* 2008; Yamada *et al.* 2012).

Basement granitoids. Granitoids of various ages (Cretaceous to Palaeogene) intrude into the basement terranes. These granitoids have been classified based on Sr–Nd isotope geochemistry into the Kitakami, North, Transitional, South and Sado zone granitoids (Fig. 9a, c; Kagami *et al.* 2000; Kagami 2005). The Kitakami zone granitoids are the most depleted isotopically, and the South zone granitoids are the most enriched. The distribution of the geochemical zoning is subparallel to the major faults in the basements, even though the zone boundaries do not follow the tectonic lines exactly (Fig. 9a). Provided that those granitoids are of lower crustal melt in origin, the spatial correlations between granitoid chemistry and the basement terranes suggest a coupled upper and lower crust (Kagami *et al.* 2000; Kagami 2005). The enriched North and South zones are extensions of the isotopic zones identified in the SW Honshu arc (Kagami 2005), and the Kitakami and Sado zones, together with the Quaternary rear arc (QRA), are plotted on the mantle array (Fig. 9c).

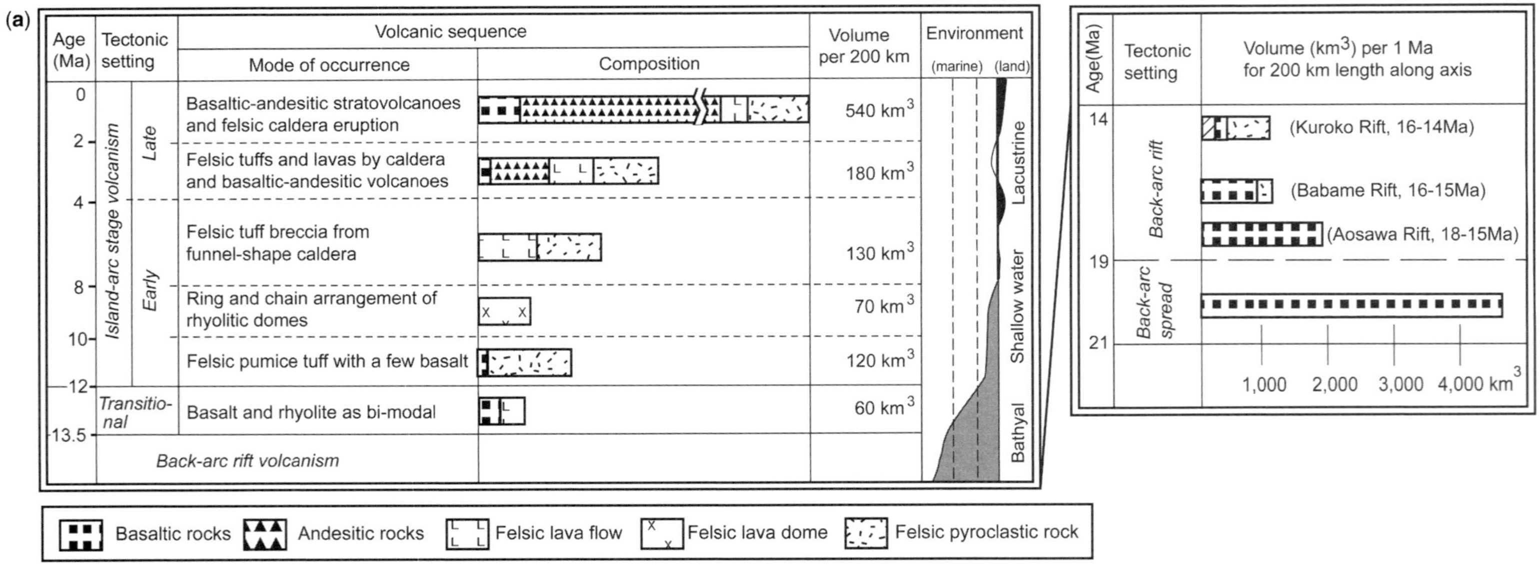

Fig. 6. (**a**) Temporal changes in the volume (km^3: per 1 Ma, for 200 km of arc length along axis) and composition of erupted magma from the back-arc to the island-arc periods (modified from Yamada & Yoshida 2011). (**b**) Temporal changes in eruptive magma volume (km^3) for stratovolcanoes (upper) and large calderas (lower) during the last 2.0 Ma along the volcanic front in the NE Japan arc. (**c**) Distribution of eruptive centres for last 2 Ma. (b, c) Modified from Umeda *et al.* (1999).

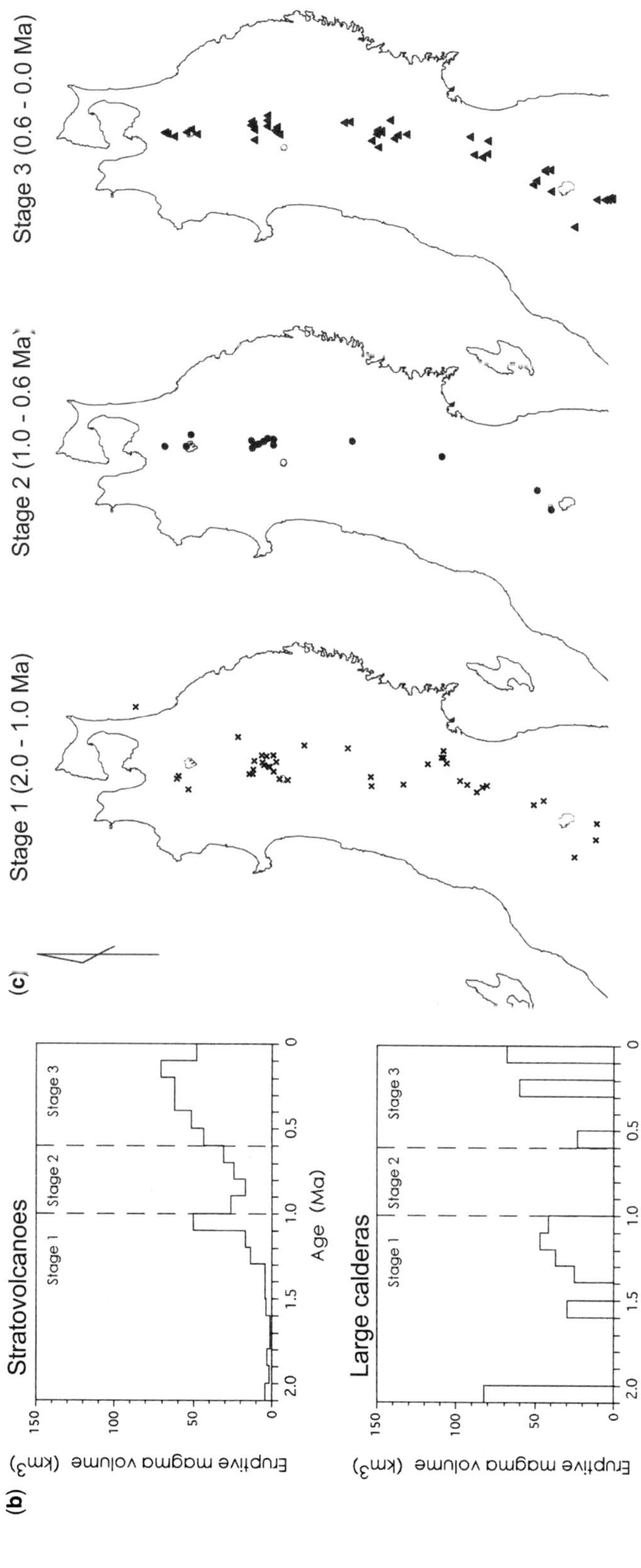

Fig. 6. *Continued.*

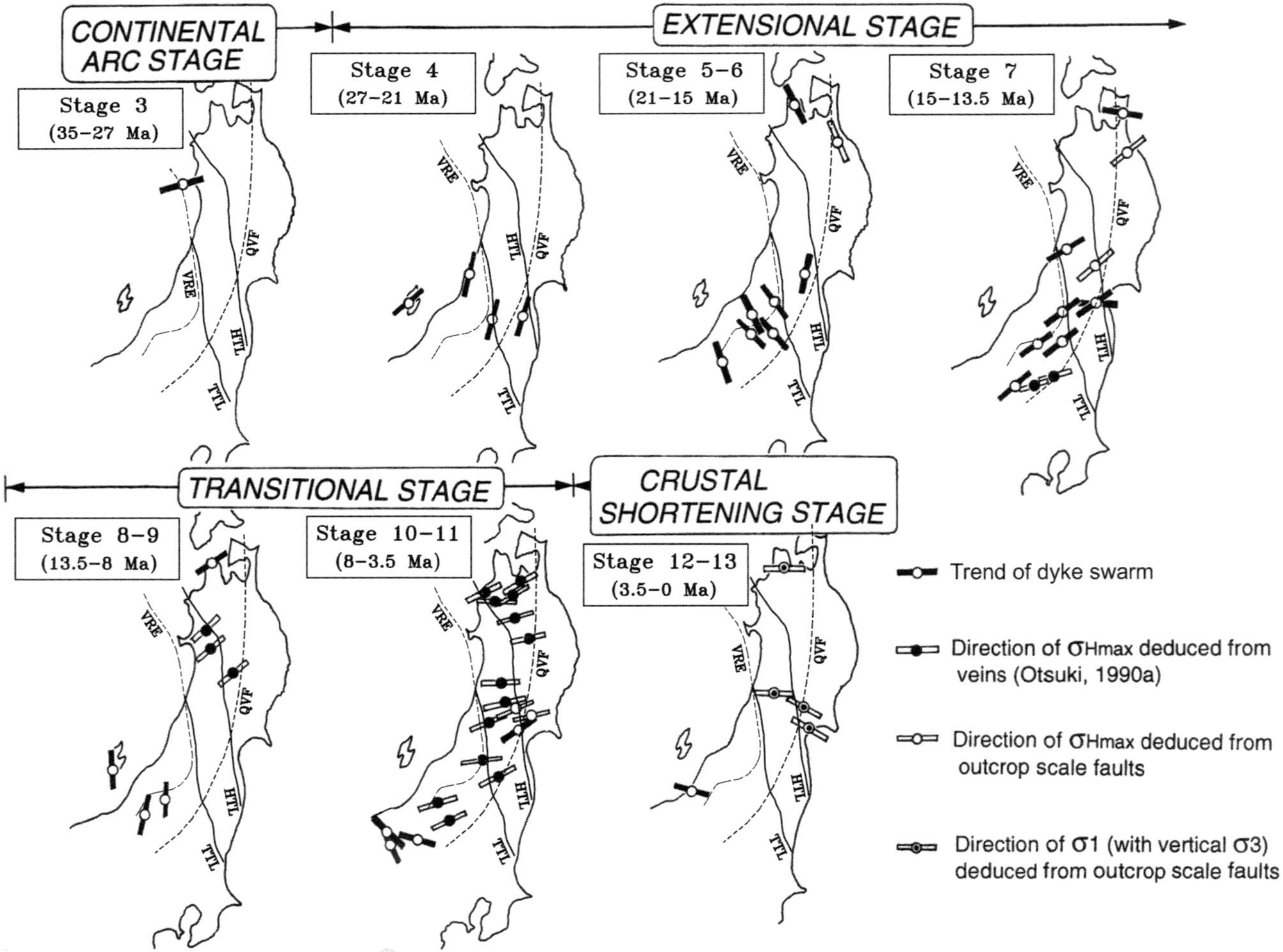

Fig. 7. Regional stress field in NE Japan since Stage 3 (modified from Sato 1994). QVF, VRE, TTL and HTL as in Figure 2.

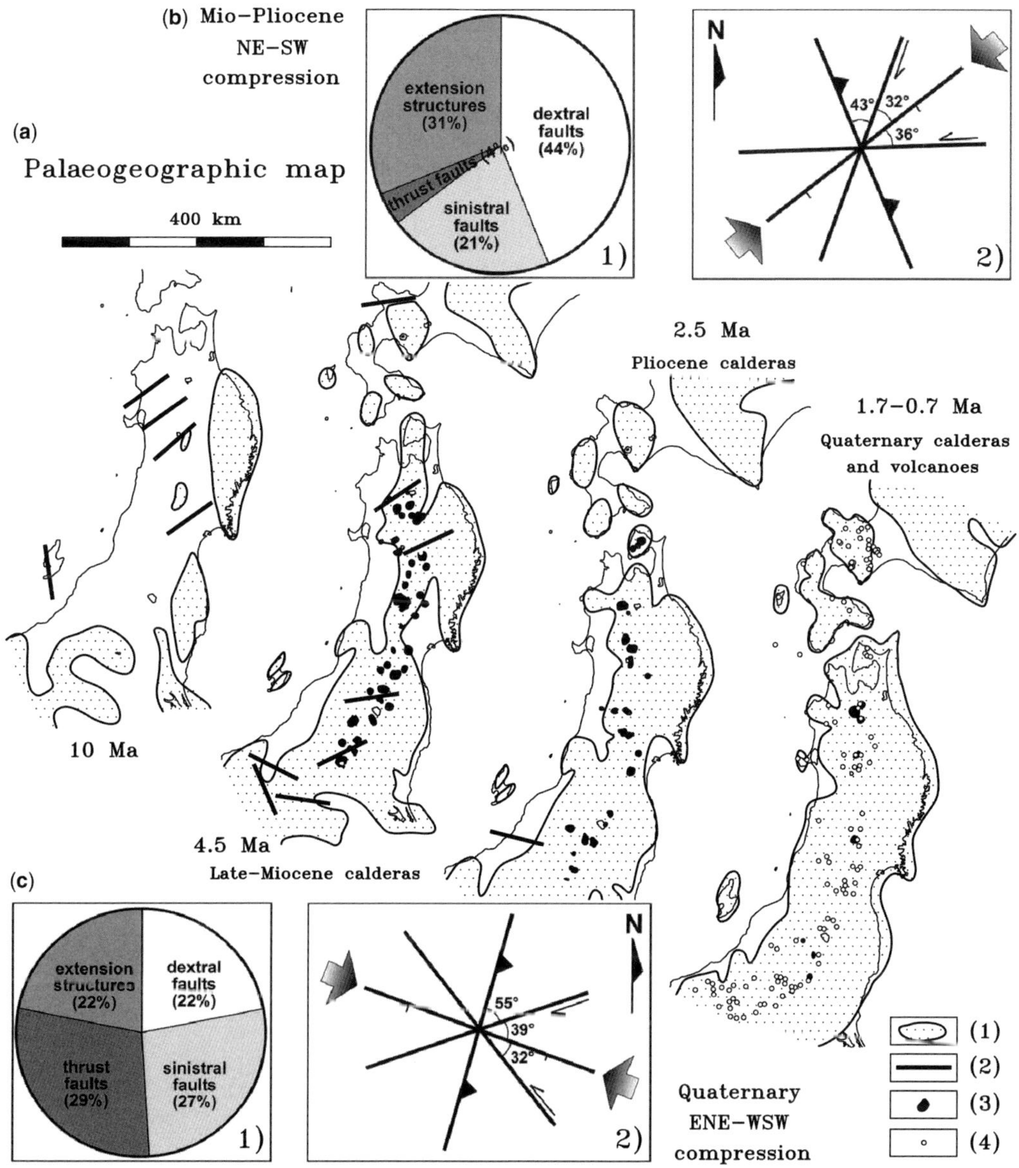

Fig. 8. (**a**) Distribution and types of volcanoes in the NE Japan arc from late Miocene to Quaternary (adapted from Yoshida *et al.* 1999*a*). (1) Hatched area represents emergent areas of the NE Japan arc in the specified time period. (2) Orientation of inferred regional maximum compression, based on direction of dyke swarms and other data (Kano *et al.* 1991; Sato 1994). (3) Calderas refer to the relative time periods. (4) Quaternary stratovolcanoes. (**b**) The late Miocene–Pliocene structural data: (1) relative abundance of the four types of fractures; (2) schematic diagram showing the mean trend of the four types of fracture, and the mutual relationships. The Mio-Pliocene deformation pattern is consistent with a NE–SW maximum compression (after Acocella *et al.* 2008). (**c**) The Quaternary structural data: (1) relative abundance of the four types of fractures; (2) schematic diagram showing the mean trend of the four types of fracture, and the mutual relationships. The Quaternary deformation pattern is consistent with a WNW–ESE (east–west) maximum compression (after Acocella *et al.* 2008).

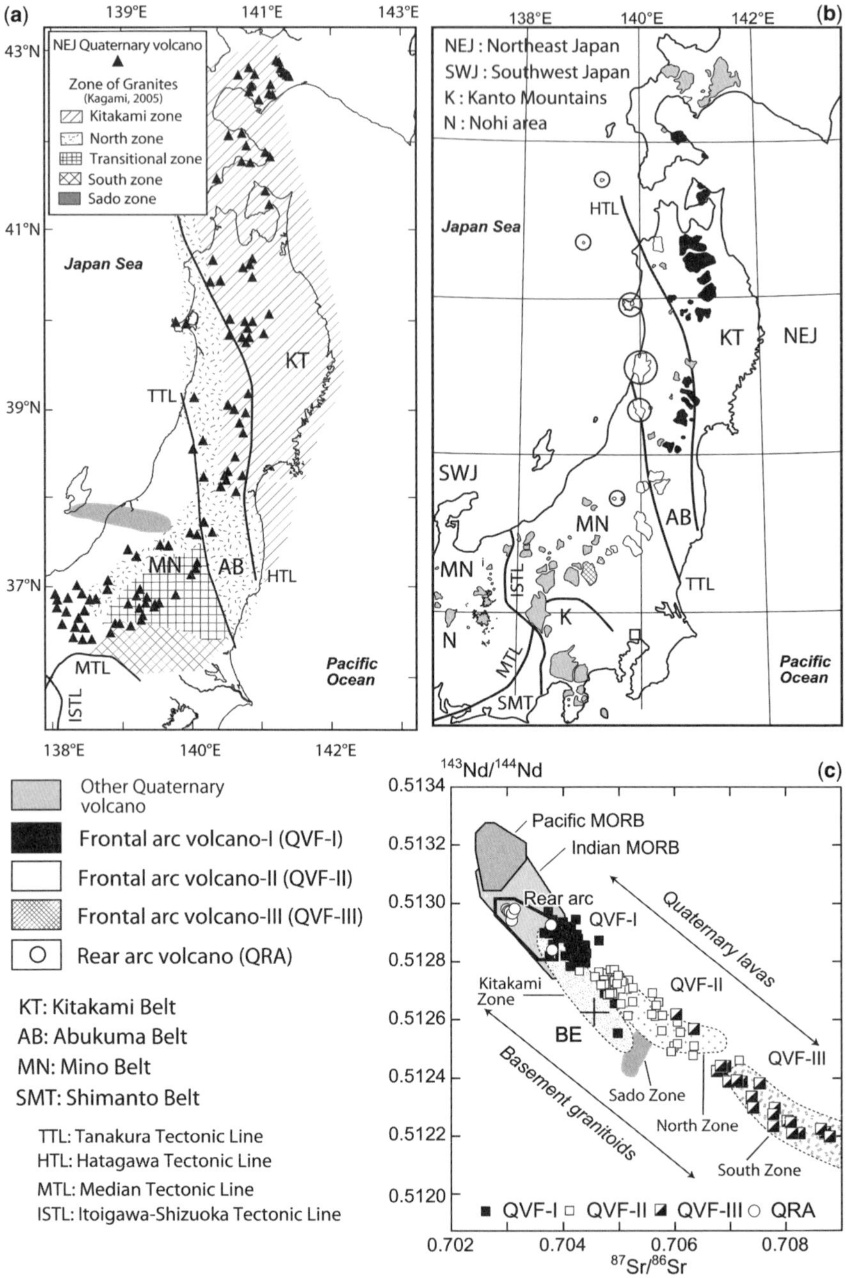

Fig. 9. (**a**) Regional isotope zoning of basement granitoids (Kitakami, North, Transitional, South, and Sado zones) by Kagami *et al.* (2000) and Kagami (2005), and tectonic boundaries of Kitakami (KT), Abukuma (AB), Mino (MN) and Shimanto (SMT) belts. These basement structures are oblique to the present NE Japan trench-arc system. (**b**) Distribution of Quaternary volcanoes (QVF-I, II, III, and QRA) in the NE Japan arc and basement tectonic boundaries (KT, AB, MN and SMT) (modified from Kimura & Yoshida 2006). (**c**) Nd–Sr isotopic compositions of basement granitoids (Kagami *et al.* 2000; Kagami 2005) and Quaternary lavas from the volcanic front and rear arc. Patterned fields show the isotopic compositions of the basement granitoids. Spatial correlations in Nd–Sr isotope composition are evident between the basement granitoids and the lavas for the frontal-arc volcanoes (Kimura & Yoshida 2006), although the rear arc lavas and the Kitakami and the Sado zone granitoids show a more depleted trend on mantle array than the frontal arc volcanics. BE, Bulk Earth.

Crustal structure

Lower crustal rocks beneath the NE Japan arc. Nishimoto *et al.* (2005) compared the V_p measurements for each xenolith from the Ichino-megata volcano with the V_p profile of the NE Japan arc given by Iwasaki *et al.* (2001). The rocks used in their study cover nearly the entire range of lithological variations of the Ichino-megata xenoliths, and are considered as representative rock samples of the lower crust and upper mantle of the back arc side of the NE Japan arc. The relatively low V_p (6.6–7.0 km s^{-1}) in the 15 km-thick lower crustal layer of the NE Japan arc is due to the high abundance of hornblende and magnetite in gabbros and amphibolites with a basic to ultrabasic composition (Fig. 10). Takahashi (1986) suggested that the thick lower crust of the NE Japan arc is composed of hornblende ($\pm$ pyroxene) gabbro and/or amphibolite. In addition, Sakuyama (1983), Fukuyama (1985) and Yoshida *et al.* (1997) suggested that the Ichino-megata basic to ultrabasic lower crustal xenoliths are highly fractionated cumulates derived from calc-alkaline magmas. Kanisawa & Yoshida (1989) showed that cumulates of low-K dacites from a Quaternary frontal volcano have the composition of hornblende gabbros (S in Fig. 10). The composition of hornblende gabbros and amphibolites of the Ichino-megata xenolith is plotted on the trend lines of the Quaternary volcanic rocks in Figure 10 Yoshida *et al.* (1997) suggested that these basic to ultrabasic cumulates comprise the lower crust of the NE Japan arc. Some amphibolite xenoliths have very similar $^{87}Sr/^{86}Sr$ ratios (0.7030–0.7032) to the host magmas (0.7030–0.7033; Fig. 11; Zashu *et al.* 1980) and a similar bulk-rock chemistry to the hornblende gabbro xenoliths. It is therefore likely that these amphibolites are originally derived from the late Cenozoic calc-alkaline magmas. Gabbroic xenoliths from Ichino-metaga volcano have higher $^{87}Sr/^{86}Sr$ ratios (0.7048–0.7035) than the host magmas (Zashu

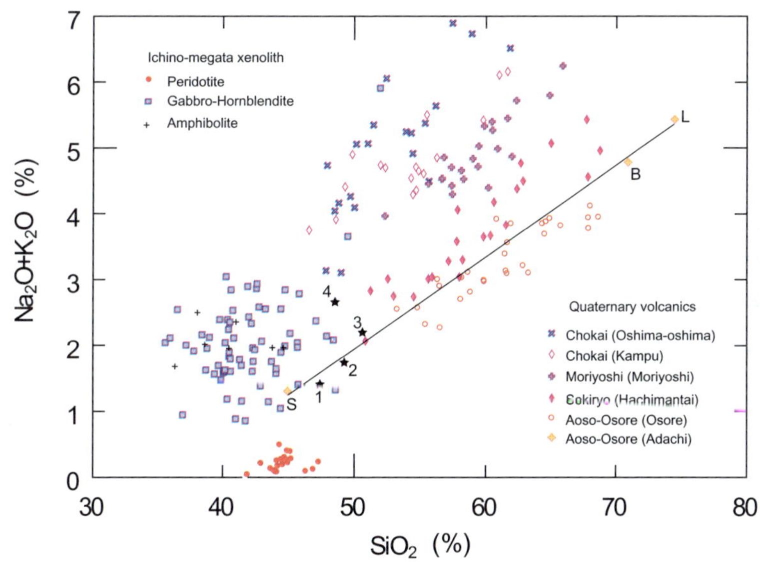

Fig. 10. Total alkali–SiO_2 diagram for the Quaternary volcanic rocks of the NE Japan arc and Ichino-megata lower crustal to mantle xenolith. Diagram includes data from Kuno (1967), Aoki (1971), Aoki & Shiba (1973), Fukuyama (1985), Aoki & Yoshida (1986), Nishimoto *et al.* (2005, 2008) for Ichino-megata xenolith, Yamamoto (1984) for Oshima–Oshima volcano of the Chokai volcanic zone, Hayashi *et al.* (1991) for Kampu volcano of the Chokai volcanic zone, Nakagawa (1983) for Moriyoshi volcano of the Moriyoshi volcanic zone, Yoshida *et al.* (1983) for Hachimantai volcano of the Sekiryo volcanic zone, Togashi (1977) for Osore volcano of the Aoso–Osore volcanic zone and Kanisawa & Yoshida (1989) for Adachi volcano of the Aoso–Osore volcanic zone. S, B and L denote calculated cumulate (S), bulk rock (B), and liquid composition (L) of the low-*K* calc-alkaline dacite of Adachi volcano (Kanisawa & Yoshida 1989). Abbreviations 2–4 show the most primitive basalt compositions of Iwate volcano (2: Kuritani *et al.* 2013), Funagata volcano (3: Kimura & Yoshida 2006) and Sanno-megata volcano (4: Yoshinaga & Nakagawa 1999). Estimated primary magma composition from Iwate primitive basalt (2) is shown in the figure as 1 (Kuritani *et al.* 2013). The compositional relation between lower crustal xenoliths and the Quaternary volcanics shows that hornblende gabbro to hornblendite and amphibolite from the lower crust of the NE Japan arc are basic to ultrabasic cumulates or restites of hydrous calc-alkaline magma. There are few mafic xenoliths directly solidified from basaltic magma.

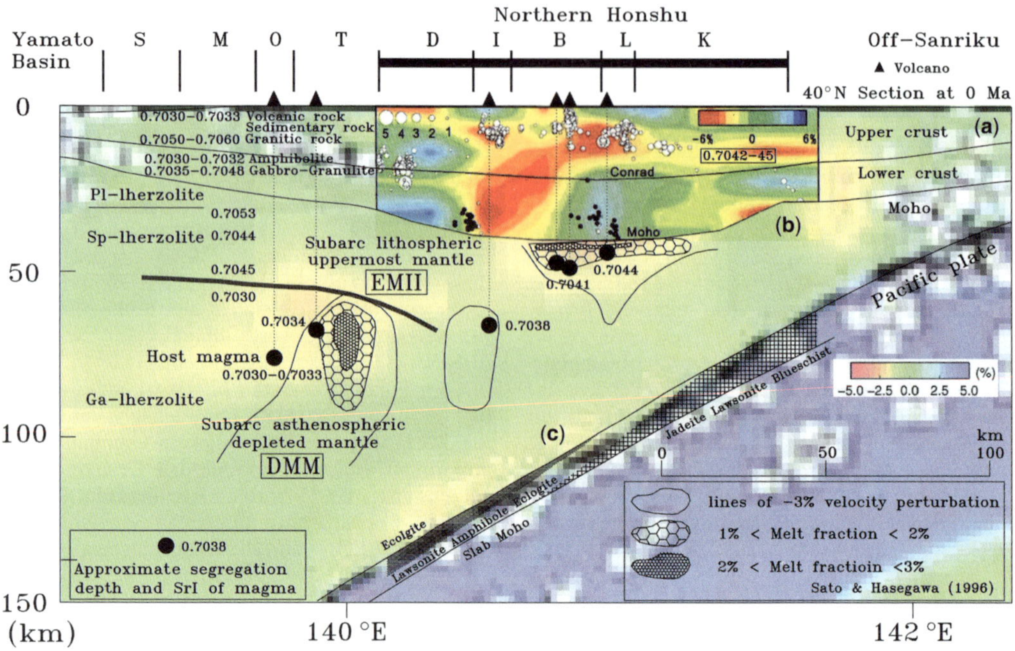

Fig. 11. Schematic cross-arc vertical section on 40°N latitude line of the NE Japan arc crust and the upper mantle, with estimates of magmatic segregation depths and SrI of Quaternary magmas inferred from seismic and petrological observations (Shibata & Ishihara 1979; Zashu *et al.* 1980; Sato & Hasegawa 1996; Yoshida *et al.* 1995, 1999*b*; Yoshida 2001), superimposed on *P*-wave velocity perturbations of the upper mantle (a: Zhao *et al.* 2012) and the crust (b: Xia *et al.* 2007). Red and blue colours denote low and high velocities, respectively. The velocity perturbation (in per cent) scale is shown in each figure. The black triangles denote Quaternary volcanoes. The black bar over a cross-section denotes land area. White dots in (**a**) denote seismicity that occurred within a 20 km width of the 40°N latitude line. (**b**) Tomographic images using both first P and PmP data (Xia *et al.* 2007). White dots denote earthquakes along the section. Black dots denote deep, low-frequency microearthquakes. Low-*V* zones are clearly visible across the entire crust beneath the active volcanoes, from the surface extending down to the Moho discontinuity with west-dipping (Xia *et al.* 2007). Abbreviations, K–S as in Figure 1. DMM, depleted MORB source mantle; EMII, enriched mantle II; SrI, initial Sr isotopic ratio. (**c**) Upper-plane seismic belt and phase boundaries of jadeite lowsonite blueschist to lawsonite amphibole eclogite, and lawsonite amphibole eclogite to eclogite. The belt of intraslab seismicity is probably caused by dehydration of eclogite-forming phase transformations in the Pacific slab crust at depths of 70–90 km (Kita *et al.* 2006).

et al. 1980). Mantle peridotites from Ichino-megata volcano also have a wide range of $^{87}Sr/^{86}Sr$ ratios (0.7030–0.7053: Zashu *et al.* 1980). The $^{87}Sr/^{86}Sr$ ratios of gabbro and peridotite cover the range of isotopic composition from the late Cenozoic continental margin to the island-arc basalts from the back-arc to the frontal region in the NE Japan arc (Fig. 11).

Lower crustal heterogeneity. Nishimoto *et al.* (2008) examined the dV_p and dV_s values from the lower crust of NE Honshu on the V_p–V_s–V_p/V_s deviation diagram, in the temperature range of 500–700°C at 0.8 GPa, and compared the data with those of typical minerals and rocks (Fig. 12c). It was concluded that: (1) the seismically high-V_p and -V_s regions beneath the Tobishima Basin (T in Fig. 1) consist of hornblende pyroxene gabbro; (2) hornblende gabbro is the predominant rock type beneath the Dewa Hills (D in Fig. 1) and the Ou Backbone range (B in Fig. 1); (3) the low-velocity anomalies in the lower crust beneath the active volcano areas may be caused by the existence of partial melts of hornblende gabbro; and (4) the low-V_p and high-V_s regions beneath the Kitakami Mountains (K in Fig. 1) consist of quartz-plagioclase-bearing rocks with intermediate compositions (granodiorite to diorite).

Tomographic images of V_p and V_s at 25 km depth beneath the eastern margin of the Japan Sea to NE Japan (Zhao *et al.* 2011) are shown in Figure 12a & b, where the nine velocity anomalies are recognized and labelled A–I. Figure 12c plots the observed average dV_p and dV_s data at a depth of 25 km (Zhao *et al.* 2011). Estimated lithologies are: (A) (high V_p, high V_s, low V_p/V_s) as peridotites;

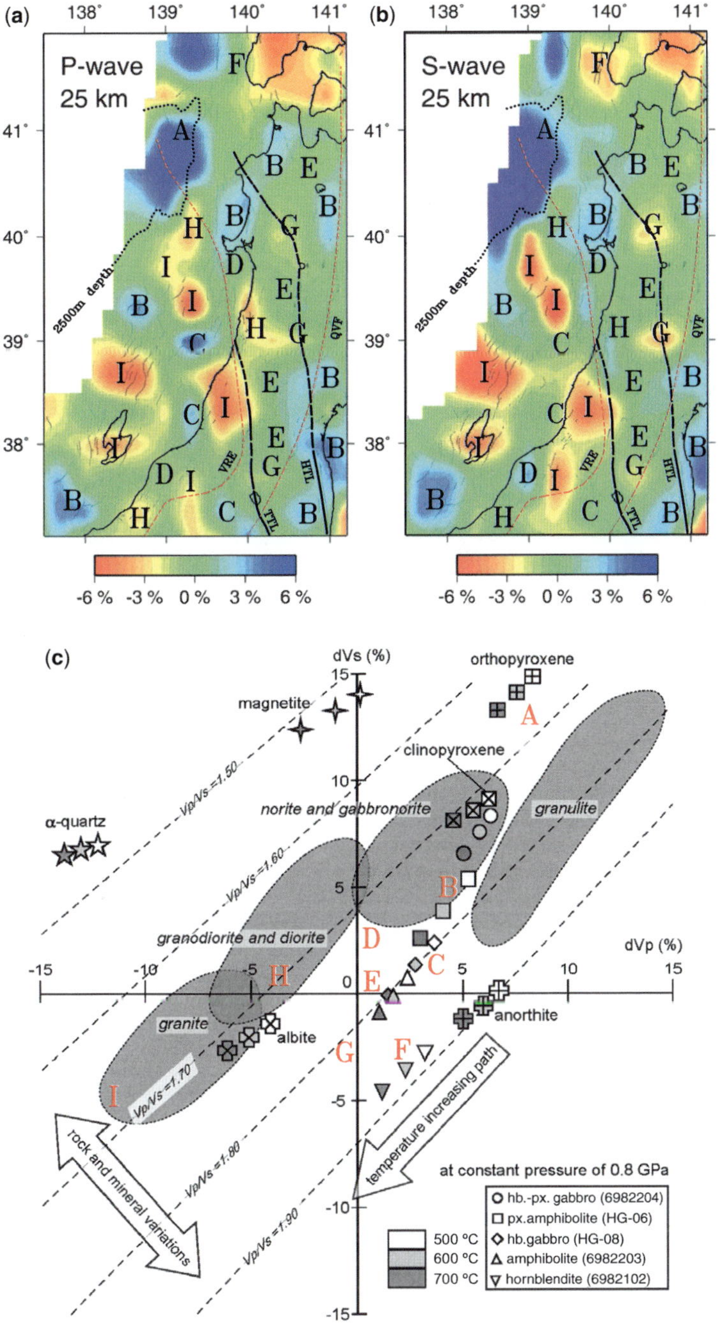

Fig. 12. (**a**, **b**) *P*-wave (a) and *S*-wave (b) velocity images at depths of 25 km in the NE Japan back-arc region (after Zhao *et al.* 2011). Red and blue colours denote low and high velocities, respectively. The velocity perturbation scales are shown at the bottom. The nine velocity anomalies discussed in the text are labelled (A–I) in (a) and (b). QVF, VRE, TTL and HTL are the same as in Figure 2. (**c**) The nine labels of velocity anomalies are plotted in the 'V_p–V_s–V_p/V_s deviation diagram', together with dV_p and dV_s values of the various minerals and rocks at 0.8 GPa (Nishimoto *et al.* 2008). Anomaly A, peridotite; B, pyroxene gabbro; C, pyroxene hornblende gabbro – amphibolite; D, gabbronorite to amphibolite; E, hornblende pyroxene gabbro to amphibolite; F, hornblende gabbro to amphibolite; G, partially molten lower crust; H, granodiorite; and I, granite.

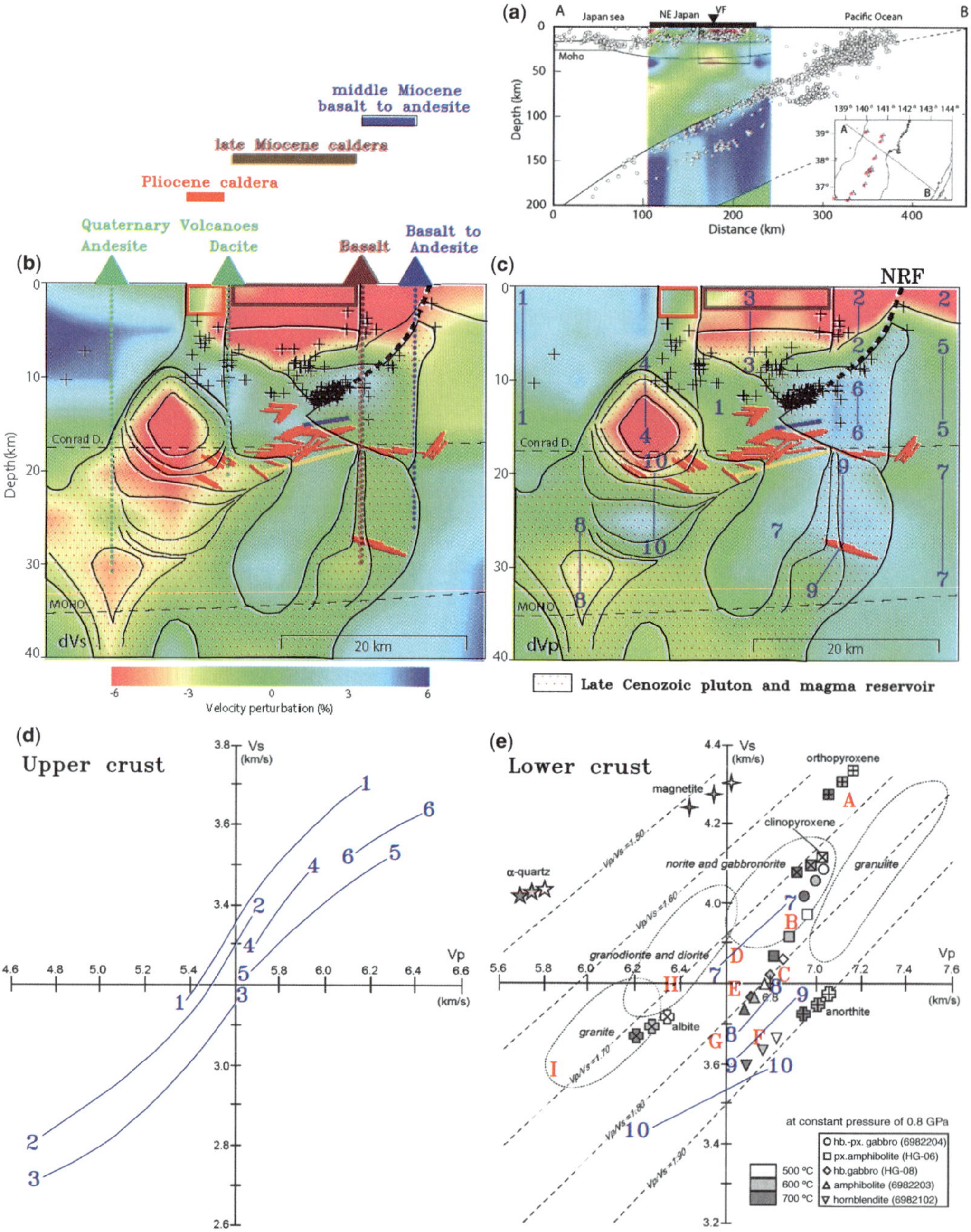

Fig. 13. (**a**) Vertical cross-section of *S*-wave velocity perturbation along the line shown in the insert map (after Nakajima *et al.* 2001). White circles and black lines show microearthquakes and seismic velocity discontinuities, respectively. The black bar and reverse triangle at the top denote the land area and the location of the volcanic front (VF), respectively. Red triangles in the insert map show Quaternary volcanoes. The rectangle indicates the area shown in (b) and (c). (**b**, **c**) Magma plumbing systems in the forearc region of the NE Japan arc superimposed on the vertical cross section of *S*-wave (b), and *P*-wave (c) velocity perturbation (after Nakajima *et al.* 2006*a*), respectively (modified from Yoshida 2009). The locality of these sections is shown in Figure 14c along the section east–west. NRF denotes the Nagamachi–Rifu fault. Red and blue colours denote low- and high-velocity perturbation. The velocity perturbation (%) scale is shown at the bottom of (b). Crosses indicate shallow microearthquakes determined by Yoshimoto *et al.* (2000)

(B) (high V_p, high V_s, medium V_p/V_s) as hornblende pyroxene gabbro; (C) (high V_p, medium V_s, high V_p/V_s) as hornblende gabbro to amphibolite; (D) (medium V_p, high V_s, low V_p/V_s) as gabbronorite to amphibolite; (E) (medium V_p, medium V_s, medium V_p/V_s) as pyroxene hornblende gabbro to amphibolite; (F) (medium V_p, low V_s, high V_p/V_s) as hornblende gabbro to hornblendite/amphibolite; (G) (medium V_p, low V_s, high V_p/V_s) as partially molten lower crust; (H) (low V_p, medium V_s, low V_p/V_s) as granodiorite; and (I) (low V_p, low V_s, low V_p/V_s) as granite.

It is likely that hornblende gabbro and amphibolite (anomalies C, E and F in Fig. 12) are the dominant lower crustal rocks, and that they are distributed along the central axis of the NE Japan arc. In the rifted back-arc basin, hornblende pyroxene gabbro to gabbronorite and/or pyroxene amphibolite (anomalies B and D) are dominant (Nishimoto *et al.* 2008). The dV_p and dV_s values of the rifted continental crust (Tamaki *et al.* 1992: anomaly I located at the eastern margin of the Japan Sea to the west of TTL) suggest that these crustal blocks consist mainly of granitic rocks. These are different to those of the granodioritic lower crust (e.g. anomaly H) and to the tonalitic lower crust of the Kitakami Mountainlands (Nishimoto *et al.* 2008). Present results show that there are regional variations of pre-Palaeogene lower crustal plutonic rocks in the NE Japan arc, from the northeastern forearc tonalite to the southwestern back-arc granite with transitional granodiorite. Lower crustal pyroxene–hornblende to hornblende gabbro in the central axis of the NE Japan arc, and hornblende pyroxene gabbro in the rifted back-arc basin, might be emplaced in the Cenozoic to the pre-Palaeogene rifted continental margin basement which consists of tonalite–granodiorite–granite.

Crustal structure and magma plumbing system. Figure 13b, c shows high-resolution tomographic images of the forearc region near Sendai in NE Japan (rectangle in Fig. 13a; Nakajima *et al.* 2006*a*). Nakajima *et al.* (2006*a*) explained the complex velocity structure using V_p, V_s and Poisson's ratio. Figure 13d, e plots the V_p and V_s data for the numbered sections 1–10 in Figure 13c. Estimated lithologies based on V_p and V_s data, using the method of Nishimoto *et al.* (2008) and geological data (Sato *et al.* 2002), are: (1) pre-Palaeogene granitic rocks; (2) Cenozoic sediments; (3) late Miocene caldera-fill pyroclastic rocks to shallow fluid-rich felsic pluton (Sato *et al.* 2002); (4) the felsic part of the Pliocene magma reservoir; (5) pre-Palaeogene basement rocks; (6) middle-crustal mafic plutonic rocks; (7) diorite to hornblende pyroxene gabbro/pyroxene amphibolite; (8) hornblende gabbro to partially molten lower crust; (9) Miocene hornblende gabbro; and (10) the mafic part of the Pliocene magma reservoir/hornblende gabbro.

Figure 14a, b shows the modelled density structure along the across-arc profile (X–Y line in Fig. 2a). Relatively denser materials are obtained in the area beneath the eastern rift border zone of the Northern Honshu rift (shown as the 'rift border zone' in Fig. 14c), which is located at the easternmost part of the Kitakami river valley (Fig. 14a). This high-density body corresponds to the Miocene hornblende gabbro (Section 9 in Fig. 13c), and it is possible that this is a mafic dyke complex which intruded into the eastern rift border zone.

The island-arc crust of the NE Japan arc is deformed under compressional stress in the direction of plate motion. The Ou Backbone range and the eastern rift border zone of the Northern Honshu rift (Fig. 14c) are two parallel seismically active zones which are NS elongated, and form the focused deformation zones (Miura *et al.* 2004). Okada *et al.* (2010) examined the deep structure of the Ou Backbone range and, by performing regional-scale seismic tomography, found that the distinct seismic low-velocity regions are continuously distributed from the mantle wedge to the middle crust just below the Ou Backbone range

Fig. 13. (*Continued*) and Umino *et al.* (2002). Red lines represent distinct *S*-wave reflectors (bright spots) detected by Hori *et al.* (2004). Blue and yellow lines denote the *S*-wave reflector identified by Umino *et al.* (2002). Thick black broken lines represent the inferred geometry of the NRF. Broken lines denote the seismic velocity discontinuities. The 10 prominent sections of velocity anomalies discussed in the text are labelled 1–10 on the lines in (c). Brown and red rectangles in (b) and (c) show the late Miocene and Pliocene calderas, respectively. Blue, brown and green triangles indicate volcanoes of middle Miocene basalt to andesite, late Miocene basalt and Quaternary dacite to andesite, respectively. The dotted line beneath each volcano denotes the magma plumbing system. Hatched area surrounded by black lines denotes the late Cenozoic plutons and magma reservoirs estimated from V_p and V_s values shown for (**d**) upper crustal rocks and (**e**) lower crustal rocks, using the same method as in Figure 12. These igneous complexes have zonal structures which are indicated by gradual changes of V_p and V. In (d) and (e), anomaly 1, pre-Cenozoic granitic rocks; 2, Cenozoic sediments; 3, late Miocene caldera-fill pyroclastic rocks to shallow fluid-rich felsic pluton; 4, felsic part of Pliocene magma reservoir; 5, pre-Cenozoic basement rocks; 6, middle-crustal mafic plutonic rocks; 7, diorite to pyroxene gabbro/amphibolite; 8, pyroxene hornblende gabbro to partially molten lower crust; 9, Miocene hornblemde gabbro; 10, mafic part of Pliocene magma reservoir/hornblende gabbro. A–I in lower crustal V_p–V_s diagram (e) are the same as in Figure 12.

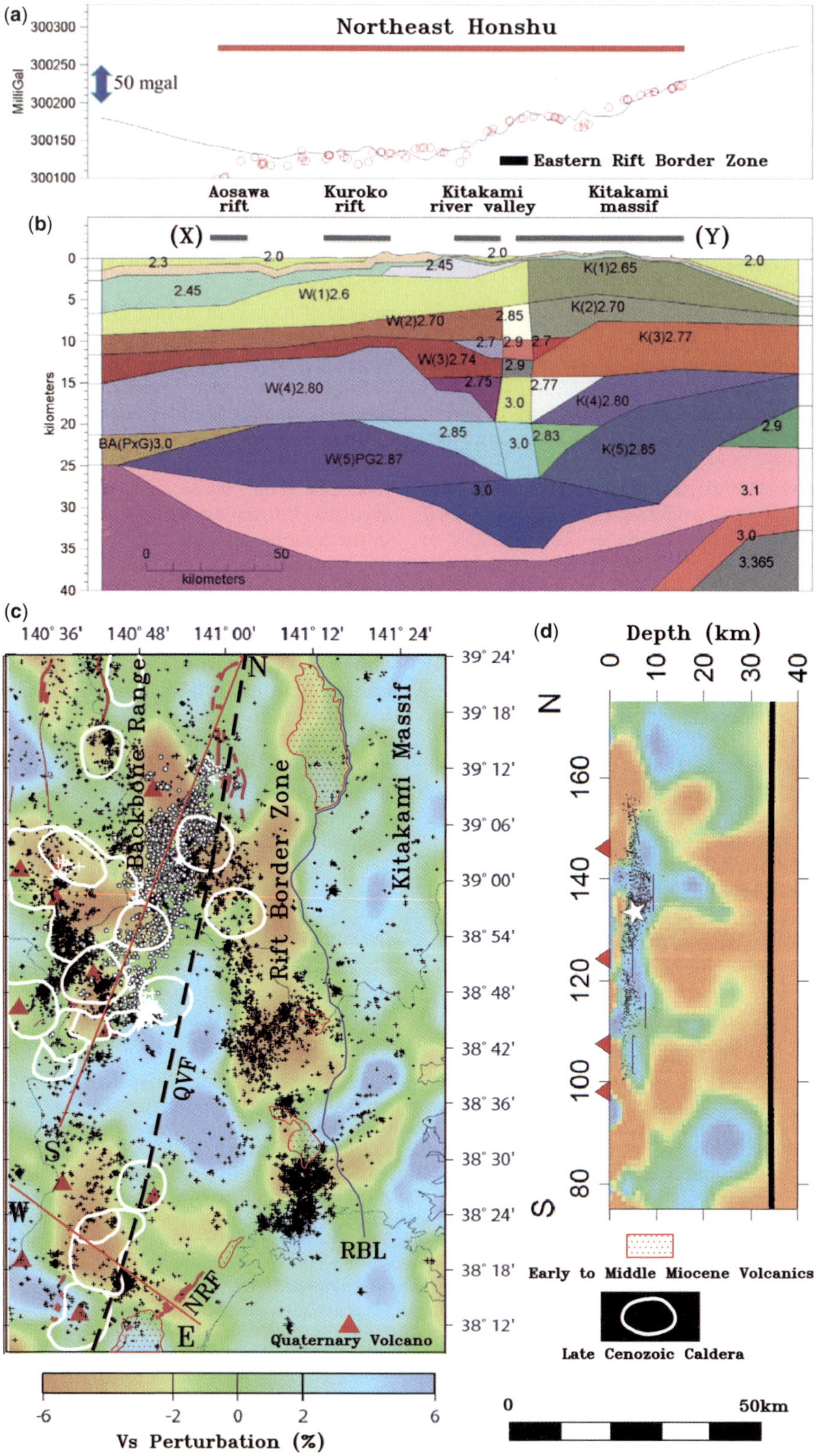
(a)
Northeast Honshu
50 mgal
MilliGal
300300
300250
300200
300150
300100
Eastern Rift Border Zone
Aosawa rift
Kuroko rift
Kitakami river valley
Kitakami massif
(b)
(X)
(Y)
kilometers
W(1)2.6
W(2)2.70
W(3)2.74
W(4)2.80
W(5)PG2.87
BA(PxG)3.0
K(1)2.65
K(2)2.70
K(3)2.77
K(4)2.80
K(5)2.85
3.365
(c)
140° 36'
140° 48'
141° 00'
141° 12'
141° 24'
39° 24'
39° 18'
39° 12'
39° 06'
39° 00'
38° 54'
38° 48'
38° 42'
38° 36'
38° 30'
38° 24'
38° 18'
38° 12'
Backbone Range
Rift Border Zone
Kitakami Massif
QVF
RBL
NRF
N
S
W
E
Quaternary Volcano
(d)
Depth (km)
Early to Middle Miocene Volcanics
Late Cenozoic Caldera
Vs Perturbation (%)
50km

(Fig. 14c). On this basis, the low-velocity zone beneath the Ou Backbone range could be interpreted as a region of partial melting, which would suggest that crustal fluids separating and upwelling from deeper depths are closely related to the active seismic zones. Around the Onikobe caldera in the middle of the Ou Backbone range, the distribution of seismic low-velocity areas, areas of high seismicity and earthquake focal mechanisms are closely related to the distribution of the late Cenozoic calderas (Onodera *et al.* 1998; Umino *et al.* 1998; Nakajima & Hasegawa 2003*b*; Yoshida *et al.* 2005). There are three levels of low-velocity bodies in the Ou Backbone range: the lower crust, middle crust and upper crust (Fig. 14d), which presumably reflect the partially molten lower crust, the middle crustal magma reservoirs and the caldera-fill pyroclastic rocks with occasionally central plutonic bodies, respectively.

Figure 15a shows the geotherms (Kushiro 1987; Tatsumi *et al.* 1994) and rock types in the crust of the NE Japan arc adapted from Yoshida (2001). The solidus of water-saturated granitic magma crosses the geotherm at a depth of 15–20 km. It is possible that the water-saturated felsic magma under the volcanic area of the Ou Backbone range could be stable in the middle to lower crustal conditions of amphibolite to hydrous granulite facies in the NE Japan arc. The brittle–ductile transition (Fig. 15b), which strongly affects the emplacement of igneous intrusions (Aizawa *et al.* 2006), is of a shallow depth near the Conrad discontinuity beneath the Ou Backbone range and the Dewa Hill, but is deeper than the Moho beneath the eastern margin of the Japan Sea.

Figure 14c shows the two low-velocity belts at depths of 24 km. As described above, the eastern low-velocity belt could be interpreted as a remnant of the middle Miocene magma plumbing system, which formed at the eastern rift border zone of the Northern Honshu rift and marked the volcanic front (RBL, rift border line in Fig. 14c) at that time. The western low-velocity belt, beneath the Ou Backbone range, reflects the present magma plumbing system from which many felsic caldera-forming magmas and stratovolcano-forming andesitic magmas have been derived. The high-resolution tomographic image around Sendai (Fig. 13b, c) confirms the existence of low *P*- and low *S*-wave velocity bodies at depth of 3–7 and 10–15 km beneath the late Miocene and Pliocene caldera complexes (Nakajima *et al.* 2006*a*), which are interpreted being still-hot, solidified pluton with a fluid-saturated top (Yoshida 2001; Sato *et al.* 2002; Nakajima *et al.* 2006*a*). The chemical composition of pumice from the late Miocene and Pliocene calderas indicates differing equilibration depths, as estimated by the An–Qz–Ab + Or–H_2O system (Luth *et al.* 1964). The estimated depths correlate well with the depth range of the low-V_p and low-V_s bodies observed in the tomographic images (Fig. 13b, c). Some Pliocene calderas have deeper magma reservoirs (10–15 km) than those of the Late Miocene (3–7 km).

Mantle structure

Seismic structure of the mantle wedge. Since the early 1990s, investigations using seismic tomography images of the sub-arc mantle wedge have been

Fig. 14. (**a**) Density structure model along the across-arc profile (X–Y line in Fig. 2a). Blue line in the gravity profile shows the computed gravity anomaly. Red circles show the observed Bouguer gravity anomaly. (**b**) Vertical cross section of density structure model along the across-arc profile is the same as in (a). Denser materials are modelled beneath the eastern rift border zone of the Northern Honshu rift (shown in (**c**) as 'rift border zone') which is located in the eastern-most part of the Kitakami river valley and is consistent with V_p, V_s and Poisson's ratio structures (Fig. 13c: Nakajima *et al.* 2006*a*) at the east–west profile shown in (c). (**c**) Two parallel seismic belts and low-velocity zones shown by V_s perturbation at a depth of 24 km elongated in a north–south direction beneath the forearc region in the Ou Backbone Range to the west of the Kitakami massif (the study area shown by the rectangle in Fig. 2a), adapted from Okada *et al.* (2010). The V_s perturbation (%) scale is shown at the bottom. Epicentre of shallow seismicity (depth <45 km) is denoted by a small black cross. White star and small circles denote the main shock and aftershocks of the 2008 earthquake, respectively. Triangles and bold red lines denote Quaternary volcanoes and active faults, respectively. The white cross denotes low-frequency micro-earthquakes. White circles denote late Cenozoic calderas (Yoshida *et al.* 1999*a*). Western zone correlates with the Quaternary volcanic belt, and the eastern zone is consistent with the distribution area of early to middle Miocene volcanics, shown by the red-dot hatched area and which correlates with Miocene plutons intruded into the eastern rift border normal fault zone (RBL: rift border line). (**d**) North–south vertical cross-sections of V_s perturbation are shown (Okada *et al.* 2010). Scale is the same as in (c). Location of the cross section is shown by the red line with S and N in (c). Triangles denote Quaternary volcanoes. Epicentre of shallow seismicity is denoted by a small black cross. White star denotes the main shock of the 2008 earthquake. Thin purple lines represent *S*-wave reflectors (Hori *et al.* 2004), and thick line denotes the Moho discontinuity (Zhao *et al.* 1990). There are three levels of low-velocity bodies in the lower crust, middle crust and upper crust, and the distribution pattern of low-velocity bodies is consistent with Figure 13, suggesting the partially molten lower crust, the middle crustal magma reservoirs and the calderas filled with pyroclastic rocks, respectively.

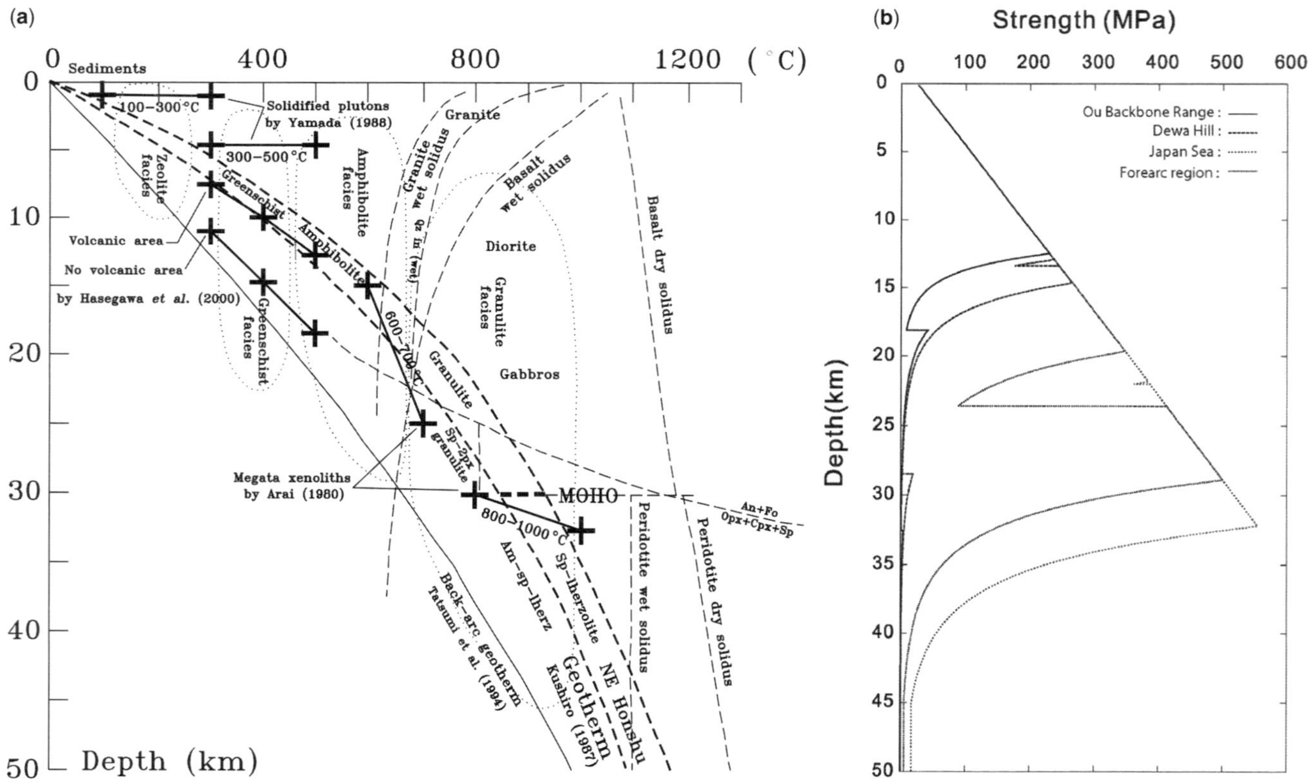

Fig. 15. (**a**) Geotherms (Kushiro 1987; Tatsumi *et al.* 1994) and rock types in the crust of the NE Japan arc, solidus of dry and wet (water-saturated) peridotite and basalt, and wet granite solidus with Qz (quartz)-in line (Wyllie 1971; Robertson & Willie 1971); adapted from Yoshida (2001). Plus signs connected by tie-lines show each temperature range estimated from: the petrologic studies of Megata xenoliths (Arai 1980); the thermodynamical studies of the Onikobe hydrothermal systems of solidified plutons (Yamada 1988); and the velocity perturbations of *P*-waves beneath volcanic and non-volcanic areas of the Ou Backbone Range (Hasegawa *et al.* 2000) in the NE Japan arc, respectively. Dotted fields show pressure and temperature conditions of the representative metamorphic facies. The estimated geotherm from Ichino-megata xenoliths is consistent with the geotherm at the present Ou Backbone Range and early Miocene back-arc rift axis, and is higher than the present back-arc geotherm (Tatsumi *et al.* 1994). (**b**) Strength profiles for the crust and upper mantle beneath: the Ou Backbone Range; the Dewa Hill; the forearc region; and the eastern margin of the Japan Sea (after Shibazaki *et al.* 2008). Shibazaki *et al.* (2008) assigned the two layers of crust (upper quartz diorite and lower wet diabase) and the upper mantle of wet olivine for modelling. Hot regions are set beneath the Ou Backbone Range and Dewa Hill based on the geothermal gradient beneath NE Japan (Yano *et al.* 1999; Tanaka *et al.* 2004).

performed (e.g. Hasegawa *et al.* 1991, 2000, 2005; Hasegawa & Nakajima 2004; Nakajima *et al.* 2001, 2005; Zhao *et al.* 1990, 1992, 2009, 2012, and references therein). The dense seismic network in NE Japan enables the world's finest images of these structures. The essential structure of the mantle wedge beneath the NE Japan arc consists of a low-V_p, low-V_s and low V_p/V_s structure lying in the middle of the mantle wedge, subparallel to the down-going Pacific Plate slab (Fig. 11a: Sato & Hasegawa 1996; Nakajima *et al.* 2001, 2005; Zhao *et al.* 2012).

The temperature (Nakajima & Hasegawa 2003*a*) and porosity (Nakajima *et al.* 2005) of the low-velocity mantle wedge have also been investigated, and a high-temperature wedge core with the presence of melts (Sato & Hasegawa 1996; Nakajima *et al.* 2005) has been assumed (Fig. 11a). The top depth of the low-velocity mantle reaches immediately below the arc Moho, and low-frequency earthquakes occur in the surrounding areas (Fig. 11b: Xia *et al.* 2007). This further suggests melt coalescence and migration from the partially molten mantle wedge (Hasegawa & Nakajima 2004). The top becomes deeper towards the rear-arc side and the low-velocity region continues beneath the Japan Sea (Zhao *et al.* 2012), although the large anomaly appears to diminish off the Japan Sea coast, across the Quaternary volcanic arc (Fig. 11). The low-velocity bodies show heterogeneities in the mantle wedge, which is thicker with a strong low-velocity amplitude beneath the Quaternary volcanic centres, and thinner and faster beneath non-volcanic areas (Hasegawa & Nakajima 2004), although a continuous low velocity region is imaged just below the Moho depth along the entire volcanic front.

Seismic structure of the slab. Using the tomographic image based on the precise analyses of the seismic locations, the dehydration and re-hydration profiles of the down-going slab and overriding forearc mantle wedge have been estimated (Kita *et al.* 2006; Tsuji *et al.* 2008). The down-going Pacific Plate slab has a relatively high V_p/V_s at a depth of about 80 km. In contrast, the overriding forearc peridotite has a lower V_p/V_s than that expected from serpentinite. The results led Tsuji *et al.* (2008) to conclude that the slab does not dehydrate significantly until a depth of *c.* 80 km depth, and that slab dehydration is largely completed between depths of 70 and 90 km. Kita *et al.* (2006) and Hasegawa *et al.* (2007) also examined slab seismicity with precise re-location of the foci. Their results support the conclusion of Tsuji *et al.* (2008), and indicate that the deepening of the top downdip seismicity surface from the slab–mantle interface, correlates with the dehydration front of the metamorphic slab that is controlled by the inverted thermal gradient of the slab (Fig. 11c).

Kawakatsu & Watada (2007) examined the receiver function structure of the slab and the mantle beneath NE Japan. They also proposed that the slab dehydrated at around 80 km deep and that the released hydrous fluids migrated upwards into the mantle wedge to form serpentinite. The hydrated serpentinite layer reaches a depth between 120 and 150 km. Together with the geodynamic model (e.g. van Keken 2003), the seismic observations beneath the NE Japan arc provide constraints to the slab–mantle thermal structure and origin of the fluids released from the slab. The finest seismic images of the mantle wedge and the slab provide strong insights into the origin of subduction volcanisms.

Along-arc mantle heterogeneity

Attenuation tomography reveals the low-attenuation (high-Q) subducting Pacific slab and the high-attenuation (low-Q) anomalies in the crust and mantle wedge (Umino & Hasegawa 1984; Tsumura *et al.* 2000). The location and geometry of low-Q zones in the mantle wedge agree with those of low-V anomalies imaged by tomography at a depth of the uppermost mantle, although they cross with the NE Japan arc at a depth of 60–80 km (Fig. 16d: Yoshida *et al.* 1999*b*; Tsumura *et al.* 2000). Tsumura *et al.* (2000) have shown that the distribution of the mantle wedge low-Q zones is different in the southwestern part compared with the northeastern part of NE Honshu. In the southwestern part (west of the TTL), low-Q zones are located beneath the active volcanoes or the volcanic front, and are continuously distributed from the crust to the mantle wedge, deepening westward. In contrast, several low-Q zones exist in the northeastern part (east of the TTL), and appear isolated from each other. They are located just beneath the active volcanoes at shallower depths, and deepen to the west of the volcanic front (Tsumura *et al.* 2000). Q-Values in the mantle wedge vary not only across the arc, but also along the arc (Tsumura *et al.* 2000). The spatial variation of Q-values is consistent with the basement structure bounded by the TTL, and suggests that the TTL has a deep mantle root (Yoshida *et al.* 1999*b*, 2005). This mantle heterogeneity along the TTL, oblique to the present volcanic front, extends to the Japan Sea side, as shown by dV_p distributions in Figure 16b, c (Zhao *et al.* 2012). This implies that the TTL extends to the Japan Sea side with a deep mantle root.

The temperature structure in the mantle wedge at a depth of 40 km inferred from seismic wave attenuation (Fig. 16e; Nakajima & Hasegawa 2003*a*) corresponds to the distribution of Quaternary

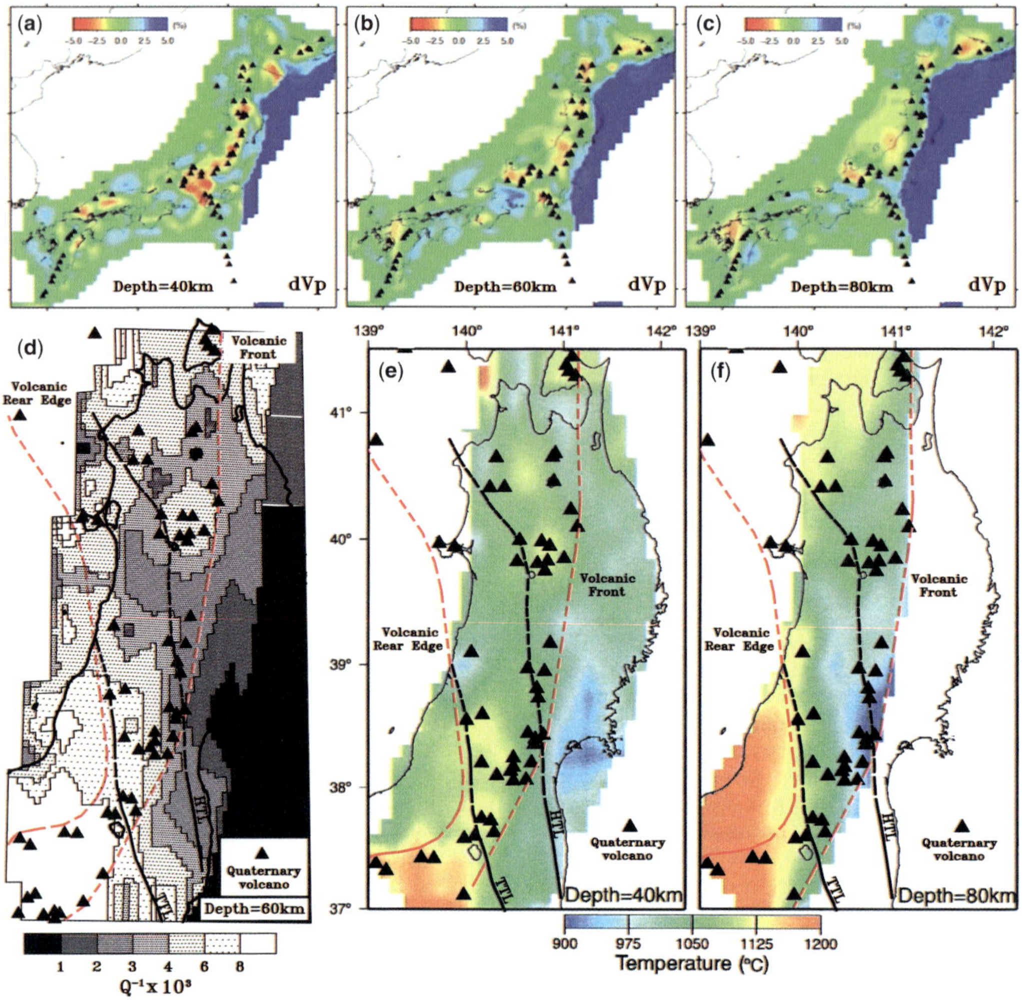

Fig. 16. (**a–c**) Map views of *P*-wave tomography at depths of (a) 40, (b) 60 and (c) 80 km. Black triangles represent the distribution of Quaternary volcanoes (after Zhao *et al.* 2012). (**d**) Seismic *Q* structure at a depth of 60 km (Tsumura *et al.* 2000; Yoshida *et al.* 1999*b*, 2005). (**e**, **f**) Relations between distribution of volcanoes and mantle temperature at depths of (e) 40 and (f) 80 km, in the NE Japan arc (Nakajima & Hasegawa 2003*a*; Yoshida *et al.* 2005). The estimated mantle temperature is higher on the SW Honshu side than on the NE Honshu side, divided by the TTL. These mantle structures are oblique to the present volcanic front and extend to the Japan Sea side, as shown by dV_p distribution in (b) and (c). TTL and HTL are the same as in Figure 1.

volcanoes. However, the temperature structure at a depth of 60–80 km (Fig. 16f) does not correspond with the distribution of the Quaternary volcanoes at the surface (Yoshida *et al.* 1999*b*; Tsumura *et al.* 2000; Nakajima & Hasegawa 2003*a*). The estimated mantle temperature is higher on the side of SW Honshu than NE Honshu divided by the TTL. If the temperature distribution in the mantle wedge is constrained only by the geodynamic models (van Keken 2003; Peacock 2003), the observed spatial variation of *Q*-values crossing the volcanic front requires explanation using other factors, including that of the heterogeneous distribution of H_2O (Karato 2003) that borders the TTL. Yoshida *et al.* (1999*b*) suggested that the counter-clockwise rotation of the continental hydrous lithosphere beneath SW Honshu, at the back-arc opening of the Japan Sea, formed this mantle heterogeneity. A high-temperature or a hydrous mantle wedge could also result in low viscosities (Honda & Yoshida 2005*a*). Honda & Yoshida (2005*a*) investigated using the inclined top-surface of a low-viscosity wedge

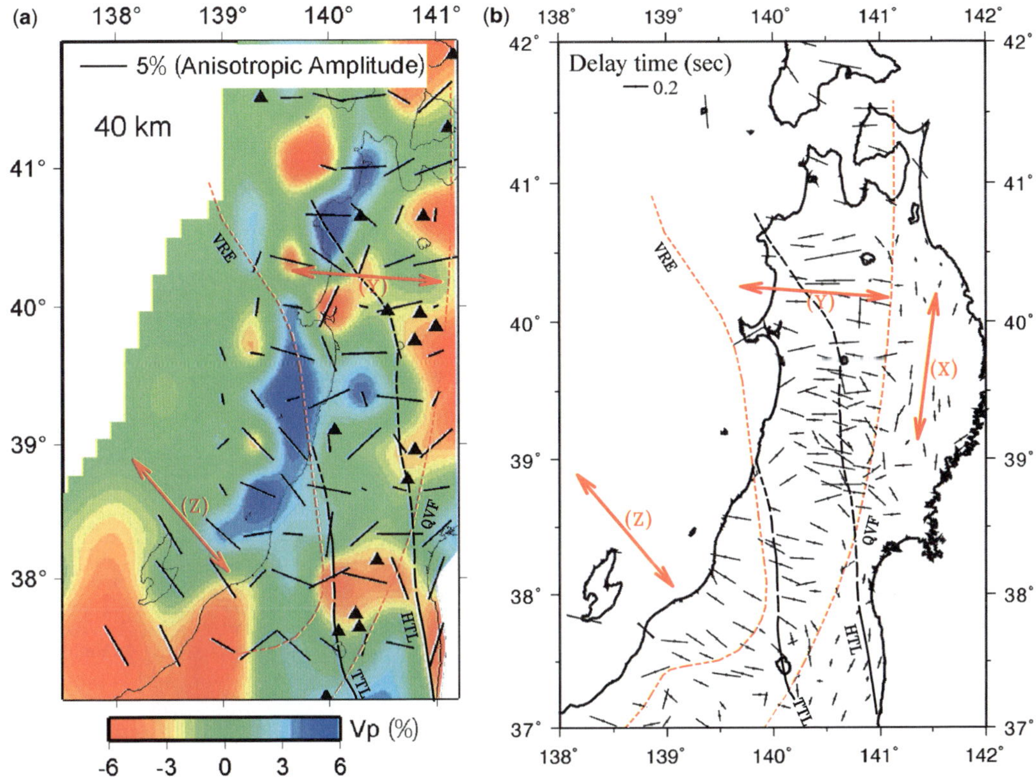

Fig. 17. (**a**) *P*-wave anisotropic tomographic images at a depth of 40 km under the eastern margin of the Japan Sea (modified from Huang *et al.* 2010). Red and blue colours denote low and high velocities, respectively, and velocity perturbation scales are shown at the bottom. The azimuth and length of bars represent the fast-velocity direction and the anisotropic amplitude, respectively. Black triangles are the Quaternary volcanoes. The fast direction of the seismic velocity anisotropy is trench-normal (*y*-direction) on the NE Honshu side, but is trench-oblique (*z*-direction) in the back-arc side of the SW Honshu side of the TTL. QVF, VRE, TTL and HTL are as in Figure 2. (**b**) Distribution of the fast *S*-wave polarization directions in the mantle wedge (modified from Nakajima & Hasegawa 2004; Nakajima *et al.* 2006*b*). Azimuth and length of each bar correspond to the average fast polarization direction and delay time at each station, respectively. The delay-time scale is shown in the upper-left corner. The pattern of their seismic anisotropy in the back-arc side is consistent with the results of Huang *et al.* (2010), and the azimuth in the forearc region is trench-parallel (*x*-direction).

similar to the low-*Q* zone in the southwestern part of the NE Japan arc (west of the TTL). The high-temperature anomaly obtained was almost slab-parallel, larger in the shallow part (volcanic front side) and weaker in the lower part (back-arc side). This result is consistent with the distribution of the Quaternary volcanoes in the southwestern part of the NE Japan arc (west of the TTL).

Heterogeneity of seismic velocity anisotropy. Shear-wave splitting analyses reveal that the low-*V* and low-*Q* zones in the mantle wedge show strong seismic anisotropies (Okada *et al.* 1995; Nakajima & Hasegawa 2004, Nakajima *et al.* 2006*b*). Anisotropic bodies are revealed in the crust, mantle wedge and the subducting Pacific slab (Hall *et al.* 2000; Nakajima & Hasegawa 2004; Nakajima *et al.* 2006*b*; Wang & Zhao 2008; Huang *et al.* 2010). Nakajima & Hasegawa (2004) have clearly shown that the azimuth in the forearc region is trench-parallel, whereas in the back-arc region the azimuth is trench-normal (Fig. 17b). The trench-normal orientation in the back-arc side might originate from the orientation of olivine by mantle wedge corner flow, associated with slab subduction (Hall *et al.* 2000). The trench parallel orientation in the forearc mantle can be explained by the preferred orientation of olivine under wet conditions (Katayama *et al.* 2004), although Wang & Zhao (2008) suggested that the trench-parallel anisotropy beneath the volcanic-front area in the mantle wedge results from the complex 3D flow in the

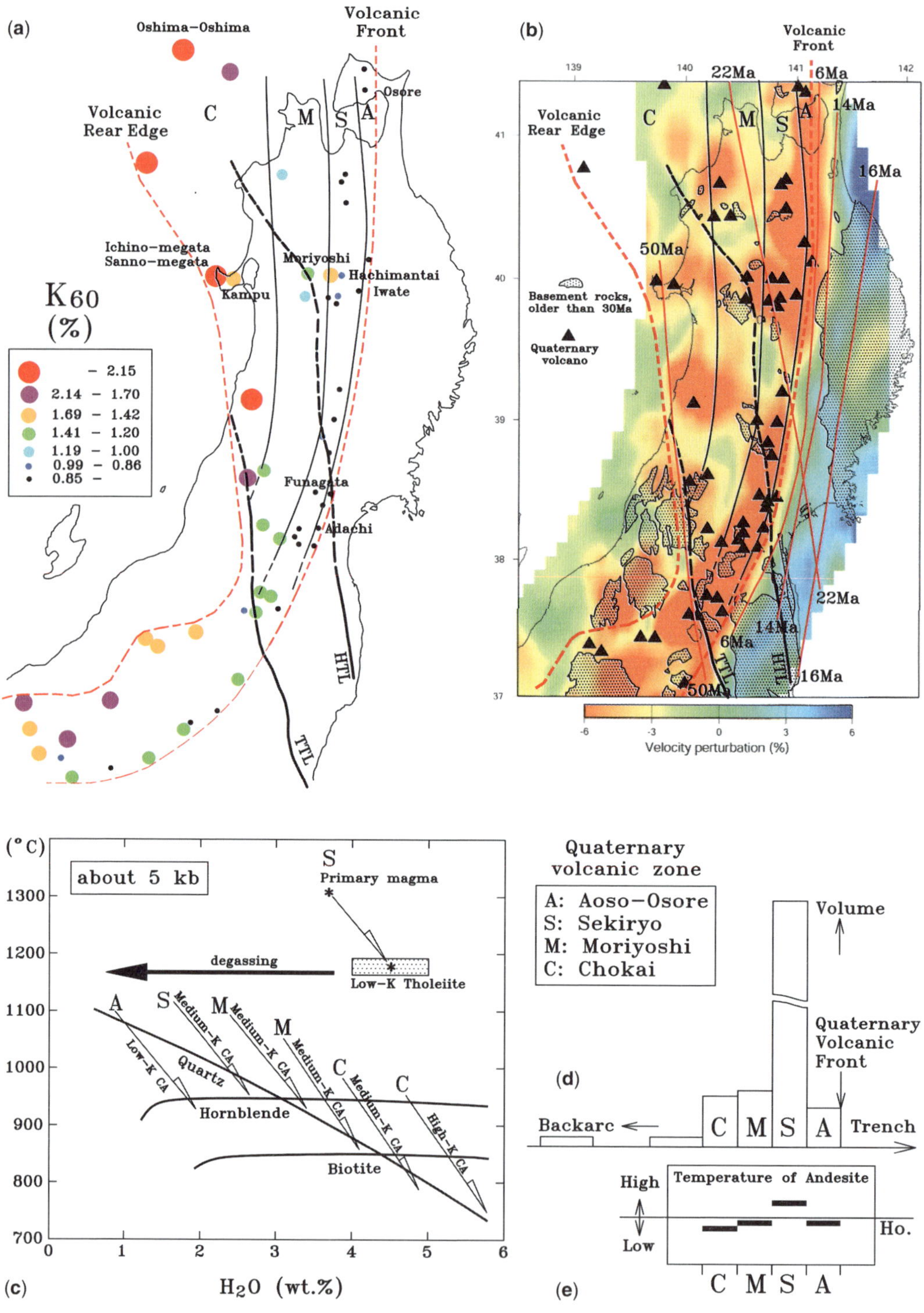

Fig. 18. (**a**) Regional distribution of K_{60} (K_2O% at 60% SiO_2) of the Quaternary volcanic rocks (modified from Yoshida *et al.* 1999*b*, 2005). The size of the circle is correlated to the range of the value in K_{60} and plotted on the location of each Quaternary volcano. The K_{60} values correlate with the boundaries of the four Quaternary volcanic zones of

mantle wedge, owing to oblique convergence or slab rollback (Hall *et al.* 2000; Kincaid & Griffiths 2003).

Huang *et al.* (2011) determined high-resolution *P*-wave isotropic and anisotropic velocity structures under the Japan Sea off NE Japan, using 175 425 high-precision *P*-wave arrival times from 2833 local earthquakes relocated with *sP* depth phase. Their results show that strong velocity variations exist in the crust and the uppermost mantle under the eastern margin of the Japan Sea. *P*-Wave azimuthal anisotropy is complex under the Japan Sea, which could also indicate the existence of a complex lithospheric structure. On the NE Honshu side, the fast direction of the seismic velocity anisotropy in the mantle wedge is trench-normal, whereas it is trench-oblique in the back-arc side of the SW Honshu side of the TTL (Fig. 17a). The east–west-trending trench-normal anisotropy shown in the NE Honshu back-arc is related to the recent slab-driven corner flow in the NE Japan mantle wedge (Hall *et al.* 2000). The NW–SE trend is parallel to the 20–15 Ma dyke direction (Fig. 7), which reveals the direction of σ_{Hmax} at that time. The NW–SE trend is also normal to the Yamato Basin axis, so this direction is consistent with the mantle wedge corner flow during the Yamato Basin opening. It is possible that the NW–SE-trending seismic velocity anisotropy beneath SW Honshu formed as trench-normal anisotropy during the Yamato Basin opening at 21–15 Ma, and this could still be maintained because of the weak mutual rotation and weak magmatism in the back-arc region to the west of the TTL after the back-arc opening.

Genesis of Quaternary magmatism in the NE Japan arc

Figure 18a shows the regional distribution of the normalized K_2O% at 60% SiO_2 (K_{60}) of Quaternary volcanic rocks (Yoshida *et al.* 1999*b*). The composition of Quaternary volcanic rocks changes across and along the arc. Distribution of K_{60} generally increases from the volcanic front to the back-arc side, and the value of K_{60} is higher in SW Honshu than in NE Honshu at the volcanic front. Typical island-arc volcanism, with clear across-arc chemical variations in the Quaternary volcanic rocks, characterizes NE Honshu to the east of the TTL. In contrast, the pattern of across-arc chemical variations is not so simple in SW Honshu to the west of the TTL, and the distribution of the Quaternary volcanoes is narrow and limited in the region near the volcanic front (Fig. 18a).

The distribution of Quaternary volcanoes is closely related to the distribution of low-V_s regions of the mantle wedge (Fig. 18b), which are interpreted as partial melting zones (Hasegawa & Nakajima 2004). This close relation suggests that the Quaternary magmatic plumbing system is directly connected from the upper mantle to the crust (Hasegawa & Nakajima 2004; Hasegawa *et al.* 2005; Acocella *et al.* 2008), and it is clear that the distribution pattern of the inclined low-*V* zones in the mantle wedge forms branches and a trunk (Fig. 18b: Kondo *et al.* 1998, 2004; Tamura *et al.* 2002; Prima *et al.* 2006).

The Quaternary volcanoes on the NE Honshu side of the TTL are divided into four across-arc volcanic zones according to their *K*-level (Fig. 18): A (Aoso-Osore volcanic zone); S (Sekiryo volcanic zone); M (Moriyoshi volcanic zone); and C (Chokai volcanic zone) (Nakagawa *et al.* 1986). The *K*-level in volcanic rocks is lowest at the volcanic front side and highest at the back-arc side (Fig. 18a). The Sekiryo volcanic zone is an axial zone of arc volcanism, in which the temperature of andesitic magmas is the highest and the eruptive materials are the most voluminous (Fig. 18d, e: Yoshida *et al.* 2005).

Sakuyama (1977) classified intermediate to felsic volcanic rocks into three types on the basis

Fig. 18. (*Continued*) Nakagawa *et al.* (1986) in the region to the eastern side of TTL. A, Aoso–Osore volcanic zone; S, Sekiryo volcanic zone; M, Moriyoshi volcanic zone; C, Chokai volcanic zone. TTL and HTL are as in Figure 1. (**b**) Boundaries (black lines) of the four Quaternary volcanic zones and migration of volcanic front (red lines: Shibata & Ishihara 1979; Ohguchi *et al.* 1989; Ohki *et al.* 1993; Yoshida *et al.* 1999*b*), superimposed on *S*-wave velocity perturbations along the inclined low-velocity zone in the mantle wedge of the NE Japan arc (Hasegawa & Nakajima 2004; Hasegawa *et al.* 2004) after Yoshida *et al.* (2005). (**c**) Schematic isobaric phase diagram at a pressure of about 5 kb (Sakuyama 1979), and differentiation courses of the Quaternary low-*K* tholeiite (Kuritani *et al.* 2013) and calc-alkaline andesites (modified from Yoshida *et al.* 2005). Lines with arrows represent differentiation courses of magma in magma chamber. Differentiation courses are estimated from the description of each volcanic zone by Yoshida (1989). Liquidus temperatures of each mineral are indicated by solid curves. CA, calc-alkaline rock series. Temperature and water content of low-K tholeiitic basalt of Iwate volcano from the Sekiryo volcanic zone are cited from Kuritani *et al.* (2013). Sekiryo calc-alkaline andesites have a lower water content than low-*K* tholeiitic basalts, so that the calc-alkaline mafic end-member must degas before forming andesite by any process. (**d**) Volumetric relation of the eruptives of the four Quaternary volcanic zones. (**e**) The temperature profile of the andesitic magmas of each Quaternary volcanic zone. Ho is the temperature of hornblende crystallization from magma. Only the andesites from the Sekiryo volcanic zone contain no hornblende phenocryst. (d, e) Adapted from Nakagawa *et al.* (1986).

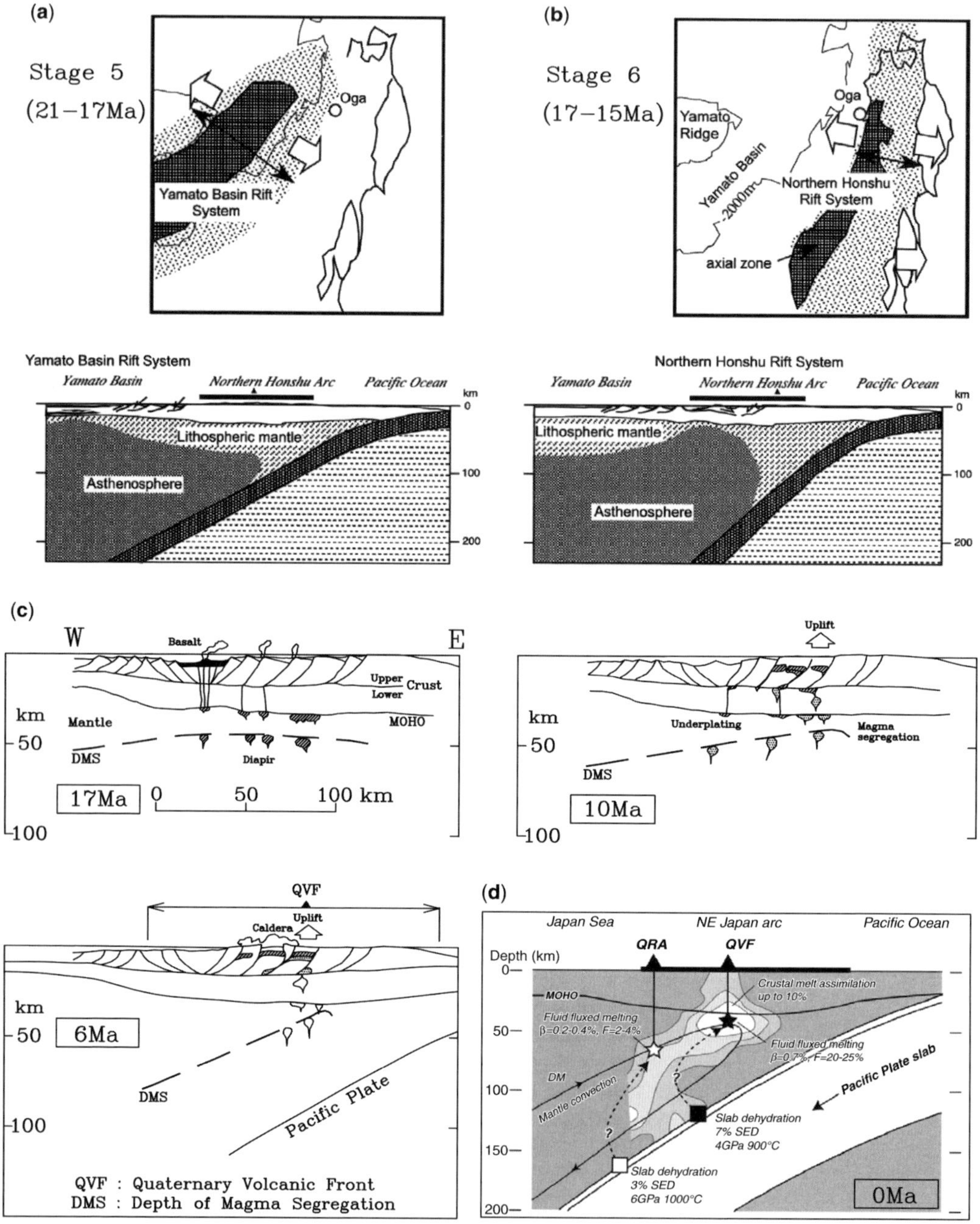

Fig. 19. (**a**, **b**) Injection of depleted asthenospheric mantle into enriched continental lithosphere during: (a) Stage 5 (21–17 Ma) to (b) Stage 6 (17–15 Ma), forming the Yamato Basin rift and the Northern Honshu rift, respectively (after Sato *et al.* 2004). (**c**) Schematic cross-sections depicting the magmatic evolution and tectonic setting of the middle to late Miocene NE Japan arc (Sato & Yoshida 1993; Yoshida *et al.* 1995). DMS, depth of magma segregation; QVF, Quaternary volcanic front. The arrow with QVF in (c) denotes the land area. (**d**) Schematic diagram showing the location of slab dehydration, fluid pathways and partial melting for frontal- (QVF) and rear-arc (QRA) basalts in the NE Japan arc, after Kimura & Yoshida (2006). A *P*-wave perturbation seismic tomography image (Nakajima *et al.* 2005) is superimposed, with lighter colours showing greater perturbation. SED, sediments; DM, depleted mantle; β and F, fluid flux and degree of melting. Black bar under QVF and QRA denotes the land area. Dashed arrows show fluid pathways from different slab source regions to the magma source regions.

of hydrous phase assemblage, and showed that the volatile contents decrease from the back-arc to the volcanic front in the Quaternary volcanoes of the NE Japan arc. Figure 18c shows a schematic isobaric phase diagram at a pressure of about 5 kbar (Sakuyama 1979; Yoshida *et al.* 2005). Volcanoes that are composed of rocks without hornblende and biotite phenocrysts are dominantly distributed in the frontal volcanic zone. Those with hornblende and no biotite phenocrysts occur mainly in the back-arc side of the volcanic front, and those with biotite and hornblende phenocrysts mostly appear in the area furthest from the trench. The most primitive Sekiryo low-*K* tholeiitic basalts, however, have a higher water content (Kuritani *et al.* 2013) than the Sekiryo calc-alkaline andesites, and this means that the calc-alkaline mafic end-member needs to degas before forming andesites by any process (Miyagi *et al.* 2012).

Across-arc geochemical variation

After establishment of the island-arc magmatism at around 13.5 Ma, a prominent across-arc variation dominated. The across-arc geochemical variation of Quaternary lavas (1.7–0 Ma) in the NE Japan arc has been examined rigorously since the 1950s. Total alkalis (Kuno 1966), K_2O (Fig. 18a) and incompatible element contents increase gradually from the volcanic front to the rear-arc lavas (Kawano *et al.* 1961; Aoki *et al.* 1981; Yoshida & Aoki 1984; Sakuyama & Nesbitt 1986; Nakagawa *et al.* 1988; Yoshida *et al.* 1997; Kimura & Yoshida 2006, and references therein). Sr and Pb are more radiogenic in the volcanic front, and Nd is less radiogenic (Fig. 9c: Shibata & Nakamura 1991; Kimura & Yoshida 2006; Takahashi *et al.* 2012). These are typical of across-arc variations commonly found in arcs (e.g. Dickinson & Hatherton 1967; Kimura & Stern 2008). Because of the relatively low slab-dip angles, the volcanic arc is wider in NE Japan than elsewhere, and the across-arc variations are more prominent compared with those of other subduction zones (Kimura & Stern 2008).

The origins of across-arc variations have been examined, and different depths of melt segregation (Kuno 1959), different degrees of partial melting in the source mantle (Miyashiro 1974), different slab flux (Shibata & Nakamura 1991) and different sources of the mantle (Nohda *et al.* 1988, 1992; Togashi *et al.* 1992; Ujike & Tsuchiya 1993) have been proposed. Kimura & Yoshida (2006) and Kimura *et al.* (2009) examined the various parameters involved in the melting of the slab liquid-fluxed mantle across the NE Japan arc, and proposed a common depleted mantle source which is fluxed by different slab fluids from the shallow slab beneath the volcanic front and from the deep slab beneath the rear-arc volcanoes. They also estimated the higher slab flux fraction beneath the volcanic front and the lower slab flux fraction beneath the rear-arc, which resulted in a larger extent of partial melting of the mantle beneath the volcanic front than the rear-arc. Magma segregation depths were also estimated to be shallower at *c.* 1 GPa beneath the volcanic front, but deeper at *c.* 2 GPa beneath the rear-arc (Fig. 19d). Kimura and co-authors concluded that seismic observations are consistent with the results of their model based on the intensive–extensive variables (e.g. slab dehydration depth, slab temperature, flux fraction to the mantle wedge, degree of partial melting, and melting pressure and temperature), derived from the geochemical mass-balance calculation package called the Arc Basalt Simulator (Kimura & Yoshida 2006; Kimura *et al.* 2009, 2010).

Along-arc geochemical variation

Along-arc isotopic variation is unusually large at the volcanic front in the Quaternary NE Japan lavas. A southward increase in Sr and Nd isotope ratios has been identified (Fig. 9b: Notsu 1983; Kersting *et al.* 1996; Kimura & Yoshida 2006; Takahashi *et al.* 2012). The spatial variation correlates closely with the basement terranes (Kersting *et al.* 1996) and the isotopic composition of basement granitoids (Kimura & Yoshida 2006), which suggests a correlation to the compositional variation of the mantle lithosphere (Kersting *et al.* 1996), the basement granitoids (Kimura & Yoshida 2006) and the lower crustal amphibolite, through the contamination of the partial melts (Takahashi *et al.* 2012). Tatsumi *et al.* (2008) and Takahashi *et al.* (2012) also found that the strong spatial variation is more prominent in the tholeiitic (TH by Miyashiro 1974) suite of basaltic andesite to andesite magmas, rather than in the calc-alkaline (CA by Miyashiro 1974) magma suite, and suggested lower crustal amphibolite origin of the TH lavas. Kimura & Yoshida (2006), however, suggested that the most primitive TH magmas along the volcanic front are upper mantle melts in origin, and that they were later affected by contamination with the lower crustal amphibolite melts.

Role of the heterogeneous lower crust

The distribution of both zones of basement terranes and basement granitoids are oblique to the alignment of the Quaternary volcanoes (Fig. 9a). For example, the Mino (south), Abukuma (middle) and Kitakami (north) belts are divided by the TTL and the HTL. An arbitrarily parallel distribution of the South, Transitional and North geochemical zones of the basement granitoids (Kagami *et al.* 2000;

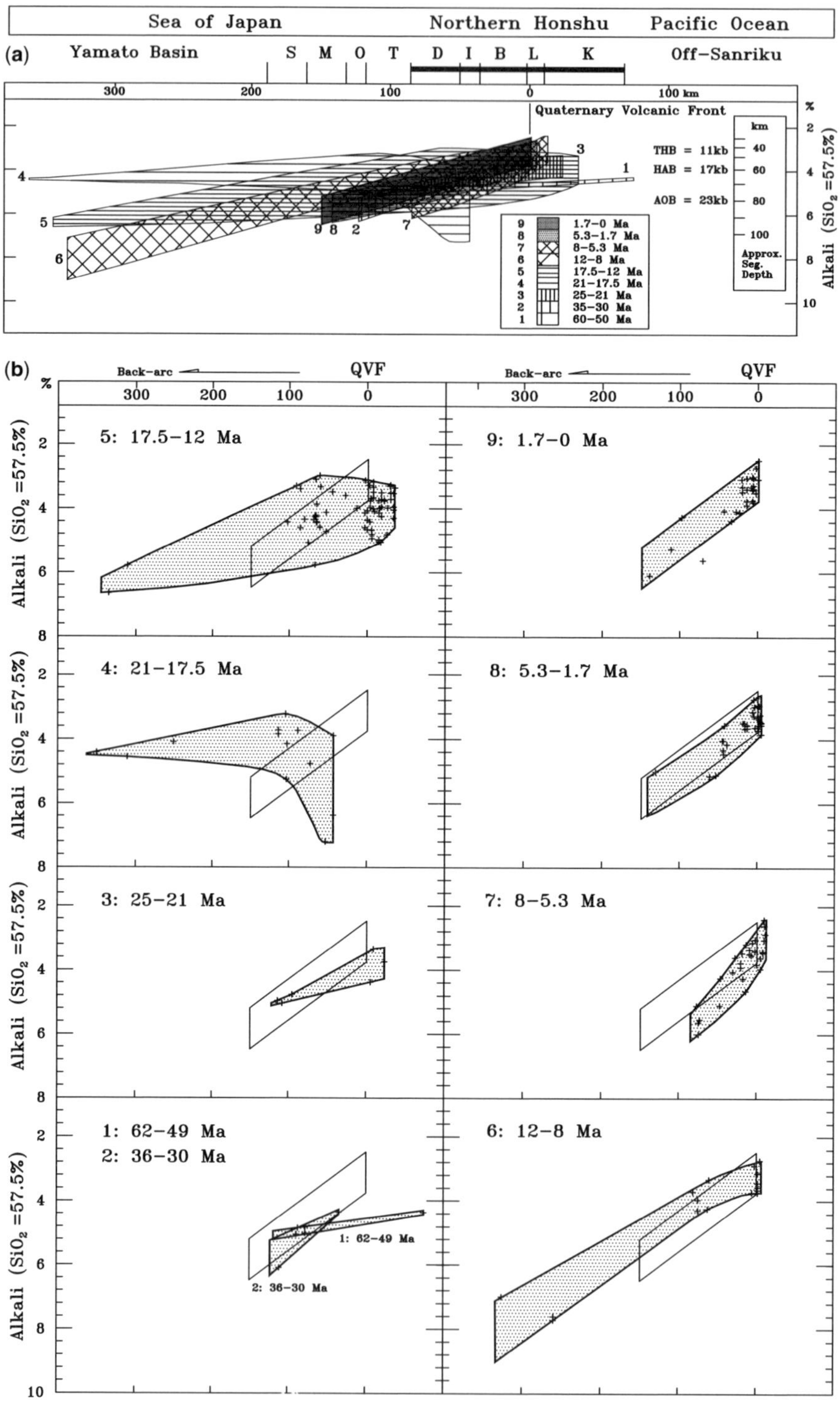

Fig. 20. (**a**) Across-arc variation of the SiO_2-normalized alkali ($Na_2O + K_2O$% at 57.5% SiO_2) level, projected to the 40°N latitude line of NE Japan, and its secular change (modified from Yoshida *et al.* 1995). Tatsumi *et al.* (1983) showed that the depth of the magma segregation beneath the NE Japan arc is shallow (1.1 GPa) beneath the volcanic front, but

Kagami 2005) is also oblique to the volcanic distribution (Fig. 9a). These reflect the changes in allocation of the arc-trench system before and after the opening of the Japan Sea; the latter was NE–SW (oblique to the present NE Japan arc) before the opening, and NNE–SSW (parallel to the present NE Japan arc) after the opening until the present (Otufuji *et al.* 1985; Ohki *et al.* 1993; Yoshida *et al.* 1999*b*). The present-day mantle structure, including the low velocity and mantle anisotropy structures (Nakajima & Hasegawa 2004; Nakajima *et al.* 2006*b*), is principally parallel to the present arc-trench system, although there are some mantle heterogeneities in the areas of SW Honshu and NE Honshu bordered by the TTL. Therefore, it is likely that the along-arc spatial variation of the Quaternary lava chemistry is mainly affected by the overlying lithosphere, and the most plausible candidate for affecting this appears to be the lower crustal melts, which can contribute to both the origin of the basement granitoids and to the Quaternary lavas.

Role of the mantle wedge

In contrast to the volcanic front lavas, the isotopic compositions of the Quaternary rear-arc lavas commonly show a contribution from a depleted source similar to those for mid-ocean ridge basalts (MORBs) or the back-arc basin basalts, irrespective of the overlying basement terranes (Kimura & Yoshida 2006). The resistance of the rear-arc lavas to crustal contamination has been ascribed to the high element concentrations in the primary mantle-derived basalts, which include Sr, Nd and Pb (Kimura & Yoshida 2006), and to a stable magma plumbing system in the thick, low-temperature, brittle crust of the back-arc region (Fig. 15b: Shibazaki *et al.* 2008). The high element abundances of incompatible elements (excluding heavy rare earth elements) could reflect a low degree of melting of the rear-arc mantle in the garnet stability field (Kimura & Yoshida 2006; Kimura *et al.* 2010), and this would be consistent with the deep top-surface of the low velocity mantle beneath the rear-arc (Fig. 11).

The common features of the rear-arc lavas, and the overall across-arc geochemical variations, correlate closely with the mantle wedge and slab structures running parallel to the arc-trench system. This indicates that the establishment of the volcanic arc was fundamentally controlled by the slab–mantle infrastructures. The proposed basic structure is a focused slab dehydration at around 80–90 km deep, and a high-temperature and melt-bearing mantle wedge developed parallel to the slab between a depth of 30 and 150 km (Fig. 19d: Kimura & Yoshida 2006; Kimura *et al.* 2009). This suggests that the contribution of slab-derived materials to the generation of frontal-arc magmas is significantly higher than the contributions to the rear-arc magmas. Moreover recent results of analysis of water contents in glass inclusions in phenocrysts suggest that frontal-arc basalts have high water contents, and that the magmatic water content does not always show an across-arc variation (Walker *et al.* 2003; Portnyagin *et al.* 2007; Kimura *et al.* 2010; Kuritani *et al.* 2013).

Development of the NE Japan arc in the late Cenozoic

The detailed images of the slab–mantle–crust seismic structures are intimately related to the spatial variation in the Quaternary magmas (e.g. Kimura & Yoshida 2006; Kimura *et al.* 2009). These enabled the most detailed discussions pertaining to the magma genesis model for magmatic arcs. The sub-arc structures and the present magmatism (the latest island-arc period), however, are evidently the products of the preceding continental margin and back-arc basin periods, as envisaged by the tectonic history and mantle–crust structure of the arc. Therefore, the developmental history of the subduction system and magmatism in these preceding periods has also been rigorously studied. This study now summarizes the results of these studies in a temporal order. The following sections deal with: (1) the drastic change in the thermal structure and source materials of the mantle in the back-arc basin period; (2) the transitional nature of the back-arc basin to island-arc periods in the early stages of the island-arc period; and (3) the youngest island-arc period.

Fig. 20. (*Continued*) deeper (2.3 GPa) beneath the back-arc, and corresponds with a change in magmatic composition from olivine tholeiite (THB) to high-alumina basalt (HAB) and then to alkali olivine basalt (AOB) at a similar temperature of about 1300 °C. This figure shows the secular changes within an estimated depth of magma segregation; the thermal structure in the mantle wedge through the Cenozoic Era. The abbreviations, K–S are the same as in Figure 1. The black bar above the cross-section denotes the land area. (**b**) Secular changes of across-arc variation of SiO_2-normalized alkali level (modified from Yoshida *et al.* 1995). The parallelograms show, for reference, the field of Quaternary (1.7–0 Ma) volcanics. SiO_2-normalized alkali contents were plotted as a function of the distance of the volcanic centre from the Quaternary volcanic front (QVF).

Secular changes in the mantle

Estimation of mantle melting conditions and thermal structure. Changes in the magmatic composition could reflect changes in: (1) the source mantle composition; (2) the slab-flux; (3) the thermal structure of the mantle; and (4) the heterogeneous overlying crust. Experimental studies have investigated the melting conditions with the assumed water content in the source. Tatsumi *et al.* (1983) conducted multiphase saturation experiments for the Quaternary basalt compositions and showed that the depths of magma segregation beneath the NE Japan arc are shallow (1.1 GPa) beneath the volcanic front, but deeper (2.3 GPa) beneath the back-arc. This corresponds with a change in magmatic composition from olivine tholeiite (THB) to high-alumina basalt (HAB) and then to alkali olivine basalt (AOB) at a similar temperature of about 1300 °C. Following the experimental results, Yoshida *et al.* (1995) formulated the relation between segregation depth and SiO_2-normalized total alkali content.

Tatsumi *et al.* (1994) also conducted experiments on the early to middle Miocene MgO-rich primitive basalt compositions in the NE Japan arc, from both the frontal-arc and the back-arc sides, and showed that the depths and temperatures of the mantle sources were constant at 1.3 ± 0.1 GPa and 1320 ± 10 °C, respectively, beneath the entire NE Japan arc. Yamashita & Fujii (1992) performed experiments on the early Miocene MgO-rich basalts erupted during the back-arc basin period, which had been sampled by drilling in the Japan Sea oceanic floor, and estimated pressure and temperature of 1.4 GPa and 1340°C. With these data, Tatsumi *et al.* (1994) concluded that the lithosphere thickness beneath the forearc had not changed at a depth of about 30–40 km (*c.* 1.1–1.3 GPa) during the past 20 Ma, but that the back-arc lithosphere has thickened through cooling at a depth of between about 40 km (*c.* 1.3 GPa) to around 70 km (*c.* 2.3 GPa; Fig. 15). The conclusion of the study suggests that asthenospheric upwelling during the back-arc basin opening period, and the subsequent deepening of the back-arc basin, may have caused cessation of volcanism in the back-arc basin, together with the thermal subsidence of the Japan Sea (Yamaji & Sato 1989). The asthenospheric deepening in the back-arc occurred through the establishment of the across-arc thermal structure of the mantle wedge in the island-arc period (Fig. 19c).

Secular changes of thermal structure in the mantle. Figure 20a summarizes the secular changes in the depth of magma segregation through the continental margin, the back-arc basin and the island-arc periods, as estimated by SiO_2-normalized total alkali content using the formula proposed by Yoshida *et al.* (1995). In the continental margin period, it is estimated that the asthenosphere was at its shallowest beneath the volcanic front, and that it deepened gently towards the back-arc. The depth of asthenosphere beneath the volcanic front, however, was deeper than today. In the back-arc basin period, the axis of shallower asthenosphere was not located beneath the volcanic front at the time, but was located along the present-day coastline of NE Japan owing to the asthenospheric upwelling during the back-arc basin opening. In the island-arc period, the asthenospheric depth steeply inclined towards the back-arc, and the depth beneath the volcanic front was shallower than in the previous period (Fig. 20b: Yoshida *et al.* 1995).

Isotopic variation of the source mantle and temporal development. Back-arc basin volcanism in the Yamato Basin started with the eruption of isotopically enriched tholeiites during 21–17 Ma (LD-group in Fig. 21a: Nohda 2009), and was followed by isotopically depleted tholeiites during 17–15 Ma (D-group in Fig. 21a: Nohda 2009). The same drastic change in Sr and Nd isotope ratios from enriched to depleted compositions, by the injection of a depleted asthenospheric mantle into an enriched continental lithosphere, was observed in: SikhoteAlin at around 35 Ma (Nohda 2009); the Oga Peninsula at 22 Ma (Kurasawa & Konda 1986); the eastern margin of the Japan Sea (Aosawa rift) at 16.5 Ma (Ujike & Tsuchiya 1993; Yagi *et al.* 2001); and the inland NE Japan (Kuroko rift) at 15 Ma (Yagi *et al.* 2001; Nohda 2009; Yamada *et al.* 2012). Nohda (2009) explained this relationship by an eastward asthenospheric flow during the continental arc to the back-arc basin periods (Fig. 19a). The back-arc basin opening ceased at around 15 Ma, and subsidence of the back-arc region began at 14.5 Ma, perhaps owing to cooling of the injected hot mantle (Yamaji 1990).

In contrast, there was little to no change in the Sr and Nd isotope ratios at the volcanic front, and enriched magmas erupted throughout the back-arc basin period, although the Hf isotope ratios appear to have become more radiogenic with time (Nohda *et al.* 1988; Tatsumi *et al.* 1988; Shuto *et al.* 1993; Hanyu *et al.* 2006). The 23 Ma high-Mg andesite at Choshi has the lowest ε_{Hf} values, while the 5 Ma Mitaki and Quaternary Iwate have the highest ε_{Hf} values, and the 16–12 Ma Ryozen, Takadate and Tomari volcanics show intermediate ε_{Hf} values (Hanyu *et al.* 2006). Hanyu *et al.* (2006) concluded that the high ε_{Hf} value of late-stage volcanism at the volcanic front can be explained by the addition of fluids from subducted sediments and the oceanic crust to the mantle wedge.

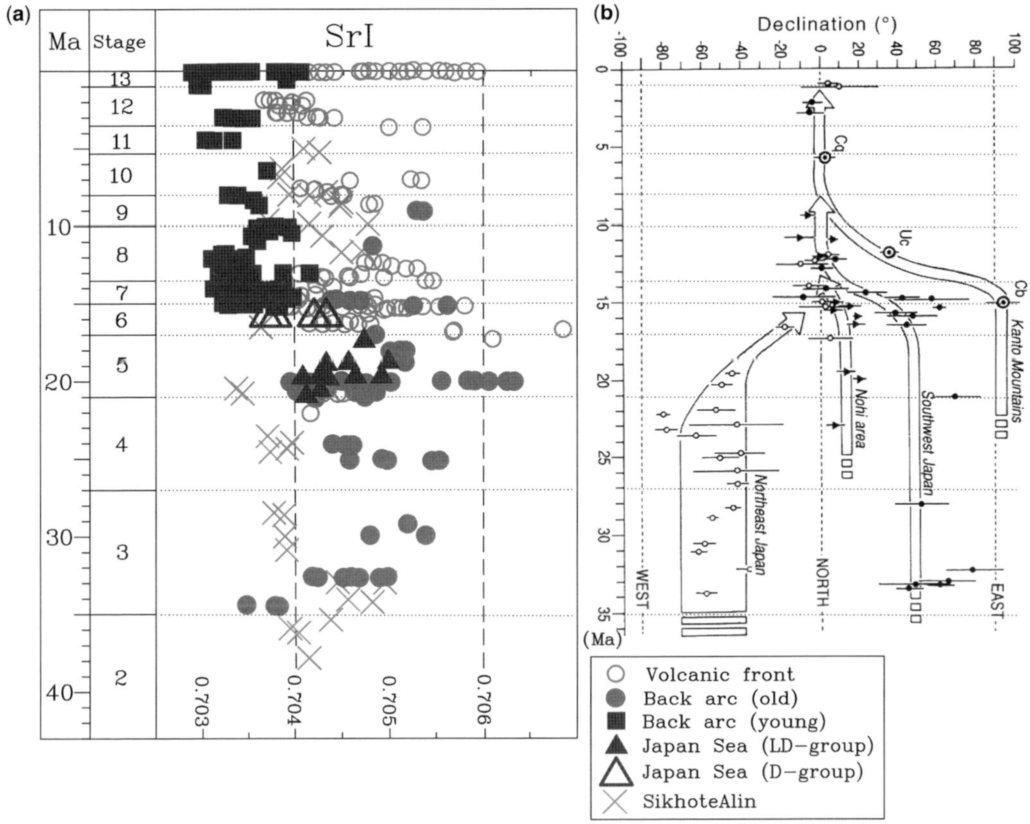

Fig. 21. (**a**) Temporal changes in initial Sr isotope ratios (SrI) in the NE Japan arc (Nohda & Wasserburg 1986; Nohda 2009; Tatsumi *et al.* 1988; Shuto *et al.* 1993, 2006; Shuto 2009). Nohda (2009) assigned a D-group to Japan Sea basalts with high $^{143}Nd/^{144}Nd$ ratios of around 0.5131, and an LD-group with lower $^{143}Nd/^{144}Nd$ ratios from 0.5130 to 0.5126. (**b**) Palaeomagnetic declination values v. age, for the Kanto Mountains, SW Japan, Nohi area and NE Japan (Takahashi 1994; Takahashi & Saito 1997). See Figure 9b for the localities. Cq, Chichibu Quatz Diorite; Uc, Uchiyama area; Cb, Chichibu Basin (Takahashi & Saito 1997). The secular change in Nd–Sr isotopic composition of the back-arc basalts has been interpreted in terms of the eastward asthenospheric flow (Nohda 2009). The timing of the remarkable SrI drop in 17–15 Ma in the back-arc region (Shuto *et al.* 1993, 2006; Shuto 2009; Nohda 2009) is consistent with the timing of rotational crustal movements associated with the back-arc basin opening in the NE Japan arc, inferred from palaeomagnetic studies (Otofuji & Matsuda 1983, 1984; Otofuji *et al.* 1985; Nohda 2009).

Thermal subsidence with lithospheric thickening. The change from extensional to neutral tectonics was associated with a rapid decrease in erupted volumes (Fig. 6a). This stage (Stage 8: oceanic island-arc stage) has been regarded as being a thermal subsidence stage under a neutral stress field (Yamaji & Sato 1989; Sato 1992, 1994). The eruption of alkali basalts in the back-arc side during this stage is generally explained by the increasing depths of magma segregation, and the decreasing degree of partial melting of parental magmas (Tatsumi *et al.* 1983, 1994; Sato & Yoshida 1993; Yoshida *et al.* 1995). The mantle isotherm steepened in the back-arc side during Stage 8. As a result, just after the end of the back-arc basin period, the thermal structure in the uppermost mantle wedge became very similar to the present one. The transition from a back-arc basin to an island-arc setting is explained by the lithospheric thickening beneath the back-arc side of the NE Honshu arc (Fig. 19c: Sato & Yoshida 1993; Tatsumi *et al.* 1994, Yoshida *et al.* 1995), owing to a temperature drop associated with a decrease in erupted volumes when the extension stopped (Yamada & Yoshida 2011). It is possible that the cooling of the back-arc upper mantle was mainly the result of a large volume of magmatic effusion in the back-arc basins. As a result of lithospheric thickening in the back-arc side, the magmatic composition changed from enriched low-*K* tholeiites to depleted

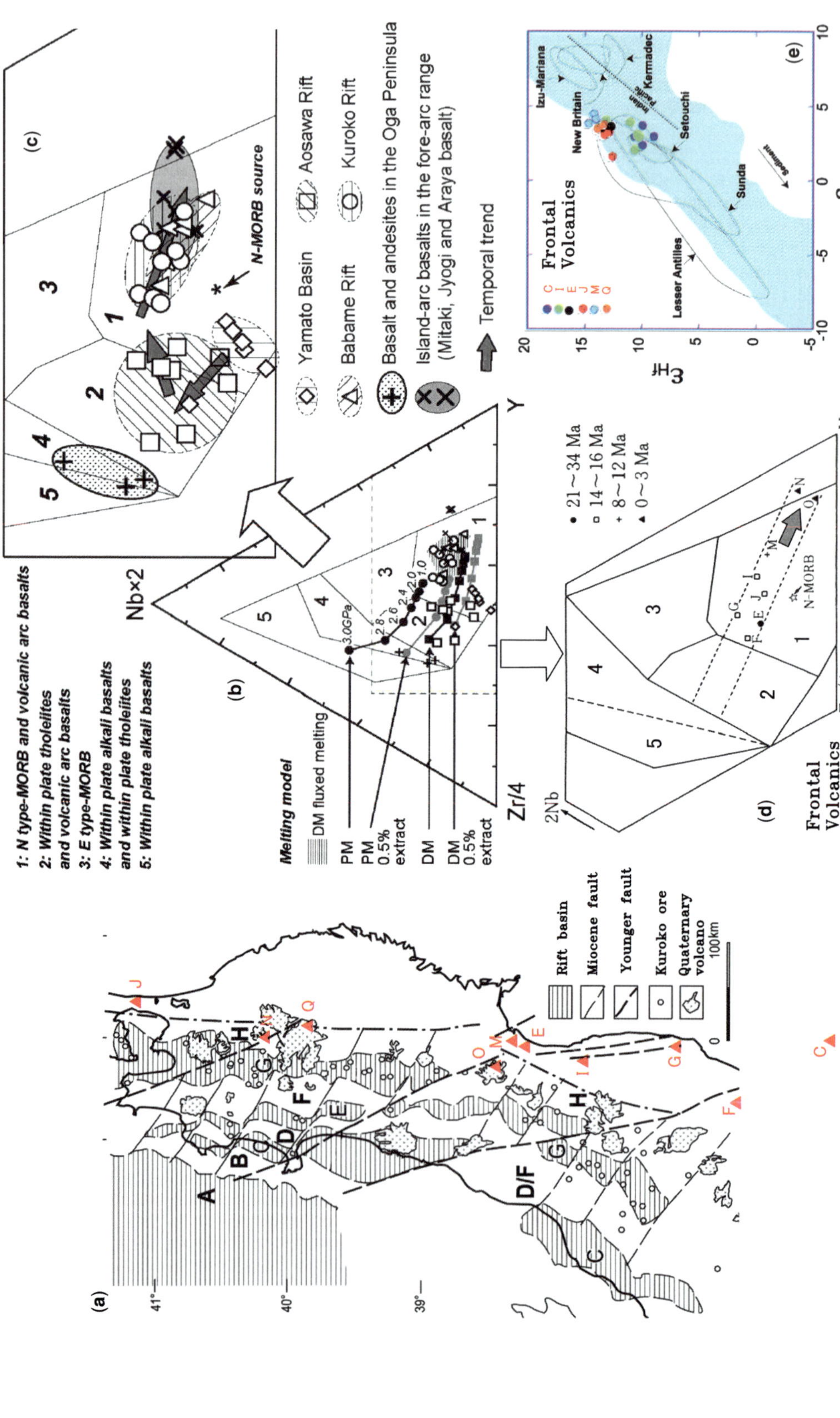

(a)
Rift basin
Miocene fault
Younger fault
Kuroko ore
Quaternary volcano
100km
1: N type-MORB and volcanic arc basalts
2: Within plate tholeiites and volcanic arc basalts
3: E type-MORB
4: Within plate alkali basalts and within plate tholeiites
5: Within plate alkali basalts
Melting model
DM fluxed melting
PM
PM 0.5% extract
DM
DM 0.5% extract
(b)
Zr/4
Nb×2
Y
(c)
N-MORB source
Yamato Basin
Aosawa Rift
Babame Rift
Kuroko Rift
Basalt and andesites in the Oga Peninsula
Island-arc basalts in the fore-arc range (Mitaki, Jyogi and Araya basalt)
Temporal trend
(d)
Frontal Volcanics
2Nb
Zr/4
Y
N-MORB
21～34 Ma
14～16 Ma
8～12 Ma
0～3 Ma
(e)
Frontal Volcanics
Izu-Mariana
New Britain
Kermadec
Indian
Pacific
Setouchi
Sunda
Lesser Antilles
Sediment
ε_{Hf}
ε_{Nd}

alkali basalts between 13.5 and 10 Ma (Fig. 21a). In the volcanic front, the magma composition evolved from high high field strength element (HFSE) magmas in the back-arc basin period to low HFSE magmas in the island-arc period, together with an increase in ε_{Hf} value (Shuto *et al.* 1992; Hanyu *et al.* 2006). Felsic magma chambers became shallower with a change from high-temperature aphyric- or plagioclase–phyric rhyolites to low-temperature plagioclase–quartz phyric rhyolites during the tectonic change from a back-arc basin to an island-arc setting (Yamada & Yoshida 2004, 2011). It is possible that this petrographical change within the felsic magmas was related to a shallowing of the emplacement level of felsic magma reservoirs from the deep crust to the shallower crust.

Temporal evolution of basaltic magma in the NE Japan arc. The chemical composition of late Cenozoic volcanic rocks from the NE Japan arc suggests the presence of three types of mantle sources: the subcontinental enriched mantle; the sub-arc lithospheric-enriched mantle (Ujike & Tsuchiya 1993); and the asthenospheric-depleted mantle (Nohda *et al.* 1988; Tatsumi *et al.* 1989; Ohki *et al.* 1994; Shuto *et al.* 1995, 2006; Yoshida *et al.* 1995). Ujike & Tsuchiya (1993) showed that the geochemical signatures of the magmas, including the Sr and Nd isotope ratios and Zr/Y ratios, changed drastically from a continental margin type to an oceanic arc type (Pearce 1983) in the Aosawa rift at around 16 Ma. Shuto *et al.* (1995) also showed that the late Cenozoic basaltic rocks in the NE Japan arc exhibit a systematic decrease in Zr/Y ratios with age (from within-plate basalts to volcanic arc basalts), from 35 Ma to the present. Yamada *et al.* (2012) showed the temporal evolution of basaltic magmas on the Nb–Zr–Y discrimination plot of basalts (Fig. 22b, c: Meschede 1986), and revealed that basalts from the Oga Peninsula of the continental margin period plot in the field of within-plate alkali basalts. After the opening of the Sea of Japan, the basaltic magmas evolved from low-Nb N-type MORB basalts (Yamato Basin) to within-plate tholeiite basalts (Aosawa Rift), and ultimately to volcanic arc basalts (Babame Rift and younger volcanic rocks).

To examine the source material and partial melting of the source material controlling the HFSE enrichments, Yamada *et al.* (2012) calculated the nature of the changes in the Zr–Nb–Y compositions of basalt melts during the adiabatic ascent of a primitive mantle (PM: Sun & McDonough 1989) and a depleted mantle source (DM: Workman & Hart 2005). Model calculations of adiabatic decompression melting (Phipps-Morgan 1999) were performed at mantle potential temperature $T_p = 1450\ ^\circ C$, with the mineralogical mode during ascent and melting based on Ghiorso *et al.* (2002). Slightly depleted PM and DM source compositions were also calculated using the ABS3 model (Kimura *et al.* 2010), assuming the extraction of a 0.5% basalt melt from PM and DM sources. The results are shown in Figure 22b.

The basalts from the PM source mantle change from within-plate alkali basalt to within-plate tholeiites and then from E-type MORB to N-type MORB, with decreasing depth of magma segregation. Similarly, the basalts from the DM source mantle follow the field of within-plate tholeiite to N-type MORB (Fig. 22b). The Oligocene Oga alkalic within-plate basalts of the continental-margin period, which have an isotopically enriched mantle source, are interpreted to have originated from the continental lithospheric mantle, and deep-level melting of a thick, enriched lithospheric

Fig. 22. (**a**) Distributions of the Miocene to Quaternary frontal basaltic volcanic rocks are shown by red triangles on the tectonic framework of the Miocene NE Japan arc, after Yamada & Yoshida (2011). J, Tomari; N, Araya; Q, Iwate; O, Funagata; M, Mitaki; E, Takadate; I, Ryozen; G, Ishimoriyama; F, Motokozawa; C, Choshi. (**b, c**) Nb–Zr–Y plot of basalts from: the Yamato Basin (area A in (a)); the Oga Ridge (area B); multiple stages of the back-arc rifting (areas C–G); and the frontal area (area H), modified after Yamada & Yoshida (2011) and Yamada *et al.* (2012). The discrimination fields are from Meschede (1986), and the N-MORB source is from Wood *et al.* (1979). Additional data are from Allan & Gorton (1992), Shuto *et al.* (1995) and Yagi *et al.* (2001). Yamada *et al.* (2012) calculated the evolution of the Zr–Nb–Y composition of basalt melts during the adiabatic ascent of a primitive mantle (PM: Sun & McDonough 1989) and a depleted mantle (DM: Workman & Hart 2005) source. They also considered slightly depleted PM and DM sources, calculated assuming the extraction of a 0.5% basalt melt from these sources, and plotted the calculated results of the evolutional melt path using the Arc Basalt Simulator model by Kimura *et al.* (2010) in (b). (**d**) Nb–Zr–Y variations of the Miocene to Quaternary frontal basaltic volcanic rocks, modified after Shuto & Gorai (1997). The discrimination fields are the same as in (b). (**e**) Nd and Hf isotope compositions of the Miocene to Quaternary frontal basaltic volcanic rocks of the NE Japan arc, modified after Hanyu *et al.* (2006). The mantle array defined by MORB, OIB and sediments is shown as the light-blue shaded area (Salters & White 1998; Vervoort *et al.* 1999). The other arc data and dividing line of the Pacific–Indian MORB fields are from Pearce *et al.* (1999), Woodhead *et al.* (2001) and Hanyu *et al.* (2002). Choshi samples (C) and some of the Ryozen samples (I) are plotted in the mantle array. However, the other samples, particularly the younger samples from Iwate (*Q*) and Mitaki (M), are plotted on the upper side of the mantle array (Hanyu *et al.* 2006).

mantle (see the 3.0–2.8 GPa symbols of PM and the 0.5%-melt-extracted PM plots in Fig. 22b) can form such alkaline basalts. In contrast, the N-type MORB of the Yamato Basin and the Aosawa within-plate tholeiite may have originated from a common DM or a 0.5%-melt-extracted DM source mantle, with shallower depths for the Yamato Basin basalts (Fig. 19a) and greater depths for the Aosawa Rift basalts (Fig. 19b).

Volcanic arc basalts from the Babame and Kuroko rifts, which lie between the DM and PM models, could originate from a shallow depth. The elevated Nb values may indicate a more fertile source mantle than that of the Yamato Basin and Aosawa Rift basalts. The most radiogenic Nd isotopic composition from the Yamato Basin basalts yields $^{143}Nd/^{144}Nd = 0.51316$ (Cousens & Allan 1992), which is comparable to values obtained for Quaternary basalts of the back-arc side, but higher than those obtained for basalts of the Babame and Kuroko rifts, suggesting a relatively fertile mantle source. It is also possible that the addition of slab melt components to the shallow (2–1 GPa) DM source mantle could result in an increased Nb value (shaded area in Fig. 22b). This fluxed melting calculation was performed using the ABS2 model (Kimura *et al.* 2010). Yamada *et al.* (2012) explained the fertile shallow mantle source of basalts for the Babame and Kuroko rifts, by a shallow DM melting induced by slab flux from dehydration of the Pacific Plate slab. This is because the tectonic setting of the Babame and Kuroko rifts was in a time of transition from the late stage of back-arc basin opening to the island-arc stage, and the compositions of basalts of the early island-arc stage are almost identical to those of the Babame and Kuroko rifts (see Fig. 22c).

Palaeomagnetic studies and rotational history. Takahashi & Saito (1997) reviewed palaeomagnetic studies and discussed Miocene intra-arc bending at the arc–arc collision zone in central Japan.

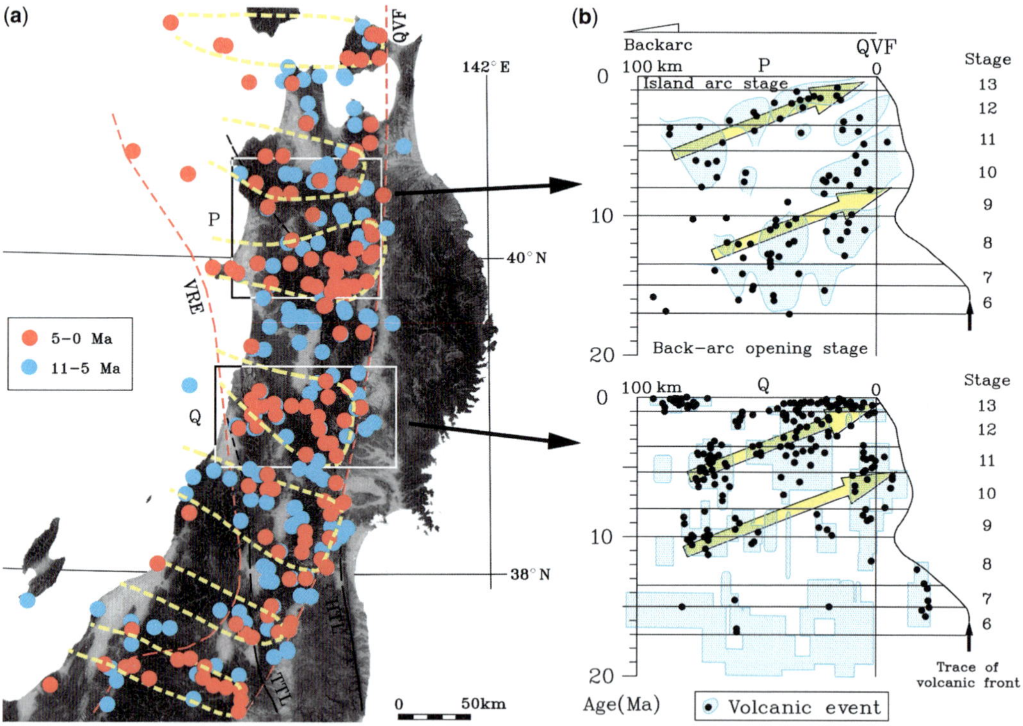

Fig. 23. (**a**) Past distribution of volcanoes in the NE Japan arc (data from Kondo *et al.* 1998). Yellow dashed lines are the surface projections of 'hot fingers' by Tamura *et al.* (2002). The data are divided into two periods: 11–5 and 5–0 Ma (modified from Honda & Yoshida 2005*a*). QVF, VRE, TTL and HTL are as in Figure 2. (**b**) Map of volcanic ages v. distance from the Quaternary volcanic front, adapted from Honda & Yoshida (2005*a*). The age data for area P (Yamada & Yoshida 2002) and area Q (Kondo *et al.* 2004) are plotted against the distance from the volcanic front to the sampling sites. Black circles are the radiometric ages, and the pale-blue-dotted hatched zone is the age ranges estimated from stratigraphic data. Each arrow shows the possible migration of volcanism from back-arc to the volcanic front side.

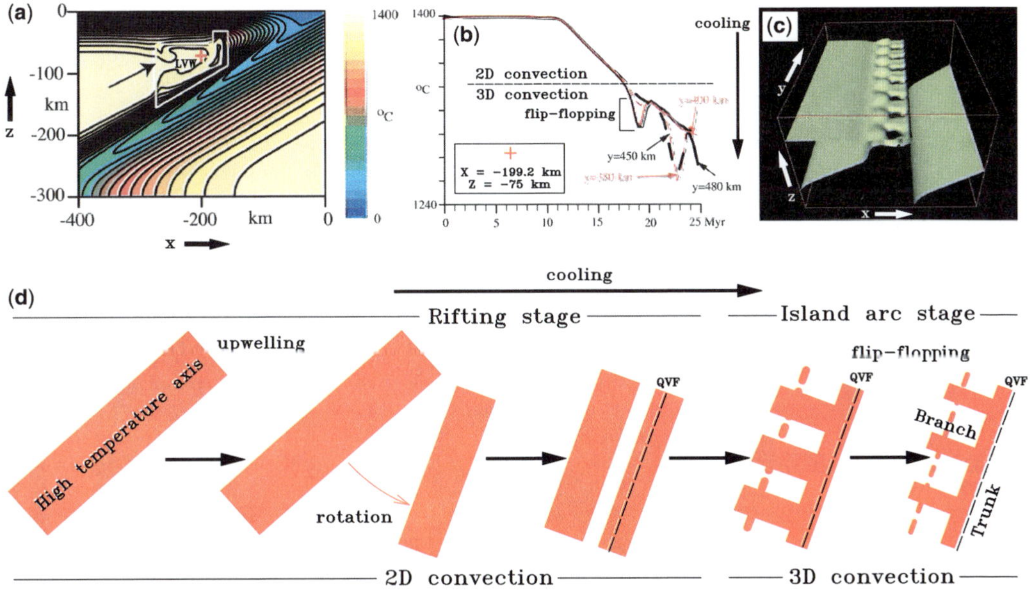

Fig. 24. (**a**) A snapshot of temperature for the small-scale convection model under the NE Japan arc, by Honda & Yoshida (2005*a*). The region within the white line is the low-viscosity wedge (LVW). The contour intervals of temperature are 70°. Arrow shows the positions of migrating cold plumes generated at the left end of the LVW. (**b**) Temporal variation of temperature at the point in the LVW. The position of point (*x*, *y* and *z*) is indicated in the 2-D section (a) by the red cross ($x = -199.2$ km, $z = -75$ km). The results reveal a temperature oscillation associated with the passage of cold plumes. The oscillation overlaps with the secular cooling of the temperature within the LVW. This figure shows the temporal variation of temperature at several places along fixed *x* and *z*. These places roughly correspond to the axes of each finger (Fig. 23a). As evident in this figure, the flip-flopping starts at around 17 Ma. (**c**) A 3D representation of the isothermal surface of temperature. This result shows the almost continuous thermal anomalies at the right edge of the LVW, which are located under the volcanic front, and the finger-like thermal anomalies. (**d**) Evolutional series of small convection from 2D to 3D at the NE Japan arc with cooling of mantle wedge. QVF, Quaternary volcanic front.

Palaeomagnetic results reveal the Miocene rotational history in central, SW and NE Japan (Fig. 21b: Takahashi & Saito 1997). As shown in Figure 21b, differential rotations in the Japanese islands took place between 20 and 10 Ma. The clockwise rotation of SW Japan occurred at *c.* 15 Ma, while the Kanto Mountains (K in Fig. 9b) rotated during the Middle Miocene (15–10 Ma). The small declinations in the Nohi area (N in Fig. 9b) indicate that differential rotation between SW Japan and the Nohi area occurred at *c.* 15 Ma. However, the palaeomagnetic deflections of NE Japan show a counter-clockwise direction during the Early Miocene. The major rotations of the SW and NE Japan arcs ceased at 14 Ma, while the Kanto Mountains rotated until the Late Miocene (Fig. 21b). The major change occurred in the stress axis in the NE Japan arc from a NNW–SSE to a NE–SW direction (Fig. 7) around 15 Ma in association with the rapid rotation of SW Japan. It is likely that this rapid rotation locked the NE Japan block to the Kuril forearc sliver, as a collision of the Kuril forearc sliver and the NE Japan arc had begun at this time (Kimura 1986; Acocella *et al.* 2008). The events at 15 Ma are consistent with the change in shear movement along the TTL, from sinistral (the back-arc opening stage) to dextral (the forearc sliver collision stage; Awaji *et al.* 2006; Yoshida 2009).

Mantle dynamics and arc magmatism

The origin of geochemical variations in volcanic rocks in late Cenozoic NE Japan arc has been discussed in the preceding sections of this paper. It is evident that the most significant event was the opening of the back-arc basin and the subsequent subduction volcanism in the island-arc period. These changes are closely related to the mantle structure, including that of: (1) the upwelling-subsidence of the asthenosphere during the back-arc basin period; and (2) the mantle wedge corner flow in the island-arc period. Such a temporal development of the mantle has been investigated by

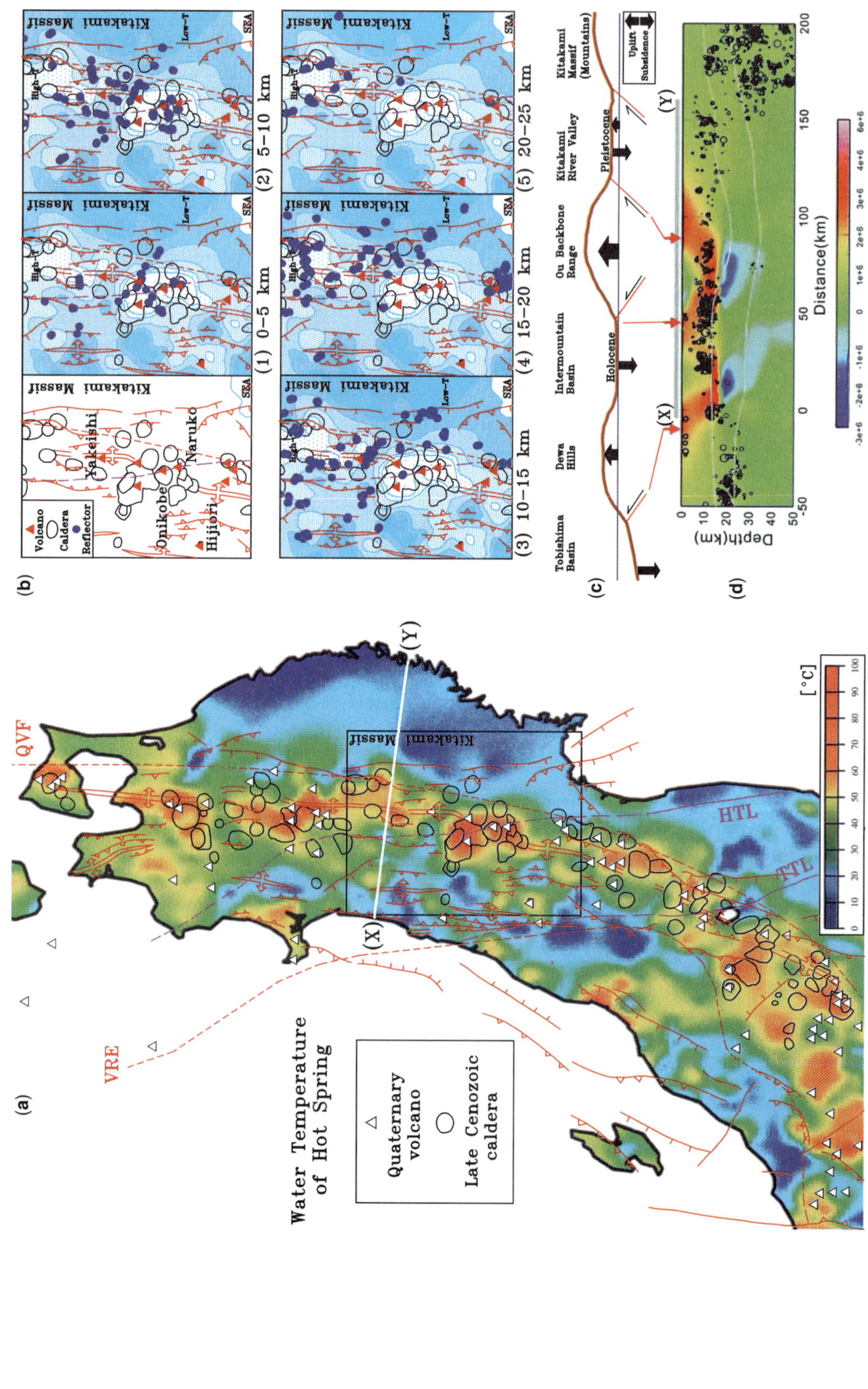
(a)
Water Temperature of Hot Spring
Quaternary volcano
Late Cenozoic caldera
VRE
QVF
Kitakami Massif
(X)
(Y)
HTL
TTL
[°C]
(b)
Volcano
Caldera
Reflector
Yakeishi
Onikobe
Naruko
Hijiori
Kitakami Massif
High-T
Low-T
SEA
(1) 0–5 km
(2) 5–10 km
(3) 10–15 km
(4) 15–20 km
(5) 20–25 km
(c)
Tobishima Basin
Dewa Hills
Intermountain Basin
Ou Backbone Range
Kitakami River Valley
Kitakami Massif (Mountains)
Holocene
Pleistocene
Uplift
Subsidence
(d)
Depth(km)
Distance(km)

several geodynamic models (Honda & Yoshida 2005*a*, *b*; Honda *et al.* 2007; Zhu *et al.* 2009).

Spatio-temporal variations of volcanisms and mantle convection. Volcanoes were distributed uniformly immediately after the cessation of the opening of the Japan Sea, but, in the island-arc period, they had a tendency to be concentrated and finally to form branches and a trunk structure (Figs 18b & 23a; Yoshida *et al.* 1995; Kondo *et al.* 1998, 2004; Yamada & Yoshida 2002; Prima *et al.* 2006). The trunk structure consists of volcanoes located along the volcanic front, and the branches coincide almost with the 'hot fingers' (Tamura *et al.* 2002). Kondo *et al.* (1998) concluded that this configuration of volcano alignment started at *c.* 8 Ma.

Honda & Yoshida (2005*a*) performed a geodynamic simulation of the mantle wedge and examined the spatial and temporal evolution of volcanism. When the ages of the volcanoes are plotted against the distance from the volcanic front (Yamada & Yoshida 2002; Kondo *et al.* 2004), the spatio-temporal plots show the migration of volcanism from back-arc to volcanic front (Fig. 23b: Honda & Yoshida 2005*a*). The most recent migration (5–0 Ma) suggests a velocity of migration of *c.* 2 cm a^{-1} (Honda & Yoshida 2005*a*), which may be correlated with dynamic movement in the mantle wedge.

Tamura *et al.* (2002) explained the distribution pattern of Quaternary volcanoes in the NE Japan arc with a 'hot fingers' model, by assuming the low-velocity mantle wedge beneath the volcanoes to have high-temperature anomalies. Honda & Yoshida (2005*a*), however, suggested that the distribution pattern of the volcanoes was not preserved over a long period, and that the finger-like structure flip-flopped at around 5 Ma. These phenomena have been explained by the existence of small-scale convections beneath the arc.

Alternatively, Zhu *et al.* (2009) simulated the 3D dynamics of hydrous thermo-chemical plumes in the oceanic subduction zones, and computed the spatial and temporal patterns of melt generation intensity above the slab, then compared results with data from the NE Japan arc. This suggested that the clustering of volcanic activity could potentially be related to the activity of thermal–chemical plumes in the mantle wedge. Even though the models vary, the important insights gained by the geodynamic model perspectives are that: (1) the mantle convection patterns can fundamentally control the magma genesis; and thus (2) control the distribution of the volcanic activity and magma production rate.

Temporal evolution of mantle convection. Temporal variations in the dynamics of the continuously cooling low-viscosity mantle wedge have been examined (Fig. 24a–c: Honda & Yoshida 2005*a*). It is

Fig. 25. (**a**) Distribution of late Cenozoic calderas and Quaternary volcanoes (Yoshida *et al.* 1999*a*) superimposed on coloured map of water temperatures of hot spring (modified from Yano *et al.* 1999). Red and blue colours denote the high and low temperatures, respectively. The temperature scale is shown at the bottom. QVF, VRE, TTL and HTL are as in Figure 2. Legends for the other symbols are shown in the upper-right box of Figure 1. The section X–Y and rectangle in (a) show the study section in (c) and (d), and the study area in (b), respectively. (**b**) Location of reflection points (blue circles) of each *S*-wave reflector in the depth ranges of: (1) 0–5 km; (2) 5–10 km; (3) 10–15 km; (4) 15–20 km; and (5) 20–25 km (Hori *et al.* 2004), at around the Onikobe caldera swarm superimposed on the map of thermal water temperatures (Yano *et al.* 1999) and modified from Yoshida *et al.* (2005). *S*-wave reflectors distributed beneath and around the caldera complex are presumed to be fracture systems filled with high temperature hydrothermal fluids. The distribution pattern of *S*-wave reflectors changes with depth, suggesting that it is controlled by the distribution of unexposed plutons under the caldera structures and the related fracture system. It may also be related to the crustal thermal structure and non-symmetrical updoming of the Ou Backbone Range. (**c**) Relationship between geomorphology, major fault systems and recent crustal movements (modified from Imaizumi 1999). Recent tectonic movements have been strongly controlled by a reverse reactivation of the Miocene basin-bounding normal faults under WNW–ESE (to east–west) compression, owing to the westward subduction of the Pacific plate since the late Pliocene (Nakamura 1992; Sato 1994; Sato *et al.* 2002; Kato *et al.* 2004). Black arrows indicate the uplift and subsidence of the each tectonic unit, estimated from geological and geomorphological data (Japan Association for Quaternary Research 1987; Imaizumi 1999; Tajikara & Ikeda 2004). (**d**) Accumulated stress (Pa) distribution after 3000 years using a finite element analysis for an X–Y section of the NE Japan arc (Shibazaki *et al.* 2008). Red and blue colours denote high and low stress concentrations. The scale (Pa) is shown at the bottom. W-shaped stress concentration (shear faulting) zones in the upper crust are created along the locations of major fault zones by the non-uniform thermal structure of the crust and uppermost mantle beneath the NE Japan arc. The W-shaped stress concentration zones arise from a shortening deformation associated with nonlinear viscous flow in the two low-viscosity regions (hot regions) beneath the Ou Backbone Range and the Dewa Hills, surrounded by the high-viscosity region in the lower crust and the uppermost mantle (see Fig. 15b). The solid grey lines indicate the inland area. Upper and lower white lines indicate the boundaries between the upper crust and the lower crust, and the Moho discontinuity, respectively. Red arrows indicate the locations of the three major active fault zones at the surface. Hypocentres of earthquakes (1 January 2003 to 31 December 2005) are also shown in the section by black circles (Shibazaki *et al.* 2008).

evident that three distinctive phases exist during cooling, including: (1) phase 1 – a steady state convection; (2) phase 2 – a constant cooling with 2D convection; and (3) phase 3 – an unstable 3D convection (Honda & Yoshida 2005*a*). The first two phases form a homogeneous thermal structure along the arc, but the final phase forms a trunk-and-branches structure (Fig. 24). The volcano distribution in the back-arc basin period of the NE Japan arc was in alignment with arc-parallel rift systems, as noted above, and thus would have related to the 2D convection. In contrast, the heterogeneous distribution of volcanoes in the late island-arc period may have related to the 3D convection (Fig. 24d).

The long-term developmental history of the mantle wedge could have been constrained by the surface manifestation of the volcanic activities, although the role of the deep-seated slab and the island-arc crust should also be considered as the driving force of the movement and production of disturbing effects to the ascending magmas, respectively. Detailed studies of erupted magmas would contribute essential knowledge towards this type of study, and would provide knowledge of additional strong constraints involved in the deep processes.

Island-arc evolution and arc magmatism

An important tectonic change associated with the change in magmatic activity occurred in the back-arc basin period and was then followed by the island-arc period (Sato 1994; Jolivet *et al.* 1994; Nakajima *et al.* 2006*c*; Yoshida 2009). Both the surface tectonic and deep mantle source conditions in the two periods were different, as noted above. The bimodal magmatism continued to the early island-arc period, and was followed by two stages of caldera-dominated magmatism, and lastly by andesite-dominated magmatism. This implies that andesite-dominated arc volcanism has been limited to the last 3.5 myr.

Tectono-magmatic evolution in the island-arc period

The island-arc period after 13.5 Ma can be divided into four phases of igneous activity: (1) the formation of oceanic island chains with submarine volcanism (13.5–8 Ma); (2) a late Miocene caldera-forming phase with weak updoming of the Ou Backbone range (8–5.3 Ma); (3) a Pliocene caldera-forming phase under weak compression (5.3–1.7 Ma); and (4) the creation of a late Quaternary highly compressional volcanic arc with andesitic stratovolcanoes (1.7–0 Ma). Nakajima *et al.* (2006*c*) argued that there were three phases of uplift in the surrounding the Ou Backbone range after the rapid subsidence (Aosawa and Kuroko rifting) in the early Middle Miocene: (1) the temporary and sudden uplift of the eastern sector of the Ou Backbone range at 12 Ma, followed by a slow deposition within the basin between 12 and 9 Ma (the stage of submarine felsic volcanism); (2) a differential uplift of the Ou Backbone range between 6.5 and 3 Ma (two caldera-forming stages); and (3) an uplift of the whole Ou Backbone range by thrusting after 3 Ma (the stage of strong compression).

Caldera-dominated magmatisms

The uplift of the present Ou Backbone Range began in the island-arc period at 10 Ma, and was associated with an increase of horizontal compression (Nakajima *et al.* 2006*c*). Between 8 and 1.7 Ma, active felsic volcanism created more than 80 calderas associated with subordinate andesites to basalts in the NE Japan arc (Figs 2 & 25a; Ito *et al.* 1989; Sato & Yoshida 1993; Sato 1994; Yoshida *et al.* 1999*a*; Prima *et al.* 2012). There are two peaks in caldera formation interrupted by a short period of dormancy at 5–4 Ma, which is related to a short transgression, and the number and size of collapse calderas decreased from the late Miocene to the Pliocene (Fig. 8a).

The calderas have an average diameter of about 10 km and an average aspect ratio of 1.24 in diameter. They are divided into three groups related to their diameter size (about 5, 10 and over 14 km), and are mainly classified into piston-cylinder type with subordinate funnel type. The spatial and size distributions of calderas are comparable with those of the Cretaceous granitic plutons from the Kitakami Mountainland in NE Honshu (Yoshida *et al.* 1999*a*). The collapse of such calderas would have formed in a neutral to weakly compressive stress field (Yoshida *et al.* 1993; Sato & Yoshida 1993; Sato 1994), and this would have resulted in the rise of felsic magmas into the crust where large intra-crustal magma chambers were formed (Aizawa & Yoshida 2000; Aizawa *et al.* 2006).

NE–SW compression to ENE–WSW (east–west) compression. It has been argued that the regional stress field controlled the volcanic activity in the NE Japan arc (Sato & Yoshida 1993; Yoshida *et al.* 1993, 1997, 1999*a*; Acocella *et al.* 2008), and that basaltic magmas derived from the mantle wedge, underplated and stagnated near the Moho, which acts as a density barrier (e.g. Ryan 1987; Takada 1989). These magmas would then have fractionated, re-mobilized or re-melted from solidified precursors or the pre-existing arc crust, to form the felsic magmas in the inland area of the NE Japan arc, with a thick crust (Fig. 19c; Sato & Yoshida 1993). Such an event is confirmed by the existence of large felsic effusives in the eastern margin of the back-arc basin rift system (Figs 4 & 5; Yamada *et al.* 2012). During the neutral stress condition between 13.5 and 10 Ma, the felsic magmas would have risen diapirically through the ductile lower crust owing to their buoyancy (Aizawa & Yoshida 2000; Aizawa *et al.* 2006), and the mode of ascent would have changed in the brittle upper crust to dyke or sheet (Fig. 15b). An increase in the compressional stress field occurred between 10 and 8 Ma, and it is likely that this increase led to the formation of sills and laccolithic shallow reservoirs in the upper crust (Fig. 19c: Sato & Yoshida 1993; Aizawa & Yoshida 2000). The regional change in the stress field was, therefore, the major control of caldera-dominated volcanism with laccolithic shallow reservoirs that occurred in the earlier half of the island-arc period. Felsic magma at this level could then have intruded along subsurface low-angle thrust sheets (Fig. 11b), and it is possible that magma migration along the thrust sheets caused the uplift of the Ou Backbone Range (Yoshida *et al.* 1993; Sato & Yoshida 1993; Sato 1994).

The clockwise rotation of SW Honshu (Otofuji & Matsuda 1983) and the collision with the Kuril forearc sliver (Kimura 1986) caused an oblique (NE–SW-trending) compression of the NE Japan arc during the Mio-Pliocene, and triggered felsic magmatism along the areas of localized extension (Acocella *et al.* 2008). After about 5 Ma, the Pacific plate accelerated (Pollitz 1986). Pollitz (1986) suggested that the change in Pacific plate motion introduced a large component of compression normal to the Japan trench. This ENE–WSW (east–west) strong compression closed the caldera-feeding systems and favoured the development of stratovolcanoes with deeper magma plumbing systems directly connected to the basaltic mantle source region. This strong coupling between the Pacific plate slab and the overriding NE Japan arc crust has continued since 3.5 Ma, perhaps with an occasional east–west stress release as envisaged by the Mw 9.0 Tohoku–Oki earthquake, which occurred on 11 March 2011 (Iio & Matsuzawa 2012), and has returned to the original NE–SW compression (Acocella *et al.* 2008; Kosuga *et al.* 2012). During the Quaternary, more andesites erupted, mainly from along-arc and east–west-aligned stratovolcanoes (Fig. 6c). The north–south thrust faults and the subordinate east–west-aligned volcanic structures produced by the east–west compression may reflect the convergence vector of the Pacific plate (Acocella *et al.* 2008).

Subcaldera plutons and S-wave reflectors. Many *S*-wave reflectors have been found along the caldera rims or around caldera complexes (Figs 13b, c & 25b) in the wide area of the NE Japan arc (Umino *et al.* 2002; Hori *et al.* 2004; Yoshida *et al.* 2005), although some of these may be related to active faults (Hori *et al.* 2004). These *S*-wave reflectors are presumed to reflect fracture systems filled with hydrothermal fluids, and seem to be generated from cooled magma reservoirs near wet solidus beneath the calderas (Fig. 25b). The shallow reflectors at depths of 0–5 km around the calderas are nearly vertical and parallel to the caldera walls, and are mainly distributed inside the caldera complexes. A number of reflectors at a depth of 5–10 km, with a gentle or flat dip, have also been found along the caldera rims or around caldera complexes. At a depth of 10–20 km, the reflectors fade away from the area of the large caldera complexes and are found on the northern to eastern sides of the caldera complexes. Those deeper than 20 km are located in an area beneath the caldera complexes, and the number of reflectors has decreased. The depth distribution of *S*-wave reflectors has two peak depths at 6 –11 and 16 km (Hori *et al.* 2004). Figure 13b, c clearly shows that many reflectors are derived from the basic part of the middle crustal magma reservoirs at around the depth condition of the wet solidus of basaltic magma (Fig. 15a) to the eastern side of reservoir. The changes in the distribution of the reflectors with depth in the wide area of the NE Japan arc could be explained by the non-symmetrical cone and ring fractures formed around the cooling magmatic reservoirs (Anderson 1936). It is possible that low angle non-symmetrical cone fractures might have formed around subcaldera middle crustal plutons in a horizontally compressional stress field. At shallow depths, pre-existing caldera structures with ring fractures would then be used as conduits of hydrothermal fluids.

Change of intra-crustal thermal structure. The intra-crustal thermal structure was affected by the intensive intrusion of magma into the crust, which was in turn controlled by the mantle thermal structure (e.g. Honda & Yoshida 2005*a*), as well as by the magma storage systems related with large calderas formation (Fig. 19c). The crustal thermal anomalies are seismologically detectable (Figs 11, 13 & 14: Sato *et al.* 2002; Nakajima *et al.* 2006*a*; Xia *et al.* 2007). The distribution of thermal waters in NE Japan (Yano *et al.* 1999) coincides well with the dense distribution of late Cenozoic calderas (Fig. 25a). Such evidence suggests that still-hot plutons, solidifying from felsic caldera-forming magma reservoirs, may exist beneath many of these calderas (Yamada 1988; Sato *et al.* 2002; Yoshida *et al.* 2005; Nakajima *et al.* 2006*a*; Takada & Fukushima 2013).

The thermal history of the NE Honshu arc, as inferred from petrological studies of the volcanic rocks (Ban *et al.* 1992; Yoshida *et al.* 1995) and studies of the thermal structure of forearc sedimentary basins (Yamaji 1994), shows that the geothermal gradient decreased from the Miocene (50–60 °C km^{-1}) to the Quaternary (20–30 °C km^{-1}). Miocene to Quaternary magma plumbing systems migrated to the back-arc side as the volcanic zone shrank, and the depths of the magma reservoirs increased from the late Miocene to the Quaternary, inducing the cooling of the shallower reservoirs. The regional westward retreat of the volcanic front (Fig. 3: Ohguchi *et al.* 1989; Yoshida *et al.* 1995), together with the geodynamic model by Honda & Yoshida (2005*a*), is consistent with the regional cooling of the NE Honshu arc crust in the island-arc period of the NE Japan arc.

Andesite-dominated magmatism

Transition in eruptive mode from felsic calderas to andesitic stratovolcanoes. In the last stage of the island-arc period, the horizontal stress increased considerably, and felsic caldera magmatism with subordinate andesitic to basaltic eruptions changed to andesite-dominated magmatism. In this stage, the mantle temperature had decreased by small-scale convection associated with mantle corner flow, induced by long-lived subduction (Honda & Yoshida 2005*a*). The crust also became more rigid as it cooled, although magmatism had converged to the volcanic frontal area (Fig. 23b).

The east–west-aligned Quaternary volcanoes are the shallowest expression of the east–west-trending hot mantle branches (Fig. 24d), suggesting that mantle–crust coupling is important for magma ascent (Acocella *et al.* 2008). Such coupling favours the extrusion of, and mixing between, deeper mafic and shallower felsic magmas (Ban & Yamamoto 2002; Ban *et al.* 2007; Hirotani *et al.* 2009). The strong compressional crustal stress then increased pressure in the deep magma reservoirs and accelerated the upwelling of the deeper mafic end-member of calc-alkaline andesites. As a result, the total erupted volume of mixed calc-alkaline andesitic magma was twice that of the felsic caldera forming stage with more ductile upper crust owing to the presence of many large shallow magma reservoirs.

Recent tectonic movements and major fault systems across the NE Japan arc. Recent tectonic movements have been strongly controlled by the reverse reactivation of the Miocene basin-bounding normal faults under a WNW–ESE (to east–west) compression, owing to the westward subduction of the Pacific plate since the late Pliocene (Nakamura 1992; Sato 1994; Sato *et al.* 2002; Kato *et al.* 2004). Figure 25c (modified from Imaizumi 1999) shows the relationship between geomorphology, major fault systems and recent crustal movements. Black arrows indicate the uplift and subsidence of the each tectonic unit estimated from geological and geomorphological data (Japan Association for Quaternary Research 1987; Imaizumi 1999; Tajikara & Ikeda 2004). The development of *S*-wave reflectors in the eastern side of the caldera complexes at a depth of 10–20 km (Figs 13b, c & 25b) suggests that a strong compression around caldera-forming middle crustal magma reservoirs caused underground thrust sheet fractures towards the trench side. The segregated magmatic fluids then moved along the sheet fractures, and this may have formed the large number of *S*-wave reflectors. Shibazaki *et al.* (2008) calculated the accumulated stress distribution using finite element analysis (Fig. 25d). W-shaped stress concentration (shear faulting) zones in the upper crust are created along the locations of the major fault zones by the non-uniform thermal structure of the crust and the uppermost mantle beneath the NE Japan arc. The W-shaped stress concentration zones arise from the shortening deformation associated with nonlinear viscous flow in the two low-viscosity regions (hot regions) beneath the Ou Backbone range and the Dewa Hills, which are surrounded by the high-viscosity region in the lower crust and uppermost mantle (see Fig. 15b). The absolute steady stress is concentrated at the bottom of the upper crust beneath the Ou Backbone range and the Dewa Hill, as this area is marked by a high temperature gradient. The formation of many *S*-wave reflectors around middle crustal magma reservoirs is consistent with this shortening deformation associated with a nonlinear, viscous flow.

Shibazaki *et al.* (2008) have shown that a large-scale upward movement occurs beneath the Ou Backbone range and the Dewa Hill, caused by a shortening deformation in the upper mantle. Age data of volcanism have shown the migration of volcanism from the back-arc area to the volcanic front (Yamada & Yoshida 2002; Kondo *et al.* 2004; Honda & Yoshida 2005*a*), and the most recent migration started at around 5 Ma in the back-arc side (Fig. 23b). If the thermal condition in the uppermost mantle controls the crustal deformation as well as controlling volcanic activity, the uplift and shortening movement from the eastern margin of the Japan Sea to the Ou Backbone range during Pliocene to Quaternary (e.g. Ikeda *et al.* 2012) could be explained by a shortening in the upper mantle associated with the migration of volcanism correlated with the dynamic movement in the mantle wedge (Honda & Yoshida 2005*a*).

Magma mixing and H_2O degassing. Miyagi *et al.* (2012) proposed that the Quaternary frontal medium-*K* series magma from Hokkaido was made by mixing of low-*K* basalts and medium-*K* rhyolites, whereby the rhyolites were partial melts of previously intruded, semi-solidified low-*K* basalts. The observed chemical variation of melts was explained by fractional crystallization and subsequent re-melting or re-mobilization of low-*K* basalt. In addition, the depleted Sr, Nd and Pb isotopic compositions of the basalt and rhyolite support the interpretation that the two rock types derived from a common source, with little or no evidence of contamination by crustal rocks. It is possible that the source of the felsic end-member of magma mixing is derived via 90% fractional crystallization, or from re-melting, or re-mobilization of a solidified or semi-solidified precursory basaltic magma. In this case, the basaltic magma chamber would have been located closely beneath the rhyolitic magma chamber, at a depth of 10–15 km or more. They also estimated that the total volume of basalt supplied beneath the volcanoes was approximately 10 times the volume of erupted rhyolite magma. The solidified basalt, with about 5 wt% H_2O, supplied a much larger amount of magmatic volatiles ('excess magmatic volatiles') to the overlying felsic magma than could be dissolved. They concluded that most of the excess magmatic volatiles were lost by degassing before the eruption.

Tectonically controlled lateral variation of phenocryst assemblage. Quaternary calc-alkaline andesites from the NE Japan arc have shown a clear lateral variation in phenocryst assemblages (Fig. 18c). Although the volatile content of the magma and its crystallization temperature are related to the magmatic alkalinity, an estimated lower H_2O content (Sakuyama 1977) in the frontal calc-alkaline andesites (Fig. 18c) suggests that the mafic end-member of the mixed calc-alkaline magma from frontal volcanoes degassed prior to producing andesite by any process. As discussed above, a large number of *S*-wave reflectors are found along caldera rims, or to the eastern side of the caldera complexes. These *S*-wave reflectors could have formed from the excess volatile degassing of the deep-seated mafic intrusives. Under strong compressional stress conditions, the excess volatiles which were degassed from the deep-seated mafic magma reservoirs could have risen directly upwards, or could have escaped to the trench side along thrust sheets, forming seismically active focused deformation zones (Okada *et al.* 2010). These processes, or other degassing processes, could have been more effective at the volcanic front to the forearc region than in the back-arc side, owing to the strong shortening deformation associated with nonlinear viscous flow (Fig. 25d).

Concluding remarks

The NE Japan arc is a typical island arc associated with a cold subduction system. Cenozoic magmatism in the NE Japan arc can be divided into three major periods of volcanic activity: continental margin, back-arc basin and island arc. Recent studies reveal secular changes in the magmatic activity, magma plumbing system, erupted volumes and magmatic composition associated with the evolution of crust–mantle structures related to the tectonic evolution of the NE Japan arc.

Each volcanic period has its unique pattern of across-arc alkali variations, which are closely related to the thermal structure and the asthenosphere–lithosphere boundary in the uppermost mantle wedge. In the back-arc basin period, the mantle isotherm was very flat, and the back-arc opening by strong extension associated with basin subsidence repeatedly expanded the area of mafic volcanism.

A drastic change from enriched to depleted magma is interpreted as being the result of an asthenospheric injection into the subcontinental lithosphere that occurred in: the continental SikhoteAlin at 35 Ma; the Oga Peninsula at 22 Ma; the Yamato Basin at 17 Ma; the Aosawa rift at 16.5 Ma; and the inland Kuroko rift area at 15 Ma. These relations are explained by the eastward asthenospheric flow (Nohda 2009). The magmatic evolution in the NE Japan arc in the back-arc basin and island-arc periods would have been related to the change in convection pattern of the mantle wedge from 2D to 3D related to the asthenosphere upwelling and subsequent cooling of the mantle. The geodynamic changes in the mantle would have caused back-arc basin basalt eruptions during the back-arc basin opening (basalt phase) followed by crustal heating and re-melting, which generated many felsic plutons and calderas (rhyolite/granite phase) in the early phase of the island-arc period. This was further followed by crustal cooling and strong trench-normal compression, which ensured vent connections and mixing between deeper mafic and shallower felsic magmas, erupting larger volumes of Quaternary calc-alkaline andesites (andesite phase).

Because of the voluminous magmatic effusions during back-arc basin opening, the temperature of the uppermost mantle beneath the back-arc decreased. Consequently, the lithosphere thickened beneath the back-arc, and the erupted magmatic composition changed from enriched low-*K* tholeiites to depleted alkali basalts at 13.5–10 Ma. In the volcanic front, the composition of magmas evolved from high HFSE magmas to low HFSE magmas, together with a change in the Hf isotope ratios. This is explained by fluxed melting, with the addition of fluids from subduction sediments and oceanic crust to the mantle wedge (Hanyu *et al.* 2006; Yamada *et al.* 2012).

During the neutral stress conditions at 13.5–10 Ma, felsic magmas ascended diapirically owing to their buoyancy within the ductile lower crust, and in the brittle upper crust the mode of magma ascent changed to that of dykes or sheets. The increase in horizontal tectonic stress at around 10–8 Ma might have formed sills or laccolithic shallow reservoirs in the upper crust followed by two peaks of late Miocene to Pliocene caldera-dominated magmatism, associated with subordinate low-*K* tholeiites. The depth of the caldera-forming magma reservoirs increased from the late Miocene to the Pliocene, and there was a regional cooling of the upper crust associated with a stress axis change from a NE–SW to an east–west compression.

It is possible that calc-alkaline andesites of the last stage of the island-arc period were formed by magma mixing or co-mingling between deep-seated mafic and shallow felsic end-members. The processes, which could possibly be responsible for producing the felsic end-member of mixed magma, are: fractional crystallization from the mafic end-member magma; melt extraction or re-melting of partially or fully crystallized precursor mafic magma in the middle crust; partial re-melting of mafic end-member material which solidified to hornblende gabbro or amphibolite; and partial re-melting of crustal basement rocks.

In the last stage of the island-arc period, strong compressional stress increased the pressure in the deep magma reservoirs and accelerated the upwelling of deeper mafic end-member of mixed calc-alkaline andesites. As a result, there was double the total erupted volume of mixed andesitic magma than that of the felsic caldera-dominated magmatism. Although phenocryst assemblages in calc-alkaline andesites suggest that the H_2O contents in magmas tended to increase from frontal-arc to the rear-arc, recent studies show a high water content in frontal primitive basalts. Under strong compressional stress conditions, the excess volatiles which had degassed from the deep-seated mafic magma reservoirs could rise directly upwards or escape trench-side along thrust sheets, forming *S*-wave reflectors and seismically active focused deformation zones. The strong degassing process of the mafic end-member would have worked more effectively at the volcanic front to the forearc region than in the back-arc side, through the shortening associated with nonlinear viscous flow in the uppermost mantle to the lower crust.

We are grateful to S. Straub, A. Gómez-Tuena and G. F. Zellmer for their editorial contributions. Constructive reviews and comments by R. Stern and T. Orozco-Esquivel are greatly appreciated. This study was partly supported by the Comparative Subductology Research Project of IFREE JAMSTEC and by grants from the Japan Society for the Promotion of Science (S11050123, 21109002 and 25610154).

References

ACOCELLA, V., YOSHIDA, T., YAMADA, R. & FUNICIELLO, F. 2008. Structural control on late Miocene to Quaternary volcanism in the NE Honshu arc, Japan. *Tectonics*, **27**, TC5008, http://dx.doi.org/10.1029/2008TC002296

ACTIVE FAULT RESEARCH GROUP 1991. *Active Faults in Japan*, revised edn. University of Tokyo Press, Tokyo [in Japanese].

AIZAWA, K. & YOSHIDA, T. 2000. Laccolithic magma reservoir – a key to the relation between collapse caldera and granitic pluton. *Earth Monthly*, **22**, 387–393 [in Japanese].

AIZAWA, K., ACOCELLA, V. & YOSHIDA, T. 2006. How the development of magma chambers affects collapse calderas: insights from an overview. *In*: TROISE, C., DE NATALE, G. & KILBURN, C. R. J. (eds) *Mechanisms of Activity and Unrest at Large Calderas*. Geological Society, London, Special Publications, **269**, 65–81.

ALLAN, J. F. & GORTON, M. P. 1992. Geochemistry of igneous rocks from Legs 127 and 128, Sea of Japan. *In*: *Proceedings of the Ocean Drilling Program, Scientific Results*. Ocean Drilling Program, Texas A&M University, College Station, TX, **127/128**, 905–929.

ANDERSON, E. M. 1936. The dynamics of the formation of cone sheets, ring dikes, and cauldron subsidence. *Proceedings of the Royal Society of Edinburgh*, **56**, 128–157.

ANDERSON, E. M. 1951. *Dynamics of Faulting and Dyke Formation*, 2nd edn. Oliver and Boyd, Edinburgh.

AOKI, K. 1971. Petrology of mafic inclusions from Itinome-gata, Japan. *Contributions to Mineralogy and Petrology*, **30**, 314–331.

AOKI, K. & SHIBA, I. 1973. Pyroxenes from lherzolite inclusions of Itinome-gata, Japan. *Lithos*, **6**, 41–51.

AOKI, K. & YOSHIDA, T. 1986. Trace element abundances in the volcanic rocks and the lower-crust derived inclusions from the Itinome-gata, Akita Prefecture, Japan. *Research Report of Laboratory for Nuclear Science, Tohoku University*, **19**, 279–287 [in Japanese].

AOKI, K., ISHIKAWA, K. & KANISAWA, S. 1981. Fluorine geochemistry of basaltic rocks from continental and oceanic regions and petrogenetic application. *Mineralogy and Petrology*, **76**, 53–59.

ARAI, S. 1980. Petrology of island arc lithosphere. *Earth Monthly*, **2**, 822–828 [in Japanese].

AWAJI, D., YAMAMOTO, D. & TAKAGI, H. 2006. Kinematic history of the Tanakura Shear Zone at brittle regime. *Journal of the Geological Society of Japan*, **112**, 222–240.

BAN, M. & YAMAMOTO, T. 2002. Petrological study of Nasu-Chausudake Volcano (ca. 16 ka to Present), northeastern

Japan. *Bulletin of Volcanology*, **64**, 100–116, http://dx.doi.org/10.1007/s00445-001-0187-9

Ban, M., Oba, Y., Ishikawa, K. & Takaoka, N. 1992. K–Ar dating of Mutsu–Hiuchidake, Osoreyama, Nanashigure, and Aoso volcanics of the Aoso – Osore volcanic zone – the formation of the present volcanic zonation of the northeast Japan arc. *Journal of Mineralogy, Petrology and Economic Geology*, **87**, 39–49 [in Japanese with English abstract].

Ban, M., Hirotani, S. *et al.* 2007. Origin of felsic magmas of some large-caldera-related stratovolcano in the central part of NE Japan – petrogenesis of the Takamatsu volcano. *Journal of Volcanology and Geothermal Research*, **167**, 100–118.

Cousens, B. L. & Allan, J. F. 1992. A Pb, Sr, and Nd isotopic study of basaltic rocks from the Sea of Japan. LEGS 127/128. *In*: *Proceedings of the Ocean Drilling Program, Scientific Results*. Ocean Drilling Program, Texas A&M University, College Station, TX, **127/128**, 805–817.

DeMets, C. 1992. Oblique convergence and deformation along the Kuril and Japanese trenches. *Journal of Geophysical Research*, **97**, 17 615–16 625.

DePaolo, D. J. 1985. Isotopic studies of processes in mafic magma chambers: I. the Kiglapait intrusion, Labrador. *Journal of Petroleum*, **26**, 925–951.

Dickinson, W. R. & Hatherton, T. 1967. Andesitic volcanism and seismicity around the Pacific. *Science*, **157**, 801–803.

Feeley, T. C., Dungan, M. A. & Frey, F. A. 1998. Geochemical constraints on the origin of mafic and silicic magmas at Cordon El Guadal, Tatara-San Pedro Complex, central Chile. *Contributions to Mineralogy and Petrology*, **131**, 393–411.

Fujinawa, A. 1988. Tholeiitic and calc-alkaline magma series at Adatara volcano, northeast Japan: 1. Geochemical constraints on their origin. *Lithos*, **22**, 135–158.

Fukuyama, H. 1985. Gabbroic inclusions of Ichinomegata tuff cone: bulk chemical composition. *Journal of the Faculty of Science – Section 2: Geology, Mineralogy, Geography, Geophysics*, **21**, 67–80.

GEOLOGICAL SURVEY OF JAPAN 1977. *Geology and Mineral Resources of Japan*, 3rd edn. Geological Survey of Japan, Kawasaki.

Ghiorso, M. S., Hirschmann, M. M., Reiners, P. W. & Kress, V. C. 2002. The pMELTS: a revision of MELTS for improved calculation of phase relations and major element partitioning related to partial melting of the mantle to 3 GPa. *Geochemistry Geophysics Geosystems*, **3**, 1030, http://dx.doi.org/10.1029/2001GC000217

Gill, J. 1981. *Orogenic Andesites and Plate Tectonics*. Springer-Verlag, Berlin.

Gordon, R. G. & Jurdy, D. M. 1986. Cenozoic global plate motions. *Journal of Geophysical Research*, **91**, 12 384–12 406.

Hall, C., Fischer, K., Parmentier, E. & Blackman, D. 2000. The influence of plate motions on three-dimensional back arc mantle flow and shear wave splitting. *Journal of Geophysical Research*, **105**, 28 009–28 033.

Hansen, J., Skjerlie, K. P., Pederson, R. B. & Rosa, R. B. 2002. Crustal melting in the lower parts of island arcs: an example from the Bremanger Granitoid Complex, west Norwegian Caledonides. *Contributions to Mineralogy and Petrology*, **143**, 316–335.

Hanyu, T., Tatsumi, Y. & Nakai, S. 2002. A contribution of slab-melts to the formation of high-Mg andesite magmas: Hf isotopic evidence from SW Japan. *Geophysical Research Letters*, **29**, 2052, http://dx.doi.org/10.1029/2002GL015856

Hanyu, T., Tatsumi, Y. *et al.* 2006. Contribution of slab melting and slab dehydration to magmatism in the NE Japan arc for last 25 Myr: constraints from geochemistry. *Geochemistry Geophysics Geosystems, 7*, **8**, 1–29, Q08002, http://dx.doi.org/10.1029/2005GC001220

Haq, B. U., Hardenbol, J. & Vail, P. R. 1988. Mesozoic and Cenozoic chronostratigraphy and eustatic cycles. *In*: Wilgus, C. K., Hastings, B. S., Kendall, C. G. St. C., Posamentier, H., Ross, C. A. & Van Wagoner, J. C. (eds) *Sea-Level Changes: An Integrated Approach*. SEPM Special Publications, Tulsa, **42**, 71–108.

Hasegawa, A. & Nakajima, J. 2004. Geophysical constraints on slab subduction and arc magmatism. *In*: Sparks, R. S. J. & Hawkesworth, C. J. (eds) *The State of the Planet: Frontiers and Challenges in Geophysics*. American Geophysical Union, Geophysical Monographs, Washington, DC, **150**, 81–94.

Hasegawa, A., Zhao, D., Hori, S., Yamamoto, A. & Horiuchi, S. 1991. Deep structure of the northeastern Japan arc and its relationship to seismic and volcanic activity. *Nature*, **352**, 683–689, http://dx.doi.org/10.1038/352683a0

Hasegawa, A., Yamamoto, A., Umino, N., Miura, S., Horiuchi, S., Zhao, D. & Sato, H. 2000. Seismic activity and deformation process of the overriding plate in the northeastern Japan subduction zone. *Tectonophysics*, **319**, 225–239, http://dx.doi.org/10.1016/S0040-1951(99)00296-6

Hasegawa, A., Nakajima, J., Umino, N., Miura, S. & Suwa, Y. 2004. Crustal deformation and shallow seismic activity beneath the Northeastern Japan Arc. *Journal of the Seismological Society of Japan*, **56**, 413–424 [in Japanese with English abstract].

Hasegawa, A., Nakajima, J., Umino, N. & Miura, S. 2005. Deep structure of the northeastern Japan arc and its implications for crustal deformation and shallow seismic activity. *Tectonophysics*, **403**, 59–75, http://dx.doi.org/10.1016/j.tecto.2005.03.018

Hasegawa, A., Nakajima, J., Kita, S., Okada, T., Matsuzawa, T. & Kirby, S. 2007. Anomalous deepening a belt of intraslab earthquakes in the Pacific slab crust under Kanto, central Japan: possible anomalous thermal shielding, dehydration reactions, and seismicity caused by shallower cold slab material. *Geophysical Research Letters*, **34**, L09305, http://dx.doi.org/10.1029/2007GL029616

Hayashi, S., Yoshida, T., Takashima, Y. & Aoki, K. 1991. Trace element abundances of the rocks from Kampu volcano, Northeastern Japan. *Research Report of Laboratory for Nuclear Science, Tohoku University*, **24**, 274–285 [in Japanese].

Hildreth, W. & Wilson, C. J. N. 2007. Compositional zoning of the Bishop tuff. *Journal of Petrology*, **48**, 951–999.

Hirotani, S. & Ban, M. 2006. Origin of silicic magma and magma feeding system of Shirataka volcano, NE Japan. *Journal of Volcanology and Geothermal Research*, **156**, 229–251.

Hirotani, S., Ban, M. & Nakagawa, M. 2009. Petrogenesis of mafic and associated silicic end-member magmas for calc-alkaline mixed rocks in the Shirataka volcano, NE Japan. *Contributions to Mineralogy and Petrology*, **157**, 709–734, http://dx.doi.org/10.1007/s00410-008-0360-7

Hoang, N., Yamamoto, T., Itoh, J. & Flower, M. F. J. 2009. Anomalous intra-plate high-Mg andesites in the Choshi area (Chiba, Central Japan) produced during early stages of Japan Sea opening? *Lithos*, **112**, 545–555.

Honda, H. & Yoshida, T. 2005*a*. Application of the model of small-scale convection under the island arc to the NE Honshu subduction zone. *Geochemistry Geophysics Geosystems*, **6**, 1–22, Q01002, http://dx.doi.org/10.1029/2004GC000785

Honda, S. & Yoshida, T. 2005*b*. Effects of oblique subduction on the 3-D pattern of small-scale convection within the mantle wedge. *Geophysical Research Letters*, **32**, 1–4, L13307, http://dx.doi.org/10.1029/2005GL023106

Honda, S., Yoshida, T. & Aoike, K. 2007. Spatial and temporal evolution of arc volcanism in the northeast Honshu and Izu–Bonin Arcs: evidence of small-scale convection under the island arc? *The Island Arc*, **16**, 2, 214–223, http://dx.doi.org/10.1111/j.1440-1738.2007.00567.x

Hori, S., Umino, N., Kono, T. & Hasegawa, A. 2004. Distinct *S*-wave reflectors (bright spots) extensively distributed in the crust and upper mantle beneath the Northeastern Japan arc. *Journal of the Seismological Society of Japan*, **56**, 435–446 [in Japanese with English abstract].

Huang, Z., Zhao, D., Umino, N., Wang, L., Matsuzawa, T., Hasegawa, A. & Yoshida, T. 2010. *P*-wave tomography, anisotropy and seismotectonics in the eastern margin of Japan Sea. *Tectonophysics*, **489**, 177–188, http://dx.doi.org/10.1016/j.tecto.2010.04.014

Huang, Z., Zhao, D. & Wang, I. 2011. Seismic heterogeneity and anisotropy of the Honshu arc from the Japan Trench to the Japan Sea. *Geophysical Journal International*, **184**, 1428–1444.

Iio, Y. & Matsuzawa, T. 2012. The generation process of the Tohoku earthquake: why did the magnitude 9 event occur? *Journal of the Geological Society of Japan*, **118**, 248–277 [in Japanese with English abstract].

Ikeda, Y., Okada, S. & Tajikara, M. 2012. Long-term strain buildup in the Northeast Japan arc-trench system and its implications for gigantic strain-release events. *Journal of the Geological Society of Japan*, **118**, 294–312.

Imaizumi, T. 1999. Geomorphological relief and distribution of active faults – some questions. *Earth Monthly*, Special issue, **27**, 113–117 [in Japanese].

Inoue, K. & Ban, M. 1996. Petrology of younger stage lavas from the Gassan volcano, northeastern Japan. *Journal of Mineralogy, Petrology and Economic Geology*, **91**, 33–47 [in Japanese with English abstract].

Isozaki, Y. 1996. Anatomy and genesis of a subduction-related orogen: a new view of geotectonic subdivision and evolution of the Japanese Islands. *The Island Arc*, **5**, 289–320.

Ito, T., Utada, M. & Okuyama, T. 1989. Mio-Pliocene calderas in the Backbone Region in northeast Japan. *Memoirs of the Geological Society of Japan*, **32**, 409–429 [in Japanese with English abstract].

Iwamori, H. & Zhao, D. 2000. Melting and seismic structure beneath the northeast Japan arc. *Geophysical Research Letters*, **27**, 425–428.

Iwasaki, T., Kato, W. *et al.* 2001. Extensional structure in northern Honshu Arc as inferred from seismic refraction/wide-angle reflection profiling. *Geophysical Research Letters*, **28**, 2329–2332.

JAPAN ASSOCIATION FOR QUATERNARY RESEARCH 1987. *Quaternary Maps of Japan*. University of Tokyo Press, Tokyo [in Japanese].

Jolivet, L., Tamaki, K. & Fournier, M. 1994. Japan Sea, opening history and mechanism: a synthesis. *Journal of Geophysical Research*, **99**, B11, 22 237–22 259, http://dx.doi.org/10.1029/93JB03463

Kagami, H. 2005. Formative periods and source materials for Cretaceous–Paleogene granitoids from Honshu arc. *Journal of the Geological Society of Japan*, **111**, 441–457.

Kagami, H., Kawano, Y. *et al.* 2000. Regional variations of initial Sr and Nd isotope ratios of late Cretaceous to Paleogene granitoid rocks, Honshu arc. *Earth Monthly*, Special issue, **30**, 185–190 [in Japanese].

Kaneoka, I., Takigami, Y., Takaoka, N., Yamashita, S. & Tamaki, K. 1992. ^{40}Ar–^{39}Ar analysis of volcanic rocks recovered from the Japan Sea floor: constraints on the age of formation of the Japan Sea. *In*: *Proceedings of the Ocean Drilling Program, Scientific Results*. Ocean Drilling Program, Texas A&M University, College Station, TX, **127/128**, 819–836.

Kanisawa, S. & Yoshida, T. 1989. Genesis of the extremely low-K tonalites from the island arc volcanism. *Bulletin of Volcanology*, **51**, 346–354.

Kano, K., Kato, H., Yanagisawa, Y. & Yoshida, F. 1991. *Stratigraphy and geologic history of the Cenozoic of Japan*. Geological Survey of Japan, Reports, **274**, 1–114.

Kano, K., Uto, K. & Ohguchi, T. 2007. Stratigraphic review of Eocene to Oligocene successions along the eastern Japan Sea: implication for early opening of the Japan Sea. *Journal of Asian Earth Sciences*, **30**, 20–23.

Kano, K., Ohguchi, T. *et al.* 2011. *Geology of the Toga and Funakawa District*. Quadrangle Series, 1: 50,000, Geological Survey of Japan, AIST [in Japanese with English abstract].

Karato, S. 2003. Mapping water content in the upper mantle. *In*: Eiler, J. (ed.) *Inside the Subduction Factory*. American Geophysical Union, Geophysical Monographs, Washington, DC, **138**, 135–152.

Katayama, I., Jung, H. & Karato, S. 2004. A new type of olivine fabric from deformation experiments at modest water content and low stress. *Geology*, **32**, 1045–1048.

Kato, N., Sato, H. *et al.* 2004. Seismic reflection profiling across the source fault of the 2003 Northern Miyagi earthquake (Mj 6.4), NE Japan: basin inversion of

Miocene back-arc rift. *Earth Planets Space*, **56**, 1369–1374.

Kawakatsu, H. & Watada, S. 2007. Seismic evidence for deep-water transportation in the mantle. *Science*, **316**, 1468.

Kawano, Y., Yagi, K. & Aoki, K. 1961. Petrography and petrochemistry of the volcanic rocks of Quaternary volcanoes of northeastern Japan. *Science Reports of the Tohoku University, Series III*, **7**, 1–46.

Kersting, A. B., Arculus, R. J. & Gust, D. A. 1996. Lithospheric contributions to arc magmatism: isotopic variations along strike in volcanoes of Honshu, Japan. *Science*, **272**, 1464–1468, http://dx.doi.org/10.1126/science.272.5267.1464

Kimura, G. 1986. Oblique subduction and collision: forearc tectonics of the Kuril arc. *Geology*, **14**, 404–407.

Kimura, J.-I. & Stern, R. J. 2008. Neogene volcanism of the Japan island arc: the K-h relationship revisted. *In*: Spencer, J. E. & Titley, S. R. (eds) *Ores and Orogenesis: Circum Pacific Tectonics, Geologic Evolution, and ore Deposits*. Arizona Geological Society Digest, Tucson, AZ, **22**, 187–202.

Kimura, J.-I. & Yoshida, T. 2006. Contribution of slab fluid, mantle wedge and crust to the origin of Quaternary Lavas in the NE Japan Arc. *Journal of Petroleum*, **47**, 2185–2232, http://dx.doi.org/10.1093/petrology/egl041

Kimura, J. I., Tanji, T., Yoshida, T. & Iizumi, S. 2001. Geology and geochemistry of lavas at Nekoma volcano: implications for origin of Quaternary low-*K* andesite in the North-eastern Honshu arc, Japan. *The Island Arc*, **10**, 116–134.

Kimura, J.-I., Yoshida, T. & Iizumi, S. 2002. Origin of low-*K* intermediate lavas at Nekoma volcano, Northeast-Honshu arc, Japan: geochemical constraints for lower-crustal melts. *Journal of Petroleum*, **43**, 631–661, http://dx.doi.org/10.1093/petrology/43.4.631

Kimura, J.-I., Hacker, B. R., van Keken, P. E., Kawabata, H., Yoshida, T. & Stern, R. J. 2009. Arc basalt simulator version 2, a simulation for slab dehydration and fluid-fluxed mantle melting for arc basalts: modeling scheme and application. *Geochemistry Geophysics Geosystems*, **10**, 1–32. Q09004, http://dx.doi.org/10.1029/2008GC002217

Kimura, J.-I., Kent, A. J. R. *et al.* 2010. Origin of cross-chain geochemical variation in Quaternary lavas from the northern Izu arc: using a quantitative mass balance approach to identify mantle sources and mantle wedge processes. *Geochemistry Geophysics Geosystems*, **11**, 24, http://dx.doi.org/10.1029/2010GC003050

Kincaid, C. & Griffiths, R. W. 2003. Laboratory models of the thermal evolution of the mantle during rollback subduction. *Nature*, **425**, 58–62.

Kita, S., Okada, T., Nakajima, J., Matsuzawa, T. & Hasegawa, A. 2006. Existence of a seismic belt in the upper plane of the double seismic zone extending in the along-arc direction at depths of 70–100 km beneath NE Japan. *Geophysical Research Letters*, **33**, L24310, http://dx.doi.org/10.1029/2006GL028239

Kobayashi, K. & Nakamura, E. 2001. Geochemical evolution of Akagi volcano, NE Japan: implications for interaction between island-arc magma and lower crust, and generation of isotopically various magmas. *Journal of Petroleum*, **42**, 2303–2331.

Kondo, H., Kaneko, K. & Tanaka, K. 1998. Characterization of spatial and temporal distribution of volcanoes since 14 Ma in the Northeast Japan arc. *Bulletin of the Volcanological Society of Japan*, **43**, 173–180.

Kondo, H., Tanaka, K., Mizouchi, Y. & Ninomiya, A. 2004. Long-term changes in distribution and chemistry of middle Miocene to Quaternary volcanism in the Chokai-Kurikoma area across the Northeast Japan Arc. *The Island Arc*, **13**, 18–46.

Kosuga, M., Watanabe, K., Hashimoto, K. & Kasai, H. 2012. Seismicity in the northern part of Tohoku district induced by the 2011 off the Pacific Coast of Tohoku earthquake. *Journal of the Seismological Society of Japan*, **65**, 69–83.

Kudo, T., Yoshida, T., Yamamoto, A., Kawamura, M. & Shichi, R. 2010. Crustal structure of the northeast Honshu arc revealed from detailed gravity anomalies. *Earth Monthly*, **32**, 373–382 [in Japanese].

Kuno, H. 1959. Origin of Cenozoic petrographic provinces of Japan and surrounding areas. *Bulletin of Volcanology*, **20**, 37–76.

Kuno, H. 1966. Lateral variation of basalt magma type across continental margins and island arcs. *Bulletin of Volcanology*, **29**, 195–222.

Kuno, H. 1967. Mafic and ultramafic nodules from Itinomegata, Japan. *In*: Wyllie, P. J. (ed.) *Ultramafic and Related Rocks*. John Wiley and Sons, New York, 337–342.

Kuno, H. 1968. Origin of andesite and its bearing on the island arc structure. *Bulletin of Volcanology*, **32**, 141–176.

Kurasawa, H. & Konda, T. 1986. Strontium isotopic ratios of Tertiary volcanic rocks of northeastern Honshu, Japan: implication for the spreading of the Japan Sea. *Journal of the Geological Society of Japan*, **92**, 205–217 [in Japanese with English abstract].

Kuritani, T., Yoshida, T., Kimura, J.-I., Hirahara, Y. & Takahashi, T. 2013. Water content of primitive low-K tholeiitic basalt magma from Iwate Volcano, NE Japan arc: implications for differentiation mechanism of frontal-arc basalt magmas. *Mineralogy and Petrology*, http://dx.doi.org/10.1007/s00710-013-0278-2

Kushiro, I. 1987. A petrological model of the mantle wedge and lower crust in the Japanese island arcs. *In*: Mysen, B. O. (ed.) *Magmatic Processes: Physicochemical Principles*. Geochemical Society, Special Publications, St Louis, MO, **1**, 165–181.

Luth, W. C., Jahns, R. H. & Tuttle, O. F. 1964. The granitic system at pressures of 4 to 10 kb. *Journal of Geophysical Research*, **69**, 759–773.

Maruyama, S. & Seno, T. 1986. Orogeny and relative plate motions: example of the Japanese islands. *Tectonophysics*, **127**, 305–329.

Meschede, M. 1986. A method of discriminating between different types of mid-ocean ridge basalts and continental tholeiite with the Nb–Zr–Y diagram. *Chemcal Geology*, **56**, 207–218.

Mimura, T. 1979. On the development of the geological structure in the southern part of Tsugaru Peninsula, Aomori Prefecture. *Journal of the Geological Society of Japan*, **85**, 719–735 [in Japanese with English abstract].

MITI (Ministry of International Trade and Industry) 1985. *Report on the regional geological structure survey, Hakkoda region during the fiscal year Showa 59.* Tokyo [in Japanese].

MITI (Ministry of International Trade and Industry) 1998. *Report on the mineral exploration technique especially for new exploration and metallurgy technique part 2, during the fiscal year Heisei 9.* Tokyo [in Japanese].

MIURA, S., SATO, T., HASEGAWA, A., SUWA, Y., TACHIBANA, K. & YUI, S. 2004. Strain concentration zone along the volcanic front derived by GPS observations in NE Japan arc. *Earth Planets Space*, **56**, 1347–1355.

MIYAGI, I., ITOH, J., HOANG, N. & MORISHITA, Y. 2012. Magma systems of the Kutcharo and Mashu volcanoes (NE Hokkaido, Japan): petrogenesis of the medium-K trend and the excess volatile problem. *Journal of Volcanology and Geothermal Research*, **231–232**, 50–60.

MIYASHIRO, A. 1974. Volcanic rock series in island arcs and active continental margins. *American Journal of Science*, **274**, 321–355.

MULLER, R. D., SDROLIAS, M., GAINA, C., STEINBERGER, B. & HEINE, C. 2008. Long-term sea-level fluctuations driven by ocean basin dynamics. *Science*, **319**, 1357–1362.

NAKAGAWA, M. 1983. Geology and petrology of Moriyoshi volcano. *Journal of the Japanese Association of Mineralogy, Petrology and Economic Geology*, **78**, 197–210 [in Japanese with English abstract].

NAKAGAWA, M., SHIMOTORI, H. & YOSHIDA, T. 1986. Aoso–Osore volcanic zone – the volcanic front of the Northeast Honshu arc, Japan. *Journal of the Japanese Association of Mineralogy, Petrology and Economic Geology*, **81**, 471–478 [in Japanese with English abstract].

NAKAGAWA, M., SHIMOTORI, H. & YOSHIDA, T. 1988. Across-arc compositional variation of the Quaternary basaltic rocks from the Northeast Japan arc. *Journal of the Japanese Association of Mineralogy, Petrology and Economic Geology*, **83**, 9–25 [in Japanese with English abstract].

NAKAJIMA, J. & HASEGAWA, A. 2003*a*. Estimation of thermal structure in the mantle wedge of northeastern Japan from seismic attenuation data. *Geophysical Research Letters*, **30**, 1760, http://dx.doi.org/10.1029/2003GL017185

NAKAJIMA, J. & HASEGAWA, A. 2003*b*. Tomographic imaging of seismic velocity structure in and around the Onikobe volcanic area, northeastern Japan: implications for fluid distribution. *Journal of Volcanology and Geothermal Research*, **127**, 1–18.

NAKAJIMA, J. & HASEGAWA, A. 2004. Shear-wave polarization anisotropy and subduction-induced flow in the mantle wedge of northeastern Japan. *Earth and Planetary Science Letters*, **225**, 365–377.

NAKAJIMA, J., MATSUZAWA, T., HASEGAWA, A. & ZHAO, D. 2001. Three-dimensional structure of Vp, Vs and Vp/Vs beneath northeastern Japan arc: implications for arc magmatism and fluids. *Journal of Geophysical Research*, **106**, 21 843–21 857.

NAKAJIMA, J., TAKEI, Y. & HASEGAWA, A. 2005. Quantitative analysis of the inclined low-velocity zone in the mantle wedge of northern Japan: a systematic change of melt-filled pore shapes with depth and its implications for melt migration. *Earth and Planetary Science Letters*, **234**, 59–70, http://dx.doi.org/10.1016/j.epsl.2005.02.033

NAKAJIMA, J., HASEGAWA, A., HORIUCHI, S., YOSHIMOTO, K., YOSHIDA, T. & UMINO, N. 2006*a*. Crustal heterogeneity around the Nagamachi–Rifu fault, northeastern Japan, as inferred from travel-time tomography. *Earth Planets Space*, **58**, 843–853.

NAKAJIMA, J., SHIMIZU, J., HORI, S. & HASEGAWA, A. 2006*b*. Shear-wave splitting beneath the southwestern Kurile arc and northeastern Japan arc: a new insight into mantle return flow. *Geophysical Research Letters*, **33**, L05305, http://dx.doi.org/10.1029/2005GL025053

NAKAJIMA, T., DANHARA, T., IWANO, H. & CHINZEI, K. 2006*c*. Uplift of the Ou Backbone Range in Northeast Japan at around 10 Ma and its implication for the tectonic evolution of the eastern margin of Asia. *Palaeogeography, Palaeoclimatology, Palaeoecology*, **241**, 28–48.

NAKAMURA, K. 1977. Volcanoes as possible indicators of tectonic stress orientation – principle and proposal. *Journal of Volcanology and Geothermal Research*, **2**, 1–16.

NAKAMURA, K. 1992. Inversion tectonics and its structural expression. *Journal of the Tectonic Research Group of Japan*, **38**, 3–45 [in Japanese with English abstract].

NISHIMOTO, S., ISHIKAWA, M., ARIMA, M. & YOSHIDA, T. 2005. Laboratory measurement of *P*-wave velocity in crustal and upper mantle xenoliths from Ichino-megata, NE Japan: ultrabasic hydrous lower crust beneath the NE Honshu arc. *Tectonophysics*, **396**, 245–259, http://dx.doi.org/10.1016/j.tecto.2004.12.010

NISHIMOTO, S., ISHIKAWA, M., ARIMA, M., YOSHIDA, T. & NAKAJIMA, J. 2008. Simultaneous high *P–T* measurements of ultrasonic compressional and shear wave velocities in Ichino-megata mafic xenoliths: their bearings on seismic velocity perturbations in lower crust of northeast Japan arc. *Journal of Geophysical Research*, **113**, B12212, http://dx.doi.org/10.1029/2008JB005587

NISHISAKA, H., SHINOHARA, M., SATO, T., HINO, R., MOCHIZUKI, K. & KASAHARA, J. 2001. Crustal structure of the Yamato basin and the margin of the northeastern Japan Sea using ocean bottom seismographs and controlled sources. *Journal of the Seismological Society of Japan*, **54**, 365–379 [in Japanese with English abstract].

NOHDA, S. 2009. Formation of the Japan Sea basin: reassessment from Ar–Ar ages and Nd–Sr isotopic data of basement basalts of the Japan Sea and adjacent regions. *Journal of Asian Earth Sciences*, **34**, 599–609.

NOHDA, S. & WASSERBURG, G. J. 1986. Trends of Sr and Nd isotopes through time near the Japan Sea in northeastern Japan. *Earth and Planetary Science Letters*, **78**, 157–167.

NOHDA, S., TATSUMI, Y., OTOFUJI, Y., MATSUDA, T. & ISHIZAKA, K. 1988. Asthenospheric injection and back-arc opening: isotopic evidence from Northeast Japan. *Chemical Geology*, **68**, 317–327.

NOHDA, S., TATSUMI, Y., YAMASHITA, S. & FUJII, T. 1992. Nd and Sr isotopic study of Leg 127 Basalts: implications for the evolution of the Japan Sea backarc basin. *In*: TAMAKI, K., SUYEHIRO, K., ALLAN, J. & MCWILLIAMS, S. (eds) *Proceedings of the Ocean Drilling Program, Scientific Results*. Ocean Drilling Program, Texas A & M University, College Station, TX, **127/128**, 899–904.

NOTSU, K. 1983. Strontium isotopic composition in volcanic rocks from the Northeast Japan arc. *Journal of Volcanology and Geothermal Research*, **18**, 531–548.

OHGUCHI, T., YOSHIDA, T. & OKAMI, K. 1989. Historical change of the Neogene and Quaternary volcanic field in the Northeast Honshu Arc, Japan. *Memoirs of the Geological Society of Japan*., **32**, 131–155 [in Japanese with English abstract].

OHKI, J., WATANABE, N., SHUTO, K. & ITAYA, T. 1993. Shifting of the volcanic fronts during Early to Late Miocene in the Northeast Japan arc. *The Island Arc*, **2**, 87–93.

OHKI, J., SHUTO, K. & KAGAMI, H. 1994. Middle Miocene bimodal volcanism by asthenospheric upwelling: Sr and Nd isotopic evidence from the back-arc region of the northeast Japan arc. *Geochemical Journal*, **28**, 473–487.

OISHI, M. & TAKAHASHI, M. 1990. Miocene formations in the Takasaki district, central Japan – with special reference to the developmental process of Niwaya Unconformity. *Contributions from the Institute of Paleontology, Tohoku University*, **92**, 1–17 [in Japanese with English abstract].

OKADA, T., MATSUZAWA, T. & HASEGAWA, A. 1995. Shear-wave polarization anisotropy beneath the north-eastern part of Honshu, Japan. *Geophysical Journal International*, **123**, 781–797.

OKADA, T., UMINO, N. & HASEGAWA, A. 2010. Deep structure of the Ou mountain range strain concentration zone and the focal area of the 2008 Iwate–Miyagi Nairiku earthquake, NE Japan – seismogenesis related with magma and crustal fluid. *Earth Planets Space*, **62**, 347–352.

OKAMURA, Y., WATANABE, M., MORIJIRI, R. & SATOH, M. 1995. Rifting and basin inversion in the eastern margin of the Japan Sea. *The Island Arc*, **4**, 166–181.

ONODERA, M., HORIUCHI, S. & HASEGAWA, A. 1998. Three-dimensional seismic velocity structure in and around the focal area of the 1996 Onikobe Earthquake based on Vp/Vs inversion. *Zisin, Series II*, **51**, 265–279 [in Japanese with English abstract].

OTOFUJI, Y. & MATSUDA, T. 1983. Paleomagnetic evidence for the clockwise rotation of Southwest Japan. *Earth and Planetary Science Letters*, **62**, 349–359.

OTOFUJI, Y. & MATSUDA, T. 1984. Timing of rotational motion of Southwest Japan inferred from paleomagnetism. *Earth and Planetary Science Letters*, **70**, 373–382.

OTOFUJI, Y., MATSUDA, T. & NOHDA, S. 1985. Paleomagnetic evidence for the Miocene counter-clockwise rotation of Northeast Japan – rifting process of the Japan Arc. *Earth and Planetary Science Letters*, **75**, 265–277.

OTSUKI, K. 1990. Neogene tectonic stress fields of northeast Honshu Arc and their implications for plate boundary condition. *Tectonophysics*, **181**, 151–164.

OTSUKI, K., NAKATA, T. & IMAIZUMI, T. 1977. Quaternary crustal movements and block model in the south-eastern region of the Northeast Japan. *Earth Science*, **31**, 1–14 [in Japanese with English abstract].

PATRIAT, P. & ARCHACHE, J. 1984. India–Asia collision chronology has implications for crustal shortening and driving mechanisms of plates. *Nature*, **311**, 615–621.

PEACOCK, S. 2003. Thermal structure and metamorphic evolution of subducting slabs. *In*: EILER, J. (ed.) *Inside the Subduction Factory*. American Geophysical Union, Geophysical Monographs, Washington, DC, **138**, 7–22.

PEARCE, J. A. 1983. Role of the sub-continental lithosphere in magma genesis at active continental margins. *In*: HAWKESWORTH, C. J. & NORRY, M. J. (eds) *Continental Basalts and Mantle Xenoliths*. Shiva, Nantwich, 230–249.

PEARCE, J. A., KEMPTON, P. D., NOWELL, G. M. & NOBLE, S. R. 1999. Hf–Nd element and isotope perspective on the nature and provenance of mantle and subduction components in western Pacific arc-basin systems. *Journal of Petroleum*, **40**, 1579–1611.

PHIPPS-MORGAN, J. 1999. Thermodynamics of pressure release melting of a veined plum pudding mantle. *Geochemistry Geophysics Geosystems*, **2**, 65, http://dx.doi.org/10.1029/2000GC000049

POLLITZ, F. R. 1986. Pliocene change in Pacific plate motion. *Nature*, **320**, 738–741.

PORTNYAGIN, M., HOERNLE, K., PLECHOV, P., MIRONOV, N. & KHUBUNAYA, S. 2007. Constraints on mantle melting and composition and nature of slab components in volcanic arcs from volatiles (H_2O, S, Cl, F) and trace elements in melt inclusions from Kamchatka Arc. *Earth and Planetary Science Letters*, **255**, 53–69.

PRIMA, O. D. A., ECHIGO, A., YOKOYAMA, R. & YOSHIDA, T. 2006. Supervised landform classification of Noetheast Honshu from DEM-derived thematic maps. *Geomorphology*, **78**, 373–386.

PRIMA, O. D. A., YOSHIDA, T., KUDO, T. & NONAKA, S. 2012. Delineation of buried caldera rims from gravity data. *Theory and Applications of GIS*, **20**, 83–93 [in Japanese with English abstract].

ROBERTSON, J. K. & WYLLIE, P. J. 1971. Rock-water systems, with special reference to the water-deficient region. *American Journal of Science*, **271**, 252–277.

RYAN, M. P. 1987. Neutral buoyancy and the mechanical evolution of magmatic systems. *In*: MYSEN, B. O. (ed.) *Magmatic Process: Physicochemical Principles*. The Geochemical Society, Special Publications, St Louis, MO, **1**, 259–287.

SAKUYAMA, M. 1977. Lateral variation of phenocryst assemblages in volcanic rocks of the Japanese islands. *Nature*, **269**, 134.

SAKUYAMA, M. 1979. Lateral variations of H2O contents in Quaternary magmas of northeastern Japan. *Earth and Planetary Science Letters*, **43**, 103–111.

SAKUYAMA, M. 1983. Phenocryst assemblages and H2O content in Circum-Pacific arc magma. *In*: HILDE, T. W. C. & UYEDA, S. (eds) *Geodynamics of the Western Pacific-Indonesian Region*. American

Geophysical Union, Geodynamics Series, Washington, DC, **11**, 143–158.

SAKUYAMA, M. & NESBITT, R. W. 1986. Geochemistry of the Quaternary volcanic rocks of the Northeast Japan arc. *Journal of Volcanology and Geothermal Research*, **29**, 413–450.

SALTERS, V. J. M. & WHITE, W. M. 1998. Hf isotope constraints on mantle evolution. *Chemical Geology*, **145**, 447–460.

SATO, H. 1986. Geological development of the Cenozoic System in central Northeast Honshu (between Sakata and Furukawa). *Contributions from the Institute of Geology and Paleontology, Tohoku University*, **88/89**, 1–32, 1–45 [in Japanese with English abstract].

SATO, H. 1992. Late Cenozoic tectonic evolution of the central part of Northern Honshu, Japan. *Bulletin of the Geological Survey of Japan*, **43**, 119–139 [in Japanese with English abstract].

SATO, H. 1994. The relationship between late Cenozoic tectonic events and stress field and basin development in northeast Japan. *Journal of Geophysical Research*, **99**, B11, 22 261–22 274.

SATO, H. & AMANO, K. 1991. Relationship between tectonics, volcanism and basin development, late Cenozoic, central part of Northern Honshu, Japan. *Sedimentary Geology*, **74**, 323–343.

SATO, H. & HASEGAWA, A. 1996. Mapping of partial melt zones in the upper mantle beneath northeastern Japan – ascent of magma in island arc. *Bulletin of the Volcanological Society of Japan*, **41**, 115–125 [in Japanese with English abstract].

SATO, H. & YOSHIDA, T. 1993. The tectonic control on the formation of large-scale collapsed calderas, late Cenozoic Northeast Japan. *Earth Monthly*, **15**, 721–724 [in Japanese].

SATO, H., OTSUKI, K. & AMANO, K. 1982. Neogene tectonic stress field of Northeast Honshu Arc. *Journal of the Tectonic Research Group of Japan*, **27**, 55–79 [in Japanese].

SATO, H., IMAIZUMI, T. & YOSHIDA, T. 2002. Tectonic evolution and deep to shallow geometry of Nagamachi–Rifu Active Fault System, NE Japan. *Earth Planets Space*, **54**, 1039–1043.

SATO, H., YOSHIDA, T., IWASAKI, T., SATO, T., IKEDA, Y. & UMINO, N. 2004. Late Cenozoic tectonic development of the back arc region of central northern Honshu, Japan, revealed by recent deep seismic profiling. *Journal of the Japanese Association for Petroleum Technology*, **69**, 145–154.

SATO, M., SHUTO, K. & YAGI, M. 2007. Mixing of asthenospheric and lithospheric mantle-derived basalt magmas as shown by along-arc variation in Sr and Nd isotopic compositions of Early Miocene basalts from back-arc margin of the NE Japan arc. *Lithos*, **96**, 453–474, http://dx.doi.org/10.1016/j.lithos.2006.11.006

SHIBATA, T. & ISHIHARA, S. 1979. Initial $^{87}Sr/^{86}Sr$ ratios of plutonic rocks from Japan. *Contributions to Mineralogy and Petrology*, **70**, 381–390.

SHIBATA, T. & NAKAMURA, E. 1991. Interaction between subducted oceanic slab and wedge mantle inferred from across-arc variations of Pb, Sr and Nd isotopic compositions in Northeastern Japan. *Proceedings of the Japan Academy*, **67**, Ser. B, 115–120.

SHIBAZAKI, B., GARATANI, K., IWASAKI, T., TANAKA, A. & IIO, Y. 2008. Faulting processes controlled by the nonuniform thermal structure of the crust and uppermost mantle beneath the northeastern Japanese island arc. *Journal of Geophysical Research*, **113**, B08415, http://dx.doi.org/10.1029/2007JB005361

SHUTO, K. 2009. *The Northeast Japan Arc – Opening of the Japan Sea and Magma Genesis*. Kyoritsu, Tokyo [in Japanese].

SHUTO, K. & GORAI, M. 1997. *Crust to Mantle-Forming Materials*. Kyoritsu, Tokyo [in Japanese].

SHUTO, K., MAKINO, A., ITAYA, T. & YASHIMA, R. 1992. K–Ar ages and petrologic characteristics of the Inase volcanic rocks from eastern part of the Kitakami district of Iwate Prefecture, Northeast Japan. *Journal of the Japanese Association of Mineralogy, Petrology and Economic Geology*, **87**, 20–34.

SHUTO, K., OHKI, J. ET AL. 1993. The relationships between drastic changes in Sr isotope ratios of magma sources beneath the NE Japan arc and the spreading of the Japan Sea back-arc basin. *Ministry of Petroleum*, **49**, 71–90.

SHUTO, K., NAKAJIMA, S., OHKI, J., UEMATSU, M., WATANABE, N. & YAMAMOTO, K. 1995. High field strength element depletions in the lithospheric mantle beneath the Northeast Japan arc. *Memoirs of the Geological Society of Japan*, **44**, 241–262 [in Japanese with English abstract].

SHUTO, K., ISHIMOTO, H. ET AL. 2006. Geochemical secular variation of magma source during Early to Middle Miocene time in the Niigata area, NE Japan: asthenospheric mantle upwelling during back-arc basin opening. *Lithos*, **86**, 1–33.

SHUTO, K., SATO, M. & OHKI, J. 2008. Interaction between sea-water and Miocene volcanic rocks recovered by deep drilling in the Niigata oil and gas fields, northeast Japan – Sr and Nd isotopic evidence from oil and gas-bearing volcanic rocks. *Journal of the Japanese Association for Petroleum Technology*, **73**, 517–530.

STERN, R. J. 2002. Subduction zones. *Reviews of Geophysics*, **40**, 1012, http://dx.doi.org/10.1029/2001RG000108

SUN, S.-S. & MCDONOUGH, W. F. 1989. Chemical and isotopic systematics of oceanic basalts: implications for mantle composition and processes. *In*: SAUNDERS, A. D. & NORRY, M. J. (eds) *Magmatism in the Ocean Basins*. Geological Society, London, Special Publications, **42**, 313–345.

TAJIKARA, M. & IKEDA, Y. 2004. Vertical crustal movement and development of basin and range topography in the middle part of the Northeast Japan arc estimated from fluvial/marine terrace data. *Quaternary Research*, **44**, 229–245 [in Japanese with English abstract].

TAKADA, A. 1989. Magma transport and reservoir formation by a system of propagating cracks. *Bulletin of Volcanology*, **52**, 118–126.

TAKADA, Y. & FUKUSHIMA, Y. 2013. Volcanic subsidence triggered by the 2011 Tohoku earthquake in Japan. *Nature Geoscience*, http://dx.doi.org/10.1038/NGEO1857

TAKAHASHI, E. 1986. Genesis of calc-alkali andesite magma in a hydrous mantle–crust boundary:

petrology of lherzolite xenoliths from the Ichinomegata crater, Oga Peninsula, northeast Japan, part II. *Journal of Volcanology and Geothermal Research*, **29**, 355–395, http://dx.doi.org/10.1016/0377-0273(86)90051-X

Takahashi, M. 1994. Miocene tectonic deformation of central Japan: paleomagnetic evidence of intra-arc bending. *Geofisica Internacional*, **33**, 25–44.

Takahashi, M. & Saito, K. 1997. Miocene intra-arc bending at an arc-arc collision zone, central Japan. *The Island Arc*, **6**, 168–182.

Takahashi, N., Kodaira, S. *et al.* 2004. Seismic structure and seismogenesis off Sanriku region, northeastern Japan. *Geophysical Journal International*, **159**, 129–145, http://dx.doi.org/10.1111/j.1365-246X.2004.02350.x

Takahashi, T., Hirahara, Y., Miyazaki, T., Senda, R., Chang, Q., Kimura, J.-I. & Tatsumi, Y. 2012. Primary magmas at the volcanic front of the NE Japan arc: coeval eruption of crustal low-*K* tholeiitic and mantle-derived medium-*K* calc-alkaline basalts at Azuma volcano. *Journal of Petrology*, **53**, http://dx.doi.org/10.1093/petrology/egs065

Takeuchi, A. 1980. Tertiary stress field and tectonic development of the southern part of the northeast Honshu arc, Japan. *Journal of Geosciences Osaka City University*, **23**, 1–64.

Tamaki, K., Suyehiro, K., Allan, J., Ingle, J. C. & Pisciotto, K. A. 1992. Tectonic synthesis and implications of Japan Sea ODP drilling. *In*: *Proceedings of the Ocean Drilling Program, Scientific Results*. Ocean Drilling Program, Texas A&M University, College Station, TX, **127/128**, 1333–1348.

Tamura, Y., Tatsumi, Y., Zhao, D., Kido, Y. & Shukuno, H. 2002. Hot fingers in the mantle wedge: new insights into magma genesis in subduction zones. *Earth and Planetary Science Letters*, **197**, 105–116.

Tanaka, A., Yano, Y. & Sasada, M. 2004. Geothermal gradient data in and around Japan. *In*: *Digital Geoscience Map DGM P-5*. Geological Survey, Japan.

Tatsumi, Y., Sakuyama, M., Fukuyama, H. & Kushiro, I. 1983. Generation of arc basalt magmas and thermal structure of the mantle wedge in subduction zones. *Journal of Geophysical Research*, **88**, 5815–5825.

Tatsumi, Y., Nohda, S. & Ishizaka, K. 1988. Secular variation of magma source compositions beneath the Northeast Japan arc. *Chemical Geology*, **68**, 309–316.

Tatsumi, Y., Otofuji, Y., Matsuda, T. & Nohda, S. 1989. Opening of the Sea of Japan back-arc basin by asthenospheric injection. *Tectonophysics*, **166**, 317–329.

Tatsumi, Y., Furukawa, Y. & Yamashita, S. 1994. Thermal and geochemical evolution of the mantle wedge in the northeast Japan arc 1. Contribution from experimental petrology. *Journal of Geophysical Research*, **99**, 22 275–22 283.

Tatsumi, Y., Takahashi, T. *et al.* 2008. New insight into andesitic genesis: the role of mantle-derived calc-alkalic and crust-derived tholeiitic melts in magma differentiation beneath Zao volcano, NE Japan. *Journal of Petroleum*, **49**, 1971–2008.

Togashi, S. 1977. Petrology of Osore-yama volcano, Japan. *Journal of the Japanese Association of Mineralogy, Petrology and Economic Geology*, **72**, 45–60 [in Japanese with English abstract].

Togashi, S., Tanaka, T., Yoshida, T., Ishikawa, K., Fujinawa, A. & Kurasawa, H. 1992. Trace elements and Nd–Sr isotopes of island arc tholeiites from frontal arc of Northeast Japan. *Geochemical Journal*, **26**, 261–277.

Toya, N., Ban, M. & Shinjo, R. 2005. Petrology of Aoso volcano, northeast Japan arc: temporal variation of the magma feeding system and nature of low-*K* amphibole andesite in the Aoso–Osore volcanic zone. *Contributions to Mineralogy and Petrology*, **148**, 566–581.

Tsuji, T., Nakajima, J. & Hasegawa, A. 2008. Tomographic evidence for hydrated oceanic crust of the Pacific slab beneath northeastern Japan: implications for water transportation in subduction zones. *Geophysical Research Letters*, **35**, L14308, http://dx.doi.org/10.1029/2008GL034461

Tsumura, N., Matsumoto, S., Horiuchi, S. & Hasegawa, A. 2000. Three-dimensional attenuation structure beneath the northeastern Japan arc estimated from spectra of small earthquakes. *Tectonophysics*, **319**, 241–260.

Tsunakawa, H. 1983. Simple two-dimensional model of propagation of magma-filled cracks. *Journal of Volcanology and Geothermal Research*, **16**, 335–343.

Tsunakawa, H. 1986. Neogene stress field of Japanese arcs and its relation to igneous activity. *Tectonophysics*, **124**, 1–22.

Ujike, O. & Tsuchiya, N. 1993. Geochemistry of Miocene basaltic rocks temporally straddling the rifting of lithosphere at the Akita–Yamagata area, northeast Japan. *Chemical Geology*, **104**, 61–74.

Umeda, K., Hayashi, S., Ban, M., Sasaki, M., Ohba, T. & Akaishi, K. 1999. Sequence of the volcanism and tectonics during the last 2.0 million years along the volcanic front in Tohoku district, NE Japan. *Bulletin of the Volcanological Society of Japan*, **44**, 233–249 [in Japanese with English abstract].

Umino, N. & Hasegawa, A. 1984. Three-dimensional Qs structure in the northeastern Japan arc. *Journal of the Seismological Society of Japan*, **37**, 217–228.

Umino, N., Matsuzawa, T., Hori, S., Nakamura, A., Yamamoto, A., Hasegawa, A. & Yoshida, T. 1998. Onikobe earthquakes and their relation to crustal structure. *Journal of the Seismological Society of Japan*, **51**, 253–264 [in Japanese with English abstract].

Umino, N., Ujikawa, H., Hori, S. & Hasegawa, A. 2002. Distinct *S*-wave reflectors (bright spots) detected beneath the Nagamachi–Rifu fault, NE Japan. *Earth Planets Space*, **54**, 1021–1026.

van Keken, P. E. 2003. The structure and dynamics of the mantle wedge. *Earth and Planetary Science Letters*, **215**, 323–338, http://dx.doi.org/10.1016/S0012-821X(03)00460-6

Vervoort, J. D., Patchett, P. J., Blichert-Toft, J. & Albarede, F. 1999. Relationships between Lu–Hf and Sm–Nd isotopic systems in the global sedimentary system. *Earth and Planetary Science Letters*, **168**, 79–99.

Walker, J. A., Roggensack, K., Patino, L. C., Cameron, B. I. & Matias, O. 2003. The water and trace element contents of melt inclusions across an active subduction zone. *Contributions to Mineralogy and Petrology*, **146**, 62–77.

WANG, J. & ZHAO, D. 2008. *P*-wave anisotropic tomography beneath Northeast Japan. *Physics of the Earth and Planetary Interiors*, **170**, 115–133, http://dx.doi.org/10.1016/j.pepi.2008.07.042

WOOD, D. A., JORON, J. L., TREUIL, M., NORRY, M. & TARNEY, J. 1979. Elemental and Sr isotope variations in basic lavas from Iceland and the surrounding ocean floor. *Contributions to Mineralogy and Petrology*, **70**, 319–339.

WOODHEAD, J. D., HERGT, J. M., DAVIDSON, J. P. & EGGINS, S. M. 2001. Hafnium isotope evidence for 'conservative' element mobility during subduction zone processes. *Earth and Planetary Science Letters*, **192**, 331–346.

WORKMAN, R. K. & HART, S. R. 2005. Major and trace element composition of the depleted MORB mantle (DMM). *Earth and Planetary Science Letters*, **231**, 53–72, http://dx.doi.org/10.1016/j.epsl.2004.12.005

WYLLIE, P. J. 1971. *The Dynamic Earth*. Wiley, New York.

XIA, S., ZHAO, D., QIU, X., NAKAJIMA, J., MATSUZAWA, T. & HASEGAWA, A. 2007. Mapping the crustal structure under active volcanoes in central Tohoku, Japan using P and PmP data. *Geophysical Research Letters*, **34**, L10309, http://dx.doi.org/10.1029/2007GL030026

YAGI, M., HASENAKA, T. *ET AL.* 2001. Transition of magmatic composition reflecting an evolution of rifting activity – a case study of the Akita–Yamagata basin in Early to Middle Miocene, Northeast Honshu, Japan. *Ganseki Kobutsu Kagaku*, **30**, 265–287 [in Japanese with English abstract], http://dx.doi.org/10.2465/gkk.30.265

YAMADA, E. 1988. Geologic development of the Onikobe caldera, northeast Japan, with special reference to its hydrothermal system. Geological Survey of Japan, Reports, **268**, 61–190.

YAMADA, R. & YOSHIDA, T. 2002. Relationships between the evolution of the Neogene volcanism and the Kuroko mineralization in the vicinity of the Hokuroku District, NE Japan – a chronological study. *Resource Geology*, **52**, 97–110 [in Japanese with English abstract].

YAMADA, R. & YOSHIDA, T. 2004. Volcanic Sequences Related to Kuroko Mineralization in the Hokuroku District, Northeast Japan. *Resource Geology*, **54**, 399–412.

YAMADA, R. & YOSHIDA, T. 2011. Relationships between Kuroko volcanogenic massive sulfide (VMS) deposits, felsic volcanism, and island arc development in the northeast Honshu arc, Japan. *Mineral Deposita*, http://dx.doi.org/10.1007/s00126-011-0362-7

YAMADA, R., YOSHIDA, T. & KIMURA, J-I. 2012. Chemical and isotopic characteristics of the Kuroko-forming volcanism. *Resource Geology*, **62**, 369–383, http://dx.doi.org/10.1111/j.1751-3928.2012.00202.x

YAMAJI, A. 1990. Rapid intra-arc rifting in Miocene northeast Japan. *Tectonics*, **9**, 365–378.

YAMAJI, A. 1994. An introduction to the Cenozoic tectonics of the Japan arc. *Geofisica International*, **33**, 21–24.

YAMAJI, A. & SATO, H. 1989. Miocene subsidence of the Northeast Honshu Arc and its mechanism. *Memoirs of the Geological Society of Japan*, **32**, 339–349 [in Japanese with English abstract].

YAMAMOTO, A., SHICHI, R. & KUDO, T. 2011. *Gravity Database of Japan* (CD-ROM). Earth Watch Safety Net Research Center, Special Publications, **1**, Chubu University, Nagoya.

YAMAMOTO, M. 1984. Origin of calc-alkaline andesite from Oshima-Oshima volcano, North Japan. *Journal of the Faculty of Science Hokkaido University, Series IV*, **21**, 77–131.

YAMAMOTO, T. 1991. Late Cenozoic dike swarms and tectonic stress field in Japan. *Bulletin of the Geological Survey of Japan*, **42**, 131–148 [in Japanese with English abstract].

YAMAMOTO, T. & HOANG, N. 2009. Synchronous Japan Sea opening Miocene fore-arc volcanism in the Abukuma Mountains, NE Japan: an advancing hot asthenosphere flow versus Pacific slab melting. *Lithos*, **112**, 575–590.

YAMASHITA, S. & FUJII, T. 1992. Experimental petrology of basement basaltic rocks from sites 794 and 797, Japan Sea. *In*: *Proceedings of the Ocean Drilling Program, Scientific Results*. Ocean Drilling Program, Texas A & M University, College Station, TX, **127/128**, 891–898.

YANO, Y., TANAKA, A., TAKAHASHI, M., OKUBO, Y., SASADA, M., UMEDA, K. & NAKATSUKA, N. 1999. *Geothermal gradient map of Japan*. Geological Survey of Japan, Tsukuba [in Japanese with English abstract].

YOKOYAMA, R., SHIRASAWA, M. & KIKUCHI, Y. 1999. Representation of topographical features by openesses. *Japan Society of Photogrammetry and Remote Sensing*, **38**, 26–34.

YOSHIDA, T. 1989. Study of the Quaternary volcanic rocks in the Northeast Honshu Arc, Japan. *Memoirs of the Geological Society of Japan*, **32**, 353–384 [in Japanese with English abstract].

YOSHIDA, T. 2001. The evolution of arc magmatism in the NE Honshu arc, Japan. *Tohoku Geophysical Journal*, **36**, 131–149.

YOSHIDA, T. 2009. Late Cenozoic Magmatism in the northeast Honshu Arc, Japan. *Earth Science*, **63**, 269–288 [in Japanese with English abstract].

YOSHIDA, T. & AOKI, K. 1984. Geochemistry of major and trace elements in the Quaternary volcanic rocks from northeast Honshu, Japan. *Science Reports of the Tohoku University, Series III*, **16**, 1–34.

YOSHIDA, T., WATANABE, H. & AOKI, K. 1983. Geochemistry of Hachimantai volcano, Nasu volcanic zone. *Research Report of Laboratory for Nuclear Science, Tohoku University*, **16**, 309–324 [in Japanese].

YOSHIDA, T., MURATA, M. & YAMAJI, A. 1993. Formation of Ishizuchi cauldron and Miocene tectonics in southwest Japan. *Memoirs of the Geological Society of Japan*, **42**, 297–349 [in Japanese with English abstract].

YOSHIDA, T., OHGUCHI, T. & ABE, T. 1995. Structure and evolution of source area of the Cenozoic volcanic rocks in Northeast Honshu arc. Japan. *Memoirs of the Geological Society of Japan*, **44**, 263–308. [in Japanese with English abstract].

YOSHIDA, T., KIMURA, J., OHGUCHI, T. & SATO, H. 1997. Structure and evolution of magma plumbing systems of island arc. *Bulletin of the Volcanological Society of Japan*, **42**, S189–S207 [in Japanese with English abstract].

YOSHIDA, T., AIZAWA, K., NAGAHASHI, Y., SATO, H., OHGUCHI, T., KIMURA, J. & OHIRA, H. 1999*a*. Geological history of Late Cenozoic island arc volcanism in Northeast Honshu arc and the formation of caldera swarm. *Earth Monthly*, Special issue, **27**, 123–129 [in Japanese].

YOSHIDA, T., TSUMURA, N., HASEGAWA, A., OKAMURA, S., ZHAO, D. & KIMURA, J. 1999*b*. Structure and evolution of upper mantle in the Northeast Honshu arc, Japan. *Earth Monthly*, **21**, 179–193 [in Japanese].

YOSHIDA, T., NAKAJIMA, J. *ET AL*. 2005. Evolution of late Cenozoic magmatism in the NE Honshu Arc and its relation to the crust–mantle structures. *Quaternary Research*, **44**, 195–216 [in Japanese with English abstract].

YOSHIMOTO, K., UCHIDA, N., SATO, H., OHTAKE, M., HIRATA, N. & OBARA, K. 2000. Microseismicity around the Nagamachi–Rifu fault, Miyagi Prefecture, Northeastern Japan. *Journal of the Seismological Society of Japan*, **52**, 407–416 [in Japanese with English abstract].

YOSHINAGA, T. & NAKAGAWA, M. 1999. Finding of primary basalt from Sannome-gata volcano, northeastern Japan, and its compositional variation. *Journal of Mineralogy, Petrology and Economic Geology*, **94**, 241–253.

ZASHU, S., KANEOKA, I. & AOKI, K. 1980. Sr isotope study of mafic and ultramafic inclusions from Itinome-gata, Japan. *Geochemical Journal*, **14**, 123–128.

ZHAO, D., HORIUCHI, S. & HASEGAWA, A. 1990. 3-D seismic velocity structure of the crust and the uppermost mantle in the northeastern Japan arc. *Tectonophysics*, **181**, 135–149.

ZHAO, D., HASEGAWA, A. & HORIUCHI, S. 1992. Tomographic imaging of P and S wave velocity structure beneath northeastern Japan. *Journal of Geophysical Research*, **97**, 19 909–19 928.

ZHAO, D., TIAN, Y., LEI, J. & LIU, L. 2009. Seismic image and origin of the Changbai intraplate volcano in East Asia: role of big mantle wedge above the stagnant Pacific slab. *Physics of the Earth and Planetary Interiors*, **173**, 197–206.

ZHAO, D., HUANG, Z., UMINO, N., HASEGAWA, A. & YOSHIDA, T. 2011. Seismic imaging of the Amur-Okhotsk plate boundary zone in the Japan Sea. *Physics of the Earth and Planetary Interiors*, **188**, 82–95, http://dx.doi.org/10.1016/j.pepi.2011.06.013

ZHAO, D., YANADA, T., HASEGAWA, A., UMINO, N. & WEI, W. 2012. Imaging the subducting slabs and mantle upwelling under the Japan Islands. *Geophysical Journal International*, **190**, 816–828.

ZHU, G., GERYA, T. V., YUEN, D. A., HONDA, S., YOSHIDA, T. & CONNOLLY, J. A. D. 2009. Three-dimensional dynamics of hydrous thermal–chemical plumes in oceanic subduction zones. *Geochemistry Geophysics Geosystems*, **10**, 11, Q11006, http://dx.doi.org/10.1029/2009GC002625

Pulling apart the Mid to Late Cenozoic magmatic record of the Gulf of California: is there a Comondú Arc?

S. E. BRYAN[1]*, T. OROZCO-ESQUIVEL[2], L. FERRARI[2,3] & M. LÓPEZ-MARTÍNEZ[4]

[1]*School of Earth, Environmental and Biological Sciences, Queensland University of Technology, GPO Box 2434, Brisbane, 4001, Australia*

[2]*Centro de Geociencias, Universidad Nacional Autonoma de Mexico, Blvd Juriquilla 3001, Querétaro, 76230, Mexico*

[3]*Instituto de Geología, Universidad Nacional Autonoma de Mexico, Circuito Investigacion Científica, Ciudad Universitaria, Mexico City, 04510, Mexico*

[4]*CICESE, Carretera Ensenada-Tijuana No 3918, Zona Playitas, Ensenada, Baja California, 22860, México*

**Corresponding author (e-mail: scott.bryan@qut.edu.au)*

Abstract: The composition of the lithosphere can be fundamentally altered by long-lived subduction processes such that subduction-modified lithosphere can survive for hundreds of millions of years. Incorrect petrotectonic interpretations result when spatial–temporal–compositional trends of, and source contributions to, magmatism are not properly considered. Western Mexico has had protracted Cenozoic magmatism developed mostly in-board of active oceanic plate subduction beneath western North America. A broad range of igneous compositions from basalt to high-silica rhyolite were erupted with intermediate to silicic compositions in particular, showing calc-alkaline and other typical subduction-related geochemical signatures. A major Oligocene rhyolitic ignimbrite 'flare-up' (>300 000 km^3) switched to a bimodal volcanic phase in the Early Miocene (*c.*100 000 km^3), associated with distributed extension and opening of numerous graben. Extension became more focused *c.*18 Ma resulting in localized volcanic activity along the future site of the Gulf of California. This localized volcanism (known as the Comondú 'arc') was dominantly effusive and andesite–dacite in composition. Past tectonic interpretations of Comondú-age volcanism may have been incorrect as these regional temporal–compositional changes are alternatively interpreted as a result of increased mixing of mantle-derived basaltic and crust-derived rhyolitic magmas in an active rift environment rather than fluid flux melting of the mantle wedge above the subducting Guadalupe Plate.

Supplementary material: References from which whole-rock geochemical and radiometric age data have been compiled in this paper are available at http://www.geolsoc.org.uk/SUP18645

Mid- to Late Tertiary magmatism along the continental margin of western Mexico (Fig. 1) is an example where interpretations on the tectonic setting of magmatism have been strongly influenced by the continent-margin position, calc-alkaline affinity, relatively primitive isotopic characteristics, the presence of andesitic or intermediate composition volcanic rocks and general association with continued subduction beneath western North America (e.g. Cameron *et al.* 1980; Lanphere *et al.* 1980; Hausback 1984; Wark *et al.* 1990; Umhoefer *et al.* 2001). A feature of the magmatic record in western Mexico is the general continuity of erupted/igneous compositions from basalt through to high-silica rhyolite (Fig. 2). This raises the question of whether the presence of calc-alkaline 'andesitic' igneous compositions offers any constraint on the tectonic setting of magmatism. Past studies have widely interpreted the mid- to Late Tertiary (*c.* 38–12 Ma) volcanism as recording a supra-subduction zone volcanic arc (e.g. Cameron *et al.* 1980; Hausback 1984; Sawlan & Smith 1984; Wark *et al.* 1990; Martín-Barajas *et al.* 1995, Ferrari *et al.* 1999; Martín *et al.* 2000; Umhoefer *et al.* 2001). However, for Oligocene to Early Miocene volcanic activity (*c.* 38–18 Ma), the overwhelming silicic composition, the eruptive scale, volume and output rate, and the rhyolite–ignimbrite-dominated character of the erupted products sourced from multiple calderas and fissures (e.g. Swanson & McDowell 1984; Aguirre-Diaz & Labarthe-Hernandez 2003; Swanson *et al.* 2006)

From: GÓMEZ-TUENA, A., STRAUB, S. M. & ZELLMER, G. F. (eds) 2014. *Orogenic Andesites and Crustal Growth.* Geological Society, London, Special Publications, **385**, 389–407.
First published online July 31, 2013, http://dx.doi.org/10.1144/SP385.8

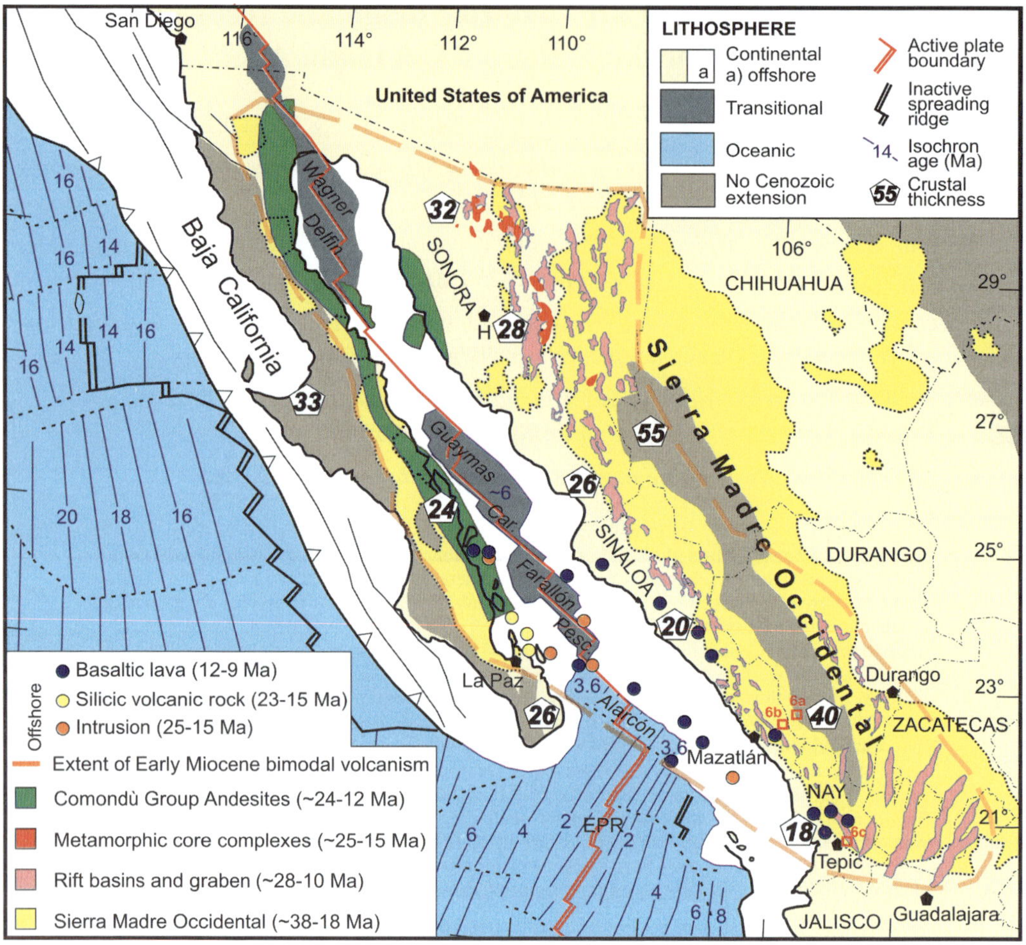

Fig. 1. Tectonic map of northwestern Mexico showing the lithospheric variation across the region, including unextended and extended continental regions and transitional to new oceanic crust formed by the propagating spreading centre in the Gulf of California. Superimposed on this tectonic map are the preserved extents of the Oligocene–Early Miocene silicic-dominant volcanic activity of the SMO (Ferrari *et al.* 2002; Bryan *et al.* 2008), and the dominantly bimodal phase during the Early Miocene that coincided with the wide development of graben and rift basins (McDowell *et al.* 1997; Ferrari *et al.* 2002) and a restricted belt of metamorphic core complexes in the state of Sonora (Nourse *et al.* 1994; Wong *et al.* 2010). Distribution of Comondú Group andesites from Umhoefer *et al.* (2001). Offshore Miocene igneous rocks from Ferrari *et al.* (2012, in press). Rift basin segments to the Gulf of California are labelled: Car, Carmen basin; Pesc, Pescadero basin; NAY, Nayarit; EPR, East Pacific Rise; H, Hermosillo. Red boxed areas near Mazatlán and Tepic refer to locations of photographs in Figure 6.

are inconsistent with modern expressions of supra-subduction zone volcanism. Equally, using narrow, silicic-magma dominated continental arc- to back-arc rifts like the Taupo Volcanic Zone (Cole 1990; Parson & Wright 1996) as analogues is also not appropriate (Bryan *et al.* 2008).

In contrast, mid-Miocene volcanism (*c.*18–12 Ma) has been reported to be predominantly intermediate in composition, with a restricted occurrence along the margins of the recently opened Gulf of California (Fig. 1). When combined with a facies comparison to stratovolcanic assemblages (Hausback 1984), this has been the basis for arguing a westward migration and re-establishment of supra-subduction zone arc volcanism along eastern Baja California before the final termination of subduction along western Mexico at *c.*12.3–12.5 Ma (Gastil *et al.* 1979; Stock & Hodges 1989; Lonsdale 1991). This interpretation is significant because: (1) prevailing models for the opening and development of the Gulf of California, despite differing in rifting kinematics, all imply Gulf extension began

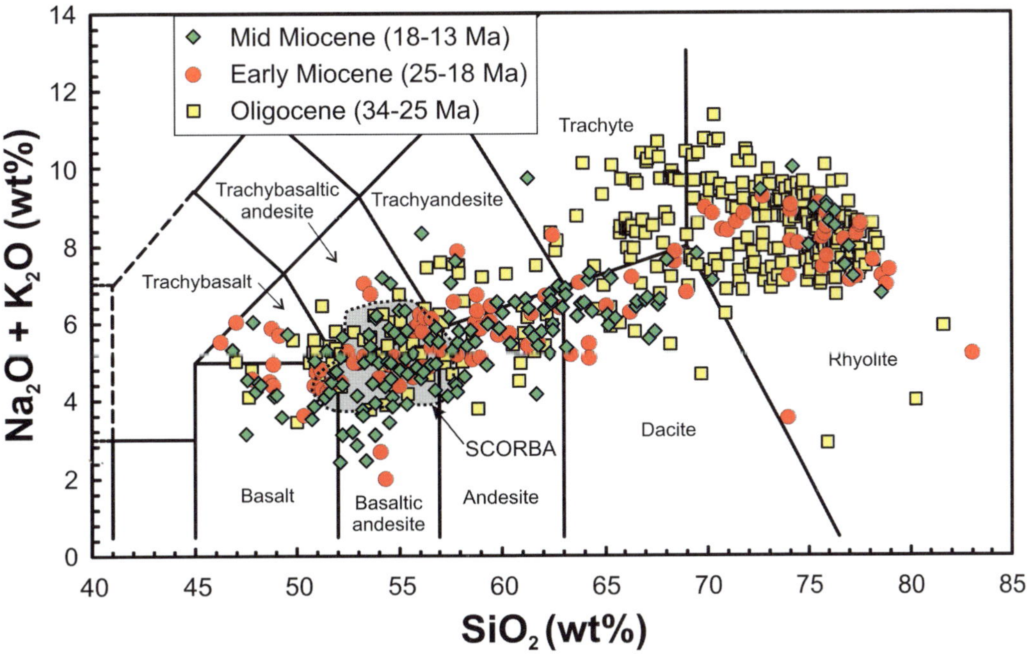

Fig. 2. Total alkali–silica diagram (TAS; Le Maitre *et al.* 1989) for Oligocene–Miocene (*c.* 34–13 Ma) igneous rocks from NW Mexico, plotted on an anhydrous basis, showing the spectrum of compositions generated during this broad interval. The field of the Southern cordillera basaltic andesites (SCORBA; Cameron *et al.* 1989) is from McDowell *et al.* (1997). Diagram contains 606 analyses.

after the termination of andesitic volcanism and subduction at 12.3 Ma (Stock & Hodges 1989; Atwater & Stock 1998; Fletcher *et al.* 2007); and (2) the Gulf of California has been considered an anomalously rapid zone of continental rupture, based on the onset of seafloor spreading interpreted to be only *c.* 6–10 myr after the cessation of subduction and arc volcanism (Umhoefer 2011). Such interpretations are predicated on extension beginning after 12 Ma because Middle Miocene andesitic volcanism has been universally considered a non-extending supra-subduction zone volcanic arc. This, however, is at odds with seismic data, well information and palaeontological data indicating rift basin development in the Gulf of California began much earlier (see Karig & Jensky 1972), and now supported by new studies (Ferrari *et al.* 2012; in press) from the Gulf region that demonstrate active extension beginning at least 18 Ma and spatially coincident with the andesitic volcanism. This is consistent with structural studies further east (e.g. Gans 1997) that indicate extension occurred mainly in the Oligo-Miocene, well before the cessation of subduction (see also Henry 1989). Several studies have demonstrated how extension can strongly influence magmatism in terms of magma generation (location, processes, rates, storage), and eruptive composition and style (e.g. Gans *et al.* 1989; Axen *et al.* 1993), such that intermediate magma compositions can be promoted by active extensional faulting (e.g. Johnson & Grunder 2000).

To understand better the origin and tectonic setting of the mid-Miocene andesitic volcanism in western Mexico, it is critical to constrain the spatial–temporal–compositional record and trends of magmatism and how this relates to the timing and location of extension across western Mexico. An important limitation on our understanding of the tectonomagmatic setting of mid- to late Tertiary volcanism in western Mexico has been that all current tectonic and petrogenetic interpretations have relied on ‘discovery-phase’ mapping begun in the 1970s and while at least 90% of the Sierra Madre Occidental (SMO) remains unmapped (Swanson *et al.* 2006); this has particularly been the case for the western margins of the province through the states of Nayarit and Sinaloa. A recent National Science Foundation (NSF) Margins focus on the opening of the Gulf of California has resulted in a closer examination by us of how magmatism transitioned from the huge SMO silicic Large Igneous Province through interpreted arc volcanism to crustal rupturing to open the Gulf of California in the Late Miocene. The examination here will begin with a review of the volcanic record of the SMO silicic Large Igneous Province that immediately preceded the mid-Miocene andesitic volcanism, referred to as the 'Comondú arc' or the middle and upper Comondú Group of Umhoefer *et al.* (2001). Constraints on the development of the Gulf of California will then be reviewed followed by an examination of the spatial–temporal–compositional trends of magmatism across this region from *c.* 40 Ma to better understand the tectonic and structural context of the mid-Miocene andesitic volcanism.

Geological background

Sierra Madre Occidental

The Sierra Madre Occidental (SMO, Fig. 1) is the largest silicic igneous province in North America (McDowell & Keizer 1977; McDowell & Clabaugh 1979; Ward 1995; McDowell & McIntosh 2012) and, although principally contained within western Mexico, is contiguous with silicic volcanism through the Basin and Range Province of western USA to the north (Lipman *et al.* 1972; Gans *et al.* 1989; Best & Christiansen 1991) and also with the ignimbrite province of the Sierra Madre Sur, south of the Trans Mexican Volcanic Belt (Martiny *et al.* 2000; Moran-Zenteno *et al.* 2007). At least 400 000 km^3 of dominantly rhyolitic ignimbrite was erupted, mostly between *c.* 38 and 18 Ma, but age dating over the last 40 years has identified two main pulses or 'flare-ups' of ignimbrite activity (Fig. 3): at *c.* 34–28 Ma and *c.* 24–18 Ma (Ferrari *et al.* 2002, 2007; Bryan *et al.* 2008; McDowell & McIntosh 2012). The Oligocene pulse is thought to be responsible for at least three-quarters of this erupted volume, whereas at least 100 000 km^3 was erupted in the Early Miocene.

Rhyolitic to dacitic ignimbrite represents at least 85–90% of the erupted volume with the remaining volume being rhyolitic lavas/domes, basaltic lavas and lesser andesitic lavas. Age dating has revealed brief but intense episodes of volcanism (*c.* 1 myr duration) emplacing kilometre-thick sections of ignimbrite across the province (e.g. McDowell & Keizer 1977; Ferrari *et al.* 2002; Swanson *et al.* 2006; McDowell & McIntosh 2012), attesting to rapid rates of silicic magma generation and eruption (Bryan *et al.* 2008). Ignimbrite sections within

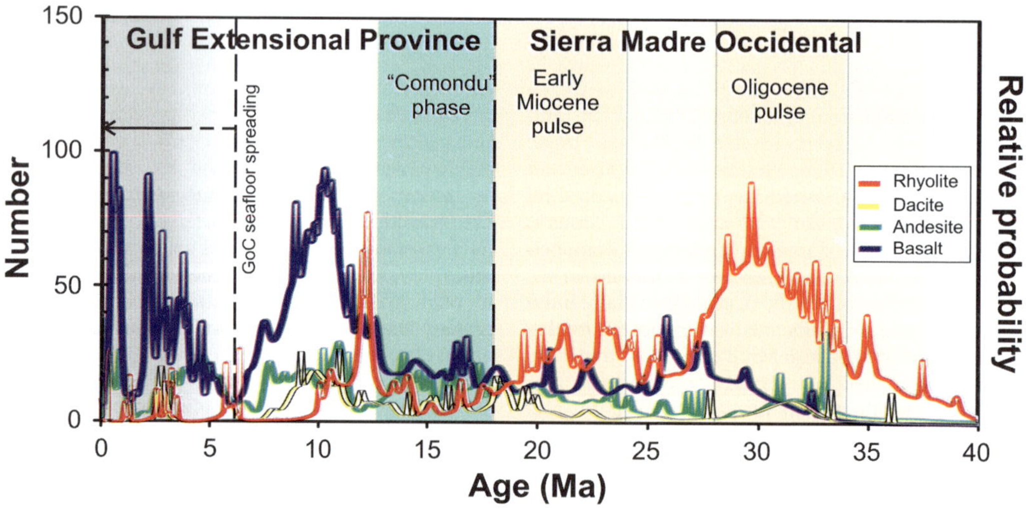

Fig. 3. Probability density plot of igneous ages from western Mexico for the period 40–0 Ma. Dated rocks have been grouped into four main compositional groupings: basalt (includes basaltic andesites and tholeiitic, calc-alkaline and alkaline varieties); andesite (includes calc-alkaline and adakitic-type compositions); dacite and rhyolite (includes high-silica rhyolites and peralkaline compositions). Important features of the diagram are: (1) the silicic dominant character of the Oligocene ignimbrite pulse; (2) the appearance of basalts during the Oligocene silicic ignimbrite pulse and increase in the frequency of basaltic eruptions up to the start of the Early Miocene pulse; these basaltic eruptions correspond to SCORBA of Cameron *et al.* (1989); (3) the bimodal character of the Early Miocene pulse; (4) the increase in andesitic compositions beginning *c.* 20 Ma until *c.* 13 Ma; (5) the abrupt decline in rhyolite magma generation and eruption beginning *c.* 19 Ma when andesite–dacite eruptions were more predominant; (6) an abrupt return to bimodal volcanism at *c.* 13 Ma with a concomitant decline in andesitic eruptions; and (7) after 10 Ma, volcanism becomes fundamentally basaltic to coincide with the onset of seafloor spreading in the GoC beginning as early as *c.* 6 Ma (Lizarralde *et al.* 2007). Diagram based on 1496 radiometric ages, and age data plotted using Isoplot (Ludwig 2003).

the central part of the province are flat-lying and define an unextended core to the SMO (Fig. 1), and this region also corresponds to the highest crustal thicknesses, up to 55 km (Bonner & Herrin 1999). In contrast, ignimbrite sections are faulted and tilted along both the western and eastern flanks of this unextended core (Henry & Aranda-Gomez 2000; Ferrari *et al.* 2002). Basaltic andesite lavas, referred to as the Southern Cordilleran Basaltic Andesite, or SCORBA (Cameron *et al.* 1989), appear to be widespread throughout the SMO, and in the northern SMO are commonly found intercalated with, or overlying, the youngest ignimbrites in each section (Swanson *et al.* 2006).

Early Miocene volcanic activity was largely superimposed on the Oligocene pulse of volcanism, except in the northern SMO, but also extended further west (Fig. 1) to be present on Baja California (e.g. Umhoefer *et al.* 2001). No apparent shift in the eastern limit of volcanism occurred for the central and southern SMO, but a westward shift is more pronounced for the northern SMO (Fig. 1). Recent dredge surveys and age dating of recovered rocks through the southern Gulf of California have confirmed the presence of Early Miocene bimodal volcanic and exhumed intrusive rocks offshore (Fig. 1) improving the pre-rift connection between Baja California and mainland Mexico (Orozco-Esquivel *et al.* 2010; Ferrari *et al.* 2012; in press). The Early Miocene phase shows significant differences from north to south, despite showing a similar north–south extent to the previous Oligocene pulse. While silicic volcanism appears to have been more volumetrically dominant in the SW part of the SMO (Ferrari *et al.* 2002; Bryan *et al.* 2008), Early Miocene volcanism was less abundant and dominantly mafic in composition across the northern SMO (McDowell *et al.* 1997; Murray *et al.* 2010, in press). In the central and southern SMO, volcanism was more bimodal, with thick rhyolitic ignimbrite packages, similar to the Oligocene sections, characterizing some areas (e.g. Espinazo del Diablo and El Salto successions, McDowell & Keizer 1977), whereas elsewhere, graben-focused bimodal volcanism is characteristic (Ferrari *et al.* 2002; Ramos Rosique 2012). Graben margins in the southern SMO are commonly defined by rhyolite domes, whereas basaltic lava packages up to 200 m thick and rhyolitic ignimbrites partly infill Early Miocene graben (Ramos Rosique *et al.* 2010; Ramos Rosique 2012). Recent zircon chronochemical studies of the Early Miocene rhyolites in the SW SMO have shown a very distinct zircon inheritance signature where the inherited zircon ages indicate remelting of silicic igneous rocks formed during the Oligocene and Early Miocene ignimbrite pulses (Bryan *et al.* 2008; Ferrari *et al.* 2012; in press; Ramos Rosique 2012). Mid to upper crustal remelting is interpreted to have been driven by basaltic magmatism invading high structural levels in the crust, which was aided by active crustal extension.

The Comondú 'Arc'

The eruption of andesitic rocks in the Early to Middle Miocene principally along eastern Baja California (the Comondú arc; Hausback 1984; Sawlan & Smith 1984; Sawlan 1991; Umhoefer *et al.* 2001) has widely been interpreted to mark the termination of the SMO and its broad zone of silicic dominant magmatism and extension beginning *c.*40 Ma, with the re-establishment of typical supra-subduction zone arc magmatism (e.g. Ferrari *et al.* 2007). This was a re-establishment because, as observed along the length of the western North American margin (McQuarrie & Oskin 2011), no well-defined frontal 'andesitic' arc had existed during the Oligocene, when instead a broad zone (up to 1000 km) of volumetrically dominant silicic volcanism occurred (e.g. Lipman *et al.* 1972; Gans *et al.* 1989; McDowell & Mauger 1994; McDowell 2007). This was the case for Baja California, which had remained attached to mainland Mexico until the Late Miocene. Along Baja California, Cretaceous rocks are generally overlain by lower Tertiary marine and local non-marine sedimentary rocks (recording this absence of frontal 'andesitic' arc volcanism), or late Oligocene to Miocene volcanic rocks above a regional unconformity (e.g. Beal 1948; Hausback 1984; Umhoefer *et al.* 2001). The Upper Oligocene to Middle Miocene volcano-sedimentary units in Baja California are known as the Comondú Group (Umhoefer *et al.* 2001), and have been widely interpreted to be forearc basin and volcanic arc deposits formed immediately prior to plate boundary reorganization and rifting that opened the Gulf of California (Hausback 1984; Umhoefer *et al.* 2001; Conly *et al.* 2005; Godinez *et al.* 2010; Umhoefer 2011). The Comondú Group is currently divided into three informal stratigraphic units (Umhoefer *et al.* 2001). A lower unit (<500 m thick, *c.* 30–19.5 Ma) is dominated by quartz sandstones and conglomerate including aeolian sandstone, and interbedded resedimented pyroclastic units (tuffaceous sandstone), whereas several rhyolitic ignimbrites and localized basaltic lavas are prominent in the upper parts. Detrital zircon U–Pb ages on aeolian sandstones suggest a maximum depositional age of *c.* 25 Ma (Godinez *et al.* 2010), and new U–Pb and Ar/Ar ages bracket ignimbrite emplacement between *c.* 24 and 19.50 ± 0.05 Ma (Drake 2005; Godinez *et al.* 2010). This lower unit is thus temporally and compositionally correlated with Early Miocene bimodal volcanism offshore in the Gulf (Orozco-Esquivel *et al.* 2010) and on mainland Mexico. These Comondú Group ignimbrites are outflow facies and generally deposited at medial to distal distances from their sources interpreted to occur to the east (Umhoefer *et al.* 2001; Drake 2005). The middle and upper 'Comondú' units (*c.* 19.5–12 Ma; *c.*1 km thick) are more intermediate in composition and dominated by sedimentary breccia with interspersed andesite–dacite lavas/domes, particularly in southern Baja California, and cross-cut by dykes and minor porphyry intrusions (Hausback 1984; Sawlan & Smith 1984; Umhoefer *et al.* 2001; Godinez *et al.* 2010). A few interbedded rhyolitic ignimbrites have also been reported (Hausback 1984; ash flow tuff D of Drake 2005). Some zircon U–Pb ages on biotite granodiorite porphyry intrusions indicate emplacement ages of 19.9 ± 0.73 and 16.3 ± 0.49 Ma (Godinez *et al.* 2010), whereas andesite–dacite lavas range in age from *c.* 19.5 to *c.* 11 Ma (Hausback 1984; Sawlan & Smith 1984; Martín-Barajas *et al.* 1995; Martín *et al.* 2000; Umhoefer *et al.* 2001; Drake 2005). The lava dome and flow units are vent to proximal facies, but critically, several stratigraphic studies have shown the andesite–dacite lavas to be interbedded with sedimentary rocks rather than forming thick stacked piles of lava, as expected in the construction of stratocones.

Gulf of California

The Gulf of California (GoC) has been one of the focus sites for the Rupturing Continental Lithosphere initiative of the NSF-funded MARGINS programme from 2004–2010. It is a *c.*1400 km long, highly sedimented, oblique rift characterized by long transform faults and short spreading centres (Lonsdale 1989; Lizarralde *et al.* 2007). Despite different models (see review in Fletcher *et al.* 2007), rifting has been considered to have developed rapidly following cessation of subduction of the Guadalupe Plate (and arc volcanism) at about *c.* 12.3–12.5 Ma (Stock & Hodges 1989; Ferrari *et al.* 2007; Fletcher *et al.* 2007; Lizarralde *et al.* 2007; Umhoefer 2011). Since rift inception, the Gulf has opened between *c.* 300 and 500 km (Oskin *et al.* 2001; Fletcher *et al.* 2007), with the Gulf progressively unzippering to the north such that active extension of continental crust is currently occurring at the northern end of the Gulf. The timing of Gulf inception and rifting at *c.* 12.3 Ma is generally consistent with the switch to more alkaline and tholeiitic basaltic volcanism, and the reappearance of more rhyolitic volcanism at *c.* 12–13 Ma (Fig. 3; e.g. Martín-Barajas *et al.* 1995; Stock *et al.* 1999; Martín *et al.* 2000; Vidal-Solano *et al.* 2007). Most dates for initiation of marine sedimentation in the Gulf region are at 8 Ma and younger (e.g. Oskin & Stock 2003), but an earlier marine incursion and the development of a seaway in the middle Miocene has been the basis for the concept of a 'Proto-gulf' (Karig & Jensky 1972). While the timing of marine incursion remains contentious and based principally on onshore exposures, comprehensive stratigraphic revision (Carreño & Smith 2007) and recent micropalaeontological studies of several deep wells drilled in the Wagner, Consag and Tiburón basins of the GoC document at least 1870 m of marine sediments deposited before 11.2 Ma (Helenes *et al.* 2009; Helenes 2012).

Seafloor-spreading centres formed at variable times from *c.*6 to 2 Ma (Umhoefer 2011), with a spreading centre in the Guaymas sub-basin (central Gulf) interpreted to have began *c.* 6 Ma (Fig. 1) based on the width of the new igneous crust observed in seismic refraction profiles (Lizarralde *et al.* 2007). The Alarcón spreading centre began forming proto-oceanic crust *c.* 3–3.5 Ma (DeMets 1995), and true seafloor spreading at present rates began at 2.4 Ma (Sutherland 2006; Umhoefer *et al.* 2008).

Summary

Two important outcomes on the Miocene history of the GoC region are apparent from the diversity of recent studies in the region. The first is that recent dating studies from the SMO, the GoC and Baja California confirm eruptive ages on rhyolites and basalts related to the Early Miocene pulse of the SMO extend to as young as *c.* 17 Ma (Hausback 1984; Martín *et al.* 2000; Umhoefer *et al.* 2001; Drake 2005; Bryan *et al.* 2008; Ramos Rosique 2012; Ferrari *et al.* 2012; in press). This continued bimodal volcanism thus overlaps the onset of interpreted arc volcanism along Baja California at *c.* 19.5 Ma (Umhoefer *et al.* 2001), whereas others have considered 'arc' volcanism began earlier in northern Baja California at *c.* 21 Ma (e.g. Martín-Barajas *et al.* 1995). This age overlap thus suggests no abrupt termination to SMO bimodal volcanism when rejuvenation of supra-subduction zone arc volcanism was apparently initiated, despite the bimodal and andesitic volcanism spatially overlapping (Fig. 1). However, regional temporal–compositional patterns in erupted magma compositions do indicate a strong compositional shift from dominant bimodal volcanism to more intermediate composition volcanism beginning *c.* 19 Ma (Figs 3 & 4). The second outcome is that biostratigraphical studies, particularly from the offshore basins are recording a more protracted history of rift basin subsidence and sediment accumulation in the GoC and that marine sedimentation began at least in the middle Miocene (*c.* 12 Ma; Helenes

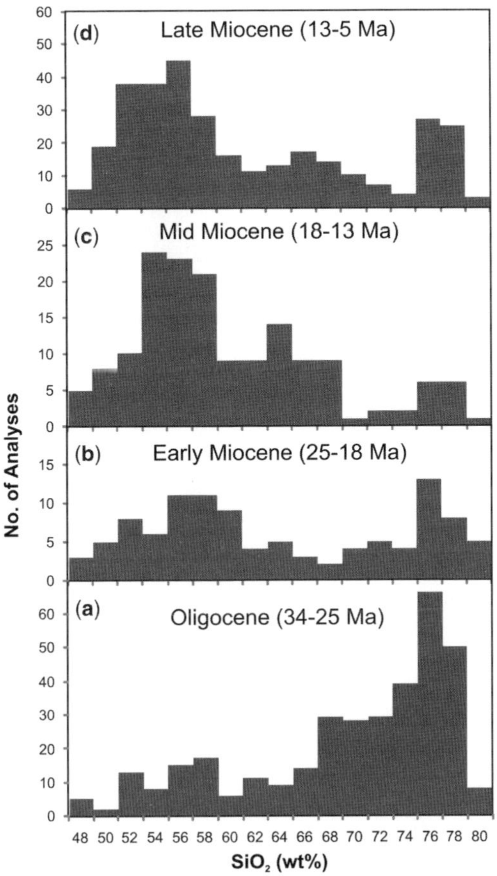

Fig. 4. SiO_2 histograms of Oligocene–Miocene igneous rocks of NW Mexico. (**a**) The Oligocene pulse of the Sierra Madre Occidental ($n = 346$ analyses) is silicic dominant and in particular, high-silica rhyolite-dominant. Many of the ignimbrites and lavas are high-silica rhyolites (>75 wt% SiO_2), and it is seldom appreciated that high-silica rhyolites, although common in intracontinental settings, are uncommon in continental margin arcs (Hildreth & Fierstein 2000). (**b**) The Early Miocene bimodal pulse of the Sierra Madre Occidental ($n = 106$ analyses). (**c**) Middle Miocene volcanism that is principally focused around the margins of the Gulf of California ($n = 159$ analyses), where in contrast to previous episodes, silicic volcanism was subdued, and basaltic andesite to dacite compositions dominant. (**d**) Late Miocene volcanism ($n = 321$ analyses) is marked by a return to significant silicic volcanism and true basaltic compositions. Note change in vertical scale for each time interval.

2012; cf. Sutherland *et al.* 2012). This would require unrealistically rapid crustal thinning and subsidence if all extension began *c.* 12 Ma. Collectively, these observations are at odds with the notion of rapid switching from bimodal volcanism across the western SMO and southern Baja California, to establishing a subduction-related andesitic magmatic arc along the eastern side of Baja California and westernmost mainland Mexico from *c.* 22–19 Ma, which then abruptly terminated at 13–12 Ma to switch back again to predominantly bimodal alkaline volcanism (Fig. 3), coincident with the onset of rifting of the GoC at this time (e.g. Stock & Hodges 1989; Umhoefer *et al.* 2001; Sutherland *et al.* 2012).

Temporal–compositional trends of Oligocene–Miocene magmatism across western Mexico

Published and unpublished whole-rock compositional and age data for 40–0 Ma igneous rocks from Western Mexico, occurring in a NW-trending belt from *c.* 22°N 103°W, to 32°N 116°W have been compiled in this study (see Supplementary material) and serves as an update to that of Ferrari *et al.* (1999). The dataset comprises three types of data: (1) whole-rock compositional data for which radiometric ages have been obtained from the same samples ($n = 461$); (2) whole-rock compositional data that have stratigraphic control and a stratigraphic age can be assigned (e.g. Oligocene, Late Miocene; $n = 1567$); and (3) samples that have radiometric age data and lithological information (e.g. basalt lava, rhyolite ignimbrite), but may lack corresponding whole-rock compositions ($n = 1520$). For samples with radiometric ages from 40–0 Ma, this includes K/Ar (whole rock, groundmass, feldspar, biotite, hornblende), $^{40}Ar/^{39}Ar$ (whole rock, groundmass, feldspar, hornblende, biotite), and zircon fission track and U–Pb ages. Of the samples with whole-rock compositional data, 1486 samples have major element analyses with the remainder having whole-rock isotopic (Sr, Pb, Nd) compositions ± some trace element abundances. Overall, trace element data are sparser, with only 288 samples having full rare earth element analyses. Most of the whole-rock geochemical data have been measured by XRF, but ICP-MS trace element data have increased over the last 10 years (e.g. Martín *et al.* 2000; Till *et al.* 2009). In the following figures, igneous compositions are grouped into the following: (1) basalts, with 45–52 wt% SiO_2; (2) basaltic andesite, 52–56 wt% SiO_2; (3) andesite, 56–63 wt% SiO_2; (4) dacite, 63–68 wt% SiO_2; and (5) rhyolite, 68–79 wt% SiO_2.

Compiled whole-rock compositional data from the SMO for which age data are available (Figs 3 & 4) illustrate the dominantly silicic compositional character of the Oligocene ignimbrite pulse, but with an increasing appearance and abundance of basaltic lavas from *c.* 30 Ma. In contrast, the Early Miocene pulse was distinctly bimodal with a paucity of volcanic compositions between 64 and 68 wt% SiO_2 (Fig. 4). Of note is the broad silicic peak for both the Oligocene and Early Miocene pulses or two compositional groupings of dacitic to low-silica rhyolite and high-silica rhyolite suites, with the latter being dominant and occurring as lavas/domes and ignimbrites.

The disappearance of rhyolite and high-silica rhyolite compositions across the region is pronounced at *c.* 19–18 Ma (Fig. 3). From 18–13 Ma, erupted magmas were more intermediate in composition, as recognized in many previous studies, with a paucity of both basalt (<52 wt% SiO_2) and rhyolitic (>68 wt% SiO_2) magmas. Basaltic andesite and low-silica andesite compositions dominate with andesite–dacite compositions also important (Fig. 4), and almost all have been emplaced as lavas. Consequently, there is not only a strong compositional change at 19–18 Ma, but also a change from mixed explosive–effusive to dominantly effusive eruptive styles. However, the period of dominant andesitic volcanism (*c.* 18–13 Ma) that defines the basis for a subduction-related andesitic magmatic arc is actually much shorter than is widely argued (Fig. 3); andesitic arc volcanism has previously been interpreted from *c.* 22–16 Ma and 22–12 Ma across northern and southern Baja California, respectively (Hausback 1984; Sawlan & Smith 1984; Sawlan 1991). The reappearance of rhyolite and high-silica rhyolite at 13–12 Ma (Fig. 3) is distinctive because these magma compositions are also peralkaline and mark the recurrence of explosive eruptions emplacing ignimbrites (Vidal-Solano *et al.* 2008) such as the 12.6 Ma alkali rhyolite Tuff of San Felipe (Stock *et al.* 1999). The alkali rhyolite eruptions coincide with the eruption of true basaltic compositions regionally that are also more alkaline in composition. From 13–5 Ma (Fig. 4) magma compositions became fundamentally more basaltic and alkalic with alkalic and tholeiitic basaltic eruptions around the margins of the GoC eventually coinciding with the onset of seafloor-spreading from 6–3.5 Ma. As with the preceding Miocene phase, dacitic compositions (64–68 wt % SiO_2) are significant between 12 and 8 Ma (Figs 3 & 4).

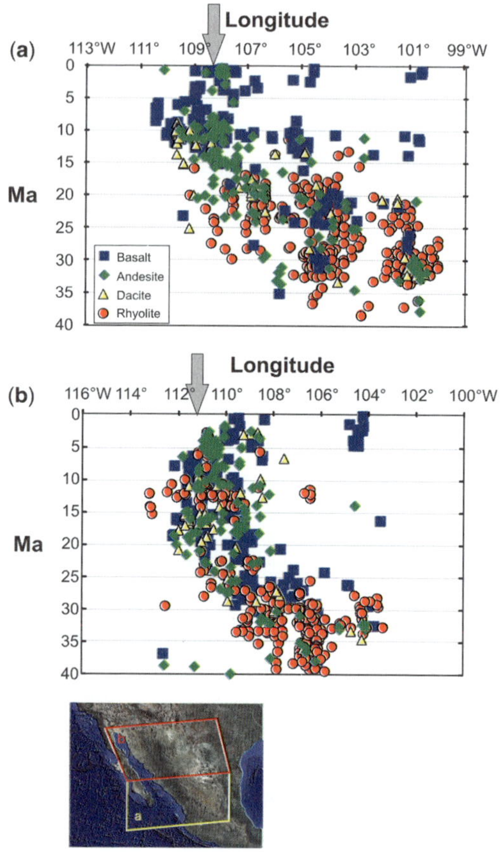

Fig. 5. Time–space plots of igneous activity across western Mexico and around the Gulf of California region. In (**a**) age data are for the region between 20° and 26°N; and in (**b**) between 26° and 30°N (see boxed areas in inset). Dated samples from Baja California have been displaced to their 'pre-rift' positions, whereby for each sample the latitude has been adjusted by 2.3°S and the longitude by 3.9°E. The approximate position of the Gulf of California in both sections is marked by the grey arrow. Igneous rocks have been grouped into four main compositional groupings: basalt (includes basaltic andesites and tholeiitic, calc-alkaline and alkaline varieties); andesite (includes calc-alkaline and adakitic-type compositions); dacite; and rhyolite (includes high-silica rhyolites and peralkaline compositions). See text for discussion. Inset map of western Mexico from Google Earth.

Spatial–temporal–compositional trends

Important insights can also be gained if the long-term temporal compositional trends discussed above can be placed in a spatial context. The age dataset for which lithological ± whole-rock compositional data are available is used here. Age data are plotted for east-west transects across the GoC for the region between 20° and 26°N and 26° and 32°N, as shown in Figure 5. Dated igneous rocks have been grouped into four compositional groupings: basalt (<56 wt% SiO_2), andesite (56–63 wt% SiO_2), dacite (63–68 wt% SiO_2) and rhyolite (>68 wt% SiO_2). Several key features are apparent from Figure 5:

(1) As observed in Figures 3 and 4, Oligocene to Early Miocene (*c.* 40–20 Ma) volcanism is silicic-dominant, and widely distributed (up to 800 km width). Across this broad zone of silicic volcanism between 40–30 Ma, low volumes of basalts and andesites were also erupted.

(2) By the Early Miocene, volcanism became strongly bimodal, but this switch occurred earlier in the northern region (*c.* 30 Ma), and basaltic magmas dominated from 30–20 Ma. In the north (Sonora-Chihuahua), basalts (SCORBA of Cameron *et al.* 1989) commonly marked the termination of silicic volcanism in an area and, since *c.* 28 Ma, were emplaced on top of rhyolitic ignimbrites and into developing extensional basins (until *c.* 15 Ma; McDowell *et al.* 1997). In the south (Jalisco, Nayarit, Sinaloa), the basalts appear to represent the triggering of the Early Miocene silicic ignimbrite pulse (Bryan *et al.* 2008).

(3) Rather than a westward migration of volcanic activity, as is widely reported from the Oligocene to Miocene (e.g. Damon *et al.* 1981; Ferrari *et al.* 1999), volcanism did extend westwards at *c.* 30 Ma, but importantly, there was no corresponding shift of the eastern limit of volcanism at this time, which would be expected if volcanic activity was intimately linked to back-arc extensional processes and driven by slab roll back (e.g. Clift *et al.* 1994). A westward shift in the location of the eastern limit of volcanism at *c.* 25 Ma is more pronounced in the northern region, and did not occur until *c.* 20 Ma in the southern SMO.

(4) The extensive Early Miocene bimodal volcanism was associated with distributed extension and the formation of numerous graben along the western and southern SMO (Fig. 1). Inception of most graben had begun by 24 Ma in the southern SMO (Ferrari *et al.* 2002; Ramos Rosique 2012) and by 27 Ma across the northern and western SMO (McDowell *et al.* 1997). Graben development occurred over a horizontal distance of up to 450 km (Fig. 1) and at least 600 km inboard of the palaeocontinental margin at that time. Despite the size and extent of graben, the

Fig. 6. Field relationships between the SMO and GoC showing the progressive tilting and deformation of the Early Miocene silicic ignimbrite successions in the central and western SMO. (**a**) view west from the Mazatlán–Durango old highway (23°39.927′N 105°43.340′W) to the flat-lying *c.* 23 Ma Espinazo–El Salto sequence of Waitt (1970) and McDowell & Keizer (1977), and part of the undeformed core to the SMO. (**b**) Tilted Early Miocene ignimbrite successions approximately 30 km to the west of A viewed to the north of the Mazatlán-Durango old highway (23°24.329′N 105°54.634′W). (**c**) Tilt blocks of Early Miocene ignimbrites 27 km east of Tepic with ignimbrites dipping up to 35°E (21°34.030′N 104°37.880′W). The ignimbrites continue to the SSE below younger lavas and where they crop out again have been dated at 20.46 Ma (Ferrari *et al.* in press). 25 km to the NW are flat-lying (post-extensional) basaltic lavas dated between 11 and 10 Ma and form part of an extensive series of flat-lying 9–12 Ma basaltic lavas along the coastal plain of Nayarit and Sinaloa (see Fig. 1). These field relationships bracket the timing of major extension in the GoC to between *c.* 18 and *c.* 12 Ma.

magnitude of Latest Oligocene–Early Miocene extension associated with them is inferred to be modest (<10% stretching, Ferrari *et al.* 2002). Soon after the onset of this wide zone of rifting, a relatively narrow belt of high magnitude extension producing metamorphic core complexes from *c.* 25–15 Ma (Nourse *et al.* 1994; Wong *et al.* 2010) also developed across the NW SMO (Fig. 1). The prominent westward shift in volcanic activity at *c.* 25 Ma and subdued silicic volcanism from *c.* 27 Ma for the northern region may in part reflect suppression of volcanism in areas of high magnitude extension (Gans & Bohrson 1998) that was occurring along the eastern limits of this volcanism.

(5) In both regions, andesite–dacite volcanism becomes more abundant beginning *c.* 19 Ma (see also Fig. 3), and in contrast to earlier episodes of andesite eruption, is much more spatially restricted to the location of the nascent GoC.

(6) Although andesite–dacite volcanism was significant from *c.* 20 to 10 Ma, as observed along Baja California, it was coeval with (basalt-dominant) bimodal volcanism occurring regionally. This is particularly the case during the period 20–16 Ma for the southern region. Rhyolitic volcanism producing ignimbrites and lavas persisted until at least 16 Ma in the southern region with rhyolites dated from Baja California (Hausback 1984; Hosack 2006), Nayarit (Ferrari *et al.* 2012; in press) and Jalisco (Nieto-Obregón *et al.* 1981; Bryan *et al.* 2008). Consequently, spatial–temporal distinction between the termination of syn-extensional bimodal volcanism across the SMO and the beginnings of more intermediate composition magmatism around the GoC is blurred.

(7) The style, composition and extent of volcanism change significantly at *c.* 13 Ma. Bimodal volcanism characterises the northern region (coastal Sonora and Baja California) and rhyolitic compositions became more peralkaline (e.g. Tuff of San Felipe, Stock *et al.* 1999; Vidal-Solano *et al.* 2007). In contrast, volcanism is more basaltic and widespread in the southern region, with few rhyolites dated younger than 13 Ma. Despite the widespread eruption of basalts across the southern region, andesite–dacite volcanism remains principally focused around the margins of the nascent GoC. Andesitic compositions are particularly spatially restricted in the northern region from 10–2 Ma, despite other magma compositions being erupted over a broader region (Figs 4 & 5).

Timing relationships to extension

An important constraint in understanding the significance of the spatial–temporal compositional trends described above is understanding how these magmatic trends correspond spatially and temporally to the onset, or changes to the rate and breadth, of extension. Most of the SMO has been affected by different episodes of dominantly extensional deformation that began in the Oligocene and potentially at the end of Eocene (Henry 1989; Aguirre-Diaz & McDowell 1993; Ferrari *et al.* 2007). However, the central zone to the SMO remains unaffected by extension (Fig. 1) where ignimbrite sections rise to high altitudes (>3000 m asl), remain flat-lying (Fig. 6) and crustal thicknesses are high (Bonner & Herrin 1999). This unextended core of the SMO represents a physiographic boundary between what has been defined as the 'Mexican Basin and Range,' to the east, and the 'Gulf Extensional Province,' to the west (Henry & Aranda-Gómez 2000). At the northern and southern ends of the SMO (i.e. northern Sonora and Chihuahua and Nayarit-Jalisco, respectively) these two provinces merge where extension has affected the entire width of the SMO (Ferrari *et al.* 2007).

The age of extension in the Gulf Extensional Province is thought to have been largely Late Miocene to Recent and associated with the termination of subduction and onset of rifting to open the GoC at *c.* 12.3 Ma (e.g. Stock & Hodges 1989; Henry & Aranda-Gomez 2000). However, an intense phase of extension resulting in up to 100% crustal thinning and producing metamorphic core complexes occurred from *c.* 25–15 Ma in Sonora (e.g. Nourse *et al.* 1994; Gans 1997; Gans *et al.* 2003; Wong *et al.* 2010). These core complexes slightly post-date the onset of a milder phase of extension producing sediment and basalt-filled graben depressions (Baucarit Formation of King 1939) through Sonora. Basaltic to andesitic lavas ranging in age between 27 and 20 Ma are common toward the base of these rift basin successions (McDowell *et al.* 1997; Paz-Moreno *et al.* 2003).

The age of extension along the southeastern margin of the GoC through Sinaloa and Nayarit has been less clear. Recent studies (Ferrari *et al.* 2012; in press) have revealed that kilometre-thick ignimbrite sections of the SMO, dated as young as 20 Ma, have been tilted by up to 35° (Fig. 6) with some blocks remaining at high elevations (*c.* 2000 m). However, along the low-relief coastal plain are extensive, flat-lying and undeformed basaltic lava fields dated between 12–9 Ma extending for at least 700 km along the eastern margin of the GoC (Fig. 1; Ferrari *et al.* 2012; in press; see also Gastil *et al.* 1979). Similar-aged basalts have also been dredged from the submerged continental margins to the southern GoC (Ferrari *et al.* 2012; in press).

Important insights also come from the Baja California sections of the Comondú Group, where interpreted vent and proximal facies lava flows and domes are buried by thick successions of medial facies bedded sedimentary conglomerate/breccia and sandstone, and distal ignimbrites (e.g. Hausback 1984; Umhoefer *et al.* 2001; Drake 2005). In particular, positive relief lava domes with thicknesses of up to 300 m have been shown to be buried by sedimentary deposits (Drake 2005). These depositional and facies relationships demonstrate that andesitic to dacitic lavas and domes of the Comondú Group were being emplaced into actively subsiding sedimentary basins.

In summary, field, stratigraphic and now new age data from the eastern margin of the GoC indicate that large-magnitude extension must have occurred between *c.* 25 and 12 Ma to the north (e.g. Gans 1997) and between *c.* 18 and 12 Ma in the central and southern Gulf region. This extension must have post-dated the final phases of bimodal and ignimbrite-dominant activity of the Early Miocene pulse of the SMO, and preceded the widespread eruption of flat-lying, (undeformed) basaltic lavas along the eastern margin of the GoC beginning *c.* 12 Ma (Fig. 1). Importantly, as recognized in previous studies, the present-day crustal thickness beneath the unextended core of the SMO is between 40 and 55 km, whereas along the margins of the GoC it is between 18 and 26 km (Gomberg *et al.* 1988; Couch *et al.* 1991; Bonner & Herrin 1999; Persaud *et al.* 2007). Consequently, most of this crustal thinning occurred prior to the termination of subduction at *c.* 12.3–12.5 Ma, and was coincident with the major period of andesitic volcanism (middle and upper Comondú Group of Umhoefer *et al.* 2001).

Discussion

Origin of the Middle Miocene Comondú Andesites

In determining the origin of the Comondú Group andesites and in particular, why there was a discrete period of dominantly intermediate volcanism between *c.* 18–13 Ma, several key points must be considered from the observations and discussion given above. First, there has been no preceding history of andesitic volcanism or existence of a well-defined frontal volcanic arc in the region since at least the Eocene, despite ongoing subduction. Andesitic volcanism occurred, but (1) was diffuse and low-volume occurring over a very broad width in-board of the margin (>600 km) where it was closely associated with silicic and basaltic volcanism (Fig. 5); (2) when eruptions of andesitic–dacitic magma compositions became dominant in the mid-Miocene, they were localized in space and time around the nascent GoC (Figs 1 & 5); (3) this andesitic phase was coincident with the major period of crustal extension through the Gulf Extensional Province with crustal thinning up to 100%; and (4) both immediately before and after, and regionally, volcanism was clearly bimodal in character (Fig. 3).

Limited petrological and geochemical data have been published on the Comondú Group andesites and much interpretation has been based on major element chemistry (Gastil *et al.* 1979; Hausback 1984; Sawlan & Smith 1984). These previous studies have established the general 'calc-alkaline' chemistry, intermediate composition and similarity to modern subduction-related arc successions for the Comondú Group andesites. However, alkalic basalts have also been reported from these suites (Martín-Barajas *et al.* 1995), which are atypical of modern, unextended arc settings. The Comondú Group andesites are commonly, moderately to very crystal-rich, and vary from hornblende to pyroxene andesites (e.g. Sawlan & Smith 1984; Martín *et al.* 2000). However, magmatic heterogeneity of these andesites is apparent from the brief petrographic descriptions with reports of resorbed olivine (Sawlan & Smith 1984), quartz and alkali feldspar (Martín *et al.* 2000); complexly zoned plagioclase (Sawlan & Smith 1984); megacrystic hornblende (Drake 2005); enclaves (Drake 2005); and granitic and upper crustal xenoliths (Martín *et al.* 2000). Recent U–Pb zircon dating of a Miocene porphyry intrusion that intrudes the Lower Comondú Group (Godinez *et al.* 2010) has revealed the presence of abundant Cretaceous zircons indicating remelting of the underlying calc-alkaline Cretaceous batholith. Collectively, these characteristics indicate the importance of open system processes, magma mixing (involving basaltic magmas) and crustal assimilation/partial melting in Comondú Group andesite genesis. Importantly, the crustal materials involved have a strong subduction heritage given the long-lived history of subduction along the western margin of North America since the Triassic. Therefore, no reliance should be placed on the whole-rock geochemical signatures for interpreting the tectonic environment in which the magmas were emplaced where crustal assimilation has demonstrably occurred (cf. Conly *et al.* 2005). Furthermore, recent studies of arc andesites are emphasizing that most if not all crystals are xenocrysts or antecrysts, being picked up by initially aphyric mafic melts on their way to the surface (Zellmer *et al.* in press) such that the bulk rock compositions will incorrectly image the mantle source and prove unreliable for tectonic discrimination. Consequently, porphyritic intermediate compositions likely represent remobilized igneous protoliths (Zellmer 2009), and much of their bulk chemical characteristics can be inherited from those source materials. As has been discussed elsewhere, there is no *a priori* requirement for I-type calc-alkaline magmatism to be directly related to active subduction processes (e.g. Roberts & Clemens 1993; Morris & Hooper 1997).

Mineralogical evidence suggests the importance of magma mixing being an important process. Previous studies have illustrated linear variation trends for major elements (e.g. Sawlan & Smith 1984; Hausback 1984). Plotting of several trace element ratio combinations also reveals linear patterns and apparent mixing trends (Fig. 7). The role of mixing is a dominant process at the regional scale, and the existence of a subduction-related signature is examined in Figure 7. The use of Ba/Nb as a subduction-related signature is similar to the approach of Till *et al.* (2009) who used Ba/Ta, but because there are few samples with Ta abundances, we use the Ba/Nb ratio here. However, elevated Ba/Nb ratios are also a crustal signature. The Ba/Nb–La/Nb data show a spectrum from 'intraplate' basaltic compositions, as represented by the Early and Late Miocene alkali basalts in the region, to more arc-like and crustal compositions with high Ba/Nb ratios. The trends for the different compositional groupings are not readily explainable by fractional crystallization processes since the mafic, intermediate and silicic igneous compositions overlap. Of note is that from the Early Miocene onwards, a distinctly high Ba/Nb suite of rocks were emplaced (Ba/Nb > 150), which are absent from the Eocene and Oligocene suites. These high Ba/Nb and La/Nb rocks are almost entirely small-volume andesite–dacite lavas geographically restricted to the northern GoC (western Sonora, and Baja California); some published whole-rock isotopic data (e.g. Mora-Klepeis & McDowell 2004) indicate relatively radiogenic Sr ($^{87}Sr/^{86}Sr = 0.70564$) and unradiogenic Nd ($^{143}Nd/^{144}Nd = 0.51258$) compositions compared to other volcanics in the region. The high Ba/Nb ratios and Sr and Nd isotopic compositions are most similar to Palaeozoic to Proterozoic silicic orthogneisses from north central Chihuahua (Cameron

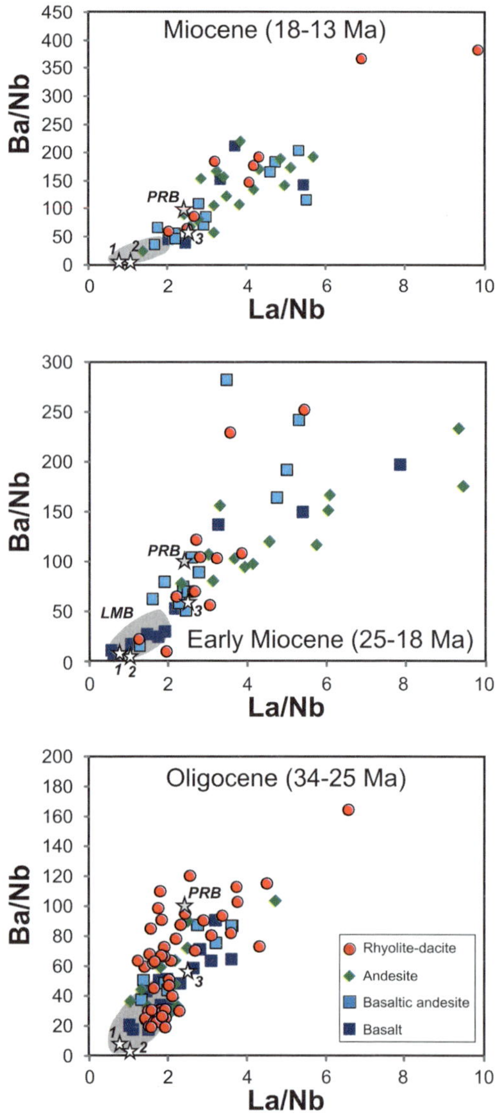

Fig. 7. Ba/Nb v. La/Nb plots for Oligocene–Miocene (34–13 Ma) igneous rocks from NW Mexico, with data organized into compositional groupings: basalts (<52 wt% SiO_2); basaltic andesite (52–56 wt% SiO_2); andesite (56–63 wt% SiO_2); and dacite–rhyolite (>63 wt% SiO_2). Stars labelled 1, 2 and 3 are E-MORB, N-MORB (Sun & McDonough 1989) and averaged calc-alkali andesite compositions (based on 2865 analyses from the GeoRoc database), respectively. PRB, is the average composition of the Peninsular Ranges Batholith from Silver & Chappell (1988). Grey shaded region labelled LMB represents compositions of Late Miocene alkali basalts from the region (e.g. Henry & Aranda-Gomez 2000; Pallares *et al.* 2007). The anomalously high Ba/Nb ratios (>200) approach values for Paleozoic–Proterozoic silicic orthogneiss and Tertiary plutonic-derived, glass-rich silicic xenoliths analysed from the La Olivina basaltic centre in northern Chihuahua by Cameron *et al.* (1992), which have Ba/Nb ratios between *c.* 90 and 371 and 100–726, respectively. Compare with Figure 12 of Ewart *et al.* (1992).

et al. 1992). We raise the possibility that these high Ba/Nb intermediate composition rocks emplaced since the Early Miocene are small-volume crustal partial melts of potentially Palaeozoic–Proterozoic crustal materials along the southwestern edge of the North American craton.

We therefore conclude that the dominant period of andesitic volcanism focused around the margins of the GoC, is at the regional scale, the product of magma mixing. Our model relieves issues around the apparent lack of geochemical discrimination of interpreted syn- and post-subduction (<14 Ma) magmatism in the region (Till *et al.* 2009) as the entire Oligocene–Miocene igneous record was unrelated to subduction. Evidence for this comes from uncontaminated Early Miocene basalts that show within-plate trace element compositions (Fig. 7). A similar asthenospheric fingerprint has also been recognized by Cameron *et al.* (1989) for Oligocene basaltic andesites (SCORBA) erupted across the northern SMO. Based on the chemical variation within the mid-Miocene (*c.* 18–13 Ma) suite, magma end-members are interpreted to be: (1) an asthenosphere-derived, within-plate alkali basalt similar to those erupted in the region during the Late Miocene to Quaternary (e.g. Luhr *et al.* 1995; Pallares *et al.* 2007), (2) crustal partial melts derived from Mesozoic–Cenozoic igneous crust, including the Jurassic–Cretaceous Peninsular Ranges Batholith (e.g. Silver & Chappell 1988) and Eocene–Oligocene batholithic rocks, as informed by inherited zircon ages (Bryan *et al.* 2008; Godinez *et al.* 2010; Ramos Rosique 2012); and (3) small-volume crustal partial melts of dacitic composition derived from Proterozoic–Palaeozoic crust (highlighted by the extreme Ba/Nb ratios, Fig. 7). Importantly, Figure 7 illustrates no apparent temporal change in basaltic magma composition from the Early to Late Miocene. Basaltic rocks with more 'arc-like' signatures are interpreted to be either crust/lithospheric mantle contaminated, or possibly derived from a previously subduction-modified lithospheric mantle, but where lithospheric mantle melting was in response to basaltic magma input from the asthenosphere and decompression due to lithospheric thinning.

How was mixing promoted? We believe that extensional faulting causing the large-magnitude crustal thinning during the Middle Miocene began actively disrupting bimodal magma systems in the GoC region (Fig. 8). Magma inputs (i.e. basaltic and rhyolitic magmas erupted during the Early Miocene) had not fundamentally changed, and regionally, away from the Gulf, basaltic and rhyolitic magmas generally continued to erupt (Fig. 5). However, locally around the GoC where extension was now being focused and was greatest, this tectonic influence had a major effect on the composition of magmas being erupted. Consequently, there was a spatial–temporal focusing of andesitic volcanism along the nascent GoC rift. Importantly, this switch from widespread Early Miocene bimodal volcanism to more localized andesitic volcanism during the Middle Miocene (Fig. 8) appears to correspond to a switch from wide to narrow rifting beginning *c.* 18 Ma. This switch had an important effect on silicic magma generation rates, which appears to have significantly decreased at *c.* 19 Ma (Fig. 3) as mafic magma inputs to the crust became more focused in the Gulf region. The Early Miocene graben that had developed across a *c.* 500 km width of western Mexico and which had significant volcanic fills, became magmatically abandoned by *c.* 18 Ma (e.g. Bolaños Graben, Ramos Rosique 2012), and continued magmatism was now largely restricted to the GoC and its immediate margins (Fig. 8). The spatial–temporal-compositional patterns of magmatism through the Miocene of western Mexico are thus providing us with an important record of changes to the extensional style, intensity and location.

Active extensional faulting, particularly during the period 18–13 Ma, modified the erupted magma compositions that became more intermediate in composition. Concomitant with this compositional change was a change in eruption styles that were dominantly effusive, producing scattered lavas and domes as well as dyking. Eruption location shifted to be concentrated around the nascent GoC rift. We therefore conclude the Comondú Group andesites are a larger-scale regional example of syn-volcanic extensional fault-driven magma mixing processes such as described by Johnson & Grunder (2000).

Concluding remarks

Magmatism is used widely as an instant tracer of the geodynamic setting, and magma composition as a proxy for a specific tectonic setting (e.g. arc/supra-subduction zone, slab melting, slab tearing, crustal rifting). In particular, calc-alkaline chemistries ± relative Nb depletions and blind use of tectonic discrimination diagrams often form the basis for interpreting an active subduction-related setting (e.g. Zhu *et al.* 2012). However, subduction leaves a strong chemical imprint on the lithosphere such that subduction-related geochemical signatures can persist for 100s of millions of years (e.g. Morris & Hooper 1997) and stamp the chemistry of new magmas produced at much younger times and unrelated or far removed from any active subduction zone. The low-Ti flood basalts of the Karoo and Paraná-Etendeka Large Igneous Provinces (LIPs) are classic examples of this (Duncan 1987; Ewart *et al.* 1998).

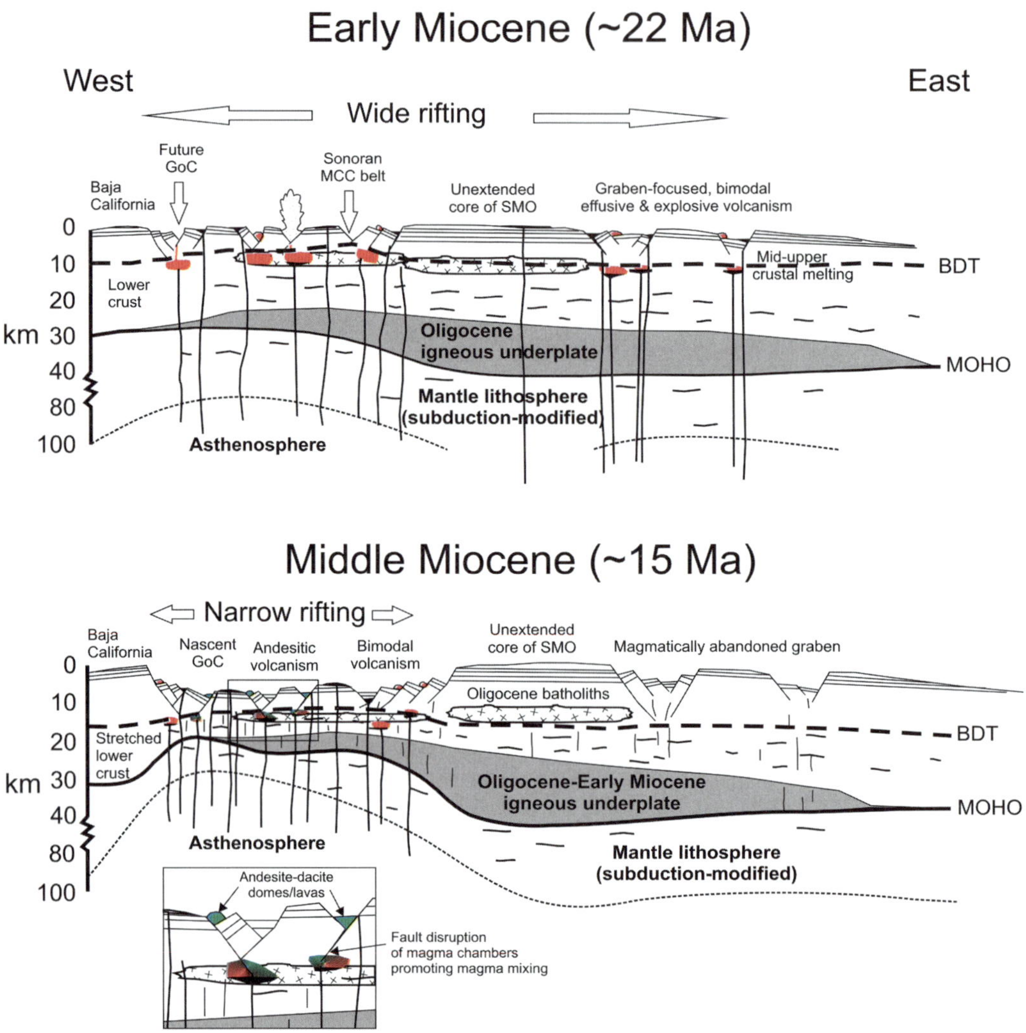

Fig. 8. A schematic model for the Early to Middle Miocene tectonomagmatic evolution of western Mexico and the initial stages of rifting and opening of the Gulf of California. The cross-sections illustrate: (1) the change from wide to narrow rifting and the focusing of magmatism into the gulf region, which occurred *c.* 18 Ma; (2) the general crustal structure, such as the development of thickened crust beneath the Sierra Madre Occidental core, and requisite thinning of the lower crust and mantle lithosphere in the gulf region beginning at least by the Early Miocene; (3) the main mafic magma source regions – the involvement of asthenosphere-derived basaltic magmas is evident from at least the Early Miocene (e.g. see Fig. 7), whereas more 'calc-alkali' or subduction signature-bearing basalts are either crust/mantle contaminated, or have been derived from subduction-modified lithospheric mantle; (4) Early Miocene silicic magmas are predominantly mid-upper crustal melts with significant material contributions from earlier formed batholithic-type granitic rocks (as informed by the zircon inheritance, Bryan *et al.* 2008; Ramos Rosique 2012; Ferrari *et al.* 2012; in press); (5) for the Middle Miocene, additional silicic magma source regions include the lower crust to account for the anomalously high Ba/Nb dacites/rhyolites (Fig. 7) where lower crustal melting may have been promoted by thinning, decompression and basaltic intrusion; and (6) active faulting interrupts bimodal magma systems (see inset) existing regionally (red, rhyolite; black, basalt) to produce the andesitic magmas (green) in the Middle Miocene around the margins of the Gulf of California.

Our study demonstrates the importance of understanding the regional tectonic setting both at the time of volcanism and for the preceding and following magmatic history in interpreting the origin of the middle Miocene Comondú Group andesites. As shown in many previous studies, when considered in isolation and focusing only on the general chemical characteristics, these andesitic rocks do resemble the products of a supra-subduction zone arc (e.g. Hausback 1984; Till *et al.* 2009). However, the andesites are spatially restricted to the nascent GoC rift, there was no preceding history of andesitic arc volcanism in this region, the region was clearly undergoing wide rifting and bimodal volcanism immediately beforehand, and their emplacement coincides in space and time with focused rifting and large magnitude (*c.* 100%) crustal thinning. The stratigraphic relationships and architecture of Middle Miocene andesite–dacite lavas and domes (proximal and vent volcanic facies) being buried by thick successions of syn-volcanic sedimentary deposits including sandstones demonstrates lava emplacement into actively subsiding rift basins. This facies architecture is more consistent with biostratigraphic and palaeontological evidence for earlier basin rifting and subsidence and a marine incursion into the GoC by 12 Ma (Helenes 2012). Andesite–dacite lavas and domes are subordinate to the volcanic-derived sedimentary rocks and isolated to scattered lavas/domes thus should not be used to define an arc. We extend the recent arguments by Castillo (2008) that the compositionally diverse suite of mafic rocks including adakitic and high-Nb basalts erupted in Baja California are unrelated to subduction or slab melting in the Late Miocene, but that asthenospheric mantle involvement in regional magmatism occurred much earlier, being prevalent in the Oligocene (Cameron *et al.* 1989) and Early Miocene (Fig. 7).

Our fundamental conclusion is that the Comondú Group does not represent a supra-subduction zone arc, and this challenges all previous studies on the Comondú Group. We caution that andesites in ancient extended continental margins may bear no relationship to active subduction, and magma chemistry cannot be solely relied on to interpret tectonic setting. Alternatively, as shown here, the andesitic volcanism and the switch from widespread bimodal to more focused andesitic volcanism can potentially provide much more important insights into the extensional history of a continental margin.

Finally, our study raises questions about genetic links between the Sierra Madre Occidental silicic volcanism and the Gulf of California, as these have previously been considered two separate phenomena. We suggest the Sierra Madre Occidental Silicic LIP is the pre- to syn-rift volcanic event to the GoC rift. This is because from *c.* 30 Ma and certainly by 24 Ma, SMO volcanism and extension were overlapping spatially and temporally. It therefore raises the question of when did rupturing of the Gulf of California really begin? Was it at the onset of wide rifting and bimodal volcanism at *c.* 30–24 Ma, or at the onset of narrow rifting at *c.* 18 Ma? Regardless, the new regional timing and stratigraphic–structural relationships evident along the eastern margin of the GoC show that the onset of rifting to open the GoC was clearly older than the *c.* 12.3 Ma end of subduction along this segment of the western North American margin. Consequently, the GoC has had a much longer history of rifting and associated magmatism before the onset of seafloor-spreading, and may not be such an anomalously rapid zone of crustal rupture (cf. Umhoefer 2011).

S.B. was supported by a Vice Chancellor's Research Fellowship from QUT. We acknowledge support of grant CONACyT 82378 to LF. The submarine samples shown on Figure 1 were collected on cruises supported by the US NSF (grants 0203348 and 0646563 to co-PIs Peter Lonsdale and Paterno Castillo), as well as grants to P. Lonsdale and J. Kluesner for the BEKL, ROCA and DANA cruises in the Gulf of California. Discussions with J. Clemens, G. Zellmer and C. Schrank on aspects of this manuscript are appreciated. We thank J. Cole, P. Castillo and S. Straub for constructive reviews of this manuscript.

References

Aguirre-Díaz, G. J. & McDowell, F. W. 1993. Nature and timing of faulting and synextensional magmatism in the southern Basin and Range, central-eastern Durango, Mexico. *Geological Society of America Bulletin*, **105**, 1435–1444.

Aguirre-Díaz, G. & Labarthe-Hernández, G. 2003. Fissure ignimbrites: fissure-source origin for voluminous ignimbrites of the Sierra Madre Occidental and its relationship with Basin and Range faulting. *Geology*, **31**, 773–776, http://dx.doi.org/10.1130/G19665.1

Atwater, T. & Stock, J. 1998. Pacific-North America plate tectonics of the Neogene southwestern United States. *International Geology Review*, **10**, 375–402.

Axen, G. J., Taylor, W. J. & Bartley, J. M. 1993. Space–time patterns and tectonic controls of Tertiary extension and magmatism in the Great Basin of the Western United States. *Geological Society of America Bulletin*, **105**, 56–76.

Beal, C. H. 1948. *Reconnaissance of the geology and oil possibilities of Baja California, Mexico*. Geological Society of America Memoir, **31**.

Best, M. G. & Christiansen, E. H. 1991. Limited extension during peakTertiary volcanism, Great Basin of Nevada and Utah. *Journal of Geophysical Research*, **96**, 13 509–13 528.

Bonner, J. L. & Herrin, E. T. 1999. Surface wave study of the Sierra Madre Occidental of northern Mexico. *Bulletin of the Seismological Society of America*, **89**, 1323–1337.

Bryan, S. E., Ferrari, L., Reiners, P. W., Allen, C. M., Petrone, C. M., Ramos Rosique, A. & Campbell, I. H. 2008. New insights into crustal contributions to large volume rhyolite generation at the mid-Tertiary

Sierra Madre Occidental Province, Mexico, revealed by U–Pb geochronology. *Journal of Petrology*, **49**, 47–77.

CAMERON, M., BAGBY, W. & CAMERON, K. L. 1980. Petrogenesis of voluminous mid-Tertiary ignimbrites of the Sierra Madre Occidental, Chihuahua, Mexico. *Contributions to Mineralogy and Petrology*, **74**, 271–284.

CAMERON, K. L., NIMZ, G. J., KUENTZ, D., NIEMEYER, S. & GUNN, S. 1989. Southern Cordilleran basaltic andesite suite, southern Chihuahua, Mexico: a link between Tertiary continental arc and flood basalt magmatism in North America. *Journal of Geophysical Research*, **94**, 7817–7840.

CAMERON, K. L., ROBINSON, J. V. *ET AL.* 1992. Contrasting styles of Pre-Cenozoic and mid-Tertiary crustal evolution in northern Mexico: evidence from deep crustal xenoliths from La Olivina. *Journal of Geophysical Research*, **97**, 17 353–17 376.

CARREÑO, A. L. & SMITH, J. T. 2007. Stratigraphy and correlation for the ancient Gulf of California and Baja California Peninsula, Mexico. *Bulletin of American Paleontology*, **371**, 158.

CASTILLO, P. R. 2008. Origin of the adakite-high Nb basalt association and its implications for post-subduction magmatism in Baja California, Mexico. *Geological Society of America Bulletin*, **120**, 451–462.

CLIFT, P. D. & ODP LEG 135 SCIENTIFIC PARTY 1994. Volcanism and sedimentation in a rifting island arc terrain: an example from Tonga, SW Pacific. *In*: SMELLIE, J. L. (ed.) *Volcanism Associated with Extension at Consuming Plate Margins*. Geological Society, London, Special Publications, **81**, 29–51.

COLE, J. W. 1990. Structural control and origin of volcanism in the Taupo Volcanic Zone, New Zealand. *Bulletin of Volcanology*, **52**, 445–459.

CONLY, A. G., BRENAN, J. M., BELLON, H. & SCOTT, S. D. 2005. Arc to rift transitional volcanism in the Santa Rosalÿa Region, Baja California Sur, Mexico. *Journal of Volcanology and Geothermal Research*, **142**, 303–341.

COUCH, R. W, NESS, G. E. *ET AL.* 1991. Gravity anomalies and crustal structure of the Gulf and Peninsular Province of the Californias. *In*: DAUPHIN, J. P. & SIMONEIT, B. R. T. (eds) *The Gulf and the Peninsular Province of the Californias*. American Association of Petroleum Geologists, Memoirs, **47**, 47–70.

DAMON, P. E., SHAFI QULLAH, M. & CLARK, K. F. 1981. Age trends of igneous activity in relation to metallogenesis in the southern Cordillera. *Arizona Geological Society Digest*, **14**, 137–154.

DEMETS, C. 1995. A reappraisal of seafloor spreading lineations in the Gulf of California: implications for the transfer of Baja California to the Pacific plate and estimates of Pacific-North America motion. *Geophysical Research Letters*, **22**, 3545–3548.

DRAKE, W. 2005. *Structural analysis, stratigraphy, and geochronology of the San José Island accommodation zone, Baja California Sur, Mexico*. MSc thesis, Northern Arizona University.

DUNCAN, A. R. 1987. The Karoo igneous province – a problem area for inferring tectonic setting from basalt geochemistry. *Journal of Volcanology and Geothermal Research*, **32**, 13–34.

EWART, A., SCHON, R. W. & CHAPPELL, B. W. 1992. The Cretaceous volcanic–plutonic province of the central Queensland (Australia) coast – a rift related 'calc-alkaline' province. *Transactions of the Royal Society of Edinburgh: Earth Sciences*, **83**, 327–345.

EWART, A., MILNER, S. C., ARMSTRONG, R. A. & DUNCAN, A. R. 1998. Etendeka volcanism of the Goboboseb Mountains and Messum Igneous complex Namibia. Part I: geochemical evidence of Early Cretaceous Tristan plume melts and the role of crustal contamination in the Paraná–Etendeka CFB. *Journal of Petrology*, **39**, 191–225.

FERRARI, L., LÓPEZ-MARTINEZ, M., AGUIRRE-DÍAZ, G. & CARRASCO-NÚÑEZ, G. 1999. Space time patterns of Cenozoic arc volcanism in central Mexico: from the Sierra Madre Occidental to the Mexican Volcanic Belt. *Geology*, **27**, 303–306.

FERRARI, L., LÓPEZ-MARTÍNEZ, M. & ROSAS-ELGUERA, J. 2002. Ignimbrite flare-up and deformation in the southern Sierra Madre Occidental, western Mexico – implications for the late subduction history of the Farallon Plate. *Tectonics*, **21**, 17-1–17-24, http://dx.doi.org/10.1029/2001TC001302

FERRARI, L., VALENCIA-MORENO, M. & BRYAN, S. 2007. Magmatism and tectonics of the Sierra Madre Occidental and its relation with the evolution of the western margin of North America. *In*: ALANIZ-ÁLVAREZ, S. A. & NIETO-SAMANIEGO, A. F. (eds) *Geology of México: Celebrating the Centenary of the Geological Society of México*. Geological Society of America Special Papers, **422**, 1–39, http://dx.doi.org/10.1130/2007.2422(01)

FERRARI, L., OROZCO-ESQUIVEL, T., LOPEZ MARTÍNEZ, M., DUQUE TRUJILLO, J., BRYAN, S. E. & CERCA, M. 2012. 25 million years to break a continent: early to middle Miocene rifting and syn-extensional magmatism in the southern Gulf of California. *Geological Society of America Abstracts with Programs*, **44**, 6.

FERRARI, L., LOPEZ MARTÍNEZ, M., OROZCO, M. T., BRYAN, S. E., DUQUE TRUJILLO, J. & LONSDALE, P. (In press) Miocene rifting and syn-extensional magmatism in the southwestern Sierra Madre Occidental (Sinaloa and Nayarit states): the early history of the Gulf of California. *Geosphere*.

FLETCHER, J. M., GROVE, M., KIMBROUGH, D., LOVERA, O. & GEHRELS, G. E. 2007. Ridge–trench interactions and the Neogene tectonic evolution of the Magdalena Shelf and southern Gulf of California; insights from detrital zircon U–Pb ages from the Magdalena Fan and adjacent areas. *Geological Society of America Bulletin*, **119**, 1313–1336, http://dx.doi.org/10.1130/B26067.1

GANS, P. 1997. Large-magnitude Oligo-Miocene extension in southern Sonora: implications for the tectonic evolution of northwest Mexico. *Tectonics*, **16**, 388–408.

GANS, P. B. & BOHRSON, W. A. 1998. Suppression of volcanism during rapid extension in the Basin and Range Province, United States. *Science*, **279**, 66–68, http://dx.doi.org/10.1126/science.279.5347.66

GANS, P., MAHOOD, G. & BORSHON, W. 1989. *Synextensional magmatism in the Basin and Range Province; a case study from the eastern Great Basin*. Geological Society of America Special Papers, **233**.

GANS, P. B., BLAIR, K. D., MACMILLAN, I., WONG, M. S. & ROLDAN-QUINTANA, J. 2003. Structural and magmatic evolution of the Sonoran rifted margin: a preliminary report. *Geological Society of America Abstracts with Programs*, **35**, 21.

GASTIL, R G., KRUMMENACHER, D. & MINCH, J. 1979. The record of Cenozoic volcanism around the Gulf of California. *Geological Society of America Bulletin*, **90**, 839–857.

GODINEZ, N. S., KIMBROUGH, D. L. & KOHEL, C. 2010. Stratigraphy and petrologic evolution of the Oligocene–Miocene Comondu Group near Bahia Concepcion and Loreto, Baja California Sur, Mexico. *Geological Society of America Abstracts with Programs*, **42**, 66.

GOMBERG, J. S., PRIESTLEY, K. F., MASTERS, G. & BRUNE, J. 1988. The structure of the crust and upper mantle in northern Mexico. *Geophysical Journal*, **94**, 1–20.

HAUSBACK, B. P. 1984. Cenozoic volcanic and tectonic evolution of Baja California Sur, Mexico. *In*: FRIZZEL, V. A., JR. (ed.) *Geology of the Baja California Peninsula*. Pacific Section, SEPM Special Publications, Tulsa, **39**, 219–236.

HELENES, J. 2012. Neogene micropaleontology and evolution of the Northern Gulf of California. *Geological Society of America Abstracts with Programs*, **44**, 18.

HELENES, J., CARREÑO, A. L. & CARRILLO, R. M. 2009. Middle to late Miocene chronostratigraphy and development of the northern Gulf of California. *Marine Micropaleontology*, **72**, 10–25, http://dx.doi.org/10.1016/j.marmicro.2009.02.003

HENRY, C. D. 1989. Late Cenozoic Basin and Range structure in western Mexico adjacent to the Gulf of California. *Geological Society of America Bulletin*, **101**, 1147–1156, http://dx.doi.org/10.1130/0016-7606(1989)101<1147:LCBARS>2.3.CO;2

HENRY, C. D. & ARANDA-GOMEZ, J. J. 2000. Plate interactions control middle–late Miocene, proto-Gulf and Basin and Range extension in the southern Basin and Range. *Tectonophysics*, **318**, 1–26.

HILDRETH, W. & FIERSTEIN, J. 2000. Katmai volcanic cluster and the great eruption of 1912. *Geological Society of America Bulletin*, **112**, 1594–1620.

HOSACK, A. 2006. *Structure and stratigraphy of the Agua Verde accommodation zone, Baja California Sur, Mexico*. MSc thesis, Northern Arizona University.

JOHNSON, J. A. & GRUNDER, A. L. 2000. The making of intermediate composition magma in a bimodal suite: Duck Butte eruptive center, Oregon, USA. *Journal of Volcanology and Geothermal Research*, **95**, 175–195.

KARIG, D. E. & JENSKY, W. 1972. The Protogulf of California. *Earth and Planetary Science Letters*, **17**, 169–174.

KING, R. E. 1939. Geological reconnaissance in northern Sierra Madre Occidental of Mexico. *Geological Society of America Bulletin*, **50**, 1625–1722.

LANPHERE, M. A., CAMERON, K. L. & CAMERON, M. 1980. Sr isotopic geochemistry of voluminous rhyolitic ignimbrites and related rocks, Batopilas area, western Mexico. *Nature*, **286**, 594–596.

LE MAITRE, R. W., BATEMAN, P. ET AL. 1989. *A Classification of Igneous Rocks and Glossary of Terms: Recommendations of the IUGS Subcommission on the Systematics of Igneous Rocks*. Blackwell Scientific, London.

LIPMAN, P. W., PROSTKA, H. J. & CHRISTIANSEN, R. L. 1972. Cenozoic volcanism and plate-tectonic evolution of the western United States, Part 1: Early and Middle Cenozoic. *Philosophical Transactions of the Royal Society of London Series A*, **271**, 217–248.

LIZARRALDE, D., AXEN, G. J. ET AL. 2007. Variable styles of rifting in the Gulf of California. *Nature*, **448**, 466–469.

LONSDALE, P. 1989. Geology and tectonic history of the Gulf of California. *In*: WINTERER, E. L., HUSSONG, D. M. & DECKER, R. W. (eds) *The eastern Pacific Ocean and Hawaii*. Geological Society of America, Geology of North America, Boulder, Colorado, v. N, 499–521.

LONSDALE, P. 1991. Structural patterns of the Pacific floor offshore of peninsular California. *In*: DAUPHIN, J. P. & SIMONEIT, B. R. T. (eds) *The Gulf and Peninsular Province of the Californias*. American Association of Petroleum Geologists, Memoirs, **47**, 87–125.

LUDWIG, K. R. 2003. *Isoplot Version 3.00: A Geochronological Tool-kit for Microsoft Excel*. Berkeley Geochronology Center, Berkeley, CA, Special Publications.

LUHR, J. F., ARANDA-GOMEZ, J. J. & HOUSH, T. B. 1995. San Quintin Volcanic field, Baja California Norte, México. Geology, petrology, and geochemistry. *Journal of Geophysical Research*, **100**, 10 353–10 380.

MARTÍN-BARAJAS, A., STOCK, J. M., LAYER, P., HAUSBACK, B., RENNE, P. & LOPEZ-MARTINEZ, M. 1995. Arc-rift transition volcanism in the Puertecitos Volcanic Province, northeastern Baja California, Mexico. *Geological Society of America Bulletin*, **107**, 407–424.

MARTÍN, A., FLETCHER, J. M., LOPEZ-MARTINEZ, M. & MENDOZA-BORUNDA, R. 2000. Waning Miocene subduction and arc volcanism in Baja California: the San Luis Gonzaga volcanic field. *Tectonophysics*, **318**, 27–51.

MARTINY, B., MARTÍNEZ–SERRANO, R. G., MORÁN–ZENTENO, D. J., MACIAS–ROMO, C. & AYUSO, R. A. 2000. Stratigraphy, geochemistry and tectonic significance of the Oligocene magmatic rocks of Western Oaxaca, Southern Mexico. *Tectonophysics*, **318**, 71–98.

MCDOWELL, F. W. 2007. Geologic transect across the northern Sierra Madre Occidental volcanic field, Chihuahua and Sonora, Mexico. *The Geological Society of America Digital Map and Chart Series*, **6**, 70, http://dx.doi.org/10.1130/2007.DMCH006

MCDOWELL, F. W. & CLABAUGH, S. E. 1979. Ignimbrites of the Sierra Madre Occidental and their relation to the tectonic history of western Mexico. *In*: CHAPIN, C. E. & ELSTON, W. E. (eds) *Ash-Flow Tuffs*. Geological Society of America Special Papers, **180**, 113–124.

MCDOWELL, F. & KEIZER, R. P. 1977. Timing of mid-Tertiary volcanism in the Sierra Madre Occidental between Durango city and Mazatlán, Mexico. *Geological Society of America Bulletin*, **88**, 1479–1487.

MCDOWELL, F. W. & MAUGER, R. L. 1994. K–Ar and U–Pb zircon chronology of Late Cretaceous and Tertiary magmatism in central Chihuahua State, Mexico. *Geological Society of America Bulletin*, **106**, 118–132, http://dx.doi.org/10.1130/0016-7606(1994)106<0118:KAAUPZ>2.3.CO;2

McDowell, F. W. & McIntosh, W. C. 2012. Timing of intense magmatic episodes in the northern and central Sierra Madre Occidental, western Mexico. *Geosphere*, **8**, 1505–1526.

McDowell, F., Roldán-Quintana, J. & Amaya-Martínez, R. 1997. Interrelationship of sedimentary and volcanic deposits associated with Tertiary extension in Sonora, Mexico. *Geological Society of America Bulletin*, **109**, 1349–1360.

McQuarrie, N. & Oskin, M. 2011. Palinspastic restoration of NAVDat and implications for the origin of magmatism in southwestern North America. *Journal of Geophysical Research*, **115**, B10401, 16, 2010, http://dx.doi.org/10.1029/2009JB006435

Mora-Klepeis, G. & McDowell, F. W. 2004. Late Miocene calc-alkalic volcanism in northwestern Mexico: an expression of rift or subduction related magmatism? *Journal of South American Earth Sciences*, **17**, 297–310, http://dx.doi.org/10.1016/j.jsames.2004.08.001

Morán–Zenteno, D. J., Alba–Aldave, L., Gonzalez–Torres, E. A., Martiny, B., Bravo–Diaz, B. A. & Sohn, E. 2007. Ignimbrite flare-up in the north–central Sierra Madre del Sur, Southern Mexico: a continuation of the Sierra Madre Occidental ignimbrite Province? *American Geophysical Union Joint Assembly Abstracts*, Acapulco, Mexico, Abstract U53A–03.

Morris, G. A. & Hooper, P. R. 1997. Petrogenesis of the Colville Igneous Complex, northeast Washington: implications for Eocene tectonics in the northern U.S. Cordillera. *Geology*, **25**, 831–834.

Murray, B., Busby, C. J., Ferrari, L. & Solari, L. 2010. Tectonic controls on Late Oligocene volcanism in the Guazapares mining district, northwestern Sierra Madre Occidental, Mexico. *Geological Society of America Abstracts with Programs*, **42**, 344.

Murray, B. P., Busby, C. J., Ferrari, F. & Solari, L. A. In press. Synvolcanic crustal extension during the mid-Cenozoic ignimbrite flare-up in the northern Sierra Madre Occidental, Mexico: evidence from the Guazapares Mining District region, western Chihuahua. *Geosphere*.

Nieto-Obregón, J., Delgado-Argote, L. & Damon, P. E. 1981. Relaciones petrológicas y geocronológicas del magmatismo de la Sierra Madre Occidental el Eje Neovolcánico en Nayarit, Jalisco y Zacatecas. *Asociación de Ingenieros de Minas, Metalurgistas y Geólogos de México, Memoria Técnica*, **XIV**, 327–361.

Nourse, J. A., Anderson, T. H. & Silver, L. T. 1994. Tertiary metamorphic core complexes in Sonora, northwestern Mexico. *Tectonics*, **13**, 1161–1182, http://dx.doi.org/10.1029/93TC03324

Orozco-Esquivel, T., López-Martínez, M., Lonsdale, P., Ferrari, L., Cornejo-Jimenez, C., Pinero-Lajas, D. & Duque-Trujillo, J. F. 2010. Tearing apart of an Early Miocene silicic igneous province in the southern Gulf of California. *Geological Society of America Abstracts with Programs*, **42**, 296.

Oskin, M. E. & Stock, J. M. 2003. Marine incursion synchronous with plate-boundary localization in the Gulf of California. *Geology*, **31**, 23–26.

Oskin, M. E., Stock, J. M. & Martín-Barajas, A. 2001. Rapid localization of Pacific–North America plate motion in the Gulf of California. *Geology*, **29**, 459–462.

Pallares, C., Maury, R. C. *et al.* 2007. Slab-tearing following ridge-trench collision: evidence from Miocene volcanism in Baja California, Mexico. *Journal of Volcanology and Geothermal Research*, **161**, 95–117, http://dx.doi.org/10.1016/j.jvolgeores.2006.11.002

Parson, L. M. & Wright, I. C. 1996. The Lau–Havre–Taupo back-arc basin: a southward- propagating, multi-stage evolution from rifting to spreading. *Tectonophysics*, **263**, 1–22.

Paz-Moreno, F. A., Demant, A., Cocheme, J.-J., Dostal, J. & Montigny, R. 2003. The Quaternary Moctezuma volcanic field: a tholeiitic to alkali basaltic episode in the central Sonora Basin and Range Province, Mexico. *In*: Johnson, S. E., Paterson, S. R., Fletcher, J. M., Girty, G. H., Kimbrough, D. L. & Martín-Barajas, A. (eds) *Tectonic evolution of northwestern Mexico and southwestern USA*. Geological Society of America Special Papers, **374**, 439–455.

Persaud, P., Pérez-Campos, X. & Clayton, R. W. 2007. Crustal thickness variations in the margins of the Gulf of California from receiver functions. *Geophysical Journal International*, **170**, 687–699, http://dx.doi.org/10.1111/j.1365-246X.2007.03412.x

Ramos Rosique, A. 2012. *Timing and evolution of Late Oligocene to Early Miocene magmatism and epithermal mineralization in the central Bolaños graben, southern Sierra Madre Occidental, México*. PhD thesis, Kingston University.

Ramos Rosique, A., Bryan, S., Ferrari, L., Allen, C., Lopez-Martinez, M. & Rankin, A. 2010. Timing and evolution of Late Oligocene to Miocene magmatism in the southern Sierra Madre Occidental silicic large igneous province: insights from zircon chronochemistry and Ar/Ar geochronology. *Geophysical Research Abstracts*, **12**, EGU2010-9788.

Roberts, M. P. & Clemens, J. D. 1993. Origin of high-potassium, calc-alkaline, I-type granitoids. *Geology*, **21**, 825–828.

Sawlan, M. G. 1991. Magmatic evolution of the Gulf of California rift. *In*: Dauphin, J. P. & Simoneit, B. R. (eds) *The Gulf and Peninsular Province of the Californias*. American Association of Petroleum Geologists, Memoirs, **47**, 301–369.

Sawlan, M. G. & Smith, J. G. 1984. Petrologic characteristics, age and tectonic setting of Neogene volcanic rocks in northern Baja California Sur, Mexico. *In*: Frizzell, V. A., Jr. (ed.) *Geology of the Baja California Peninsula*. SEPM Special Publications, Tulsa, **39**, 237–251.

Silver, L. T. & Chappell, B. W. 1988. The Peninsular Ranges Batholith: an insight into the evolution of the Cordilleran batholiths of southwestern North America. *Transactions of the Royal Society of Edinburgh, Earth Sciences*, **79**, 105–121.

Stock, J. M. & Hodges, K. V. 1989. Pre-Pliocene extension around the Gulf of California and the transfer of Baja California to the Pacific plate. *Tectonics*, **8**, 99–115.

Stock, J. M., Lewis, C. J. & Nagy, E. A. 1999. The Tuff of San Felipe: an extensive middle Miocene pyroclastic flow deposit in Baja California, Mexico. *Journal of Volcanology and Geothermal Research*, **93**, 53–74.

Sun, S. & McDonough, W. F. 1989. Chemical and isotopic systematics of oceanic basalts: implications for mantle composition and processes. *In*: Saunders, A. D. & Norry, M. J. (eds) *Magmatism in the Ocean Basins*. Geological Society, London, Special Publications, **42**, 313–345.

Sutherland, F. H. 2006. *Continental rifting across the southern Gulf of California*. PhD thesis, University of California at San Diego.

Sutherland, F. H., Kent, G. M. *et al.* 2012. Middle Miocene to early Pliocene oblique extension in the southern Gulf of California. *Geosphere*, **8**, 752–770, http://dx.doi.org/10.1130/GES00770.1

Swanson, E. R. & McDowell, F. W. 1984. Calderas of the Sierra Madre Occidental volcanic field western Mexico. *Journal of Geophysical Research*, **89**, 8787–8799.

Swanson, E. R., Kempter, K. A., McDowell, F. W. & McIntosh, W. C. 2006. Major ignimbrites and volcanic centers of the Copper Canyon area: a view into the core of Mexico's Sierra Madre Occidental. *Geosphere*, **2**, 125, http://dx.doi.org/10.1130/GES00042.1.

Till, C. B., Gans, P. B., Spera, F. J., MacMillan, I. & Blair, K. D. 2009. Perils of petrotectonic modeling: a view from southern Sonora, Mexico. *Journal of Volcanology & Geothermal Research*, **186**, 160–168, http://dx.doi.org/10.1016/j.jvolgeores.2009.06.014

Umhoefer, P. J. 2011. Why did the Southern Gulf of California rupture so rapidly? Oblique divergence across hot, weak lithosphere along a tectonically active margin. *GSA Today*, **21**, 4–10.

Umhoefer, P. J., Dorsey, R. J., Willsey, S., Mayer, L. & Renne, P. 2001. Stratigraphy and Geochronology of the Comondú Group near Loreto, Baja California Sur, Mexico. *Sedimentary Geology*, **44**, 125–147.

Umhoefer, P. J., Sutherland, F. *et al.* 2008. Synchronous changes in rift-margin basins and initiation of the Alarcón spreading ridge and related transform fault, Southwestern Gulf of California. *Geological Society of America Abstracts with Programs*, **40**, 151.

Vidal-Solano, J. R., Paz-Moreno, F. A., Demant, A. & Lopez-Martinez, M. 2007. Ignimbritas hiperalcalinas del Mioceno medio en Sonora Central: revaluación de la estratigrafía y significado del volcanismo terciario. *Revista Mexicana de Ciencias Geológicas*, **24**, 47–67.

Vidal-Solano, J. R., Demant, A., Paz-Moreno, F. A., Lapierre, H., Ortega-Rivera, M. A. & Lee, J. K. W. 2008. Insights into the tectonomagmatic evolution of NW Mexico: geochronology and geochemistry of the Miocene volcanic rocks from the Pinacate area, Sonora. *Geological Society of America Bulletin*, **120**, 691–708.

Waitt, R. B. 1970. *Ignimbrites of the Sierra Madre Occidental between Durango and Mazatlán, Mexico*. MA thesis, University of Texas at Austin.

Ward, P. L. 1995. Subduction cycles under western North America during the Mesozoic and Cenozoic eras. *In*: Miller, D. M. & Busby, C. (eds) *Jurassic Magmatism and Tectonics of the North American Cordillera*. Geological Society of America Special Papers, **299**, 1–45.

Wark, D. A., Kempter, K. A. & McDowell, F. W. 1990. Evolution of waning, subduction-related magmatism, northern Sierra Madre Occidental, Mexico. *Geological Society of America Bulletin*, **102**, 1555–1564.

Wong, M. S., Gans, P. B. & Scheier, J. 2010. The $^{40}Ar/^{39}Ar$ thermochronology of core complexes and other basement rocks in Sonora, Mexico: implications for Cenozoic tectonic evolution of northwestern Mexico. *Journal of Geophysical Research*, **115**, http://dx.doi.org/10.1029/2009JB007032

Zellmer, G. F. 2009. Petrogenesis of Sr-rich adakitic rocks at volcanic arcs: insights from global variations of eruptive style with plate convergence rates and surface heat flux. *Journal of the Geological Society, London*, **166**, 725–734, http://dx.doi.org/10.1144/0016-76492008-0721

Zellmer, G. F., Sakamoto, N., Iizuka, Y., Miyoshi, M., Tamura, Y., Hsieh, H.-H. & Yurimoto, H. 2013. Crystal uptake into aphyric arc melts: insights from two-pyroxene pseudo-decompression paths, plagioclase hygrometry, and measurement of hydrogen in olivines from mafic volcanics of SW Japan. *In*: Gómez-Tuena, A., Straub, S. M. & Zellmer, G. F. (eds) *Orogenic Andesites and Crustal Growth*. Geological Society, London, Special Publications, **385**. First published online June 11, 2013, http://dx.doi.org/10.1144/SP385.3

Zhu, D.-C., Zhao, Z. D. *et al.* 2012. Cambrian bimodal volcanism in the Lhasa Terrane, southern Tibet: record of an early Paleozoic Andean-type magmatic arc in the Australian proto-Tethyan margin. *Chemical Geology*, **328**, 290–308, http://dx.doi.org/10.1016/j.chemgeo.2011.12.024

Index

Page numbers in *italic* denote figures. Page numbers in **bold** denote tables.